International Health and Safety at Work

nebosh

Endorsed by NEBOSH

This publication is endorsed by NEBOSH as offering high quality support for the delivery of NEBOSH qualifications. NEBOSH endorsement does not imply that this publication is essential to achieve a NEBOSH qualification, nor does it mean that this is the only suitable publication available to support NEBOSH qualifications. No endorsed material will be used verbatim in setting any NEBOSH examination and all responsibility for the content remains with the publisher. Copies of official specifications for all NEBOSH qualifications may be found on the NEBOSH website – **www.nebosh.org.uk**

Third Edition

International Health and Safety at Work

For the NEBOSH
International General Certificate in Occupational
Health and Safety

Phil Hughes MBE, MSc, CFIOSH

Chairman NEBOSH 1995–2001. President of IOSH 1990–1991

Ed Ferrett PhD, BSc (Hons Eng), CEng, MIMechE, MIET, CMIOSH

Vice Chairman NEBOSH 1999–2008

Endorsed by NEBOSH

nebosh
Endorsed by NEBOSH

Routledge
Taylor & Francis Group
LONDON AND NEW YORK

Third edition first published 2016
by Routledge
2 Park Square, Milton Park, Abingdon, Oxon OX14 4RN

and by Routledge
711 Third Avenue, New York, NY 10017

Routledge is an imprint of the Taylor & Francis Group, an informa business

First edition first published by Elsevier 2010
Second edition first published by Routledge 2013

British Library Cataloguing-in-Publication Data
A catalogue record for this book is available from the British Library

Library of Congress Cataloging-in-Publication Data
Hughes, Phil (Phillip W.), 1946- , author.
International health and safety at work : for the NEBOSH international general certificate in
occupational health and safety / Phil Hughes and Ed Ferrett. -- Third edition.
p. ; cm.
Includes bibliographical references and index.
ISBN 978-1-138-83130-8 (pbk. : alk. paper) -- ISBN 978-1-315-73301-2 (ebook)
I. Ferrett, Ed, author. II. Title.
[DNLM: 1. National Examination Board in Occupational Safety and Health. 2. Occupational
Health--Great Britain--Examination Questions. 3. Safety Management--Great Britain--Examination
Questions. WA 18.2]
T55
363.11--dc23
2015009678

ISBN: 978-1-138-83130-8 (pbk)
ISBN: 978-1-315-73301-2 (ebk)

Typeset in Univers LT by
Servis Filmsetting Ltd, Stockport, Cheshire

Printed by Bell and Bain Ltd, Glasgow

Contents

List of illustrations viii
Preface xv
Acknowledgements xvii
About the authors xviii
How to use this book and what it covers xix
List of principal abbreviations xxv
Safety signs xxviii

1. Foundations in health and safety 1
1.1 The scope and nature of occupational health and safety 2
1.2 The moral, social and economic reasons for maintaining and promoting good standards of health and safety in the workplace 4
1.3 The role of national governments and international bodies in formulating a framework for the regulation of health and safety 9
1.4 Further information 16
1.5 Practice revision questions 17
Appendix 1.1 Scaffolds and ladders 18

2. Health and safety management systems – PLAN 19
2.1 The key elements of a health and safety management system 20
2.2 The purpose and importance of setting policy for health and safety 29
2.3 The key features and appropriate content of an effective health and safety policy 30
2.4 Further information 36
2.5 Practice revision questions 36
Appendix 2.1 Health and Safety Policy checklist 38

3. Health and safety management systems – Organising – DO 1 41
3.1 Organisational health and safety roles and responsibilities of employers, directors, managers, workers and other relevant parties 43
3.2 Concept of health and safety culture and its significance in the management of health and safety in an organisation 59
3.3 Human factors which influence behaviour at work 62

3.4 How health and safety behaviour at work can be improved 68
3.5 Further information 79
3.6 Practice revision questions 79
Appendix 3.1 Leadership actions for directors and board members....................................82
Appendix 3.2 Detailed health and safety responsibilities 84
Appendix 3.3 Checklist for supply chain health and safety management...........................86
Appendix 3.4 Safety culture questionnaire 87

4. Health and safety management systems – Risk assessment and controls – DO 2 89
4.1 Principles and practice of risk assessment..........90
4.2 General principles of control and hierarchy of risk reduction measures....................................108
4.3 Sources of health and safety information 109
4.4 Safe systems of work 110
4.5 Role and function of a permit-to-work system 115
4.6 Emergency procedures and arrangements for contacting the emergency services....................120
4.7 Requirements for, and effective provision of, first-aid in the workplace...................................123
4.8 Further information 125
4.9 Practice revision questions 125
Appendix 4.1 Procedure for risk assessment and management (European Commission) 127
Appendix 4.2 Hazard checklist.................................127
Appendix 4.3 Risk assessment example 1: Hairdressing salon.................................129
Appendix 4.4 Risk assessment example 2: Office cleaning...131
Appendix 4.5 Asbestos examples of safe systems of work...133
Appendix 4.6 Emergency numbers in some countries worldwide 135

5. Health and safety management systems – Monitoring, Investigation and recording – CHECK...139
5.1 Active and reactive monitoring 140
5.2 Investigating incidents 150
5.3 Recording and reporting incidents 158

Contents

5.4 Further information 163
5.5 Practice revision questions 163
Appendix 5.1 Workplace inspection exercises 165
Appendix 5.2 ILO Code of Practice: Annex
H: Classification of industrial accidents
according to type of accident 167
Appendix 5.3 ILO Code of Practice: Annex
I: Classification of industrial accidents
according to agency 168
Appendix 5.4 ILO Code of Practice: Annex B:
Proposed list of occupational diseases 169

6. Health and safety management systems – Audit and review – ACT 171
6.1 Health and safety auditing 172
6.2 Review of health and safety performance 176
6.3 Further information 179
6.4 Practice revision questions 179

7. Workplace hazards and risk control 181
7.1 Health, welfare and work environment requirements 182
7.2 Violence at work 187
7.3 Substance misuse at work 191
7.4 Safe movement of people in the workplace 192
7.5 Working at height 197
7.6 Hazards and control measures for works of a temporary nature 210
7.7 Construction activities 212
7.8 Further information 221
7.9 Practice revision questions 221
Appendix 7.1 Scaffolds and ladders 223
Appendix 7.2 Inspection recording form with timing and frequency chart 224
Appendix 7.3 Checklist of typical scaffolding faults 226
Appendix 7.4 Recommendations for excavation work in the ILO Code of Practice 'Safety and Health in Construction' 227

8. Transport hazards and risk control 229
8.1 Safe movement of vehicles in the workplace 230
8.2 Driving at work 236
8.3 Further information 240
8.4 Practice revision questions 241
Appendix 8.1 Advice on driving and the use of taxis 242

9. Musculoskeletal hazards and risk control 243
9.1 Work-related upper limb disorders 244
9.2 Manual handling hazards and control measures 248
9.3 Manually operated load handling equipment 254
9.4 Powered load handling equipment 256
9.5 Further information 266
9.6 Practice revision questions 267
Appendix 9.1 A typical UK risk assessment for the use of lifting equipment 269

Appendix 9.2 Examples of manually operated load handling equipment 270

10. Work equipment hazards and risk control 273
10.1 General requirements for work equipment 274
10.2 Hazards and controls for hand-held tools 284
10.3 Mechanical and non-mechanical hazards of machinery 290
10.4 Control measures for reducing risks from machinery hazards 295
10.5 Further information 307
10.6 Practice revision questions 308

11. Electrical safety 311
11.1 Principles, hazards and risks associated with the use of electricity at work 312
11.2 Control measures when working with electrical systems or using electrical equipment in all workplace conditions 320
11.3 Further information 329
11.4 Practice revision questions 330

12. Fire hazards and risk control 331
12.1 Fire initiation, classification, spread and some legal standards 332
12.2 Fire risk assessment 340
12.3 Fire prevention and prevention of fire spread 344
12.4 Fire alarm system and fire-fighting arrangements 354
12.5 Evacuation of a workplace 361
12.6 Further information 367
12.7 Practice revision questions 367
Appendix 12.1 Fire risk assessment checklist as recommended in Fire Safety Guides published by the UK Department for Communities and Local Government in 2006 369
Appendix 12.2 Typical fire notice 371

13. Chemical and biological health hazards and risk control 373
13.1 Forms of, classification of, and health risks from hazardous substances 374
13.2 Assessment of health risks 379
13.3 Occupational exposure limits 387
13.4 Control measures 389
13.5 Specific agents 400
13.6 Safe handling and storage of waste 411
13.7 Further information 415
13.8 Practice revision questions 415
Appendix 13.1 GHS Hazard (H) Statements (Health only) 418
Appendix 13.2 Hazardous properties of waste 419
Appendix 13.3 Different types of protective gloves 420

14. Physical and psychological health hazards and risk control 421
14.1 Noise 422

14.2 Vibration 428
14.3 Radiation 434
14.4 Stress...442
14.5 Further information 444
14.6 Practice revision questions 444

15. Summary of ILO, OSH Conventions, legal frameworks and country examples 447
15.1 ILO International Conventions on OSH referenced in the NEBOSH International General Certificate syllabus 448
15.2 Typical OSH legal frameworks in the USA, EU and UK...471
15.3 National implementing legislation 476
15.4 Common themes in national legislation...........527
Appendix 15.1 Seoul Declaration on Safety and Health at Work 533
Appendix 15.2 ILO – C155 Occupational Safety and Health Convention, 1981 534

16. Study skills.................................539
16.1 Introduction..540
16.2 Find a place to study 540
16.3 Time management....................................540
16.4 Blocked thinking......................................541
16.5 Taking notes..541
16.6 Reading for study.....................................541
16.7 Free learning resources from the Open University...541

16.8 Organising for revision 542
16.9 Organising information..................................542
16.10 Being aware of your learning style..................544
16.11 How does memory work? 544
16.12 How to deal with exams 545
16.13 The examiners' reports 546
16.14 Conclusion 547
16.15 Further information 547

17. Specimen answers to practice questions 549
17.1 Introduction...550
17.2 The written examinations 550
17.3 Unit GC3 – Health and Safety Practical Application 555
Appendix 17.1 Practical application report...............559
Appendix 17.2 Practical application observation sheets 563

18. International sources of information and guidance 569
18.1 Introduction...570
18.2 How to search the internet effectively 570
18.3 Some useful websites 572
18.4 Health and safety forms 574

Index 609

List of illustrations

Figures

1.1 At work (© Beci Phipps) 5
1.2 Economic reasons for good health and safety management 7
1.3 Insured and uninsured costs (© Beci Phipps).....8
1.4 Diagrammatic view of 'reasonably practicable' (© Beci Phipps) 12
1.5 Typical health and safety legal framework (© Beci Phipps) 14
1.6 The inspector inspects (© ndoeljindoel Shutterstock) 14
1.7 Good standards prevent harm and save money (© Beci Phipps) 15
2.1 The Plan, Do, Check, Act management cycle (© Beci Phipps) 21
2.2 This chapter covers the PLAN stage of the management cycle (© Beci Phipps)..................22
2.3 Key elements of ILO-OSH 2001 (Source: ILO) (© Beci Phipps) 24
2.4 Well-presented policy documents (© Beci Phipps) 30
2.5 Part of a policy commitment (© Beci Phipps)...31
2.6 SMART performance standards or Objectives (© Beci Phipps) 32
2.7 (a) (b) and (c) Good information, training and working with employees is essential (© Beci Phipps) 34
2.8 The policy might be good but is it put into practice – unsafe use of ladder (© mikeledray / Shutterstock.com) 35
2.9 Emergency procedures (© Shutterstock) 38
2.10 Responsibilities (© Beci Phipps) 38
2.11 Reach truck in warehouse (© Corepics VOF Shutterstock) 39
3.1 DO part of the management cycle involves Risk Profiling (Chapter 4), Organising and Implementing plans 42
3.2 Everyone from senior managers down has health and safety responsibilities (© Beci Phipps) 43
3.3 Safety practitioner at the front line (© ndoeljindoel Shutterstock) 50
3.4 NEBOSH in control here 51
3.5 Typical supply chain (© Beci Phipps) 52

3.6 Inadequate chair – it should have five feet and an adjustable backrest – take care when buying especially if second-hand 53
3.7 Contractors at work from a MEWP (Mobile Elevating Work Platform) 57
3.8 Safety investment............................60
3.9 Heinrich's accidents/incidents ratios (© Beci Phipps) 62
3.10 Well-designed workstation for sitting or standing (Source: © HSE) 64
3.11 Poor working conditions 64
3.12 Motivation and activity 66
3.13 Visual perception: (a) Are the lines the same length? (b) Face or vase? (c) Face or saxophone player? 66
3.14 Types of human failure (Source: HSE) (© Beci Phipps) 67
3.15 Effective consultation is essential 73
3.16 Health and safety training needs and opportunities (© Beci Phipps) 74
3.17 Internal influences on safety culture (© Beci Phipps) 77
3.18 External influences on safety culture (© Beci Phipps) 78
4.1 Risk Assessment or Profiling is covered by the DO part of the management cycle (© Beci Phipps) 90
4.2 Reducing the risk – finding an alternative safer method when fitting a wall-mounted boiler (Source: © HSE)............................92
4.3 Accident at work............................92
4.4 F. E. Bird's well-known accident triangle (© Beci Phipps) 93
4.5 Five steps to risk assessment (© Beci Phipps) 94
4.6 Proper control of gases, vapours and biological agents in a laboratory (© emin kuliyev Shutterstock) 98
4.7 Colour categories and shapes of signs (© Stocksigns) 98
4.8 Examples of pictorial signs (© Stocksigns).......99
4.9 Falling objects and construction site entrance signs (© Stocksigns) 99

4.10 Wet floor signs (© Stocksigns) 99

4.11 Examples of chemical warning signs (© Stocksigns) 100

4.12 Examples of fire safety signs (© Stocksigns) 100

4.13 Examples of fire action signs (© Stocksigns) 100

4.14 Examples of first-aid signs (© Stocksigns) 100

4.15 LPG sign (© Stocksigns) 100

4.16 Smoke-free – no smoking signs (© HM Gov) 100

4.17 Fragile roof signs (© Stocksigns) 101

4.18 Emergency shower and eye wash station where there is a serious risk of contamination (© Shutterstock) 102

4.19 Good dust control for a chasing operation. A dust mask is still required for complete protection 103

4.20 Respiratory protection and disposable overalls are needed when working in high levels of asbestos dust 103

4.21 A lone worker – special arrangements required (© istockphoto) 107

4.22 When controls break down (© Pictureguy – Shutterstock) 108

4.23 Recommended format for a Material Safety Data Sheet (MSDS) 109

4.24 Health risk – checking on the contents...........110

4.25 (a) and (b) Multipadlocked hasp for locking off an isolation valve – each worker puts on their own padlock 110

4.26 Multihasp used on an electrical isolator 111

4.27 Lone workers need special consideration (© Shutterstock) 115

4.28 A hot work permit is usually essential for welding, cutting and burning except in designated areas like a welding shop (© Shutterstock) 117

4.29 Entering a confined space – using a rescue tripod with hoist, harness, breathing apparatus, and person outside to keep watch and call for further assistance if required 119

4.30 Emergency services at work 121

4.31 (a) First-aid and stretcher sign; (b) first-aid sign (© Stocksigns) 124

5.1 CHECK involves measuring performance and investigating incidents (© Beci Phipps) 142

5.2 Effective risk control (Source: © HSE)...........143

5.3 (a) Poor conditions – inspection needed; (b) inspection in progress (© Shutterstock); (c) Poor conditions in offices can cause accidents (trek and shoot © Shutterstock) 144

5.4 Using a checklist (© Beci Phipps) 147

5.5 An accident at work – hit by a forklift truck...........................149

5.6 A dangerous occurrence – fire (© Jason Salmon Shutterstock) 150

5.7 Reconstruction of a fatal accident with the author lying where the deceased person was found under the ladder after trying to lengthen the extending ladder (© Phil Hughes)...........152

5.8 (a) Accident; (b) near miss; (c) undesired circumstance or unsafe condition (© HSE).....153

5.9 F. E. Bird's well-known accident ratio/triangle 153

5.10 Appropriate levels of investigation (Source: © HSE)...........154

5.11 Questions to be asked in an investigation (© Beci Phipps) 156

5.12 Office.................................165

5.13 Road repair...........................165

5.14 Workshop 166

5.15 Roof repair/unloading...........166

5.16 Construction site...........167

6.1 ACT part of the health and safety management system (© Beci Phipps) 172

6.2 The Audit Process (© Beci Phipps) 173

6.3 Using the audit questions for interviews and collecting information (© Beci Phipps) 174

6.4 The audit report should be reviewed by senior managers with an action plan and follow-up (© Beci Phipps) 176

6.5 Review of performance (© Beci Phipps) 177

6.6 Continual improvement part of the health and safety management process (© Beci Phipps) 178

7.1 Welfare washing facilities: washbasin large enough for people to wash their forearms (© bikeriderlondon Shutterstock) 183

7.2 Natural ventilation in a building (Source: © HSE)...........184

7.3 A well-lit workplace...........184

7.4 The heat equation (Source: © Occupational Safety and Health Administration, USA).........186

7.5 Security access and surveillance CCTV camera (© Shutterstock)...........188

7.6 It takes a healthy liver about one hour to break down and remove one unit of alcohol. A unit is equivalent to 8 mg or 10 ml (1 cl) of pure alcohol...........192

7.7 Tripping hazards (Source: © HSE)...........192

7.8 Cleaning must be done carefully to prevent slipping (Source: © HSE) 193

7.9 Falling from a height – tower scaffold with inadequate handrails (no centre rail) and should never be moved when in use 195

7.10 Typical pedestrian/vehicle crossing area (Source: © HSE)...........196

7.11 Walkway and traffic rate separated with kerb and partial rails. Traffic movement stopped during factory visit 197

List of illustrations

7.12 (a) A mobile elevating work platform being used on a fragile roof to replace a roof sheet (Source: © HSE)....................199

7.12 (b) Workmen wearing harnesses attached to a work positioning line which is fitted to the staging (Source: © HSE) 200

7.13 Fall arrest harness and device 201

7.14 Ladder showing correct 1 in 4 angle (means of securing omitted for clarity) (Source: © HSE)....................203

7.15 (a) Ladder tied at top stiles (correct for working on, not for access); (b) tying part way down; (c) tying at base; (d) securing at the base (Source: © HSE) 204

7.16 Working with stepladders (© Bratladders)......205

7.17 (a) Typical independent tied scaffold; (b) fixed scaffold left in place to fit the gutters (Source: © HSE) 206

7.18 Typical pre-fabricated tower scaffold with outriggers 207

7.19 Mobile elevating work platforms: (a) scissor lift; (b) hydraulic access platform in Belgium (© Phil Hughes)....................207

7.20 (a) and (b) Secure site access gates (© Phil Hughes)....................213

7.21 Demolition in progress (© Phil Hughes) 213

7.22 Barriers around excavation by footpath (Source: © HSE)....................214

7.23 Portable toilet, washing, eating and locker facilities for a building site 215

7.24 Contractors at work in market square, Bruges (© Phil Hughes) 219

7.25 Timbered excavation with ladder access and supported services (guard removed on one side for clarity) (Source: © HSE) 220

8.1 Industrial counterbalanced lift truck 231

8.2 Telescopic materials handler (© Phil Hughes)....................231

8.3 Various construction plant with driver protection (© Phil Hughes) 233

8.4 Excavator digging out a swimming pool in France (© Phil Hughes) 234

8.5 Professional driver at work on vehicle recovery (© RH Kerkham) 236

8.6 Occupational road risk – animals can be very unpredictable on busy roads with major potential consequences when elephants are involved (© Phil Hughes)..........237

8.7 Specialised training is required for heavy goods vehicles and other similar categories 238

8.8 Road accidents are a significant risk when vehicles are overloaded – hay bales in transit in Morocco (© Phil Hughes)................238

9.1 Handling roofing slates onto a roof using a teleporter lift truck (© Phil Hughes)................244

9.2 A tilted worktable. The distance between the operator and the work can be reduced by putting the table at a more vertical angle. The table is adjustable in height and angle to suit the particular job (Source: © HSE)........245

9.3 Feed material into a machine from a bulk bin or drum by suction or pumping to save workers having to climb steps carrying loads. This reduces the risk of trips and falls as well as avoiding manual handling and improves the efficiency of your process. Consider this solution wherever you have to transfer powders, granules and liquids 246

9.4 Workstation design 247

9.5 Manual handling: there are many potential hazards....................249

9.6 Main injury sites caused by manual handling accidents 249

9.7 UK HSE guidance for manual lifting – recommended weights (Source: © HSE) 251

9.8 The main elements of a good lifting technique....................253

9.9 A pallet truck (Source: © HSE) 255

9.10 Mechanical aids to lift patients in hospital 255

9.11 Conveyor systems: (a) roller conveyor (these may have powered and free running rollers); (b) an overhead conveyor handling wheels. Other designs of overhead conveyor are useful for transferring components and garments between workstations in, for example, manufacture of machines or clothing; (c) a slat conveyor in use in a food factory 257

9.12 A brick and tile elevator (Source: © HSE) 258

9.13 Reach truck – designed so that the load retracts within the wheelbase to save space . 258

9.14 Manoeuvring a yacht using a large overhead travelling gantry and slings in a Turkish marina 260

9.15 Hoists for lifting cars in a high quality French repair garage 262

9.16 Specially designed safety hooks (Source: © HSE)....................262

9.17 Lifting equipment in the forest 263

9.18 Truck-mounted lifting equipment....................263

10.1 (a) CE mark 276

10.1 (b) Division of responsibility for the safety of machinery (© Beci Phipps) 276

10.2 Typical Declaration of Conformity (© Dewalt) 277

10.3 A typical bench-mounted abrasive wheel (© Draper) 279

10.4 Typical electrically powered compressor with air receiver tank attached (© Draper)......281

10.5 Emergency stop button (© Praphan Jampala Shutterstock) 282

10.6 Non-powered set of engineering hand tools (© Shutterstock) 285

10.7 Range of hand-held power tools (© Dewalt)....................286

10.8   Electric drill with percussion hammer action to drill holes in masonry 288

10.9   Disc sander with dual hand holds.................289

10.10 (a) Rotary drum floor sander; (b) orbital finishing sander.................................289

10.11 Range of mechanical hazards 291

10.12 Range of fixed guards 296

10.13 Adjustable guard for a rotating shaft, such as a pedestal drill 297

10.14 Self-adjusting guard on a wood saw..............297

10.15 Typical interlocking guards: (a) sliding and (b) hinged 297

10.16 Schematic diagram of a telescopic trip device fitted to a radial drill 298

10.17 Pedestal-mounted free-standing two-handed control device.................................299

10.18 Typical multifunction printer/photocopier 301

10.19 Typical office paper shredder (© Fellowes) 301

10.20 Typical bench-mounted grinder (© Draper).....302

10.21 Typical pedestal drill. Guard should be adjusted close to the work-piece to cover the rotating drill (© Draper) 302

10.22 (a) Typical large cylinder mower (© Chelle129 – Shutterstock)..........................303

10.22 (b) Tractor-mounted cylinder mower for hedge cutting (© Shutterstock) 303

10.23 Typical brush cutter (© STIHL).......................303

10.24 Typical chainsaw with rear handle. The rear handle projects from the back of the saw. It is designed to always be gripped with both hands, with the right hand on the rear handle. It may be necessary to have a range of saws with different guide bar lengths available. As a general rule, choose a chainsaw with the shortest guide bar suitable for the work. 1 – hand guard with integral chain brake; 2 – exhaust outlet directed to the right-hand side away from the operator; 3 – chain breakage guard at bottom of rear handle; 4 – chain designed to have low-kickback tendency; 5 – rubber anti-vibration mountings; 6 – lockout for the throttle trigger; 7 – guide bar (should be protected when transporting chainsaw); 8 – bottom chain catcher; 9 – PPE hand/eye/ear defender signs; 10 – on/off switch (© HSE) 304

10.25 Kevlar gloves, overtrousers and overshoes providing protection against chainsaw cuts; helmet, ear and face shield protect the head. Apprentice under training – first felling (© Phil Hughes) 304

10.26 Typical retail compactor (© Pakawaste Ltd)...305

10.27 Typical retail checkout conveyor (© Alamy)....305

10.28 Small concrete/cement mixer (© Winget Ltd)...305

10.29 Diesel concrete mixer with hopper 305

10.30 Typical bench-mounted circular saw (© HSE)..306

11.1   Electrical hazard warning sign 312

11.2   A poster about electric shock (© Stocksigns) 315

11.3   Keep 18 m clear of high-voltage lines............316

11.4   Over 25% of fires are caused by electrical malfunction 317

11.5   Modern European multi-plug – France (© Louis Phipps) 317

11.6   Earth bonding of containers to prevent a static discharge – anti-static outer clothing and conductive footwear should be used (© Phil Hughes)...318

11.7   Portable hand-held electric power tools (© Dewalt) 319

11.8   Typical 240 volt mini circuit breaker (© Evgeny Tomeev Shutterstock) 322

11.9   A variety of electrical equipment: (a) flush mounted exterior socket with RCD; (b) lockable electrical isolator (© Phil Hughes)...................................322–323

11.10 Multi-plug extension lead and special plugs and sockets: (a) reduced voltage usually 110 volt; (b) 240 volt mains extension lead with cut out (© Phil Hughes) 323

11.11 Double insulation sign.....................................324

11.12 Precautions for overhead lines: (a) 'goalpost' crossing points beneath lines to avoid contact by plant; (b) diagram showing normal dimensions for 'goalpost' crossing points and barriers (© HSE)..325

11.13 Using a cable detector (© Phil Hughes)..........326

12.1   Fire is still a significant risk in many workplaces (© Shutterstock) 332

12.2   Average number of fire deaths per 100 fires (Source: CTIF World Fire Statistics) (© Beci Phipps) 332

12.3   Economic-statistical evaluation of 'costs' of fire for 14 countries (Source: CTIF World Fire Statistics) (© Beci Phipps) 333

12.4   Fire triangle..335

12.5   (a) Transport of flammable solid sign; (b) GHS packaging sign 335

12.6   (a) Transport of flammable liquid sign; (b) GHS packaging sign 336

12.7   (a) Transport of flammable gas sign; (b) GHS packaging sign 336

12.8   (a) Transport of oxidising agent sign; (b) GHS packaging sign 336

12.9   Principles of heat transmission......................337

12.10 Fire and smoke spread in buildings 338

12.11 UK causes of fire in recent years (© Beci Phipps)..339

12.12 UK accidental fires – sources of ignition in recent years (© Beci Phipps) 339

12.13 Distribution of fires by fire origin (Source: CTIF World Fire Statistics) (© Beci Phipps) 340

12.14 (a) Before risk assessment; (b) after risk assessment (© HM Government) 342

12.15 Partly blocked fire exit door 345

12.16 (a)–(c) Various storage arrangements for highly flammable liquids 350

12.17 Steel structures can collapse in the heat of a fire (© Phil Hughes) 352

12.18 Insulated core panels 352

12.19 Firebreak wall between dwellings 353

12.20 Earth bonding of containers to prevent a static discharge – anti-static outer clothing and conductive footwear should be used (© Phil Hughes) 354

12.21 Simple electrical fire alarm system components (© Beci Phipps) 355

12.22 Marking layout for an extinguisher under ISO 7165:2009 (© ISO/DIS 7165) 359

12.23 Use-code symbols for extinguishers under ISO 7165:2009. Class A: Ordinary solid material fires. Class B: Flammable liquid fires. Class C: Gas and vapour fires. Class D: Combustible metal fires. Fires involving energised electrical conductors. Class F: Cooking oil fires 359

12.24 Various items of fire-fighting equipment: (a) fire blanket; (b) water hose reel; (c) ISO labelled fire extinguisher (© Phil Hughes) 359

12.25 Various sprinkler heads designed to fit into a high-level water pipe system and spray water at different angles onto a fire below.....360

12.26 External fire escape from a multi-storey building. Extreme caution would be needed in winter in a cold climate, and enclosure of the staircase may be essential 361

12.27 International Fire Escape pictorial 362

12.28 (a) Fire evacuation diagram in the UK 364

12.28 (b) Fire evacuation plan in four languages in Bruges, Belgium (© Phil Hughes) 364

12.29 Special addition to the fire notice for people with a disability (© Stocksigns) 366

12.30 Stair evacuation chair for people with a disability 366

13.1 (a) Use of GHS symbols (© Phil Hughes) 377

13.1 (b) How the European packaging symbols relate to the new GHS labels (© Phil Hughes).......................377

13.2 Paint spraying – risk of sensitising, particulary if isocyanate-based paint used and inadequate local exhaust ventilation (© Phil Hughes).......................379

13.3 Hazardous substances – principal routes of entry into the human body 379

13.4 The (a) upper and (b) lower respiratory system.......................380

13.5 The nervous system 381

13.6 The cardiovascular system 381

13.7 Parts of the urinary system...........381

13.8 The skin – main structures in the dermis........382

13.9 (a) Typical symbols and (b) product label on containers 386

13.10 Hand pump and stain detectors (© Draeger) .. 386

13.11 Chemical storage in France which needs to comply with European standards (© Phil Hughes) 388

13.12 (a) A LEV system; (b) welding with an adjustable LEV system to remove dust and fumes (© Auto Extract Systems Ltd).............392

13.13 (a) Local exhaust ventilation systems can quickly corrode and need regular maintenance of fans and filters (b) Large roof level system with access platform for maintenance of the filters and fans (c) Paint-spraying inside a special room with general ventilation to remove fumes 394

13.14 Personal protective equipment (PPE) at work (© Shutterstock) 394

13.15 Types of respiratory protective equipment: (a) filtering half-mask; (b) half-mask reusable with filters; (c) compressed air-line breathing apparatus with full-face mask fitted with demand valve (©HSE) 396

13.16 Variety of eye protection goggles (© Draper) 397

13.17 Damaged asbestos lagging on pipework (© PA Group Ltd) 404

13.18 High hazard vacuum cleaner to clear up asbestos material...........405

13.19 Example of dermatitis due to wet cement (© VrisPhuket Shutterstock) 408

13.20 Example of heavy industrial air pollution (© Phil Hughes)...........412

13.21 (a) Water pollution from oil spillage; (b) water pollution from plastic and other solid waste...........412

13.22 Commercial waste collection (© Shutterstock) 413

13.23 A designated waste collection area with two types of skip commonly used for waste collection. Heavy materials would be transported in the smaller skip. Sizes of skip range from about 4 cu metres (small skip shown) to about 35 cu metres (large skip shown)...........414

13.24 Electronic waste under EU WEEE (© Alamy).....................415

14.1 It is better to control noise at source rather than wear ear protection (© Phil Hughes) 422

14.2 Passage of sound waves: (a) the ear with cochlea uncoiled; (b) summary of transmission 423

14.3 Typical ear protection zone sign 426

14.4 Noise paths found in a workplace: (a) the quiet area is subjected to reflected noise from a machine elsewhere in the building;

(b) the correct use of roof absorption will reduce the reflected noise; (c) segregation of the noisy operation will benefit the whole workplace (© HSE) 427

14.5 Injuries which can be caused by hand–arm vibration (© HSE) 430

14.6 Breaker mounted on an excavator's jib to reduce vibrations (© Phil Hughes) 431

14.7 (a) Vibrating roller with risk of whole-body vibration (© Phil Hughes) 433
(b) Remote control vibrating plate weighing 1.2 tons with compaction in excess of a 7-ton roller which eliminates the risk of whole-body vibration. The operator is protected from vibrations, noise and dust. The machine can only be operated if line of sight is intact. In case of a loss of control the proximity recognition sensor keeps the operator safe (© Waker neuson) 433

14.8 Typical ionising radiation warning sign............435

14.9 X-ray generator cabinet (© Balteau NDT)........435

14.10 Radon monitoring equipment 436

14.11 A proper Class 4 laser area setup for defeatable access control. Curtain design and layout varies with environment. Class 4 laser areas in a research or university environment usually run long-term experiments that require unattended operation. In such cases, defeatable entryway controls are appropriate. By design, persons who have been properly trained and given the keypad access code may momentarily 'defeat' the interlock to enter and exit the laser controlled area protected by a magnetically locked door 437

14.12 Metal furnace – source of infrared heat (© Alamy).............................438

14.13 Specialised eye protection for work with lasers 439

14.14 Breakdown of mental health cases by type of event which precipitated stress between 2010 and 2012 (Source: HSE Stress and Psychological Disorders Great Britain 2013, © HSE).............................443

15.1 ILO's strategic approach to strengthening national OSH systems (Source: ILO – Introductory Report: Decent Work – Safe Work) 450

15.2 Fatal work-related accident rates worldwide. Estimated rate of work-related fatal accidents per 100,000 workers (Source: ILO – Introductory Report: Decent Work – Safe Work) (© Beci Phipps)...............478

15.3 Disease fatalities attributed to work. The 2.2 million work-related deaths are broken down as shown (Source: ILO – World Day for Safety and Health at Work 2005: A Background Paper) (© Beci Phipps)...............478

15.4 Some national and international OSH development and activities, 2004–2005 (Source: ILO – Decent Work – Safe Work Programme).............................479

16.1 Revision notes (© Liz Hughes) 542

16.2 Mind map – report writing (© Liz Hughes) 543

17.1 Select a competent and experienced person to carry out a risk assessment (© sima Shutterstock).............................550

17.2 Glass-blowing factory 553

17.3 Flat roof protection: (a) using handrails, and toe boards; (b) using a harness and proprietary anchor.............................554

17.4 Asbestos pipe lagging (© dedek Shutterstock) 555

M1 General health & safety risk assessment example 1 575

M2 Risk assessment report form example 2.......576

M3 Contractors' risk assessment example for confined spaces.............................577

M4 Contractors' risk assessment example for work on fragile roofs.............................578

M5 Workplace inspection report form 579

M6 Workplace inspection checklist 580–581

M7 Job safety analysis.............................582

M8 Essential elements – permit to work 583

M9 Witness statement form.............................584

M10 Accident/incident report.............585–586

M11 First aid treatment and accident record ..587–588

S1 Machinery risk assessment 589

S2(a) General work permit.............................590

S2(b) General work permit: back of form................591

H1a COSHH assessment example 592

H1 COSHH assessment 1 (blank form) 593

H2 COSHH assessment 2: details of substances used or stored; assessment of a substance.............................594–595

H3 Example of a workstation self assessment checklist.............................596–597

H4 Example of a noise assessment record form 598

H5 Manual handling of loads: assessment checklist.............................599

H6 Manual handling risk assessment: employee checklist.............................600

F1 Fire safety maintenance checklist 601–602

F2 Fire risk assessment record – significant findings 603

C1 Construction inspection report 604

C2 Example risk assessment for contract bricklayers.............................605–606

C3 Example risk assessment for woodwork.............................607–608

Tables

1.1 Numbers of global work-related adverse events 3

1.2 Distribution of fatal occupational injuries and incidence rates around the world in recent years 5

1.3 Global estimates of work-related fatalities caused by occupational accidents and work-related diseases in recent years 7

1.4 Causes of global work-related fatalities 8

2.1 Essential elements of any national occupational health and safety management system 21

2.2 Location and contents of the key elements of a health and safety management system in chapters 2, 3, 4, 5 and 6 23

4.1 Typical contents of first-aid box – low risk 124

4.2 Number of first-aid personnel 125

7.1 Typical workplace lighting levels 185

7.2 Trend in physical assaults and threats at work, 2006–2014 (based on working adults of working age) 188

9.1 Examples of low-cost ergonomic and other improvements resulting from an ILO initiative 246

9.2 Safe driving of lift trucks 259

11.1 Standard wiring colours 320

12.1 Classification for fire extinguishers used in Australia 358

12.2 Classification for fire extinguishers used in the UK under EN3:7 358

12.3 Classification for fire extinguishers used in the USA 358

12.4 Maintenance and testing of fire equipment....360

12.5 Maximum travel distances............................362

13.1 Examples of the new hazard warning (H) and precautionary statements (P) 378

13.2 Examples of workplace exposure limits (WELs) 389

13.3 The hazards and types of PPE for various parts of the body................................395

14.1 Some typical sound pressure levels (SPL) (dB(A) values)................................424

14.2 Typical noise levels at wood working machines 424

14.3 Simple observations to determine the need for a noise risk assessment............................425

14.4 Examples of vibration exposure values measured by HSE on work equipment...........429

14.5 Machines which could produce significant whole-body vibration................................430

14.6 The change in exposure times as vibration increases................................432

14.7 Typical radiation dose limits............................434

15.1 Essential elements of a national OSH system................................452

15.2 Estimates of work-related occupational accidents and diseases – 2001 data mainly....477

15.3 Summary of obligations (with relevant section) 481

15.4 Summary of obligations under the Factory Acts................................493

15.5 Common themes in national OSH legislation................................531

16.1 Terminology used in NEBOSH exams 545

Boxes

2.1 Example of objectives................................32

5.1 Key data for medium level of investigation.....158

5.2 ILO Code of Practice – types of incidents which should be recorded for national statistics................................160

5.3 Information for insurance/compensation purposes following accident or incident 162

12.1 Extract from ILO Code of Practice 'Safety in the Use of Chemicals at Work' 347

Preface to the third edition

International Health and Safety at Work has quickly established itself as the foremost textbook for students taking the NEBOSH International General Certificate (IGC) and has now been endorsed by NEBOSH as a recommended textbook for the IGC course.

The International General Certificate in Occupational Health and Safety is suitable for managers, supervisors and staff based outside the UK from all types of organisations making day-to-day decisions at work who need a broad understanding of health and safety issues and to be able to manage risks effectively. Over 50,000 people have achieved this qualification since it was introduced in 2004.

The NEBOSH International General Certificate is also suitable for those embarking on a career in health and safety, providing a valuable foundation for further professional study (such as the NEBOSH International Diploma in Occupational Health and Safety). The IGC is modelled on the NEBOSH National General Certificate in Occupational Health and Safety, the most widely recognised health and safety qualification of its kind in the UK. The key difference between the two qualifications is in the applicability of legal requirements. Rather than being guided by a specifically UK framework, the IGC takes a risk management approach based on best practice and international standards, such as International Labour Organisation (ILO) codes of practice, with special reference to the model proposed in the ILO's 'Guidelines on Occupational Safety and Health Management Systems' (ILO-OSH 2001). Local laws and cultural factors form part of the study programme where relevant and appropriate.

Many larger organisations choose the NEBOSH qualifications as a key part of their supervisors' or management development programme. By ensuring that line managers have a sound understanding of the principles of risk management they build an effective safety culture in the organisation. Smaller organisations, operating in lower risk environments, often choose the NEBOSH International General Certificate in Occupational Health and Safety as the appropriate qualification for the manager taking the lead on health and safety issues.

This book has been produced as a textbook for students studying the NEBOSH IGC course and closely follows the latest NEBOSH syllabus. The qualification is divided into three distinct units each of which is assessed separately. This development offers the opportunity for additional and more flexible course formats and students may now study parallel courses (in, say, fire and construction) without repeating the management unit. Students who decide to take individual units will, on passing, receive a Unit Certificate.

The publication of the amended HSG65 – Managing for health and safety, recommends a new model for health and safety management based on the 'Plan, Do, Check, Act' principle replacing the 'Policy, Organising, Planning, Measuring performance, Auditing and Review (POPMAR)' model. This has produced a significant change to IGC1 – the management unit. This change produced a very large Do element which we have split into two chapters – Do1 that covers 'organising' and Do2 that covers 'risk assessment and controls' resulting in an extra chapter in this edition.

The International General Certificate and the National General Certificate in Occupational Health and Safety now have a common hazards syllabus – GC2. The most significant change to this syllabus is the introduction in Element 1 of the hazards and control measures for works of a temporary nature. This replaces some of the more specific construction items and concentrates on the effects all types of temporary works can have on health and safety in the workplace. The other major change is the replacement of CHIP4 with the CLP Regulation. The tutor references for each element have been updated and there are two separate reference lists for the National General Certificate and the International General Certificate. There are also several minor changes which have been included in this edition.

Since the first edition of this book was published, NEBOSH has allowed us to use past NEBOSH examination questions at the end of each chapter. Over the last few years, it has become evident that a small number of candidates have memorised these questions and the contents of the accompanying examiner reports. As a result of this problem, NEBOSH has withdrawn permission to use past examination questions and changed the format of examiner reports. We have, therefore, included our own questions at the end of each chapter using the NEBOSH format. Candidates that can successfully answer these questions unaided should have no problems in the examinations.

NEBOSH is anxious to dispel the myths surrounding their examinations and have introduced regional meetings for course providers to introduce changes to the syllabuses and to answer any queries so that their students get the best possible preparation for the assessment tasks. The NEBOSH website is also a very useful channel of communication with course providers and students.

We have tried to reference best practice in each of the chapters to the Conventions, Recommendations and Codes of Practice of the International Labour Organisation (ILO), although we have also used good practice examples from the UK Health and Safety Executive and the European Union. We hope that this use of ILO material will help to publicise the excellent work that the ILO has done in furthering the occupational health and safety of workers throughout the world.

Finally, one of the objectives of the book is to provide a handbook for the use of any person who has health and safety as part of his/her responsibilities. Chapter 15 gives a useful summary of the principal ILO Occupational Health and Safety Conventions and some examples of legal frameworks in over 20 countries around the world. Chapter 18 gives some useful advice on searching the internet and information sources plus a range of forms for health and safety management. The forms can be accessed on the website and downloaded for use by readers. We thought that it would also be useful to add a few topics which are outside the syllabus, such as some additional environmental matters at the end of Chapter 13.

We hope that you find this new edition to be useful.

Phil Hughes
Ed Ferrett

Acknowledgements

The authors' grateful thanks go to Liz Hughes and Jill Ferrett for proof reading and patience and their administrative help during the preparation of this edition. The authors are particularly grateful to Liz for the excellent study guide that she has written for all NEBOSH students, which is included at the end of this book, and for the section on report writing in Chapter 5. Liz gained an honours degree in psychology at the University of Warwick, later going on to complete a Master's degree at the same university. She taught psychology in further and higher education, where most of her students were either returning to education after a gap of many years, or were taking a course to augment their existing professional skills. She went on to qualify as a social worker specialising in mental health, and later moved into the voluntary sector where she managed development for a number of years. Liz then helped to set up and manage training for the National Schizophrenia Fellowship (now called Rethink) in the Midlands.

We would also like to acknowledge the additional contribution made by Jill Ferrett for the help that she gave during the research for the book and with some of the word processing. Given her background in economics and higher education, her advice on certain legal and economic issues has been particularly valuable.

We would like to thank Teresa Budworth, the Chief Executive of NEBOSH, for her support during this third edition and various NEBOSH and HSE staff for their generous help and advice. Finally we would like to thank Stephen Vickers, the immediate past Chief Executive of NEBOSH, for his encouragement at the beginning of the project and Sadé Lee and all the production team at Routledge who have worked hard to translate our dream into reality.

About the authors

Phil Hughes MBE is a well-known UK safety professional with over 40 years worldwide experience as Head of Environment, Health and Safety at two large multinationals: Courtaulds and Fisons. Phil started work in health and safety in the Factory Inspectorate at the Derby District, UK, in 1969 and moved to Courtaulds in 1974. He joined IOSH in that year and became Chairman of the Midland Branch, then National Treasurer and was President in 1990–1991.

Phil was very active on the NEBOSH Board for over 10 years and served as Chairman from 1995 to 2001. He was also a Professional Member of the American Society of Safety Engineers for many years and has lectured widely throughout the world. Phil received the RoSPA Distinguished Service Award in May 2001 and was a Director and Trustee of RoSPA from 2003 to 2010. He received an MBE in the New Year Honours List 2005 for services to Health and Safety. Phil is a Chartered Fellow of IOSH.

Ed Ferrett is an experienced health and safety consultant who has practised for over 25 years. With a PhD and an honours degree in Mechanical Engineering from Nottingham University, Ed spent 30 years in Higher and Further education, retiring as the Head of the Faculty of Technology of Cornwall College in 1993. Since then he has been an independent consultant to several public and private sector organisations including construction businesses, the Regional Health and Safety Adviser for the Government Office (West Midlands), and was Chair of West of Cornwall Primary Care NHS Trust for 6 years until 2006.

Ed was a member of the NEBOSH Board from 1995 until 2010 and Vice Chair from 1999 to 2008. He has delivered many health and safety courses and has been a lecturer on NEBOSH courses for various course providers. He has been an External Examiner for an MSc course and BSc course in Health and Safety at two UK Universities, a Reporting Inspector for Independent Further and Higher Education with the British Accreditation Council and a NEBOSH Ambassador. Ed is a Chartered Engineer and a Chartered Member of IOSH.

How to use this book and what it covers

International Health and Safety at Work, third edition, is basically designed to:

1. cover the syllabus of the NEBOSH International General Certificate in Occupational Health and Safety (IGC1 and GC2) 2015 editions and other level 3 OSH awards;
2. go beyond the NEBOSH syllabus in covering some construction and environmental aspects of health and safety;
3. provide a good basis in OSH for students who wish to progress to the NEBOSH Diploma or a University first or second degree;
4. provide a text which more than covers the IOSH Managing Safely syllabus or other similar awards;
5. show the reader how some major OSH frameworks cover the subject including the USA, Europe and the UK;
6. give summaries of relevant ILO Conventions and national OSH legislation from 23 countries particularly where there are the most candidates for the IGC courses;
7. help students study, revise and sit the examinations;
8. provide guidance to students who carry out the practical assessment;
9. provide guidance for searching the internet and supplying a range of significant websites;
10. provide a good updated reference text for managers with OSH responsibilities and OSH practitioners in industry and commerce.

We expect the book to be used as a basis for training, and as further reference when students are back in their own workplaces. We believe that all questions can be answered from the material in the book but we would also urge students to study some of the documents given as reference sources at the end of each chapter. Also it would be advantageous to visit some of the websites where further detailed guidance is available.

Poor

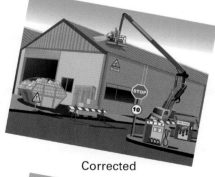

Corrected

Poor

Corrected

Animated inspection exercises

The websites featured in the text were found to be correct at the time of writing in January 2015.

There is a companion website http://www.routledge.com/cw/hughes/ where can be found: animated versions of the workplace inspection exercises in Chapter 5 showing corrected versions with labels on what has been corrected; copies of the forms in Chapter 18 in Word; a range of slides; many of the illustrations which are also available for downloading and use in training materials; and a range of multiple answer quizzes for revision purposes.

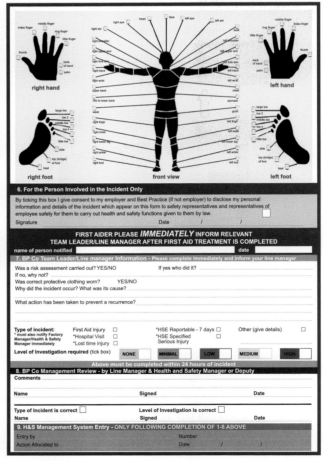

Usable images

Forms in Word that can be downloaded for use at work (see Chapter 18 for full set)

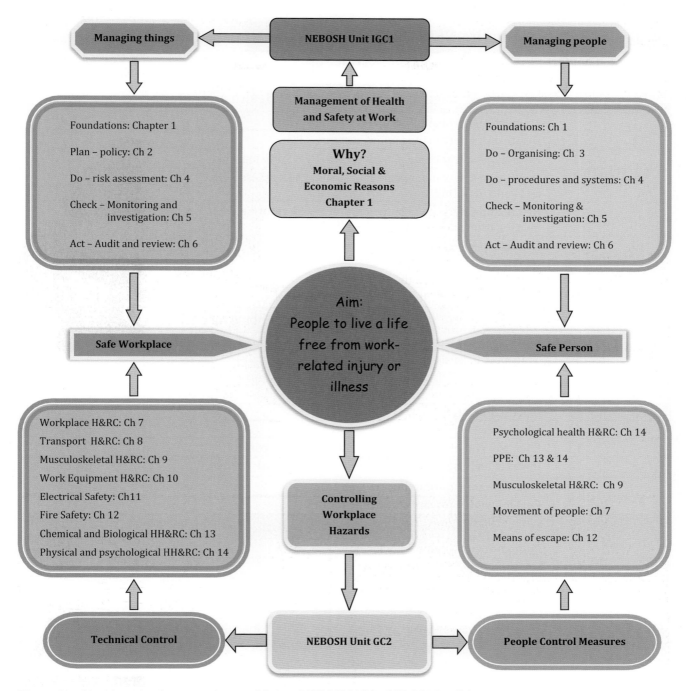

Figure X1 Health and safety overview and link to NEBOSH IGC, GC2 2015 syllabus

Element No	Chapter	Title	Recommended Study Hours
\multicolumn{4}{c}{**Syllabus for the NEBOSH International General Certificate**}			
\multicolumn{4}{c}{Unit IGC1 Management of International Health and Safety}			
1	1	Foundations in health & safety	7
2	2	Health and safety management systems – Plan	3
3	3	Health and safety management systems – Organising – Do1	7
3	4	Health and safety management systems – Risk assessment and controls – Do2	10
4	5	Health and safety management systems – Monitoring, investigation and recording – Check	5
5	6	Health and safety management systems – Audit and review – Act	4
		Minimum total tuition time for Unit IGC1	36
		Recommended private study time for IGC1	23
\multicolumn{4}{c}{Unit GC2 Control of international workplace hazards}			
1	7	Workplace hazards and risk control	8
2	8	Transport hazards and risk control	4
3	9	Musculoskeletal hazards and risk control	6
4	10	Work equipment hazards and risk control	6
5	11	Electrical safety	3
6	12	Fire safety	6
7	13	Chemical and biological health hazards and risk control	6
8	14	Physical and psychological health hazards and risk control	3
		Minimum total tuition time for Unit GC2	42
		Recommended private study time for GC2	26
\multicolumn{4}{c}{Unit GC3 The Practical Application}			
		Minimum total tuition time for Unit GC3	2
		Recommended private study time for GC3	4
		Minimum total tuition time	**80**
		Recommended private study time	**53**

Table X.1 The NEBOSH IGC syllabus is divided into three units. Each of the first two units (IGC1 and GC2) is further divided into a number of elements. For more details see the NEBOSH syllabus guide at www.nebosh.org.uk.

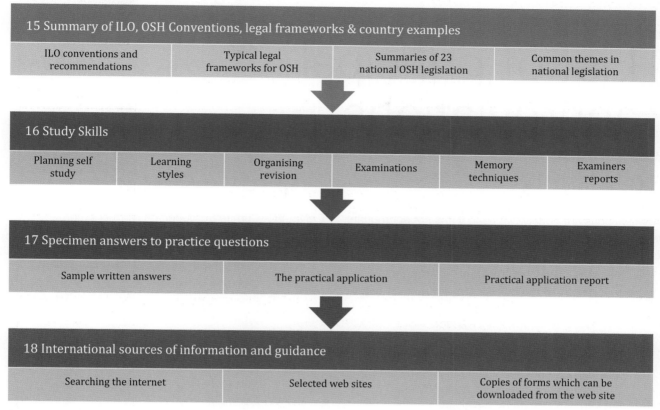

Figure X2 Chapters 15–18. The extra chapters in Figure X2 are designed to help the student understand their own OSH legislation. There is information on how to study, the standard for NEBOSH answers, how to research the internet and essential websites for OSH information, plus a range of form templates which can be freely used by readers.

List of principal abbreviations

Most abbreviations are defined within the text. Abbreviations are not always used if it is not appropriate within the particular context of the sentence. The most commonly used ones are as follows:

ACGIH	American Conference of Governmental Industrial Hygienists
ACL	Approved Carriage List
ACM	Asbestos Containing Material
ACoP	Approved Code of Practice
ADR	Accord dangereux routier (European agreement concerning the international carriage of dangerous goods by road)
AFNOR	French Standards Association
AFSSET	French Agency for Environmental and Occupational Health Safety
AIB	Asbestos Insulation Board
ALARP	As low as reasonably practicable
AND	European provisions concerning the international carriage of dangerous goods by inland waterways
ANSI	American National Standards Institute
ASCC	Australian Safety and Compensation Council
ASEAN	Association of Southeast Asian Nations
ASSE	American Society of Safety Engineers
ASTM	American Society for Testing and Materials (now ASTM International)
ATSDR	Agency for Toxic Substances and Disease Registry (USA)
ATEX	Atmosphere Explosive (used in the context of two European Directives, 94/9/EC and 1999/92/EC)
BA	Breathing apparatus
BAT	Best available techniques
BEBOH	British Examining Board in Occupational Hygiene
BIOH	British Institute of Occupational Hygiene
BLR	Blue light radiation
BPM	Best practicable means
BRE	Building Research Establishment
BSI	British Standards Institution
CAR	Control of Asbestos Regulations (UK)
CAS	Chemical Abstracts Service (USA)
CBI	Confederation of British Industry
CD	Consultative document
CDM	Construction (Design and Management) Regulations (UK)

CEN	Comité Européen de Normalisation
CENELEC	Comité Européen de Normalisation Électrotechnique
CIB	Chartered Institute of Building
CIRC	Centre International de Recherche sur le Cancer (France)
CIS	International Occupational Safety and Health Information Centre
CISDOC	International Labour Organisation database available on OSHROM
CISPR	Comité International Spécial des Perturbations Radioélectriques
CLAW	Control of Lead at Work Regulations
CLP	Classification Packaging and Labelling
CNS	Comité de Normalisation de la Soudure (France)
CO	Carbon monoxide
CONIAC	Construction Industry Advisory Committee
COSHH	Control of Substances Hazardous to Health Regulations (UK)
CPR	Cardiopulmonary resuscitation
CSA	Canadian Standards Association
CTS	Carpal tunnel syndrome
CVD	Cardiovascular disease
dB	Decibels
DB	Dry bulb
dB(A)	Decibel (A-weighted)
dB(C)	Decibel (C-weighted)
DSE	Display Screen Equipment
DSEAR	Dangerous Substances and Explosive Atmospheres Regulations (UK)
E&W	England and Wales
EA	Environment Agency
EAV	Exposure Action Value
EC	European Commission
EEF	Engineering Employers Federation
ELV	Exposure Limit Value
EMAS	Employment Medical Advisory Service
EPA	Environmental Protection Act 1990 (UK)
EU	European Union
EU-OSH	European Agency for Safety and Health at Work
FAO	Food and Agriculture Organisation of the United Nations
FOPS	Falling-Object Protective Structure(s)
FPO	Fire Prevention Officer

GATT General Agreement on Tariffs and Trade
GHGB Good Health is Good Business
GHS Globally Harmonised System of Classification and Labelling of Chemicals
GTAW Gas tungsten arc welding
HACCP Hazard analysis critical control point
HAI Hospital acquired infections
HAM Hand-held monitor
HASAC Health and Safety Advice Centre
HAV Hand–Arm Vibration
HGV Heavy Goods Vehicle
HOPE Healthcare, Occupational and Primary for Employees
HSCER Health and Safety (Consultation with Employees) Regulations (UK)
HSE Health and Safety Executive (UK)
HSG Health and Safety Guidance Booklet
HSW Act Health and Safety at Work etc. Act 1974 (UK)
HWL Healthy Working Lives
IAC Industry Advisory Committee
IChemE Institution of Chemical Engineers
IEA International Ergonomics Association
IEC International Electrotechnical Commission
IEE Institution of Electrical Engineers
ILO International Labour Organisation
INDG Industry Guidance
IOH Institution of Occupational Hygienists
IOSH Institution of Occupational Safety and Health
IPCS International Programme on Chemical Safety
IPMS Institution of Professionals, Managers and Specialists
IPPC Integrated pollution prevention and control
IPPR Institute for Public Policy Research
IPR Integrated pollution regulation
IRM Institute of Risk Management
IRSM International Institute of Risk and Safety Management
ISBN International Standard Book Number(ing)
ISO International Organisation for Standardisation
LD50 Lethal dose fifty
LDLo Lethal dose low
LEA Local Enterprise Agency
LEAL Lower Exposure Action Level
LEL Lower exposure limit
Leq Equivalent continuous sound level
Leq(8)hr Equivalent continuous sound level (normalised to 8 hours)
LEV Local exhaust ventilation
LNG Liquefied natural gas
LOLER Lifting Operations and Lifting Equipment Regulations (UK)
LPG Liquefied petroleum gas
MEL Maximum exposure limit
mg/m^3 Milligrams per cubic metre
MHOR Manual Handling Operations Regulations (UK)

MHSW Management of Health and Safety at Work Regulations (UK)
MORR Management of Occupational Road Risk
MOT Ministry of Transport (still used for vehicle tests in UK)
MSD Musculoskeletal disorder
MSDS Material Safety Data Sheet(s)
NEBOSH National Examination Board in Occupational Safety and Health
NIOSH National Institute for Occupational Safety and Health (USA)
NVQ National Vocational Qualification
OECD Organisation for Economic Cooperation and Development
OEL Occupational exposure limit
OES Occupational exposure standard
OHS Occupational Health Service
OHSAS Occupational Health and Safety Assessment Series
OHSLB Occupational Health and Safety Lead Body
OIAC Oil Industry Advisory Committee
OSH Occupational Safety and Health or Occupational Health and Safety
OSHA Occupational Safety and Health Administration (USA)
PPE Personal Protective Equipment
ppm Parts per million
PTFE Polytetrafluoroethylene
PUWER Provision and Use of Work Equipment Regulations (UK)
PVC Polyvinyl chloride
RCD Residual current device
RCE Report of the Committee of Experts
REACH Registration Evaluation and Authorisation and Restriction of Chemicals
RIDDOR Reporting of Injuries, Diseases and Dangerous Occurrences Regulations (UK)
ROES Representative(s) of Employee Safety
RoSPA Royal Society for the Prevention of Accidents
RPE Respiratory protective equipment
RRFSO Regulatory Reform Fire Safety Order (UK)
RTA Road traffic accident
SaHW Safe and Healthy Working
SFAIRP So far as is reasonably practicable
SMEs Small and medium-sized enterprises
SPL Sound Pressure Level
STEL Short-term Exposure Limit
SWL Safe working load
SWP Safe working pressure
TLV Threshold limit value
TUC Trades Union Congress
TWA Time-weighted average
UEAL Upper Exposure Action Level
UK United Kingdom
ULD Upper Limb Disorder
UNEP United Nations Environment Programme

List of principal abbreviations

UNESCO   United Nations Educational, Scientific and Cultural Organisation

VAWR   Vibration at Work Regulations (UK)

WAHR   Work at Height Regulations (UK)

WBV   Whole-Body Vibration

WEL   Workplace Exposure Limit

WHO   World Health Organisation

WRULD   Work-related Upper Limb Disorder

See ILO for more information on abbreviations and acronyms at: http://www.ilo.org/legacy/english/protection/safework/cis/products/safetytm/acronym.htm

Safety signs

PROHIBITION SIGNS

 No smoking

 No open flame; Fire, open ignition source and smoking prohibited

 No access for unauthorised persons

 No access for pedestrians

 Not drinking water

 No access for fork lift trucks and other industrial vehicles

 Do not extinguish with water

 Do not touch

WARNING SIGNS

 General danger

 Explosive material

 Radioactive material

 Laser beam

 Non-ionising radiation

 Magnetic field

 Obstacles

 Drop (fall)

Biological hazard

 Low temperature

 Electricity

 Fork lift trucks and other industrial vehicles

 Overhead load

 Toxic material

 Risk of fire/ flammable materials

 Corrosive substance

 Oxidizing substance

MEANS OF ESCAPE AND EMERGENCY EQUIPMENT (SAFE CONDITION) SIGNS

 Emergency exit (left hand)

 Emergency exit (right hand)

 First aid

 Emergency telephone

 Eyewash station

 Safety shower

 Stretcher

FIRE SAFETY SIGNS

 Fire extinguisher

 Fire hose reel

 Ladder

 Fire alarm call point

 Emergency fire telephone

MANDATORY ACTION SIGNS

 General mandatory action sign

 Wear safety footwear

 Wear protective gloves

 Wear protective clothing (overalls symbol)

 Wear ear protection

 Wear eye protection (not opaque image)

 Wear face shield

 Wear head protection

 Wear respiratory protection

 Wear safety harness

Pedestrians must use this route

SIGNS FOR GLOBALLY HARMONISED SYSTEM (GHS) OF CLASSIFICATION AND LABELLING OF CHEMICALS

 Warning

 Fatal if swallowed/ inhaled/in contact with skin

 May cause damage to organs/genetic defects/cancer/ damage to fertility or the unborn child

 Flammable

 May cause or intensify fire; oxidizer (oxidising image)

 Contains gas under pressure

 Heating may cause explosion

 May be corrosive to metals

Hazardous to the environment

HSE coded hand signals

A. General signals

START
Both arms are extended horizontally with the palms facing forwards

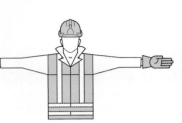

STOP
End of movement the right arm points upwards with the palm facing forwards

END
Of the operation both hands are clasped at chest height

B. Vertical movements

RAISE
The right arm points upwards with the palm facing forward and slowly makes a circle

LOWER
The right arm points downwards with the palm facing inwards and slowly makes a circle

VERTICAL DISTANCE
The hands indicate the relevant distance

C. Horizontal movements

MOVE FORWARDS
Both arms are bent with the palms facing upwards, and the forearms make slow movements towards the body

MOVE BACKWARDS
Both arms are bent with the palms facing downwards, and the forearms make slow movements away from the body

RIGHT (of the signalman)
The right arm is extended more or less horizontally with the palm facing downwards and slowly makes small movements to the right

LEFT (of the signalman)
The left arm is extended more or less horizontally with the palm facing downwards and slowly makes small movements to the left

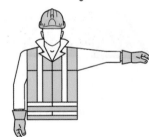

HORIZONTAL DISTANCE
The hands indicate the relevant distance

D. Danger

DANGER
Emergency stop – both arms point upwards with the palms facing forwards

Artwork from www.hse.gov.uk/pubns/priced/164.pdf

CHAPTER 1

Foundations in health and safety

1.1 The scope and nature of occupational health and safety ▶ 2

1.2 The moral, social and economic reasons for maintaining and promoting good standards of health and safety in the workplace ▶ 4

1.3 The role of national governments and international bodies in formulating a framework for the regulation of health and safety ▶ 9

1.4 Further information ▶ 16

1.5 Practice revision questions ▶ 17

Appendix 1.1 Scaffolds and ladders ▶ 18

This chapter covers the following NEBOSH learning objectives:

1. Outline the scope and nature of occupational health and safety
2. Explain the moral, social and economic reasons for maintaining and promoting good standards of health and safety in the workplace
3. Explain the role of national governments and international bodies in formulating a framework for the regulation of health and safety

1.1 The scope and nature of occupational health and safety

1.1.1 Introduction

Occupational health and safety is relevant to all branches of industry, business and commerce including traditional industries, information technology companies, hospitals, care homes, schools, universities, leisure facilities and offices.

The purpose of this chapter is to introduce the foundations on which appropriate health and safety management systems may be built. Occupational health and safety affects all aspects of work. In a low hazard organisation, health and safety may be supervised by a single competent manager. In a high hazard manufacturing plant, many different specialists, such as engineers (electrical, mechanical and civil), lawyers, medical doctors and nurses, trainers, work planners and supervisors, may be required to assist the professional health and safety practitioner in ensuring that there are satisfactory health and safety standards within the organisation.

There are many obstacles in the way of achieving good standards. The pressure of production or performance targets, financial constraints and the complexity of the organisation are typical examples of such obstacles. However, there are some powerful incentives for organisations to strive for high health and safety standards. These incentives are moral, legal and economic.

Corporate responsibility, a term used extensively in the 21st century world of work, covers a wide range of issues. It includes the effects that an organisation's business has on the environment, human rights and Third World poverty. Health and safety in the workplace is an important corporate responsibility issue.

Corporate responsibility has various definitions. However, broadly speaking, it covers the ways in which organisations manage their core business to add social, environmental and economic value in order to produce a positive, sustainable impact on both society and the business itself. Terms, such as 'corporate social responsibility', 'socially responsible business' and 'corporate citizenship', all refer to this concept.

The UK Health and Safety Executive's (HSE) mission is to ensure that the risks to health and safety of workers are properly controlled. In terms of corporate responsibility, it is working to encourage organisations to:

▶ improve health and safety management systems to reduce injuries and ill-health;
▶ demonstrate the importance of health and safety issues at board level;
▶ report publicly on health and safety issues within their organisation, including their performance against targets.

The HSE believes that effective management of health and safety:

▶ is vital to employee well-being;
▶ has a role to play in enhancing the reputation of businesses and helping them achieve high-performance teams;
▶ is financially beneficial to business.

The need for a global approach to occupational health and safety management was recognised as a logical and necessary response to increasing economic globalisation, while the benefits of systematic models of managing occupational health and safety became apparent as a result of the impact of ISO standards for quality and the environment. Current management science theories suggest that performance is better in all areas of business, including occupational health and safety, if it is measured and continuous improvement sought in an organised fashion. Successful management of health and safety is a top priority throughout the world and for this reason a comparison of the three major occupational health and safety management systems is covered in this chapter.

The International Labour Organisation (ILO), the World Health Organisation (WHO) and United Nations (UN) have estimated that there are 270 million occupational accidents and 160 million occupational diseases every year throughout the world – and these are recognised as relatively conservative estimates due to probable under-reporting. The ILO estimates that 2 million people die each year as a result of occupational accidents and work-related diseases. Table 1.1 shows the global trends over recent years in more detail.

In the USA in one year, approximately 2 million workers were victims of workplace violence. In the UK, in the same year, 1.7% of working adults (357,000 workers) were the victims of one or more incidents of workplace violence.

Table 1.1 Numbers of global work-related adverse events

Event	Average (daily)	Annually
Work-related death	5,000	2,000,000
Work-related deaths to children	60	22,000
Work-related accidents	740,000	270,000,000
Work-related disease	438,000	160,000,000
Hazardous substance deaths	1,205	440,000
Asbestos-related deaths	274	100,000

Source: ILO.

Ten per cent of all skin cancers are estimated to be attributable to workplace exposure to hazardous substances. Thirty seven per cent of miners in Latin America have silicosis, rising to 50% among miners over 50 years of age. In India 54.6% of slate pencil workers and 36.2% of stone cutters have silicosis.

In the course of the 20th century, industrialised countries saw a clear decrease in serious injuries, not least because of real advances in making the workplace healthier and safer. The challenge is to extend the benefits of this experience to the whole working world. However, 1984 saw the worst chemical disaster ever when 2,500 people were killed, and over 200,000 injured, in the space of a few hours at Bhopal. This affected not only the workers, but also their families, their neighbours and whole communities. More than thirty years later many people are still affected by the disaster and are dying as a result. The rusting remains of a once-magnificent plant exist as a reminder of the disaster.

Experience has shown that a strong safety culture is beneficial for workers, employers and governments alike. Various prevention techniques have proved themselves effective, both in avoiding workplace accidents and illnesses and improving business performance. Today's high standards in some countries are a direct result of long-term policies encouraging tripartite social dialogue, collective bargaining between trade unions and employers, and effective health and safety legislation backed by potent labour inspection. The ILO believes that safety management systems like ILO-OSH 2001 provide a powerful tool for developing a sustainable safety and health culture at the enterprise level and mechanisms for the continual improvement of the working environment.

The size of the health and safety 'problem' in terms of numbers of work-related fatalities and injuries and incidence of ill-health varies from country to country. However, these figures should be available from the statistics branch of the national regulator, similar to their availability in the UK from the annual report on health and safety statistics from the Health and Safety Executive (HSE).

Finally, a knowledge of international health and safety issues is important to international business travellers. In 2010, over 3.5 million employees undertook overseas trips and one-quarter of these trips were to high- or extreme-risk locations.

1.1.2 Some basic definitions

Before a detailed discussion of health and safety issues can take place, some basic occupational health and safety definitions are required.

▶ **Health** – The protection of the bodies and minds of people from illness resulting from the materials, processes or procedures used in the workplace.

▶ **Safety** – The protection of people from physical injury. The borderline between health and safety is ill-defined and the two words are normally used together to indicate concern for the physical and mental well-being of the individual at the place of work.

▶ **Welfare** – The provision of facilities to maintain the health and well-being of individuals at the workplace. Welfare facilities include washing and sanitation arrangements, the provision of drinking water, heating, lighting, accommodation for clothing, seating (when required by the work activity or for rest), eating and rest rooms. First-aid arrangements are also considered as welfare facilities.

▶ **Occupational or work-related ill-health** – This is concerned with those illnesses or physical and mental disorders that are either caused or triggered by workplace activities. Such conditions may be induced by the particular work activity of the individual or by activities of others in the workplace. They may be either physiological or psychological or a combination of both. The time interval between exposure and the onset of the illness may be short (e.g. asthma attacks) or long (e.g. deafness or cancer).

▶ **Environmental protection** – These are the arrangements to cover those activities in the workplace which affect the environment (in the form of flora, fauna, water, air and soil) and, possibly, the health and safety of employees and others. Such activities include waste and effluent disposal and atmospheric pollution.

▶ **Accident** – This is defined by the UK HSE as 'any unplanned event that results in injury or ill-health of people, or damage or loss to property, plant, materials or the environment or a loss of a business opportunity'. Other authorities define an accident more narrowly by excluding events that do not involve injury or ill-health. This book will always use the HSE definition. It is important to note that work-related accidents may not always occur at the place of work. *Commuting accidents* occur during work-related travel (usually by road).

▶ **Near miss** – This is any incident that could have resulted in an accident. Knowledge of near misses is very important as research has shown that, approximately, for every 10 'near miss' events at a particular location in the workplace, a minor accident will occur.

▶ **Dangerous occurrence** – This is a 'near miss' which could have led to serious injury or loss of life. Specified dangerous occurrences are always reportable to the enforcement authorities. Examples include the collapse of a scaffold or a crane or the failure of any passenger-carrying equipment.

▶ **Hazard and risk** –

A **hazard** is something with the *potential* to cause harm (this includes articles, substances, plant or machinery, methods of work, the working environment and other aspects of work organisation). Hazards take many forms including, for example, chemicals, electricity and noise. A hazard can be ranked relative to other hazards or to a possible level of danger.

▷ A **risk** is the *likelihood* of potential harm from that hazard being realised. Risk (or strictly the level of risk) is also linked to the severity of its consequences. This can involve the likelihood of a substance, activity or process to cause harm together with its resulting severity. A risk can be reduced and the hazard can be eliminated or controlled by good management.

It is very important to distinguish between a **hazard** and a **risk** – the two terms are often confused and activities such as construction work are frequently called **high risk** when they are **high hazard**. Although the hazard will continue to be high, the risks will be reduced as controls are implemented. The level of risk remaining when controls have been adopted is known as the **residual risk**. There should only be high residual risk where there is poor health and safety management and inadequate control measures.

1.2 The moral, social and economic reasons for maintaining and promoting good standards of health and safety in the workplace

The reasons for establishing good occupational health and safety standards are frequently identified as moral, social (and/or legal) and economic. Each will be discussed in turn.

1.2.1 Moral reasons

The moral reasons are supported by the occupational accident and disease rates. The ILO estimates that globally some 2.3 million people die from work-related

accidents or contract work-related diseases every year. There are around 270 million occupational accidents and 160 million victims of work-related illnesses annually.

According to the ILO, deaths due to work-related accidents and illnesses represent 3.9% of all deaths and 15% of the world's population suffers a minor or major occupational accident or work-related disease in any one year. A large number of the unemployed – up to 30% – report that they suffer from an injury or disease dating from the time at which they were employed. The number of fatal occupational accidents, especially in Asia and Latin America, is increasing.

The main (preventable) factors for accidents are:

▶ lack of a preventative safety and health culture;
▶ poor management systems;
▶ poor supervision and enforcement by the government.

The International Labour Organisation reckons that globally every year:

▶ 2 million people die from work-related diseases;
▶ 350,000 people die from occupational accidents;
▶ there are around 300 million occupational accidents, many of these resulting in extended absences from work; and
▶ 160 million people contract an occupational disease.

The safety and health conditions at work are very different between countries, economic sectors and social groups. Deaths and injuries take a particularly heavy toll in developing countries, where a large part of the population is engaged in hazardous activities, such as agriculture, construction, fishing and mining. Throughout the world, the poorest and least protected – often women, children and migrants – are among the most affected.

However, these figures are only the tip of an iceberg; the ILO believes that the unofficial number of dead, injured and ill workers is much higher.

Accident rates

An employee should not have to risk injury, ill-health or death at work, nor should others associated with the work environment. Accidents at work can lead to serious injury and even death. Although accident and work-illness rates are discussed in greater detail in later chapters, some trends are shown in Tables 1.2–1.4. Statistics are collected on all people who are injured at places of work, not just employees. A certain amount of caution must be used when quoting global workplace accident/incident data due to the significant level of under-reporting in many countries.

Table 1.2 shows the distribution of fatal occupational injuries and fatality incidence rates around the major global economic geographical areas in recent years.

Figure 1.1 At work

Whilst the figures will almost certainly have changed from one year to the next, the order of magnitude and the accident rates are still relevant.

Table 1.3 compares the number of fatalities due to work-related accidents and diseases for the same economic areas. Table 1.4 shows the top eight causes of work-related fatalities across the world.

It can be seen from these tables that the fatality rates are much lower in the developed or established market economies than in underdeveloped or emerging economies of Asia and South America. It is, however, important to stress that many of the hazardous industries have relocated from the developed to the emerging economies.

The statistics published by the International Labour Organisation tend also to include within work-related fatal accidents the number of deaths through accidents that occur while the victims are commuting to or from work. Various researchers have estimated that approximately 30% of all global work-related accidents are due to commuting accidents. A study for the ILO has estimated that there are approximately 350,000 work-related fatalities annually, of which 160,000 are due to fatal commuting accidents. It has been estimated

that a significant proportion of the work-related accidents recorded by the ILO across the globe annually in fact take place while commuting, and not on work premises at all.

Table 1.2 Distribution of fatal occupational injuries and incidence rates around the world in recent years

Region	Percentage share of fatal injuries	Fatal accidents per 100, 000 workers
Established market economies	5	4.5
Former socialist economies	5	13.0
India	11	11.0
China	26	10.0
Other Asia and islands	22	20.5
Sub-Saharan Africa	15	21.0
Latin-America and Caribbean	11	15.0
Middle Eastern crescent	5	17.0
WORLD	100	

Source: ILO.

Disease rates

Diseases related to work cause the most deaths among workers. Globally, as recorded by the ILO, around 2 million deaths per year are due to work-related disease – which corresponds to nearly 80% of *all* work-related deaths. However, the estimated figure for the annual number of sufferers from work-related disease across the planet is 160 million. This estimate is reasonable for the 2.8 billion global workforce, if non-recorded, part-time, child and other informal sector workers are taken into account.

Work-related ill-health and occupational disease can lead to absence from work and, in some cases, to death. Hazardous substances kill about 438,000 workers annually; asbestos alone claims 100,000 lives. Most of the other deaths are due to various forms of cancer. Another major killer is silicosis, which affects 37% of miners in Latin America.

Accident rates amongst female workers are different to those of male workers. ILO research appears to show that even in the same jobs, women tend to adopt more preventative and protective ways of carrying out work. On the other hand, with large numbers of women working in agriculture in developing countries, they are particularly vulnerable to communicable diseases, such as work-related malaria, hepatitis, schistosomiasis (infection by a water-borne parasite) and other bacterial, viral and vector-borne diseases. While men are more likely to be involved in fatal accidents and other work-related deaths, the everyday burden of musculoskeletal disorders (MSDs), stress and violence hits women hard, but the outcome may often be long-term disabilities rather than death.

In many countries, accident figures are reasonably well reported but the same cannot be said for occupational or work-related diseases.

In the UK, a study during 2007/08 showed that there were an estimated 2.1 million people suffering from work-related illness, of whom 563,000 were new cases in that year. This led to 28 million working days lost, compared to 6 million due to workplace injury. The largest groups of illness were vibration white finger, carpal tunnel syndrome and respiratory diseases. Some 8.8 million working days were lost due to musculoskeletal disorders causing each sufferer to have, on average, 21 days off work; 13.5 million working days were lost due to stress, depression and anxiety causing each sufferer to have, on average, 31 days off work. Recent research has shown that, in the UK, one in five people who are on sickness leave from work for six weeks will stay off work permanently and leave paid employment.

The World Health Organisation (WHO) has estimated that 37% of low back pain, 16% of hearing loss, 13% of chronic obstructive pulmonary disease, 11% of asthma and 8% of injuries are related to workplace activities. Several countries are in the process of developing or implementing policies that target emerging ill-health issues such as stress and musculoskeletal disorders, and promoting best remedial practices.

1.2.2 Social reasons

In all countries, employers owe a duty of care to each of their employees and others who might be affected by their undertaking, such as contractors and members of the public. This duty must not be assigned to others, even if a consultant is employed to advise on health and safety matters or if the employees are sub-contracted to work with another employer. This duty may be sub-divided into five groups. Employers must:

1. provide a safe place of work, including access and egress;
2. provide safe plant and equipment;
3. provide a safe system of work;
4. provide safe and competent fellow employees; and
5. provide adequate levels of supervision, information, instruction and training.

The requirement to provide competent fellow employees includes the provision of adequate supervision, instruction and training. In many countries, employers are responsible for the actions of their employees (known as **vicarious liability**) provided that the action in question took place during the normal course of their employment.

Employer responsibilities (see 1.3.5) are often mirrored in national legislation and apply even if the employee is working at a third-party premises or if he/she has been hired by the employer to work for another employer. These employer responsibilities indicate clear social reasons for sound health and safety management systems to protect employees, members of the public and, in some cases, the general environment.

Occupational health and safety requirements may be reinforced in national civil law and/or criminal law as many countries accept that without the extra 'encouragement' of potential regulatory action or litigation, many organisations would not act upon their implied moral obligations.

1.2.3 Financial reasons

Poor occupational health and safety performance results in additional costs to both public and private sectors of the economy of a country.

In addition to all of the personal suffering involved, this situation also has quite an impact on the global economy: the ILO estimates that about 4% of the global gross domestic product – that is, around US$2.8 trillion – is lost due to work accidents and occupational diseases.

Table 1.3 Global estimates of work-related fatalities caused by occupational accidents and work-related diseases in recent years

Region	Economically active population	Fatal occupational accidents	Fatal work-related diseases	Total work-related fatalities
Established market economies	419,732,002	15,879	281,364	297,243
Former socialist economies	183,089,714	17,416	148,194	165,610
India	443,860,000	40,133	261,891	302,024
China	740,703,800	90,295	386,645	476,940
Other Asia and islands	415,527,598	76,886	178,786	255,672
Sub-Saharan Africa	279,680,390	53,292	211,262	264,554
Latin-America and Caribbean	219,083,179	39,372	108,195	147,567
Middle Eastern crescent	135,220,721	17,977	120,725	138,702
WORLD	2,836,897,404	351,251	1,697,061	2,048,312

Source: ILO.

'Prevention is paying not only in human terms but also in better performance by business and national economic strength. Together we can make sure that decent work is safe work.'

(Thaksin Shinawatra, Former Prime Minister of Thailand)

Figure 1.2 Economic reasons for good health and safety management

Costs of accidents

The human cost of this daily adversity is vast and the economic burden of poor occupational safety and health practices is estimated at 4% of global GDP each year. The UK Institution of Occupational Safety and Health (IOSH) says too many employers are failing to appreciate the financial benefits of worker protection and are placing it way down on their list of priorities. According to IOSH, the Irish economy is losing €3.2 billion per year and the cost of health and safety failures per worker in Ireland is €1,711 because of workplace injury and ill-health. This equates to one million lost working days each year.

The cost of work-related injuries and illnesses in the United States has been estimated at some $300 billion (£200 billion) a year. In 2011, according to final fatality data from the Bureau of Labor Statistics, 4,693 workers were killed on the job – an average of 13 a day and corresponding to an incidence rate of 3.5 per 100,000 workers. This is almost six times the UK rate of 0.6 per 100,000 workers. In addition, an estimated 50,000 people died from occupational diseases and more than 3.8 million work-related injuries and illnesses were reported. By comparison, in the UK in 2011/12 there were 1.1 million cases of work-related illness and 323,000 reportable injuries.

Any accident or incidence of ill-health will cause both direct and indirect costs and incur an insured and an uninsured cost. It is important that all of these costs are taken into account when the full cost of an accident is calculated. In a study undertaken by the UK HSE, it was shown that indirect costs or hidden costs could be 36 times greater than direct costs of an accident. In other words, the direct costs of an accident or disease represent the tip of the iceberg when compared to the overall costs (Figure 1.3).

The European Union estimated that the cost of occupational accidents among the member states (15 at that time) was €55 billion each year – this did not cover the costs of work-related diseases. This estimate was probably an underestimate and is undoubtedly much higher now since the expansion of the EU. The study suggested that the costs due to work-related diseases may be twice as high as that caused by accidents at work. The ILO estimated that there were 120,000 annual deaths in the EU in the year in which the estimate was made, caused by work-related diseases, compared to 6,000 fatal occupational accidents during the same time frame.

Direct costs —predetable

These are costs which are directly related to the accident and may be insured or uninsured.

Insured direct costs normally include:

▶ claims on employers and public liability insurance;
▶ damage to buildings, equipment or vehicles;
▶ any attributable production and/or general business loss;
▶ the absence of employees.

Uninsured direct costs include:

▶ fines resulting from prosecution by the enforcement authority;
▶ sick pay;
▶ some damage to product, equipment, vehicles or process not directly attributable to the accident (e.g. caused by replacement staff);
▶ increases in insurance premiums resulting from the accident;

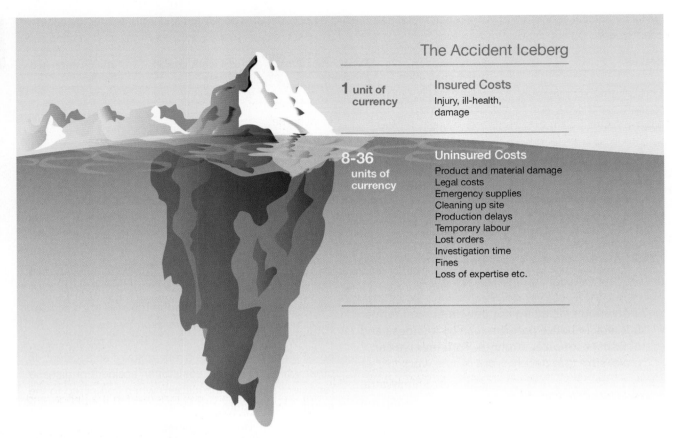

Figure 1.3 Insured and uninsured costs

▶ any compensation not covered by the insurance policy due to an excess agreed between the employer and the insurance company;

▶ legal representation following any compensation claim.

Table 1.4 Causes of global work-related fatalities

Fatality cause	Percentage share %
Cancer	32
Cardio-vascular	23
Accidents and violence	19
Contagious diseases	17
Respiratory system	7
Psychological disorders	1
Digestive system	1
Uro-genital system	0.4

Source: ILO.

Indirect costs ~ unpredictable

These are costs which may not be directly attributable to the accident but may result from a series of accidents. Again these may be insured or uninsured.

Insured indirect costs include:

▶ a cumulative business loss;

▶ product or process liability claims;

▶ recruitment of replacement staff.

Uninsured indirect costs include:

▶ loss of goodwill and a poor corporate image;

▶ accident investigation time and any subsequent remedial action required;

▶ production delays;

▶ extra overtime payments;

▶ lost time for other employees, such as first-aid staff, who tend to the needs of the injured person;

▶ the recruitment and training of replacement staff;

▶ additional administration time incurred;

▶ first-aid provision and training;

▶ lower employee morale possibly leading to reduced productivity.

Some of these items, such as business loss, may be uninsurable or too prohibitively expensive to insure. Therefore, insurance policies can never cover all of the costs of an accident or disease because either some items are not covered by the policy or the insurance excess is greater than the particular item cost.

Employers' liability insurance

In many countries, employers are required to take out employers' liability insurance to cover their liability in the event of accidents and work-related ill-health to employees and others who may be affected by their operations. This ensures that any employee, who successfully sues his/her employer following an accident, is assured of receiving compensation irrespective of the financial position of the employer.

Many countries have either fault or no-fault compensation schemes for workers involved in accidents. Knowledge of these schemes is important for those who work in more than one country.

Fault and no-fault injury compensation

In the UK, compensation for an injury following an accident is achieved by means of a successful legal action in a Civil Court. In such cases, injured employees sue their employer for negligence and the employer is found liable or at fault. This approach to compensation is adversarial, costly and can deter injured individuals of limited means from pursuing their claim. In a recent medical negligence claim in Ireland, costs were awarded against a couple who were acting on behalf of their disabled son, and they were faced with a bill for £3 million.

The spiralling cost of insurance premiums to cover the increasing level and number of compensation awards, despite the Woolf reforms in the UK (see Chapter 5), has led to another debate on the introduction of a no-fault compensation system. It has been estimated that in medical negligence cases, it takes, on average, six years to settle a claim and only 10% of claimants ever see any compensation.

No-fault compensation systems are available in many parts of the world, in particular New Zealand and some states of the USA. In these systems, amounts of compensation are agreed centrally at a national or state level according to the type and severity of the injury. The compensation is often in the form of a structured continuous award rather than a lump sum and may be awarded in the form of a service, such as nursing care, rather than cash.

The no-fault concept was first examined in the 1930s in the USA to achieve the award of compensation quickly in motor car accident claims without the need for litigation. It was introduced first in the state of Massachusetts in 1971 and is now mandatory in nine other states, although several other states have either repealed or modified no-fault schemes.

In 1978, the Pearson Commission in the UK rejected a no-fault system for dealing with clinical negligence even though it acknowledged that the existing tort system was costly, cumbersome and prone to delay. Its principal reasons for rejection were given as the difficulties in reviewing the existing tort liability system and in determining the causes of injuries.

In New Zealand, there was a general dissatisfaction with the workers' compensation scheme, which was similar to the adversarial fault-based system used in Australia and the UK. In 1974, a no-fault accident compensation system was introduced and administered by the Accident Compensation Corporation (ACC). This followed the publication of the Woodhouse Report in 1966 which advocated 24-hour accident cover for everybody in New Zealand. The Woodhouse Report suggested the following five principles for any national compensation system:

▶ community responsibility;
▶ comprehensive entitlement irrespective of income or job status;
▶ complete rehabilitation for the injured party;
▶ real compensation for the injured party; and
▶ administrative efficiency of the compensation scheme.

The proposed scheme was to be financed by channelling all accident insurance premiums to one national organisation (the ACC).

The advantages of a no-fault compensation scheme include:

1. Accident claims are settled much quicker than in fault schemes.
2. Accident reporting rates will improve.
3. Accidents become much easier to investigate because blame is no longer an issue.
4. Normal disciplinary procedures within an organisation or through a professional body are unaffected and can be used if the accident resulted from negligence on the part of an individual.
5. More funds are available from insurance premiums for the injured party and less used in the judicial and administrative process.

The possible disadvantages of a no-fault compensation system are:

1. There is often an increase in the number of claims, some of which may not be justified.
2. There is a lack of direct accountability of managers and employers for accidents.
3. Mental injury and trauma are often excluded from no-fault schemes because of the difficulty in measuring these conditions.
4. There is more difficulty in defining the causes of many injuries and industrial diseases than in a fault scheme.
5. The monetary value of compensation awards tends to be considerably lower than those in fault schemes (although this can be seen as an advantage).

No-fault compensation schemes also exist in countries such as Canada and the Scandinavian countries. However, an attempt to introduce such a scheme in New South Wales, Australia, was defeated in the legislative assembly.

1.3 The role of national governments and international bodies in formulating a framework for the regulation of health and safety

1.3.1 The role and function of the International Labour Organisation (ILO)

The ILO is a specialised agency of the United Nations that seeks to promote social justice through establishing and safeguarding internationally recognised human and labour rights. It was founded in 1919 by the Treaty of Versailles at the end of the First World War.

The motivation behind the creation of such an organisation was primarily humanitarian. Working conditions at the time were becoming unacceptable to a civilised society. Long hours, unsafe, unhygienic and dangerous conditions were common in low-paid manufacturing careers. Indeed, in the wake of the Russian Revolution, there was concern that such working conditions could lead to social unrest and even other revolutions. The ILO was created as a tripartite with governments, employers and workers represented on its governing body.

The ILO formulates international labour standards and attempts to establish minimum rights including freedom of association, the right to organise, collective bargaining, abolition of forced labour, equality of opportunity and treatment and other standards that regulate conditions across all work-related activities.

Representatives of all ILO member states meet annually in Geneva for the International Labour Conference, acting as a forum where social and labour questions of importance to the entire world are discussed. At this conference, labour standards are adopted and decisions made on policy and future programmes of work.

The ILO has 178 member states but if a country is not a member, the ILO still has influence as a source of guidance when social problems occur.

The main principles on which the ILO is based are:

1. labour is not a commodity;
2. freedom of expression and of association are essential to sustained progress;
3. poverty anywhere constitutes a danger to prosperity everywhere;
4. the 'war against want' is required to be carried out with unrelenting vigour within each nation, and by continuous and concerted international effort in which the representatives of workers and employers, enjoying equal status with those of governments, join with them in free discussion and democratic decision with a view to the promotion of the common welfare.

A campaign launched by the ILO has been to seek to eliminate child labour throughout the world. In particular, the ILO is concerned about children who work in hazardous working conditions, bonded child labourers and extremely young working children. It is trying to create a worldwide movement to combat the problem by:

- ▶ implementing measures which will prevent child labour;
- ▶ withdrawing children from dangerous working conditions;
- ▶ providing alternatives; and
- ▶ improving working conditions as a transitional measure towards the elimination of child labour.

The developed countries, for example the USA, Canada, Australia and those in the EU have health and safety legislation and standards in place. Unfortunately, the developing countries do not have similar cultures or safety standards as the recent Bangladesh clothes factory collapse indicated.

1.3.2 ILO Conventions and Recommendations

The international labour standards were developed for four reasons. The main motivation was to improve working conditions with respect to health and safety and career advancement. The second motivation was to reduce the potential for social unrest as industrialisation progressed. Third, the member states want common standards so that no single country has a competitive advantage over another due to poor working conditions. Finally, the union of these countries creates the possibility of a lasting peace based on social justice.

International labour standards are adopted by the International Labour Conference. They take the form of Conventions and Recommendations. At the present time, there are 187 Conventions and 198 Recommendations, some of which date back to 1919. See Chapter 15 for more information on the background of the ILO Conventions and Recommendations.

International labour standards contain flexible measures to take into account the different conditions and levels of development among member states. However, a government that ratifies a Convention must comply with all of its articles. Standards reflect the different cultural and historical backgrounds of the member states as well as their diverse legal systems and levels of economic development.

ILO occupational safety and health standards can be divided into four groups, and an example is given in each case:

1. Guiding policies for action – The Occupational Safety and Health Convention, 1985 (No. 155) and its accompanying Recommendation (No. 164) emphasise the need for preventative measures and a coherent national policy on occupational safety and health. They also stress employers' responsibilities and the rights and duties of workers.
2. Protection in given branches of economic activity – The Safety and Health in Construction Convention, 1988 (No. 167) and its accompanying Recommendation (No. 175) stipulate the basic principles and measures to promote safety and health of workers in construction.
3. Protection against specific risks – The Asbestos Convention, 1986 (No. 162) and its accompanying Recommendation (No. 172) gives managerial, technical and medical measures to protect workers against asbestos dust.

4. Measures of protection – Migrant Workers (Supplementary Provisions) Convention, 1975 (No. 143) aims to protect the safety and health of migrant workers.

ILO **Conventions** are international treaties signed by ILO member states and each country has an obligation to comply with the standards that the Convention establishes.

In contrast, ILO **Recommendations** are non-binding instruments that often deal with the same topics as Conventions. Recommendations are adopted when the subject, or an aspect of it, is not considered suitable or appropriate at that time for a Convention. Recommendations guide the national policy of member states so that a common international practice may develop and be followed by the adoption of a Convention.

ILO standards are the same for every member state and the ILO has consistently opposed the concept of different standards for different regions of the world or groups of countries.

The standards are modified and modernised as needed. The governing body of the ILO periodically reviews individual standards to ensure their continuing relevance.

Supervision of international labour standards is conducted by requiring the countries that have ratified Conventions to periodically present a report with details of the measures that they have taken, in law and practice, to apply each ratified Convention. In parallel, employers' and workers' organisations can initiate contentious proceedings against a member state for its alleged non-compliance with a convention it has ratified. In addition, any member country can lodge a complaint against another member state which, in its opinion, has not ensured, in a satisfactory manner, the implementation of a Convention which both of them have ratified. Moreover, a special procedure exists in the field of freedom of association to deal with complaints submitted by governments or by employers' or workers' organisations against a member state, whether or not the country concerned has ratified the relevant Conventions. Finally, the ILO has systems in place to examine the enforcement of international labour standards in specific situations.

The ILO also publishes Codes of Practice, guidance and manuals on health and safety matters. These are often used as reference material by either those responsible for drafting detailed Regulations or those who have responsibility for health and safety within an organisation. They are more detailed than either Conventions or Recommendations and suggest practical solutions for the application of ILO standards. Codes of Practice indicate 'what should be done'. They are developed by tripartite meetings of experts and the final publication is approved by the ILO governing body.

For example, the construction industry has a Safety and Health in Construction Convention, 1988 (No. 167) that obliges signatory ILO member states to comply with the construction standards laid out in the Convention – the Convention is a relatively brief statement of those standards. The accompanying Recommendation (No. 175) gives additional information on the Convention statements. The Code of Practice gives more detailed information than the Recommendation. This can best be illustrated by contrasting the coverage of scaffolds and ladders by the three documents shown in Appendix 1.1.

The ILO Codes of Practice and guidelines on health and safety matters that are relevant to the International General Certificate are:

▶ Safety and Health in Construction (ILO Code of Practice);
▶ Ambient factors in the workplace (ILO Code of Practice);
▶ Safety in the Use of Chemicals at Work (ILO Code of Practice);
▶ Recording and Notification of Occupational Accidents and Diseases;
▶ Ergonomic Checkpoints;
▶ Work Organisation and Ergonomics;
▶ Occupational Safety and Health Management Systems (ILO Guidelines).

Important ILO Conventions (C) and Recommendations (R) in the field of occupational safety and health include:

▶ C 115 Radiation Protection and (R114), 1960;
▶ C 120 Hygiene (Commerce and Offices) and (R120), 1964;
▶ C 139 Occupational Cancer and (R147), 1974;
▶ C 148 Working Environment (Air Pollution, Noise and Vibration) and (R156), 1977;
▶ C 155 Occupational Safety and Health and (R164), 1981;
▶ C 161 Occupational Health Services and (R171), 1985;
▶ C 162 Asbestos and (R172), 1986;
▶ C 167 Safety and Health in Construction and (R175), 1988;
▶ C 170 Chemicals and (R177), 1990;
▶ C 174 Prevention of Major Industrial Accidents and (R181), 1993;
▶ C 176 Safety and Health in Mines and (R176), 1995;
▶ C 184 Safety and Health in Agriculture and (R192), 2001;
▶ C 187 Promotional Framework for Occupational Safety and Health and (R197), 2006;
▶ R97 Protection of Workers' Health Recommendation, 1953;
▶ R102 Welfare Facilities Recommendation, 1956;
▶ R31 List of Occupational Diseases Recommendation, 2002.

See Chapter 15 for more details on occupational health and safety Conventions and Recommendations. ILO member states are expected in due course either to pass these as new national law or to adjust the wording of existing national law to conform to them.

The ILO Conventions, Recommendations and Codes of Practice often use the terms 'practicable' and 'reasonably practicable' when describing duties or recommendations. These two terms together with 'absolute' form a hierarchy or levels of duty which also appear in the health and safety laws of several countries. These levels of duty have particular and precise meanings.

1.3.3 The role of the European Union (EU)

The European Commission has presented a new strategic framework on Health and Safety at Work 2014–2020 to better protect the more than 217 million workers in the EU from work-related accidents and diseases.

The strategic framework identifies three major health and safety at work challenges:

- to improve implementation of existing health and safety rules, in particular by enhancing the capacity of micro and small enterprises to put in place effective and efficient risk prevention strategies;
- improve the prevention of work-related diseases by tackling new and emerging risks without neglecting existing risks; and
- the ageing of the EU's workforce.

The strategic framework proposes to address these challenges with a range of actions under seven key strategic objectives:

- further consolidating national health and safety strategies;
- providing practical support to small and micro enterprises to help them to better comply with health and safety rules;
- improving enforcement by member states, for example by evaluating the performance of national labour inspectorates;
- simplifying existing legislation where appropriate to eliminate unnecessary administrative burdens, while preserving a high level of protection for workers' health and safety;
- addressing the ageing of the European workforce and improving prevention of work-related diseases to tackle existing and new risks such as nanomaterials, green technology and biotechnologies;
- improving statistical data collection to have better evidence and developing monitoring tools; and
- reinforcing coordination with international organisations (such as the International Labour Organisation (ILO), the World Health Organisation (WHO) and the Organisation for Economic Cooperation and Development (OECD)) and partners to contribute to reducing work accidents and

occupational diseases and to improving working conditions worldwide.

An EU study has shown that the development of the health and safety legislative system in the UK has produced one of the best health and safety performances in the EU.

1.3.4 Levels of duty

The three levels of duty are absolute, practicable and reasonably practicable.

Absolute duty

This is the highest level of duty and, often, occurs when the risk of injury is so high that injury is inevitable unless safety precautions are taken. Many health and safety management requirements contained in national health and safety law place an absolute duty on the employer. Examples of this include the need for written safety policies, risk assessments, information and training.

Practicable

This level of duty is more often used than the absolute duty as far as the provision of safeguards is concerned and, in many ways, has the same effect. A duty that 'the employer ensures, so far as is practicable, that any control measure is maintained in an efficient state' means that if the duty is technically possible or feasible then it must be done irrespective of any difficulty, inconvenience or cost.

Reasonably practicable

This is the most common level of duty and means that if the risk of injury is very small compared to the cost, time and effort required to reduce it, no action is necessary. It is important to note that money, time and trouble must 'grossly outweigh', not balance, the risk (see Figure 1.4). This duty requires judgement on the part of the employer (or their adviser) and clearly needs a risk assessment to be undertaken with conclusions

Figure 1.4 Diagrammatic view of 'reasonably practicable'

noted. Continual monitoring is also required to ensure that risks do not increase.

1.3.5 Employers' duties and responsibilities

The principal general duties of employers under the ILO Recommendation 164 are:

(a) to provide and maintain workplaces, machinery and equipment, and use work methods, which are as safe and without risk to health as is reasonably practicable;

(b) to give necessary instruction and training that takes into account the functions and capabilities of different categories of workers;

(c) to provide adequate supervision of work practices ensuring that proper use is made of relevant occupational health and safety measures;

(d) to institute suitable occupational health and safety management arrangements appropriate to the working environment, the size of the undertaking and the nature of its activities; and

(e) to provide, without any cost to the worker, adequate personal protective clothing and equipment which are reasonably necessary when workplace hazards cannot be otherwise prevented or controlled.

A comprehensive and detailed list of employers' duties is given in chapters 3 and 15.

1.3.6 Workers' rights and responsibilities

Workers' rights

In 1998, ILO member states adopted the *Declaration on Fundamental Principles and Rights at Work* and agreed to uphold a set of core labour standards. These are human rights and form basic workers' rights. The ILO is actively campaigning for improvements in the areas covered by the Declaration.

The Declaration covers four areas:

1. **Freedom of Association** – The right of workers and employers to form and join organisations of their choice is an integral part of a free and open society and is linked to the recognition of the right to collective bargaining.
2. **Forced Labour** – The ILO is pressing for effective national laws and stronger enforcement mechanisms, such as legal sanctions and vigorous prosecution against those who exploit forced labourers.
3. **Discrimination** – Hundreds of millions of people suffer from discrimination in the world of work. Discrimination stifles opportunities, wasting the human talent needed for economic progress and accentuating social tensions and inequalities.
4. **Child Labour** – There are more than 200 million children working throughout the world, many

full-time. They are deprived of adequate education, good health and basic freedoms. Of these, 126 million – or one in every 12 children worldwide – are exposed to hazardous forms of child labour, work that endangers their physical, mental or moral well-being.

The rights of workers are also contained in the ILO Code of Practice – 'Ambient factors in the workplace'. The code specifies that workers and their representatives should have the right to:

(a) be consulted regarding any hazards or risks to health and safety from hazardous factors at the workplace;

(b) enquire into and receive information from the employer regarding any hazards or risks to health and safety from hazardous factors in the workplace. This information should be provided in forms and languages easily understood by the workers;

(c) take adequate precautions, in cooperation with their employer, to protect themselves and other workers against hazards or risks to their health and safety;

(d) request and be involved in the assessment of hazards and risks to health and safety by the employer and/or the competent authority, and in any subsequent control measures and investigations;

(e) be involved in the inception and development of workers' health surveillance, and participate in its implementation;

(f) be informed in a timely, objective and comprehensible manner:
 (i) of the reasons for any examinations and investigations relating to the health hazards involved in their workplace;
 (ii) individually of the results of medical examinations, including pre-assignment medical examinations, and of the subsequent assessment of health.

In accordance with national laws and regulations, workers should have the right:

(a) to bring to the attention of their representatives, employer or competent authority any hazards or risks to health and safety at the workplace;

(b) to appeal to the competent authority if they consider that the measures taken and the means employed by the employer are inadequate for the purposes of ensuring health and safety at work;

(c) to remove themselves from a hazardous situation when they have good reason to believe that there is an imminent and serious risk to their health and safety and inform their supervisor immediately;

(d) in the case of a health condition, such as sensitisation, to be transferred to alternative work that does not expose them to that hazard, if such work is available and if the workers concerned have the qualifications or can reasonably be trained for such alternative work;

(e) to compensation if the case referred to in (d) above results in loss of employment;

(f) to adequate medical treatment and compensation for occupational injuries and diseases resulting from hazards at the workplace;

(g) to refrain from using any equipment or process or substance which can reasonably be expected to be hazardous, if relevant information is not available to assess the hazards or risks to health and safety.

The ILO code recommends that workers should receive training and, where necessary, retraining in the most effective methods which are available for minimising risks to health and safety from hazards at work. Female workers should have the right, in the case of pregnancy or during lactation, to work that is not hazardous to the health of the unborn or nursing child where such work is available, and to return to their previous jobs at the appropriate time.

Workers responsibilities

Employees or workers have specific responsibilities under the ILO Convention 187, which are to:

(a) take reasonable care for their own safety and that of other persons who may be affected by their acts or omissions at work;

(b) comply with instructions given for their own health and safety and those of others and with health and safety procedures;

(c) use safety devices and protective equipment correctly and not to render them inoperative;

(d) report forthwith to their immediate supervisor any situation which they have reason to believe could present a hazard and which they cannot themselves correct; and

(e) report any accident or injury to health which arises in the course of or in connection with work.

1.3.7 Role of enforcement agencies

The legal framework

The framework for regulating health and safety will vary across the world, for example European countries use the EU framework, the Pacific Rim countries tend to use the USA framework, whereas the Caribbean countries follow the UK framework. The course provider should be able to describe the legal and regulatory framework appertaining to any particular country.

Most legislation is driven by a framework of Acts, Regulations and support material including Codes of Practice and Standards, as illustrated in Figure 1.5. Within Europe there is another layer of legislation known as Directives, above the member states' own legislation. These are legally binding on each member state. The US system of federal and individual state legislation is very similar.

Figure 1.5 Typical health and safety legal framework

Regulatory authorities and safety management systems

The role of the national regulatory authority is crucial to the successful implementation of an occupational health and safety management system. In many parts of the world, such as Southeast Asia, formal adoption of a recognised management system is required with third party auditing by government-approved auditors.

Figure 1.6 The inspector inspects

In the USA, organisations with approved management systems may be exempted from normal inspections by the Occupational Safety and Health Administration.

In the UK, there has been a movement from prescriptive legislation to risk assessment by the employer and this movement is now occurring in many parts of the world including the EU. A management system is an essential tool to achieve this movement and such a system is implied in the UK Management of Health and Safety at Work Regulations. Countries such as Canada, Australia, New Zealand and Norway have developed occupational health and safety management systems as an encouragement for such self-regulation. The ILO-OSH 2001 system has been adopted by Germany, Sweden, Japan, Finland, Korea, China, Mexico, Costa Rica, Brazil, Indonesia, Vietnam, Malaysia, India, Thailand, the Czech Republic, Poland and Russia.

It is likely that more countries and multinational companies will expect occupational health and safety management systems to be adopted either as a legal duty or an implied duty following rulings from local courts of law.

Consequences of non-compliance

The direct consequences of non-compliance are similar for most countries and include the following:

▶ loss of competitive advantage;
▶ inability to compete for certain contracts;
▶ fines; and
▶ imprisonment.

There are other indirect consequences and these include:

▶ a high staff turnover;
▶ poor levels of health and safety competence;
▶ a high sickness, ill-health and absentee rate among the workforce;
▶ a lack of compliance with relevant health and safety law and the safety rules and procedures of the organisation; and
▶ higher insurance premiums.

Figure 1.7 Good standards prevent harm and save money

1.3.8 **International standards**

As has already been described under 1.3.2, the International Labour Organisation (ILO) has many Conventions, Recommendations and Codes of Practice. It also has a health and safety management system (ILO-OSH 2001) which is described in Chapter 2.

Another international standards organisation is the ISO. ISO (International Organisation for Standardisation) is a **network** of the national standards institutes of 162 countries with a Central Secretariat based in Geneva, Switzerland. The standards produced by the ISO attempt to ensure that desirable characteristics of products and services such as quality, environmental friendliness, safety, reliability, efficiency and interchangeability, are maintained by suppliers. These standards should make the development, manufacture and supply of products and services more efficient, safer and cleaner. They should also safeguard consumers (and users in general) of those products and services.

ISO has more than 18,500 International Standards of which the vast majority are specific to a particular product, material or process. ISO's work programme includes standards for traditional activities, such as agriculture and construction, mechanical engineering, manufacturing and distribution, transport, medical devices, information and communication technologies, and standards for good management practice and services.

However, ISO 9001 (quality) and ISO 14001 (environment) are 'generic management system standards' meaning that the same standard can be applied to any organisation, large or small, whatever its product or service, in any sector of activity, and whether it is a business enterprise, a public administration or a government department. ISO 9001 contains a generic set of requirements for implementing a quality management system and ISO 14001 for an environmental management system.

Standards ensure important characteristics of products and services such as quality, environmental friendliness, safety, reliability, efficiency and interchangeability – and at an economic cost. When products, systems, machinery and devices work well and safely, it is often because they meet standards. Thus ISO standards:

▶ make the development, manufacturing and supply of products and services more efficient, safer and cleaner;
▶ facilitate trade between countries and make it fairer;
▶ provide governments with a technical base for health, safety and environmental legislation, and conformity assessment;
▶ share technological advances, good management practice and disseminate innovation;
▶ safeguard consumers, and users, of products and services; and

▶ make life simpler by providing solutions to common problems.

Many organisations are also implementing an occupational health and safety management system (OHSMS) as part of their risk management strategy to address changing legislation and to protect their workforce. Such a system, discussed in detail in Chapter 2, promotes a safe and healthy working environment by providing a framework that allows an organisation to consistently identify and control its health and safety risks, reduce the possibility of accidents, aid legislative compliance and improve overall performance. ISO introduced OHSAS 18001 in 1999 as an internationally recognised assessment specification for occupational health and safety management systems. It was developed by a selection of leading trade bodies, international standards and certification bodies to address a gap where no third-party certifiable international standard existed. OHSAS 18001 provides an approved 'best-practice' framework for delivering a practical workable solution to reduce risk across an organisation. It was designed to be compatible with ISO 9001 and ISO 14001, and addressed the following key areas:

▶ planning for hazard identification, risk assessment and risk control;
▶ OHSAS management programme;
▶ structure and responsibility;
▶ training, awareness and competence;
▶ consultation and communication;
▶ operational control;
▶ emergency preparedness and response; and
▶ performance measuring, monitoring and improvement.

The most recent update to this standard occurred in 2007 (BS OHSAS 18001:2007) and the principal changes were a much greater emphasis on 'health' rather than just 'safety', and significantly improved alignment to ISO 14001:2004 (to enable organisations to develop 'integrated management systems'). ISO 45001 replaces OHSAS 18001 as the definitive occupational health and safety standard in 2016.

ISO standards provide technological, economic and societal benefits for:

▶ businesses – to develop and offer products and services meeting specifications that have wide international acceptance in their sectors;
▶ governments – the technological and scientific bases that underpin occupational health and safety legislation;
▶ developing countries – the characteristics that products and services will be expected to meet on export markets; and
▶ consumers – conformity of products and services to international standards provides assurance about their quality, safety and reliability.

Certification by a body, such as ISO, can also provide other financial benefits, such as helping to reduce insurance costs. However, one of the biggest benefits from a commercial perspective is that OHSAS 18001 certification will demonstrate credibility and confidence to support the retention of existing clients and help win new business.

1.3.9 Sources of information on national standards

There are several sources of information on national standards including:

▶ health and safety legislation;
▶ HSE publications, such as Approved Codes of Practice, guidance documents, leaflets, journals, books and their websites;
▶ European and British Standards;
▶ International Labour Organisation (ILO);
▶ Occupational Safety and Health Administration (USA);
▶ European Agency for Safety and Health (EU);
▶ Worksafe (Western Australia);
▶ health and safety magazines and journals;
▶ information published by trade associations, employer organisations and trade unions;
▶ specialist technical and legal publications;
▶ information and data from manufacturers and suppliers; and
▶ the internet and encyclopedias.

1.4 Further information

ILO Guidelines on Occupational Safety and Health Management Systems (ILO-OSH 2001) http://www.ilo.org/global/publications/ilo-bookstore/order-online/books/WCMS_PUBL_9221116344_EN/lang--en/index.htm

ILOLEX (ILO database of International Law) http://www.ilo.org/ilolex/index.htm

Occupational Health and Safety Assessment Series (OHSAS 18000): Occupational Health and Safety Management Systems OHSAS 18001:2007 ISBN 978-0-5805-9404-5 OHSAS18002:2008 ISBN 978-0-5806-2686-9

Occupational Safety and Health Convention (C155), ILO http://www.ilo.org/dyn/normlex/en/f?p=NORMLEXPUB:12100:0::NO:12100:P12100_ILO_CODE:C155

Occupational Safety and Health Recommendation (R164), ILO http://www.ilo.org/dyn/normlex/en/f?p=NORMLEXPUB:12100:0::NO:12100:P12100_INSTRUMENT_ID:312502

1.5 Practice revision questions

1. In relation to occupational health and safety, **explain**, using an example in **EACH** case, the meaning of the following terms:
   (a) hazard
   (b) risk
   (c) welfare
   (d) work-related ill-health
   (e) near miss.

2. (a) **Explain** the reasons that organisations are required to have employer liability insurance.
   (b) **Outline SIX** possible direct **AND SIX** possible indirect costs to an organisation following a serious accident in the workplace. Identify those costs that are insurable and those that are not.
   (c) **Identify** possible costs to an organisation resulting from poor health and safety standards.

3. **Outline** the ways in which the International Labour Organisation (ILO) formulates a framework of international standards in occupational health and safety. *Pg.9*

4. **Explain**, giving an example in each case, the differences between the levels of duty – absolute, practicable and reasonably practicable.

5. (a) **Outline** the principal duties and responsibilities of employers to safeguard the health and safety of workers in their organisation.
   (b) **Outline** the main health and safety responsibilities and rights of the workers in the organisation.

6. (a) **Outline** reasons why national or state governments have health and safety laws.
   (b) **Outline** ways in which national/state governments try to help ensure organisations comply with health and safety laws.
   (c) **Identify** the actions that a national health and safety enforcement agency may take if a workplace is deemed to be unsafe.

7. **Identify THREE** sources of information on national/state health and safety standards.

8. (a) **Identify** the possible consequences of an accident to:
   (i) the injured workers;
   (ii) their employer.
   (b) **Outline** actions management may take to prevent similar accidents.

APPENDIX 1.1 Scaffolds and ladders

1.1 Convention (Safety and Health in Construction) (167)

Article 14

1. Where work cannot be done safely on or from the ground or from part of a building or other permanent structure, a safe and suitable scaffold shall be provided and maintained, or other equally safe and suitable provision shall be made.
2. In the absence of alternative safe means of access to elevated working places, suitable and sound ladders shall be provided. They shall be properly secured against inadvertent movement.
3. All scaffolds and ladders shall be constructed and used in accordance with national laws and regulations.
4. Scaffolds shall be inspected by a competent person in such cases and at such times as shall be prescribed by national laws or regulations.

1.2 Recommendation (Safety and Health in Construction) (175)

Scaffolds

16. Every scaffold and part thereof should be of suitable and sound material and of adequate size and strength for the purpose for which it is used and be maintained in a proper condition.
17. Every scaffold should be properly designed, erected and maintained so as to prevent collapse or accidental displacement when properly used.

18. The working platforms, gangways and stairways of scaffolds should be of such dimensions and so constructed and guarded as to protect persons against falling or being endangered by falling objects.
19. No scaffold should be overloaded or otherwise misused.
20. A scaffold should not be erected, substantially altered or dismantled except by or under the supervision of a competent person.
21. Scaffolds as prescribed by national laws or Regulations should be inspected, and the results recorded, by a competent person:
    (i) before being taken into use;
    (ii) at periodic intervals thereafter;
    (iii) after any alteration, interruption in use, exposure to weather or seismic condition or any other occurrence likely to have affected their strength or stability.

1.3 Code of Practice – Safety and Health in Construction

The Code of Practice covers scaffolds and ladders under the following topics over five pages:

1. general provisions
2. materials
3. design and construction
4. inspection and maintenance
5. lifting appliances on scaffolds
6. prefabricated scaffolds
7. use of scaffolds
8. suspended scaffolds.

CHAPTER 2

Health and safety management systems – PLAN

2.1 The key elements of a health and safety management system ▶ 20

2.2 The purpose and importance of setting policy for health and safety ▶ 29

2.3 The key features and appropriate content of an effective health and safety policy ▶ 30

2.4 Further information ▶ 36

2.5 Practice revision questions ▶ 36

Appendix 2.1 Health and Safety Policy checklist ▶ 38

> This chapter covers the following NEBOSH learning objectives:
>
> 1 Outline the key elements of a health and safety management system
> 2 Explain the purpose and importance of setting policy for health and safety
> 3 Describe the key features and appropriate content of an effective health and safety policy

2.1 The key elements of a health and safety management system

2.1.1 Introduction

Every organisation should have a policy for the systematic management of health and safety so that health and safety risks may be effectively addressed and controlled. A good health and safety policy will indicate the goals that the adopted health and safety management system hopes to achieve. The health and safety policy and management system will also complement those policies in areas such as quality, the environment and human resources. As for those areas, for the health and safety policy and its associated management system to be successful, it must have realistic goals and the active support and involvement of all levels of management within the organisation.

Planning to implement the health and safety policy is the key element of any health and safety management system. An effective planning system for health and safety requires organisations to set up, operate and maintain a management system that can detect, eliminate and control hazards and risks. This process is particularly important when there are long latency periods before a health problem becomes apparent, such as asbestosis or lung cancer caused by asbestos fibre inhalation.

Since the main objective of any health and safety plan is to prevent people being harmed at work, workplace precautions must be provided and maintained at the point of risk. Risks are created in the workplace as resources and information are used to create products and services in the manufacturing process (see Figure 2.1). This business activity model also applies to other industries including construction, education, leisure, hospitals and local authorities. Workplace precautions to match the hazards and risks are needed at each stage of the business activity. They can include machine guards, electrical safeguards, flammable liquid storage, dust controls, safety instructions and systems of work.

This chapter is concerned with the planning for managing occupational health and safety within an organisation. The preparation of an effective health and safety policy is an essential first step in the formulation of the health and safety management system. The policy will only remain as words on paper, however good the intentions, until there is an effective organisation set up to implement and monitor its requirements. Some policies are written so that most of the wording concerns strict requirements laid on employees and only a few vague words cover managers' responsibilities. For a policy to be effective, it must be honoured in the spirit as well as the letter. A good health and safety policy together with an effective occupational health and safety management system will also enhance the performance of the organisation in areas other than health and safety, help with the personal development of the workforce and reduce financial losses. The chapter begins with a description of the key elements of such a management system.

2.1.2 The key elements of all health and safety management systems

The latest ILO Convention No. 187 (2003) and its accompanying Recommendation (No. 197) in 2006 applies a similar approach to the management of national occupational health and safety systems to ensure they are improved through a continuous cycle of policy review, evaluation and action for improvement. The essential elements of a national occupational health and safety management system are shown in Table 2.1. The principles are based on two fundamental concepts:

▶ to develop a preventative safety and health culture; and
▶ to apply a systems approach to managing occupational health and safety nationally.

See Chapter 15 for more information on legal frameworks around the world.

Most of the key elements required for effective health and safety management are very similar to those required for good quality finance and general business management. Commercially successful organisations usually have good health and safety management systems in place. The principles of good and effective management provide a sound basis for the improvement of health and safety performance.

In the publication *Successful health and safety management – HSG65*, the UK HSE recommend a four-step approach to occupational health and safety management known as:

▶ **PLAN** – establish standards for health and safety management based on risk assessment and legal requirements.
▶ **DO** – implement plans to achieve objectives and standards.

Table 2.1 Essential elements of any national occupational health and safety management system

> - Relevant occupational health and safety legislation
> - One or more authorities responsible for occupational health and safety
> - Regulatory compliance mechanisms, including systems of inspection
> - A national advisory body to advise on occupational health and safety issues
> - A national occupational health and safety information and advisory service
> - National occupational health services
> - An organisation for the collection and analysis of data on occupational injuries and diseases
> - Provision of relevant insurance or social security schemes covering occupational injuries and diseases
> - Research budgets for occupational health and safety topics
> - Formal systems for the provision of occupational health and safety training.

> - CHECK – measure progress with plans and compliance with standards.
> - ACT – review against objectives and standards and take appropriate action.

The *Plan, Do, Check, Act* for occupational health and safety management forms the basis of the three occupational health and safety management systems HSG65, OHSAS 18001 and ILO-OSH 2001.

All recognised occupational health and safety management systems have some basic and common elements. These are:

> - a planning phase;
> - a performance phase;
> - a performance assessment phase; and
> - a performance improvement phase.

The planning phase – PLAN

The planning phase always includes a **policy statement** which outlines the health and safety aims, objectives and commitment of the *organisation* and

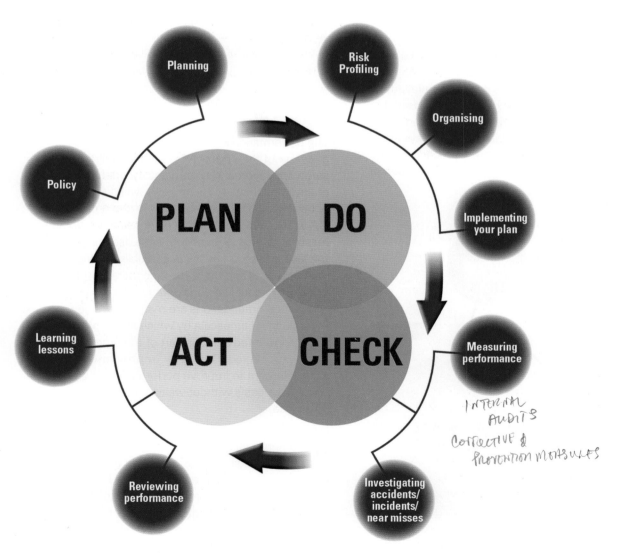

Figure 2.1 The Plan, Do, Check, Act management cycle

lines of responsibility. **Hazard identification** and **risk assessment** takes place during this phase and the significant hazards may well be included in the policy statement. It is important to note that in some reference texts, in particular those in languages other than English, the whole process of hazard identification, risk determination and the selection of risk reduction or control measures is termed 'risk assessment'. However, all three occupational health and safety management systems described in this chapter refer to the individual elements of the process separately and use the term 'risk assessment' for the determination of risk only.

At the **planning stage**, emergency procedures should be developed and relevant health and safety legal requirements and other standards identified together with appropriate benchmarks from similar industries. An **organisational structure** must be defined so that health and safety responsibilities are allocated at all levels of the organisation and issues such as competent persons and health and safety training are addressed. Realistic targets should be agreed within the organisation and be published as part of the policy.

The performance phase – DO

The performance phase will only be successful if there is good **communication** at and between all levels of the organisation. This implies employee participation as both *worker representatives* and on **safety committees**. Effective communication with the workforce, for example with clear **safe systems of work** and other health and safety procedures, will not only aid the implementation and operation of

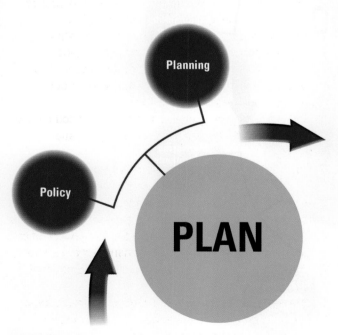

Figure 2.2 This chapter covers the PLAN stage of the management cycle

the plan but also produce continual improvement of performance – a key requirement of all occupational health and safety, quality and environmental management systems. There should also be effective communication with other stakeholders, such as regulators, contractors, customers and trade unions. The performance phase must be *monitored* on a regular basis since this will indicate whether there is an effective occupational health and safety management system and a good *health and safety culture* within the organisation.

The performance assessment phase – CHECK

The performance assessment phase may be either **active** or **reactive** or, ideally, a mixture of both. Active assessment includes work-based **inspections and audits**, regular health and safety committee meetings, feedback from training sessions and a constant review of risk assessments. Reactive assessment relies on records of accidents, work-related injuries and ill-health as well as near miss and any enforcement notices. Any recommended remedial or preventative actions, following an investigation, must be implemented immediately and monitored regularly. Strictly speaking Audit comes in CHECK but the NEBOSH syllabus places it in ACT. Therefore, Audit is discussed in detail in Chapter 6.

The performance improvement phase – ACT

The performance improvement phase involves a **review** of the effectiveness of the health and safety management system and the identification of any weaknesses. The review, which should be undertaken by the management of the organisation, will assess whether targets have been met and the reasons for any underperformance. Issues such as the level of resources made available, the vigilance of supervisors and the level of cooperation of the workforce should be considered at the review stage. When recommendations are made, the review process must define a timescale by which any improvements are implemented and this part of the process must also be monitored. **Continual improvement** implies a commitment to improve performance on a proactive continuous basis without waiting for a formal review to take place.

Most management systems include an **audit** requirement, which may be either internal or external, or both. The audit process examines the effectiveness of the whole management process and may act as a control on the review process. Many inquiry reports into health and safety management issues have asserted that health and safety performance should be subject to audit in the same way that financial performance must be audited.

Table 2.2 Location and contents of the key elements of a health and safety management system in chapters 2, 3, 4, 5 and 6

Plan, Do, Check, Act	Topics covered	Chapter
PLAN	Key elements of a health and safety management system	2
	Purpose and importance of setting a policy for health and safety	
	Key features and appropriate content of an effective health and safety policy	
DO	Organisational health and safety roles and responsibilities of employers, directors, managers and supervisors	3 & 4
	Concept of health and safety culture and its significance in the management of health and safety in an organisation	
	Human factors which influence behaviour at work	
	How health and safety behaviour at work can be improved	
	Importance of planning	
	Principles and practice of risk assessment	
	General principles of control and hierarchy of risk reduction measures	
	Sources of health and safety information	
	Safe systems of work	
	Permits to work	
	Emergency procedures and arrangement for contacting the emergency services	
	Requirements for, and effective provision of, first-aid in the workplace	
CHECK	Active and reactive monitoring	5
	Investigating incidents	
	Recording and reporting incidents	
ACT	Health and safety auditing	6
	Review of health and safety performance	

2.1.3 Major occupational health and safety management systems

There are three major occupational health and safety management systems that are in use globally:

▶ ILO-OSH 2001 was developed by the ILO after an extensive study of many occupational health and safety management systems used across the world. It was established as an international system following the publication of *Guidelines on occupational safety and health management systems* in 2001. It is very similar to OHSAS 18001.

▶ OHSAS 18001:2007 has been developed in conjunction with the ISO 9000 series for quality management and the ISO 14000 series for environmental management.

▶ HSG65, which has been developed by the UK HSE. The syllabuses for the International General Certificate and the chapter headings in this textbook have followed the elements of HSG65.

ILO-OSH 2001

The ILO has a considerable influence on the development of employment law in many countries throughout the world. As companies have become more international in terms of both markets and production bases, this influence of the ILO has increased and its working standards have become accepted in many parts of the world. Many of the ILO standards cover health and safety issues and are used in the International General Certificate course. As mentioned earlier, the ILO-OSH 2001 guidelines offer a recommended occupational health and safety management system based on an ILO survey of several contemporary schemes including HSG65 and OHSAS 18001. There are, therefore, many common elements between the three schemes. The guidelines are not legally binding and are not intended to replace national laws, Regulations or accepted standards.

At the national level, the guidelines should be used to establish a national framework for occupational health and safety management systems, preferably supported by national laws and Regulations. They should also provide guidance for the development of voluntary arrangements to strengthen compliance with Regulations and standards leading to continual improvement in occupational health and safety performance.

The ILO recognises that a management system can usually only be successful in a country if there is some form of national policy on health and safety and occupational health and safety management systems. The ILO recommends, therefore, that the following general principles and procedures be established:

1. the implementation and integration of occupational health and safety management systems as part of the overall management of an organisation;
2. the introduction and improvement of voluntary arrangements for the systematic identification, planning, implementation and improvement of occupational health and safety activities at both national and organisational levels;
3. the promotion of worker participation in occupational health and safety management at organisational level;
4. the implementation of continual improvement without unnecessary bureaucracy and cost;
5. the encouragement of national labour inspectors to support the arrangements at organisational level for a health and safety culture within the framework of an occupational health and safety management system;
6. the evaluation of the effectiveness of the national policies at regular intervals;
7. the evaluation of the effectiveness of occupational health and safety management systems within those organisations which are operating within the country;

8. the inclusion of all those affected by the organisation, such as contractors, members of the public and temporary workers, at the same level of health and safety provision as employees.

Figure 2.3 shows the key elements of the ILO-OSH 2001 occupational health and safety management system. Each element will be discussed in turn but since it has been developed from HSG65 and OHSAS 18001 among other systems, there are many similarities. The elements are:

1. **Policy** – There is a more specific emphasis on worker participation, which is seen as an essential element of the management system and should be referenced in the policy statement. It is expected that workers and their health and safety representatives should have sufficient time and resources allocated to them so that they can participate actively in each element of the management system. The formation of a health and safety committee is also a recommended part of the system. The occupational health and safety management system should be compatible with or integrated with other management systems operating in the organisation.

2. **Organising** – There is much in common with both HSG65 and OHSAS 18001, with responsibility, accountability, competence, training and communication being key parts of this element. There is a specific responsibility to provide effective supervision to ensure the protection of the health and safety of workers and to establish prevention and health promotion programmes. Workers should have access to any records, such as accident and monitoring records, which are relevant to their occupational health and safety while respecting the need for confidentially. Health and safety training must be available to all members of the organisation and be provided during normal working hours at no cost to employees.

3. **Planning and implementation** – Unlike other systems, the ILO system combines planning and implementation. Following an initial review of any existing health and safety management system, a plan should be developed to remedy any deficiencies found. The plan should support compliance with national laws and Regulations and include continual improvement of health and safety performance. It should contain measurable objectives which are realistic and achievable and, as with the other occupational health and safety management systems, hazard identification and risk assessment. There should also be an adequate provision of resources and technical support. The plan must be capable of accommodating the impact on health and safety of any internal changes in the organisation, such as new processes, new technologies and amalgamations with other organisations, or

external changes due, for example, to changes in national laws or Regulations. As with OHSAS 18001, emergency and procurement arrangements and detailed arrangements for the selection and supervision of contractors must be included in the health and safety plan.

4. **Evaluation** – This is very similar to the performance measurement phase of HSG65 with a greater emphasis on the health and welfare of the worker. The recommendations concerning the investigation of work-related injuries, ill-health, diseases and incidents are identical to those for the management review element of OHSAS 18001.

5. **Action for improvement** – Arrangements should be introduced and maintained for any preventative and corrective action to be undertaken identified by performance monitoring, audits and management reviews of the health and safety management system. Arrangements should also be in place for the continual improvement of the management system. More details on the factors to be considered for continual improvement are given later in this chapter.

6. **Audit** – The ILO recommends that an audit should be performed by competent and trained personnel at agreed and regular intervals. It should cover all elements of the management system including worker participation, communication, procurement, contracting and continual improvement. The audit conclusions must state whether the health and safety management system is effective in meeting the organisational health and safety policy and objectives and promotes full worker participation. The audit should also check that there is compliance with national laws and regulations and that there has been a satisfactory response to earlier audit findings.

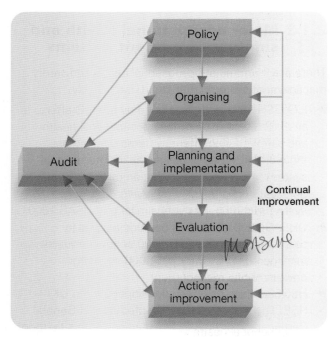

Figure 2.3 Key elements of ILO-OSH 2001

Additional features of ILO-OSH 2001

1. **Worker participation** – This is an essential element of the occupational health and safety management system in the organisation. The employer should ensure that workers and their health and safety representatives are consulted, informed and trained on all aspects of occupational health and safety, including emergency arrangements, associated with their work. Arrangements should be made for workers and their health and safety representatives to have the time and resources to participate actively in the processes of organising, planning and implementation, evaluation and action for improvement of the occupational health and safety management system. Finally, the employer should ensure, as appropriate, the establishment and efficient functioning of a health and safety committee and the recognition of workers' health and safety representatives, in accordance with national laws and practice.

2. **Documentation** – Depending upon the size and activities of the organisation, occupational health and safety management system documentation should be established and maintained, and may cover:
   (a) the health and safety policy and objectives of the organisation;
   (b) the allocated key health and safety management roles and responsibilities for the implementation of the occupational health and safety management system;
   (c) the significant occupational health and safety hazards/risks arising from the organisation's activities, and the arrangements for their prevention and control; and
   (d) arrangements, procedures, instructions or other internal documents used within the framework of the occupational health and safety management system.

The occupational health and safety management system documentation should be:

(a) clearly written and presented in a way that is understood by those who have to use it; and
(b) periodically reviewed, revised as necessary, communicated and readily accessible to all appropriate or affected members of the organisation.

Occupational health and safety records should be established, managed and maintained locally and according to the needs of the organisation. They should be identifiable and traceable, and their retention times should be specified. Workers should have the right to access records relevant to their working environment and health, while respecting the need for confidentiality. Occupational health and safety records may include:

(a) records arising from the implementation of the occupational health and safety management system;
(b) records of work-related injuries, ill-health, diseases and incidents;
(c) records arising from national laws or regulations dealing with occupational health and safety;
(d) records of workers' exposures, surveillance of the working environment and workers' health; and
(e) the results of both active and reactive monitoring.

OHSAS 18001:2007

The OHSAS 18001 occupational health and safety management system elements can be linked to the Plan, Do, Check, Act model as follows:

- ▶ policy (PLAN)
- ▶ planning (PLAN)
- ▶ implementation and operation (DO)
- ▶ checking and corrective action (CHECK)
- ▶ management review (ACT)
- ▶ continual improvement (ACT)

OHSAS 18001 provides an approved 'best-practice' framework for delivering a practical workable solution to reduce risk across an organisation. Those organisations that create, implement and comply with a health and safety management system compliant to OHSAS 18001 will have a structure in place that provides reassurance that all necessary policies, procedures and controls are in place to continuously improve health and safety.

ISO 45001 replaces OHSAS 18001 as the definitive occupational health and safety standard in 2016. Its purpose is to enable an organisation to proactively improve its occupational health and safety performance in preventing injury and ill-health. As with HSG65 it applies the Plan, Do, Check, Act management model. It places more emphasis on seeking continual improvement by addressing occupational health and safety risks and other initiatives such as health, education and training. It also stresses the importance of risk management and continuing risk assessment. A new area of focus of ISO 45001 is the context of an organisation, for example its supply chains and local communities.

HSG65

The HSE document HSG65, *Managing for health and safety*, describes the occupational health and safety management system used extensively in the UK. In 2013, the HSE's guidance in HSG65 moved from using the POPMAR model (Policy, Organising, Planning, Measuring performance, Auditing and Review) to the 'Plan, Do, Check, Act' approach as mentioned earlier.

The HSE have argued that the move towards Plan, Do, Check, Act achieves a better balance between the systems and behavioural aspects of management. It also treats health and safety management as an integral part of good management generally, rather than as a stand-alone system.

It is clear that there is much that is common between the three management systems discussed and any differences are in emphasis. HSG65 was the first of the three systems and the emphasis was on legal compliance. OHSAS 18001 introduced the concept of continual improvement, integration with other management systems and certification. ILO-OSH 2001 combines the features of HSG65 and OHSAS 18001 and stresses the importance of worker participation for the system to be effective.

2.1.4 Planning a health and safety management system

The planning (**PLAN**) of a health and safety management system involves the development and implementation of suitable management arrangements to control risks by the introduction of workplace preventative measures. Such measures must be proportionate and appropriate to the hazards and risks in the organisation. Any management system must be able to accommodate changes in the activities of the organisation, its personnel or legislative requirements.

There are three key questions that need to be answered during the planning process:

▶ Where is the organisation at the moment in terms of the management of health and safety?
▶ Where does it want to be?
▶ How does it get there?

The answers to these questions at site level will be different to those at the headquarters in a large multi-site organisation. Therefore they should be asked at all levels in an organisation to obtain a complete understanding of the effectiveness of health and safety management across the whole organisation. For planning to be effective it must be properly coordinated throughout the organisation to avoid overlaps and omissions.

The health and safety planning process comprises three sections:

▶ correct information about the existing situation;
▶ suitable benchmarks against which to make comparisons (see later in this chapter);
▶ competent people to carry out the analysis and make judgements.

The organisation has to compare the current situation against both the health and safety management framework described earlier (Figure 2.1) and specific legal requirements. The key requirements of this framework are as follows:

1. **A clear health and safety policy** – Evidence shows that a well-considered policy contributes to business efficiency and continual improvement throughout the organisation. It helps to minimise financial losses arising from avoidable accidents and demonstrates

to the workforce that accidents are not necessarily the fault of any individual member of the workforce. Such a management attitude could lead to an increase in workforce cooperation, job satisfaction and productivity. This demonstration of senior management involvement offers evidence to all stakeholders that responsibilities to people and the environment are taken seriously by the organisation. A good health and safety policy helps to ensure that there is a systematic approach to risk assessment and sufficient resources, in terms of people and money, have been allocated to protect the health and safety and welfare of the workforce. It can also support quality improvement programmes which are aimed at continual improvement.

2. **A well-defined health and safety organisation** – The shared understanding of the organisation's values and beliefs, at all levels of the organisation, is an essential component of a positive health and safety culture. For a positive health and safety culture to be achieved, an organisation must have clearly defined health and safety responsibilities so that there is always management control of health and safety throughout the organisation. The formal organisational structure should be such that the promotion of health and safety becomes a collaborative activity between the workforce, safety representatives and the managers. An effective organisation is essential for the **DO** stage of the management system and will be noted for good staff involvement and participation, high-quality communications, the promotion of competency and the empowerment of all employees to make informed contributions to the work of the organisation.

3. **A clear health and safety plan** – This involves the setting and implementation of performance standards and procedures using an effective occupational health and safety management system. The plan is based on risk assessment methods to decide on priorities and set objectives for controlling or eliminating hazards and reducing risks. Measuring success requires the establishing of performance standards against which achievements can be identified.

4. **The measurement of health and safety performance** – This includes both active (sometimes called proactive) and reactive monitoring to see how effectively the occupational health and safety management system is working. It forms part of the **CHECK** stage of the management system. Active monitoring involves looking at the premises, plant and substances plus the people, procedures and systems. Reactive monitoring discovers through investigation of accidents and incidents why controls have failed. It is also important to measure the organisation against its own long-term goals and objectives.

3. REASONS FOR DOING / NOT DOING THINGS   (TO PROMOTE GOOD H&S)

1. MORAL
2. SOCIAL
3. LEGAL
4. FINANCIAL

5. **Reviewing performance** – The results of monitoring and independent audits should be systematically reviewed to evaluate the performance of the management system against the objectives and targets established by the health and safety policy. It is at the review or **ACT** stage of the management system that the objectives and targets set in the health and safety policy may be changed. Changes in the health and safety environment in the organisation, such as an accident, should also trigger a performance review – this is discussed in more detail later in this chapter. Performance reviews are part of any organisation's commitment to continuous improvement. Comparisons should be made with internal performance indicators and the external performance indicators of similar organisations with exemplary practices and high standards.

6. **Auditing** – An independent and structured audit of all parts of the health and safety management system reinforces the review process. Such audits may be internal or external (the differences are discussed in Chapter 6). The audit assesses compliance with the health and safety management arrangements and procedures. If the audit is to be really effective, it must assess both the compliance with stated procedures and the performance in the workplace. It will identify weaknesses in the health and safety policy and procedures and identify unrealistic or inadequate standards and targets.

2.1.5 Other aspects of health and safety management systems

Any occupational health and safety management system will fail unless there is a positive health and safety culture within the organisation and the active involvement of internal and external stakeholders.

A structured and well-organised occupational health and safety management system is essential for the maintenance of high health and safety standards within all organisations and countries. Some systems, such as OHSAS 18001, offer the opportunity for integration with quality and environmental management systems. This enables a sharing of resources although it is important that technical activities, such as health and safety risk assessment, are only undertaken by persons trained and competent in that area.

For an occupational health and safety management system to be successful, it must address workplace risks and be 'owned' by the workforce. It is, therefore, essential that the audit process examines shop floor health and safety behaviour to check that it mirrors that required by the health and safety management system.

Finally, whichever system is adopted, there must be continual improvement in health and safety performance if the application of the occupational health and safety management system is to succeed in the long term.

2.1.6 Other key characteristics of a health and safety management system

The four basic elements common to all occupational health and safety management systems, as described earlier in this chapter, contain the different activities of the system together with the detailed arrangements and activities required to deliver those activities. However, there are four key characteristics of a *successful* occupational health and safety management system:

▶ a positive health and safety culture;
▶ the involvement of all stakeholders;
▶ an effective audit; and
▶ continual improvement.

A positive health and safety culture

In Chapter 3, the essential elements for a successful health and safety culture are detailed and discussed. In summary, they are:

▶ leadership and commitment to health and safety throughout the organisation;
▶ an acceptance that high standards of health and safety are achievable;
▶ the identification of all significant hazards facing the workforce and others;
▶ a detailed assessment of health and safety risks in the organisation and the development of appropriate control and monitoring systems;
▶ a health and safety policy statement outlining short- and long-term health and safety objectives. Such a policy should also include national codes of practice and health and safety standards;
▶ relevant communication and consultation procedures and training programmes for employees at all levels of the organisation;
▶ systems for monitoring equipment, processes and procedures and the prompt rectification of any defects found;
▶ the prompt investigation of all incidents and accidents and reports made detailing any necessary remedial actions.

Some of these essential elements form part of the health and safety management system but unless all are present within the organisation, it is unlikely that occupational health and safety will be managed successfully no matter which system is introduced. The chosen management system must be effective in reducing risks in the workplace or else it will be nothing more than a paper exercise.

The involvement of stakeholders

There are a number of internal and external stakeholders of the organisation who will have an interest and influence on the introduction and development of the occupational health and safety management system.

The **internal stakeholders** include:

▶ **Directors and trustees of the organisation** – Following several national and international reports on corporate governance in recent years (such as the UK's Combined Code of Corporate Governance 2003), the measurement of occupational health and safety performance and the attainment of health and safety targets have been recognised as being as important as other measures of business performance and targets. A report on health and safety performance should be presented at each board meeting and be a periodic agenda item for sub-committees of the Board such as audit and risk management.

▶ **The workforce** – Without the full cooperation of the workforce, including contractors and temporary employees, the management of health and safety will not be successful. The workforce is best qualified to ensure and provide evidence that health and safety procedures and arrangements are actually being implemented at the workplace. So often this is not the case even though the occupational health and safety managements system is well designed and documented. Worker representatives can provide useful evidence on the effectiveness of the management system at shop floor level and they can also provide a useful channel of communication between the senior management and the workforce. One useful measure of the health and safety culture is the enthusiasm with which workers volunteer to become, and continue to be, Occupational Health and Safety worker representatives. For a representative to be really effective, some training is essential. Organisations which have a recognised trade union structure will have fewer problems with finding and training worker representatives. More information on the duties of worker representatives is given in Chapter 3.

▶ **Health and safety professionals** – Such professionals will often be appointed by the organisation to manage the occupational health and safety management system and monitor its implementation. They will, therefore, have a particular interest in the development of the system, the design of objectives and the definition of targets or goals. Unless they have a direct-line management responsibility, which is not very common, they can only act as advisers, but still have a positive influence on the health and safety culture of the organisation. The appointed health and safety professional also liaises with other associated competent persons, such as for electrical appliance and local exhaust ventilation testing.

The **external stakeholders** include:

▶ **Insurance companies** – As compensation claims increase, insurance companies are requiring more and more evidence that health and safety is being effectively managed. There is increasing evidence that insurance companies are becoming less prepared to offer cover to organisations with a poor health and safety record and/or occupational health and safety management system.

▶ **Investors** – Health and safety risks will, with other risks, have an effect on investment decisions. Increasingly, investment organisations require evidence that these risks are being addressed before investment decisions are made.

▶ **Regulators** – In many parts of the world, particularly in the Far East, national regulators and legislation require certification to a recognised international occupational health and safety management system standard. In many countries, regulators use such a standard to measure the awareness of a particular organisation of health and safety issues.

▶ **Customers** – Customers and others within the supply chain are increasingly insisting on some form of formal occupational health and safety management system to exist within the organisation. The construction industry is a good example of this trend. Much of this demand is linked to the need for corporate social responsibility and its associated guidelines on global best practice.

▶ **Neighbours** – The extent of the interest of neighbours will depend on the nature of the activities of the organisation and the effect that these activities have on them. The control of noise, and dust and other atmospheric contaminants are examples of common problem areas which can only be addressed on a continuing basis using a health, safety and environmental management system.

▶ **International organisations** – The United Nations, the ILO, the International Monetary Fund, the World Bank and the World Health Organisation are all examples of international bodies which have shown a direct or indirect interest in the management of occupational health and safety. In particular, the ILO is keen to see minimum standards of health and safety established around the world. The ILO works to ensure for everyone the right to work in freedom, dignity and security – which includes the right to a safe and healthy working environment. More than 70 ILO Conventions and Recommendations relate to questions of safety and health. In addition, the ILO has issued more than 30 Codes of Practice on Occupational Health and Safety. For more information see their website www.ilo.org/safework.

There is a concern of many of these international organisations that as production costs are reduced by relocating operations from one country to another, there is also a lowering in occupational health and safety standards. The introduction of internationally recognised occupational health and safety management systems will help to alleviate such fears.

An effective audit

An effective audit is the final step in the occupational health and safety management system control cycle. The use of audits enables the reduction of risk levels and the effectiveness of the occupational health and safety management system to be improved. The auditing process is discussed in detail in Chapter 6 as is the important choice that must be made between the use of internal or external auditors.

Continual improvement

Continual improvement is recognised as a vital element of all occupational health and safety management systems if they are to remain effective and efficient as internal and external changes affect the organisation. Internal changes may be caused by business reorganisation, such as a merger, new branches and changes in products, or by new technologies, employees, suppliers or contractors. External changes could include new or revised legislation, guidance or industrial standards, new information regarding hazards or campaigns by regulators.

Continual improvement is part of the **ACT** stage of the management system and need not necessarily be done at high cost or add to the complexity of the management system. Continual improvement is discussed in detail in Chapter 6.

2.1.7 The benefits and problems associated with occupational health and safety management systems

Occupational health and safety management systems have many benefits, of which the principal ones are:

▶ it is much easier to achieve and demonstrate legal compliance. Enforcement authorities have more confidence in organisations that have a health and safety management system in place;
▶ they ensure that health and safety is given the same emphasis as other business objectives, such as quality and finance. They will also aid integration, where appropriate, with other management systems;
▶ they enable significant health and safety risks to be addressed in a systematic manner;
▶ they can be used to show legal compliance with terms such as 'practicable' and 'so far as is reasonably practicable';
▶ they indicate that the organisation is prepared for an emergency;
▶ they illustrate that there is a genuine commitment to health and safety throughout the organisation.

There are, however, several problems associated with occupational health and safety management systems, although most of them are solvable because they are caused by poor implementation of the system. The main problems are:

▶ the arrangements and procedures are not apparent at the workplace level and the audit process is only concerned with a desktop review of procedures;
▶ the documentation is excessive and not totally related to the organisation due to the use of generic procedures;
▶ other business objectives, such as production targets, lead to ad hoc changes in procedures;
▶ integration, which should really be a benefit, can lead to a reduction in the resources and effort applied to health and safety;
▶ a lack of understanding by supervisors and the workforce leads to poor system implementation;
▶ the performance review is not implemented seriously thus causing cynicism throughout the organisation.

2.2 The purpose and importance of setting policy for health and safety

2.2.1 Introduction

A clear policy for the management of health and safety allows everybody associated with the organisation to be aware of its health and safety aims and objectives and how they are to be achieved. For a policy to be effective, it must be honoured in the spirit as well as the letter. A health and safety policy contributes to business efficiency and continuous improvement throughout the operation. The demonstration of senior management involvement provides evidence to all stakeholders that responsibilities to people and the environment are taken seriously. The policy should state the intentions of the business in terms of clear aims, objectives, organisation, arrangements and targets for all health and safety issues.

Some policies are written so that most of the wording concerns strict requirements laid on employees and only a few vague words cover managers' responsibilities. Generally, such policies do not meet the recommendations of the ILO that require an effective policy with a robust organisation and arrangements to be set up. A good health and safety policy will also enhance the performance of the organisation in areas other than health and safety, help with the personal development of the workforce and reduce financial losses.

2.2.2 ILO Recommendations

The ILO recommends in the ILO-OSH 2001 management system that the employer, in consultation with workers and their representatives, should set out in writing a health and safety policy, which should be:

(a) specific to the organisation and appropriate to its size and the nature of its activities;

(b) concise, clearly written, dated and made effective by the signature or endorsement of the employer or the most senior accountable person in the organisation;

(c) communicated and readily accessible to all persons at their place of work;

(d) reviewed for continuing suitability; and

(e) made available to relevant external interested parties, as appropriate.

Figure 2.4 Well-presented policy documents

The ILO further recommends that the health and safety policy should include, as a minimum, the following key principles and objectives to which the organisation is committed:

(a) protecting the safety and health of all members of the organisation by preventing work-related injuries, ill-health, diseases and incidents;

(b) complying with relevant occupational health and safety national laws and regulations, voluntary programmes, collective agreements on occupational health and safety and other requirements to which the organisation subscribes;

(c) ensuring that workers and their representatives are consulted and encouraged to participate actively in all elements of the occupational health and safety management system; and

(d) continually improving the performance of the occupational health and safety management system.

2.3 The key features and appropriate content of an effective health and safety policy

2.3.1 Policy statement of intent

The health and safety policy statement of intent is often referred to as the health and safety policy statement or simply (and incorrectly) as the health and safety policy. It should contain the aims (which are not measurable) and objectives (which are measurable) of the organisation or company. Aims will probably remain unchanged during policy revisions whereas objectives will be reviewed

and modified or changed every year. The statement should be written in clear and simple language so that it is easily understandable. It should also be fairly brief and broken down into a series of smaller statements or bullet points.

The statement should be signed and dated by the most senior person in the organisation. This will demonstrate management commitment to health and safety and give authority to the policy. It will indicate where ultimate responsibility lies and the frequency with which the policy statement is reviewed.

The most senior manager is normally the Chief Executive Officer (CEO) or the Managing Director. It is the responsibility of the CEO (or equivalent) to ensure that the health and safety policy is developed and communicated to all employees in the organisation. He/she will need to ensure the following:

▶ key functions of health and safety management, such as monitoring and audit, accident investigation and training, are included in the organisational structure;

▶ adequate resources are available to manage health and safety effectively;

▶ the production of various health and safety arrangements in terms of rules and procedures;

▶ arrangements for the welfare of employees;

▶ the regular review and, if necessary, updating of the health and safety policy.

The policy statement should be written by the organisation and not by external consultants, as it needs to address the specific health and safety issues and hazards within the organisation. In large organisations, it may be necessary to have health and safety policies for each department and/or site with an overarching general policy incorporating the individual policies. Such an approach is often used by local authorities and multinational companies.

The following points should be included or considered when a health and safety policy statement is being drafted:

▶ the aims, which should cover health and safety, welfare and relevant environmental issues;

▶ the position of the senior person in the organisation or company who is responsible for health and safety (normally the Chief Executive Officer);

▶ the names of the Health and Safety Adviser and any safety representatives or other competent health and safety persons;

▶ a commitment to the basic requirements of access, egress, risk assessments, safe plant and systems of work, use, handling, transport and handling of articles and substances, information, training and supervision;

▶ a commitment to the additional requirements for emergency procedures, health surveillance and employment of competent persons;

- duties towards the wider general public and others (contractors, customers, students, etc.);
- the principal hazards in the organisation;
- specific policies of the organisation (e.g. smoking policy, violence to staff, etc.);
- a commitment to employee consultation possibly using a safety committee or plant council;
- duties and rights of employees;
- specific health and safety performance targets for the immediate and long-term future;
- a commitment to provide the necessary resources to achieve the objectives outlined in the policy statement.

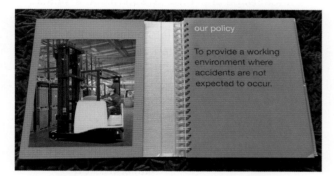

our policy

To provide a working environment where accidents are not expected to occur.

Figure 2.5 Part of a policy commitment

Health and safety **performance targets** are an important part of the statement of intent because:

- they indicate that there is management commitment to improve health and safety performance;
- they motivate the workforce with tangible goals resulting, perhaps, in individual or collective rewards;
- they offer evidence during the monitoring, review and audit phases of the management system.

The type of target chosen depends very much on the areas that need the greatest improvement in the organisation. The following list, which is not exhaustive, shows common health and safety performance targets:

- a specific reduction in the number of accidents, incidents (not involving injury) and cases of work-related ill-health (perhaps to zero);
- a reduction in the level of sickness absence;
- a specific increase in the number of employees trained in health and safety;
- an increase in the reporting of minor accidents and 'near miss' incidents;
- a reduction in the number of civil claims;
- no enforcement notices from the Enforcement Agency;
- a specific improvement in health and safety audit scores;
- the achievement of a nationally recognised health and safety management standard such as OHSAS 18001.

In the ILO-OSH 2001 management system, the ILO recommends that measurable health and safety objectives, consistent with the health and safety policy and based on any initial or subsequent reviews, should be established, which are:

(a) specific to the organisation, and appropriate to and according to its size and nature of activity;

(b) consistent with the relevant and applicable national laws and regulations, and the technical and business obligations of the organisation with regard to health and safety;

(c) focused towards continually improving workers' health and safety protection to achieve the best health and safety performance;

(d) realistic and achievable;

(e) documented, and communicated to all relevant functions and levels of the organisation; and

(f) periodically evaluated and if necessary updated.

The policy statement of intent should be posted on prominent notice boards throughout the workplace and brought to the attention of all employees at induction and refresher training sessions. It can also be communicated to the workforce during team briefing sessions, at 'toolbox' talks which are conducted at the workplace or directly by email, intranet, newsletters or booklets. It should be a permanent item on the agenda for health and safety committee meetings where it and its related targets should be reviewed at each meeting.

2.3.2 Setting health and safety objectives

A health and safety plan is necessary to guide the organisation – setting out the objectives for a specified time period. The speed with which the objectives will be achieved will depend on the resources available, the current state of health and safety compliance and the policy aims of the organisation. If the policy is to achieve excellence, the objectives will need to be tougher than those wishing simply to comply with minimum legal standards. It is essential to decide on priorities, with the emphasis on effective and adequate workplace precautions that meet legal requirements. High hazard/risk activities should receive priority. In some cases short-term measures may be needed quickly to minimise risks while longer-term solutions are designed and implemented.

There are two basic methods of setting objectives within an organisation – from the top or from the bottom – and either has its merits. Many policies are set from the boardroom and it is here that the resources for health and safety and the associated standards can be agreed. This is the top-down approach.

However, at times directors may be somewhat remote from the workplaces where most accidents occur. In some world-leading businesses, objectives are set at the workplace by those who are exposed to the significant risks. These are then approved at higher levels and coordinated with other workplaces within the organisation. The overall aims may be set centrally and then the details organised at site or workplace levels. This is the 'bottom-up' approach and helps to achieve commitment from people at the workplace.

Whatever approach is used, it is vitally important to ensure that safety representatives and/or representatives of employee safety are fully involved through consultation when setting the objectives and during their implementation. The objectives should be properly documented and may differ in detail at each functional level.

Health and safety objectives need to be specific, measurable, achievable, agreed with those who deliver them, realistic and set against a suitable timescale (SMART). Both short- and long-term objectives should be set and prioritised against business needs. Objectives at different levels or within different parts of an organisation should be aligned so they support the overall policy objectives. Personal targets can also be agreed with individuals to secure the attainment of objectives of the organisation and an example of such objectives is given in Box 2.1.

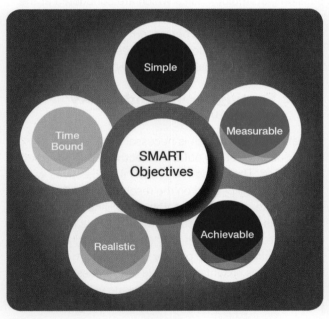

Figure 2.6 SMART performance standards or Objectives

There are three complementary outputs from the planning process:

▶ health and safety plans with objectives for developing, maintaining and improving the health and safety management system, such as:
  ▷ requiring each site of a multi-site organisation to have an annual health and safety plan and an accident and incident investigation system;
  ▷ establishing a reliable risk assessment process for COSHH;
  ▷ involving employees in preparing workplace precautions;

Box 2.1 Example of objectives

An organization may wish to minimize the risk to workers of working at height and aim to prevent all serious injury from falls when working at height. The objectives at each functional level may be as follows:

1. Board room – Objective is to eliminate all serious injuries from working at height within one year. The Board may set one of its members as the champion to make this happen and provide advice from in-house expertise or external consultants. The budget to be approved within the first month.

2. The Board champion may set the following objectives/plan to achieve the Board's aim in consultation with the organisation's central health and safety steering committee:
   a) All sites to reduce serious accidents by 50% within 6 months and 100% within 12 months.
   b) All site managers to audit existing work at height operations; determine what can be done to avoid working at height; and report in 1 month with a draft budget.
   c) Safety adviser to establish existing accident performance; produce best practice guidance with safety representatives for those work at height operations which are considered unavoidable; identify suitable levels of competency and training requirements for those working at height.

3. Site managers may set the following in consultation with safety representatives and the site health and safety committee:
   a) All departmental supervisors to ensure immediately that work at height operations are inspected and risk assessed before work starts, all equipment to be assessed for suitability and checked for condition.
   b) All people working at height to be assessed for competence and specific training given as needed, for example working in mobile elevating work platforms.
   c) Toolbox talks on general work at height awareness to be provided.

  ▷ completing all manual handling assessments by the end of the current year;
  ▷ providing a new guard for a particular machine;
▶ specifications for management arrangements, risk control and workplace precautions; and
▶ performance standards for implementing the health and safety management system, identifying the contribution of individuals to implementing the

system (this is essential to building a positive health and safety culture).

Health and safety targets will need to be set so that the agreed performance standards are approached and met. Since there may well be many standards, it will be necessary to prioritise important standards and associated targets. These prioritised targets are known as **Key Performance Indicators (KPIs)**. The targets must be realistic and a procedure known as benchmarking will help to ascertain whether this is the case.

Benchmarking

Benchmarking compares the performance of the organisation with that of similar organisations. A small construction company will benchmark itself against the performance of the UK construction industry perhaps using statistics supplied by the HSE. Such benchmarks, or examples of good practice, are defined by comparison with the health and safety performance of other parts of the organisation or the national performance of the occupational group of the organisation. The HSE publish an annual report, statistics and a bulletin, all of which may be used for this purpose. Typical benchmarks include accident rates per employee and accident or disease causation. It is important that the correct benchmarks are chosen and the organisation is assured of the accuracy of the data. The advantages of benchmarking are that:

▶ the key performance indicators for a particular organisation may be easily identified;
▶ it helps with continuous improvement;
▶ it focuses attention on weaker performance areas;
▶ it gives confidence to various stakeholders; and
▶ it is useful feedback for Boards, Chief Executives and managers.

The relationship of health and safety performance to other business activities

The best health and safety policies do not separate health and safety and human resources management since people are a key resource. Organisations want a fit, competent and committed workforce rather than simply preventing accidents and ill-health. This integrated approach will extend to people outside the organisation with policies for the control of environmental pollution and product safety. Organisations often fail to manage health and safety effectively because they see it as something distinct from other managerial tasks. The principles and approach to managing health and safety are exactly the same as those required for managing quality or the environment.

Well-managed health and safety is important in business terms because stakeholders (such as shareholders, customers and the general public)

are influenced by any bad publicity resulting from poor health and safety standards. The identification, assessment and control of health and safety and other risks is a managerial responsibility and of equal importance to production and quality. There is good evidence that good health and safety management also produces good quality products at a reasonable cost. However, the practical implications of health and safety must be carefully thought through to avoid conflict between the demands of policy and other operational requirements. Insufficient attention to health and safety can lead to problems with production – for example, work schedules that fail to take into account the problems of fatigue or inadequate resources allocated to training.

The considerable advances in information technology have been very helpful in the development of health and safety management systems. It has enabled the rapid identification of data that is critical to the management of health and safety (e.g. accident and ill-health trends, health and safety performance variations between departments within the organisation). It has also simplified the collection and analysis of essential data.

Research by the HSE found that 'in many organisations the visible leadership and emphasis on continual improvement with respect to health and safety lagged behind that for quality of a product or service'.

2.3.3 Organisation of health and safety

This section of the policy defines the names, positions and duties of those within the organisation or company who have a specific responsibility for health and safety. Therefore, it identifies those health and safety responsibilities and the reporting lines through the management structure. This section will include the following groups together with their associated responsibilities:

▶ directors and senior managers (responsible for setting policy, objectives and targets);
▶ supervisors (responsible for checking day-to-day compliance with the policy);
▶ health and safety advisers (responsible for giving advice during accident investigations and on compliance issues);
▶ other specialists, such as an occupational nurse, chemical analyst and an electrician (responsible for giving specialist advice on particular health and safety issues);
▶ health and safety representatives (responsible for representing employees during consultation meetings on health and safety issues with the employer);
▶ employees (responsible for taking reasonable care of the health and safety of themselves and others who may be affected by their acts or omissions);

▶ fire marshals (responsible for the safe evacuation of the building in an emergency);

▶ first-aiders (responsible for administering first-aid to injured persons).

For smaller organisations, some of the specialists mentioned above may well be employed on a consultancy basis.

For the health and safety organisation to work successfully, it must be supported from the top (preferably at board level) and some financial resource made available.

It is also important that responsibility for certain key functions are included in the organisation structure. These include:

▶ accident investigation and reporting;

▶ health and safety training and information;

▶ health and safety monitoring and audit;

▶ health surveillance;

▶ monitoring of plant and equipment, their maintenance and risk assessment;

▶ liaison with external agencies;

▶ management and/or employee safety committees – the management committee will monitor day-to-day problems and any concerns of the employee health and safety committee.

The role of the health and safety adviser is to provide specialist information to managers in the organisation and to monitor the effectiveness of health and safety procedures. The adviser is not 'responsible' for health and safety or its implementation; that is the role of the line managers.

Finally the job descriptions, which define the duties of each person in the health and safety organisational structure, must not contain responsibility overlaps or blur chains of command. Each individual must be clear about his/her responsibilities and the limits of those responsibilities.

2.3.4 Arrangements for health and safety

The arrangements section of the health and safety policy gives details of the specific systems and procedures used to assist in the implementation of the policy statement. This will include health and safety rules and procedures and the provision of facilities such as a first-aid room and washrooms. It is common for risk assessments (including those for hazardous substances, manual handling and personal protective equipment (PPE) assessments) to be included in the arrangements section, particularly for those hazards referred to in the policy statement. It is important that arrangements for fire and other emergencies and for information, instruction, training and supervision are also covered. Local codes of practice (e.g. for fork-lift truck drivers) should be included.

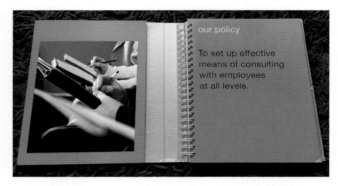

Figure 2.7 (a) (b) (c) Good information, training and working with employees is essential

The following list covers the more common items normally included in the arrangements section of the health and safety policy:

▶ employee health and safety code of practice;

▶ accident and illness reporting and investigation procedures;

▶ emergency procedures, first-aid;

▶ fire drill procedure;

▶ procedures for undertaking risk assessments;

▶ control of exposure to specific hazards (noise, vibration, radiation, manual handling, hazardous substances, etc.);

▶ machinery safety (including safe systems of work, lifting and pressure equipment);

▶ electrical equipment (maintenance and testing);

▶ maintenance procedures;

▶ permits to work procedures;

▶ use of PPE;

▶ monitoring procedures including health and safety inspections and audits;

- procedures for the control and safety of contractors and visitors;
- provision of welfare facilities;
- training procedures and arrangements;
- catering and food hygiene procedures;
- arrangements for consultation with employees;
- terms of reference and constitution of the safety committee;
- procedures and arrangements for waste disposal.

The three sections of the health and safety policy are usually kept together in a health and safety manual and copies distributed around the organisation.

2.3.5 Review of health and safety policy

It is important that the health and safety policy is monitored and reviewed on a regular basis. For this to be successful, a series of benchmarks needs to be established. Such benchmarks, or examples of good practice, are defined by comparison with the health and safety performance of other parts of the organisation or the national performance of the occupational group of the organisation. Many national health and safety enforcement agencies publish an annual occupational accident and disease report, statistics and a bulletin, all of which may be used for this purpose. Typical benchmarks include accident rates per employee and accident or disease causation.

There are several reasons to review the health and safety policy. The more important reasons are:

- significant organisational changes may have taken place;
- there have been changes in key personnel;
- there have been changes in legislation and/or guidance;
- new work methods have been introduced;
- there have been alterations to working arrangements and/or processes;
- there have been changes following consultation with employees;
- the monitoring of risk assessments or accident/ incident investigations indicates that the health and safety policy is no longer totally effective;
- information from manufacturers has been received;
- advice from an insurance company has been received;
- the findings of an external health and safety audit;
- enforcement action has been taken by the national health and safety enforcement agency; and
- a sufficient period of time has elapsed since the previous review.

A positive promotion of health and safety performance will achieve far more than simply preventing accidents and ill-health. It will:

- support the overall development of personnel;
- improve communication and consultation throughout the organisation;
- minimise financial losses due to accidents and ill-health and other incidents;
- directly involve senior managers in all levels of the organisation;
- improve supervision, particularly for young persons and those on occupational training courses;
- improve production processes;
- improve the public image of the organisation or company.

It is apparent, however, that some health and safety policies appear to be less than successful. There are many reasons for this. The most common are:

- the statements in the policy and the health and safety priorities are not understood by or properly communicated to the workforce;
- minimal resources made available for the implementation of the policy;
- too much emphasis on rules for employees and too little on management policy;
- a lack of parity with other activities of the organisation (such as finance and quality control) due to mistaken concerns about the costs of health and safety and the effect of those costs on overall performance. This attitude produces a poor health and safety culture;
- a lack of senior management involvement in health and safety, particularly at board level;

Figure 2.8 The policy might be good but is it put into practice – unsafe use of ladder

▶ employee concerns that their health and safety issues are not being addressed or that they are not receiving adequate health and safety information. This can lead to low morale among the workforce and, possibly, high absenteeism;

▶ high labour turnover;

▶ inadequate or no PPE;

▶ unsafe and poorly maintained machinery and equipment;

▶ a lack of health and safety monitoring procedures.

In summary, a successful health and safety policy is likely to lead to a successful organisation or company. A checklist for assessing any health and safety policy has been produced by the UK HSE and has been reproduced in Appendix 2.1.

2.3.6 Standards and guidance relating to health and safety policy

The three health and safety management systems discussed earlier in this chapter give specific guidance on health and safety policy. Guidance is also available from specialist health and safety organisations including the UK HSE and the ILO. The ILO recommends (in ILO-OSH 2001) that a health and safety policy should be:

(a) specific to the organisation and appropriate to its size and the nature of its activities;

(b) concise, clearly written, dated and made effective by the signature or endorsement of the employer or the most senior accountable person in the organisation;

(c) communicated and readily accessible to all persons at their place of work;

(d) reviewed for continuing suitability; and

(e) made available to relevant external interested parties, as appropriate.

Further, the ILO recommends that the policy should include, as a minimum, the following key principles and objectives:

(a) the protection of the health and safety of all members of the organisation by the prevention of work-related injuries, ill-health, diseases and incidents;

(b) the compliance with relevant national laws and regulations, voluntary programmes and collective agreements involving occupational health and safety;

(c) consultation with the workers and their representatives and encouragement for them to participate actively in all elements of the health and safety management system; and

(d) the continual improvement of the performance of the health and safety management system.

2.4 Further information

ILO Guidelines on Occupational Safety and Health Management Systems (ILO-OSH 2001). ISBN 978-0-580-37805-5 http://www.ilo.org/global/publications/ilo-bookstore/order-online/books/WCMS_PUBL_9221116344_EN/lang--en/index.htm

Occupational Health and Safety Assessment Series (OHSAS 18000): Occupational Health and Safety Management Systems OHSAS 18001:2007 ISBN 978-0-5805-9404-5 OHSAS18002:2008 ISBN 978-0-5806-2686-9

2.5 Practice revision questions

1. (a) **Outline** the key elements of a health and safety management system.
   (b) **Identify** documents that may be examined when reviewing an organisation's health and safety management system.

2. (a) **Outline** factors that should be considered during the 'planning' stage of a safety management system.
   (b) Review (evaluation) is a key element of a health and safety management system. **Identify THREE** parts of this element **AND give** an example in **EACH** case.

3. (a) **Identify** the economic benefits that an organisation may obtain by implementing a successful health and safety management system.

   (b) **Discuss** the possible results of a poor implementation of a health and safety management system.

4. (a) **Outline** the role of the health and safety policy in decision making in an organisation.
   (b) **Identify** the three main sections of a health and safety policy document and explain the purpose and general content of each section.
   (c) **Outline** the responsibilities of a Chief Executive in relation to the health and safety policy.

5. (a) **Explain** the purpose of the 'statement of intent' section of a health and safety policy.
   (b) **Outline** the **SIX** topics that may be addressed in this section of the health and safety policy.

(c) **Explain** why a health and safety policy should be signed by the most senior person in an organisation.

(d) **Outline** the indications that the objectives within the 'statement of intent' of the health and safety policy are not reflected in the health and safety performance of an organisation.

6. (a) **Outline** the key functions that should be included in the health and safety organisational structure of a company.

(b) **Identify** the categories of persons that would be found in the health and safety organisational structure.

(c) **Identify SIX** employee responsibilities which could be included in the 'organisation' section of a health and safety policy.

7. The 'arrangements' section is an important part of a health and safety policy.

(a) **Identify** the information that should be contained in it.

(b) **Outline** the key areas that should be addressed within it.

8. (a) **Outline** the occasions when a health and safety policy should be reviewed.

(b) **Describe FOUR** external **AND FOUR** internal influences that might initiate a health and safety policy review.

9. (a) **Explain** why it is important for an organisation to set performance targets in terms of its health and safety performance.

(b) **Outline SEVEN** types of performance target that an organisation might typically set in relation to health and safety.

(c) **Outline** the importance of benchmarking health and safety performance and identify an example of information that could be used for benchmarking.

10. (a) **Outline** why it is important to maintain and promote good standards of health and safety within an organisation.

(b) **Identify** sources of information and guidance that may be used to maintain and promote good standards of health and safety in the organisation.

(c) **Outline** the actions that the Board of Directors can take to promote good health and safety standards within an organisation.

APPENDIX 2.1 Health and Safety Policy checklist

The following checklist is intended as an aid to the writing and review of a health and safety policy. It is derived from UK HSE Information.

General policy and organisation

▶ Does the statement express a commitment to health and safety and are your obligations towards your employees made clear?

▶ Does it say which senior manager is responsible for seeing that it is implemented and for keeping it under review, and how this will be done?

▶ Is it signed and dated by you or a partner or senior director?

▶ Have the views of managers and supervisors, safety representatives and the safety committee been taken into account?

▶ Were the duties set out in the statement discussed with the people concerned in advance, and accepted by them, and do they understand how their performance is to be assessed and what resources they have at their disposal?

▶ Does the statement make clear that cooperation on the part of all employees is vital to the success of your health and safety policy?

▶ Does it say how employees are to be involved in health and safety matters, for example by being consulted, by taking part in inspections and by sitting on a safety committee?

▶ Does it show clearly how the duties for health and safety are allocated and are the responsibilities at different levels described?

▶ Does it say who is responsible for the following matters (including deputies where appropriate)?
  ▷ Reporting investigations and recording accidents.
  ▷ Fire precautions, fire drill and evacuation procedures.
  ▷ First-aid.
  ▷ Safety inspections.
  ▷ The training programme.
  ▷ Ensuring that legal requirements are met, for example regular testing of lifts and notifying accidents to the health and safety inspector.

Arrangements that need to be considered

▶ Keeping the workplace, including staircases, floors, ways in and out, washrooms, etc., in a safe and clean condition by cleaning, maintenance and repair;

▶ The requirements of the Work at Height Regulations;

▶ Any suitable and sufficient risk assessments.

Figure 2.9 Emergency procedures

our policy

To ensure that employees at all levels, have a clear understanding of their responsibilities and accountabilities for environment and health & safety.

Figure 2.10 Responsibilities

Plant and substances

▶ Maintenance of equipment such as tools, ladders, etc. – are they in a safe condition?

▶ Maintenance and proper use of safety equipment such as helmets, boots, goggles, respirators, etc.;

▶ Maintenance and proper use of plant, machinery and guards;

▶ Regular testing and maintenance of lifts, hoists, cranes, pressure systems, boilers and other dangerous machinery, emergency repair work, and safe methods of carrying out these functions;

▶ Maintenance of electrical installations and equipment;

▶ Safe storage, handling and, where applicable, packaging, labelling and transport of flammable and/or hazardous substances;

▶ Controls of work involving harmful substances such as lead and asbestos;

▶ The introduction of new plant, equipment or substances into the workplace by examination, testing and consultation with the workforce;

▶ Exposure to non-ionising and ionising radiation.

Figure 2.11 Reach truck in warehouse

Other hazards

▶ Noise problems – wearing of hearing protection, and control of noise at source;

▶ Vibration problems – hand–arm and whole-body control techniques and personal protection;

▶ Preventing unnecessary or unauthorised entry into hazardous areas;

▶ Lifting of heavy or awkward loads;

▶ Protecting the safety of employees against assault when handling or transporting the employer's money or valuables;

▶ Special hazards to employees when working on unfamiliar sites, including discussion with site manager where necessary;

▶ Control of works transport, for example fork-lift trucks, by restricting use to experienced and authorised operators or operators under instruction (which should deal fully with safety aspects);

▶ Driving on public roads while at work.

Emergencies

▶ Ensuring that fire exits are marked, unlocked and free from obstruction;

▶ Maintenance and testing of fire-fighting equipment, fire drills and evacuation procedures;

▶ First-aid, including name and location of person responsible for first-aid and deputy, and location of first-aid box.

Communication

▶ Giving employees information about the general duties under relevant national legislation and specific legal requirements relating to their work;

▶ Giving employees necessary information about substances, plant, machinery and equipment with which they come into contact;

▶ Discussing with contractors, before they come on site, how they plan to do their job, whether they need any equipment from your organisation to help them, whether they can operate in a segregated area or only when part of the plant is shut down and, if not, what hazards they may create for your employees and vice versa.

Training

▶ Training employees, supervisors and managers to enable them to work safely and to carry out their health and safety responsibilities efficiently.

Supervising

▶ Supervising employees so far as necessary for their safety – especially young workers, new employees and employees carrying out unfamiliar tasks.

Keeping check

▶ Regular inspections and checks of the workplace, machinery appliances and working methods.

CHAPTER 3

Health and safety management systems – Organising – DO 1

3.1 Organisational health and safety roles and responsibilities of employers, directors, managers, workers and other relevant parties ▶ 43

3.2 Concept of health and safety culture and its significance in the management of health and safety in an organisation ▶ 59

3.3 Human factors which influence behaviour at work ▶ 62

3.4 How health and safety behaviour at work can be improved ▶ 68

3.5 Further information ▶ 79

3.6 Practice revision questions ▶ 79

Appendix 3.1 Leadership actions for directors and board members ▶ 82

Appendix 3.2 Detailed health and safety responsibilities ▶ 84

Appendix 3.3 Checklist for supply chain health and safety management ▶ 86

Appendix 3.4 Safety culture questionnaire ▶ 87

This chapter covers the following NEBOSH learning objectives:

1. Outline the health and safety roles and responsibilities of employers, managers, supervisors, workers and other relevant parties
2. Explain the concept of health and safety culture and its significance in the management of health and safety in an organisation
3. Outline the human factors which influence behaviour at work in a way that can affect health and safety
4. Explain how health and safety behaviour at work can be improved

Introduction

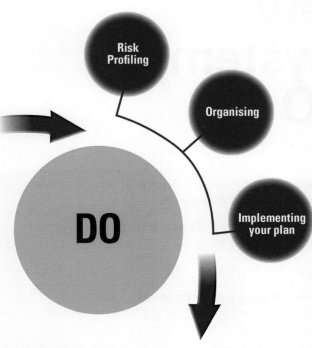

Figure 3.1 DO part of the management cycle involves Risk Profiling (Chapter 4), Organising and Implementing plans

This chapter is about managers in businesses, or other organisations, setting out clear responsibilities and lines of communications for everyone in the enterprise. The chapter also covers the ILO Conventions and Recommendations, advises legal responsibilities that exist between people who control premises and those who use them, and between contractors and those who hire them; and the duties of suppliers, manufacturers and designers of articles and substances for use at work. The policy, described in Chapter 2, is an essential first step and sets the direction for health and safety within the enterprise and forms the written intentions of the principals or directors of the business. The occupational health and safety policy of any organisation needs to be clearly communicated and people need to know what they are responsible for in the day-to-day operations. A vague statement that 'everyone is responsible for health and safety' is misleading and fudges the real issues.

Everyone is responsible for health and safety, but this is particularly true for management. The policy will only remain as words on paper, however good the intentions, until there is an effective organisation set up to implement and monitor its requirements. Occupational health and safety management systems such as ILO-OSH 2001 and OHSAS 18001:2007 also require that an effective organisation is established to implement the policy.

There is generally no equality of responsibility under law between those who provide direction and create policy and those who are employed to follow. Principals or employers, generally, have substantially more responsibility than employees.

Some policies are written so that most of the wording concerns strict requirements laid on employees and only a few vague words cover managers' responsibilities. Generally, such policies do not meet the ILO recommendations and the occupational health and safety law of many countries, which usually require an effective policy with a robust organisation and arrangements to be set up. See Chapter 15 for more details.

In the UK in 1972, a Government Inquiry Report (known as the Robens report) recognised that the introduction of health and safety management systems was essential if the ideal of self-regulation of health and safety by industry was to be realised. It further recognised that a more active involvement of the workforce in such systems was essential if self-regulation was to work. Self-regulation and the implicit need for health and safety management systems and employee involvement were incorporated into the Health and Safety at Work (HSW) Act.

Since the introduction of the HSW Act, health and safety standards have improved considerably in the UK but there have been some catastrophic failures. One of the worst was the fire on the offshore oil platform, Piper Alpha, in 1988, when 167 people died. At the subsequent enquiry, the concept of a safety culture was defined by the Director General of the Health and Safety Executive (HSE) at that time, J. R. Rimington. This definition has remained as one of the key points for a successful health and safety management system.

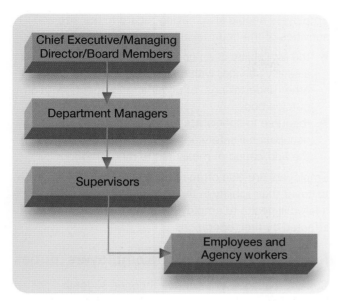

Figure 3.2 Everyone from senior managers down has health and safety responsibilities

3.1 Organisational health and safety roles and responsibilities of employers, directors, managers, workers and other relevant parties

3.1.1 Control of the health and safety organisation

Like all management functions, establishing control and maintaining it day in day out is crucial to effective health and safety management. Managers, particularly at senior levels, must take proactive responsibility for controlling issues that could lead to ill-health, injury or loss. A nominated senior manager at the top of the organisation needs to oversee policy implementation and monitoring. The nominated person will need to report regularly to the most senior management team and will be a director or principal of the organisation.

Health and safety responsibilities will need to be assigned to line managers and expertise must be made available, either inside or outside the enterprise, to help them achieve their legal requirements. The purpose of the health and safety organisation is to harness the collective enthusiasm, skills and effort of the entire workforce with managers taking key responsibility and providing clear direction. The prevention of accidents and ill-health through management systems of control becomes the focus rather than looking for individuals to blame after the incident occurs.

The control arrangements should be part of the written health and safety policy. Performance standards will need to be agreed and objectives set which link the outputs required to specific tasks and activities for which individuals are responsible. For example, the objective could be to carry out a workplace inspection

once a week to an agreed checklist and rectify faults within three working days. The periodic, say annual, audit would check to see if this was being achieved and, if not, what the reasons for non-compliance with the objective were.

People should be held accountable for achieving the agreed objectives through existing or normal procedures such as:

- ▶ job descriptions, which include health and safety responsibilities;
- ▶ performance appraisal systems, which look at individual contributions;
- ▶ arrangements for dealing with poor performance;
- ▶ where justified, the use of disciplinary procedures.

Such arrangements are only effective if health and safety issues achieve the same degree of importance as other key management concerns, and a good performance is considered to be an essential part of career and personal development.

3.1.2 Organisational health and safety responsibilities – employers

Employers must protect the health, safety and welfare of employees and others who might be affected by their work activities including:

- ▶ other workers – agency, temporary or casual
- ▶ trainees
- ▶ contractors
- ▶ visitors
- ▶ neighbours and the general public.

Health and safety is about sensible, proportionate actions that protect people – not unnecessary bureaucracy and paperwork.

This 10-point list shows some of the key actions required by law that apply to nearly every organisation:

1. Obtain Employers' Liability Compulsory Insurance and display the certificate.
2. Ensure that the organisation has competent health and safety advice available. This does not have to be an external consultant.
3. Develop a health and safety policy that outlines the health and safety management system.
4. Undertake risk assessments on the business activities that include details of the required control measures.
5. Ensure that relevant actions are taken following the risk assessment to prevent accidents and ill-health.
6. Provide basic welfare facilities, such as toilets, washing facilities and drinking water.
7. Provide free health and safety training for workers.
8. Consult workers on health and safety.
9. Display the health and safety law poster or give workers a leaflet with the information.
10. Report certain work-related accidents, diseases and dangerous occurrences.

Failure to comply with these requirements can have serious consequences – for both organisations and individuals. As has already been stated, sanctions include fines, imprisonment and disqualification.

Employers are required to do a risk assessment of the work activities carried out by homeworkers. It may be necessary for employers to visit the home of the worker to carry out a risk assessment, although homeworkers can also help in identifying the hazards for their employers themselves.

ILO Recommendations

The ILO Convention 155 and Recommendations 164 place primary responsibilities on employers (see Chapter 15 for a summary). This concept has been followed by many countries although a few still put responsibilities on factory occupiers. The general duties of employers under the ILO Recommendations 164 are:

▶ to provide and maintain workplaces, machinery and equipment, and use work methods, which are as safe and without risk to health as is reasonably practicable;

▶ to give necessary instructions and training, taking account of the functions and capacities of different categories of workers;

▶ to provide adequate supervision of work, of work practices and of application and use of occupational safety and health measures;

▶ to institute organisational arrangements regarding occupational safety and health and the working environment adapted to the size of the undertaking and the nature of its activities;

▶ to provide, without any cost to the worker, adequate personal protective clothing and equipment which are reasonably necessary when hazards cannot be otherwise prevented or controlled;

▶ to ensure that work organisation, particularly with respect to hours of work and rest breaks, does not adversely affect occupational safety and health;

▶ to take all reasonably practicable measures with a view to eliminating excessive physical and mental fatigue;

▶ to undertake studies and research or otherwise keep abreast of the scientific and technical knowledge necessary to comply with the foregoing clauses;

▶ to consult with workers or their representatives and form occupational health and safety committees;

▶ to set out their occupational health and safety policy in writing (where the size and nature of the organisation makes this appropriate);

▶ to keep records (such as incidences of accidents and ill-health at work) as required by local legislation, and report these as required to the authorities concerned;

▶ to verify implementation of applicable standards, e.g. by environmental monitoring and systematic safety audits.

Employers should also, having regard to the size and activities of the undertaking, make provision for:

▶ the availability of an occupational health service and a safety service, within the undertaking, jointly with other undertakings, or under arrangements with an outside body;

▶ recourse to specialists to advise on particular occupational safety or health problems or supervise the application of measures to meet them.

Employers are also often required to take out insurance to cover their liability in the event of accidents and work-related ill-health to employees and others who may be affected by their operations.

Recent extensions of the employers' duties

Current management science theories suggest that performance is better in all areas of business, including occupational health and safety, if it is measured and continuous improvement sought in an organised fashion. Drawing from the principles defined in the ILO Guidelines on Occupational Safety and Health Management Systems 2001, the latest Convention 187 (2003) and its accompanying Recommendation 197 (in 2006) apply a similar approach to the management of national occupational health and safety systems to ensure they are improved through a continuous cycle of policy review, evaluation and action for improvement.

The principles are based on two fundamental concepts, namely to:

▶ develop a preventative safety and health culture; and

▶ apply a systems approach to managing occupational health and safety nationally.

A national preventative safety and health culture is one in which the right to a safe and healthy working environment is respected at all levels. It is also one where governments, employers, workers and other interested stakeholders actively participate in securing a safe and healthy working environment through a system of defined rights, responsibilities and duties, and where the principle of prevention is accorded the highest priority. The issues surrounding a health and safety culture are covered in detail later in this chapter. Employers' duties have therefore often been extended in modern occupational health and safety legislation to include:

▶ the need to conduct and record risk assessments;

▶ requirements to include continual improvement and develop a preventative occupational health and safety culture;

▶ setting up a systems approach to occupational health and safety management by following the ILO-OSH 2001 or the OHSAS 18001:2007 approaches which are largely identical.

Visitors and general public

Organisations usually have a duty to ensure the health and safety of the public while on their premises, even if the individuals concerned, like children, are not supposed to be there. Two cases in the UK reported in the Autumn of 2008 have highlighted how far this liability can extend.

UK Case studies

Two companies pleaded guilty to charges of breaching the HSW Act 1974 following the drowning of a 9-year-old girl, who was playing with other children in the company car park when they strayed onto nearby reservoirs. At the time of the accident, in 2004, the main gates to the factory were off their hinges because work was being carried out on the site and a second gate, which led to the reservoir, was only secured with a nylon rope.

In a separate case, a construction company has been found guilty of failing to prevent unauthorised persons, including children, from gaining access to an area where construction material and equipment were stored. A child was seriously injured by falling paving stones while playing on a partly built housing estate where materials were being stored during construction work.

Misunderstandings regarding who is responsible for monitoring and protecting contractors and sub-contractors – leading to the failure to carry out responsibilities – is often at the heart of high-profile health and safety cases. But situations such as these could be avoided with a clearer understanding of employers' and contractors' responsibilities.

Visitors to a site whether authorised or not are often more at risk than employees because:

- they are unfamiliar with the workplace processes, the hazards and associated risks they present;
- they may not have the appropriate personal protective equipment (PPE);
- they will have a lack of knowledge of the site or premises layout;
- walkways are often inadequate, unsigned or poorly lit;
- they are not familiar with the emergency procedures or means of escape;
- they may be particularly vulnerable if they suffer from a disability or are very young.

Many of these problems with visitors can be overcome by, for example:

- visitors signing in and being provided with a site escort;

- providing appropriate PPE and identity badges;
- providing simple induction procedures with a short video and information on site rules, hazards and emergency procedures;
- clear marking of walkways and areas where unauthorised people are not permitted.

Night working

Employers should ascertain whether they employ people who would be classified as night workers. If so, they should check:

- whether there are specific local restrictions on night working;
- who may work at night – some vulnerable people like young people under 18 years of age and pregnant women are excluded;
- how much time night workers normally work;
- if night workers work more than say eight hours per day on average, whether the number of hours can be reduced and if any exceptions apply;
- that a health assessment is made on each night worker;
- that a regular programme of health checks is made on each night worker;
- that proper records of night workers are maintained, including details of health assessments (where required);
- that night workers are not involved in work which is particularly hazardous.

Temporary workers Directive adopted by EU Parliament

The European Parliament has adopted a Directive on temporary agency work which enables temporary workers to be treated equally with those of the employer company. However, following an agreement reached in 2008, agency workers should receive the same pay and conditions as permanent staff after being employed for 12 weeks.

The agreement has been made between the UK Government, the CBI and the TUC, and the following points have been agreed:

- After 12 weeks in a given job, there will be entitlement to equal treatment.
- Equal treatment will be defined to mean 'at least the basic working and employment conditions that would apply to the workers concerned if they had been recruited directly by that undertaking to occupy the same job'. This does not cover occupational social security schemes.
- The government will consult regarding the implementation of the Directive, in particular how disputes will be resolved regarding the definition of equal treatment and how both sides of the industry can reach appropriate agreements on the treatment of agency workers.

▶ The new arrangements will be reviewed at an appropriate point 'in the light of experience'.

Migrant workers

Migrant workers are employed throughout the world and there have been concerns at the accident rate of migrant workers in several countries. The following suggestions have been made by a global health and safety think tank (organised by the UK Health and Safety Laboratory and hosted by L'Oréal in Paris) for managing health and safety in a culturally diverse workforce:

▶ Have a clear, transparent health and safety strategy against which company values and expectations can be anchored.
▶ Specify the targets that need to be achieved, but allow how these are to be achieved to vary with local culture or languages.
▶ Embed company standards by periodically rotating line management across different regions, but ensure that they have the 'inter-cultural' competencies to remain responsive to local issues.
▶ Use local stories and pictures to help get health and safety messages across.
▶ Overcome language barriers by using pictograms rather than words, by providing interpretation and translation services, and by developing company dictionaries for technical terms.

The following suggestions were made to manage the health and safety of a transient workforce:

▶ Give temporary workers a probationary period with site supervisors and use induction events and meetings to educate contractors around company health and safety standards and expectations.
▶ Ensure that the temporary workforce is integrated into health and safety management, by providing ways for workers to communicate with their managers.
▶ Include temporary workers within company health and safety statistics.
▶ Take a proportionate approach to risk particularly where transient workers may be confronted by major hazards.
▶ Involve other key players (such as shareholders) who can also champion the importance of good health and safety standards.
▶ Manage health and safety information to maximise its impact positively with the workforce.

3.1.3 Organisational health and safety responsibilities – directors

In the UK, the HSE and the Institute of Directors have published INDG417. The guide is based on a plan that will deliver, monitor and review the management concept and the following information is closely based on this guide.

Effective health and safety performance comes from the top; members of the Board have both collective and individual responsibility for health and safety. Directors and boards need to examine their own behaviours, both individually and collectively, against the guidance given by national enforcement agencies. If they fall short of the standards it sets them, then they must change so that they become effective leaders in health and safety.

The following are quotations from health and safety leaders in the public and private sectors in the UK:

> 'Health and safety is integral to success. Board members who do not show leadership in this area are failing in their duty as directors and their moral duty, and are damaging their organisation.'

> 'An organisation will never be able to achieve the highest standards of health and safety management without the active involvement of directors. External stakeholders viewing the organisation will observe the lack of direction.'

For many organisations, health and safety is a corporate governance issue. The Board should integrate health and safety into the main governance structures, including Board sub-committees, such as for risk, remuneration and audit. The Turnbull guidance on the Combined Code on Corporate Governance requires companies listed on the London Stock Exchange to have robust systems of internal control, covering not just 'narrow' financial risks but also risks relating to the environment, business reputation and health and safety.

Directors and Board members need to take action so that:

▶ the health and safety of employees and others, such as members of the public, may be protected;
▶ risk management includes health and safety risks and becomes a key business risk in board decisions; and
▶ health and safety duties imposed by legislation are followed.

The UK Institute of Directors and the HSE state that:

> 'Protecting the health and safety of employees or members of the public who may be affected by the activities of an organisation is an essential part of risk management and must be led by the Board.'

The starting points are the following essential principles. These principles are intended to underpin the actions in this guidance and so lead to good health and safety performance.

▶ strong and active leadership from the top;
  ▷ visible, active commitment from the Board;
  ▷ establishing effective 'downward' communication systems and management structures;
  ▷ integration of good health and safety management with business decisions;

- worker involvement;
  - ▷ engaging the workforce in the promotion and achievement of safe and healthy conditions;
  - ▷ effective 'upward' communication;
  - ▷ providing high quality training;
- assessment and review;
  - ▷ identifying and managing health and safety risks;
  - ▷ accessing (and following) competent advice; and
  - ▷ monitoring, reporting and reviewing performance.

The UK HSE in its guidance on Directors and Board responsibility for health and safety recommends the following five action points:

- The Board needs to accept formally and publicly its collective role in providing health and safety leadership in its organisation.
- Each member of the Board needs to accept their individual role in providing health and safety leadership for their organisation.
- The Board needs to ensure that all board decisions reflect its health and safety intentions, as articulated in the health and safety policy statement.
- The Board needs to recognise its role in engaging the active participation of workers in improving health and safety.
- The Board needs to ensure that it is kept informed of, and alert to, relevant health and safety risk management issues. The HSE recommends that boards appoint one of their number to be the 'Health and Safety Director'.

Directors need to ensure that the Board's health and safety responsibilities are properly discharged. To achieve this, the Board will need to:

- carry out an annual review of health and safety performance;
- keep the health and safety policy statement up to date with current board priorities and review the policy at least every year;
- ensure that there are effective management systems for monitoring and reporting on the organisation's health and safety performance;
- ensure that any significant health and safety failures and their investigation are communicated to Board members;
- ensure that when decisions are made the health and safety implications are fully considered;
- ensure that regular audits are carried out to check that effective health and safety risk management systems are in place.

By appointing a 'Health and Safety Director' there will be a Board member who can ensure that these health and safety risk management issues are properly addressed, both by the Board and more widely throughout the organisation.

The Chairman and/or Chief Executive have a critical role to play in ensuring risks are properly managed and that the Health and Safety Director has the

necessary competence, resources and support of other Board members to carry out their functions. Indeed, some boards may prefer to see all the health and safety functions assigned to their Chairman and/ or Chief Executive. As long as there is clarity about the health and safety responsibilities and functions, and the Board properly addresses the issues, this is acceptable.

The health and safety responsibilities of all Board members should be clearly articulated in the organisation's statement of health and safety policy and arrangements. It is important that the role of the Health and Safety Director should not detract either from the responsibilities of other directors for specific areas of health and safety risk management or from the health and safety responsibilities of the Board as a whole. Some form of health and safety training will probably be required for directors.

There are four elements that boards need to incorporate into their management of health and safety. These are:

- planning the direction of health and safety;
- delivering the plan for health and safety;
- monitoring health and safety performance; and
- reviewing health and safety performance.

Details of these four elements are given in Appendix 3.1

Some Australian states have introduced positive director duties – a director must exercise 'due diligence' to ensure that the company complies with its health and safety duties. Failure to undertake such due diligence is an offence in and of itself, and does not require the commission of an offence by a company. Depending on the category of offence, imprisonment for five years can be imposed in the event of conviction.

In the Republic of Ireland, there is a reverse burden of proof on directors, which requires them to show, in the event of a breach by the company, that the offence is not attributable to their consent, connivance or neglect.

Either the 'due diligence' approach, or the admittedly less attractive 'reverse burden' approach, and the associated increase in personal liability for directors, would likely have the effect of forcing them to do more to ensure health and safety compliance 'within their organisations'.

3.1.4 Organisational health and safety responsibilities – managers

In addition to the legal responsibilities on management, there are many specific responsibilities imposed by each organisation's health and safety policy. A summary of the organisational responsibilities for health and safety of typical line managers and their accountability of each level of the line organisation is given below. Many organisations will not fit this exact structure but most will have those who direct, those who manage or supervise and those who have no line responsibility, but have responsibilities to themselves and fellow workers. A more detailed list of the responsibilities is given in Appendix 3.2.

Managing Directors/Chief Executives

Managing Directors/Chief Executives are responsible for the health, safety and welfare of all those who work or visit the organisation. In particular, they:

1. are responsible and accountable for health and safety performance within the organisation;
2. must ensure that adequate resources are available for the health and safety requirements within the organisation including the appointment of a senior member of the senior management with specific responsibility for health and safety;
3. appoint one or more competent persons and adequate resources to provide assistance in meeting the organisation's health and safety obligations including specialist help where appropriate;
4. establish, implement and maintain a formal, written health and safety programme for the organisation that encompasses all areas of significant health and safety risk;
5. approve, introduce and monitor all site health and safety policies, rules and procedures;
6. review annually the effectiveness and, if necessary, require revision of the health and safety programme.

Departmental managers

The principal departmental managers may report to the Site Manager, Managing Director or Chief Executive. In particular, they:

1. are responsible and accountable for the health and safety performance of their department;
2. are responsible for the engagement and management of contractors and that they are properly supervised;
3. must ensure that any machinery, equipment or vehicles used within the department are maintained, correctly guarded and meet agreed health and safety standards. Copies of records of all maintenance, statutory and insurance inspections must be kept by the Departmental Manager;
4. develop a training plan that includes specific job instructions for new or transferred employees and follow up on the training by supervisors. Copies of records of all training must be kept by the Departmental Manager;
5. personally investigate all lost workday cases and dangerous occurrences and report to their line manager. Progress any required corrective action.

Managers and supervisors

Managers and supervisors are essential communication links between the Board and the workforce. They should ensure that the agreed health and safety standards are communicated to the workers and adherence to those standards is monitored effectively. They have a responsibility to deliver the duty of care of the employer to the workforce.

Therefore managers should:

▶ familiarise themselves with the health and safety management system of the organisation;
▶ ensure that there is an adequate and appropriate level of supervision for all workers;
▶ ensure that supervisors are aware of:
  ▷ the health and safety standards of the organisation;
  ▷ the specific hazards within their area of supervision;
  ▷ the need to set a good example on health and safety issues;
  ▷ the need to monitor the health and safety performance of their workforce; and
  ▷ the training needs, including induction training, of their workforce;
▶ ensure that sufficient resources are available to allow tasks to be completed safely and without risks to health; and
▶ communicate to the Chief Executive and the Board the adherence or otherwise of the health and safety standards agreed by the Board.

Effective supervision of the workforce is one of the important keys to successful health and safety management. It will enable managers to monitor the effectiveness of the training that workers have received, and whether they have the necessary capacity and competence to do the job assigned to them. Supervisors should lead by example in health and safety matters (for example in the wearing of PPE).

Hence supervisors should ensure that:

▶ workers in their department understand the risks associated with their workplace and the measures available to control them;
▶ the risk control measures are up to date and are being properly used, maintained and monitored;
▶ particular attention is paid to new, inexperienced or young people and those whose first language is not English or the language commonly used in the region;
▶ workers are encouraged to raise concerns over any shortcomings in health and safety provision;
▶ arrangements are in place to supervise the work of contractors.

Supervisors are responsible to and report to their Departmental Manager. In particular, they:

1. are responsible and accountable for their team's health and safety performance;
2. enforce all safe systems of work procedures that have been issued by the Departmental Manager;

3. instruct employees in relevant health and safety rules, make records of this instruction and enforce all health and safety rules and procedures;

4. supervise any contractors that are working within their area of supervision; and

5. enforce personal protective equipment requirements, make spot checks to determine that protective equipment is being used and periodically appraise condition of equipment. Record any infringements of the personal protective equipment policy.

3.1.5 Role and functions of health and safety practitioners and other advisers

Competent person

The ILO Convention requires that there should be an occupational health service and a safety service available within the organisation, jointly with other organisations or from an outside body. Some countries require this by law; others are not specific on the issue. In Europe a person must be appointed to assist managers to implement the legislation. The essential point is that managers should have access to expertise to help them fulfil their legal requirements. However, they will always remain as advisers and will not assume responsibility in law for health and safety matters. This responsibility always remains with line managers and cannot be delegated to an adviser whether inside or outside the organisation. The appointee could be:

▶ the employer themselves if they are competent. This may be appropriate in a small, low-hazard business;

▶ one or more employees, provided that they have sufficient time and other resources to undertake the task properly;

▶ a person(s) from outside the organisation who has sufficient expertise to help.

When there is an employee with the ability to do the job, it is better for them to be appointed than to use outside specialists. Many health and safety issues can be tackled by people with an understanding of current best practice and the ability to judge and solve problems. Some specialist help is needed on a permanent basis whilst other help may only be required for a short period. There is a wide range of specialists available for different types of health and safety problem, these include:

▶ engineers for specialist ventilation, specialist machinery or chemical processes;

▶ occupational hygienists for assessment and practical advice on exposure to chemical (dust, gases, fumes, etc.), biological (viruses, fungi, etc.) and physical (noise, vibration, etc.) agents;

▶ occupational health professionals for medical examinations and diagnosis of work-related disease, pre-employment and sickness advice, health education;

▶ ergonomists for advice on suitability of equipment, comfort, physical work environment, work organisation;

▶ physiotherapists for prevention and treatment of musculoskeletal disorders;

▶ radiation protection advisers for advice on radiation issues;

▶ health and safety practitioners for general advice on implementation of legislation, health and safety management, risk assessment, control measures and monitoring performance.

Health and Safety Practitioner

Status and **competence** are essential for the role of health and safety practitioners and other advisers. They must be able to advise management and employees or their representatives with authority and independence. They need to be able to advise on:

▶ creating and developing health and safety policies. These will be for existing activities in addition to new acquisitions or processes;

▶ the promotion of a positive health and safety culture. This includes helping managers to ensure that an effective health and safety policy is implemented;

▶ health and safety planning. This will include goal-setting, deciding priorities and establishing adequate systems and performance standards. Short- and long-term objectives need to be realistic;

▶ day-to-day implementation and monitoring of policy and plans. This will include accident and incident investigation, reporting and analysis;

▶ performance reviews and audit of the whole health and safety management system.

To do this properly, health and safety practitioners need to:

▶ have proper training and be suitably qualified – for example NEBOSH Diploma or competence-based IOSH membership, relevant degree or higher degree and, where appropriate, a Chartered Safety and Health Practitioner or a NEBOSH Certificate in small to medium-sized low-hazard premises, like offices, call centres, warehouses and retail stores;

▶ keep up-to-date information systems on such topics as civil and criminal law, health and safety management and technical advances;

▶ know how to interpret the law as it applies to their own organisation;

▶ actively participate in the establishment of organisational arrangements, systems and risk control standards relating to hardware and human performance. Health and safety practitioners will need to work with management on matters such as legal and technical standards;

49

▶ undertake the development and maintenance of procedures for reporting, investigating, recording and analysing accidents and incidents;

▶ develop and maintain procedures to ensure that senior managers get a true picture of how well health and safety is being managed (where a benchmarking role may be especially valuable). This will include monitoring, review and auditing;

▶ be able to present their advice independently and effectively.

Figure 3.3 Safety practitioner at the front line

Relationships within the organisation

Health and safety practitioners:

▶ support the provision of authoritative and independent advice;

▶ report directly to directors or senior managers on matters of policy and have the authority to stop work if it contravenes agreed standards and puts people at risk of injury;

▶ are responsible for professional standards and systems.

They may also have line management responsibility for other health and safety practitioners, in a large group of companies or on a large and/or high-hazard site.

Relationships outside the organisation

Health and safety practitioners also have a function outside their own organisation. They provide the point of liaison with a number of other agencies including the following:

▶ local occupational health and safety enforcement officers and licensing officials;

▶ architects and consultants;

▶ local fire and rescue authorities;

▶ local police authorities;

▶ local authorities;

▶ insurance companies;

▶ contractors;

▶ clients and customers;

▶ the public;

▶ equipment suppliers;

▶ the media;

▶ professional occupational health and safety associations;

▶ other occupational health and safety specialists and services.

Many safety advisers may think it unlikely they will ever be investigated and if they are, they will be covered by insurance. This may not necessarily be the case. If a safety adviser is being investigated, the normal procedure is to claim under the insurance cover that is in place, either the professional indemnity (PI) if self- employed, or under the employer's policy if employed in-house. It is not unusual to find that cover for advice and representation in criminal investigations and proceedings is excluded from these policies. Therefore, health and safety advisers, particularly those that are self-employed, should always check their insurance cover.

3.1.6 Duties and responsibilities of employees and others

Employees or workers have specific responsibilities under the ILO Convention, which are to:

▶ take reasonable care for their own safety and that of other persons who may be affected by their acts or omissions at work;

▶ comply with instructions given for their own safety and health and those of others and with safety and health procedures;

▶ use safety devices and protective equipment correctly and do not render them inoperative;

▶ report forthwith to their immediate supervisor any situation which they have reason to believe could present a hazard and which they cannot themselves correct; and

▶ report any accident or injury to health which arises in the course of or in connection with work.

Where a worker complains, in good faith, about what they consider is a breach of legal requirements or a serious inadequacy in measures taken by the employer in occupational health and safety or working environment matters, no measures prejudicial to the worker should be taken.

Persons in control of premises

In some situations the employer is not the only person responsible for the safety of people in a workplace. Certain duties are placed on, for example, 'Persons in control of [usually *non-domestic*] premises', to take such steps as are reasonable in their position to ensure that there are no risks to the health and safety of people who are not employees but use the premises. Examples would be where there are several workplaces or companies occupying a large building owned and controlled by a landlord. This duty may extend to:

- people entering the premises to work;
- people entering the premises to use machinery or equipment, for example a launderette;
- access to and exit from the premises; and
- corridors, stairs, lifts and storage areas.

Those in control of premises need to take a range of steps depending on the likely use of the premises and the extent of their control and knowledge of the actual use of the premises. For example they may need to provide fire alarms and escape routes which are common to all occupiers of the premises.

Figure 3.4 NEBOSH in control here

The self-employed

The duties imposed on the self-employed are often non-existent or fairly limited. Because the work of self-employed people can affect the safety of others they should be responsible for:

- their own health and safety;
- ensuring that others who may be affected are not exposed to risks to their health and safety;
- undertaking relevant risk assessment;
- cooperating with other people who work in the premises and, where necessary, in the appointment of a health and safety coordinator; and
- the provision of comprehensible information to other peoples' employees working in the same workplace.

3.1.7 Duties and responsibilities of manufacturers and others in the supply chain

Introduction

Market leaders in every industry are increasing their grip on the chain of supply. They do so by monitoring rather than managing, and also by working more closely with suppliers. The result of this may be that suppliers or contractors are absorbed into the culture of the dominant firm, while avoiding the costs and liabilities of actual management. Powerful procurement departments emerge to define and impose the necessary quality standards and guard the lists of preferred suppliers.

The trend in many manufacturing businesses is to involve suppliers in a greater part of the manufacturing process so that much of the final production is the assembly of pre-fabricated subassemblies. This is particularly true of the automotive and aircraft industries. This is good practice as it:

- involves the supplier in the design process;
- reduces the number of items being managed within the business;
- reduces the number of suppliers; and
- improves quality management by placing the onus on suppliers to deliver fully checked and tested components and systems.

In retail, suppliers are even given access to daily sales figures and forecasts of demand which would normally be considered as highly confidential information. In the process, the freedom of local operating managers to pick and choose suppliers is reduced. Even though the responsibility to do so is often retained, it is strongly qualified by centrally imposed rules and lists, and assistance or oversight.

Suppliers have to be:

- trusted;
- treated with fairness in a partnership; and
- given full information to meet the demands being placed on them.

Under these conditions, suppliers and contractors looking for business with major firms need greater flexibility and wider competence than earlier. This often implies increased size and perhaps mergers, though in principle bids could be, and perhaps are, made by loose partnerships of smaller firms organised to secure such business.

Advantages of good supply chain management

Reduction of waste

This is an important objective of any business and involves not only waste of materials but also that of time. Examples of waste are:

- unwanted materials due to over ordering, damage or incorrect specifications;
- extraneous activities like double handling, for example between manufacturer, builders merchant and the site;
- re-working and re-fitting due to poor quality, design, storage or manufacture; and
- waste of time such as waiting for supplies due to excessive time from ordering to delivery or early delivery long before they are needed.

Faster reaction

A well-managed supply chain should be able to respond rapidly to changing requirements. Winter conditions require very different materials than those used during warmer or drier seasons. A contractor may have to modify plans rapidly and suppliers may need to ramp up or change production at short notice.

Reduction in accidents

A closer relationship between client, designers, principal contractors and suppliers of services and products can result not only in a safer finished product but, in construction, a safer method of erection. If more products are pre-assembled in ideal factory conditions and then fixed in place on site, it is often safer than utilising a full assembly approach in poor weather conditions on temporary work platforms. Examples are made-up roof trusses and pre-fabricated doors and windows already fitted to their frames.

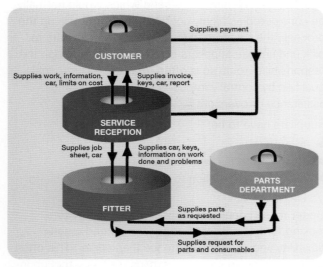

Figure 3.5 Typical supply chain

Legislation and standards

Legislation in a number of countries places a duty on everyone in the supply chain, from the designer to the final installer, of articles of plant or equipment for use at work. However, supply side legislation often has two purposes and is separate from much of the workplace occupational health and safety law. First, it provides a quality standard to protect users and consumers which includes their health and safety. Second, it can provide a barrier to trade by insisting that all imported goods must comply with a particular country's unique set of supply requirements.

Within the countries of the European Union many of the agreed standards for plant and equipment have been promulgated to remove barriers to trade between member states. A piece of equipment with a CE mark must be accepted as complying with all requirements across the EU and can be sold throughout the EU. See Chapter 10 for more details of this EU safeguard.

There are global examples of standards which are primarily designed to protect the health and safety of users. The best example of this is the ILO-sponsored Globally Harmonised System of Classification and Labelling of Chemicals, which is due to be fully operational by June 2015. This is an international agreement and requires local legislation in all trading blocs/countries to implement it globally. See Chapter 13 for more information.

ILO-OSH 2001 requires that procedures should be established and maintained to ensure that:

▶ compliance with safety and health requirements for the organisation is identified, evaluated and incorporated into purchasing and leasing specifications;

▶ national laws and regulations and the organisation's own occupational health and safety requirements are identified prior to the procurement of goods and services; and

▶ arrangements are made to achieve conformity to the requirements prior to their use.

Good practice in local legislation should:

▶ ensure that the article will be safe and without risk to health at all times when it is being set, used, cleaned or maintained;

▶ carry out any necessary testing and examination to ensure that it will be safe;

▶ provide adequate information about its safe setting, use, cleaning, maintenance, dismantling and disposal;

▶ ensure that there is an obligation on designers or manufacturers to do any research necessary to prove safety in use. Erectors or installers have special responsibilities to make sure, when handed over, that the plant or equipment is safe to use; and

▶ place similar duties on manufacturers and suppliers of substances for use at work to ensure that the substance is safe when properly used, handled, processed, stored or transported and to provide adequate information and do any necessary research, testing or examining.

Where articles or substances are imported, the suppliers' obligations outlined above attach to the importer, whether they are a separate importing business or the user personally who does the importing.

Often items are obtained through hire purchase, leasing or other financing arrangements with the ownership of the items being vested with the financing organisation. Where the financing organisation's only function is to provide the money to pay for the goods, the supplier's obligations do not attach to them.

Information for customers

The quality movement has drawn attention to the need to ensure that there are processes in place which ensure quality, rather than just inspecting and removing defects when it is too late. In much the same way, organisations need to manage health and safety rather than acting when it is too late.

Customers need information and specifications from the manufacturer or supplier – especially where there is a potential risk involved for them. When deciding what the supplier needs to pass on, careful thought is required about the health and safety factors associated with any product or service.

This means focusing on four key questions and then framing the information supplied so that it deals with each one. The questions are as follows:

▶ Are there any inherent dangers in the product or service being passed on – what could go wrong?
▶ What can the manufacturer or supplier do while working on the product or service to reduce the chance of anything going wrong later?
▶ What can be done at the point of handover to limit the chances of anything going wrong?
▶ What steps should customers take to reduce the chances of something going wrong? What precisely would they need to know?

Buying problems

Examples of problems that may arise when purchasing include:

▶ second-hand equipment which does not conform to current safety standards such as an office chair which does not provide adequate back support or have five feet/castors;
▶ starting to use new substances which do not have safety data sheets;
▶ machinery which, while well-guarded for operators, may pose risks for a maintenance engineer.

Figure 3.6 Inadequate chair – it should have five feet and an adjustable backrest – take care when buying especially if second-hand

A risk assessment should be done on any new product, taking into account the likely life expectancy (delivery, installation, use, cleaning, maintenance, disposal, etc.). The supplier should be able to provide the information needed to do this. This will help the purchaser make an informed decision on the total costs because the risks will have been identified as will the precautions needed to control those risks.

A note on CE marking

A risk assessment will still be needed for a CE-marked product. The CE marking signifies the manufacturer's declaration that the product conforms to relevant European Directives. Declarations from reputable manufacturers will normally be reliable. However, purchasers should be alert to fake or inadequate declarations and technical standards which may affect the health and safety of the product despite the CE mark. The risk assessment is still necessary to consider how and where the product will be used, what effect it might have on existing operations and what training will be required.

Employers have some key duties when buying plant and equipment:

▶ They must ensure that work equipment is safe and suitable for its purpose and complies with the relevant legislation. This applies equally to equipment which is adapted to be used in ways for which it was not originally designed.
▶ When selecting work equipment, they must consider existing working conditions and health and safety issues.
▶ They must provide adequate health and safety information, instructions, training and supervision for operators. Manufacturers and suppliers are required by law to provide information that will enable safe use of the equipment, substances, etc. and without risk to health.

Some of the issues that will need to be considered when buying in product or plant include:

▶ ergonomics – risk of work-related upper limb disorders (WRULDs);
▶ manual handling needs;
▶ access/egress;
▶ storage, for example of chemicals;
▶ risk to contractors when decommissioning old plant or installing new plant;
▶ hazardous materials – provision of extraction equipment or personal protective equipment;
▶ waste disposal;
▶ safe systems of work;
▶ training;
▶ machinery guarding;

▶ emissions from equipment/plant, such as noise, heat or vibration.

Appendix 3.3 gives a useful checklist for supply chain health and safety management.

Supply chains involving multinational companies

There has been an enormous increase in globalisation over the past three decades with technological improvements in information technology, communication and container shipping.

Multinational companies require management systems that can meet the different legislative requirements of countries in which they operate.

There are three main health and safety challenges that multinationals face in ensuring consistently safe ways of working across all the areas in which they operate, and throughout their supply chains. These challenges are:

▶ cultural diversity (due to the large numbers of nationalities in the workforce);
▶ workforce transience; and
▶ persuading senior managers to deal more seriously with health and safety issues.

All these challenges need to be addressed if a uniformly positive safety culture is to be established throughout an organisation and its supply chain. Cultural differences caused by a reliance on migrant workers can also lead to more specific challenges caused by language barriers, a limited knowledge of a host country's health and safety requirements, deference to authority in a new country, and an eagerness to please.

3.1.8 The relationship between the client and the contractor

A contractor is anyone who is brought in to work and is not an employee. People on work experience or on labour-only contracts or temporary staff are normally considered to be employees. Contractors are used for maintenance, repairs, installation, construction, demolition, computer work, cleaning, security, health and safety and many other tasks. Sometimes there are several contractors on site at any one time. Clients need to think about how their work may affect each other and how they interact with the normal site occupier.

The following points should be considered when contractors are employed:

▶ Health and safety must be included in the contract specification.
▶ All significant hazards must be included in the contract specification.
▶ The contractor must be selected with safety in mind.
▶ Prior to the start of work, health and safety policies should be exchanged.

▶ The contractor must be given basic site and health and safety information, such as welfare and first-aid arrangements, significant hazards, safe storage of chemicals and the name of the contract supervisor.
▶ Where appropriate, the contractor should supply risk assessments and method statements.
▶ For construction work, the contractor should be aware of the position of buried/overhead services and the arrangements for the disposal of waste.
▶ The contractor should be monitored during the progress of the contract by the contract supervisor.
▶ The contract supervisor should check that the work has been completed safely at the end of the contract.

Legal considerations

Most contracted work is governed by a legally binding contract and it is, therefore, very important that the contract covers all parts of the work – for example, competent workers, safe access when working at height, fire precautions and safe waste disposal.

All parties to a contract normally have specific responsibilities under local health and safety law, and these cannot be passed on to someone else. For example:

▶ employers are responsible for protecting people from harm caused by work activities. This includes the responsibility not to harm contractors and sub-contractors on site;
▶ employees and contractors have to take care not to endanger themselves, their colleagues or others affected by their work;
▶ contractors also have to comply with local health and safety legislation. Clearly, when contractors are engaged, the activities of different employers do interact. So cooperation and communication are needed to make sure all parties can meet their obligations;
▶ employees have to cooperate with their employer on health and safety matters, and not do anything that puts them or others at risk;
▶ employees must be trained and clearly instructed in their duties;
▶ self-employed people must not put themselves in danger, or others who may be affected by what they do;
▶ suppliers of chemicals, machinery and equipment have to make sure their products or imports are safe, and provide information on this.

Good practice encourages employers to take a more systematic approach to dealing with health and safety by:

▶ assessing the risks which affect employees and anyone who might be affected by the site occupier's work, including contractors;
▶ setting up emergency procedures;

▶ providing training;
▶ cooperating with others on health and safety matters, for example contractors who share the site with an occupier;
▶ providing temporary workers, such as contractors, with health and safety information.

The principles of cooperation, coordination and communication between organisations in a construction project are explained next.

ILO-OSH 2001 requires that arrangements should be established and maintained for ensuring that the organisation's safety and health requirements, or at least the equivalent, are applied to contractors and their workers.

Arrangements for contractors working on site should:

▶ include occupational health and safety criteria in procedures for evaluating and selecting contractors;
▶ establish effective ongoing communication and coordination between appropriate levels of the organisation and the contractor prior to commencing work. This should include provisions for communicating hazards and the measures to prevent and control them;
▶ include arrangements for reporting of work-related injuries, ill-health, diseases and incidents among the contractors' workers while performing work for the organisation;
▶ provide relevant workplace safety and health hazard awareness and training to contractors or their workers prior to commencing work and as work progresses, as necessary;
▶ regularly monitor occupational health and safety performance of contractor activities on site; and
▶ ensure that on-site occupational health and safety procedures and arrangements are followed by the contractor(s).

Construction projects

Businesses often engage contractors for construction projects to build plant, convert or extend premises and demolish buildings. Good practice on construction projects requires the following details.

Clients must:

▶ check the competence of all their appointees;
▶ ensure there are suitable management arrangements for the project;
▶ allow sufficient time and resources for all stages; and
▶ provide pre-construction information to designers and contractors.

Designers must:

▶ eliminate hazards and reduce risks during design; and
▶ provide information about remaining risks.

Contractors must:

▶ plan, manage and monitor their own work and that of employees;
▶ check the competence of all their appointees and employees;
▶ train their own employees;
▶ provide information to their employees;
▶ comply with the requirements for health and safety on site detailed in local legislation; and
▶ ensure there are adequate welfare facilities for their employees.

Everyone must:

▶ assure their own competence;
▶ cooperate with others and coordinate work so as to ensure the health and safety of construction workers and others who may be affected by the work; and
▶ report obvious risks; take account of the general principles of prevention in planning or carrying out construction work; and comply with the requirements of local legislation for any work under their control.

For even small projects, clients should ensure that contractors provide:

▶ information regarding the contractor's health and safety policy;
▶ information on the contractor's health and safety organisation detailing the responsibilities of individuals;
▶ information on the contractor's procedures and standards of safe working;
▶ the method statements for the project in hand;
▶ details on how the contractor will audit and implement its health and safety procedure; and
▶ procedures for investigating incidents and learning the lessons from them.

Smaller contractors may need some guidance to help them produce suitable method statements. While they do not need to be lengthy, they should set out those features essential to safe working, for example access arrangements, PPE and control of chemical risks.

Copies of relevant risk assessments for the work to be undertaken should be requested. These need not be very detailed but should indicate the risk and the control methods to be used.

3.1.9 The principles of assessing and managing contractors

Scale of contractor use

The use of contractors is increasing as many companies turn to outside resources to supplement their own staff and expertise. Contractors are often used for information technology projects, maintenance, repairs, installation, construction, demolition and many other jobs. Organisations have increasingly used contractors over recent years rather than employ more full-time employees. This may be to supplement their staff particularly for specialist tasks, or to undertake non-routine activities. There may also be other reasons for the increased use of contractors:

▶ Demand for products or services is uncertain.
▶ Contractors can be used when more flexibility is required.
▶ Contractors usually supply their own tools and equipment associated with the contract.
▶ There are no permanent staff available to perform the work.
▶ The financial overheads and legal employment obligations are lower.
▶ Most of the costs associated with increasing and reducing employee numbers as product demand varies do not relate to contractors.
▶ Permanent staff can concentrate on the core business of the organisation.

The advantage of the flexibility of contractors in uncertain economic times is the main reason for the increased use of them over recent years. There are, however, some disadvantages in the use of independent contractors. These include:

▶ Contractors/sub-contractors may cost more than the equivalent daily rate for employing a worker.
▶ By relying on contractors, the skills of permanent staff are not developed.
▶ There is less control over contractors than permanently employed staff. This can be a significant problem if the contractor sub-contracts some of the work.
▶ The control of the contractor and the quality of the work is crucially dependent on the terms of the contract.

International contract work

The UK Institution of Occupational Safety and Health has produced a useful advice document *'Global best practices in contractor safety'*.

It recommends the following key elements of an international company's health and safety programme for contractors:

▶ involving senior level management;
▶ providing an on-site safety manager, certified in the relevant country; and
▶ involving local health and safety personnel.

While the project is in preparation, the following should be considered:

▶ contractor qualifications, including a background check and insurance review;
▶ pre-employment procedures, including drug and alcohol screening and site safety orientation, plus specific task orientation with a middle management foreman;
▶ employee training;
▶ first-aid and emergency procedures;

▶ workers to organise their tasks effectively by identifying hazards and prepare a plan to control them; and
▶ fall protection for work at height and working procedures in confined spaces.

Emergency plans must be established with strategies in place.

The document identifies the following key issues for all international contracts:

▶ an acceptable balance between corporate and local values, standards and cultures at a typical worksite (involving international and/or national employers) with a potentially multinational workforce;
▶ the availability and use of internationally agreed standards, for example from the International Labour Organisation, the International Organisation for Standardisation, the World Bank, the International Monetary Fund and the United Nations;
▶ variations in type of contract and size of organisation – for example those working on a small sub-contract have much less freedom of action.

For both clients and contractors, good practice management solutions include:

▶ the publication of clear corporate value statements covering ethics, human rights and health and safety risk acceptance;
▶ being willing to challenge constructively local health and safety standards where they are lower than corporate ones;
▶ at all levels of the organisation implement a health and safety management system based on a recognised international standard;
▶ the inclusion of all contractors in safety improvement targets;
▶ setting realistic, balanced health and safety key performance indicators, including both leading and lagging indicators, and defining clear accountabilities;
▶ publishing and sharing results and lessons learned, both internally and through external bodies including poor results as well as good ones.

For developing economies, the key issues for international contracts are:

▶ communication – the need to be able to speak and listen to local people in their own language, not via interpreters;
▶ the local availability of suitably safe and reliable equipment; and
▶ community and/or government expectations for infrastructure development and education or training as part of the contract.

For both clients and contractors, good practice solutions include:

1. For management personnel:
   ▶ a core, highly skilled, expatriate management team supplementing this with skilled local people, training them elsewhere if necessary

to ensure that good practices are more readily transferred;

▶ the setting of high personal standards by management team members;

▶ involving local government bodies to increase their long-term ownership of health and safety results, infrastructure, training and education; and

▶ being visible on site to engage with, praise and police the workforce.

2. For the local workforce:

▶ the provision of brief, practical and relevant training related to daily hazards;

▶ the implementation of an agreed education and training strategy;

▶ the provision of family welfare facilities for employees;

▶ the protection of local workers from hazards that they may not be aware of.

Contractor selection

The selection of the right contractor for a particular job is probably the most important element in ensuring that the risks to the health and safety of all involved in the activity and people in the vicinity are reduced as far as possible. Ideally, selection should be made from a list of approved contractors who have demonstrated that they are able to meet the client's requirements

The selection of a contractor has to be a balanced judgement with a number of factors taken into account. Fortunately, a contractor who works well and meets the client's requirements in terms of the quality and timeliness of the work is also likely to have a better-than-average health and safety performance. Cost, of course, will have to be part of the judgement but may not provide any indication of which contractor is likely to give the best performance in health and safety terms. In deciding which contractor should be chosen for a task, the following should be considered:

▶ Do they have an adequate health and safety policy?

▶ Can they demonstrate that the person responsible for the work is competent?

▶ Can they demonstrate that competent safety advice will be available?

▶ Do they monitor the level of accidents at their work site(s)?

▶ Do they have a system to assess the hazards of a job and implement appropriate control measures?

▶ Will they produce a method statement which sets out how they will deal with all significant risks?

▶ Do they have guidance on health and safety arrangements and procedures to be followed?

▶ Do they have effective monitoring arrangements?

▶ Do they use trained and skilled staff who are qualified where appropriate? (Judgement will be

→ Provide a SAFETY METHODS STATEMENT

required, as many construction workers have had little or no training except training on the job.)

▶ Can the company demonstrate that the employees or other workers used for the job have had the appropriate training and are properly experienced and, where appropriate, qualified?

▶ Can they produce good references indicating satisfactory performance?

Figure 3.7 Contractors at work from a MEWP (Mobile Elevating Work Platform)

Management and authorisation of contractors

The management of contractors is essential since accidents tend to happen more easily when the work of contractors is not properly supervised or the hazards of their job have not been identified and suitable controls introduced. It is important that contractors are made aware of:

▶ the health and safety procedures and rules of the organisation;

▶ the hazards on your site particularly those associated with the project;

▶ any special equipment or personal protective equipment that they need to use;

▶ the emergency procedures and the sound of the alarm; and

▶ the safe disposal of waste.

Good communication and supervision of contractors is essential with a named point of contact from the organisation known to the contractors. Regular liaison is important so that the progress of the contract to a satisfactory completion may be monitored.

It is important that the activities of the contractors within the organisation of the client are effectively planned and coordinated so there is minimal interference with the normal activities of the organisation and the health and safety of its staff is not put at risk. This means that staff should be made aware of the contractor activity and any particular hazards associated with it, such as dust and noise.

Contractors, their employees, and sub-contractors and their employees, should not be allowed to commence work on any client's site without authorisation signed by the company contact. The authorisation should clearly define the range of work that the contractor can carry out and set down any special requirements, for example protective clothing, fire exits to be left clear, and isolation arrangements.

Permits will be required for operations such as hot work. All contractors should keep a copy of their authorisation at the place of work. A second copy of the authorisation should be kept at the site and be available for inspection.

The company contact signing the authorisation will be responsible for all aspects of the work of the contractor. The contact will need to check as a minimum the following:

- that the correct contractor for the work has been selected;
- that the contractor has made appropriate arrangements for supervision of staff;
- that the contractor has received and signed for a copy of the contractor's safety rules;
- that the contractor is clear what is required, the limits of the work and any special precautions that need to be taken;
- that the contractor's personnel are properly qualified for the work to be undertaken.

The company contact should check whether sub-contractors will be used. They will also require authorisation, if deemed acceptable. It will be the responsibility of the company contact to ensure that sub-contractors are properly supervised.

Appropriate supervision will depend on a number of factors, including the risk associated with the job, experience of the contractor and the amount of supervision the contractor will provide. The responsibility for ensuring there is proper supervision lies with the person signing the contractor's authorisation.

The company contact will be responsible for ensuring that there is adequate and clear communication between different contractors and company personnel where this is appropriate.

Safety rules for contractors

In the conditions of contract, there should be a stipulation that the contractor and all of their employees adhere to the contractor's safety rules. Contractors' safety rules should contain as a minimum the following points:

- **Health and safety** – that the contractor operates to at least the minimum legal standard and conforms to accepted industry good practice;
- **Supervision** – that the contractor provides a good standard of supervision of their own employees;
- **Sub-contractors** – that they may not use sub-contractors without prior written agreement from the organisation;
- **Authorisation** – that each employee must carry an authorisation card issued by the organisation at all times while on site.

Example of rules for contractors

Contractors engaged by the organisation to carry out work on its premises will:

- familiarise themselves with so much of the organisation's health and safety policy as affects them and will ensure that appropriate parts of the policy are communicated to their employees, and any sub-contractors and employees of sub-contractors who will do work on the premises;
- cooperate with the organisation in its fulfilment of its health and safety duties to contractors and take the necessary steps to ensure the like cooperation of their employees;
- comply with their legal and moral health, safety and food hygiene duties;
- ensure the carrying out of their work on the organisation's premises in such a manner as not to put either themselves or any other persons on or about the premises at risk;
- ensure that where they wish to avail themselves of the organisation's first-aid arrangements/facilities while on the premises, written agreement to this effect is obtained prior to first commencement of work on the premises;
- supply a copy of their statement of policy, organisation and arrangements for health and safety;
- abide by all relevant provisions of the organisation's safety policy, including compliance with health and safety rules and relevant national legislation;
- ensure that on arrival at the premises, they and any other persons who are to do work under the contract report to reception or their designated organisation contact.

Without prejudice to the requirements stated above, contractors, sub-contractors and employees of contractors and sub-contractors will, to the extent that such matters are within their control, ensure:

► the safe handling, storage and disposal of materials brought onto the premises;
► that the organisation is informed of any hazardous substances brought onto the premises and that any relevant national legislation has been complied with;
► that fire prevention and fire precaution measures are taken in the use of equipment which could cause fires;
► that steps are taken to minimise noise and vibration produced by their equipment and activities;
► that scaffolds, ladders and other such means of access, where required, are erected and used in accordance with relevant national legislation and good working practice;
► that any welding or burning equipment brought onto the premises is in safe operating condition and used in accordance with all safety requirements;
► that any lifting equipment brought onto the premises is adequate for the task and has been properly tested/certified;
► that any plant and equipment brought onto the premises is in safe condition and used/operated by competent persons;
► that for vehicles brought onto the premises, any speed, condition or parking restrictions are observed;
► that compliance is made with the relevant requirements of national electrical safety law;
► that connection(s) to the organisation's electricity supply is from a point specified by its management and is by proper connectors and cables;
► that they are familiar with emergency procedures existing on the premises;
► that welfare facilities provided by the organisation are treated with care and respect;
► that access to restricted parts of the premises is observed and the requirements of food safety legislation are complied with;
► that any major or lost-time accident or dangerous occurrence on the organisation's premises is reported as soon as possible to their site contact;
► that where any doubt exists regarding health and safety requirements, advice is sought from the site contact.

The foregoing requirements do not exempt contractors from their statutory duties in relation to health and safety, but are intended to assist them in attaining a high standard of compliance with those duties.

3.1.10 Joint occupation of premises

The occupational health and safety legislation of many countries covers the cooperation required between two or more employers who share a workplace, whether on a temporary or a permanent basis. In this situation, each employer should:

► cooperate with other employers;
► take reasonable steps to coordinate between other employers to comply with legal requirements;
► take reasonable steps to inform other employers where there are risks to health and safety.

All employers and self-employed people involved should satisfy themselves that the arrangements adopted are adequate. Where a particular employer controls the premises, the other employers should help to assess the shared risks and coordinate any necessary control procedures. Where there is no controlling employer, the organisations present should agree joint arrangements to meet regulatory obligations, such as appointing a health and safety coordinator.

3.2 Concept of health and safety culture and its significance in the management of health and safety in an organisation

3.2.1 Definition of a health and safety culture

The health and safety culture of an organisation may be described as the development stage of the organisation in health and safety management at a particular time. HSG65 gives the following definition of a health and safety culture:

'The safety culture of an organisation is the product of individual and group values, attitudes, perceptions, competencies and patterns of behaviour that determine the commitment to, and the style and proficiency of, an organisation's health and safety management.

Organisations with a positive safety culture are characterised by communications founded on mutual trust, by shared perceptions of the importance of safety and by confidence in the efficacy of preventive measures.'

There is concern among some health and safety professionals that many health and safety cultures are developed and driven by senior managers with very little input from the workforce. Others argue that this arrangement is sensible because the legal duties are placed on the employer. A positive health and safety culture needs the involvement of the whole workforce just as a successful quality system does. There must be a joint commitment in terms of attitudes and values. The workforce must believe that the safety measures put in place will be effective and followed even when financial and performance targets may be affected.

3.2.2 ILO perspective on health and safety culture

The ILO defines a health and safety culture in its Convention C187 (Promotional Framework for Occupational Safety and Health Convention, 2006) as:

'A culture in which the right to a safe and healthy working environment is respected at all levels, where government, employers and workers actively participate in securing a safe and healthy working environment through a system of defined rights, responsibilities and duties, and where the principle of prevention is accorded the highest priority.'

It further recommends in R197 (Promotional Framework for Occupational Safety and Health Recommendation, 2006), that in promoting a national preventative health and safety culture member states should seek to:

▶ raise workplace and public awareness on occupational health and safety through national campaigns linked with, where appropriate, workplace and international initiatives;

▶ promote mechanisms for delivery of occupational health and safety education and training, in particular for management, supervisors, workers and their representatives, and government officials responsible for health and safety;

▶ introduce occupational health and safety concepts and, where appropriate, competencies, in educational and vocational training programmes;

▶ facilitate the exchange of occupational health and safety statistics and data among relevant authorities, employers, workers and their representatives;

▶ provide information and advice to employers and workers and their respective organisations and to promote or facilitate cooperation among them with a view to eliminating or minimising, so far as is reasonably practicable, work-related hazards and risks;

▶ promote, at the level of the workplace, the establishment of safety and health policies and joint health and safety committees and the designation of workers' occupational health and safety

representatives, in accordance with national law and practice; and

▶ address the constraints of micro-enterprises and small and medium-sized enterprises and contractors in the implementation of occupational health and safety policies and regulations, in accordance with national law and practice.

3.2.3 Safety culture and safety performance

The relationship between health and safety culture and health and safety performance

The following elements are the important components of a **positive** health and safety culture:

▶ leadership and commitment to health and safety throughout and at all levels of the organisation, which is demonstrated in a genuine and visible way;

▶ acceptance that high standards of health and safety are achievable as part of a long-term strategy formulated by the organisation requiring sustained effort and interest;

▶ a detailed assessment of health and safety risks in the organisation and the development of appropriate control and monitoring systems;

▶ a health and safety policy statement that conveys a sense of optimism and outlines short- and long-term health and safety objectives. Such a policy should also include codes of practice and required health and safety standards;

▶ relevant employee training programmes and communication and consultation procedure to ensure ownership and participation in health and safety throughout the organisation;

▶ systems for monitoring equipment, processes and procedures and the prompt rectification of any defects;

▶ the prompt investigation of all incidents and accidents and reports made detailing any necessary remedial actions.

If the organisation adheres to these elements, then a basis for a good performance in health and safety will have been established. However, to achieve this level of performance, sufficient financial and human resources must be made available for the health and safety function at all levels of the organisation.

Managers at all levels of the organisation must show their commitment to health and safety by leading through example and ensuring that adequate resources in terms of time, money and people are allocated to health and safety management. Good managers appear regularly on the shop floor, talk about health and safety and visibly demonstrate their commitment by their actions – such as stopping production to resolve issues. If employees perceive that managers are not seriously committed to high health and safety standards, they will assume that they are expected to put commercial

"Safety is, without doubt, the most crucial investment we can make. And the question is not what it costs us, but what it saves."

Robert E McKee, Chairman and Managing Director, Conoco (UK) Ltd

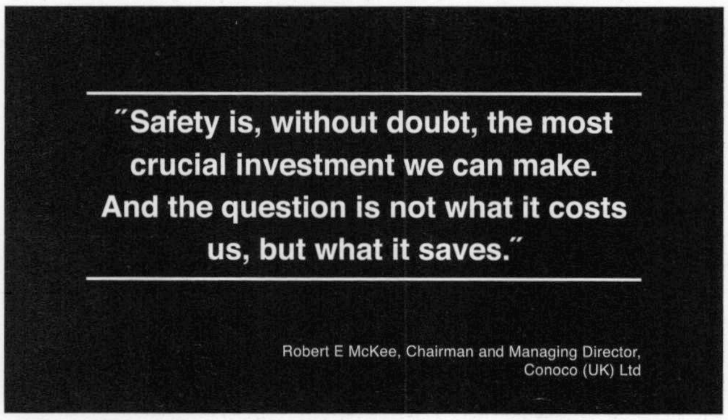

Figure 3.8 Safety investment

interests first, and safety initiatives or programmes will be undermined as a result.

All managers, supervisors and members of the governing body (e.g. directors) should receive training in health and safety and be made familiar during training sessions with the health and safety targets of the organisation. The depth of training undertaken will depend on the level of competence required of the particular manager. Managers should be accountable for health and safety within their departments and be rewarded for significant improvements in health and safety performance. They should also be expected to discipline employees within their departments who infringe health and safety policies or procedures.

Important indicators of a health and safety culture

As the safety culture of an organisation develops, increasing use should be made of specific leading indicators and the setting and monitoring of specific goals. The main indicators for the development of health and safety must:

▶ be objective and easy to measure and collect;
▶ be relevant to the organisation;
▶ provide immediate and reliable indication of the performance level;
▶ be cost-effective in terms of equipment, personnel and additional technology required to gather the information; and
▶ be understood and owned by the organisation.

The goals must be relevant and improve safety performance over a reasonable time period and be specific to the key risks and activities of the organisation.

There are several outputs or indicators of the state of the health and safety culture of an organisation. The most important are the numbers of accidents, near misses and occupational ill-health cases occurring within the organisation.

Although the number of accidents may give a general indication of the health and safety culture, a more detailed examination of accidents and accident statistics is normally required. A calculation of the rate of accidents enables health and safety performance to be compared between years and organisations.

The simplest measure of accident rate is called the incidence rate and is defined as:

$$\frac{\text{Total number of accidents}}{\text{Number of persons employed}} \times 1,000$$

or the total number of accidents per 1,000 employees.

A similar measure (per 100,000) is used by the UK HSE in its annual report on national accident statistics and enables comparisons to be made within an organisation between time periods when employee numbers may

change. It also allows comparisons to be made with the national occupational or industrial group relevant to the organisation.

There are four main problems with this measure which must be borne in mind when it is used. These are:

▶ there may be a considerable variation over a time period in the ratio of part-time to full-time employees;
▶ the measure does not differentiate between major and minor accidents and takes no account of other incidents, such as those involving damage but no injury (although it is possible to calculate an incidence rate for a particular type or cause of accident);
▶ there may be significant variations in work activity during the periods being compared;
▶ under-reporting of accidents will affect the accuracy of the data.

Some industries prefer the accident frequency rate per million hours worked. This method, by counting hours worked rather than the number of employees, avoids distortions by including part-time as well as full-time employees and overtime worked.

The calculation is:

$$\frac{\text{Number of accidents in the period}}{\text{Total hours worked in the period}} \times 1,000,000$$

There is more information on these two measures in Chapter 5.

Indications of a poor health and safety culture

Subject to the above limitations, an organisation with a high accident incidence rate is likely to have a negative or poor health and safety culture.

There are other indications of a poor health and safety culture or climate. These include:

▶ a high sickness, ill-health and absentee rate among the workforce;
▶ the perception of a blame culture;
▶ high staff turnover leading to a loss of momentum in making health and safety improvements;
▶ no resources (in terms of budget, people or facilities) made available for the effective management of health and safety;
▶ management decisions that consistently put production or cost before health and safety considerations;
▶ a lack of compliance with relevant health and safety law and the safety rules and procedures of the organisation;
▶ poor selection procedures and management of contractors;
▶ poor levels of communication, cooperation and control;

▶ a weak health and safety management structure;

▶ either a lack or poor levels of health and safety competence;

▶ high insurance premiums.

A positive health and safety culture is not enforceable through legislation. However, there can be enforcement that addresses the outcomes of a poor culture. For example if a company has unsuccessfully relied on procedural controls to avoid major accidents, there could be enforcement action to ensure compliance with relevant health and safety legislation or for the provision of appropriate safeguards.

In summary, a poor health and safety performance within an organisation is an indication of a negative health and safety culture.

Factors affecting a health and safety culture

The most important factor affecting the health and safety culture of an organisation is the commitment to health and safety from the top of an organisation. This commitment may be shown in many different ways. It needs to have a formal aspect in terms of an organisational structure, job descriptions and a health and safety policy, but it also needs to be apparent during crises or other stressful times. The health and safety procedures may be circumvented or simply forgotten when production or other performance targets are threatened.

Structural reorganisation or changes in market conditions will produce feelings of uncertainty among the workforce which, in turn, will affect the health and safety culture.

Poor levels of supervision, health and safety information and training are very significant factors in reducing health and safety awareness and, therefore, the culture.

Finally, the degree of consultation and involvement with the workforce in health and safety matters is crucial for a positive health and safety culture. Most of these factors may be summed up as human factors.

A safety culture questionnaire is given in Appendix 3.4.

3.3 Human factors which influence behaviour at work

3.3.1 Human factors

Over the years, there have been several studies undertaken to examine the link between various accident types, graded in terms of their severity, and near misses. One of the most interesting was conducted in the USA by H. W. Heinrich in 1950. He looked at over 300 accidents/incidents and produced the ratios illustrated in Figure 3.9.

This study indicated that for every 10 near misses, there will be an accident. Although the accuracy of

this study may be debated and other studies have produced different ratios, it is clear that if near misses are continually ignored, an accident will result. Further, the UK HSE Accident Prevention Unit has suggested that 90% of all accidents are due to human error and 70% of all accidents could have been avoided by earlier (proactive) action by management. It is clear from many research projects that the major factors in most accidents are human factors.

Figure 3.9 Heinrich's accidents/incidents ratios

While Heinrich's Triangle is useful in many ways, it does not suggest that minor and major incidents have the same causes as has been suggested by some technical papers over the years. Neither does it imply that the causes of incident frequency are the same as the causes of severe (major) injuries. The causes of severe injuries are often different to those of the causes of minor injuries. It is not reasonable, therefore, to assume that the majority of major injuries or incidents will occur where the majority of incidents take place. The hazards in a construction company's main office will not present the same major hazards as those at the company's various construction sites.

The UK HSE has defined human factors as, 'environmental, organisational and job factors, and human and individual characteristics which influence behaviour at work in a way which can affect health and safety'.

In simple terms, in addition to the environment, the health and safety of people at work are influenced by:

▶ the organisation;

▶ the job;

▶ individual (personal) factors.

These are known as human factors as they each have a human involvement. The personal factors which differentiate one person from another are only one part of those factors – and not always the most important.

Each of these elements will be considered in turn.

The organisation

The organisation is the company or corporate body and has the major influence on health and safety. It must have its own positive health and safety culture and produce an environment in which it:

▶ manages health and safety throughout the organisation, including the setting and publication of a health and safety policy and the establishment of a health and safety organisational structure;

▶ measures the health and safety performance of the organisation at all levels and in all departments. The performance of individuals should also be measured. There should be clear health and safety targets and standards and an effective reporting procedure for accidents and other incidents so that remedial actions may be taken;

▶ motivates managers within the organisation to improve health and safety performance in the workplace in a proactive rather than reactive manner.

An organisation needs to provide the following elements within its management system so that a positive health and safety culture may be developed:

▶ a clear and evident commitment from the most senior manager downwards, which provides a climate for safety in which management's objectives and the need for appropriate standards are communicated and in which constructive exchange of information at all levels is positively encouraged;

▶ an analytical and imaginative approach identifying possible routes to human factor failure. This may well require access to specialist advice;

▶ procedures and standards for all aspects of critical work and mechanisms for reviewing them;

▶ effective monitoring systems to check the implementation of the procedures and standards;

▶ incident investigation and the effective use of information drawn from such investigations;

▶ adequate and effective supervision with the power to remedy deficiencies when found.

It is important to recognise that there are often reasons for these elements not being present, resulting in weak management of health and safety. The most common reason is that individuals within the management organisation do not understand their roles – or their roles have never been fully explained to them. The higher a person is within the structure, the less likely it is that he/she has received any health and safety training. Such training at Board level is rare.

Objectives and priorities may vary across and between different levels in the structure, leading to disputes which affect attitudes to health and safety. For example, a warehouse manager may be pressured to block walkways so that a large order can be stored prior to dispatch.

Motivations can also vary across the organisation, which may cause health and safety to be compromised. The production controller will require that components of a product are produced as near simultaneously as possible so that their final assembly is performed as quickly as possible. However, the health and safety adviser will not want to see safe systems of work compromised.

To address some of these problems, it is important to define the safety duties of company directors. Each director and the Board, acting collectively, must provide health and safety leadership in the organisation. The Board needs to ensure that all its decisions reflect its health and safety intentions and that it engages the workforce actively in the improvement of health and safety. The Board will also be expected to keep itself informed of changes in health and safety risks.

The following simple checklist may be used to check any organisational health and safety management structure. Does the structure have:

▶ an effective health and safety management system?
▶ a positive health and safety culture?
▶ arrangements for the setting and monitoring of standards?
▶ adequate supervision?
▶ effective incident reporting and analysis?
▶ learning from experience?
▶ clearly visible health and safety leadership?
▶ suitable team structures?
▶ efficient communication systems and practices?
▶ adequate staffing levels?
▶ suitable work patterns?

HSG48 *Reducing Error and Influencing Behaviour* gives the following causes for failures in organisational and management structures:

▶ poor work planning leading to high work pressure;
▶ lack of safety systems and barriers;
▶ inadequate responses to previous incidents;
▶ management based on one-way communications;
▶ deficient coordination and responsibilities;
▶ poor management of health and safety;
▶ poor health and safety culture.

Organisational factors play a significant role in the health and safety of the workplace. However, this role is often forgotten when health and safety is being reviewed after an accident or when a new process or piece of equipment is introduced.

The job

Jobs may be highly dangerous or present only negligible risk of injury. Health and safety is an important element during the design stage of the job and any equipment, machinery or procedures associated with the job. Method study helps to design the job in the most cost-effective way and ergonomics helps to design the job with health and safety in mind. Ergonomics is the science of matching equipment, machines and processes to people rather than the other way round. An ergonomically designed machine will ensure that control levers, dials, meters and switches are sited in a convenient and comfortable position for the machine operator. Similarly, an ergonomically designed workstation will be designed for the comfort and health of the operator. Chairs, for example, will be designed to support the back properly throughout the working day (Figure 3.10).

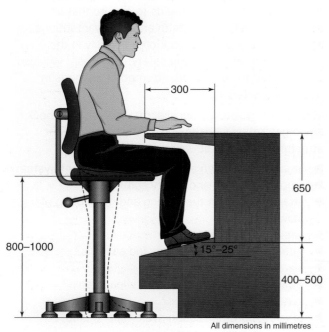

All dimensions in millimetres

Workstation where workers can sit or stand

Figure 3.10 Well-designed workstation for sitting or standing

Physically matching the job and any associated equipment to the person will ensure that the possibility of human error is minimised. It is also important to ensure that there is mental matching of the person's information and decision-making requirements. A person must be capable, either through past experience or through specific training, of performing the job with the minimum potential for human error.

The major considerations in the design of the job, which would be undertaken by a specialist, are as follows:

▶ the identification and detailed analysis of the critical tasks expected of individuals and the appraisal of any likely errors associated with those tasks;
▶ evaluation of the required operator decision making and the optimum (best) balance between the human and automatic contributions to safety actions (with the emphasis on automatic whenever possible);
▶ application of ergonomic principles to the design of man–machine interfaces, including displays of plant and process information, control devices and panel layout;

▶ design and presentation of procedures and operating instructions in the simplest terms possible;
▶ organisation and control of the working environment, including the workspace, access for maintenance, lighting, noise and heating conditions;
▶ provision of the correct tools and equipment;
▶ scheduling of work patterns, including shift organisation, control of fatigue and stress and arrangements for emergency operations;
▶ efficient communications, both immediate and over a period of time.

For some jobs, particularly those with a high risk of injury, a job safety analysis should be undertaken to check that all necessary safeguards are in place. All jobs should carry a job description and a safe system of work for the particular task. The operator should have sighted the job description and be trained in the safe system of work before commencing the job. More information on both these latter items is given in Chapter 4.

The following simple checklist may be used to check that the principal health and safety considerations of the job have been taken into account:

▶ Have the critical parts of the job been identified and analysed?
▶ Have the employee's decision-making needs been evaluated?
▶ Has the best balance between human and automatic systems been evaluated?
▶ Have ergonomic principles been applied to the design of equipment displays, including displays of plant and process information, control information and panel layouts?
▶ Has the design and presentation of procedures and instructions been considered?
▶ Has the guidance available for the design and control of the working environment, including the workspace, access for maintenance, lighting, noise and heating conditions, been considered?
▶ Have the correct tools and equipment been provided?
▶ Have the work patterns and shift organisation been scheduled to minimise their impact on health and safety?

Figure 3.11 Poor working conditions

▶ Has consideration been given to the achievement of efficient communications and shift handover?

HSG48 gives the following causes for failures in job health and safety:

▶ illogical design of equipment and instruments;
▶ constant disturbances and interruptions;
▶ missing or unclear instructions;
▶ poorly maintained equipment;
▶ high workload;
▶ noisy and unpleasant working conditions.

It is important that health and safety monitoring of the job is a continuous process. Some problems do not become apparent until the job is started. Other problems do not surface until there is a change of operator or a change in some aspect of the job.

It is very important to gain feedback from the operator on any difficulties experienced because there could be a health and safety issue requiring further investigation.

Individual (personal) factors

Individual or personal factors, which affect health and safety, may be defined as any condition or characteristic of an individual which could cause or influence him/her to act in an unsafe manner. They may be physical, mental or psychological in nature. Individual factors, therefore, include issues such as attitude, motivation, training and human error and their interaction with the physical, mental and perceptual capability of the individual.

Typical individual factors could include:

▶ self-interest in the case of incentive or bonus schemes to maximise earnings;
▶ loss of/or poor memory;
▶ loss of/or poor hearing;
▶ language issues and communication problems;
▶ physical health;
▶ experience and competence;
▶ personality and attitude;
▶ age;
▶ the position in the team; and
▶ training and competence.

Language and communication issues may be particular problems for some migrant workers. The law requires that employers provide workers with comprehensible and relevant information about risks and about the procedures they need to follow to ensure they can work safely and without risk to health. This does not have to be in English but can be in the language common to the region. The employer may make special arrangements, which could include translation, using interpreters or replacing written notices with clearly understood symbols or diagrams. Any health and safety training provided must take into account the worker's capabilities, including language skills.

The following **negative** factors can also affect the individual:

▶ low skill and competence levels;
▶ tired colleagues;
▶ bored or disinterested colleagues;
▶ individual medical or mental problems;
▶ complacency from repetitive tasks and lack of awareness training;
▶ inexperience especially if the employee is new or a young person; and
▶ peer pressure to conform to the 'group' or an individual's perception of how a task should be completed.

These factors have a significant effect on health and safety. Some of them, normally involving the personality of the individual, are unchangeable but others, involving skills, attitude, perception and motivation, can be changed, modified or improved by suitable training or other measures. In summary, the person needs to be matched to the job.

Studies have shown that the most common personal factors which contribute to accidents are low skill and competence levels, tiredness, boredom, low morale and individual medical problems. The setting of realistic timescales and work task schedules will lead to considerable improvements in these factors. However, the key issue is effective and relevant training. This can help to improve attitudes and the understanding of risks and safe working. It will also enhance or improve skills.

It is difficult to separate all the physical, mental or psychological factors because they are interlinked. However, the three most common factors are psychological factors: attitude, motivation and perception.

Attitude is the tendency to behave in a particular way in a certain situation. One of the principal aims of a good safety culture is to change behaviour. But, if this change is to become permanent, then attitude will have to change. However, good intentions are not enough and attitudes and behaviour do not always coincide. Attitudes are influenced by the prevailing health and safety culture within the organisation, the commitment of the management, the experience of the individual and the influence of the peer group. Peer group pressure is a particularly important factor among young people and health and safety training must be designed with this in mind by using examples or case studies that are relevant to them. Behaviour may be changed by training, the formulation and enforcement of safety rules and meaningful consultation – attitude change often follows.

Motivation is the driving force behind the way a person acts or the way in which people are stimulated to act. Involvement in the decision-making process in a meaningful way will improve motivation as will the use of incentive schemes. However, there are other

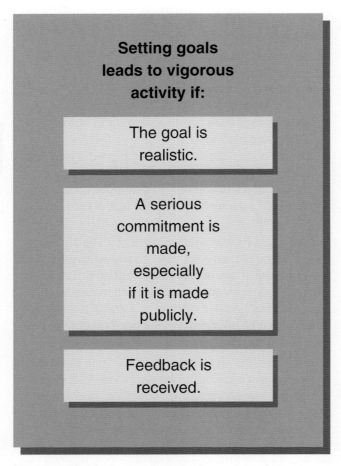

Setting goals leads to vigorous activity if:

The goal is realistic.

A serious commitment is made, especially if it is made publicly.

Feedback is received.

Figure 3.12 Motivation and activity

important influences on motivation such as recognition and promotion opportunities, job security and job satisfaction. Self-interest, in all its forms, is a significant motivator and personal factor.

Perception is the way in which people interpret the environment or the way in which a person believes or understands a situation (Figure 3.13). In health and safety, the perception of hazards is an important concern. Many accidents occur because people do not perceive that there is a risk. There are many common examples of this, including the use of personal protective equipment (such as hard hats) and guards on drilling machines and the washing of hands before meals. It is important to understand that when perception leads to an increased health and safety risk, it is not always caused by a conscious decision of the individual concerned. The stroboscopic effect caused by the rotation of a drill at certain speeds under fluorescent lighting will make the drill appear stationary. It is a well-known phenomenon, especially among illusionists, that people will often see what they expect to see rather than reality. Routine or repetitive tasks will reduce attention levels, leading to the possibility of accidents.

Other individual factors which can affect health and safety include physical stature, age, experience, health, hearing, intelligence, language skills, level of competence and qualifications.

Memory is an important personal factor, as it is influenced by training and experience. The efficiency of memory varies considerably between people and during the lifetime of an individual. The overall health of a person can affect memory as can personal crises. Owing to these possible problems with memory, important safety instructions should be available in written as well as verbal form.

Finally, it must be recognised that some employees do not follow safety procedures either due to peer pressure or a wilful disregard of those procedures.

The following checklist, given in HSG48, may be used to check that the relevant personal factors have been covered to help minimise human error:

► Has the job specification been drawn up and does it include age, physique, skill, qualifications, experience, aptitude, knowledge, intelligence and personality?
► Have the skills and aptitudes been matched to the job requirements?
► Have the personnel selection policies and procedures been set up to select appropriate individuals?

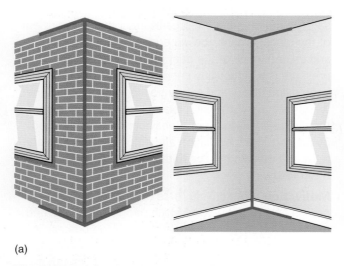

(a)

(b)

(c)

Figure 3.13 Visual perception: (a) Are the lines the same length? (b) Faces or vase? (c) Face or saxophone player?

- ▶ Has an effective training system been implemented?
- ▶ Have the needs of special groups of employees been considered?
- ▶ Have the monitoring procedures been developed for the personal safety performance of safety critical staff?
- ▶ Have fitness for work and health surveillance been provided where it is needed?
- ▶ Have counselling and support for ill-health and stress been provided?

Individual factors are the attributes that employees bring to their jobs and may be strengths or weaknesses. Negative personal factors cannot always be neutralised by improved job design. It is, therefore, important to ensure that personnel selection procedures should match people to the job. This will reduce the possibility of accidents or other incidents.

3.3.2 Human errors and violations

In 1985, a report by the UK HSE found that senior management failures resulted in 61% of all accidents, and 47% of those accidents had human causes. Human failures in health and safety are classified either as errors or violations. An error is an unintentional deviation from an accepted standard, whereas a violation is a deliberate deviation from the standard.

Human errors

Human errors fall into three groups – slips, lapses and mistakes, which can be further sub-divided into rule-based and knowledge-based mistakes.

Slips and lapses

Slips and lapses are very similar in that they are caused by a momentary memory loss often due to lack of attention or loss of concentration. They are not related to levels of training, experience or motivation and they can usually be reduced by re-designing the job or equipment or minimising distractions.

Slips are failures to carry out the correct actions of a task. Examples include the use of the incorrect switch, reading the wrong dial or selecting the incorrect component for an assembly. A slip also describes an action taken too early or too late within a given working procedure.

Lapses are failures to carry out particular actions which may form part of a working procedure. A fork-lift truck driver leaving the keys in the ignition lock of his truck is an example of a lapse, as is the failure to replace the petrol cap on a car after filling it with petrol. Lapses may be reduced by re-designing equipment so that, for example, an audible horn indicates the omission of a task. They may also be reduced significantly by the use of detailed checklists.

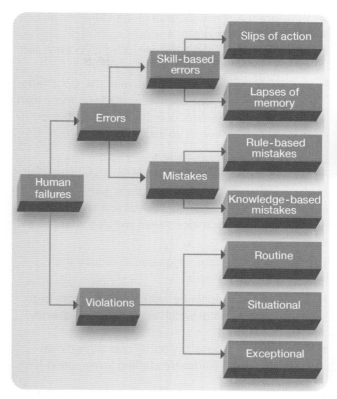

Figure 3.14 Types of human failure

Mistakes

Mistakes occur when an incorrect action takes place but the person involved believes the action to be correct. A mistake involves an incorrect judgement. There are two types of mistake – rule-based and knowledge-based.

Rule-based mistakes occur when a rule or procedure is remembered or applied incorrectly. These mistakes usually happen when, due to an error, the rule that is normally used no longer applies. For example, a particular job requires the counting of items into groups of ten followed by the adding together of the groups so that the total number of items may be calculated. If one of the groups is miscounted, the final total will be incorrect even though the rule has been followed.

Knowledge-based mistakes occur when well-tried methods or calculation rules are used inappropriately. For example, the depth of the foundations required for a particular building was calculated using a formula. The formula, which assumed a clay soil, was used to calculate the foundation depth in a sandy soil. The resultant building was unsafe.

The following points have been suggested by the UK HSE when the potential source of human errors is to be identified:

- ▶ What human errors can occur with each task? There are formal methods available to help with this task.
- ▶ What influences are there on performance? Typical influences include time pressure, design of controls, displays and procedures, training and experience, fatigue and levels of supervision.

67

▶ What are the consequences of the identified errors?

▶ What are the significant errors?

▶ Are there opportunities for detecting each error and recovering it?

▶ Are there any relationships between the identified errors? Could the same error be made on more than one item of equipment due, for example, to the incorrect calibration of an instrument?

Errors and mistakes can be reduced by the use of instruction, training and relevant information. However, communication can also be a problem, particularly at shift handover times. Environmental and organisational factors involving workplace stress will also affect error levels.

The following steps are suggested to reduce the likelihood of human error:

▶ Examine and reduce the workplace stressors (e.g. noise, poor lighting) which increase the frequency of errors.

▶ Examine and reduce any social or organisational stressors (e.g. insufficient staffing levels, peer pressure).

▶ Design plant and equipment to reduce error possibilities (e.g. poorly designed displays, ambiguous instructions).

▶ Ensure that there are effective training arrangements.

▶ Simplify any complicated or complex procedures.

▶ Ensure that there is adequate supervision, particularly for inexperienced or young trainees.

▶ Check that job procedures, instructions and manuals are kept up to date and are clear.

▶ Include the possibility of human error when undertaking the risk assessment.

▶ Isolate the human error element of any accident or incident and introduce measures to reduce the risk of a repeat.

▶ Monitor the effectiveness of any measures taken to reduce errors.

Violations

There are three categories of violation: routine, situational and exceptional.

Routine violation occurs when the breaking of a safety rule or procedure is the normal way of working. It becomes routine not to use the recommended procedures for tasks. An example of this is the regular high speed use of fork-lift trucks in a warehouse so that orders can be fulfilled on time.

There are many reasons given for routine violations; for example:

▶ taking short cuts to save time and energy;

▶ a belief that the rules are unworkable or too restrictive;

▶ lack of knowledge of the procedures;

▶ perception that the rules are no longer applied;

▶ poor supervision and a lack of enforcement of the rules;

▶ new workers thinking that routine violations are the norm and not realising that this was not the safe way of working.

Finally, it must be recognised that there are some situations where peer pressure or simply a wilful disregard for procedures or other people's safety may result in routine violations. Routine violations can be reduced by regular monitoring, ensuring that the rules are actually necessary, or re-designing the job.

The following features are very common in many workplaces and often lead to routine violations:

▶ poor working posture due to poor ergonomic design of the workstation or equipment;

▶ equipment difficult to use and/or slow in response;

▶ equipment difficult to maintain or pressure on time available for maintenance;

▶ procedures unduly complicated and difficult to understand;

▶ unreliable instrumentation and/or warning systems;

▶ high levels of noise and other poor aspects to the environment (fumes, dusts, humidity);

▶ associated PPE either inappropriate, difficult and uncomfortable to wear or ineffective due to lack of maintenance.

Situational violations occur when particular job pressures at particular times make rule compliance difficult. They may happen when the correct equipment is not available or weather conditions are adverse. A common example is the use of a ladder rather than a scaffold for working at height to replace window frames in a building. Situational violations may be reduced by improving job design, the working environment and supervision.

Exceptional violations rarely happen and usually occur when a safety rule is broken to perform a new task. Good examples of this are the violations which can occur during the operations of emergency procedures such as for fires or explosions. These violations should be addressed in risk assessments and during training sessions for emergencies (e.g. fire training).

Everybody is capable of making errors. It is one of the objectives of a positive health and safety culture to reduce them and their consequences as much as possible.

3.4 How health and safety behaviour at work can be improved

Safe behaviour should be encouraged and strengthened through positive reinforcement by good example and supervision. Workers should be encouraged to set their own safety targets. Positive feedback on behaviour can sometimes reverse normal peer pressure so that the safe working becomes the accepted social norm. However, safe methods of working need to be identified and communicated to workers. This requires a significant amount of time and resources, but will result in an improvement in health and safety performance. The success of behavioural change is very dependent on the commitment of management.

No single section or department of an organisation can develop a positive health and safety culture on its own. There needs to be commitment by the management, the promotion of health and safety standards, identifying and keeping up to date with legal requirements, effective communication within the organisation, cooperation from and with the workforce and an effective and developing training programme. Each of these topics will be examined in turn to show their effect on improving the health and safety culture in the organisation.

3.4.1 Commitment from management

As mentioned earlier, there needs to be a commitment from the very top of the organisation and this commitment will, in turn, produce higher levels of motivation and commitment throughout the organisation. Probably the best indication of this concern for health and safety is shown by the status given to health and safety and the amount of resources (money, time and people) allocated accordingly. The management of health and safety should form an essential part of a manager's responsibility and they should be held to account for their performance on health and safety issues. Specialist expertise should be made available when required (e.g. for noise assessment), either from within the workforce or by the employment of external contractors or consultants. Health and safety should be discussed on a regular basis at management meetings at all levels of the organisation. If the organisation employs sufficient people to make direct consultation with all employees difficult, there should be a health and safety committee at which there is employee representation. In addition, there should be recognised routes for anybody within the organisation to receive health and safety information or have their health and safety concerns addressed.

The health and safety culture is enhanced considerably when senior managers appear regularly at all levels of an organisation, whether it be the shop floor, the hospital ward or the general office, and are willing to discuss health and safety issues with staff. A visible management is very important for a positive health and safety culture.

Finally, the positive results of management commitment to health and safety will be the active involvement of all employees in health and safety, the continuing improvement in health and safety standards and the subsequent reduction in accident and occupational ill-health rates. This will lead, ultimately, to a reduction in the number and size of compensation claims.

HSG48 makes some interesting suggestions to managers on the improvements that may be made to health and safety which will be seen by the workforce as a clear indication of their commitment. The suggestions are:

▶ review the status of the health and safety committees and health and safety practitioners. Ensure that any recommendations are acted upon or implemented;
▶ ensure that senior managers receive regular reports on health and safety performance and act on them;
▶ ensure that any appropriate health and safety actions are taken quickly and are seen to have been taken;
▶ any action plans should be developed in consultation with employees, based on a shared perception of hazards and risks, be workable and continually reviewed.

3.4.2 The promotion of health and safety standards

For a positive health and safety culture to be developed, everyone within the organisation needs to understand the standards of health and safety expected by the organisation and the role of the individual in achieving and maintaining those standards. Such standards are required to control and minimise health and safety risks.

Standards should clearly identify the actions required of people to promote health and safety. They should also specify the competencies needed by employees and should form the basis for measuring health and safety performance.

Health and safety standards cover all aspects of the organisation. Typical examples include:

▶ the design and selection of premises;
▶ the design and selection of plant and substances (including those used on site by contractors);
▶ the recruitment of employees and contractors;
▶ the control of work activities, including issues such as risk assessment;
▶ competence, maintenance and supervision;
▶ emergency planning and training;
▶ the transportation of the product and its subsequent maintenance and servicing.

Having established relevant health and safety standards, it is important that they are actively promoted within the organisation by all levels of management. The most effective method of promotion is by leadership and example. There are many ways to do this such as:

▶ the involvement of managers in workplace inspections and accident investigations;
▶ the use of PPE (e.g. goggles and hard hats) by all managers and their visitors in designated areas;
▶ ensuring that employees attend specialist refresher training courses when required (e.g. first-aid and fork-lift truck driving);

▶ full cooperation with fire drills and other emergency training exercises;
▶ comprehensive accident reporting and prompt follow-up on recommended remedial actions.

The benefit of good standards of health and safety will be shown directly in less lost production, accidents and compensation claims, and lower insurance premiums. It may also be shown in higher product quality and better resource allocation.

An important and central necessity for the promotion of high health and safety standards is health and safety competence. What is meant by 'competence'?

Competence

The word 'competence' is often used in health and safety literature. One definition, made during a civil case in 1962 in the UK, stated that a competent person is:

> *'a person with practical and theoretical knowledge as well as sufficient experience of the particular machinery, plant or procedure involved to enable them to identify defects or weaknesses during plant and machinery examinations, and to assess their importance in relation to the strength and function of that plant and machinery.'*

This definition concentrates on a manufacturing rather than service industry requirement of a competent person. Occupational health and safety competence includes education, work experience and training, or a combination of these.

The ILO has defined competence more broadly in the ILO-OSH 2001 management system in terms of an authority, an institution and a person as follows:

▶ **Competent institution** – A government department or other body with the responsibility to establish a national policy and develop a national framework for health and safety management systems in organisations and to provide relevant guidance.
▶ **Competent authority** – A minister, government department or other public authority having the power to issue regulations, orders or other instructions having the force of law.
▶ **Competent person** – A person possessing adequate qualifications, such as suitable training and sufficient knowledge, experience and skill, for the safe performance of the specific work. The competent authorities may define appropriate criteria for the designation of such persons and may determine the duties to be assigned to them.

The ILO further recommends that the necessary health and safety competence requirements should be defined by the employer, and arrangements established and maintained to ensure that all persons are competent to carry out the health and safety aspects of their duties and responsibilities. The employer should have, or have access to, sufficient health and safety competence to identify and eliminate or control work-related hazards and risks, and to implement the health and safety management system. Competence training programmes should:

▶ cover all members of the organisation, as appropriate;
▶ be conducted by competent persons;
▶ provide effective and timely initial and refresher training at appropriate intervals;
▶ include participants' evaluation of their comprehension and retention of the training;
▶ be reviewed periodically. The review should include the safety and health committee, where it exists, and the training programmes, modified as necessary to ensure their relevance and effectiveness; and
▶ be documented, as appropriate and according to the size and nature of activity of the organisation.

Finally the ILO recommends that training should be provided to all participants at no cost and should take place during working hours, if possible.

Competent persons are required to assist the employer in meeting their obligations under the particular national health and safety law. This may mean a health and safety adviser in addition to, say, an electrical engineer, an occupational nurse and a noise assessment specialist. The number and range of competent persons will depend on the nature of the business of the organisation.

It is recommended that competent employees are used for advice on health and safety matters rather than external specialists (consultants). However, if employees competent in health and safety are not available in the organisation, then an external service may be enlisted to help. The key is that management and employees need access to health and safety expertise.

Employers should appoint health and safety advisers who should have:

▶ a knowledge and understanding of the work involved, the principles of risk assessment and prevention and current health and safety applications;
▶ the capacity to apply this to the task required by the employer in the form of problem and solution identification, monitoring and evaluating the effectiveness of solutions and the promotion and communication of health and safety and welfare advances and practices.

Such competence does not necessarily depend on the possession of particular skills or qualifications. It may only be required to understand relevant current best practice, be aware of one's own limitations in terms of experience and knowledge and be willing to supplement existing experience and knowledge. However, in more complex or technical situations, membership of a relevant professional body and/or the possession of an appropriate qualification in health and safety may be necessary. It is important that any competent person employed to help with health and safety has evidence of relevant knowledge, skills and experience for the tasks involved. The appointment of a competent person as an adviser to an employer does not normally absolve the employer from his/her responsibilities under relevant national statutory health and safety legislation.

Finally, it is worth noting that the requirement to employ competent workers is not restricted to those having a health and safety function but covers the whole workforce.

Competent workers must have sufficient training, experience, knowledge and other qualities to enable them to properly undertake the duties assigned to them.

3.4.3 Identifying and keeping up to date with legal requirements

The comparisons with local legal standards and requirements together with ILO Conventions is an essential part of the health and safety planning process. This is important even if the organisation wishes to achieve more than minimum legal compliance. The organisation should have systems to identify local legal requirements and keep up to date with changes to the law. This might be achieved by:

▶ regular checking of a specific country's OSH organisations' websites for updates;
▶ checking international sources of information such as the ILO (www.ilo.org/global/standards/lang–en/index.htm), WHO (www.who.int/about/en);
▶ checking the OSH World website (www.oshworld.com);
▶ reading health and safety periodicals from Occupational Safety and Health (OSH) organisations;
▶ many trade associations also help members keep up to date through their journals or websites.

There is more information on OSH websites and guidance on searching the internet in Chapter 18. Today there are so many ways of keeping up to date through the internet that the comment 'ignorance of the law is no excuse' is more poignant than ever.

3.4.4 Effective communication

Many problems in health and safety arise due to poor communication. It is not just a problem between management and workforce; it is often a problem the other way or indeed at the same level within an organisation. It can arise from ambiguities or, even, accidental distortion of a message.

There are three basic methods of communication in health and safety: verbal, written and graphic.

Verbal communication is the most common. It is communication by speech or word of mouth. Verbal communication should only be used for relatively simple pieces of information or instruction. It is most commonly used in the workplace, during training sessions or at meetings.

There are several potential problems associated with verbal communication. The speaker needs to prepare the communication carefully so that there is no confusion about the message. It is very important that the recipient is encouraged to indicate their understanding of the communication. There have been many cases of accidents occurring because a verbal instruction has not been clearly understood. There are several barriers to this understanding from the point of view of the recipient, including language and dialect, the use of technical language and abbreviations, background noise and distractions, hearing limitations, ambiguities in the message, mental weaknesses and learning disabilities, and lack of interest and attention.

Having described some of the limitations of verbal communication, it does have some merits. It is less formal, enables an exchange of information to take place quickly and the message to be conveyed as near to the workplace as possible. Training or instructions that are delivered in this way are called toolbox talks and can be very effective.

Written communication takes many forms from a simple memo to a detailed report.

A memo should contain one simple message and be written in straightforward and clear language. The title should accurately describe the contents of the memo. In recent years, emails have largely replaced memos, as it has become a much quicker method to ensure that the message gets to all concerned (although a recent report has suggested that many people are becoming overwhelmed by the number of emails which they receive!). The advantage of memos and emails is that there is a record of the message after it has been delivered. The disadvantage is that they can be ambiguous or difficult to understand or, indeed, lost within the system.

Reports are more substantial documents and cover a topic in greater detail. The report should contain a detailed account of the topic and any conclusions or recommendations. The main problem with reports is that they are often not read properly due to the time constraints on managers. It is important that all reports have a summary attached so that the reader can decide whether it needs to be read in detail (see Chapter 5 for further discussion on report writing).

The most common way in which written communication is used in the workplace is the **notice board**. For a notice board to be effective it needs to be well positioned within the workplace and there needs to be a regular review of the notices to ensure that they are up to date and relevant. The use of notice boards as a means of communicating health and safety information to employees has some limitations that include:

▶ the information may not be read;
▶ the notice boards may not be accessible;
▶ the information may become outdated or defaced;
▶ some employees may not be able to read while others may not understand what they have read;
▶ there may be language barriers;
▶ the information is mixed in with other non-health and safety information; and
▶ there is no opportunity offered for feedback.

The following alternative methods could be used for the communication of essential health and safety information:

▶ memos, emails and company intranet;
▶ toolbox talks and team briefings;
▶ induction training and any further back-up training sessions;
▶ newsletters, bulletins and payslips;
▶ digital video media including DVDs;
▶ a staff handbook; and
▶ through safety committees, safety representatives, and representatives of employee safety.

The following types of health and safety information could be displayed on a workplace notice board:

▶ a copy of the Employer's Liability Insurance Certificate;
▶ details of first-aid arrangements;
▶ emergency evacuation and fire procedures;
▶ minutes of the last health and safety committee meeting;
▶ details of health and safety targets and performance against them;
▶ health and safety posters and campaign details.

There are many other examples of written communications in health and safety, such as employee handbooks, company codes of practice, minutes of safety committee meetings and health and safety procedures.

Graphic communication is communication by the use of drawings, photographs or DVDs. It is used to impart either health and safety information (e.g. fire exits) or health and safety propaganda. The most common forms of health and safety propaganda are the poster and the DVD. Both can be used very effectively as training aids, as they can retain interest and impart a simple message. Their main limitation is that they can become out of date fairly quickly or, in the case of posters, become largely ignored.

The ILO in the ILO-OSH 2001 management system recommends that arrangements and procedures should be established and maintained for:

▶ receiving, documenting and responding appropriately to internal and external communications related to health and safety;

▶ ensuring the internal communication of health and safety information between relevant levels and functions of the organisation; and
▶ ensuring that the concerns, ideas and inputs of workers and their representatives on health and safety matters are received, considered and responded to.

There are many sources of health and safety information which may need to be consulted before an accurate communication can be made. These include national legislation, Codes of Practice, guidance, ILO and European standards, periodicals, case studies and publications (for example, from the ILO and the UK HSE).

3.4.5 Consultation with the workforce

General

It is important to gain the cooperation of all employees if a successful health and safety culture is to become established. This cooperation is best achieved by consultation. Joint consultation can help businesses be more efficient and effective by reducing the number of accidents and work-related ill-health and also to motivate staff by making them aware of health and safety problems.

The ILO Recommendations R164 require that:

'The measures taken to facilitate the cooperation referred to in Article 20 of the Convention C155 should include, where appropriate and necessary, the appointment, in accordance with national practice, of workers' safety delegates, of workers' safety and health committees, and/or of joint safety and health committees. In joint safety and health committees workers should have at least equal representation with employers' representatives. Workers' safety delegates, workers' safety and health committees, and joint safety and health committees or, as appropriate, other workers' representatives should:'

▶ be given adequate information on safety and health matters, to enable them to examine factors affecting safety and health, and be encouraged to propose measures on the subject;
▶ be consulted when major new safety and health measures are envisaged and before they are carried out, and seek to obtain the support of the workers for such measures;
▶ be consulted in planning alterations of work processes, work content or organisation of work, which may have safety or health implications for the workers;
▶ be given protection from dismissal and other measures prejudicial to them while exercising their functions in the field of occupational safety and health as workers' representatives or as members of safety and health committees;

▶ be able to contribute to the decision-making process at the level of the undertaking regarding matters of safety and health;

▶ have access to all parts of the workplace and be able to communicate with the workers, including contractors, on safety and health matters during working hours at the workplace;

▶ be free to contact labour inspectors;

▶ be able to contribute to negotiations in the undertaking on occupational safety and health matters;

▶ have reasonable time during paid working hours to exercise their safety and health functions and to receive training related to these functions;

▶ have recourse to specialists to advise on particular safety and health problems.

It follows, therefore, that the role of workers' safety delegates includes to:

▶ inform the employer of health and safety concerns of the workforce;

▶ inform the employer of potential hazards and dangerous occurrences in the workplace;

▶ inform the employer of any general matters that affect the health and safety of the workforce; and

▶ speak on behalf of the workforce to health and safety enforcement inspectors.

Employers should consult on the following issues:

▶ new processes or equipment or any changes in them;

▶ the appointment arrangements for the health and safety competent person;

▶ the results of any risk assessments;

▶ the arrangements for the management of health and safety training; and

▶ the introduction of new technologies.

Safety committees

In medium and large organisations, the easiest and often the most effective method of consultation is the health and safety committee. It will realise its full potential if its recommendations are seen to be implemented and both management and employee concerns are freely discussed. It will not be so successful if it is seen as a talking shop.

The committee should have stated objectives which mirror the objectives in the organisation's safety and health policy statement, and its own terms of reference. Terms of reference should include the following:

▶ the study of accident and notifiable disease statistics to enable reports to be made of recommended remedial actions;

▶ the examination of occupational health and safety audits and statutory inspection reports;

▶ the consideration of reports from the external enforcement agency;

▶ the review of new legislation and guidance and their effect on the organisation;

▶ the monitoring and review of all occupational health and safety training and instruction activities in the organisation;

▶ the monitoring and review of occupational health and safety publicity and communication throughout the organisation;

▶ development of safe systems of work and safety procedures;

▶ reviewing risk assessments;

▶ considering reports from safety representatives;

▶ continuous monitoring of arrangements for occupational health and safety and revising them whenever necessary.

In some countries there are fixed rules on the composition of the health and safety committee, which must be followed. Good practice would require that:

▶ it should be representative of the whole organisation;

▶ it should have representation from the workforce and the management (probably in equal numbers) including at least one senior manager (other than the Health and Safety Adviser);

▶ manager and worker safety delegates should agree who should chair the meetings, how often meetings should be held and what they hope to achieve.

Figure 3.15 Effective consultation is essential

Accident and ill-health investigations

Properly investigated accidents and cases of ill-health can reveal weaknesses that need to be remedied. A joint investigation with safety delegates is more likely to inspire confidence with workers so that they cooperate fully with the investigation, as in many cases those involved may be concerned about being blamed for the accident.

Safety delegates are entitled to contact labour inspectors. If this is just for information, they can be contacted directly. If it is a formal complaint against the employer, the labour inspector will need to know if the employer has been informed. The inspectors can be contacted anonymously. They will normally keep the person's identity secret in such circumstances.

Training, facilities and assistance

Safety delegates are legally entitled to paid leave for training. Training course topics often include:

▶ the role and functions of the safety delegate;
▶ health and safety legislation;
▶ how to identify and minimise hazards;
▶ how to carry out a workplace inspection and accident investigation;
▶ employer's health and safety arrangements, including emergency procedures, risk assessments and health and safety policies;
▶ further information on training courses.

The employer should also provide facilities and assistance for safety delegates. Depending on the circumstances, these could include:

▶ notice board;
▶ telephone;
▶ lockable filing cabinet;
▶ access to an office to meet workers in private;
▶ camera;
▶ key occupational health and safety information;
▶ access to specialist assistance and support in understanding technical issues.

There are several benefits that accrue to organisations that consult with their employees. These include:

▶ healthier and safer workplaces because employees can help to identify hazards and develop relevant procedures to eliminate, reduce or control risks;
▶ a stronger commitment to implementing procedures and decisions because employees have been involved proactively in the decision-making process;
▶ the development of better work practices;
▶ a reduction in workplace accidents; and
▶ greater cooperation and trust between employers and employees so that the overall performance of the organisation improves.

Consultation does not remove the employer's right and duty to manage since the employer will always make the final decision following the consultation. But talking to employees and involving them in the decision-making process is a vital part of successfully managing health and safety.

3.4.6 Health and safety training

The provision of information and training for employees will develop their awareness and understanding of the specific hazards and risks associated with their jobs and working environment. It will inform them of the control measures that are in place and any related safe procedures that must be followed. Apart from satisfying legal obligations, several benefits will accrue to the employer by the provision of sound information and training to employees. These benefits include:

▶ a reduction in accident severity and frequency;
▶ a reduction in injury and ill-health related absence;
▶ a reduction in compensation claims and, possibly, insurance premiums;
▶ an improvement in the health and safety culture of the organisation;
▶ improved staff morale and retention.

Health and safety training is a very important part of the health and safety culture and it is often a legal requirement for an employer to provide such training. Training should be provided on recruitment, at induction or on being exposed to new or increased risks due to:

▶ being transferred to another job or given a change in responsibilities;
▶ the introduction of new work equipment or a change of use in existing work equipment;
▶ the introduction of new technology;
▶ the introduction of a new system of work or the revision of an existing system of work;
▶ an increase in the employment of more vulnerable employees (young or disabled persons);
▶ particular training required by the organisation's insurance company (e.g. specific fire and emergency training).

Additional training may well be needed following a single or series of accidents or near misses, the introduction of new legislation, the issuing of an enforcement notice or as a result of a risk assessment or safety audit.

It is important during the development of a training course that the target audience is taken into account. If the audience are young persons, the chosen approach must be capable of retaining their interest and any illustrative examples used must be within their experience. The trainer must also be aware of external influences, such as peer pressure, and use them to their

Figure 3.16 Health and safety training needs and opportunities

advantage. For example, if everybody wears PPE then it will be seen as the thing to do. Levels of literacy and numeracy are other important factors.

The way in which the training session is presented by the use of DVDs, Powerpoint slides, case studies, lectures or small discussion groups needs to be related to the material to be covered and the backgrounds of the trainees. Supplementary information in the form of copies of slides and additional background reading is often useful.

The environment used for the training sessions is also important in terms of room layout and size, lighting and heating.

Attempts should be made to measure the effectiveness of the training by course evaluation forms issued at the time of the session, by a subsequent refresher session and by checking for improvements in health and safety performance (such as a reduction in specific accidents).

There are several different types of training; these include induction, job specific, supervisory and management, and specialist. Informal sessions held at the place of work are known as toolbox talks. Such sessions should only be used to cover a limited number of issues. They can become a useful route for employee consultation.

Induction training

Induction training should always be provided to new employees, trainees and contractors. While such training covers items such as pay, conditions and quality, it must also include health and safety. It is useful if the employee signs a record to the effect that training has been received. This record may be required as evidence should there be a subsequent legal claim against the organisation.

New workers are most vulnerable to accidents during their first six months at a new workplace. In the construction industry in the UK, eight out of 16 fatal accidents happened during the first ten days on site, half of them on the very first day. The causes of these accidents were:

▶ lack of experience of working in a new industry or workplace;
▶ lack of familiarity with the job and the work environment;
▶ reluctance to raise concerns (or not knowing how to); and
▶ eagerness to impress workmates and managers.

They may not recognise hazards as a potential source of danger, possibly due to a lack of familiarity with the workplace, and may ignore warning signs and rules, or cut corners.

A key requirement for all new workers is adequate supervision, particularly during the first few weeks. They must know how and with whom they can raise any concerns about any health and safety issues. It is

important that they have understood the information, instruction and training needed to work safely. They must also have understood any emergency arrangements or procedures. Young persons and migrant workers are two groups of potentially vulnerable workers who may require particular attention at induction.

Most induction training programmes would include the following topics:

▶ the health and safety policy of the organisation including a summary of the organisation and arrangements including employee consultation;
▶ a brief summary of the health and safety management system including the name of the employee's direct supervisor, safety representative and source of health and safety information;
▶ the employee responsibility for health and safety including any general health and safety rules (e.g. smoking prohibitions);
▶ the accident reporting procedure of the organisation, the location of the accident book and the location of the nearest first-aider;
▶ the fire and other emergency procedures including the location of the assembly point;
▶ the hazards that are specific to the workplace;
▶ a summary of any relevant risk assessments and safe systems of work;
▶ the location of welfare, canteen facilities and rest rooms;
▶ procedures for reporting defects or possible hazards and the name of the responsible person to whom the report should be made;
▶ details of the possible disciplinary measures that may be enacted for non-compliance with health and safety rules and procedures.

Additional items which are specific to the organisation may need to be included such as:

▶ internal transport routes and pedestrian walkways (e.g. fork-lift truck operations);
▶ the correct use of PPE and maintenance procedures;
▶ manual handling techniques and procedures;
▶ details of any hazardous substances in use and any procedures relating to them (e.g. health surveillance).

There should be some form of follow-up with each new employee after three months to check that the important messages have been retained. This is sometimes called a refresher course, although it is often better done on a one-to-one basis.

It is very important to stress that the content of the induction course should be subject to constant review and updated following an accident investigation, new legislation, changes in the findings of a risk assessment or the introduction of new plant or processes. A well-designed induction programme may help to encourage a positive health and safety culture and reduce the effect of negative peer pressure.

The induction training of **migrant workers** is of particular importance. A worker may have enough language to gain employment, but may not be able to understand safety instructions. Many migrant workers' reading skills lag their spoken language ability. Even if the range of languages spoken on site is established, translated manuals and posters may be of little value. A qualified interpreter should be used to satisfactorily induct new migrant workers. Another option is to use a multilingual DVD which can reach multiple sites easily and consistently. Finally, it is important to check that if acronyms are used then it does not have a different meaning in the languages of the migrant workers.

Job-specific training

Job-specific training ensures that employees undertake their job in a safe manner. Such training, therefore, is a form of skill training and is often best done 'on the job' – sometimes known as 'toolbox training'. Details of the safe system of work or, in more hazardous jobs, a permit-to-work system, should be covered. In addition to normal safety procedures, emergency procedures and the correct use of PPE also need to be included. The results of risk assessments are very useful in the development of this type of training. It is important that any common causes of human errors (e.g. discovered as a result of an accident investigation), any standard safety checks or maintenance requirements, are addressed.

It is common for this type of training to follow an operational procedure in the form of a checklist which the employee can sign on completion of the training. The new employee will still need close supervision for some time after the training has been completed.

Supervisory and management training

Supervisory and management health and safety training follows similar topics to those contained in an induction training course but will be covered in more depth. There will also be a more detailed treatment of health and safety law. There has been considerable research over the years into the failures of managers that have resulted in accidents and other dangerous incidents. These failures have included:

▶ lack of health and safety awareness, enforcement and promotion (in some cases, there has been encouragement to circumvent health and safety rules);
▶ lack of consistent supervision of and communication with employees;
▶ lack of understanding of the extent of the responsibility of the supervisor.

It is important that all levels of management, including the Board, receive health and safety training. This will not only keep everybody informed of health and safety legal requirements, accident prevention techniques and changes in the law, but also encourage everybody

to monitor health and safety standards during visits or tours of the organisation.

Specialist training

Specialist health and safety training is normally needed for activities that are not related to a specific job but more to an activity. Examples include first-aid, fire prevention, fork-lift truck driving, overhead crane operation, scaffold inspection and statutory health and safety inspections. These training courses are often provided by specialist organisations and successful participants are awarded certificates. Details of two of these courses will be given here by way of illustration. Details of other types of specialist training appear elsewhere in the text.

Fire prevention training courses include the causes of fire and fire spread, fire and smoke alarm systems, emergency lighting, the selection and use of fire extinguishers and sprinkler systems, evacuation procedures, high-risk operations and good housekeeping principles.

A fork-lift truck drivers' course would include the general use of the controls, loading and unloading procedures, driving up or down an incline, speed limits, pedestrian awareness (particularly in areas where pedestrians and vehicles are not segregated), security of the vehicle when not in use, daily safety checks and defect reporting, refuelling and/or battery charging and emergency procedures.

Training is a vital part of any health and safety programme and needs to be constantly reviewed and updated. In many countries, health and safety legislation requires specific training (e.g. manual handling, PPE and display screens). Additional training courses may be needed when there is a major reorganisation, a series of similar accidents or incidents, or a change in equipment or a process. Finally, the methods used to deliver training must be continually monitored to ensure that they remain effective.

3.4.7 Internal influences

There are many influences on health and safety standards, some are positive and others negative (Figure 3.17). No business, in particular a small business, is totally divorced from its suppliers, customers and neighbours. This section considers the internal influences on a business, including management commitment, production demands, communication, competence and employee relations.

Management commitment

Managers, particularly senior managers, can give powerful messages to the workforce by what they do for health and safety. Managers can achieve the level of health and safety performance that they demonstrate they want to achieve. Employees soon get

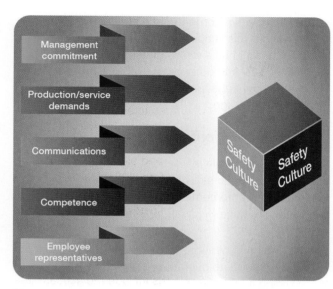

Figure 3.17 Internal influences on safety culture

the negative message if directors disregard safety rules and ignore written policies to get urgent production to the customer or to avoid personal inconvenience. It is what they do that counts, not what they say. Depending on the size and geography of the organisation, senior managers should be personally involved in:

▶ health and safety inspections or audits;
▶ meetings of the central health and safety or joint consultation committees;
▶ the investigation of accidents, ill-health and incidents. The more serious the incident, the more senior the manager who takes an active part in the investigation.

Production/service demands

Managers need to balance the demands placed by customers with the action required to protect the health and safety of their employees. How this is achieved has a strong influence on the standards adopted by the organisation. The delivery driver operating to near-impossible delivery schedules and the manager agreeing to the strident demands of a large customer regardless of the risks to employees involved, are typical dilemmas facing workers and managers alike. The only way to deal with these issues is to be well organised and agree with the client/employees ahead of the crisis as to how they should be prioritised.

Rules and procedures should be intelligible, sensible and reasonable. They should be designed to be followed under normal production or service delivery conditions. If following the rules involves very long delays or impossible production schedules, they should be revised rather than ignored by workers and managers alike until it is too late and an accident occurs. Sometimes the safety rules are simply used in a court of law, in an attempt to defend the company concerned, following an accident.

Managers who plan the impossible schedule or ignore a safety rule to achieve production or service demands are held responsible for the outcomes. It is acceptable to balance the cost of the action against the level of risk being addressed but it is never acceptable to ignore safety rules or standards simply to get the work done. Courts do not condone the action of managers who put profit considerations ahead of safety requirements.

Communication

Communication was covered in depth earlier in the chapter and clearly will have significant influence on health and safety issues. This will include:

▶ poorly communicated procedures that will not be understood or followed;
▶ poor verbal communication which will be misunderstood and will demonstrate a lack of interest by senior managers;
▶ missing or incorrect signs that may cause accidents rather than prevent them;
▶ managers who are nervous about face-to-face discussions with the workforce on health and safety issues, which will have a negative effect.

Managers and supervisors should plan to have regular discussions to learn about the problems faced by employees and discuss possible solutions. Some meetings, like that of the safety committee, are specifically planned for safety matters, but this should be reinforced by discussing health and safety issues at all routine management meetings. Regular one-to-one talks should also take place in the workplace, preferably to a planned theme or safety topic, to get specific messages across and get feedback from employees.

Competence

Competent people, who know what they are doing and have the necessary skills to do the task correctly and safely, will make the organisation effective. Competence can be brought in through recruitment or consultancy but it is often much more effective to develop it among employees. It demonstrates commitment to health and safety and a sense of security for the workforce. The loyalty that it creates in the workforce can be a significant benefit to safety standards. Refer back to earlier in this chapter for more details on achieving competence.

Employee representation

Given the resources and freedom to fulfil their function effectively, enthusiastic, competent employee safety representatives can make a major contribution to good health and safety standards. They can provide the essential bridge between managers and employees. People are more willing to accept the restrictions that some precautions bring if they are consulted and feel involved, either directly in small workforces or through their safety representatives.

3

3.4.8 External influences

The role and function of some external organisations are set out in Chapter 1. In this chapter, the influence of external organisations is briefly discussed, including societal expectations, legislation and enforcement, insurance companies, trade unions, economics and commercial stakeholders.

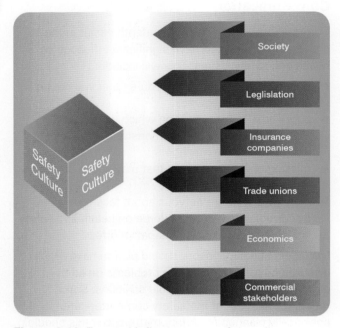

Figure 3.18 External influences on safety culture

Societal expectations

Societal expectations are not static and tend to rise over time. For example, the standards of safety accepted in a motor car 50 years ago would be considered to be totally inadequate at the beginning of the 21st century. We expect safe, quiet, comfortable cars that do not break down and which retain their appearance for many thousands of miles. Industry should strive to deliver these same high standards for the health and safety of employees or service providers. The question is whether societal expectations are as great an influence on workplace safety standards as they are on product safety standards. Society can influence standards through:

▶ people only working for good employers. This is effective in times of low unemployment;
▶ national and local news media highlighting good and bad employment practices;
▶ schools teaching good standards of health and safety;
▶ the purchase of fashionable and desirable safety equipment, such as trendy crash helmets for mountain bikes;
▶ buying products only from responsible companies. The difficulty of defining what is responsible has

been partly overcome through ethical investment criteria but this is possibly not widely enough understood to be a major influence;
▶ watching TV and other programmes which improve safety knowledge and encourage safe behaviour from an early age.

Legislation and enforcement

Good comprehensible legislation should have a positive effect on health and safety standards. Taken together, legislation and enforcement can affect standards by:

▶ providing a level to which every employer has to conform;
▶ insisting on minimum standards which also enhances people's ability to operate and perform well;
▶ providing a tough, visible threat of getting shut down or a heavy fine;
▶ stifling development by being too prescriptive;
▶ providing well-presented and easily read guidance for specific industries at reasonable cost or free.

On the other hand, a weak enforcement regime can have a powerful negative effect on standards.

Insurance companies

Insurance companies can influence health and safety standards mainly through financial incentives. Employers' liability insurance is a legal requirement in the UK and therefore all employers have to obtain this type of insurance cover. Insurance companies can influence standards through:

▶ discounting premiums to those in the safest sectors or best individual companies;
▶ insisting on risk reduction improvements to remain insured. This is not very effective where competition for business is fierce;
▶ encouraging risk reduction improvements by bundling services into the insurance premium;
▶ providing guidance on standards at reasonable cost or free.

Trade unions

Trade unions can influence standards by:

▶ providing training and education for members;
▶ providing guidance and advice cheaply or free to members;
▶ influencing governments to regulate, enhance enforcement activities and provide guidance;
▶ influencing employers to provide high standards for their members. This is sometimes confused with financial improvements, with health and safety getting a lower priority;
▶ encouraging members to work for safer employers;
▶ helping members to get proper compensation for injury and ill-health if it is caused through their work.

Economics

Economics can play a major role in influencing health and safety standards. The following ways are the most common:

▶ lack of orders and/or money can cause employers to try to ignore health and safety requirements;

▶ if employers were really aware of the actual and potential cost of accidents and fires, they would be more concerned about prevention. The UK HSE believes that the ratio between insured and uninsured costs of accidents is between 1:8 and 1:36;

▶ perversely, when the economy is booming activity increases and, particularly in the building industry, accidents can sharply increase. The pressures to perform and deliver for customers can be safety averse;

▶ businesses that are only managed on short-term performance indicators seldom see the advantage of the long-term gains that are possible with a happy, safe and fit workforce.

The cost of accidents and ill-health, in both human and financial terms, needs to be visible throughout the organisation so that all levels of employee are encouraged to take preventative measures.

Commercial stakeholders

A lot can be done by commercial stakeholders to influence standards. This includes:

▶ insisting on proper arrangements for health and safety management at supplier companies before they tender for work or contracts;

▶ checking on suppliers to see if the workplace standards are satisfactory;

▶ encouraging ethical investments;

▶ considering ethical as well as financial standards when banks provide funding;

▶ providing high-quality information for customers;

▶ insisting on high standards to obtain detailed planning permission (where this is possible);

▶ providing low-cost guidance and advice.

3.5 Further information

ILO Guidelines on Occupational Safety and Health Management Systems (ILO-OSH 2001) http://www.ilo.org/global/publications/ilo-bookstore/order-online/books/WCMS_PUBL_9221116344_EN/lang--en/index.htm

ILOLEX (ILO database of International Law) http://www.ilo.org/ilolex/index.htm

Occupational Health and Safety Assessment Series (OHSAS 18000): Occupational Health and Safety Management Systems OHSAS 18001:2007 ISBN 978-0-5805-9404-5 OHSAS18002:2008 ISBN 978-0-5806-2686-9

Occupational Safety and Health Convention (C155) 2003, ILO http://www.ilo.org/dyn/normlex/en/f?p=NORMLEXPUB:12100:0::NO:1210 0:P12100_ILO_CODE:C155

Occupational Safety and Health Recommendation (R164) 2006, ILO http://www.ilo.org/dyn/normlex/en/f?p=NORMLEXPUB:12100:0::NO:12100:P12100_INSTRUMENT_ID:312502

Reducing Error and Influencing Behaviour (HSG48), HSE Books, ISBN 978-0-7176-2452-2 http://www.hse.gov.uk/pubns/priced/hsg48.pdf

ILO Safety and Health in Construction, Code of Practice, ILO Geneva, ISBN 92-2-107104-9 http://www.ilo.org/wcmsp5/groups/public/---ed_protect/---protrav/---safework/documents/normativeinstrument/wcms_107826.pdf

ILO Training package on workplace risk assessment and management for small and medium size enterprises, 2013, ILO Geneva, ISBN 978-92-2-127065-2 (web pdf) http://www.ilo.org/wcmsp5/groups/public/---ed_protect/---protrav/---safework/documents/instructionalmaterial/wcms_215344.pdf

ILO Code of Practice Ambient factors in the workplace (Chapter 3 – General principles of prevention and control) http://www.ilo.org/wcmsp5/groups/public/---ed_protect/---protrav/---safework/documents/normativeinstrument/wcms_107729.pdf

3.6 Practice revision questions

1. (a) **Describe** the four elements that boards of directors should take into account when overseeing the management of health and safety within their organisation.

   (b) **Outline** the responsibilities that departmental managers have to ensure the health, safety and welfare of those working in the department.

   (c) **Outline** the factors that will determine the level of supervision that a new employee should receive during their initial period of employment within an organisation.

   (d) **Outline** why it is important that everyone is aware of their roles and responsibilities for health and safety in an organisation.

2. (a) **Explain** the meaning of the term 'competent person' when related to health and safety.

   (b) **Outline** the factors that the employer should consider when selecting an individual for the role of health and safety competent person within the organisation.

(c) **Outline** the organisational factors that may cause a person to work unsafely even though they are competent.

(d) **Identify FOUR** types of specialist help that may be required to support the health and safety function.

3. (a) **Outline** the responsibilities of employers towards workers in respect of health and safety at work.

(b) **Outline** the rights and responsibilities of workers in respect of health and safety.

4. (a) **Outline** the reasons that visitors to a workplace may be more at risk from injury than employees.

(b) **Outline** the controls that should be in place if night workers are employed by an organisation.

(c) **Identify THREE** principal responsibilities that each employer has in a jointly occupied workplace.

(d) **Identify** the possible health and safety issues that may arise in the management of a supply chain.

5. Several construction contractors are to be employed by a printing company to refurbish its building.

(a) **Outline** the health and safety responsibilities of the self-employed.

(b) **Outline** the main issues to be considered during the selection of the contractors by the printing company.

(c) **Outline** the matters to be addressed by the company while supervising the contractors and the contractors in supervising its employees.

6. (a) **Explain** the meaning of the term *'health and safety culture'*.

(b) **Describe FIVE** requirements of a positive health and safety culture.

(c) **Identify** ways in which the health and safety culture of an organisation might be improved.

(d) **Outline EIGHT** possible indications of a poor health and safety culture in an organisation.

7. (a) The number of accidents is a measure of the health and safety culture in a company. **State** the definition of 'accident incidence rate'.

(b) **Describe** other measures of health and safety culture. **Outline** how information on accidents could be used to promote health and safety in the workplace.

(c) **Identify SIX** indicators of management commitment to health and safety in the workplace.

8. (a) **Explain**, using an example, the meaning of the term 'attitude'.

(b) **Outline THREE** influences on the attitude towards health and safety of employees within an organisation.

9. (a) **Explain** the meaning of the term 'motivation'.

(b) Other than lack of motivation, **outline FOUR** reasons why employees may fail to comply with safety procedures at work.

(c) **Outline** ways in which employers may motivate their employees to comply with health and safety procedures.

10. (a) **Explain** the meaning of the term 'perception'.

(b) **Outline** the factors relating to the individual that may influence a person's perception of an occupational risk.

(c) **Outline** ways in which employees' perceptions of hazards in the workplace might be improved.

11. The human factors which influence behaviour in the workplace are the organisation, the job and individual factors.

(a) **Describe** the essential elements in the organisation and the job that will ensure good health and safety behaviour.

(b) **Outline** personal or individual factors that may contribute to human errors occurring at work.

12. (a) **Explain** the meaning of the term 'human error'.

(b) **Describe** the difference between human errors and violations.

(c) **Describe**, using practical examples, **THREE** types of human error and **THREE** types of violation that can lead to accidents in the workplace.

(d) **Outline** ways of reducing the likelihood of human error in the workplace.

13. (a) **Identify FIVE** topics for which there would be health and safety standards in a company.

(b) **Outline** the possible actions that a Chief Executive and senior managers could take to improve the health and safety standards in the workplace.

14. Following a significant increase in accidents, a health and safety campaign is to be launched within a manufacturing company to encourage safer working by employees.

(a) **Outline** how the company might ensure that the nature of the campaign is effectively communicated to, and understood by, employees.

(b) **Identify** a range of methods that an employer can use to provide health and safety information directly to individual employees.

(c) **Outline** reasons why the safety procedures may not have been followed by all employees.

15. **Outline** the advantages and disadvantages of communicating health and safety messages to employees:

(a) Verbally
(b) In writing
(c) Graphically.

16. Notice boards are often used to display health and safety information in the workplace.

(a) **Outline** the limitations of relying on this method to communicate to employees and how these limitations may be addressed.

(b) **Identify FOUR** types of health and safety information that might usefully be displayed on a notice board.

(c) **Identify FOUR** other methods which the employer could use to communicate essential health and safety information to their employees.

17. (a) **Explain** the difference between consulting and informing workers on health and safety issues.

(b) **Outline** why it is important for an organisation to consult with its workers on health and safety issues.

(c) **Outline** how arrangements for consultation with workers may be made more effective.

(d) **Outline** the reasons why a health and safety committee may prove to be ineffective in practice.

18. (a) **Outline** the importance of induction training for new employees in reducing the number of accidents in a workplace.

(b) **Outline** reasons for additional refresher health and safety training at a later stage of employment within a workplace.

(c) **Identify** the reasons for a review of the health and safety training programme within an organisation.

19. A building contractor is to undertake building maintenance work in the roof space of a busy warehouse. **Outline** the issues that should be covered in an induction programme for the contractor's employees.

20. Following an accident at a manufacturing company, an investigation concluded that there had been a poor record of specialist and supervisory training.

(a) **Identify THREE** types of specialist training that may have been relevant to the company.

(b) **Outline** the possible supervisory failures that could have caused an accident.

21. **Describe TWO** internal and **TWO** external influences on the health and safety culture of an organisation.

APPENDIX 3.1 Leadership actions for directors and board members

This guidance, taken from INDG 417, sets out an agenda for the effective leadership of health and safety by all directors, governors, trustees, officers and their equivalents in the private, public and third sectors. It applies to organisations of all sizes.

There are four elements that boards need to incorporate into their management of health and safety. These are:

▶ planning the direction of health and safety;
▶ delivering the plan for health and safety;
▶ monitoring health and safety performance; and
▶ reviewing health and safety performance.

1. Planning the direction of health and safety

The Board should set the direction for effective health and safety management. Board members need to establish a health and safety policy that is much more than a document – it should be an integral part of the organisation's culture, its values and performance standards.

All board members should take the lead in ensuring the communication of health and safety duties and benefits throughout the organisation. Executive directors must develop the policy since they have responsibility for the day-to-day operation of the organisation and must respond quickly where difficulties arise or new risks are introduced; non-executive directors must make sure that health and safety is properly addressed.

Essential actions

For a policy to be effective, all board members must be aware of the significant risks faced by their organisation. The policy should include the Board's role and that of individual board members in leading the management of health and safety in the organisation. The Board needs to 'own' and understand the key issues involved and decide on the best methods to communicate, promote and champion health and safety.

The health and safety policy is a 'living' document and it should evolve over time, for example, during major organisational changes such as restructuring or a significant acquisition.

Good practice guide

▶ Health and safety should appear regularly on the agenda for board meetings.
▶ The Chief Executive should give the clear leadership, but it may also be useful to name one of their members as the health and safety 'champion'.

▶ The presence on the Board of a Health and Safety Director gives a strong signal that the issue is being taken seriously and that its strategic importance is understood.
▶ The setting of targets helps the Board to measure performance.
▶ A non-executive director can act as a scrutiniser to ensure that robust processes are in place to support boards that face significant health and safety risks.

2. Delivering the plan for health and safety

Delivery depends on an effective management system to ensure, so far as is reasonably practicable, the health and safety of employees, customers and members of the public. Organisations should aim to protect people by introducing management systems and practices that ensure risks are dealt with sensibly, responsibly and proportionately.

Essential actions

To take responsibility and 'ownership' of health and safety, members of the Board must ensure that:

▶ health and safety arrangements are adequately resourced;
▶ they obtain competent health and safety advice;
▶ risk assessments are carried out;
▶ employees or their representatives are involved in decisions that affect their health and safety.

The Board should consider the health and safety implications of introducing new processes, working practices or personnel and dedicate adequate resources to the task and seek competent advice when necessary. All board decisions, particularly when implementing change, must be made in the context of the organisation's health and safety policy.

Good practice guide

▶ Leadership is more effective if visible – board members can reinforce health and safety policy by being seen on the 'shop floor', following all safety measures themselves and addressing any breaches immediately.
▶ Health and safety should be considered when senior management appointments are made.
▶ Having good procurement standards for goods, equipment and services should prevent the introduction of expensive health and safety hazards.
▶ The health and safety arrangements of partners, key suppliers and contractors should be assessed

regularly since their performance could adversely affect a director's own performance.

▶ A separate risk management or health and safety sub-committee of the Board, chaired by a senior executive, should ensure that key issues are addressed and guard against time and effort being wasted on trivial risks and unnecessary bureaucracy.

▶ Health and safety training to some or all board members will promote understanding and knowledge of the key health and safety issues in the organisation.

▶ Worker involvement in health and safety, above the legal duty to consult worker representatives, will improve participation and indicate the commitment of senior management.

3. Monitoring health and safety performance

Monitoring and reporting are vital parts of a health and safety system. Management systems must allow the Board to receive both specific (e.g. incident-led) and routine reports on the performance of health and safety policy. Much day-to-day health and safety information need be reported only at the time of a formal review. But only a strong system of monitoring can ensure that the formal review will proceed as planned – and that relevant events in the interim are brought to the Board's attention.

Core actions

The Board should ensure that:

▶ appropriate weight is given to reporting both preventative information (such as progress of training and maintenance programmes) and incident data (such as accident and sickness absence rates);

▶ periodic audits of the effectiveness of management structures and risk controls for health and safety are carried out;

▶ the impact of changes such as the introduction of new procedures, work processes or products, or any major health and safety failure, is reported as soon as possible to the Board;

▶ there are procedures to implement new and changed legal requirements and to consider other external developments and events.

Good practice guide

▶ Effective monitoring of sickness absence and workplace health will alert the Board to underlying problems that could seriously damage

performance or result in accidents and long-term illness.

▶ The collection of workplace health and safety data will allow the Board to benchmark the organisation's performance against others in its sector.

▶ Appraisals of senior managers should include an assessment of their contribution to health and safety performance.

▶ Boards should receive regular reports on the health and safety performance and actions of contractors.

▶ Organisations will find that they win greater support for health and safety from workers by involving them in the monitoring process.

4. Reviewing health and safety performance

A formal boardroom review of health and safety performance is essential. It allows the Board to establish whether the essential health and safety principles – strong and active leadership, worker involvement, and assessment and review – have been embedded in the organisation. It tells senior managers whether their system is effective in managing risk and protecting people.

Core actions

The Board should review health and safety performance at least once a year. The review process should:

▶ examine whether the health and safety policy reflects the organisation's current priorities, plans and targets;

▶ examine whether risk management and other health and safety systems have been effectively reported to the Board;

▶ report health and safety shortcomings, and the effect of all relevant board and management decisions;

▶ decide actions to address any weaknesses and a system to monitor their implementation;

▶ instigate immediate reviews in the light of major shortcomings or events.

Good practice guide

▶ Performance on health and safety and well-being should be reported in the organisations' annual report to investors and stakeholders.

▶ Board members should make extra 'shop floor' visits to gather information for the formal review.

▶ Good health and safety performance should be celebrated at central and local level.

APPENDIX 3.2 Detailed health and safety responsibilities

Managing Directors/Chief Executives

1. are responsible and accountable for health and safety performance at the organisation;
2. develop a strong, positive health and safety culture throughout the company and communicate it to all managers. This should ensure that all managers have a clear understanding of their health and safety responsibilities;
3. provide guidance and leadership on health and safety matters to their management team;
4. establish minimum acceptable health and safety standards within the organisation;
5. ensure that adequate resources are available for the health and safety requirements within the organisation and authorise any necessary major health and safety expenditures;
6. evaluate, approve and authorise health and safety related projects developed by the organisation's health and safety advisers;
7. review and approve health and safety policies, procedures and programmes developed by the organisation's managers;
8. ensure that a working knowledge of the areas of health and safety that are regulated by various governmental agencies are maintained;
9. ensure that health and safety is included as an agenda topic at all formal senior management meetings;
10. review and act upon major recommendations submitted by outside loss prevention consultants and insurance companies;
11. ensure that health and safety is included in any tours such as fire inspections of the organisation's sites and note any observed acts or conditions that fall short of or exceed agreed health and safety standards;
12. ensure that all fatalities, major property losses, serious lost workday injuries and dangerous occurrences are investigated;
13. establish, implement and maintain a formal, written health and safety programme for the organisation that encompasses all areas of significant health and safety risk;
14. establish controls to ensure uniform adherence to the health and safety programme across the organisation;
15. attend the health and safety committee meetings at the organisation;
16. review, on a regular basis, all health and safety activity reports and performance statistics;
17. review health and safety reports submitted by outside agencies and determine that any agreed actions have been taken;
18. review annually the effectiveness of, and, if necessary, require revision of, the site health and safety programme;
19. appraise the performance of the health and safety advisers and provide guidance or training where necessary;
20. monitor the progress of managers and others towards achieving their individual health and safety objectives.

Departmental managers

1. are responsible and accountable for the health and safety performance of their department;
2. contact each supervisor frequently (daily) to monitor the health and safety standards in the department;
3. hold departmental health and safety meetings for supervisors and employee representatives at least once a month;
4. ensure that any machinery, equipment or vehicles used within the department are maintained and correctly guarded and meet agreed health and safety standards. Copies of records of all maintenance, statutory and insurance inspections must be kept by the departmental manager;
5. ensure that all fire and other emergency equipment is properly maintained on a regular basis with all faults rectified promptly and that all departmental staff are aware of fire and emergency procedures;
6. ensure that there is adequate first-aid cover on all shifts and all first-aid boxes are adequately stocked;
7. ensure that safe systems of work procedures are in place for all jobs and that copies of all procedures are submitted to the site Managing Director for approval;
8. review all job procedures on a regular basis and require each supervisor to check that the procedures are being used correctly;
9. approve and review, annually, all departmental health and safety risk assessments, rules and procedures, maintain strict enforcement and develop plans to ensure employee instruction and re-instruction;
10. ensure that all health and safety documents (such as the organisation's health and safety manual, risk assessments, rules and procedures) are easily accessible to all departmental staff;
11. establish acceptable housekeeping standards, defining specific areas of responsibility, and assign areas to supervisors; make a weekly spot check across the department, hold a formal

inspection with supervisors at least monthly and submit written reports of the inspections to the health and safety adviser with deadlines for any required actions;

12. authorise purchases of tools and equipment necessary to attain compliance with the organisation's specifications and relevant statutory Regulations;

13. develop a training plan that includes specific job instructions for new or transferred employees and follow up on the training by supervisors. Copies of records of all training must be kept by the departmental manager;

14. review the health and safety performance of their department each quarter and submit a report to the Managing Director/Chief Executive;

15. personally investigate all lost workday cases and dangerous occurrences and report to the Managing Director/Chief Executive. Progress any required corrective action;

16. adopt standards for assigning PPE to employees, insist on strict enforcement and make spot field checks to determine compliance;

17. evaluate the health and safety performances of supervisors;

18. develop in each supervisor strong health and safety attitudes and a clear understanding of their specific duties and responsibilities;

19. instil, by action, example and training, a positive health and safety culture among all departmental staff;

20. instruct supervisors in site procedures for the care and treatment of sick or injured employees;

21. ensure that the names of any absentees, written warnings and all accident reports are submitted to the human resources manager.

Supervisors

1. are responsible and accountable for their team's health and safety performance;

2. conduct informal health and safety meetings with their employees at least monthly;

3. enforce all safe systems of work procedures that have been issued by the departmental manager;

4. report to the departmental manager any weaknesses in the safe system of work procedures or any actions taken to revise such procedures. These weaknesses may be revealed by either health and safety risk assessments or observations;

5. report any jobs that are not covered by safe systems of work procedures to the departmental manager;

6. review any unsafe acts and conditions and either eliminate them or report them to the departmental manager;

7. instruct employees in relevant health and safety rules, make records of this instruction and enforce all health and safety rules and procedures;

8. make daily inspections of assigned work areas and take immediate steps to correct any unsafe or unsatisfactory conditions, report to the departmental manager those conditions that cannot be immediately corrected and instruct employees on housekeeping standards;

9. instruct employees that tools/equipment are to be inspected before each use and make spot checks of tools'/equipment's condition;

10. instruct each new employee personally on job health and safety requirements in assigned work areas;

11. provide on-the-job instruction on safe and efficient performance of assigned jobs for all employees in the work area;

12. report any apparent employee health problems to the departmental manager;

13. enforce PPE requirements; make spot checks to determine that protective equipment is being used and periodically appraise condition of equipment. Record any infringements of the PPE policy;

14. in the case of a serious injury, ensure that the injured employee receives prompt medical attention, isolate the area and/or the equipment as necessary and immediately report the incident to the departmental manager. In case of a dangerous occurrence, the supervisor should take immediate steps to correct any unsafe condition and, if necessary, isolate the area and/or the equipment. As soon as possible, details of the incident and any action taken should be reported to the departmental manager;

15. investigate all accidents, serious incidents and cases of ill-health involving employees in assigned work areas. Immediately after an accident, complete an accident report form and submit it to the departmental manager for onward submission to the health and safety adviser. A preliminary investigation report and any recommendations for preventing a recurrence should be included on the accident report form;

16. check for any changes in operating practices, procedures and other conditions at the start of each shift/day and before relieving the 'on duty' supervisor (if applicable). A note should be made of any health and safety related incidents that have occurred since their last working period;

17. at the start of each shift/day, make an immediate check to determine any absentees. Report any absentees to the departmental manager;

18. make daily spot checks and take necessary corrective action regarding housekeeping,

unsafe acts or practices, unsafe conditions, job procedures and adherence to health and safety rules;

19. attend all scheduled and assigned health and safety training meetings;

20. act on all employee health and safety complaints and suggestions;

21. maintain, in their assigned area, health and safety signs and notice boards in a clean and legible condition.

Employees and agency workers

1. are responsible for their own health and safety;
2. ensure that their actions will not jeopardise the safety or health of other employees;
3. obey any safety rules, particularly regarding the use of PPE or other safety equipment;
4. learn and follow the operating procedures and health and safety rules and procedures for the safe performance of the assigned job;
5. must correct, or report to their supervisor, any observed unsafe practices and conditions;
6. maintain a healthy and safe place to work and cooperate with managers in the implementation of health and safety matters;
7. make suggestions to improve any aspect of health and safety;
8. maintain an active interest in health and safety;
9. follow the established procedures if accidents occur by reporting any accident to the supervisor;
10. report any absence from the company caused by illness or an accident.

APPENDIX 3.3 Checklist for supply chain health and safety management

This checklist is taken from the UK's HSE leaflet INDG368 *Working Together: Guidance on Health and Safety for Contractors and Suppliers 2003*. It is a reminder of the topics that might need to be discussed with people with whom individual contractors may be working.

It is not intended to be exhaustive and not all questions will apply at any one time, but it should help people to get started.

1. Responsibilities

▶ What are the hazards of the job?
▶ Who is to assess particular risks?
▶ Who will coordinate action?
▶ Who will monitor progress?

2. The job

▶ Where is it to be done?
▶ Who with?
▶ Who is in charge?
▶ How is the job to be done?
▶ What other work will be going on at the same time?
▶ How long will it take?
▶ What time of day or night?
▶ Do you need any permit to do the work?

3. The hazards and risk assessments

Site and location

Consider the means of getting into and out of the site and the particular place of work – are they safe? And:

▶ Will any risks arise from environmental conditions?
▶ Will you be remote from facilities and assistance?
▶ What about physical/structural conditions?
▶ What arrangements are there for security?

Substances

▶ What suppliers' information is available?
▶ Is there likely to be any microbiological risk?
▶ What are the storage arrangements?
▶ What are the physical conditions at the point of use? Check ventilation, temperature, electrical installations, etc.
▶ Will you encounter substances that are not supplied, but produced in the work, for example fumes from hot work during dismantling plant? Check how much, how often, for how long, method of work, etc.
▶ What are the control measures? For example, consider preventing exposure, providing engineering controls, using personal protection (in that order of choice).
▶ Is any monitoring required?
▶ Is health surveillance necessary, for example for work with sensitisers? (Refer to health and safety data sheet.)

Plant and equipment

▶ What are the supplier/hirer/manufacturer's instructions?
▶ Are any certificates of examination and test needed?
▶ What arrangements have been made for inspection and maintenance?
▶ What arrangements are there for shared use?
▶ Are the electrics safe to use? Check the condition of power sockets, plugs, leads and equipment.

(Don't use damaged items until they have been repaired.)

▶ What assessments have been made of noise levels?

4. People

▶ Is information, instruction and training given, as appropriate?
▶ What are the supervision arrangements?
▶ Are members of the public/inexperienced people involved?
▶ Have any disabilities/medical conditions been considered?

5. Emergencies

▶ What arrangements are there for warning systems in case of fire and other emergencies?

▶ What arrangements have been made for fire/emergency drills?
▶ What provision has been made for first-aid and fire-fighting equipment?
▶ Do you know where your nearest fire exits are?
▶ What are the accident reporting arrangements?
▶ Are the necessary arrangements made for availability of rescue equipment and rescuers?

6. Welfare: Who will provide

▶ shelter?
▶ food and drinks?
▶ washing facilities?
▶ toilets (male and female)?
▶ clothes changing/drying facilities?

There may be other pressing requirements which make it essential to re-think health and safety as the work progresses.

APPENDIX 3.4 Safety culture questionnaire

The HSE have suggested that the following questionnaire can be used to inspect the safety culture of an organisation.

1. Management commitment

Where is safety perceived to be in management's priorities (senior/middle/first line)?
How do they show this?
How often are they seen in the workplace?
Do they talk about safety when in the workplace and is this visible to the workforce?
Do they 'walk the talk'?
Do they deal quickly and effectively with safety issues raised?
What balance do their actions show between safety and production?
Are management trusted over safety?

2. Communication

Is there effective two-way communication about safety?
How often are safety issues discussed;
    ▷ With line manager/subordinate?
    ▷ With colleagues?
What is communicated about the safety programme of the company?
How open are people about safety?

3. Employee involvement

How are people (all levels, especially operators) involved in safety?

How often are individual employees asked for their input to safety issues?
How often do operators report unsafe conditions or near misses, etc.?
Is there active, structured operator involvement, e.g. workshops, projects, safety circles?
Is there a continuous improvement/total quality approach?
Whose responsibility is safety regarded to be?
Is there genuine cooperation over safety – a joint effort between all in the company?

4. Training/information

Do employees feel confident that they have all the training that they need?
How accurate are employees' perceptions of hazards and risks?
How effective is safety training in meeting needs (including managers!)?
How are needs identified?
How easily available is safety information?

5. Motivation

Do managers give feedback on safety performance (and how)?
Are they likely to notice unsafe acts?
Do managers (all levels – senior/middle/first line) always confront unsafe acts?
How do they deal with them?
Do employees feel they can report unsafe acts?
How is discipline applied to safety?
What do people believe are the expectations of managers?

Do people feel that this is a good place to work (why/why not)?

Are they proud of their company?

6. Compliance with procedures

What are written procedures used for?

What decides whether a particular task will be captured in a written procedure?

Are they read? • Are they helpful?

What other rules are there?

Are there too many procedures and rules?

How well are people trained in them?

Are they audited effectively?

Are they written by users?

Are they linked to risks?

7. Learning organisation

Does the company really learn from accident history, incident reporting, etc.?

Do employees feel confident in reporting incidents or unsafe conditions?

Do they report them?

Do reports get acted upon?

Do they get feedback?

CHAPTER 4

Health and safety management systems – Risk assessment and controls – DO 2

4.1 Principles and practice of risk assessment ▶ 90

4.2 General principles of control and hierarchy of risk reduction measures ▶ 108

4.3 Sources of health and safety information ▶ 109

4.4 Safe systems of work ▶ 110

4.5 Role and function of a permit-to-work system ▶ 115

4.6 Emergency procedures and arrangements for contacting the emergency services ▶ 120

4.7 Requirements for, and effective provision of, first-aid in the workplace ▶ 123

4.8 Further information ▶ 125

4.9 Practice revision questions ▶ 125

Appendix 4.1 Procedure for risk assessment and management (European Commission) ▶ 127

Appendix 4.2 Hazard checklist ▶ 127

Appendix 4.3 Risk assessment example 1: Hairdressing salon ▶ 129

Appendix 4.4 Risk assessment example 2: Office cleaning ▶ 131

Appendix 4.5 Asbestos examples of safe systems of work ▶ 133

Appendix 4.6 Emergency numbers in some countries worldwide ▶ 135

> **This chapter covers the following NEBOSH learning objectives:**
> 1. Explain the principles and practice of risk assessment
> 2. Explain the general principles of control and a general hierarchy of risk reduction measures
> 3. Identify the key sources of health and safety information
> 4. Explain what factors should be considered when developing and implementing a safe system of work for general activities
> 5. Explain the role and function of a permit-to-work system
> 6. Outline the need for emergency procedures and the arrangements for contacting emergency services
> 7. Outline the requirements for, and effective provision of, first-aid in the workplace

Introduction

Risk assessment and minimum levels of risk prevention or control are regulated by health and safety legal requirements. Risk assessment is an essential part of the DO stage of any health and safety management system (see Figure 4.1). Risk assessment methods are used to decide on priorities and to set objectives for eliminating hazards and reducing risks. Wherever possible, risks should be eliminated through the selection and design of facilities, equipment and processes. If risks cannot be eliminated, they should be minimised by the use of physical controls or, as a last resort, through systems of work and personal protective equipment. The control of risks is essential to secure and maintain a healthy and safe workplace, which complies with the relevant legal requirements. In this chapter, hazard identification and risk assessment are covered together with appropriate risk control measures. A hierarchy of control methods is discussed that gives a preferred order of approach to risk control.

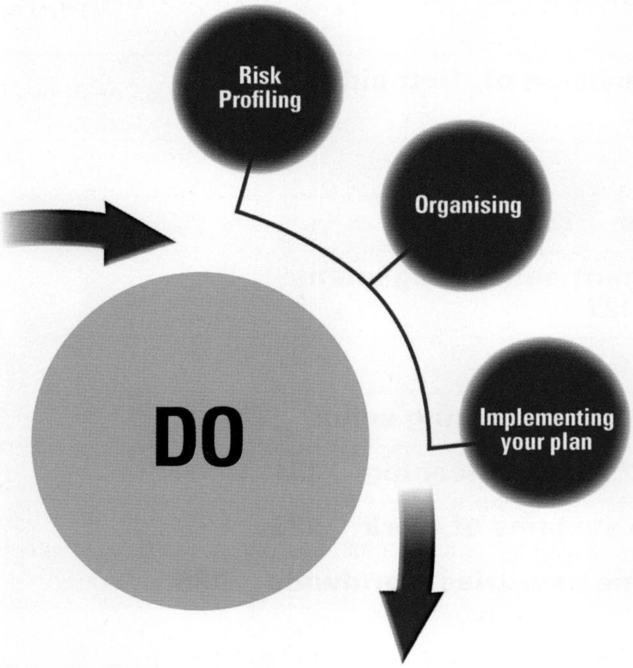

Figure 4.1 Risk Assessment or Profiling is covered by the DO part of the management cycle

This chapter also concerns the principles of prevention that should be adopted when deciding on suitable measures to eliminate or control both acute and chronic risks to the health and safety of people at work. The principles of control can be applied to both health risks and safety risks although health risks have some distinctive features that require a special approach. The chapter concludes with emergency procedures and the treatment of first-aid.

4.1 Principles and practice of risk assessment

4.1.1 Introduction

There are several descriptions of the meaning of 'risk assessment'. NEBOSH defines risk assessment as:

> 'the identification of preventative and protective measures by the evaluation of the risk(s) arising from a hazard(s), taking into account the adequacy of any existing controls, and deciding whether or not the risk(s) is acceptable.'

The ILO acknowledges in its guidelines on occupational safety and health management systems (ILO-OSH 2001) that the impact on occupational health and safety of internal changes (such as those in staffing or due to new processes, working procedures, organisational structures or acquisitions), and of external changes (for example as a result of amendments of national laws and regulations, organisational mergers, and developments in occupational health and safety knowledge and technology), should be evaluated and appropriate preventative steps taken prior to the introduction of changes. To achieve this, a workplace hazard identification and risk assessment should be carried out before any modification or introduction of new work methods, materials, processes or machinery. Such assessment should be done in consultation with and involving workers and their representatives, and the health and safety committee, where appropriate.

Risk assessment is an essential part of the planning stage of any health and safety management system. The UK HSE, in the publication HSG65 *Successful Health and Safety Management,* states that the aim of the planning process is to minimise risks.

'Risk assessment methods are used to decide on priorities and to set objectives for eliminating hazards and reducing risks. Wherever possible, risks are eliminated through selection and design of facilities, equipment and processes. If risks cannot be eliminated, they are minimised by the use of physical controls or, as a last resort, through systems of work and personal protective equipment.'

4.1.2 International requirements for risk assessment

Many countries, particularly those who have updated their health and safety legislation in the last 20 years or so, have incorporated a requirement for risk assessment. This has been driven by the ILO and, within Europe, the European Commission.

In the Code of Practice 'Ambient factors in the workplace', the ILO recommends that employers should make periodic assessments of the hazards and risks to health and safety from hazardous ambient and other factors at each permanent or temporary workplace, and implement the control measures required to prevent those hazards and risks, or to reduce them to the lowest reasonable and practicable level. If a new source of hazard is introduced, the assessment should be made before workers are exposed to the hazard. The assessment should gather information on the hazards present at the workplace, the degree of exposure and risk, appropriate control measures, health surveillance, and training and information.

In its guidance document *Guidance on risk assessment at work* published in 1996, the European Commission advises that:

1. The employer at each workplace has a general duty to ensure the health and safety of workers in every aspect related to work. The purpose of carrying out a risk assessment is to enable the employer to effectively take the measures necessary for the health and safety protection of workers. These measures include:
   ▶ prevention of occupational risks;
   ▶ provision of information to workers;
   ▶ provision of training to workers;
   ▶ organisation and means to implement the necessary measures.
2. Where elimination of risk is not realised, then the risks should be reduced and the residual risk controlled. At a later stage, as part of the review programme, such residual risks should be reassessed and reduced further.

A flow chart outlining the risk assessment procedure recommended by the European Commission is shown in Appendix 4.1 and incorporates the elements of risk management.

The general duties of employers to their employees, for example in Section 2 of the UK HSW Act 1974,

imply the need for risk assessment. This duty was also extended by Section 3 of the Act to anybody else affected by activities of the employer – contractors, visitors, customers or members of the public. However, the UK Management of Health and Safety at Work Regulations are much more specific concerning the need for risk assessment.

'the risk assessment shall be 'suitable and sufficient' and cover both employees and non-employees affected by the employer's undertaking (e.g. contractors, members of the public, students, patients, customers); every self-employed person shall make a 'suitable and sufficient' assessment of the risks to which they or those affected by the undertaking may be exposed;

any risk assessment shall be reviewed if there is reason to suspect that it is no longer valid or if a significant change has taken place;

where there are five or more employees, the significant findings of the assessment shall be recorded and any specially at risk group of employees identified. (This does not mean that employers with four or less employees need not undertake risk assessments.)'

The term '**suitable and sufficient**' is important as it defines the limits to the risk assessment process. A suitable and sufficient risk assessment should:

▶ identify the significant risks and ignore the trivial ones;
▶ identify and prioritise the measures required to comply with any relevant statutory provisions;
▶ remain appropriate to the nature of the work and valid over a reasonable period of time;
▶ identify the risk arising from or in connection with the work. The level of detail should be proportionate to the risk.

The significant findings that should be recorded include a detailed statement of the hazards and risks, the preventative, protective or control measures in place and any further measures required to reduce the risks and present proof that a suitable and sufficient assessment has been made.

4.1.3 Forms of risk assessment

There are two basic forms of risk assessment.

A **quantitative** risk assessment attempts to measure the risk by relating the probability of the risk occurring to the possible severity of the outcome and then giving the risk a numerical value. This method of risk assessment is used in situations where a malfunction could be very serious (e.g. aircraft design and maintenance or the petrochemical industry).

The more common form of risk assessment is the **qualitative** assessment, which is based purely on personal judgement and is normally defined as high,

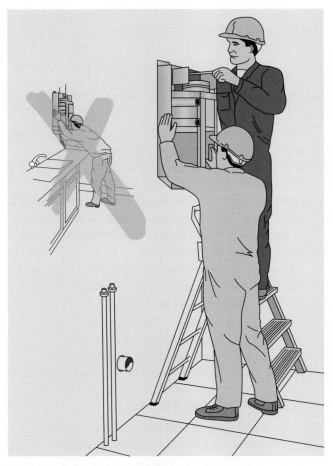

Figure 4.2 Reducing the risk – finding an alternative safer method when fitting a wall-mounted boiler

medium or low. Qualitative risk assessments are usually satisfactory as the definition (high, medium or low) is normally used to determine the time frame over which further action is to be taken.

The term '**generic**' risk assessment is sometimes used, and describes a risk assessment which covers similar activities or work equipment in different departments, sites or companies. Such assessments are often produced by specialist bodies, such as trade associations. If used, they must be appropriate to the particular job and they will need to be extended to cover additional hazards or risks.

4.1.4 Some definitions

Some basic definitions were introduced in Chapter 1 and those relevant to risk assessment are reproduced here.

Hazard and risk

A hazard is something with the **potential** to cause harm (this can include articles, substances, plant or machines, methods of working, the working environment and other aspects of work organisation). Hazards take many forms including, for example, chemicals, electricity or noise. A hazard can be ranked relative to other hazards or to a possible level of danger.

A risk is the **likelihood** of potential harm from that hazard being realised. Risk (or strictly the level of risk) is also linked to the severity of its consequences. A risk can be reduced and the hazard controlled by good management.

It is very important to distinguish between a hazard and a risk – the two terms are often confused and activities often called high risk are in fact high hazard. There should only be high residual risk where there is poor health and safety management and inadequate control measures.

Electricity is an example of a high hazard as it has the potential to kill a person. The risk associated with electricity – the likelihood of being killed on coming into contact with an electrical device – is, hopefully, low. However, the risk is of course much higher if the person is an electrical engineer and dismantles a piece of electrical equipment.

Occupational or work-related ill-health

This is concerned with those acute and chronic illnesses or physical and mental disorders that are either caused or triggered by workplace activities. Such conditions may be induced by the particular work activity of the individual or by activities of others in the workplace. The time interval between exposure and the onset of the illness may be short (e.g. acute asthma attacks) or long (e.g. chronic deafness or cancer).

Accident

Occupational Accident is defined in the ILO Code of Practice as: '*An occurrence arising out of or in the course of work which results in:*

(a) *fatal occupational injury;*
(b) *non-fatal occupational injury.*'

An accident is defined by the UK HSE in HSG65 as: '*any undesired circumstances which give rise to ill-health*

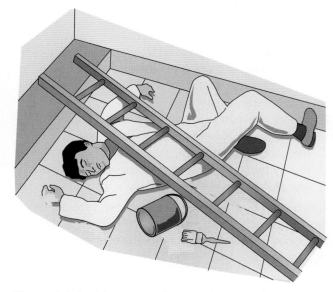

Figure 4.3 Accident at work

or injury; damage to property, plant, products or the environment; production losses or increased liabilities'. Other authorities define an accident more narrowly by excluding events that do not involve injury or ill-health. This book will use the HSE definition.

Incident and near miss

The ILO define this as: *'An unsafe occurrence arising out of or in the course of work where no personal injury is caused, or where personal injury requires only first-aid treatment.'*

A near miss would be an incident where there is no personal injury. Knowledge of near misses is very important as research has shown that, approximately, for every 10 'near miss' events at a particular location in the workplace, a minor accident will occur. See Chapter 5 for further explanation.

Dangerous occurrence

This is a 'near miss' or 'damage incident' which could have led to serious injury or loss of life. Dangerous occurrences are defined in the ILO Code of Practice as a: *'Readily identifiable event as defined under national laws and regulations, with potential to cause an injury or disease to persons at work or the public.'*

As an example, within the UK dangerous occurrences are defined in the Reporting of Injuries, Diseases and Dangerous Occurrences Regulations 1995 (often known as RIDDOR) and are always reportable to the enforcement authorities. Examples include the collapse of a scaffold or a crane or the failure of any passenger-carrying equipment.

In 1969, F. E. Bird collected a large quantity of accident data and produced a well-known triangle (Figure 4.4). It can be seen that damage and near miss incidents occur much more frequently than injury accidents and are, therefore, a good indicator of risks. The study also shows that most accidents are predictable and avoidable.

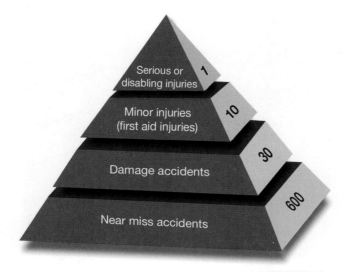

Figure 4.4 F. E. Bird's well-known accident triangle

Occupational disease

Occupational disease is defined by the ILO as: *'A disease contracted as a result of an exposure to risk factors arising from work activity.'*

Annex B to the ILO Code of Practice proposes a list of occupational diseases. See Chapter 5 for more details of this list.

4.1.5 The objectives of risk assessment

The *main* objective of risk assessment is to determine the measures required by the organisation to comply with relevant health and safety legislation and, thereby, reduce the level of occupational injuries and ill-health. The purpose is to help the employer or self-employed person to determine the measures required to comply with national legal obligations. The risk assessment will need to cover all those who may be at risk, such as customers, contractors and members of the public. In the case of shared workplaces, an overall risk assessment may be needed in partnership with other employers.

In Chapter 1, the moral, legal and financial arguments for health and safety management were discussed in detail. The important distinction between the direct and indirect costs of accidents is reiterated here.

Any accident or incidence of ill-health will cause both direct and indirect costs and incur an insured and an uninsured cost. It is important that all of these costs are taken into account when the full cost of an accident is calculated. In a study undertaken by the UK HSE, it was shown that indirect or hidden costs could be 36 times greater than direct costs of an accident. In other words, the direct costs of an accident or disease represent the tip of the iceberg when compared with the overall costs (see Figure 1.3).

Direct costs are costs that are directly related to the accident. They may be insured (claims on employers' and public liability insurance, damage to buildings, equipment or vehicles) or uninsured (fines, sick pay, damage to product, equipment or process).

Indirect costs may be insured (business loss, product or process liability) or uninsured (loss of goodwill, extra overtime payments, accident investigation time, production delays).

There are many reasons for the seriousness of a hazard not to be obvious to the person exposed to it. It may be that the hazard is not visible (radiation, certain gases and biological agents) or has no short-term effect (work-related upper limb disorders). Some common causes of accidents include lack of attention, lack of experience, not wearing appropriate PPE, sensory impairment and inadequate information, instruction and training.

4.1.6 Accident classification

The principal classification of industrial accidents according to type of accident in the ILO Code of Practice is as follows:

1. Falls of persons;
2. Struck by falling objects;
3. Stepping on, striking against or struck by objects excluding falling objects;
4. Caught in or between objects;
5. Over exertion or strenuous movements;
6. Exposure to or contact with extreme temperatures;
7. Exposure to or contact with electric current;
8. Exposure to or contact with harmful substances or radiations;
9. Other types of accident, not elsewhere classified, including accidents not classified for lack of sufficient data.

See Chapter 5 for more details.

4.1.7 Health risks

Risk assessment is not only concerned with injuries in the workplace but also needs to consider the possibility of occupational ill-health. Health risks fall into the following four categories:

1. chemical (e.g. paint solvents, exhaust fumes);
2. biological (e.g. bacteria, pathogens);
3. physical (e.g. noise, vibrations);
4. psychological (e.g. occupational stress).

There are two possible health effects of occupational ill-health.

They may be **acute**, which means that they occur soon after the exposure and are often of short duration, although in some cases emergency admission to hospital may be required.

They may be **chronic**, which means that the health effects develop with time. It may take several years for the associated disease to develop and the effects may be slight (mild asthma) or severe (cancer).

Health risks are discussed in more detail in chapters 13 and 14.

4.1.8 The management of risk assessment

Risk assessors

It is important that the risk assessment team is selected on the basis of its competence to assess risks in the particular areas under examination in the organisation. The Team Leader or Manager should have health and safety experience and relevant training in risk assessment. It is sensible to involve the appropriate line manager, who has responsibility for the area or activity being assessed, as a team member. Other members of the team will be selected on the basis of their experience, their technical and/or design

knowledge and any relevant standards or Regulations relating to the activity or process. At least one team member must have communication and report writing skills. A positive attitude and commitment to the risk assessment task are also important factors. It is likely that team members will require some basic training in risk assessment.

The approach to risk assessment in the United Kingdom

Risk assessment is part of the planning and performance stages of the health and safety management system recommended by the UK HSE in its publication HSG65. All aspects of the organisation, including health and safety management, need to be covered by the risk assessment process. This will involve the assessment of risk in areas such as maintenance procedures, training programmes and supervisory arrangements. A general risk assessment of the organisation should reveal the significant hazards present and the general control measures that are in place. Such a risk assessment should be completed first and then followed by more specific risk assessments that examine individual work activities.

The UK HSE has produced a free leaflet entitled 'Risk assessment – A brief guide to controlling risks in the workplace', INDG163 (Figure 4.5). It gives practical advice on assessing risks and recording the findings, and is aimed at small and medium-sized companies in the service and manufacturing sectors. There are five steps in risk assessment. These are:

1. look for the hazards;
2. decide who might be harmed, and how;
3. evaluate the risks and decide whether existing precautions are adequate or more should be done;
4. record the significant findings;
5. review the assessment and revise it if necessary.

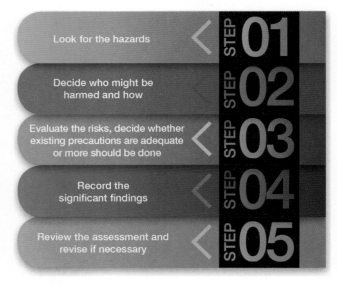

Figure 4.5 Five steps to risk assessment

This simple approach to risk assessment is similar to that adopted by the EU (see Appendix 4.1) and the ILO.

The approach of the ILO to risk assessment

Guidance on the approach of the ILO to risk assessment is given in its guidelines on occupational safety and health management systems ILO-OSH 2001 and in the Code of Practice 'Ambient factors in the workplace'. The assessment should be carried out in consultation with workers and/or their representatives by the employers or by persons acting on their behalf who are competent and have the necessary information, instruction and training.

Where the outcome of the assessment indicates a potential injury or risk to health and safety, the results of the assessment should be recorded and made available for inspection by the competent authority, and to workers exposed to the hazards and the workers' representatives. The record of the assessment should be retained for a period of time as may be specified by the competent authority.

The **first stage of the assessment** should include inspection of the workplace in order to identify:

(a) the hazards that are present or likely to occur, including chemical, biological, physical and psychological hazards and those associated with the work organisation;
(b) the activities that are likely to expose workers and others to the hazards identified, including those using maintenance, cleaning and emergency procedures.

The **second stage of the assessment** should consist of the collection of information about the hazards that are present or likely to occur so that the magnitude and significance for health and safety of any hazard or risk may be determined. This information should include that provided by suppliers and that available to the public. The relevance of work organisation and the practicability of various methods of control should be considered at this stage. Determining the magnitude of the hazard or risk should include determining the exposure of the workers to the hazard, unless other information is adequate for quantification of the risk. The exposure levels should be compared, when possible, with those exposure limits or standards prescribed by the competent authority. Where there are no such limits or standards, other national or internationally recognised standards should be used for comparison. In either case, account should be taken of the criteria on which those limits have been based.

The **third stage of the assessment** should establish whether hazards or risks to health and safety can be eliminated. If they cannot be eliminated, the employer should plan how they can be reduced to the lowest reasonable level, or to a level which, in the light of currently available national and international knowledge and data, would not lead to injury if exposure continued for a working lifetime.

As part of the assessment, the employer should:

(a) determine what instructions, training and information need to be given to the workers and, where appropriate, to their representatives, and others likely to be exposed to the hazards;
(b) determine what measures are needed to ensure that the information is kept up to date;
(c) plan for necessary training to be given to new or transferred workers;
(d) ensure that a programme for review of the assessment, including future monitoring of exposure levels, is established.

The assessment should be **reviewed** whenever there has been a significant change in the work processes to which it relates or when there is reason to suspect that it is no longer valid. The review should be incorporated into a system of management accountability which ensures that control action shown to be necessary by the initial assessment is in fact taken.

The reasons indicating that an assessment might no longer be valid include:

(a) complaints by workers of adverse health effects or the detection of health impairment by health surveillance;
(b) an accident, dangerous occurrence or incident leading to exposure to a level of hazard or risk that is different from that quantified in the initial assessment;
(c) subsequent measurement of exposure levels to hazardous substances;
(d) availability of updated or new information on the particular hazards described in the assessment;
(e) plant modification, including engineering control measures, changes in the process or methods of work and in the volume or rate of production which lead to a change in the hazards present in the workplace.

The review should reconsider all parts of the initial assessment, and in particular whether it is now:

(a) practicable to eliminate any hazards;
(b) possible to control at source and minimise hazards or risks which had previously required personal protective equipment.

The review should also consider the results of the programme for monitoring of exposure level, and whether:

(a) exposure levels previously considered to be acceptable should now be regarded as too high in the light of available and updated information on the hazards and risks;
(b) any control action needs to be taken;
(c) the frequency and type of monitoring previously decided is still appropriate.

The results of the review should be recorded and made available in the same way as the initial assessment.

4.1.9 The practice of risk assessment

The UK HSE approach to risk assessment (five steps) will be used to discuss the process of risk assessment.

Step 1 – Look for the hazards

The essential first step in risk assessment is to seek out and identify hazards. Relevant sources of information include:

▶ legislation and supporting codes of practice which give practical guidance and include basic minimum requirements;
▶ process information;
▶ product information;
▶ relevant international standards;
▶ industry or trade association guidance;
▶ the personal knowledge and experience of managers and employees;
▶ accident, ill-health and incident data from within the organisation, from other organisations or from central sources;
▶ expert advice and opinion and relevant research.

There should be a critical appraisal of all routine and non-routine business activities. People exposed may include not just employees, but also others such as members of the public, contractors and users of the products and services. Employees and safety representatives can make a useful contribution in identifying hazards.

In the simplest cases, hazards can be identified by observation and by comparing the circumstances with the relevant information (e.g. single-storey premises will not present any hazards associated with stairs). In more complex cases, measurements such as air sampling or examining the methods of machine operation may be necessary to identify the presence of hazards from chemicals or machinery. In the most complex or high-risk cases (for example in the chemical or nuclear industry) special techniques and systems may be needed such as hazard and operability studies (HAZOPS) and hazard analysis techniques such as event or fault-tree analysis. Specialist advice may be needed to choose and apply the most appropriate method.

Only significant hazards, which could result in serious harm to people, should be identified. Trivial hazards are a lower priority.

A tour of the area under consideration by the risk assessment team is an essential part of hazard identification as is consultation with the relevant section of the workforce.

A review of accident, incident and ill-health records will also help with the identification. Other sources of information include safety inspection, survey and audit reports, job or task analysis reports, manufacturers' handbooks or data sheets and Approved Codes of Practice and other forms of guidance.

Hazards will vary from workplace to workplace but the checklist in Appendix 4.2 shows the common hazards that are significant in many workplaces. Many questions in the NEBOSH examinations involve several common hazards found in most workplaces.

It is important that unsafe conditions are not confused with hazards, during hazard identification. Unsafe conditions should be rectified as soon as possible after observation. Examples of unsafe conditions include missing machine guards, faulty warning systems and oil spillage on the workplace floor.

Step 2 – Decide who might be harmed and how

Employees and contractors who work full time at the workplace are the most obvious groups at risk and it will be a necessary check that they are competent to perform their particular tasks. However, there may be other groups who spend time in or around the workplace. These include young workers, trainees, new and expectant mothers, cleaners, contractor and maintenance workers and members of the public. Members of the public will include visitors, patients, students or customers as well as passers-by.

The risk assessment must include any additional controls required due to the vulnerability of any of these groups, perhaps caused by inexperience or disability. It must also give an indication of the numbers of people from the different groups who come into contact with the hazard and the frequency of these contacts.

Step 3 – Evaluating the risks and the adequacy of current controls

This step is really two steps –evaluating the risks and evaluating the adequacy of current controls.

Evaluating the risks

During most risk assessments it will be noted that some of the risks posed by the hazard have already been addressed or controlled. The purpose of the risk assessment, therefore, is to reduce the remaining risk. This is called the **residual risk**.

The goal of risk assessment is to reduce all residual risks to as low a level as reasonably practicable. In a relatively complex workplace, this will take time so that a system of ranking risk is required – the higher the risk level the sooner it must be addressed and controlled.

For most situations, a **qualitative** risk assessment will be perfectly adequate. (This is certainly the case for NEBOSH International Certificate candidates and is suitable for use during the practical assessment.) During the risk assessment, a judgement is made as to whether the risk level is high, medium or low in terms of the risk of somebody being injured. This designation defines a timetable for remedial actions to be taken thereby reducing the risk. High-risk activities should normally be addressed in days, medium risks in weeks and low risks in months or in some cases no action will be required. It will usually be necessary for risk assessors to receive some training in risk level designation.

A **quantitative** risk assessment attempts to quantify the risk level in terms of the likelihood of an incident and its subsequent severity. Clearly the higher the likelihood and severity, the higher the risk will be. The likelihood depends on such factors as the control measures in place, the frequency of exposure to the hazard and the category of person exposed to the hazard. The severity will depend on the magnitude of the hazard (voltage, toxicity, etc.). The UK HSE suggests in HSG65 a simple 3 × 3 matrix to determine risk levels.

Likelihood of occurrence	Likelihood level
Harm is certain or near certain to occur	High 3
Harm will often occur	Medium 2
Harm will seldom occur	Low 1
Severity of harm	**Severity level**
Death or major injury (as defined by RIDDOR)	Major 3
3-day injury or illness (as defined by RIDDOR)	Serious 2
All other injuries or illnesses	Slight 1
Risk = Severity × Likelihood	

Likelihood	Severity		
	Slight 1	**Serious 2**	**Major 3**
Low 1	Low 1	Low 2	Medium 3
Medium 2	Low 2	Medium 4	High 6
High 3	Medium 3	High 6	High 9

It is possible to apply such methods to organisational risk or to the risk that the management system for health and safety will not deliver in the way in which it was expected or required. Such risks will add to the activity or occupational risk level. In simple terms, poor supervision of an activity will increase its overall level of risk. A risk management matrix has been developed which combines these two risk levels, as shown below.

RISK MANAGEMENT MATRIX	Occupational risk levels		
	Low	Medium	High
Organisational risk level — Low	Low	Low	Medium
Organisational risk level — Medium	Medium	Medium	High
Organisational risk level — High	High	High	Unsatisfactory

Whichever type of risk evaluation method is used, the level of risk simply enables a timetable of risk reduction to an acceptable and tolerable level to be formulated. The legal duty normally requires that all risks should be reduced to as low as is reasonably practicable, or words to this effect.

In established workplaces, some control of risk will be in place already. The effectiveness of these controls needs to be assessed so that an estimate of the residual risk may be made. Many hazards have had specific legislation or other recognised standards developed to reduce associated risks. Examples of such hazards may be fire, electricity, lead and asbestos. The relevant legislation and any accompanying codes of practice or guidance should be consulted first and any recommendations implemented. Advice on control measures may also be available from trade associations, trade unions or employers' organisations.

Where there are existing preventative measures in place, it is important to check that they are working properly and that everybody affected has a clear understanding of the measures. It may be necessary to strengthen existing procedures, for example by the introduction of a permit-to-work system. More details on the principles of control are contained later in this chapter.

Evaluating the controls

1. Hierarchy of risk control

When assessing the adequacy of existing controls or introducing new controls, a hierarchy of risk controls should be considered. The health and safety management system ISO 45001 (to replace OHSAS 18001 in 2016) states that the organisation shall establish a process for achieving risk reduction based upon the following hierarchy:

(a) eliminate the hazard;
(b) substitute with less hazardous materials, processes, operations or equipment;
(c) use engineering controls;
(d) use safety signs, markings and warning devices and administrative controls;
(e) use personal protective equipment.

The organisation shall ensure that the Occupational Health and Safety risks and determined controls are taken into account when establishing, implementing and maintaining its Occupational Health and Safety management system.

The hierarchy reflects that risk elimination and risk control by the use of physical engineering controls and safeguards can be more reliably maintained than those which rely solely on people. These concepts are written into the ILO Code of Practice 'Ambient factors in the workplace'.

Where a range of control measures is available, it will be necessary to weigh up the relative costs of each against the degree of control each provides, both in the short and long term. Some control measures, such as eliminating a risk by choosing a safer alternative substance or machine, provide a high degree of control and are reliable. Physical safeguards such as guarding a machine or enclosing a hazardous process need to be maintained. In making decisions about risk control, it will therefore be necessary to consider the degree of control and the reliability of the control measures along with the costs of both providing and maintaining the measure. The ISO 45001 hierarchy will now be discussed in detail.

(a) Elimination of the hazard

The best and most effective way of reducing risks is by avoiding a hazard and its associated risks. For example, avoid working at height by using a long-handled tool to clean windows; avoid entry into a confined space by, for example, using a sump pump in a pit which is removed by a lanyard for maintenance; eliminate the fire risks from tar boilers by using bitumen which can be applied cold.

(b) Substitution

Substitution describes the use of a less hazardous form of a substance or process. There are many examples of substitution such as the use of water-based rather than solvent-based paints; the use of asbestos substitutes; the use of compressed air as a power source rather than electricity to reduce both electrical and fire risks; and the use of mechanical excavators instead of hand digging.

In some cases it is possible to change the method of working so that risks are reduced. For example use rods to clear drains instead of strong chemicals; use a mobile elevating work platform instead of climbing a ladder. Sometimes the pattern of work can be changed so that people can do things in a more natural way, for example when placing components for packing consider whether people are right- or left-handed; encourage people in offices to take breaks from computer screens by getting up to photocopy, fetch files or print documents.

Care must be taken to consider any additional hazards which may be involved, and thereby introduce additional risks, as a result of a substitution.

(c) Engineering controls

This describes the control of risks by means of engineering design rather than a reliance on preventative actions by the employee. There are several ways of achieving such controls:

1. Control the risks at the source (e.g. the use of more efficient dust filters or the purchase of less noisy equipment).
2. Control the risk of exposure by:
   - ▶ *isolating* the equipment by the use of an enclosure, a barrier or guard;
   - ▶ *insulating* any electrical or temperature hazard;
   - ▶ *ventilating* away any hazardous fumes or gases either naturally or by the use of extractor fans and hoods (Figure 4.6).

(d) Safety signs, markings and warning devices and administrative controls

Safety signs, markings and warning devices

All general health and safety signs used in the workplace should include a pictorial symbol categorised

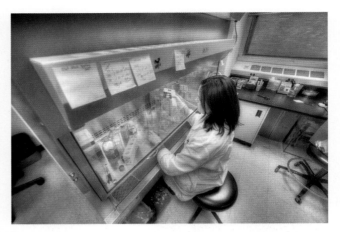

Figure 4.6 Proper control of gases, vapours and biological agents in a laboratory

by shape, colour and graphic image. A picture may be worth a thousand words, but when it comes to graphical symbols for safety-related information, misunderstanding the message may have serious consequences. The use of an International Standard aims to improve the understanding and proper application of safety signs worldwide, regardless of language and culture (Figures 4.7 and 4.8).

Prohibition
A red circular band with diagonal crossbar on a white background, the symbol within the circle to be black denoting a safety sign that indicates that a certain behaviour is prohibited.

Warning
A yellow triangle with black border and symbol within the yellow area denoting a safety sign that gives warning of a hazard.

Mandatory
A blue circle with white symbol denoting a sign that indicates that a specific course of action must be taken.

Safe condition
A green oblong or square with symbol or text in white denoting a safety sign providing information about safe conditions.

Fire equipment
A red oblong or square with symbol in white denoting a safety sign that indicates the location of fire-fighting equipment.

Figure 4.7 Colour categories and shapes of signs

Overhead load

Safety helmet
must be worn

Safety harness
must be worn

Not drinkable

No access for
pedestrians

Figure 4.8 Examples of pictorial signs

Published by ISO (International Organisation for Standardisation), ISO 7010:2003, *Graphical symbols – Safety colours and safety signs – Safety signs used in workplaces and public areas*, provides a method of communicating safety information through a collection of signs designed for use in any workplace, location and sector where safety-related questions may be posed. ISO 7010 sets out to guarantee that, wherever in the world, a manufacturer of safety signs for workplaces and public areas uses exactly the same pattern. The standard seeks, moreover, to give guidance to the designers of safety signs and get them to use ISO 7010 with a view to obtaining greater overall consistency and, thereby, better universal public recognition.

The collection of safety signs contained in ISO 7010 is not a mere 'collection' of more or less randomly sampled proposals. The signs included have given evidence, after year-long use in different countries, that they will also be globally understood.

The standard covers 32 safety signs designed for use in accident prevention, fire protection and emergency evacuation. Each is displayed by a visual illustration together with the image content, function, field and format of application. Geometric shape and colour are also indicated as prescribed by ISO 3864-1:2002, *Graphical symbols – Safety colours and safety signs – Part 1: Design principles for safety signs in workplaces and public areas*.

All workplaces need to display safety signs of some kind but deciding what is required can be confusing. Here are the basic suggestions for the majority of small premises or sites like small construction sites, canteens, shops, small workshop units and offices. This does not cover any signs which food hygiene law may require.

Many national regulatory regimes require signs to be displayed where a risk has not been controlled by other means. For example if a wet area of floor is cordoned off, a warning sign will not be needed, because the barrier will keep people out of the danger area. Signs are generally not needed where the sign would not reduce the risk or the risk is insignificant.

The following signs are typical of some of the ones most likely to be needed in these premises. Others may be necessary, depending on the hazards and risks present. The wording will of course be in the relevant local language.

(i) Overhead obstacles, construction site and prohibition notices (Figure 4.9)

Figure 4.9 Falling objects and construction site entry signs

(ii) Wet floors

These need to be used wherever a slippery area is not cordoned off. Lightweight stands holding double-sided signs are readily available (Figure 4.10).

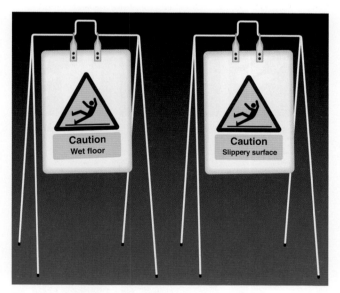

Figure 4.10 Wet floor signs

(iii) Chemical storage

Where hazardous cleaning chemicals are stored, apart from keeping the store locked, a suitable warning notice should be posted if it is considered this would help to reduce the risk of injury (Figure 4.11).

Toxic   Corrosive

Figure 4.11 Examples of chemical warning signs

(iv) Fire safety signs

These are needed to indicate emergency routes and emergency exits (Figure 4.12).

Fire exit

Figure 4.12 Examples of fire safety signs

(v) Fire action signs

These are needed to show actions necessary in an emergency such as sounding a fire alarm, location of fire extinguishers or hose reels (Figure 4.13).

Hose reel   Fire extinguisher   Fire alarm call point

Figure 4.13 Examples of fire action signs

(vi) First-aid

Signs showing the location of first-aid facilities will be needed. Advice on the action to take in the case of electric shock is no longer a legal requirement but is recommended (Figure 4.14).

First-aid post   First-aid stretcher   Eyewash

Figure 4.14 Examples of first-aid signs

(vii) Gas pipes and LPG cylinder stores

LPG cylinder stores should have the sign shown in Figure 4.15.

Figure 4.15 LPG sign

(viii) No smoking

Where smoking is not permitted the no-smoking sign is required. In many EU member states under smoke-free legislation substantially enclosed areas should have a sign (Figure 4.16).

No smoking

Figure 4.16 Smoke-free – no smoking signs

(ix) Fragile roofs

Signs should be erected at roof access points and at the top of outside walls where ladders may be placed (Figure 4.17).

Figure 4.17 Fragile roof signs

(x) Obstacles or dangerous locations

For example low head height, tripping hazard, etc. – alternating yellow and black stripes.

(xi) Other signs and posters

There may be other specific signs or posters required by national legislation; here are examples from the UK:

▶ UK Health and Safety Law – What you should know (there is a legal requirement to display this poster or distribute an equivalent leaflet).

▶ Certificate of Employer's Liability Insurance (there is a legal requirement to display this).

▶ Scalds and burns are common in kitchens. A poster showing recommended action is advisable, for example 'First-Aid for Burns'.

(xii) Sign checklist

Existing signs should be checked to ensure that:

▶ they are correct and up to date;

▶ they carry the correct warning symbol where appropriate;

▶ they are relevant to the hazard;

▶ they are easily understood;

▶ they are suitably located and not obscured;

▶ they are clean, durable and weatherproof where necessary;

▶ illuminated signs have regular lamp checks;

▶ they are used when required (e.g. 'Caution: wet floor' signs);

▶ they are obeyed and effective.

Administrative controls

Reduced time exposure

This involves reducing the time during the working day that the employee is exposed to the hazard, by giving the employee either other work or rest periods. It is normally only suitable for the control of health hazards associated with, for example, noise, vibration, excessive heat or cold, display screens and hazardous substances. However, it is important to note that, for many hazards, there are short-term exposure limits as well as normal occupational exposure limits (OELs) over an 8-hour period (see Chapter 13). Short-term limits must not be exceeded during the reduced time exposure intervals.

It cannot be argued that a short time of exposure to a dangerous part of a machine is acceptable. However, it is possible to consider short bouts of intensive work with rest periods when employees are engaged in heavy labour, such as manual digging when machines are not permitted due to the confines of the space or buried services.

Isolation/segregation

Controlling risks by isolating them or segregating people and the hazard is an effective control measure and is used in many instances; for example separating vehicles and pedestrians on factory sites, providing separate walkways for the public on road repairs, providing warm rooms on sites or noise refuges in noisy processes.

The principle of isolation is usually followed with the storage of highly flammable liquids or gases which are put into open, air-ventilated compounds away from other hazards such as sources of ignition, or away from people who may be at risk from fire or explosion.

Safe systems of work

Operating procedures or safe systems of work are probably the most common form of control measure used in industry today and may be the most economical and, in some cases, the only practical way of managing a particular risk. They should allow for methodical execution of tasks. The development of safe operating procedures should address the hazards that have been identified in the risk assessment. The system of work describes the safe method of performing the job or activity. A safe system of work is often a requirement of national legislation and is dealt with in detail later.

If the risks involved in the task are high or medium, the details of the system should be in writing and should be communicated to the employee formally in a training session. Details of systems for low-risk activities may be conveyed verbally. There should be records that the employee (or contractor) has been trained or instructed in the safe system of work and that they understand it and will abide by it.

Training

Training helps people acquire the skills, knowledge and attitudes to make them competent in the health and safety aspects of their work. There are generally two types of safety training:

▶ Specific safety training (or on-the-job training) which aims at tasks where training is needed due to the specific nature of such tasks. This is usually a job for supervisors, who, by virtue of their authority and close daily contact, are in a position to convert safety generalities to the everyday safe practice procedures that apply to individual tasks, machines, tools and processes;

▶ Planned training, such as general safety training, induction training, management training, skill training or refresher courses that are planned by the organisation, and relate to managing risk through policy, legislative or organisational requirements that are common to all employees.

Before any employee can work safely, they must be shown safe procedures for completing their tasks. The purpose of safety training should be to improve the safety awareness of employees and show them how to perform their jobs while employing acceptable, safe behaviour.

See Chapter 3 for more detail on health and safety training.

Information

Organisations need to ensure that they have effective arrangements for identifying and receiving relevant health and safety information from outside the organisation including:

▶ ensuring that pertinent health and safety information is communicated to all people in the organisation who need it;

▶ ensuring that relevant information is communicated to people outside the organisation who require it;

▶ encouraging feedback and suggestions from employees on health and safety matters.

Anyone who is affected by what is happening in the workplace will need to be given safety information. This does not only apply to staff. It can also apply to visitors, members of the public and contractors.

Information to be provided for people in a workplace includes:

▶ who is at risk and why;

▶ how to carry out specific tasks safely;

▶ correct operation of equipment;

▶ emergency action;

▶ accident and hazard reporting procedures;

▶ the safety responsibilities of individual people.

Information can be provided in a variety of ways. These include safety signs, posters, newsletters, memos, emails, personal briefings, meetings, toolbox talks, formal training, written safe systems of work and written health and safety arrangements.

Welfare

Welfare facilities include general workplace ventilation, lighting and heating and the provision of drinking water, sanitation and washing facilities. There is also a requirement to provide eating and rest rooms.

Risk control may be enhanced by the provision of good hand and forearm washing facilities and eye washing and shower facilities for use after certain accidents

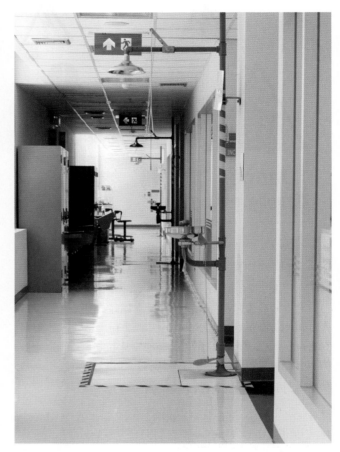

Figure 4.18 Emergency shower and eye wash station where ther is a serious risk of contamination

(Figure 4.18).

Good housekeeping is a very cheap and effective means of controlling risks. It involves keeping the workplace clean and tidy at all times and maintaining good storage systems for hazardous substances and other potentially dangerous items. The risks most likely to be influenced by good housekeeping are fire and slips, trips and falls.

See Chapter 7 for more information on the work environment.

Monitoring and supervision

All risk control measures, whether they rely on engineered or human behavioural controls, must be monitored for their effectiveness, with supervision to ensure that they have been applied correctly. Competent people who have a sound knowledge of the equipment or process should undertake this monitoring. Checklists are useful to ensure that no significant factor is forgotten. Any statutory inspection or insurance company reports should be checked to see whether any areas of concern were highlighted and if any recommendations were implemented. Details of any accidents, illnesses or other incidents will give an indication on the effectiveness of the risk control measures. Any emergency arrangements, including first-aid provision, should be tested during the monitoring process.

It is crucial that the operator should be monitored to ascertain that all relevant procedures have been understood and followed. The operator may also be able to suggest improvements to the equipment or system of work. The supervisor is an important source of information during the monitoring process.

Where the organisation is involved with shift work, it is essential that the risk controls are monitored on all shifts to ensure the uniformity of application.

The effectiveness and relevance of any training or instruction given should be monitored.

Periodically the risk control measures should be reviewed to ensure that they are correct and achieving what was intended. Monitoring and other reports are crucial for the review to be useful. Reviews often take place at safety committee and/or at management meetings. A serious accident or incident should always lead to an immediate review of the risk control measures in place.

(e) Personal protective equipment

Personal protective equipment (PPE) can provide very limited protection and should only be used as a last resort, when all other control measures have been considered and used or discarded. There are many reasons for this. The most important limitations are that PPE:

▶ only protects the person wearing the equipment, not others nearby;

▶ relies on people wearing the equipment at all times;

▶ must be used properly;

▶ must be replaced when it no longer offers the correct level of protection. This last point is particularly relevant when respiratory protection is used.

However, PPE does have some benefits which are:

▶ it gives immediate protection to allow a job to continue while more effective engineering controls are put in place;

▶ in an emergency it can be the only practicable way of effecting rescue or shutting down plant in hazardous atmospheres;

▶ it can be used to carry out work, for example, in a confined space, where alternatives are impracticable. But it should never be used to allow people to work in dangerous atmospheres, which are, for example, enriched with oxygen or potentially explosive.

See Chapter 13 for more details on PPE.

Figure 4.19 Good dust control for a chasing operation. A dust mask is still required for complete protection

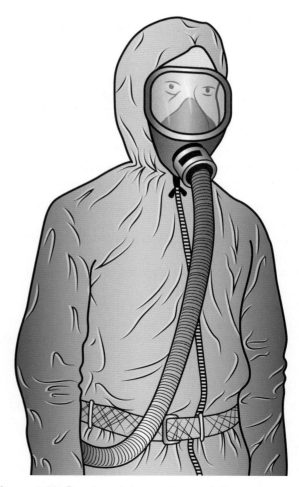

Figure 4.20 Respiratory protection and disposable overalls are needed when working in high levels of asbestos dust

2. Other hierarchies of risk control

There are several other similar hierarchies of risk control, which have been used over many years. A typical example is as follows:

▶ elimination:
▶ substitution;
▶ changing work methods/patterns;
▶ reduced or limited time exposure;
▶ engineering controls (e.g. isolation, insulation and ventilation);
▶ good housekeeping;
▶ safe systems of work;
▶ training and information;
▶ personal protective equipment;
▶ welfare;
▶ monitoring and supervision; and
▶ review.

In a similar manner, the ILO advises (in the ILO-OSH 2001 management system guidelines) that hazards and risks to workers' health and safety should be identified and assessed on an ongoing basis. Preventative and protective measures should be implemented in the following order of priority:

(a) eliminate the hazard/risk;
(b) control the hazard/risk at source, through the use of engineering controls or organisational measures;
(c) minimise the hazard/risk by the design of safe work systems, which include administrative control measures; and
(d) where residual hazards/risks cannot be controlled by collective measures, the employer should provide for appropriate personal protective equipment, including clothing, at no cost, and should implement measures to ensure its use and maintenance.

Hazard prevention and control procedures or arrangements should be established and should:

(a) be adapted to the hazards and risks encountered by the organisation;
(b) be reviewed and modified if necessary on a regular basis;
(c) comply with national laws and regulations, and reflect good practice; and
(d) consider the current state of knowledge, including information or reports from organisations, such as labour inspectorates, occupational health and safety services, and other services as appropriate.

3. Prioritisation of risk control

The prioritisation of the implementation of risk control measures will depend on the risk rating (high, medium and low), but the timescale in which the measures are introduced will not always follow the ratings. It may be convenient to deal with a low-level risk at the same time as a high-level risk or before a medium-level risk. It may also be that work on a high-risk control system is delayed due to a late delivery of an essential component – this should not halt the overall risk reduction work. It is important to maintain a continuous programme of risk improvement rather than slavishly following a predetermined priority list.

Step 4 – Recording significant findings

It is very useful to keep a written record of the risk assessment even if there are less than five employees in the organisation. For an assessment to be 'suitable and sufficient', only the significant hazards and conclusions need be recorded. The record should also include details of the groups of people affected by the hazards and the existing control measures and their effectiveness. The conclusions should identify any new controls required and a review date. The UK HSE booklet *Five Steps to Risk Assessment* provides a very useful guide and examples of the detail required for most risk assessments.

There are many possible layouts which can be used for the risk assessment record. Examples are given in Appendices 4.3 and 4.4 and Chapter 18. Also see the UK HSE's website which has further risk assessment examples at: http://www.hse.gov.uk/risk/index.htm.

It should be noted that in Appendices 4.3 and 4.4, the initial qualitative risk level at the time of the risk assessment is given – the residual risk level when all the additional controls have been implemented will be 'low'. This should mean that an annual review will be sufficient.

The written record provides excellent evidence to a health and safety enforcement officer of compliance with the law. It is also useful evidence if the organisation should become involved in a civil action.

The record should be accessible to employees and a copy kept with the safety manual containing the safety policy and arrangements.

Step 5 – Monitoring and review

As mentioned earlier, the risk controls should be reviewed periodically. This is equally true for the risk assessment as a whole. Review and revision may be necessary when conditions change as a result of the introduction of new machinery, processes or hazards. There may be new information on hazardous substances or new legislation. There could also be changes in the workforce, for example the introduction of trainees. The risk assessment needs to be revised only if significant changes have taken place since the last assessment was done. An accident or incident or a series of minor ones provides a good reason for a review of the risk assessment. This is known as the post-accident risk assessment.

Examples of completed risk assessments are given in Appendices 4.3 and 4.4. They show how small and medium-sized businesses have approached risk assessment.

4.1.10 Cost-benefit analysis

In recent years, risk assessment has been accompanied by a cost-benefit analysis that attempts to evaluate the costs and benefits of risk control and reduction. The costs could include capital investment, maintenance and training and produce benefits such as reduced insurance premiums, higher productivity and better product quality. The pay-back period for most risk reduction projects (other than the most simple) has been shown to be between two and five years. Although the benefits are often difficult to quantify, cost-benefit analysis does help to justify the level of expenditure on a risk reduction project.

4.1.11 Special cases

There are several groups of persons who require an additional risk assessment due to their being more 'at risk' than other groups. Five such groups will be considered – young persons, expectant and nursing mothers, workers with a disability, lone workers and business travellers.

Young persons

There are about 20 fatalities of young people at work each year in the UK. Therefore, any risk assessment involving young people needs to consider the particular vulnerability of young persons in the workplace. Young workers clearly have a lack of experience and awareness of risks in the workplace, a tendency to be subject to peer pressure and a willingness to work hard. Many young workers will be trainees or on unpaid work experience. Young people are not fully developed and are more vulnerable to physical, biological and chemical hazards than adults.

The following key elements should be covered by the risk assessment:

- details of the work activity, including any equipment or hazardous substances;
- details of any prohibited equipment or processes;
- details of health and safety training to be provided;
- details of supervision arrangements.

The extent of the risks identified in the risk assessment will determine whether employers should restrict the work of the people they employ. Except in special circumstances, young people should not be employed to do work which:

- is beyond their physical or psychological capacity;
- exposes them to substances chronically harmful to human health, for example toxic or carcinogenic substances, or effects likely to be passed on genetically or likely to harm an unborn child;
- exposes them to radiation;
- involves a risk of accidents which they are unlikely to recognise because of, for example, their lack of experience, training or insufficient attention to safety;

- involves a risk to their health from extreme heat, noise or vibration.

These restrictions will not apply in *special circumstances* where young people *over* the minimum school leaving age are doing work necessary for their training, under proper supervision by a competent person, and providing the risks are reduced to the lowest level, so far as is reasonably practicable. Under no circumstances can children of compulsory school age do work involving these risks, whether they are employed or under training such as work experience.

Induction training is important for young workers and such training should include site rules, restricted areas, prohibited machines and processes, fire precautions, emergency procedures, welfare arrangements and details of any further training related to their particular job. At induction, they should be introduced to their mentor and given close supervision, particularly during the first few weeks of their employment.

The ILO has also examined occupational health and safety among younger and older workers. Younger and older workers are particularly vulnerable. The ageing population in developed countries means that an increasing number of older persons are working and need special consideration.

According to ILO estimates for younger and older workers:

- Young workers aged 15–24 are much more likely to suffer non-fatal but serious accidents at work compared to their older colleagues. In the European Union, for example, the incidence rate for non-fatal accidents is at least 50% higher among workers aged 18–24 than in any other age category.
- Young workers also appear to be more vulnerable to certain types of risk than their older colleagues. For example, in Australia, fatal injuries involving electricity are twice as common amongst younger workers than amongst their older colleagues, according to the National Occupational Health & Safety Commission.

On the other hand, workers aged 55 years and over seem to be more likely to suffer fatal injuries at work compared to their younger colleagues.

Younger and older workers are covered by the Minimum Age Convention, 1973 (No. 138), the Worst Forms of Child Labour Convention, 1999 (No. 182) and their associated Recommendations, and the Older Workers Recommendation, 1980 (No. 162).

Expectant and nursing mothers

If any type of work could present a particular risk to expectant or nursing mothers, the risk assessment must include an assessment of such risks. Should these risks be unavoidable, then the woman's working conditions or hours must be altered to avoid the risks. The alternatives for her are to be offered other work or be suspended from work on full pay. The woman must notify the employer in writing that she is pregnant, or has given birth within the previous six months and/or is breastfeeding.

Pregnant workers should not be exposed to chemicals, such as pesticides and lead, or to biological hazards, such as hepatitis. Female agricultural workers, veterinary workers or farmers' wives who are pregnant, should not assist with lambing so that any possible contact with ovine chlamydia is avoided.

Other work activities that may present a particular risk to pregnant women at work are radiography, involving possible exposure to ionising radiation, and shop work when long periods of standing are required during shelf filling or stock-taking operations.

Typical factors which might affect such women are:

▶ manual handling, especially later in pregnancy;
▶ chemical or biological agents (e.g. lead and the rubella virus);
▶ passive smoking;
▶ lack of rest room facilities;
▶ temperature variations;
▶ ergonomic issues related to prolonged standing, sitting or the need for awkward body movement;
▶ issues associated with the use and wearing of personal protective equipment;
▶ working excessive hours;
▶ night working; and
▶ stress and violence to staff.

The ILO has introduced the Maternity Protection Convention 2000 (No. 183) and its accompanying Maternity Protection Recommendation 2000 (No. 191) to cover the needs of expectant and nursing mothers in the workplace.

The Convention expects member states to adopt appropriate measures to ensure that pregnant or breastfeeding mothers are not obliged to perform work which has been determined by the competent authority to be prejudicial to the health of the mother or the child, or where an assessment has established a significant risk to the mother's health or that of her child. On production of a medical certificate or other appropriate certification stating the presumed date of childbirth, a woman to whom the Convention applies is entitled to a period of maternity leave of not less than 14 weeks. The Convention states that it is unlawful for an employer to terminate the employment of a woman during her pregnancy or absence on leave or during a period following her return to work to be prescribed by national laws or regulations, except on grounds unrelated to the pregnancy or birth of the child and its consequences or nursing. The burden of proving that the reasons for dismissal are unrelated to pregnancy or childbirth and its consequences or nursing rests on the employer.

The Convention guarantees a woman the right to return to the same position or an equivalent position paid at the same rate at the end of her maternity leave. A woman shall be provided with the right to one or more daily breaks or a daily reduction of hours of work to breastfeed her child. The period during which nursing breaks or the reduction of daily hours of work are allowed, their number, the duration of nursing breaks and the procedures for the reduction of daily hours of work shall be determined by national law and practice. These breaks or the reduction of daily hours of work shall be counted as working time and remunerated accordingly.

The Maternity Protection Recommendation R191 recommends that member states extend the period of maternity leave referred to in the Convention to at least 18 weeks. If the current work of the woman presents a risk to her or her unborn child, as indicated by an appropriate medical certificate, an alternative to such work in the form of:

(a) elimination of risk;
(b) an adaptation of her conditions of work;
(c) a transfer to another post, without loss of pay, when such an adaptation is not feasible; or
(d) paid leave, in accordance with national laws, regulations or practice, when such a transfer is not feasible.

The following types of work could be hazardous for pregnant women:

(a) arduous work involving the manual lifting, carrying, pushing or pulling of loads;
(b) work involving exposure to biological, chemical or physical agents which represent a reproductive health hazard;
(c) work requiring special equilibrium;
(d) work involving physical strain due to prolonged periods of sitting or standing, to extreme temperatures, noise or to vibration.

A pregnant or nursing woman should not be obliged to do night work if a medical certificate declares such work to be incompatible with her pregnancy or nursing. She should retain the right to return to her job or an equivalent job as soon as it is safe for her to do so. She should also be allowed to leave her workplace, if necessary, after notifying her employer, for the purpose of undergoing medical examinations relating to her pregnancy.

On production of a medical certificate or other appropriate certification as determined by national law and practice, breastfeeding mothers should be allowed breaks, the frequency and length of which should be adapted to particular needs. Where practicable and with the agreement of the employer and the woman concerned, it should be possible to combine the time allotted for daily nursing breaks to allow a reduction of hours of work at the beginning or at the end of the working day. Where practicable, provision should be made for the establishment of facilities for nursing under adequate hygienic conditions at or near the workplace.

Detailed guidance is available in *New and Expectant Mothers at Work,* HSG122, HSE Books.

Workers with a disability

Organisations have been encouraged for many years to employ workers with disabilities and to ensure that their premises provide suitable access for such people. From a health and safety point of view, it is important that workers with a disability are covered by special risk assessments so that appropriate controls are in place to protect them. For example, employees with a hearing problem will need to be warned when the fire alarm sounds or a fork-lift truck approaches. Special vibrating signals or flashing lights may be used. Similarly workers in wheelchairs will require a clear, wheelchair-friendly route to a fire exit and onwards to the assembly point. Safe systems of work and welfare facilities need to be suitable for any workers with disabilities.

The special risk assessment should identify:

▶ the jobs with particular health and fitness requirements;

▶ the types of disability that would make certain jobs unsuitable;

▶ the staff whose disabilities would exclude them from undertaking those jobs safely;

▶ and screen staff against these criteria – this may have the effect of excluding those with a certain disability from doing these jobs (e.g. fork-lift truck driving).

It is important to recognise that there may be other employment and anti-discrimination legislation that should be considered before the findings of the risk assessment are finalised.

Lone workers

People who work alone, like those in small workshops, remote areas of a large site, social workers, sales personnel or mobile maintenance staff, should not be at more risk than other employees (Figure 4.21). Lone workers are a group of workers who are especially vulnerable in certain situations. If, for example, the worker is in contact with members of the public, they may be at risk of assault from people who could be violent. People who work alone in confined spaces could also be at risk if they have an accident or become ill. There are, however, no absolute restrictions on working alone; it will depend on the findings of a risk assessment.

When a risk assessment shows that it is not possible for the work to be done safely by a lone worker, arrangements for providing help or back-up should be put in place. Where a lone worker is working at the premises of another employer, that employer should inform the employer of the lone worker of any risks and the precautions that should be taken. A risk assessment is, therefore, essential for all instances of lone working.

It is important to consider whether the risks of the job can be properly controlled by one person. Other considerations in the risk assessment include:

▶ does the particular workplace present a special risk to someone working alone?

▶ is there safe entrance and exit from the workplace?

▶ can all the equipment and substances be safely handled by one person?

▶ is violence from others a risk?

▶ would women and young persons be specially at risk?

▶ is the worker medically fit and suitable for working alone?

▶ are special training and supervision required?

▶ has the worker access to first-aid?

For details of further precautions for lone workers with regard to safe systems of work, see 4.4.9.

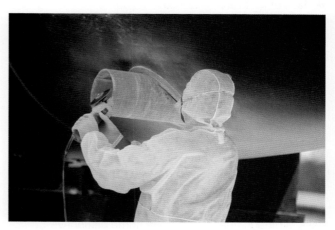

Figure 4.21 A lone worker – special arrangements required

Pre-travel risk assessments

Despite great strides in communication technology, business travel has not lost its appeal and in 2010 over 3.5 million employees undertook overseas trips. One-quarter of the overseas trips undertaken in 2010 were to high or extreme risk locations.

With recent political unrest around the world, as well as natural disasters in Japan and New Zealand, many organisations are addressing the hazards associated with sending employees overseas on business. The following points should be considered when making a pre-travel risk assessment:

▶ the political, medical and security risks of the countries involved;

▶ infrastructure and contacts in the countries involved;

▶ cultural awareness and training;

▶ travel planning and vaccination schedule;
▶ personal safety and security training;
▶ communications arrangements;
▶ details of accommodation;
▶ travel within the country, including driving; and
▶ contingency and emergency strategy and response.

4.2 General principles of control and hierarchy of risk reduction measures

4.2.1 Introduction

The control of risks is essential to secure and maintain a healthy and safe workplace which complies with the relevant legal requirements. Hazard identification and risk assessment are covered earlier in this chapter and these together with appropriate risk control measures form the core of the Occupational Safety and Health Management Systems such as ILO-OSH 2001, OHSAS 45001 (previously 18001:2007) and UK, HSG65.

Today safety is controlled through a combination of **engineered measures** such as the provision of safety protection (e.g. guarding and warning systems), and **operational measures** in training, safe work practices, operating procedures and method statements, along with **management supervision**.

These measures (collectively) are commonly known in health and safety terms as **control measures**. Some of these more common measures will be explained in more detail later.

This section concerns the principles that should be adopted when deciding on suitable measures to eliminate or control both acute and chronic risks to the health and safety of people at work. The principles of control can be applied to both health risks and safety risks, although health risks have some distinctive features that require a special approach.

Chapters 7–14 deal with specific workplace hazards and controls, subject by subject. The principles of prevention now enshrined in EU Directives, OHSAS 45000 (previously 18001:2007) and ILO OSH-2001 should be

used jointly with the hierarchy of control methods which give the preferred order of approach to risk control.

When risks have been analysed and assessed, decisions can be made about workplace precautions.

All final decisions about risk control methods must take into account the relevant national legal requirements, which establish minimum levels of risk prevention or control. Sometimes the duties imposed by national statutory provisions are **absolute** and must be complied with. Many requirements are, however, qualified by phrases such as: suitable and sufficient, proportional to the risk, **so far as is reasonably practicable,** or **so far as is practicable**. These require an assessment of cost, along with information about relative costs, effectiveness and reliability of different control measures. Further guidance on the meaning of these four expressions is provided in Chapter 1.

4.2.2 General principles of prevention

The general principles of prevention cover technical, behavioural and procedural controls. They are set out in Article 6(2) of the European Council Directive 89/391/EEC but are just as relevant outside the EU. These general principles require that where an employer implements any preventative measures they shall do so on the basis of these principles which are not a hierarchy or order of priority. The general principles of prevention are as follows:

1. **Avoiding risks**
   This means, for example, trying to stop doing the task or using different processes or doing the work in a different, safer way.
2. **Evaluating the risks which cannot be avoided**
   This requires a risk assessment to be carried out.
3. **Combating the risks at source**
   This means that risks, such as a dusty work atmosphere, are controlled by removing the cause of the dust rather than providing special protection; or that slippery floors are treated or replaced rather than putting up a sign.
4. **Adapting the work to the individual**
   This involves the design of the workplace, the choice of work equipment and the choice of working and production methods, with a view, in particular, to alleviating monotonous work and work at a predetermined work rate and to reducing their effect on health.

   This will involve consulting those who will be affected when workplaces, methods of work and safety procedures are designed. The control individuals have over their work should be increased, and time spent working at predetermined speeds and in monotonous work should be reduced where it is reasonable to do so.
5. **Adapting to technical progress**
   It is important to take advantage of technological and technical progress, which often gives designers

Figure 4.22 When controls break down

and employers the chance to improve both safety and working methods. With the Internet and other international information sources available, very wide knowledge, going beyond what is happening in a particular nation state or region, is likely to be expected by the enforcing authorities and the courts.

6. **Replacing the dangerous by the non-dangerous or the less dangerous**
   This involves substituting, for example, equipment or substances with non-hazardous or less hazardous substances.

7. **Developing a coherent overall prevention policy**
   This covers technology, organisation of work, working conditions, social relationships and the influence of factors relating to the working environment.
   Health and safety policies should be prepared and applied by reference to these principles.

8. **Giving collective protective measures priority over individual protective measures**
   This means giving priority to control measures which make the workplace safe for everyone working there, giving the greatest benefit, for example removing hazardous dust by exhaust ventilation rather than providing a filtering respirator to an individual worker. This is sometimes known as a 'Safe Place' approach to controlling risks.

9. **Giving appropriate instruction to employees**
   This involves making sure that employees are fully aware of company policy, safety procedures, good practice, official guidance, any test results and legal requirements. This is sometimes known as a 'Safe Person' approach to controlling risks where the focus is on individuals. A properly set-up health and safety management system should cover and balance both a Safe Place and Safe Person approach.

4.3 Sources of health and safety information

When anybody, whether a health and safety professional, a manager or an employee, is confronted with a health and safety problem, they will need to consult various items of published information to ascertain the scale of the problem and its possible remedies. The sources of this information may be internal to the organisation and/or external to it.

4.3.1 Internal sources which should be available within the organisation include

▶ accident and ill-health records and investigation reports;
▶ absentee records;
▶ inspection and audit reports undertaken by the organisation and by external organisations including enforcement agencies;

▶ OSH manuals and procedures;
▶ maintenance, risk assessment and training records;
▶ documents which provide information to workers;
▶ any equipment examination or test reports.

4.3.2 External sources, which are available outside the organisation, are numerous and include

▶ health and safety legislation;
▶ Enforcement agencies guidance documents, leaflets, journals, books and their websites;
▶ International (e.g. International Labour Organisation, ILO), European and National Standards;
▶ health and safety magazines and journals;
▶ information published by trade associations, employer organisations and trade unions;
▶ specialist technical and legal publications;
▶ information and data from manufacturers and suppliers such as MSDS (Material Safety Data Sheets);
▶ Occupational Safety and Health Administration (OSHA) USA
▶ Worksafe, Western Australia
▶ dedicated and company websites.

Recommended Format for Material Safety Data Sheets (MSDSs)

By following this recommended format, the information of greatest concern to workers is featured at the beginning of the data sheet, including information on chemical composition and first aid measures. More technical information that addresses topics such as the physical and chemical properties of the material and toxicological data appears later in the document. The 16-section MSDS is becoming the international norm. The 16 sections are:

▶ Identification
▶ Hazard(s) identification
▶ Composition/information on ingredients
▶ First-aid measures
▶ Fire-fighting measures
▶ Accidental release measures
▶ Handling and storage
▶ Exposure controls/personal protection
▶ Physical and chemical properties
▶ Stability and reactivity
▶ Toxicological information
▶ Ecological information
▶ Disposal considerations
▶ Transport information
▶ Regulatory information
▶ Other information

Figure 4.23 Recommended format for a Material Safety Data Sheet (MSDS)

See Chapter 18 for more information and website addresses. Many of these sources of information will be referred to throughout this book.

Figure 4.24 Health risk – checking on the contents

4.4 Safe systems of work

4.4.1 What is a safe system of work?

A safe system of work has been defined as:

'*The integration of personnel, articles and substances in a laid out and considered method of working which takes proper account of the risks to employees and others who may be affected, such as visitors and contractors, and provides a formal framework to ensure that all of the steps necessary for safe working have been anticipated and implemented.*'

In simple terms, a safe system of work is a defined method for doing a job in a safe way. It takes account of all foreseeable hazards to health and safety and seeks to eliminate or minimise these. Safe systems of work are normally formal and documented, for example in written operating procedures but, in some cases, they may be verbal.

The particular importance of safe systems of work stems from the recognition that most accidents are caused by a combination of factors (plant, substances, lack of training and/or supervision, etc.). Hence prevention must be based on an integral approach and not one which only deals with each factor in isolation. The adoption of a safe system of work provides this integral approach because an effective safe system:

▶ is based on looking at the job as a whole;
▶ starts from an analysis of all foreseeable hazards, for example physical, chemical, health;
▶ brings together all the necessary precautions, including design, physical precautions, training, monitoring, procedures and PPE.

It follows from this that the use of safe systems of work is in no way a replacement for other precautions, such as good equipment design, safe construction and the use of physical safeguards. However, there are many situations where these will not give adequate protection in themselves, and then a carefully thought-out and properly implemented safe system of work is especially important. The best example is maintenance and repair work, which will often involve, as a first-stage, dismantling the guard or breaking through the containment, which exists for the protection of the ordinary process operator. In some of these operations, a permit-to-work procedure will be the most appropriate type of safe system of work.

The operations covered may be simple or complex, routine or unusual.

Whether the system is verbal or written, and whether the operation it covers is simple or complex, routine or unusual, the essential features are forethought and planning – to ensure that all foreseeable hazards are identified and controlled. In particular, this will involve scrutiny of:

▶ the sequence of operations to be carried out;
▶ the equipment, plant, machinery and tools involved;
▶ chemicals and other substances to which people might be exposed in the course of the work;
▶ the people doing the work – their skill and experience;

Figure 4.25 (a) and (b) Multipadlocked hasp for locking off an isolation valve – each worker puts on their own padlock

▶ foreseeable hazards (health, safety, environment), whether to the people doing the work or to others who might be affected by it;

▶ practical precautions which, when adopted, will eliminate or minimise these hazards (Figure 4.25);

▶ the training needs of those who will manage and operate under the procedure;

▶ monitoring systems to ensure that the defined precautions are implemented effectively.

4.4.2 Legal requirements

Article 10(a) of the ILO Recommendation R164 requires employers, 'to provide and maintain workplaces, machinery and equipment, and use work methods, which are as safe and without risk to health as is reasonably practicable'.

In addition 10(b) requires employers, 'to give necessary instructions and training, taking account of the functions and capacities of different categories of workers'.

Many national regulations require information and instruction to be provided to employees and others. In effect, this is also a more specific requirement to provide safe systems of work. Many of these safe systems, information and instructions will need to be in writing.

There may also be a need for employers to provide a safe system of work to fulfil their civil law/common law duty of care.

Figure 4.26 Multihasp used on an electrical isolator

4.4.3 Assessment of what safe systems of work are required

Requirement

It is the responsibility of the management in each organisation to ensure that its operations are assessed to determine where safe systems of work need to be developed.

This assessment must, at the same time, decide the most appropriate form for the safe system; that is:

▶ Is a written procedure required?

▶ Should the operation only be carried out under permit to work?

▶ Is an informal system sufficient?

Factors to be considered

It is recognised that each organisation must have the freedom to devise systems that match the risk potential of their operations and which are practicable in their situation. However, they should take account of the following factors in making their decision:

▶ types of risk involved in the operation;

▶ magnitude of the risk, including consideration of the worst foreseeable loss;

▶ complexity of the operation;

▶ past accident and loss experience;

▶ requirements and recommendations of the relevant health and safety authorities;

▶ the type of documentation needed;

▶ resources required to implement the safe system of work (including training and monitoring).

4.4.4 Development of safe systems

Role of competent person

A competent person or safety and/or health adviser may be required to be appointed under national legislation. A competent person is normally considered to be someone who has appropriate practical and theoretical knowledge with the necessary qualifications and experience to perform a job properly. This could be as an employee or an external consultant. Article 13 of ILO-R164 requires that:

'As necessary in regard to the activities of the undertaking and practicable in regard to size, provision should be made for:

(a) the availability of an occupational health service and a safety service, within the undertaking, jointly with other undertakings, or under arrangements with an outside body;

(b) recourse to specialists to advise on particular occupational safety or health problems or supervise the application of measures to meet them.'

Appointed competent persons should assist managers to draw up guidelines for safe systems of work.

111

This will include, where necessary, particularly in construction work, method statements. The competent person should prepare suitable forms and should advise management on the adequacy of the safe systems produced.

Role of managers

Primarily management is responsible for providing safe systems of work, as they will know the detailed way in which the task should be carried out.

Management is responsible to ensure that employees are adequately trained in a specific safe system of work and are competent to carry out the work safely. Managers need to provide sufficient supervision to ensure that the system of work is followed and the work is carried out safely. The level of supervision will depend on the experience of the particular employees concerned and the complexity and risks of the task.

When construction work is involved, principal or main contractors will need to monitor sub-contractors to check that they are providing suitable safe systems of work, have trained their employees and are carrying out the tasks in accordance with the safe systems.

Role of employees/consultation

Many people operating a piece of machinery or a manufacturing process are in the best position to help with the preparation of safe systems of work. Consultation with those employees who will be exposed to the risks, either directly or through their representatives, is required and may also be a legal requirement. The importance of discussing the proposed system with those who will have to work under it, and those who will have to supervise its operation, cannot be emphasised enough.

Employees have a responsibility to follow the safe system of work.

Analysis

The safe system of work should be based on a thorough analysis of the job or operation to be covered by the system. The way this analysis is done will depend on the nature of the job/operation.

If the operation being considered is a new one involving high loss potential, the use of formal hazard analysis techniques such as hazard and operability (HAZOP) study, fault-tree analysis (FTA) or failure modes and effects analysis should be considered.

However, where the potential for loss is lower, a more simple approach, such as job safety analysis (JSA), will be sufficient. This will involve three key stages:

- identification of the key steps in the job/operation – What activities will the work involve?
- analysis and assessment of the risks associated with each stage – What could go wrong?

- definition of the precautions or controls to be taken – What steps need to be taken to ensure the operation proceeds without danger, either to the people doing the work, or to anyone else?

The results of this analysis are then used to draw up the safe operating procedure or method statement. (See Chapter 18 for a suitable form.)

Introducing controls

There are a variety of controls that can be adopted in safe systems of work. They can be split into the following three basic categories:

(i) **Technical** – these are engineering or process type controls which engineer out or contain the hazard so that the risks are acceptable. For example exhaust ventilation, a machine guard, dust respirator.

(ii) **Procedural** – these are ways of doing things to ensure that the work is done according to the procedure, legislation or cultural requirements of the organisation. For example a supervisor must be involved, the induction course must be taken before the work commences, a particular type of form or a person's signature must be obtained before proceeding, the names of the workforce must be recorded.

(iii) **Behavioural** – these are controls which require a certain standard of behaviour from individuals or groups of individuals. For example no smoking is permitted during the task, hard hats must be worn, all lifts are to be in tandem between two workers.

4.4.5 Preparation of safe systems

A checklist for use in the preparation of safe systems of work is set out as follows:

- What is the work to be done?
- What are the potential hazards?
- Is the work covered by any existing instructions or procedures? If so, to what extent (if any) do these need to be modified?
- Who is to do the work?
- What are their skills and abilities – is any special training needed?
- Under whose control and supervision will the work be done?
- Will any special tools, protective clothing or equipment (e.g. breathing apparatus) be needed? Are they ready and available for use?
- Are the people who are to do the work adequately trained to use the above?
- What isolations and locking-off will be needed for the work to be done safely?
- Is a permit to work required for any aspect of the work?
- Will the work interfere with other activities? Will other activities create a hazard to the people doing the work?

- Have other departments been informed about the work to be done, where appropriate?
- How will the people doing the work communicate with each other?
- Have possible emergencies and the action to be taken been considered?
- Should the emergency services be notified?
- What are the arrangements for handover of the plant/ equipment at the end of the work? (For maintenance/project work, etc.)
- Do the planned precautions take account of all foreseeable hazards?
- Who needs to be informed about or receive copies of the safe system of work?
- What arrangements will there be to see that the agreed system is followed and that it works in practice?
- What mechanism is there to ensure that the safe system of work stays relevant and up to date?

4.4.6 Documentation

Safe systems of work should be properly documented.

Wherever possible, they should be incorporated into normal process operating procedures. This is so that:

- health and safety are seen as an integral part of, and not add-on to, normal production procedures;
- the need for operators and supervisors to refer to separate manuals is minimised.

Whatever method is used, all written systems of work should be signed by the relevant managers to indicate approval or authorisation. Version numbers should be included so that it can quickly be verified that the most up-to-date version is in use. Records should be kept of copies of the documentation, so that all sets are amended when updates and other revisions are issued.

As far as possible, systems should be written in a non-technical style and should specifically be designed to be as intelligible and user-friendly as possible. It may be necessary to produce simple summary sheets which contain all the key points in an easy-to-read format.

4.4.7 Communication and training

People doing work or supervising work must be made fully aware of the laid-down safe systems that apply. The preparation of safe systems will often identify a training need that must be met before the system can be implemented effectively.

In addition, people should receive training in how the system is to operate. This applies not only to those directly involved in doing the work but also to supervisors/managers who are to oversee it.

In particular, the training might include:

- why a safe system is needed;
- what is involved in the work;
- the hazards which have been identified;

- the precautions which have been decided and, in particular:
  - ▷ the isolations and locking-off required, and how this is to be done;
  - ▷ details of the permit-to-work system, if applicable;
  - ▷ any monitoring (e.g. air testing) which is to be done during the work, or before it starts;
  - ▷ how to use any necessary PPE;
  - ▷ emergency procedures.

4.4.8 Monitoring safe systems

Safe systems of work should be monitored to ensure that they are effective in practice. This will involve:

- reviewing and revising the systems themselves, to ensure they stay up to date;
- inspecting to identify how fully they are being implemented.

In practice, these two things go together, as it is likely that a system that is out of date will not be fully implemented by the people who are intended to operate it.

All organisations are responsible for ensuring that their safe systems of work are reviewed and revised as appropriate. Monitoring of implementation is part of all line managers' normal operating responsibilities, and should also take place during health and safety audits.

4.4.9 Definition of and specific examples of safe systems of work

Confined spaces

A confined space is defined as: *'any place, including any chamber, tank, vat, silo, pit, trench, pipe, sewer, flue, well or other similar space in which, by virtue of its enclosed nature, there arises a reasonably foreseeable specified risk'.*

These specified risks mean a risk to a worker of:

- serious injury arising from a fire or explosion;
- loss of consciousness arising from an increase in body temperature;
- loss of consciousness or asphyxiation arising from gas, fume, vapour or the lack of oxygen;
- drowning arising from an increase in the level of liquid;
- asphyxiation arising from a free-flowing solid or because of entrapment by it.

Therefore, confined spaces include chambers, tanks (sealed and open-top), vessels, furnaces, ducts, sewers, manholes, pits, flues, excavations, boilers, reactors and ovens.

The principal hazards associated with a confined space are the difficult access and egress, which can make escape and rescue more difficult. Other hazards associated with confined spaces include:

- asphyxiation due to oxygen depletion;
- poisoning by toxic substance or fumes;
- explosions due to gases, vapours and dust;
- fire due to flammable liquids;
- fall of materials leading to possible head injuries;
- free-flowing solid such as grain in a silo;
- electrocution from unsuitable equipment;
- difficulties of rescuing injured personnel;
- drowning due to flooding; and
- fumes from plant or processes entering confined spaces.

Since work in confined spaces is a high risk work activity, a risk assessment is essential. The following items should be included in the assessment:

- the task;
- the working environment;
- working materials and tools;
- the suitability of those carrying out the task; and
- arrangements for emergency rescue.

If the assessment identifies risks of serious injury from work in the confined space, then the following key duties apply. These duties are:

- if possible, avoid entry to confined space by doing the work outside the space (e.g. remote cameras can be used for internal inspection of vessels);
- if entry to a confined space is unavoidable, follow a safe system of work; and
- ensure that adequate emergency arrangements are in place before the work starts.

If entry into a confined space cannot be avoided, then a safe system of work for working inside the space must be developed that uses the results of the risk assessment to identify the controls needed to reduce the risk of injury. These controls will depend on the nature of the confined space, the associated risk and the work involved. It is important that the agreed safe system of work is fully implemented and everyone who is to work in the confined space must be trained and instructed on the controls required.

The following topics need to be addressed in a safe system of work for a confined space:

- the appointment of a supervisor;
- the competence and experience required of the workers in the confined space;
- the isolation of mechanical and electrical equipment in an emergency;
- the size of the entrance must enable rapid access and exit in an emergency;
- the provision of adequate ventilation;
- the testing of the air inside the space to ensure that it is fit to breathe. If the air inside the space is not fit to breathe, then breathing apparatus will be essential;
- the provision of special tools and lighting, such as extra low voltage equipment, non-sparking tools and specially protected lighting;

- the emergency arrangements to cover the necessary equipment, training, practice drills and the raising of the alarm; and
- adequate communications arrangements to enable communication between people inside and outside the confined space and to summon help in an emergency.

The provision of suitable rescue and resuscitation equipment will depend on the likely emergencies identified. Where such equipment is provided for rescuers to use, training in the correct operation of the equipment is essential. Rescuers should also be trained in all aspects of the emergency procedures and relevant first-aid procedures.

For particularly hazardous confined space working a permit to work may be required, this is discussed later in 4.6.4.

Lone workers

Section 4.1.11 covered the need and contents of a risk assessment for lone workers. In this section the controls required to protect lone workers are discussed.

People who work by themselves without close or direct supervision are found in many work situations. In some cases they are the sole occupant of small workshops or warehouses; they may work in remote sections of a large site; they may work out of normal hours, like cleaners or security personnel; they may be working away from their main base as installers, or maintenance people; they could be people giving a service, like domiciliary care workers, drivers and estate agents.

There is no general legal reason why people should not work alone, but there may be special risks which require two or more people to be present; for example during entry into a confined space in order to effect a rescue. See 4.5.4 for more details on confined spaces. It is important to ensure that a lone worker is not put at any higher risk than other workers. This is achieved by carrying out a specific risk assessment and introducing special protection arrangements for their safety. People particularly at risk, like young people or women, should also be considered. People's overall health and suitability to work alone should be taken into account. It is important to ascertain whether the work should be performed alone particularly where there is a possibility of a serious risk, such as violence, being confronted by the worker.

Typical control procedures may include:

- documented records of the location or itineraries of the lone workers;
- periodic visits from the supervisor to observe what is happening;
- regular voice contact between the lone worker and the supervisor;

▶ automatic warning devices to alert others if a specific signal is not received from the lone worker;

▶ other devices to raise the alarm, which are activated by the absence of some specific action;

▶ checks that the lone worker has returned safely home or to their base;

▶ special arrangements for first-aid to deal with minor injuries – this may include mobile first-aid kits;

▶ arrangements for emergencies – these should be established and employees trained.

One of the largest increases in lone working has been in work-related driving. Incidents of 'road-rage' and violent assaults are becoming more common as roads become busier.

All lone workers at risk of violence should receive training to help them recognise and anticipate violence and difficult situations. Such training should develop communication skills to help the workers remain diplomatic and non-confrontational when facing potentially violent circumstances.

The UK Institution of Occupational Safety and Health (IOSH) have produced a very useful document on lone working that contains an audit checklist for remote working.

Figure 4.27 Lone workers need special consideration

4.5 Role and function of a permit-to-work system

4.5.1 Introduction

Safe systems of work are crucial in work such as the maintenance of chemical plant where the potential risks are high and the careful coordination of activities and precautions is essential to safe working. In this situation and others of similar risk potential, the safe system of work is likely to take the form of a permit-to-work procedure.

The permit-to-work procedure is a specialised type of safe system of work for ensuring that potentially very dangerous work (e.g. entry into process plant and other confined spaces) is done safely.

Although this procedure has been developed and refined by the chemical industry, the principles of the permit-to-work procedure are equally applicable to the management of complex risks in other industries.

The fundamental principle is that certain defined operations are prohibited without the specific permission of a responsible manager, this permission being only granted once stringent checks have been made to ensure that all necessary precautions have been taken and that it is safe for work to go ahead.

The people doing the work take on responsibility for following and maintaining the safeguards set out in the permit, which will define the work to be done (no other work being permitted) and the timescale in which it must be carried out.

To be effective, the permit system requires the training needs of those involved to be identified and met, and the monitoring procedures must ensure that the system is operating as intended.

4.5.2 The principles that apply to permits to work

Permit systems must adhere to the following eight principles:

1. Wherever possible, and especially with routine jobs, hazards should be eliminated so that the work can be done safely without requiring a permit to work.
2. Although the Site Manager may delegate the responsibility for the operation of the permit system, the overall responsibility for ensuring safe operation rests with him/her.
3. The permit must be recognised as the master instruction which, until it is cancelled, overrides all other instructions.
4. The permit applies to everyone on site, including contractors.
5. Information given in a permit must be detailed and accurate. It must state:
   (a) which plant/equipment has been made safe and the steps by which this has been achieved;

(b)   what work may be done;

(c)   the time at which the permit comes into effect.

6. The permit remains in force until the work has been completed and the permit is cancelled by the person who issued it or by the person nominated by management to take over the responsibility (e.g. at the end of a shift or during absence).

7. No work other than that specified is authorised. If it is found that the planned work has to be changed, the existing permit should be cancelled and a new one issued.

8. Responsibility for the plant must be clearly defined at all stages.

4.5.3 Permit-to-work procedures

The permit-to-work procedure is a specialised type of safe system of work, under which certain categories of high risk potential work may only be done with the specific permission of an authorised manager. This permission (in the form of the permit to work) will be given only if the laid-down precautions are in force and have been checked.

The permit document should specify the following key items of information:

▶ the date, time and duration of the permit;

▶ a description and assessment of the task to be performed and its location;

▶ the plant/equipment involved, and how it is identified;

▶ the authorised persons to do the work;

▶ the steps which have already been taken to make the plant safe;

▶ potential hazards which remain, or which may arise as the work proceeds;

▶ the precautions to be taken against these hazards;

▶ the action to be taken prior to the task being started, such as:

▷ the isolation of sources of energy and outlets;

▷ emergency procedures and equipment;

▷ ensuring the competency of those involved;

▷ communication arrangements; and

▷ reference to any other relevant documents;

▶ the equipment to be released to those who are to carry out the work.

In accepting the permit, the person in charge of doing the authorised work normally undertakes to take/ maintain whatever precautions are outlined in the permit, such as:

▶ isolation of the area;

▶ carrying out atmospheric monitoring;

▶ the provision and use of personal protective equipment;

▶ the provision of suitable equipment including lighting and tools;

▶ ensuring an adequate level of supervision; and

▶ arrangements for any extension to or handover of the permit.

The permit will also include spaces for:

▶ a signature certifying that the work is complete; and

▶ a signature confirming re-acceptance of the plant/ equipment.

See Appendix 4.5.

4.5.4 Work requiring a permit

The nature of permit-to-work procedures will vary in their scope depending on the job and the risks involved. However, a permit-to-work system is unlikely to be needed where, for example:

(a) the assessed risks are low and can be controlled easily;

(b) the system of work is very simple;

(c) other work being done nearby cannot affect the work concerned in say a confined space entry, or a welding operation.

However, where there are high risks and the system of work is complex and other operations may interfere, a formal permit to work should be used.

The main types of permit and the work covered by each are identified below. See Chapter 18 for an illustration of the essential elements of a permit-to-work form with supporting notes on its operation and an example of a general work permit.

General permit

The general permit should be used for work such as:

▶ alterations to or overhaul of plant or machinery where mechanical, toxic or electrical hazards may arise. This is particularly important for:

▷ large machines where visual contact between workers is difficult;

▷ where work has to be done near dangerous parts of the machine;

▷ where there are multiple isolations for energy sources and/or dangerous substances;

▷ where dangerous substances are being used in confined areas with poor ventilation;

▶ work on or near overhead crane tracks;

▶ work on pipelines with hazardous contents;

▶ repairs to railway tracks, tippers, conveyors;

▶ work with asbestos-based materials;

▶ work involving ionising radiation;

▶ roof work and other hazardous work at height. This is particularly important where there are:

▷ no permanent work platforms with fixed handrails on flat roofs;

▷ on sloping or fragile roofs;

▷ where specialist access equipment, like rope hung cradles, is required;

▷ where access is difficult;

▶ excavations to avoid underground services.

Typical work tasks that might require a permit to work

Hot work

Hot work is potentially hazardous because:

▶ It may act as a source of ignition in any plant in which flammable materials are handled;

▶ It may act as a cause of fires in all processes, regardless of whether flammable materials are present.

Hot work includes cutting, welding, brazing, soldering and any process involving the application of a naked flame. Drilling and grinding should also be included where a flammable atmosphere is potentially present.

Figure 4.28 A hot work permit is usually essential for welding, cutting and burning except in designated areas like a welding shop

Hot work should therefore be done under the terms of a hot work permit, the only exception being where hot work is done in a designated maintenance area suitable for the purpose. Typical controls include:

▶ a suitable fire extinguisher nearby;

▶ prompt removal of flammable waste material; and

▶ the damping down of nearby wooden structures such as floors.

Work on high-voltage apparatus (including testing)

Work on high-voltage apparatus (over about 600V) is potentially high risk. Hazards include:

▶ possibly fatal electric shock/burns to the people doing the work;

▶ electrical fires/explosions;

▶ consequential danger from disruption of power supply to safety-critical plant and equipment.

In view of the risk, this work must only be done by suitably trained and competent people acting under the terms of a high-voltage permit. The most important control is to ensure that the necessary isolation is provided. Other controls for this type of work are covered in Chapter 11.

Confined spaces

The safe systems of work required for confined spaces were discussed under 4.4.9. If the work and/or confined space is very hazardous, then a permit to work will be required. This will ensure that a formal check is undertaken to make sure all the elements of a safe system of work are in place before people are allowed to enter or work in the confined space. It will also be a means of communication between site management, supervisors and those carrying out the hazardous work. The essential features of such a permit to work are:

▶ the clear identification of who may authorise particular jobs (and any limits to their authority) and who, including any contractors, are responsible for specifying the necessary precautions (e.g. isolation, air testing and emergency arrangements);

▶ any particular training and instruction that may be required; and

▶ monitoring and auditing of the permit-to-work system.

Many fatal accidents have occurred where inadequate precautions were taken before and during work involving entry into confined spaces (Figure 4.29). Two hazards are the potential presence of toxic or other dangerous substances, and the absence of adequate oxygen. In addition, there may be mechanical hazards (entanglement on agitators), ingress of fluids, risk of engulfment in a free-flowing solid like grain or sugar, and raised temperatures. The work to be carried out may itself be especially hazardous when done in a confined space, for example cleaning using solvents, or cutting/welding work. Should the person working in a confined space get into difficulties for whatever reason, getting help in and getting the individual out may prove difficult and dangerous.

Stringent preparation, isolation, air testing and other precautions are therefore essential, and experience shows that the use of a confined space entry permit is essential to confirm that all the appropriate precautions have been taken.

Precaution

The hazards identified will normally indicate what precautions are needed to reduce the risks to people entering confined spaces. Where necessary, full isolation of substances and mechanical devices should be covered using a dedicated or Standard Isolation Sheet or a Safe System of Work. Refer to these on the permit.

At least one watcher, who should normally be a trained first-aider capable of performing resuscitation, should be in the immediate vicinity of the entrance at all times while someone is in a confined space. If the watcher has to leave the area for any reason, all people in the space must come out of it before the watcher leaves. The watcher's duty is to look for any signs of problems or abnormal behaviour in the space and in this event to order evacuation of the space. If someone collapses, the watcher should summon help and start rescue procedures if possible. Remember, if the atmosphere has suddenly deteriorated, you may only have four minutes to complete a rescue before the victim dies. In some cases rescuers will need to be standing by with equipment ready for use.

People working in confined spaces should be given regular breaks if the working conditions are hot or very cramped. Typically for work in cool conditions, but with restricted room, a ten-minute break every half-hour is reasonable. Adjust these times depending on the severity of the working environment. In large vessels, where it is possible to walk around, normal working is permissible.

With some confined space working, it may be necessary to erect barriers and toe boards around the opening to prevent objects falling on people in the space.

Personal protective equipment

The crucial decision is whether or not breathing apparatus is required. Base this on inspection of the equipment and atmosphere tests. If in doubt, err on the safe side.

Many entries into confined spaces will require other PPE to be worn, for example gloves, special overalls, safety shoes and helmets. Do not forget that the watcher(s) may also be exposed to hazardous substances and may also need protective equipment.

When arc welding is being carried out in a confined space, the watchers must wear eye protection against arc eye (caused by ultra-violet light).

Atmosphere tests

These should always be carried out so that the state of the atmosphere is known. This applies even when breathing apparatus is to be used. Always avoid entry into a space that is at more than 10% of the explosive limit. If it is impossible to clean and ventilate the space to below this level the atmosphere should be made inert with nitrogen or carbon dioxide. Entry will then require self-contained breathing apparatus with full rescue facilities and people immediately outside the space at all times.

Use an Atmosphere Test Record Sheet to record the results and enter the sheet number on the permit. For entry without breathing apparatus, the space should be ventilated to below the 8-hour safe exposure standard (OEL, TLV, etc.) for the substance(s). Use analytical equipment that measures the materials themselves, not indirect equipment such as explosimeters, which depend on normal oxygen levels to work and are inaccurate at safe breathing concentrations for many solvents and gases.

If sludge or deposits have to be removed by hand, consider whether breathing apparatus or continuous monitoring is required. Use of any solvent-based substance, for example paint or glue, in the space requires similar consideration.

Breathing apparatus or some other suitable type of supplied air respiratory equipment should always be used if the oxygen level is likely to be depleted below the normal atmosphere levels of 20.8%. Work should cease if levels get down to 19.5% (USA: OSHA standard) when people are not wearing suitable supplied air respiratory protection.

Some welding operations can enhance oxygen levels, greatly increasing the possibility and severity of fire. If the oxygen concentration rises above 23.5% (OSHA standard) evacuate the space immediately and ventilate to below 23.5%.

Ventilation equipment

If special ventilation is needed, specify this on the permit. Even if breathing apparatus is to be used, there are advantages in providing ventilation. Breathing apparatus is not infallible and it is much safer if the atmosphere has some capacity to support life. In addition, where flammable materials are involved, ventilation may well prevent the Lower Explosive Limit being reached.

Forced ventilation using air movement fans and/or air ventilation pipes should be used whenever possible. In small spaces, the air should be directed to the base of the equipment, i.e. near the feet of the person in the space. In large spaces, the air should be directed to the area where people are working.

Note that blowing in air from compressed air pipes, while better than nothing, is not very effective. Air movement fans based on four- or six-inch flexible ducting are much more effective. Ensure the source of the air is clean and fresh. If flammable materials are present then the fan should be electrically protected to a suitable national standard.

Rescue equipment

There are several types of person-operated hoists available and these are preferred to simple ropes or pulley systems. With hoists, two people can easily carry out a vertical rescue and one person can do so in many cases. To carry out a vertical rescue with just a simple rope through a manhole, four or five people are needed. In all cases, anyone entering the space should wear a full harness with a lifting point at neck level to give a vertical lift. Whenever possible, people entering a confined space should have a lifeline to the outside.

Powered hoists should never be used. If a limb becomes trapped during the rescue a powered hoist may tear it off. For a horizontal rescue, dragging is only possible if the floor of the space is smooth and free from obstructions. If the floor is obstructed, a stretcher will be required and the rescuers must wear breathing apparatus if the atmosphere could be hazardous.

When considering arrangements for carrying out a confined space rescue, it is important to ensure that there is no temptation for inadequately protected people to go into the space to attempt to rescue a workmate who has collapsed. This is the commonest cause of multiple fatalities associated with confined spaces entry.

When breathing apparatus is used for an entry, the watchers and rescuers should also have, and be trained in the use of, breathing apparatus. Even when the entry is to be done without breathing apparatus, it is good practice to have it available at the job. Be sure that it is possible to get into the space while breathing from the breathing apparatus.

There may be specific national regulations for entry into confined spaces which detail the controls that are necessary when people enter confined spaces.

A triple fatal accident occurred in an African factory when a man went inside a tank to clean up sludge. He was not wearing breathing apparatus and soon got into difficulties with breathing due to the movement of the sludge. The worker collapsed and the outside rescue man also went inside to save him, again without breathing apparatus.

Two other men hearing the shouts went to help. They saw two men collapsed and one went in to help. He immediately got into difficulties but managed to get out and the fourth man went inside the tank.

By this time the site fire team arrived at the tank. They put on their rescue suits but their breathing apparatus would not fit inside the suits and the instructions were all in German which they did not understand. A rescue team member eventually went inside the tank with his suit half open to allow the use of his breathing apparatus. By this time all three men inside the tank were dead.

The rescue team, although classed as the site fire brigade, had never been trained in the use of the rescue equipment. This is a classic tragedy but all too typical of a confined space accident.

Machinery maintenance

All work equipment should be maintained in an efficient state and in good repair. Any maintenance operations on the equipment should be carried out safely. The frequency and nature of maintenance should be determined through risk assessment, taking full account of:

▶ the manufacturer's recommendations;
▶ the intensity and frequency of use;
▶ operating environment (e.g. the effect of temperature, corrosion, weathering);

Figure 4.29 Entering a confined space – using a rescue tripod with hoist, harness, breathing apparatus, and person outside to keep watch and call for further assistance if required

▶ user knowledge and experience; and
▶ the risk to health and safety from any foreseeable failure or malfunction.

Maintenance work should only be undertaken by those who are competent to do the work, who have been provided with sufficient information, instruction and training. For high-risk equipment, positive means of disconnecting the equipment from the energy source may be required (e.g. isolation), along with means to prevent inadvertent reconnection (e.g. by locking off). Such procedures are best controlled using permits to work. Where possible, equipment should normally be shut down and any residual/stored energy safely released (e.g. pneumatic pressure dumped, parts with gravitational/rotational energy stopped or brought to a safe position). In some cases, it may not be possible to avoid particular significant hazards during the maintenance of work equipment so appropriate measures should be taken to protect people and minimise the risk. These may include:

▶ physical measures, such as temporary guarding, slow speed hold-to-run control devices, safe means of access and personal protective equipment;
▶ management issues, including safe systems of work, supervision and monitoring; and
▶ personnel competence (training, skill, awareness and knowledge of risk).

Work at height

A permit to work may be required for some hazardous work at height such as roof work to ensure that a fall arrest strategy is in place. This is particularly important where there are:

▶ no permanent work platforms with fixed handrails on flat roofs;

▶ on sloping or fragile roofs;
▶ where specialist access equipment, like rope hung cradles, is required; and
▶ where access is difficult.

4.5.5 Responsibilities

The effective operation of the permit system requires the involvement of many people. The following specific responsibilities can be identified:

(Note: all appointments, definitions of work requiring a permit, etc. must be in writing. All the categories of people identified below should receive training in the operation of the permit system as it affects them.)

Site manager

▶ has overall responsibility for the operation and management of the permit system;
▶ appoints a senior manager (normally the Chief Engineer) to act as a senior authorised person.

Senior authorised person

▶ is responsible to the Site Manager for the operation of the permit system;
▶ defines the work on the site which requires a permit;
▶ ensures that people responsible for this work are aware that it must only be done under the terms of a valid permit;
▶ appoints all necessary authorised persons;
▶ appoints a deputy to act in his/her absence.

Authorised persons

▶ issue permits to competent persons and retain copies;
▶ personally inspect the site to ensure that the conditions and proposed precautions are adequate and that it is safe for the work to proceed;
▶ accompany the competent person to the site to ensure that the plant/equipment is correctly identified and that the competent person understands the permit;
▶ cancel the permit on satisfactory completion of the work.

Competent persons

▶ receive permits from authorised persons;
▶ read the permit and make sure they fully understand the work to be done and the precautions to be taken;
▶ signify their acceptance of the permit by signing both copies;
▶ comply with the permit and make sure those under their supervision similarly understand and implement the required precautions;
▶ on completion of the work, return the permit to the authorised person who issued it.

Operatives

▶ read the permit and comply with its requirements, under the supervision of the competent person.

Specialists

A number of permits require the advice/skills of specialists in order to operate effectively. Such specialists may include chemists, electrical engineers, health and safety advisers and fire officers. Their role may involve:

▶ isolations within his/her discipline – for example electrical work;
▶ using suitable techniques and equipment to monitor the working environment for toxic or flammable materials, or for lack of oxygen;
▶ giving advice to managers on safe methods of working.

Specialists must not assume responsibility for the permit system. This lies with the Site Manager and the senior authorised person.

Engineers (and others responsible for work covered by permits)

▶ ensure that permits are raised as required.

Contractors

The permit system should be applied to contractors in the same way as to direct employees.

The contractor must be given adequate information and training on the permit system, the restrictions it imposes and the precautions it requires.

4.6 Emergency procedures and the arrangements for contacting the emergency services

4.6.1 Introduction

Emergency procedures involve control procedures and equipment to limit the damage to people and property caused by an incident. Local fire and rescue authorities will often be involved and are normally prepared to give advice to employers.

Under ILO-R164 Article 3(q) there is a requirement to establish emergency plans. This requirement is also reflected in many national occupational health and safety legislation. Procedures should be established and set in motion when necessary to deal with serious and imminent danger to persons at work. Necessary links should be maintained with local authorities, particularly with regard to first-aid, emergency medical care and rescue work.

Under the ILO Ambient Factors Code of Practice, emergency prevention, preparedness and response arrangements should be established and maintained.

Figure 4.30 Emergency services at work

These arrangements should identify the potential for accidents and emergency situations, and address the prevention of occupational health and safety risks associated with them. The arrangements should be made according to the size and nature of activity of the organisation. They should:

▶ ensure that the necessary information, internal communication and coordination are provided to protect all people in the event of an emergency at the worksite;

▶ provide information to, and communication with, the relevant competent authorities, and the neighbourhood and emergency response services;

▶ address first-aid and medical assistance, fire-fighting and evacuation of all people at the worksite; and

▶ provide relevant information and training to all members of the organisation, at all levels, including regular exercises in emergency prevention, preparedness and response procedures.

Emergency prevention, preparedness and response arrangements should be established in cooperation with external emergency services and other bodies where applicable.

Although fire is the most common emergency likely to be faced, there are many other possibilities which should be considered, including:

▶ gas explosion;

▶ electrical burn or electrocution;

▶ escape of toxic gases or fumes;

▶ discovery of dangerous dusts like asbestos in the atmosphere;

▶ terrorist threat;

▶ large vehicle crashing into the premises;

▶ aircraft crash if near a flight path;

▶ spread of highly infectious disease; and

▶ severe weather with high winds and flooding.

For fire emergencies, see Chapter 12.

4.6.2 Points to include in emergency procedures

Emergency procedures form an essential part of any health and safety management system. A quick and effective response to an emergency may well help to ease the situation and reduce the consequences. In any emergency, people are more likely to respond quickly and effectively if they:

▶ are well trained and competent;

▶ take part in regular and realistic practice exercises; and

▶ have clearly agreed, recorded and rehearsed plans, actions and responsibilities.

The following points should be considered when developing an emergency procedure:

▶ decide on the possible nature of any emergency that might occur and the potential consequences to the organisation;

▶ consider how the alarm will be raised. This may be particularly important when the organisation is closed at weekends or holiday times;

▶ decide on how the emergency services will be contacted – again including times when the organisation is closed (if there are 25 tonnes or more of dangerous substances on the premises – the exact figure may vary from country to country – the fire and rescue service should be notified and warning signs erected);

▶ decide on an emergency assembly point for staff;

▶ ensure there are enough emergency exits for everyone to escape quickly, and keep emergency doors and escape routes unobstructed and clearly marked;

▶ nominate competent people to take control in the event of an emergency;

▶ decide which other key people are needed, such as activators of emergency equipment, a nominated incident controller, and first-aiders;

▶ plan essential actions such as emergency plant shutdown, isolation or making processes safe. Clearly identify important items such as shut-off valves and electrical isolators;

▶ ensure that everyone is included in emergency procedures including people with disabilities and vulnerable workers; and finally

▶ decide on the arrangements for liaising with the media during and after the event.

4.6.3 Supervisory duties

A member of the site staff should be nominated to supervise and coordinate all emergency arrangements. This person should be in a senior position or at least have direct access to a senior manager. Senior members of the staff should be appointed as departmental fire/emergency procedure wardens, with deputies for every occasion of absence, however brief.

They should ensure that the following precautions are taken:

- everyone on site can be alerted to an emergency;
- everyone on site knows what signal will be given for an emergency and knows what to do;
- someone who has been trained in what to do is on site and ready to coordinate activities;
- emergency routes are kept clear, signed and adequately lit;
- there are arrangements for calling the fire and rescue services and to give them special information about high-risk work, for example in tunnels or confined spaces;
- there is adequate access to the site for the emergency services and this is always kept clear;
- suitable arrangements for treating and recovering injured people are set up;
- someone is posted to the site entrance to receive and direct the emergency services.

4.6.4 Assembly and roll-call

Assembly points should be established for use in the event of evacuation. They should be at a position, preferably undercover, that is unlikely to be affected at the time of emergency. In some cases, it may be necessary to make mutual arrangements with the client or occupiers of nearby premises.

In the case of small sites, a complete list of the names of all staff should be maintained so that a roll-call can be made if evacuation becomes necessary.

In those premises where the number of staff would make a single roll-call difficult, each area warden should maintain a list of the names of employees and contractors in their area. Roll-call lists must be updated regularly.

4.6.5 Contacting the emergency services

Emergency services should be called as soon as possible after the start of an emergency. The system for calling out the services will vary between countries. The following information and Appendix 4.6 gives emergency information for a number of countries worldwide. Many emergency services prefer to be called early as this gives them the best chance to contain the emergency and save lives. Within Europe such services are generally provided at taxpayers' cost and there is no direct charge at the point of use. This encourages people to use the services. This may not be the case in all countries, and it is then even more important that the organisation has a well-defined policy to call the emergency services in good time, so that individuals are not personally penalised for calling the services.

Many countries' public telephone networks have a single emergency telephone number, sometimes known as the universal emergency telephone number or occasionally the emergency services number, that allows a caller to contact local emergency services for assistance. The emergency telephone number may differ from country to country. It is typically a three-digit number so that it can be easily remembered and dialled quickly. Some countries have a different emergency number for each of the different emergency services; these often differ only by the last digit.

An emergency telephone number call may be answered by either a telephone operator or an emergency service dispatcher. The nature of the emergency (police, fire, medical) is then determined. If the call has been answered by a telephone operator, they then connect the call to the appropriate emergency service, which then dispatches the appropriate help. In the case of multiple services being needed on a call, the most urgent need must be determined, with other services being called in as needed.

Emergency dispatchers are trained to control the call in order to provide help in an appropriate manner. The emergency dispatcher may find it necessary to give urgent advice in life-threatening situations. Some dispatchers have special training in telling people how to perform first-aid or **cardiopulmonary resuscitation** (CPR) in an effort to return life to a person in cardiac arrest.

The emergency dispatcher will need the following information:

- The caller's telephone number.
- The exact location of the incident, e.g. the road name and any important details about approaching and accessing the site.
- The type and seriousness of the incident, i.e. major fire, robbery taking place.
- Details of any hazards, e.g. gas leak or fire.

In many parts of the world, an emergency service can identify the telephone number that a call has been placed from. This is normally done using the system that the telephone company uses to bill calls, making the number visible even for users who have unlisted numbers or who block caller ID. For an individual fixed landline telephone, the caller's number can often be associated with the caller's address and therefore their location. However, with mobile phones and business telephones, the address may be a mailing address rather than the caller's location. The latest 'enhanced' systems, such as Enhanced 911 in the USA, are able to provide the physical location of mobile telephones. This is often specifically mandated in a country's legislation.

Europe

The most common European emergency number is *112 (following Directive 2002/22/EC – Universal Service Directive)* and is also standard on *GSM mobile phones.* 112 is used in Austria, Belgium, Bulgaria, Croatia,

Cyprus, Czech Republic, Denmark, Estonia, Finland, France, Germany, Greece, Hungary, Iceland, Ireland, Italy, Latvia, Liechtenstein, Lithuania, Luxembourg, Republic of Macedonia, Malta, Netherlands, Norway, Poland, Portugal, Romania, Serbia, Slovakia, Slovenia, Spain, Sweden, Switzerland, Ukraine and the United Kingdom in addition to their other emergency numbers.

If people struggle with the language, or are just not confident of getting their message across in an emergency situation, then the **pan-European Union number 112** is worth keeping to hand.

The number allows people to speak to the emergency services in all member states of the EU, although some people claim almost a third of callers get a poor response, or none at all.

Anyone can call the single European emergency call number 112 from all telephones, mobile and fixed line, even if they have no money, calling credit card or even a SIM card.

4.6.6 Testing and training for emergencies

It is important that the emergency procedures are covered during the induction training. Separate training courses should also be given to staff on a regular basis on the following topics:

▶ use of fire-fighting equipment;
▶ regular refresher first-aid training;
▶ suitable training (and competency assessment) for all those allocated particular roles in an emergency; and
▶ regular timed fire drills.

Emergency lighting should be regularly tested and fire-fighting equipment examined. In high hazard plants, there should be an occasional full-scale rehearsal of the emergency procedure in action.

4.7 Requirements for, and effective provision of, first-aid in the workplace

4.7.1 Introduction

People at work can suffer injuries or fall ill. It does not matter whether the injury or the illness is caused by their work or not. It is important that they receive immediate attention and that an ambulance is called in serious cases. First-aid at work covers the arrangements that employers must make to achieve this. It can save lives and prevent minor injuries becoming major ones.

ILO-R164 requires the employer to provide first-aid treatment. The UK's HSE guide on the provision of first-aid at work (L74) is used as the basis for the guidance given here.

Employers should provide adequate and appropriate equipment, facilities and personnel to enable first-aid to be given to employees if they are injured or become ill at work.

What is adequate and appropriate will depend on the circumstances in a particular workplace.

The recommended minimum first-aid provision on any work site is:

▶ a suitably stocked first-aid box;
▶ an appointed person to take charge of first-aid arrangements; and
▶ information for all employees giving details of first-aid arrangements.

It is also important to remember that accidents can happen at any time. First-aid provision must be available at all times to people who are at work.

Many small firms will only need to make the minimum first-aid provision. However, there are factors which might make greater provision necessary. The following checklist covers the points that should be considered.

4.7.2 First-aid provision checklist

The risk assessments should show whether there are any specific risks in the workplace requiring particular first-aid provision. The following should be considered:

▶ Are there hazardous substances, dangerous tools and equipment, dangerous manual handling tasks, electrical shock risks, dangers from neighbours or animals?
▶ Are there different levels of risk in parts of the premises or site?
▶ What is the history of accidents and ill-health, and the type and location of incidents?
▶ What is the total number of persons likely to be on site?
▶ Are there young people, pregnant or nursing mothers on site, and employees with disabilities or special health problems?
▶ Are the working facilities widely dispersed within several buildings or compact in a multi-storey building?
▶ What is the pattern of working hours? Does it involve night work?
▶ Is the site remote from emergency medical services?
▶ Do employees travel or work alone for a considerable amount of time?
▶ Do any employees work at sites occupied by other employers?
▶ Are members of the public regularly on site?

4.7.3 Impact on first-aid provision if risks are significant

Qualified first-aiders may need to be appointed if risks are significant. This will involve a number of factors which must be considered, including:

- training for first-aiders;
- additional first-aid equipment and the contents of the first-aid box;
- siting of first-aid equipment to meet the various demands in the premises; for example provision of equipment in each building or on several floors;
- the need for first-aid provision at all times during working hours;
- informing local medical services of the site and its risks;
- any special arrangements that may be needed with the local emergency services.

Any first-aid room provided must be easily accessible to stretchers and to other equipment needed to convey patients to and from the room. They must be signposted according to International standard ISO 7010.

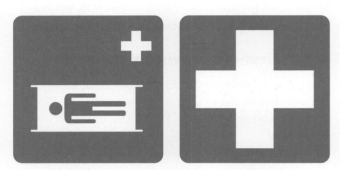

Figure 4.31 (a) First-aid and stretcher sign; (b) first-aid sign

If employees travel away from the site, the employer needs to consider:

- issuing personal first-aid kits and providing training;
- issuing mobile phones to employees;
- making arrangements with employers on other sites.

Although there may be no legal responsibilities for non-employees, it is strongly recommended that they are included in any first-aid provision.

4.7.4 Contents of the first-aid box

There is no standard list of items to put in a first-aid box. It depends on what the employer assesses the needs to be. Where there is no special risk in the workplace, a minimum recommended stock of first-aid items is listed in Table 4.1.

Tablets or medicines should not be kept in the first-aid box. Table 4.1 shows a suggested contents list only; equivalent but different items will be considered acceptable.

4.7.5 Appointed persons

An appointed person is someone who is appointed by management to:

- take charge when someone is injured or falls ill. This includes calling an ambulance if required;

Table 4.1 Typical contents of first-aid box – low risk

Stock for up to 50 persons:	
A leaflet giving general guidance on first-aid, for example HSE leaflet *Basic advice on first-aid at work.*	
Medical adhesive plaster	40
Sterile eye pads	4
Individually wrapped triangular bandages	6
Safety pins	6
Individually wrapped: medium sterile unmedicated wound dressings	8
Individually wrapped: large sterile unmedicated wound dressings	4
Individually wrapped wipes	10
Paramedic shears	1
Pairs of latex gloves	2
Sterile eyewash if no clean running water	2

- provide emergency cover where a first-aider is absent due to unforeseen circumstances;
- look after the first-aid equipment, for example keeping the first-aid box replenished;
- keep records of treatment given.

Appointed persons should never attempt to give first-aid for which they are not competent. Short emergency first-aid training courses are often available locally. An appointed person should be available at all times when people are at work on site – this may mean appointing more than one. The training should be repeated about every three years to keep up to date.

4.7.6 A first-aider

A first-aider is someone who has undergone a recognised training course with, for example, the Red Cross/Red Crescent Society in administering first-aid at work and holds a current certificate. Many countries will have local lists of training organisations. The training should be repeated every three years to maintain a valid certificate and keep the first-aider up to date.

It is not possible to give definitive rules on when or how many first-aiders or appointed persons might be needed. This will depend on the circumstances of each particular organisation or work site. Table 4.2 offers suggestions on how many first-aiders or appointed persons might be needed in relation to categories of risk and number of employees. The details in the table are suggestions only; they are not definitive.

Employees should be informed of the first-aid arrangements. Putting up notices telling staff who and where the first-aiders or appointed persons are and where the first-aid box is will usually be enough. Special arrangements will be needed for employees with reading or language difficulties.

To ensure cover at all times when people are at work and where there are special circumstances, such as remoteness from emergency medical services, shift work or sites with several separate buildings, there may need to be more first-aid personnel than set out in Table 4.2.

Table 4.2 Number of first-aid personnel

Category of risk	Numbers employed at any location	Suggested number of first-aid personnel
Lower risk		
For example shops and offices, libraries	Fewer than 50	At least one appointed person
	50–100	At least one first-aider
	More than 100	One additional first-aider for every 100 employees
Medium risk		
For example light engineering and assembly work, food processing, warehousing	Fewer than 20	At least one appointed person
	20–100	At least one first-aider for every 50 employed (or part thereof)
	More than 100	One additional first-aider for every 100 employees
Higher risk		
For example most construction, slaughterhouses, chemical manufacture, extensive work with dangerous machinery or sharp instruments	Fewer than 5	At least one appointed person
	5–50	At least one first-aider
	More than 50	One additional first-aider for every 50 employees

4.8 Further information

ILO Guidelines on Occupational Safety and Health Management Systems (ILO-OSH 2001) http://www.ilo. org/global/publications/ilo-bookstore/order-online/books/ WCMS_PUBL_9221116344_EN/lang--en/index.htm

ILOLEX (ILO database of International Law) http://www. ilo.org/ilolex/index.htm

Occupational Health and Safety Assessment Series (OHSAS 18000): Occupational Health and Safety Management Systems OHSAS 18001:2007 ISBN 978-0-5805-9404-5 OHSAS18002:2008 ISBN 978-0-5806-2686-9

Occupational Safety and Health Convention (C155) 2003, ILO http://www.ilo.org/dyn/normlex/ en/f?p=NORMLEXPUB:12100:0::NO:1210 0:P12100_ILO_CODE:C155

Occupational Safety and Health Recommendation (R164) 2006, ILO http://www.ilo.org/dyn/normlex/ en/f?p=NORMLEXPUB:12100:0::NO:1210 0:P12100_INSTRUMENT_ID:312502

ILO Safety and Health in Construction, Code of Practice, ILO Geneva, ISBN 92-2-107104-9 http:// www.ilo.org/wcmsp5/groups/public/---ed_protect/--- protrav/---safework/documents/normativeinstrument/ wcms_107826.pdf

ILO Training package on workplace risk assessment and management for small and medium size enterprises, 2013, ILO Geneva, ISBN 978-92-2-127065-2 (web pdf) http://www.ilo.org/wcmsp5/groups/public/---ed_ protect/---protrav/---safework/documents/ instructionalmaterial/wcms_215344.pdf

Five Steps to Risk Assessment (INDG163), HSE Books, ISBN 978-0-7176-6440-5

Example risk assessments, HSE http://www.hse.gov. uk/risk/casestudies/index.htm

ILO Code of Practice Ambient factors in the workplace (Chapter 3 – General principles of prevention and control) http://www.ilo.org/wcmsp5/groups/public/---ed_ protect/---protrav/---safework/documents/ normativeinstrument/wcms_107729.pdf

ISO 7010:2011/Amd5:2014, Graphical symbols – Safety colours and safety signs – Registered Safety Signs

Safe Work in Confined Spaces (ACoP, Regulations and guidance), L101, HSE Books, ISBN 978-0-7176-6233-3 www.hse.gov.uk/pubns/priced/l101.pdf

Guidance on permit-to-work systems. A guide for the petroleum, chemical and allied industries, HSE Books, ISBN 978-0-7176-2943-5 www.hse.gov.uk/pubns/priced/ hsg250.pdf

4.9 Practice revision questions

1. (a) **Explain** the criteria that must be met for the assessment to be deemed 'suitable and sufficient'.
   (b) **Outline** the **FIVE** key stages of the risk assessment process, identifying the issues that would need to be considered at **EACH** stage.
   (c) **Outline THREE** reasons for reviewing a risk assessment.

2. A plastics manufacturing company has moved to a new larger site and a programme of general risk assessments needs to be undertaken.
   (a) **Outline** the factors that should be considered when planning the risk assessment programme.
   (b) **Identify** the important characteristics required of the risk assessment team to ensure that it has the necessary competence.

(c) **Outline** the content of a training course for the risk assessment team.

3. (a) **Outline** the hazards that might be encountered in a busy hotel kitchen.

(b) **Identify SIX** hazards that might be considered when assessing the risk to the health and safety of a multi-storey car park attendant.

(c) **Outline** the hazards that might be encountered by a gardener employed by a local authority parks department.

4. (a) **Define** the meaning of the term 'young person' as used in health and safety legislation.

(b) **Identify FOUR** 'personal' factors that may place young persons at a greater risk from workplace hazards.

(c) **Outline** the factors to be taken into account when undertaking a risk assessment on young persons who are to be employed in the workplace.

5. (a) **Identify** work activities that may present a particular risk to pregnant women at work giving an example of **EACH** type of activity.

(b) **Outline** the factors that the employer should consider when undertaking a specific risk assessment in relation to a pregnant employee.

6. **Outline** the issues to be considered to ensure the health and safety of disabled workers in the workplace.

7. **Outline** the issues that should be considered to ensure the health and safety of cleaners employed in a school out of normal working hours.

8. (a) **Identify** the personal factors that may increase the risks to an individual who is required to work alone away from their workplace.

(b) **Describe** the controls that employers could implement to help minimise the risk to lone workers.

9. (a) **List SIX** of the general principles of prevention of risk control.

(b) **Outline,** with examples, the standard hierarchy of control that should be applied when controlling health and safety risks in the workplace.

10. **Outline** the possible consequences of poor standards of housekeeping on health and safety in the workplace.

11. **Identify** the shape and colour, and give a relevant example, of each of the following types of safety sign:

(a) prohibition

(b) warning

(c) mandatory

(d) safe condition.

12. **Explain** why personal protective equipment (PPE) should be considered as a last resort in the control of occupational health hazards.

13. **Outline FOUR** sources of internal and **FOUR** sources of external information that may be consulted when dealing with a health and safety issue in the workplace.

14. (a) **Explain** the meaning of the term 'safe system of work'.

(b) **State** the legal requirements for an employer to provide a safe system of work to an employee.

(c) **Outline** the factors that should be considered when developing a safe system of work.

(d) **Outline SIX** issues that should be addressed during the preparation of a safe system of work.

15. (a) **Define** the term 'permit-to-work system'.

(b) **Outline FIVE** types of work situation that may require a permit-to-work system, giving reasons in **EACH** case for the requirement.

(c) **Outline** the details that a permit to work should specify.

16. With reference to confined space:

(a) **Explain** the meaning of 'confined space', giving **FOUR** workplace examples.

(b) **Outline** specific hazards associated with working in confined spaces.

(c) **Outline** the controls that should be in place to ensure the safety of employees undertaking maintenance work in an underground storage vessel.

17. (a) **Identify FOUR** types of emergency procedure that an organisation might need to have in place.

(b) **Outline** reasons why workplace emergency procedures should be practised.

(c) **Explain** why visitors to a workplace should be informed of the emergency procedures.

18. (a) **Identify** the two main functions of first-aid treatment.

(b) **Outline** the factors to consider when making a risk assessment of first-aid provision in a workplace.

APPENDIX 4.1 Procedure for risk assessment and management (European Commission)

1. Establish a programme of risk assessment at work.
2. Structure the assessment – decide on approach (site- or process-based).
3. Collect information (jobs, environment, management methods, past experience).
4. Identify hazards.
5. Identify those at risk.
6. Identify the patterns of exposure of those at risk.
7. Evaluate significant risks (probability of harm/ severity of harm in actual circumstances).
8. Check whether existing control measures are adequate to control the risks.
9. Investigate options for eliminating or controlling the risks.
10. Prioritise action and decide on suitable control measures.
11. Implement controls.
12. Record the risk assessment.
13. Measure the effectiveness of the controls.
14. Review the risk assessment either periodically or if changes have been introduced. If the assessment is still valid no further action is required, otherwise revision is needed.
15. Monitor the programme of risk assessment.

Note: The content and extent of each step depend on the condition in the workplace (e.g. number of workers, accident and ill-health records and work equipment and process).

APPENDIX 4.2 Hazard checklist

The following checklist may be helpful.

1. Equipment/mechanical

entanglement
friction/abrasion
cutting
shearing
stabbing/puncturing
impact
crushing
drawing-in
air or high-pressure fluid injection
ejection of parts
pressure/vacuum
display screen equipment
hand tools

2. Transport

works vehicles
mechanical handling
people/vehicle interface

3. Access

slips, trips and falls
falling or moving objects
obstruction or projection
working at height
confined spaces
excavations

4. Handling/lifting

manual handling
mechanical handling

5. Electricity

fixed installation
portable tools and equipment

6. Chemicals

dust/fume/gas
toxic
irritant
sensitising
corrosive
carcinogenic
nuisance

7. Fire and Explosion

flammable materials/gases/liquids
explosion
means of escape/alarms/detection

8. Particles and dust

inhalation
ingestion
abrasion of skin or eye

9. Radiation

ionising
non-ionising

10. Biological

bacterial
viral
fungal

11. Environmental

noise
vibration

light
humidity
ventilation
temperature
overcrowding

12. The individual

individual not suited to work
long hours
high work rate
violence to staff
unsafe behaviour of individual
stress

pregnant/nursing women
young people

13. Other factors to consider

people with a disability
lone workers
poor maintenance
lack of supervision
lack of training
lack of information
inadequate instruction
unsafe systems

APPENDIX 4.3 Risk assessment example: Hairdressing salon

Name of Company: His and Hers Hairdressers Date of Assessment: 11 January 2014
Name of Assessor: A. R. Smith Date of Review: 12 July 2014

Hazards	Persons affected	Risks	Initial risk level	Existing controls	Additional controls	Action by whom?	Action by when?	Done
Hairdressing products and chemicals Various bleaching and cleansing products, in particular: Lightening (bleach) product Hydrogen peroxide Oxidative colourants	Staff and customers	Eye and/or skin irritation Possible allergic reaction	Medium	• COSHH assessment completed • Non-latex gloves are provided for staff when using products • Customers are protected with single use towels • Only non-dusty bleaches used • Staff report any allergies at induction	• Needs to be reviewed • Eye baths to be purchased for treatment of eye splashes • Repeat allergy checks every 3 months and records kept	Manager Owner Manager	12/2/14 12/2/14 14/2/14 then every 3 months	8/2/14 24/1/14 12/2/14
				• Store-room and salon well ventilated • Products stored as per manufacturer's recommendations • Staff are specifically trained in the correct use of products • Staff check whether customers have allergies to any products	• Storage in salon kept to 1 day requirement • Refresher training every 3 months and records kept • Records kept of customer allergies and simple patch tests introduced	Manager Manager Staff Manager	18/1/14 11/4/14 25/4/14	16/1/14 8/4/14 7/5/14
Sharp instruments	Staff and customers	Cuts, grazes and blood-borne infections	Medium	• All sharp instruments sterilised after use • Sterilising liquid changed daily • Sharps box available for disposable blades • First aid box available	• Regular recorded checks that sterilisation procedures correctly followed • Contents of first aid box checked weekly	Manager Manager	18/1/14 then monthly 18/1/14 then weekly	18/1/14 18/1/14
Fire	Staff and customers	Smoke inhalation and burns	Low	• Fire risk assessment completed	• No flammable products will be displayed in the windows or near heater	Manager	18/1/14	18/1/14

(Continued)

4

Hazards	Persons affected	Risks	Initial risk level	Existing controls	Additional controls	Action by whom?	Action by when?	Done
Electricity	Staff and customers	Electrical shocks and burns. Also a risk of fire	Low	• Any damaged cables, plugs or electrical equipment is reported to the manager	• An appropriate fire extinguisher (carbon dioxide) should be available	Owner	25/1/14	23/1/14
				• All portable electrical equipment and thermostats checked every 6 months by a competent person	• Make a visual check of electrical equipment, cables and sockets every month and record findings	Manager	25/1/14 then every month	25/1/14
				• All electrical equipment purchased from a reliable supplier • Staff shown how to use and store hairdryers, etc. safely and isolate electrical supply	• All internal wiring should be checked by a qualified electrician	Owner	25/4/14	3/4/14
Standing for long periods	Staff	Back pain and pain in neck, shoulders, legs and feet – musculoskeletal injuries	Medium	• Staff given regular breaks • Stools available for staff for use while trimming hair • Customer chairs are adjust-able in height	• Develop a formal rota system for breaks	Manager	25/1/14	21/1/14
Wet hand work	Staff	Skin sensitisation, dry skin, dermatitis	Medium	• Staff are trained to wash and dry hands thoroughly between hair washes • Non-latex gloves are provided for staff • Moisturising hand cream is provided for staff	• Ensure that a range of glove sizes are available for staff	Manager	25/1/14	31/1/14
					• Staff will always wear gloves for all wet work	Manager	25/1/14	1/2/14
Slips and trips	Staff and customers	Bruising, lacerations and possible fractures	Low	• Cut hair is swept up regularly • Staff must wear slip-resistant footwear • No trailing leads on floor • Any spills are cleaned immediately • Door mat provided at shop entrance	• Organise repair of worn floor covering	Manager	8/2/14	5/2/14

APPENDIX 4.4: Risk assessment example: Office cleaning

Name of Company:	Apex Cleaning Company	Date of Assessment:	14 May 2014
Name of Assessor:	T.W.James	Date of Review:	14 November 2014

Hazards	Persons affected	Risks	Initial risk level	Existing controls	Additional controls	Action by whom?	Action by when?	Done
Machine cleaning of floors	Staff and others	Injury to ankles due to incorrect use of machinery	Low	• Machine supplied for the job is suitable • Machine is maintained regularly and examined by a competent person • Cleaners trained in the safe use of the machine	• Maintenance and inspections to be documented • Training to be documented	Manager Manager	30/5/14 30/5/14	21/5/14 28/5/14
Electrical	Staff	Electric shock and burns	Low	• Staff trained in visual inspection of plugs, cables and switches before use on each shift • Staff inform manager of any defects • All portable electrical equipment is regularly tested by a competent person • Staff trained not to splash water near machines or wall sockets	• Training to be documented • Defect forms and a procedure to be developed and communicated to staff • All electrical equipment to be listed and results of portable appliance testing (PAT) tests recorded • Training to be documented	Manager Manager Manager	30/5/14 13/6/14 20/6/14	28/5/14 18/6/14 18/6/14
Lone working	Staff	Accident, illness or attack by intruder	Medium	• Staff sign in and out at either end of shift with security staff • Security staff visit staff regularly • All staff issued with mobile telephone number of manager for emergency use	• Check that staff are aware that security staff are trained first aiders • Check on regularity of these visits • If manager cannot respond, staff to be told to ring emergency services	Manager Manager Manager	22/5/14 22/5/14 22/5/14	20/5/14 20/5/14 20/5/14
Work at height	Staff	Bruising, sprains, lacerations and fractures	Low	• Staff issued with long-handled equipment for work at high level • Staff told not to stand on chairs or stepladders	• Staff to be trained in safe system of work for cleaning stairways, escalators and external windows	Manager	11/6/14	3/6/14

(Continued)

4

131

Hazards	Persons affected	Risks	Initial risk level	Existing controls	Additional controls	Action by whom?	Action by when?	Done
Contact with bleach and other cleaning chemicals	Staff	Skin irritation. Possible allergic reaction and eye injuries from splashes	Low	• COSHH assessment completed	• COSHH assessment to be reviewed	Manager	27/6/14	18/6/14
				• Staff trained in safe use and storage of cleaning chemicals	• Records of the staff training to be kept and updated	Manager	30/5/14	28/5/14
				• Staff induction questionnaire includes details of any allergies and skin problems	• Check whether products marked 'harmful' or 'irritant' can be substituted with milder alternatives	Manager	20/6/14	25/6/14
				• Impervious rubber gloves are issued for use with any chemical cleaners	• Staff to report any health problems	Staff	Ongoing	
				• Equipment used with chemicals is regularly serviced and/or cleaned				
Musculoskeletal disorders and injuries	Staff	Musculoskeletal injuries to the back, neck, shoulders, legs and feet	Medium	• Staff trained in correct lifting technique	• Check that staff are not lifting heavy objects or furniture or unduly stretching while cleaning	Manager	28/5/14	26/5/14
				• Long-handled equipment used to prevent stooping	• Check whether more up-to-date equipment is available (long-handled wringers and buckets on wheels)	Manager	28/5/14	28/5/14
				• Each floor is provided with all necessary cleaning equipment				
				• Staff trained not to overfill buckets				
Slips, trips and falls	Staff and others	Bruising, sprains, lacerations and possible fractures	Medium	• Warning signs placed on wet floors	• Introduce a wet and dry mops cleaning system for floors	Manager	28/5/14	26/5/14
				• Staff use nearest electrical socket to reduce trip hazards				
				• Staff told to wear slip-resistant footwear				
				• Wet floor work restricted to less busy times at client offices				
				• Client encouraged in a good housekeeping policy				
Fire	Staff and others	Smoke inhalation and burns	Low	• Staff trained in client's emergency fire procedures including assembly point	• On each floor, a carbon dioxide fire extinguisher is available for cleaning staff	Manager	6/6/14	2/6/14

APPENDIX 4.5 Asbestos examples of safe systems of work

Standards required

All work involving asbestos in any form will be carried out in accordance with the current Control of Asbestos Regulations and Approved Code of Practice. Any asbestos removal must be done in accordance with the Asbestos Licensing Regulations. The use of new asbestos-containing materials is prohibited.

Planning procedures

All work will be tendered for or negotiated in accordance with the approved standards.

The Contracts Manager will ascertain at an early stage whether asbestos in any form is likely to be present or used on the site. If details provided by the client are inconclusive, then an occupational hygiene specialist will be asked to take and analyse samples.

Method statements will be prepared by the Contracts Manager in conjunction with an occupational hygiene specialist, and, where necessary, a licensed asbestos removal contractor will be selected to carry out the work.

The Contracts Manager will ensure that any requirement to give notice of the work to the Health and Safety Executive is complied with.

Where any work involving asbestos materials not subject to the licensing requirements is to be carried out by employees, the working methods, precautions, safety equipment, protective clothing, special tools, etc. will be arranged by the Contracts Manager.

Supervision

Before work starts, all information on working methods and precautions agreed will be issued to site supervision by the Contracts Manager in conjunction with the Safety Adviser/Officer.

The Site Supervisor in conjunction with management will ensure that the licensed contractor contracted to carry out the removal work has set up operations in accordance with the agreed method statement and that the precautions required are fully maintained throughout the operation so that others not involved are not exposed to risk.

Where necessary, smoke testing of the enclosure and monitoring of airborne asbestos fibre concentrations outside the removal enclosure will be carried out by an occupational hygiene specialist.

The Site Supervisor will ensure that when removal operations have been completed, no unauthorised person enters the asbestos removal area until clearance samples have been taken by an occupational hygiene specialist and confirmation received that the results are satisfactory.

Where employees are required to use or handle materials containing asbestos not subject to the Licensing Regulations, the Site Supervisor will ensure that the appropriate safety equipment and protective clothing is provided and that the agreed safe working procedures are understood by employees and complied with.

All warning labels will be left in place on any asbestos materials used on site.

Examples

1. Painting undamaged asbestos insulating boards

Description

This task guidance sheet can be used where undamaged asbestos insulating boards need to be painted. This may be to protect them, or for aesthetic reasons.

It is not appropriate where the material is damaged. Use a specialist contractor licensed by the HSE.

Carry out this work only if you are properly trained.

PPE

▶ Disposable overalls fitted with a hood.
▶ Boots without laces (laced boots can be difficult to decontaminate).
▶ Disposable particulate respirator (FFP3).

Equipment

▶ 500-gauge polythene sheeting and duct tape.
▶ Warning tape and notices.
▶ Type H vacuum cleaner to BS5415 (if dust needs to be removed from the asbestos insulating board).
▶ Paint conforming to the original specification, for example fire resistant. Select one low in hazardous constituents, for example solvents.
▶ Low-pressure spray or roller/brush.
▶ Bucket of water and rags.
▶ Suitable asbestos waste container, for example a labelled polythene sack.
▶ Appropriate lighting.

Preparing the work area

This work may be carried out at height; if so, the appropriate precautions MUST be taken.

▶ Carry out the work with the minimum number of people present.
▶ Restrict access, for example close the door and/or use warning tape and notices.
▶ Use polythene sheeting, secured with duct tape, to cover surfaces within the segregated area, which could become contaminated.
▶ Ensure adequate lighting.

Painting

▶ Never prepare surfaces by sanding.
▶ Before starting, check there is no damage.
▶ Repair any minor damage.
▶ If dust needs to be removed, use a Type H vacuum cleaner or rags.
▶ Preferably use the spray to apply the paint.
▶ Spray using a sweeping motion.
▶ Do not concentrate on one area as this could cause damage.
▶ Alternatively, apply the brush/roller lightly to avoid abrasion/damage.

Cleaning

▶ Use wet rags to clean the equipment.
▶ Use wet rags to clean the segregated area.
▶ Place debris, used rags, polythene sheeting and other waste in the waste container.

Personal decontamination

▶ Use a suitable personal decontamination procedure.

Clearance procedure

▶ Visually inspect the area to make sure that it has been properly cleaned.
▶ Clearance air sampling is not normally required.

2. Removal of asbestos cement sheets, gutters, etc.

Description

This task guidance sheet can be used where asbestos cement sheets, gutters, drains and ridge caps, etc. need to be removed.

For the large-scale removal of asbestos cement, for example demolition, read Working with Asbestos Cement HSG189/2, HSE Books, 1999, ISBN 978 0 7176 1667 1.

It is not appropriate for the removal of asbestos insulating board.

Carry out this work only if you are properly trained.

PPE

▶ Use disposable overalls fitted with a hood.
▶ Waterproof clothing may be required outside.
▶ Boots without laces (laced boots can be difficult to decontaminate).
▶ Use disposable particulate respirator (FFP3).

Equipment

▶ 500- and 1,000-gauge polythene sheeting and duct tape.
▶ Warning tape and notices.
▶ Bolt cutters.
▶ Bucket of water, garden type spray and rags.
▶ Suitable asbestos waste container, for example a labelled polythene sack.
▶ Lockable skip for larger quantities of asbestos cement.

▶ Asbestos warning stickers.
▶ Appropriate lighting.

Preparing the work area

This work may be carried out at height; if so, the appropriate precautions to prevent the risk of fails MUST be taken.

▶ Carry out the work with the minimum number of people present.
▶ Restrict access, for example close the door and/or use warning tape and notices.
▶ Use 500-gauge polythene sheeting, secured with duct tape, to cover any surface within the segregated area, which could become contaminated.
▶ It is dangerous to seal over exhaust vents from heating units in use.
▶ Ensure adequate lighting.

Overlaying

▶ Instead of removing asbestos cement roofs, consider overlaying with a non-asbestos material.
▶ Attach sheets to existing purlings but avoid drilling through the asbestos cement.
▶ Note the presence of the asbestos cement so that it can be managed.

Removal

▶ Avoid breaking the asbestos cement products.
▶ If the sheets are held in place with fasteners, dampen and remove – take care not to create a risk of slips.
▶ If the sheets are bolted in place, use bolt cutters avoiding contact with the asbestos cement. Remove bolts carefully.
▶ Unbolt or use bolt cutters to release gutters, drain pipes and ridge caps, avoiding contact with the asbestos cement.
▶ Lower the asbestos cement to the ground. Do not use rubble chutes.
▶ Check for debris in fasteners or bolt holes. Clean with wet rags.
▶ Single asbestos cement products can be double wrapped in 1,000-gauge polythene sheeting (or placed in waste containers if small enough). Attach asbestos warning stickers.
▶ Where there are several asbestos cement sheets and other large items, place in a lockable skip.

Cleaning

▶ Use wet rags to clean the equipment.
▶ Use wet rags to clean the segregated area.
▶ Place debris, used rags, polythene sheeting and other waste in the waste container.

Personal decontamination

▶ Use a suitable personal decontamination system.

Clearance procedure

▶ Visually inspect the area to make sure that it has been properly cleaned.

▶ Clearance air sampling is not normally required.

3. Personal decontamination system

Description

This guidance sheet explains how you should decontaminate yourself after working with asbestos materials.

If you do not decontaminate yourself properly, you may take asbestos fibres home on your clothing. You or your family and friends could be exposed to them if they were disturbed and became airborne.

It is important that you follow the procedures given in the task guidance sheets and wear PPE such as overalls correctly; this will make cleaning easier.

Removing and decontaminating PPE

▶ Remove your respirator last.

▶ Clean your boots with wet rags.

▶ Where available, use a Type H vacuum cleaner to clean your overalls.

▶ Otherwise use a wet rag – use a 'patting' action – rubbing can disturb fibres.

▶ Where two or more workers are involved they can help each other by 'buddy' cleaning.

▶ Remove overalls by turning inside out – place in suitable asbestos waste container.

▶ Use wet rags to clean waterproof clothing.

▶ Disposable respirators can then be removed and placed in a suitable asbestos waste container.

Personal decontamination

▶ Site-washing facilities can be used but restrict access during asbestos work.

▶ Wash each time you leave the work area.

▶ Use wet rags to clean washing facilities at the end of the job.

▶ Clean facilities daily if the job lasts more than a day.

▶ Visually inspect the facilities once the job is finished.

▶ Clearance air sampling is not normally required.

Further information

These examples are taken from Asbestos Essentials Task Manual HSG210 (now revised), HSE Books, 2008, ISBN 978 0 7176 6263 0. Many more examples are contained in the publication including equipment and method guidance sheets. Obtainable from HSE Books.

APPENDIX 4.6 Emergency numbers in some countries worldwide

Africa

Country	Police	Medical	Fire	Notes
Egypt	122	123	180	Tourist Police – 126; Traffic Police – 128; Electricity Emergency – 121; Natural Gas Supply Emergency – 129.
Nigeria	199	199	199	199 for any of the 3 services.
South Africa	10111	10177	10111	112 from mobile phones (soon also from fixed line phones).

Asia

Country	Police	Medical	Fire	Notes
China	Patrol: 110 Traffic: 122	120	119	Traffic accident – 122 999 for private ambulance service in Beijing, along with government owned ambulance service 120.
India	100	102,108,104	101	Traffic police – 103 112 from any GSM handsets are redirected to the local emergency number. Central Govt of India designate 108 as the national emergency contact number for Police, Medical and Fire emergencies. Central Govt of India designate 104 as the Andhra Pradesh state health advice emergency contact number for Medical emergencies.
Indonesia	110	118/119	113	Search and rescue team – 115; Natural disaster – 129; Electricity – 123; Mobile phone and satellite phone emergency number – 112.
Japan	110	119		Emergency at sea – 118.
South Korea	112	119		National security hotline – 111; Reporting spies – 113; Reporting a child, mentally handicapped, or elderly person wandering – 182 (missing child report hotline); 114 connects to the phone service provider.
Malaysia	999			
Mongolia	102	103	101	
Oman	9999			
Pakistan	15/1122	115	16	15/1122 can be used to redirect to any service. 112 from any GSM handset will forward to the local emergency number.

Philippines	117			**112** and **911** redirect to 117. 112 and 911 can be dialled from mobile phones. 117 may also be texted from mobile phones. **136** for motorist assistance (Metro Manila only), **163** for child abuse (Bantay Bata).
Saudi Arabia	999	997	998	Traffic police – **993**; Rescue emergency – **911**, **112** or **08**.
Thailand	191	1669	199	Bangkok EMS Command Center – **1646** (Bangkok only), Tourist Police '1155' (English-speaking emergency and routine assistance).
United Arab Emirates	999 or 112	998 or 999	997	
Bangladesh	999			For the cities of Dhaka and Chittagong only (Dhaka Metropolitan Police – '999' and Chittagong Metropolitan Police – '999').

Europe

EU

The most common European emergency number is 112 (following Directive **2002/22/EC – Universal Service Directive)** and is also standard on GSM mobile phones. 112 is used in Austria, Belgium, Bulgaria, Croatia, Cyprus, Czech Republic, Denmark, Estonia, Finland, France, Germany, Greece, Hungary, Iceland, Ireland, Italy, Latvia, Liechtenstein, Lithuania, Luxembourg, Republic of Macedonia, Malta, Netherlands, Norway, Poland, Portugal, Romania, Serbia, Slovakia, Slovenia, Spain, Sweden, Switzerland, Ukraine and the United Kingdom in addition to their other emergency numbers.

Non-EU

Country	Police	Medical	Fire	Notes
Kazakhstan	112			Police – **102**; Ambulance – **103**; Fire – **101**; Gas leaks – **104**.
Latvia	112			Police – **02**; Ambulance – **03**; Fire – **01**; Gas leaks – **04**.
Moldova	902	903	901	112
Norway	112	113	110	Police (non-urgent) – **02800**.
Russia	112			Police (until 2016) – **02**; Ambulance (until 2016) – **03**; Fire (until 2016) – **01**; Gas leaks – **04**. After 2016, 112 for all emergency services.
Serbia	112			Police – **92**; Ambulance – **94**; Fire – **93**.
Switzerland	112			Police – **117**; Ambulance – **144**; Fire – **118**; Poison – **145**; Road emergency – **140**; Psychological support (free and anonymous) – **143**; Psychological support for teens and children (free and anonymous) – **147**; Helicopter air-rescue (Rega) –**1414** or by radio on 161.300 MHz; Air rescue (Air Glaciers) (in Valais only) – **1415**.
Turkey	155	112	110	Gendarmerie – **156**; Coast Guard – **158**.
Ukraine	112			Police – **102**; Ambulance – **103**; Fire – **101**; Gas leaks – **104**.

Australia and Oceania

Country	Police	Medical	Fire	Notes
Australia	000			On a mobile phone, dial **112**, **000**, remembering to tell the operator your exact location. If you have a textphone/TTY, you can use the National Relay Service on **106**. SES units in the Australian Capital Territory, Victoria, New South Wales, Queensland and South Australia can be contacted on **132 500**. In Western Australia, the number is **1300 130 039**. In Tasmania and Northern Territory, you will have to call the individual units. The number **131 444** is used for non-emergency police. For reporting crimes, Crime Stoppers can be called on 1800 333 000 from all internal states and territories. Threats to national security can be reported on 1800 123 400. 911 may also be dialled in an emergency situation from mobile phones ONLY; however, the call will be redirected to **000**.
New Zealand	111			Urgent but not emergency police/traffic number ***555** (from mobile phones only). Redirect connects many popular foreign emergency numbers. From mobile phones, the international emergency numbers **112**, **911** and **08** also work. The **0800 161616** TTY and **0800 161610** fax numbers are operated by the police for all three services.

North America

Country	Police	Medical	Fire	Notes
Canada	**911**			Non-emergency **311** in certain areas. Some rural areas still lack 911 service. Also **112** is being redirected to 911 on GSM mobile phones.
Mexico	**066**, **060**, or **080**			Some regions redirect 911 calls to the proper number.
United States of America	**911**			Non-emergency **311** in certain areas. A few rural areas still lack 911 service. Also **112** is being redirected to 911 on GSM mobile phones.

Central America and the Caribbean

Country	Police	Medical	Fire	Notes
Jamaica	**119**	**110**		
Trinidad and Tobago	**999**	**990**		

South America

Country	Police	Medical	Fire	Notes
Brazil	**190**	**192**	**193**	Federal highway police **191**; federal police **194**; civil police **197**; state highway police **198**; civil defence **199**; human rights **100**; emergency number for Mercosularea **128**; **112** will be redirected to 190 when dialled from mobile phones and **911** will also be redirected to the police number (**190**).

CHAPTER 5

Health and safety management systems – Monitoring, investigation and recording – CHECK

5.1 **Active and reactive monitoring** ▶ 140

5.2 **Investigating incidents** ▶ 150

5.3 **Recording and reporting incidents** ▶ 158

5.4 **Further information** ▶ 163

5.5 **Practice revision questions** ▶ 163

Appendix 5.1 **Workplace inspection exercises** ▶ 165

Appendix 5.2 **ILO Code of Practice: Annex H: Classification of industrial accidents according to type of accident** ▶ 167

Appendix 5.3. **ILO Code of Practice: Annex I: Classification of industrial accidents according to agency** ▶ 168

Appendix 5.4 **ILO Code of Practice: Annex B: Proposed list of occupational diseases** ▶ 169

This chapter covers the following NEBOSH learning objectives:

1. Outline the principles, purpose and role of active and reactive monitoring
2. Explain the purpose of, and procedures for, investigating incidents (accidents, cases of work-related ill-health and other occurrences)
3. Describe the legal and organisational requirements for recording and reporting incidents

5.1 Active and reactive monitoring

5.1.1 Introduction

This chapter concerns the checking or monitoring of health and safety performance, including both active measures such as inspections, and reactive measures like injury statistics. It is concerned with the recording of incidents and accidents at work, their investigation, the legal reporting requirements and simple analysis of incidents to help managers benefit from the investigation and recording process.

Measurement is a key step in any management process and forms the basis of continuous improvement. If measurement is not carried out correctly, the effectiveness of the health and safety management system is undermined and there is no reliable information to show managers how well the health and safety risks are controlled.

Managers should check by asking key questions to ensure that arrangements for health and safety risk control are in place, comply with the law as a minimum, and operate effectively.

There are two basic types of checking or monitoring:

Active monitoring, by taking the initiative before things go wrong, involves routine inspections and checks to make sure that standards and policies are being implemented and that controls are working.

Reactive monitoring, after things go wrong, involves looking at historical events to learn from mistakes and see what can be put right to prevent a recurrence.

The ILO guidelines on occupational safety and health management systems (ILO-OSH 2001) require the following under performance monitoring and measurement.

(a) *Procedures to monitor, measure and record OSH performance on a regular basis should be developed, established and periodically reviewed. Responsibility, accountability and authority for monitoring at different levels in the management structure should be allocated.*

(b) *The selection of performance indicators should be according to the size and nature of activity of the organisation and the OSH objectives.*

(c) *Both qualitative and quantitative measures appropriate to the needs of the organisation should be considered. These should:*

(i) *be based on the organisation's identified hazards and risks, the commitments in the OSH policy and the OSH objectives; and*

(ii) *support the organisation's evaluation process, including the management review.*

(d) *Performance monitoring and measurement should:*

(i) *be used as a means of determining the extent to which OSH policy and objectives are being implemented and risks are controlled;*

(ii) *include both active and reactive monitoring, and not be based only upon work-related injury, ill-health, disease and incident statistics; and*

(iii) *be recorded.*

(e) *Monitoring should provide:*

(i) *feedback on OSH performance;*

(ii) *information to determine whether the day-to-day arrangements for hazard and risk identification, prevention and control are in place and operating effectively; and*

(iii) *the basis for decisions about improvement in hazard identification and risk control, and the OSH management system.*

(f) *Active monitoring should contain the elements necessary to have a proactive system and should include:*

(i) *monitoring of the achievement of specific plans, established performance criteria and objectives;*

(ii) *the systematic inspection of work systems, premises, plant and equipment;*

(iii) *surveillance of the working environment, including work organisation;*

(iv) *surveillance of workers' health, where appropriate, through suitable medical monitoring or follow-up of workers for early detection of signs and symptoms of harm to health in order to determine the effectiveness of prevention and control measures; and*

(v) *compliance with applicable national laws and regulations, collective agreements and other commitments on OSH to which the organisation subscribes.*

(g) *Reactive monitoring should include the identification, reporting and investigation of:*

(i) work-related injuries, ill-health (including monitoring of aggregate sickness absence records), diseases and incidents;

(ii) other losses, such as damage to property;

(iii) deficient safety and health performance, and OSH management system failures; and

(iv) workers' rehabilitation and health-restoration programmes.

The UK Health and Safety Executive's (HSE's) experience is that organisations find health and safety performance measurement a difficult subject. They struggle to develop health and safety performance measures which are not based solely on injury and ill-health statistics.

5.1.2 The traditional approach to measuring health and safety performance

Senior managers often measure company performance by using, for example, percentage profit, return on investment or market share. A common feature of these measures would be that they are generally positive in nature, which demonstrates achievement, rather than negative, which demonstrates failure.

Yet, if senior managers are asked how they measure their companies' health and safety performance, it is likely that the only measure would be accident or injury statistics. Although the general business performance of an organisation is subject to a range of positive, active measures, for health and safety it too often comes down to one negative, reactive measure of failure.

Health and safety differs from many subjects measured by managers because:

▶ improvement in performance means **fewer** outcomes from the measure (injuries or ill-health) rather than more;

▶ a low injury or ill-health rate trend over years is still no guarantee that risks are being controlled and that incidents will not happen in the future. This is particularly true in organisations when major hazards are present, where, fortunately, there is a low probability of an incident occurring but it would be catastrophic if it happened.

There is no single reliable measure of health and safety performance. What is required is a 'basket' of measures, providing information on a range of health and safety issues.

There are some significant problems with the use of injury/ill-health statistics in isolation:

▶ there may be under-reporting — focusing on injury and ill-health rates as a measure, especially if a reward system is involved, can lead to non-reporting to keep up performance;

▶ it is often a matter of chance whether a particular incident causes an injury, and it may not show whether or not a hazard is under control. Luck, or a reduction in the number of people exposed, may produce a low injury/accident rate rather than good health and safety management;

▶ an injury is the particular consequence of an incident and often does not reflect the potential severity. For example an unguarded machine could result in a cut finger or an amputation;

▶ people can be absent from work for reasons which are not related to the severity of the incident;

▶ there is evidence to show that there is little relationship between 'occupational' injury statistics (e.g. slips, trips and falls) and the reasons for the lack of control of major accident hazards (e.g. loss of containment of flammable or toxic material);

▶ a small number of accidents may lead to complacency;

▶ injury statistics demonstrate outcomes not causes.

Because of the potential shortcomings related to the use of accident/injury and ill-health data as a single measure of performance, more active or 'upstream' measures are required. These require a systematic approach to deriving positive performance standards and how they link to the overall risk control process, rather than a quick-fix based on things that can easily be counted, such as the numbers of training courses or numbers of inspections carried out, which has limited value. The resultant data would provide no information on how the figure was arrived at, whether it is 'acceptable' (i.e. good/bad) or the quality and effectiveness of the activity. A more disciplined approach to health and safety performance standards measurement is required. This needs to develop as the health and safety management system develops.

5.1.3 Why measure performance?

You can't manage what you can't measure.

– Drucker

Measurement is an accepted part of the 'Plan, Do, Check, Act' management process discussed in Chapter 2. Measuring performance is as much part of a health and safety management system as financial, production or service delivery management.

Figure 5.1 shows where measuring performance standards fits within the overall health and safety management system.

The main purpose of measuring health and safety performance is to provide information on the progress and current status of the strategies, processes and activities employed to control health and safety risks. Effective measurement not only provides information on what the levels are but also why they are at this level, so that corrective action can be taken.

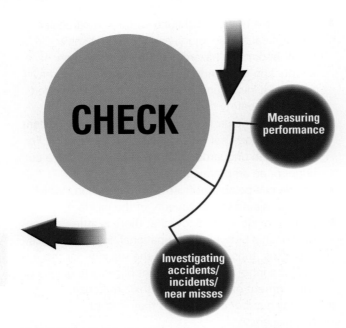

Figure 5.1 CHECK involves measuring performance and investigating incidents

Answering questions

Active checking or monitoring of health and safety performance standards should seek to answer such questions as the following:

▶ Where is the position relative to the overall health and safety aims and objectives?
▶ Where is the position relative to the control of hazards and risks?
▶ How does the organisation compare with others?
▶ What is the reason for the current position?
▶ Is the organisation getting better or worse over time?
▶ Is the management of health and safety doing the right things?
▶ Is the management of health and safety doing things right consistently?
▶ Is the management of health and safety proportionate to the hazards and risks?
▶ Is the management of health and safety efficient?
▶ Is an effective health and safety management system in place across all parts of the organisation?
▶ Is the culture supportive of health and safety, particularly in the face of competing demands?

These questions should be asked at all management levels throughout the organisation. The aim of checking or monitoring performance standards should be to provide a complete picture of how well an organisation's health and safety risks are being controlled.

Decision making

The checking or monitoring of performance standards helps in deciding:

▶ where the organisation is in relation to where it wants to be;

▶ what progress is necessary and reasonable in the circumstances;
▶ how that progress might be achieved against particular restraints (e.g. resources or time);
▶ priorities – what should be done first and what is most important;
▶ effective use of resources.

Addressing different information needs

Information from the monitoring of performance standards is needed by a variety of people. These will include directors, senior managers, line managers, supervisors, health and safety professionals and employees/safety representatives. They each need information appropriate to their position and responsibilities within the health and safety management system.

For example what the Chief Executive Officer of a large organisation needs to know from the performance standards measurement system will differ in detail and nature from the information needs of the manager of a particular location.

A coordinated approach is required so that individual monitoring activities fit within the general performance standards measurement framework.

Although the primary focus for performance standards measurement is to meet the internal needs of an organisation, there is an increasing need to demonstrate to external stakeholders (regulators, insurance companies, shareholders, suppliers, contractors, members of the public, etc.) that arrangements to control health and safety risks are in place, operating correctly and effectively.

5.1.4 What to measure

In order to achieve an outcome of no injuries or work-related ill-health, and to satisfy stakeholders, health and safety risks need to be controlled. Effective risk control is founded on an effective health and safety management system. This is illustrated in Figure 5.2.

Effective risk control

The health and safety management system comprises three levels of control (see Figure 5.2):

Level 3 – effective workplace precautions provided and maintained to prevent harm to people who are exposed to the risks;

Level 2 – risk control systems (RCSs): the basis for ensuring that adequate workplace precautions are provided and maintained;

Level 1 – the key elements of the health and safety management system: the management arrangements (including plans and objectives) necessary to organise, plan, control and monitor the design and implementation of RCSs.

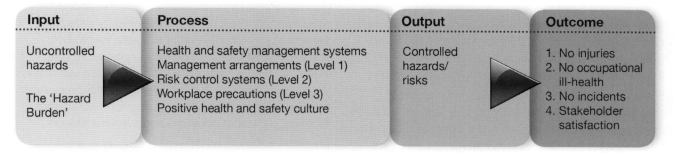

Figure 5.2 Effective risk control

The health and safety culture must be positive to support each level.

Performance measurement should cover all elements of Figure 5.2 and be based on a balanced approach which combines:

Input: monitoring the scale, nature and distribution of hazards created by the organisation's activities – measures of the hazard burden.

Process: active monitoring of the adequacy, development, implementation and deployment of the health and safety management system and the activities to promote a positive health and safety culture – measures of success.

Outcomes: reactive monitoring of adverse outcomes resulting in injuries, ill-health, loss and accidents with the potential to cause injuries, ill-health or loss – measures of failure.

5.1.5 Active monitoring – how to measure performance

Introduction and types of inspection

Active monitoring is a means of verifying the adequacy of, and the degree of compliance with, the risk control systems that have been established. It is intended to identify deficiencies in the safety management system for subsequent remedial action, and thus prevent accidents by eradicating their potential causes. It also enables a company to gain a picture of its health and safety performance and chart the progress of improvement.

Active monitoring employs several complementary methods which address differing aspects and areas of the organisation. These methods may be usefully categorised as follows:

(i) The **Safety Management System (SMS) Audit** see Chapter 6 for more information on auditing.

(ii) The **Safety Survey** is usually a detailed assessment of one aspect of an organisation's SMS, e.g. the organisation's training arrangements.

(iii) The **Safety Inspection** consists of a formal assessment of workplace safety, and the identification of hazardous conditions or practices, for subsequent remedial action. It is normally

carried out by first line managers referring to, and completing, a checklist.

(iv) The **Safety Tour** is not a detailed inspection but a way of demonstrating management commitment and interest and to see obvious examples of good or bad performance. They can be planned to cover the whole site or operation progressively or to focus attention on current priorities in the overall safety effort. The safety tour addresses the 'people' aspects of workplace safety, and by discussions with a range of staff, establishes their familiarity with safety procedures and requirements. It is normally carried out by middle and senior management, as one means of demonstrating their commitment to safety. A questionnaire is frequently used.

(v) **Safety sampling** is a useful technique that encourages organisations to concentrate on one particular area or subject at a time. A specific area is chosen which can be inspected in about 30 minutes. A checklist is drawn up to facilitate the inspection, looking at specific issues. These may be different types of hazard: they may be unsafe acts or conditions noted; they may be proactive, good behaviour or practices noted. The inspection team or person then carries out the sampling at the same time each day or week in the specified period. The results are recorded and analysed to see if the changes are good or bad over time. Of course, defects noted must be brought to the notice of the appropriate person for action on each occasion.

Management roles

Performance should be measured at each management level from directors downwards. It is not sufficient to monitor by exception, where, unless problems are raised, it is assumed to be satisfactory. Senior managers must satisfy themselves that the correct arrangements are in place and working properly. Responsibilities for both active and reactive monitoring must be laid down and managers need to be personally involved in making sure that plans and objectives are met and compliance with standards is achieved. Although systems may be set up with the guidance of safety professionals, managers should be personally involved and given sufficient training to be competent to make informed judgements about monitoring performance.

Other people, such as safety representatives, will also have the right to inspect the workplace. Each employee should be encouraged to inspect their own workplace frequently to check for obvious problems and rectify them if possible or report hazards to their supervisors.

Specific statutory (or thorough) examinations of, for example, lifting equipment or pressure vessels have to be carried out at intervals laid down in written schemes by competent persons – usually specially trained and experienced inspection/insurance company personnel.

This frequency of monitoring and inspections will depend on the level of risk and any statutory inspection requirement. Directors may be expected to examine the premises formally at an annual audit, whereas departmental supervisors may be expected to carry out inspection every week. Senior managers should regularly monitor the health and safety plan to ensure that objectives are being met and to make any changes to the plan as necessary.

Data from reactive monitoring should be considered by senior managers at least once a month. In most organisations serious events would be closely monitored as they happen.

Depending on the size and complexity of the organisation there should be arrangements for active monitoring and comparing the results of such monitoring, checking that appropriate actions are being taken. It is at this point that decisions may be required on resources (e.g. investment in a new machine which does not spill oil) and priorities, and that consultation with employees and/or their representatives is desirable (see also chapter 3 on Consultation). It is important that all places and activities are included because it is often at the margins of responsibility that hazards are overlooked.

The measurement process can gather information through:

▶ direct observation of conditions and of people's behaviour (sometimes referred to as unsafe acts and unsafe conditions monitoring);
▶ talking to people to elicit facts and their experiences as well as gauging their views and opinions;
▶ examining written reports, documents and records.

These information sources can be used independently or in combination. Direct observation includes inspection activities and the monitoring of the work environment (e.g. temperature, dust levels, solvent levels, noise levels) and people's behaviour. Each risk control system (RCS) should have a built-in monitoring element that will define the frequency of monitoring; these can be combined to form a common inspection system.

5.1.6 Safety inspection programme

Within any active monitoring programme there should be a system for workplace safety inspections. It often forms part of the preventative maintenance scheme for plant and equipment. However, they may

also be covered by legal examination and inspection requirements. Equipment in this category includes steam boilers, other pressure vessels, hoists, lifts, cranes, chains, ropes, lifting tackle, power presses, scaffolds, trench supports and local exhaust ventilation. Workplace inspections should include other precautions, such as those covering the use of premises, welfare facilities, behaviour of people and systems of work.

Inspections should be done by people who are competent to identify the relevant hazards and risks and who can assess the conditions found. Preferably they should mainly be local staff who understand the

(a)

(b)

(c)

Figure 5.3 (a) Poor conditions – inspection needed; (b) inspection in progress; (c) poor conditions in offices can cause accidents

problems and can immediately present solutions. It is essential that they do not, in any way, put themselves or anyone else at risk during the inspection. Particular care must be taken with regard to safe access. In carrying out inspections, the safety of people's actions should be considered, in addition to the safety of the conditions they are working in – a ladder might be in perfect condition but it has to be used properly too.

A properly thought-out approach to inspection will include:

- well-designed inspection forms to help plan and initiate remedial action by requiring those doing the inspection to rank any deficiencies in order of importance;
- summary lists of remedial action with names and deadlines to track progress on implementing improvements;
- periodic analysis of inspection forms to identify common features or trends which might reveal underlying weaknesses in the system;
- information to aid judgements about any changes required in the frequency or nature of the inspection program.

The inspection programme should be properly targeted whilst considering all risks in the premises. For example, low risks could be covered by monthly or bi-monthly general inspections involving a wide range of workplace safety issues such as the general condition of premises, floors, passages, stairs, heating, ventilation, lighting, welfare facilities and first-aid. Higher risks need more frequent and detailed inspections; perhaps weekly or even, in extreme cases, daily or before using a piece of equipment like an injection moulding machine or mobile plant.

The daily, weekly or monthly inspections will be aimed at checking conditions in a specified area against a fixed checklist drawn up by local management. It will cover specific items, such as the guards at particular machines, whether access/agreed routes are clear, whether fire extinguishers are in place, etc. The checks should be carried out primarily by competent staff of the department who should sign off the checklist. They should not last more than half an hour, perhaps less. This is not a specific hazard-spotting operation, but there should be a space on the checklist for the inspectors to note down any particular problems encountered.

In order to get maximum value from inspection checklists, they should be designed so that they require objective rather than subjective judgements of conditions. For example asking the people undertaking a general inspection of the workplace to rate housekeeping as good or bad begs questions as to what does good and bad mean, and what criteria should be used to judge this. If good housekeeping means there is no rubbish left on the floor, all waste bins are regularly emptied and not overflowing, floors are swept each day and cleaned once a week,

decorations should be in good condition with no peeling paint, then this should be stated. Adequate expected standards should be provided in separate notes so that those inspecting know the standards that are required.

The checklist or inspection form should facilitate:

- the planning and initiation of remedial action, by requiring those doing the inspection to rank deficiencies in priority order (those actions which are most important rather than those which can easily be done quickly);
- identifying those responsible for taking remedial actions, with sensible timescales to track progress on implementation;
- periodic monitoring to identify common themes which might reveal underlying problems in the system;
- management information on the frequency or nature of the monitoring arrangements.

The checklist could be structured using the 'four Ps' (note that the examples are not a definitive list. See chapter 18 for an example):

Premises, including:

- work at height
- access
- working environment
- welfare
- services
- fire precautions

Plant and substances, including:

- work equipment
- manual and mechanical handling
- vehicles
- dangerous/flammable substance
- hazardous substance

Procedures, including:

- risk assessments
- safe systems of work
- permits to work
- personal protective equipment
- contractors
- notices, signs and posters

People, including:

- health surveillance
- people's behaviour
- training and supervision
- appropriate authorised persons
- violence
- those especially at risk.

Key points in becoming a good observer

To improve health and safety performance, managers and supervisors must eliminate unsafe acts by observing them, taking immediate corrective action and following up to prevent recurrence. To become a good observer, they must improve their observation skills and must learn how to observe effectively. Effective observation includes the following key points:

▶ be selective
▶ know what to look for
▶ practice
▶ keep an open mind
▶ guard against habit and familiarity
▶ do not be satisfied with general impressions
▶ record observations systematically.

Observation techniques

In addition, to become a good observer, a person must:

▶ stop for 10–30 seconds before entering a new area to ascertain where employees are working;
▶ be alert for unsafe practices that are corrected as soon as you enter an area;
▶ observe activity – do not avoid the action;
▶ remember ABBI – look above, below, behind, inside;
▶ develop a questioning attitude to determine what injuries might occur if the unexpected happened and how the job might be accomplished more safely. Ask 'why?' and 'what could happen if …?';
▶ use all senses: sight, hearing, smell, touch;
▶ maintain a balanced approach. Observe all phases of the job;
▶ be inquisitive;
▶ observe for ideas – not just to determine problems;
▶ recognise good performance.

Some commonly observed poor work practices include:

▶ using machinery or tools without authority
▶ operating at unsafe speeds or in some other violation of a safe working practice
▶ removing guards or other safety devices, or rendering them ineffective
▶ using defective tools or equipment or using tools or equipment in unsafe ways
▶ using hands or body instead of tools or push sticks
▶ overloading, crowding, or failing to balance materials or handling materials in other unsafe ways, including improper lifting
▶ repairing or adjusting equipment that is in motion, under pressure, or electrically charged
▶ failing to use or maintain, or improperly using, personal protective equipment or safety devices
▶ creating unsafe, unsanitary, or unhealthy conditions by improper personal hygiene, by using compressed air for cleaning clothes, by poor housekeeping, or by smoking in unauthorised areas
▶ standing or working under suspended loads, scaffolds, shafts, or open hatches.

Additional principles for conducting inspections:

▶ Draw attention to the presence of any immediate danger – other items can await the final report.
▶ Shut down and 'lock out' any hazardous items that cannot be brought to a safe operating standard until repaired.

▶ Do not operate equipment. Ask the operator for a demonstration. If the operator of any piece of equipment does not know what dangers may be present, this is cause for concern. Items should never be ignored because the observer did not have the knowledge to make an accurate assessment of safety.
▶ Clearly describe each hazard and its exact location in rough notes. Allow 'on-the-spot' recording of all findings before they are forgotten. Record what hazards have or have not been examined in case the inspection is interrupted.
▶ Ask questions, but do not unnecessarily disrupt work activities. This may interfere with efficient assessment of the job function and may also create a potentially hazardous situation.
▶ Consider the static (stop position) and dynamic (in motion) conditions of the item you are inspecting. If a machine is shut down, consider postponing the inspection until it is functioning again.
▶ Discuss as a group, 'Can any problem, hazard or accident generate from this situation when looking at the equipment, the process or the environment?' Determine what corrections or controls are appropriate.
▶ Do not try to detect all hazards simply by relying on human senses or by looking at them during the inspection. Monitoring equipment may have to be used to measure the levels of exposure to chemicals, noise, radiation or biological agents.
▶ Take photographs to help with any description of a particular situation. Make sure it is safe to operate a camera in the workplace concerned.

Reports from inspections

Some of the items arising from safety inspections will have been dealt with immediately; other items will require action by specified people. Where there is some doubt about the problem, and what exactly is required, advice should be sought from the safety adviser or external expert. A brief report of the inspection and any resulting action list should be submitted to the safety committee. While the committee may not have the time available to consider all reports in detail, it will want to be satisfied that appropriate action is taken to resolve all matters; it will be necessary for the committee to follow up the reports until all matters are resolved.

Essential elements of a report are:

▶ identification of the organisation, workplace, inspector and date of inspection;
▶ list of observations;
▶ priority or risk level;
▶ actions to be taken;
▶ timescale for completion of the actions.

Appendix 5.1 gives examples of poor workplaces which can be used for practice exercises. Chapter 17 shows a specimen workplace inspection form which can be

Figure 5.4 Using a checklist

adapted for use at many workplaces. There is more information on report writing in general in Section 5.1.7.

Chapter 18 also provides a workplace inspection checklist which can be used to assist in drawing up a specific checklist for any particular workplace or as an aide-memoire for the workplace inspection report form.

Advantages of using a checklist

The advantages of using a checklist for inspection are:

► the opportunity given for prior preparation and planning in order to ensure that the inspection is both structured and systematic;
► the degree of consistency obtained first by those carrying out the inspection and second in the areas and issues to be covered;
► it can be easily adapted or customised for different areas;
► it results in an immediate record of findings;
► it provides an easy method of comparison and audit.

Checklists can be used to improve safety performance by:

► checking adherence to standards;
► allowing an analysis of trends to be made;
► identifying problems which if quickly rectified can boost the morale of the workforce;
► demonstrating the commitment of management; and
► the involvement of employees which will increase the ownership of health and safety within the organisation.

There are a few downsides with checklists, including:

► the lack of comments so that reasons for passing or failing are not given;
► the tick-box mentality when it is all too easy to merely tick a box than provide any original thinking;
► people may shut their eyes to items not on the list.

These can be overcome by requiring inspectors to fill in the comments section with each item and providing

space for other issues with perhaps a few prompts of additional hazards which could be present.

5.1.7 Effective report writing

There are three main aims to the writing of reports and they are all about communication. A report should aim to:

► get a message through to the reader;
► make the message and the arguments clear and easy to understand;
► make the arguments and conclusions persuasive.

Communication starts with trying to get into the mind of the reader, imagining what would most effectively catch the attention, what would be most likely to convince, and what will make this report stand out among others.

A vital part of this is presentation; so while a handwritten report is better than nothing if time is short, a well-organised, typed report is very much clearer. To the reader of the report, who may well be very busy with a great deal of written information to wade through, a clear, well-presented report will produce a positive attitude from the outset, with instant benefit to the writer.

Five factors which help to make reports effective are:

► structure;
► presentation of arguments;
► style;
► presentation of data;
► how the report itself is presented.

Structure

The structure of a report is the key to its professionalism. Good structuring will:

► help the reader to understand the information and follow the arguments contained in the report;
► increase the writer's credibility;
► ensure that the material contained in the report is organised to the best advantage.

The following list shows a frequently used method of producing a report, but always bear in mind that different organisations use different formats:

1. title page
2. summary
3. contents list
4. introduction
5. main body of the report
6. conclusions
7. recommendations
8. appendices
9. references.

In the case of an inspection report, there should be an introduction, an executive summary and main body recommendations. For all reports it is important to check with the organisation requesting them in case their in-house format differs.

1. Title page

This will contain:

- a title and often a subtitle;
- the name of the person or organisation to whom the report is addressed;
- the name of the writer(s) and their organisation;
- the date on which the report was submitted.

As report writing is about communication, it is a good idea to choose a title that is eye-catching and memorable as well as being informative, if this is appropriate to the subject.

2. Summary

Limit the summary to between 150 and 500 words. Do not include any evidence or data. This should be kept for the main report. Include the main conclusions and principal recommendations and place the summary near the front of the report.

3. Contents list

Put the contents list near the beginning of the report. Short reports do not need a list, but if there are several headings it does help the reader to grasp the overall content of the report in a short time.

4. Introduction

The introduction should contain the following:

- information about who commissioned the report and when;
- the reason for the report;
- objectives of the report;
- terms of reference;
- preparation of the report (type of data, research undertaken, subjects interviewed, etc.);
- methodology used in any analysis;
- problems and the methods used to tackle them;
- details about consultation with clients, employees, etc.

There may be other items that are specific to the report.

5. The main body of the report

This part of the report should describe, in detail, what was discovered (the facts), and the significance of these discoveries (analysis) and their importance (evaluation). Specific technical terms should be explained where required. Graphs, tables and charts are often used at this stage in the report. These should have the function of summarising information rather than giving large amounts of detail. The more detailed graphics should be made into appendices.

To make it more digestible, this part of the report should be divided into sections, using numbered headings and sub-headings. Very long and complex reports will need to be broken down into chapters.

When writing an inspection report, the recommendations should be made in this section

(according to the examiners' report for IGC1 in 2011). Priorities should be noted and recommendations made clearly for each action, with a proposed timescale. There should be a comparison made with previous inspections to indicate whether standards have improved or deteriorated. The report should acknowledge strengths, for example where a good standard of compliance has been found.

6. Conclusions

The concluding part of the report should be a reasonably detailed 'summing up'. It should give the conclusions arrived at by the writer and explain why the writer has reached these conclusions.

7. Recommendations

The use of this section depends on the requirements of the person commissioning the report. If recommendations are required, provide as few as possible, to retain a clear focus. Report writers are often asked only to provide the facts. In the case of an inspection report, the recommendations should be made in the main body of the report (see no. 5 above).

8. Appendices

This part of the report should contain sections that may be useful to a reader who requires more detail. Examples would be the detailed charts, graphs and tables, any questionnaires used in constructing the evidence mentioned in no. 5, forms, case studies and so on. The appendices are the background material of the report.

9. References

If any books, papers or journal articles have been used as source material, this should be acknowledged in a reference section. There are a number of accepted referencing methods used by academics.

Because the reader is likely to be a person with some degree of expertise in the subject, a report must be reliable, credible, relevant and thorough. It is therefore important to avoid emotional language, opinions presented as facts and arguments that have no supporting evidence. To make a report more persuasive, the writer needs to:

- present the information clearly;
- provide reliable evidence;
- present arguments logically;
- avoid falsifying, tampering with or concealing facts.

In conclusion

Expertise in an area of knowledge means that distortions, errors and omissions will quickly be spotted by the discerning reader and the presence of any of these will cast doubt on the credibility of the whole report.

Reports are usually used as part of a decision-making process. If this is the case, clear, unembellished facts are needed. Exceptions to this would be where the report is a proposal document or where a recommendation is specifically requested. Unless this is the case, it is better not to make recommendations.

A report should play a key role in organising information for the use of decision makers. It should review a complex and/or extensive body of information and make a summary of all the important issues.

It is relatively straightforward to produce a report, as long as the writer keeps to a clear format. Using the format described here, it should be possible to tell the reader as clearly as possible:

▶ what happened and why;
▶ who was involved;
▶ what it cost if appropriate;
▶ what the result was.

There may be a request for a special report and this is likely to be longer and more difficult to produce. Often it will relate to a 'critical incident' and the decision makers will be looking for information to help them:

▶ decide whether this is a problem or an opportunity;
▶ decide whether to take action;
▶ decide what action, if any, to take.

Finally, report writing should be kept simple. Nothing is gained in the use of long, complicated sentences, jargon and official-sounding language. When the report is finished, it is helpful to run through it with the express intention of simplifying the language and making sure that it says what was intended in a clear and unambiguous way.

5.1.8 Measuring failure – reactive monitoring

So far, this chapter has concentrated on measuring activities designed to prevent the occurrence of injuries and work-related ill-health (active monitoring). Failures in risk control also need to be measured (reactive monitoring), to provide opportunities to check performance, learn from failures and improve the health and safety management system.

Reactive monitoring arrangements include systems to identify and report:

▶ injuries and work-related ill-health (details of the incidence rate calculation are given in Chapter 3);
▶ other losses such as damage to property;
▶ incidents, including those with the potential to cause injury, ill-health or loss (near misses);
▶ hazards and faults;
▶ weaknesses or omissions in performance standards and systems, including complaints from employees and enforcement action by the authorities.

Each of the above provides opportunities for an organisation to check performance, learn from mistakes,

Figure 5.5 An accident at work – hit by a forklift truck

and improve the health and safety management system and risk control systems. In some cases, the organisation must send a report of the circumstances and causes of incidents such as major injuries to the appropriate enforcing authority. In some countries like the UK statutorily appointed safety representatives are entitled to investigate.

The 'corporate memory' is enhanced by these events and information obtained by investigations can be used to reinforce essential health and safety messages. Safety committees and safety representatives should be involved in discussing common features or trends. This will help the workforce to identify jobs or activities which cause the most serious or the greatest number of injuries where remedial action may be most beneficial. Investigations may also provide valuable information if an insurance claim is made or legal action taken.

Most organisations should not find it difficult to collect data on serious injuries and ill-health. However, learning about minor injuries, other losses, incidents and hazards can prove more challenging. There is value in collecting information on all actual and potential losses to learn how to prevent more serious events. Accurate reporting can be promoted by:

▶ training which clarifies the underlying objectives and reasons for identifying such events;
▶ a culture which emphasises an observant and responsible approach and the importance of having systems of control in place before harm occurs;
▶ open, honest communication in a just environment, rather than a tendency merely to allocate 'blame';
▶ cross-referencing and checking first-aid treatments, health records, maintenance or fire reports and insurance claims to identify any otherwise unreported events.

Guidance on recording, investigating and analysing these incidents is given later in this chapter.

5.2 Investigating incidents

5.2.1 Function of incident investigation

Incidents and accidents rarely result from a single cause and many turn out to be complex. Most incidents involve multiple, interrelated causal factors. They can occur whenever significant deficiencies, oversights, errors, omissions or unexpected changes occur. Any one of these can be the precursor of an accident or incident. There is a value in collecting data on all incidents and potential losses as it helps to prevent more serious events.

Incidents and accidents, whether they cause damage to property or more serious injury and/or ill-health to people, should be properly and thoroughly investigated to allow an organisation to take the appropriate action to prevent a recurrence (Figure 5.6). Good investigation is a key element to making improvements in health and safety performance.

Incident investigation is considered to be part of a reactive monitoring system because it is triggered after an event.

Incident/accident investigation is based on the logic that:

▶ all incidents/accidents have causes – eliminate the cause and eliminate future incidents;
▶ the direct and indirect causes of an incident/accident can be discovered through investigation;
▶ corrective action indicated by the causation can be taken to eliminate future incidents/accidents.

Investigation is not intended to be a mechanism for apportioning blame. There are often strong emotions associated with injury or significant losses. It is all too easy to look for someone to blame without considering the reasons why a person behaved in a particular way. Often short cuts to working procedures that may have contributed to the accident give no personal advantage to the person injured. The short cut may have been taken out of loyalty to the organisation or ignorance of a safer method.

Figure 5.6 A dangerous occurrence – fire

Valuable information and understanding can be gained from carrying out accident/incident investigations. These include:

▶ an understanding of how and why problems arose which caused the accident/incident;
▶ an understanding of the ways people are exposed to substances or situations which can cause them harm;
▶ a snapshot of what really happens, for example why people take short cuts or ignore safety rules;
▶ identifying deficiencies in the control of risks in the organisation.

5.2.2 National/international requirements

Here are some of the requirements for incident investigations laid down by the International Labour Organisation (ILO).

▶ One of the essential elements of a National OSH system contained in the Promotional Framework for ILO-OSH Convention No. 187 (see Table 15.1) is: *'A mechanism for the collection and analysis of data on occupational injuries and diseases.'*
▶ ILO-OSH Convention 155 Article 11 (c) and (d) requires national authorities to ensure that the following are carried out (see Appendix 15.2):
   '(c) the establishment and application of procedures for the notification of occupational accidents and diseases, by employers and, when appropriate, insurance institutions and others directly concerned, and the production of annual statistics on occupational accidents and diseases;
   (d) the holding of inquiries, where cases of occupational accidents, occupational diseases or any other injuries to health which arise in the course of or in connection with work appear to reflect situations which are serious';
▶ Under ILO Recommendation R164 paragraph 15, employers are required to:
   'Keep such records relevant to occupational safety and health and the working environment as are considered necessary by the competent authority or authorities; these might include records of all notifiable occupational accidents and injuries which arise in the course of or in connection with work, records of authorisation and exemptions under laws or regulations to supervision of the health of workers in the undertaking, and data concerning exposure to specified substances and agents'
▶ Paragraph 2.1.4 of the ILO Code of Practice 'Recording and notification of occupational accidents and diseases, 1996', states that:
   'The competent authority should establish and implement progressively a national system for the recording, notification and investigation of

occupational accidents, occupational diseases, commuting accidents, dangerous occurrences and incidents for all branches of economic activity and all enterprises, and for all workers, regardless of their status in employment.'

▶ Paragraph 4.3 of the 1996 Code of Practice states that arrangements should include:

'(a) the provision of information by workers, workers' representatives, physicians and other appropriate persons on occupational accidents, occupational diseases, dangerous occurrences and incidents in the enterprise, and commuting accidents;

(b) the identification of a competent person, where appropriate:

(c) measures to ensure the confidentiality of personal and medical data in the employer's possession, in accordance with national laws and regulations, conditions and practice.'

The ILO Conventions and Recommendations are very clear on:

▶ the need for reporting undesired incidents, such as accidents and injuries to health, to national authorities;

▶ the need for a system of reporting and analysis to be set up by national authorities;

▶ the need for employers to record incidents; and

▶ the need for authorities to investigate serious incidents, for example where there have been multiple fatalities, extensive damage to property or the environment or significant leaks of toxic materials;

▶ the need for safety representatives to be allowed to investigate incidents or participate in investigations.

This is reflected in many countries' national legislation. However, the requirement for employers to conduct investigations is much more of an open issue. It is left more to good practice, to ILO Code of Practice and to OSH management systems to specify the requirements. The following extract from ILO-OSH 2001 clearly covers the issue of investigation at the level of an organisation. Despite the lack of specific legal requirements, it can be safely stated that unless incidents are investigated properly the organisation will not know if they were operating in compliance with their legal obligations, as they relate to a particular incident.

ILO-OSH 2001 covers in section 3.12 the 'Investigation of work-related injuries, ill-health, diseases and incidents, and their impact on safety and health performance'. This requires that:

(a) The investigation of the origin and underlying causes of work-related injuries, ill-health, diseases and incidents should identify any failures in the OSH management system and should be documented.

(b) Such investigations should be carried out by competent persons, with the appropriate participation of workers and their representatives.

(c) The results of such investigations should be communicated to the safety and health committee, where it exists, and the committee should make appropriate recommendations.

(d) The results of investigations, in addition to any recommendations from the safety and health committee, should be communicated to appropriate persons for corrective action, included in the management review and considered for continual improvement activities.

(e) The corrective action resulting from such investigations should be implemented in order to avoid repetition of work-related injuries, ill-health, diseases and incidents.

(f) Reports produced by external investigative agencies, such as inspectorates and social insurance institutions, should be acted upon in the same manner as internal investigations, taking into account issues of confidentiality.

5.2.3 Benefits

There are many **benefits** from investigating accidents/incidents. These include:

▶ the prevention of similar events occurring again. Where the outcomes are serious injuries, the enforcing authorities are likely to take a tough stance if previous warnings have been ignored;

▶ the prevention of business losses due to disruption immediately after the event, loss of production, loss of business through a lowering of reputation or inability to deliver, and the costs of criminal and legal actions;

▶ improvement in employee morale and general attitudes to health and safety, particularly if they have been involved in the investigations;

▶ improving management skills to improve health and safety performance throughout the organisation.

The case for investigating near misses and undesired circumstances may not be so obvious but it is just as useful and much easier as there are no injured people to deal with. There are no demoralised people at work or distressed families and seldom a legal action to answer. Witnesses will be more willing to speak the truth and help with the investigation.

Managers need to:

▶ communicate the type of accident and incident that needs to be reported;

▶ provide a system for reporting and recording;

▶ check that proper reports are being made;

▶ make appropriate records of accidents and incidents;

▶ investigate incidents and accidents reported;

▶ analyse the events routinely to check for trends in performance and the prevalence of types of incident or injury;

151

▶ monitor the system to make sure that it is working satisfactorily.

5.2.4 Types of incident or adverse events

Occupational or work-related ill-health

This is concerned with those acute and chronic illnesses or physical and mental disorders that are either caused or triggered by workplace activities. Such conditions may be induced by the particular work activity of the individual or by activities of others in the workplace. The time interval between exposure and the onset of the illness may be short (e.g. acute asthma attacks) or long (e.g. chronic deafness or cancer).

Accident

Occupational accident is defined in the ILO Code of Practice as: *'An occurrence arising out of or in the course of work which results in:*

(a) *fatal occupational injury;*
(b) *non-fatal occupational injury.'*

An accident is defined by the UK HSE in HSG65 as *'any undesired circumstances which give rise to ill-health or injury; damage to property, plant, products or the environment; production losses or increased liabilities'*. Other authorities define an accident more narrowly by excluding events that do not involve injury or ill-health. This book will use the UK HSE definition.

Figure 5.7 Reconstruction of a fatal accident with the author lying where the deceased person was found under the ladder after trying to lengthen the extending ladder

Incident and near miss

The ILO define this as: *An unsafe occurrence arising out of or in the course of work where no personal injury is caused, or where personal injury requires only first-aid treatment.*

A near miss would be an incident where there is no personal injury. Knowledge of near misses is very important as research has shown that, approximately,

for every 10 'near miss' events at a particular location in the workplace, a minor accident will occur.

Figure 5.8 demonstrates the difference between an accident, near miss and undesired circumstances or unsafe condition.

Dangerous occurrence

This is a 'near miss' or 'damage incident' which could have led to serious injury or loss of life. Dangerous occurrences are defined in the ILO Code of Practice as a: *Readily identifiable event as defined under national laws and regulations, with potential to cause an injury or disease to persons at work or the public.*

As an example, within the UK dangerous occurrences are defined in the Reporting of Injuries, Diseases and Dangerous Occurrences Regulations 2013 (often known as RIDDOR) and are always reportable to the enforcement authorities. Examples include the collapse of a scaffold or a crane or the failure of any passenger-carrying equipment.

Occupational disease

Occupational disease is defined by the ILO as: *'A disease contracted as a result of an exposure to risk factors arising from work activity.'*

Annex B to the ILO Code of Practice proposes a list of occupational diseases. See Appendix 5.4 for more details of this list.

Each type of incident gives the opportunity to:

▶ check performance;
▶ identify underlying deficiencies in management systems and procedures;
▶ learn from mistakes and add to the corporate memory;
▶ reinforce key health and safety messages;
▶ identify trends and priorities for prevention;
▶ provide valuable information if there is a claim for compensation;
▶ help meet legal requirements for reporting certain incidents to the authorities.

5.2.5 Accident triangles and their limitation

In 1969, F. E. Bird collected a large quantity of accident data and produced a well-known triangle (Figure 5.9). It can be seen that damage and near miss incidents occur much more frequently than injury accidents and, therefore, may be a good indicator of risks. The study also shows that most accidents are predictable and avoidable.

There are other accident ratios/triangles which have been produced, for example Heinrich (see Chapter 3) and the HSE. They are all very similar despite different actual numbers but the concept has limitations as

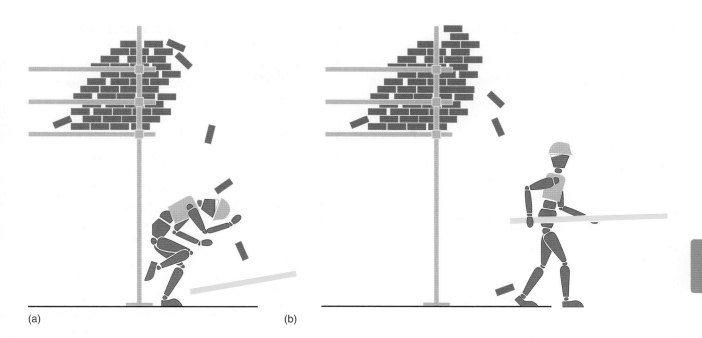

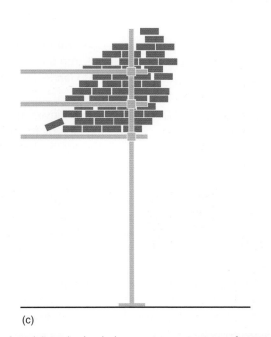

Figure 5.8 (a) Accident; (b) near miss; (c) undesired circumstances or unsafe condition

Figure 5.9 F. E. Bird's well-known accident ratio/triangle

recent research in the USA by Behavioral Science Technology (BST) has shown.

The safety triangle, or pyramid, states that reducing minor injuries or near misses (the large number at the bottom) leads to a proportionate reduction in severe injuries and deaths (the small number at the top). However, statistics over the past 10 years or so have shown that minor injuries have steadily declined while the number of serious injuries and deaths has not changed. The reasons for this are suggested as:

▶ Similar injuries may have completely different causes;
▶ Some incidents have the potential for serious injury yet may only result in a minor one. For example falling off a step ladder may only cause a bruise but could be very serious;

- ▶ Serious injuries have different underlying causes to minor injuries. For example missing controls, poor procedures, badly designed equipment create high-risk situations that may lead to a major incident;
- ▶ Many injuries are musculoskeletal sprains and strains which could not have ended up as fatal injuries. Whereas falling from height often results in serious injury or death;
- ▶ Insurance companies may concentrate more on frequent minor injuries because serious injuries, although individually expensive, are very rare;
- ▶ The solutions to major injury risks may be expensive and difficult to design or effect, for example changes to aircraft controls to prevent an accident like Virgin Galactic.

It is clear that a good safety record based solely on few lost time injuries or minor accidents does not guarantee a safe workplace. Companies must systematically examine their procedures, designs, supervision and the standards implemented in the workplace to identify the potential for serious injuries and major incidents, as well as paying attention to the causes of minor injuries.

5.2.6 Which incidents/accidents should be investigated?

The types of incident which may need to be investigated and the depth of the investigation are usually determined by their outcomes or consequences. Should every accident be investigated or only those that lead to serious injury? In fact the main determinant is the potential of the accident to cause harm rather than the actual harm resulting. For example, a slip can result in an embarrassing flailing of arms or, just as easily, a broken leg. The frequency of occurrence of the accident type is also important — for example a stream of minor cuts from paper needs investigating.

As it is not possible to determine the potential for harm simply from the resulting injury, the only really sensible solution is to investigate all accidents. The amount of time and effort spent on the investigation should, however, vary depending on the level of risk (severity of potential harm and frequency of occurrence). The most effort should be focused on significant events involving serious injury, ill-health or losses and events which have the potential for multiple or serious harm to people or substantial losses. These factors should become clear during the accident investigation and be used to guide how much time should be taken.

Figure 5.10 has been developed by the UK HSE to help to determine the level of investigation which is appropriate. The potential worst injury consequences in any particular situation should be considered when using the table. A particular incident like a scaffold collapse may not have caused an injury but had the potential to cause major or fatal injuries.

In a **minimal-level** investigation, the relevant supervisor will look into the circumstances of the accident/incident and try to learn any lessons which will prevent future incidents.

A **low-level** investigation will involve a short investigation by the relevant supervisor or line manager into the circumstances and immediate underlying and root causes of the accident/incident, to try to prevent a recurrence and to learn any general lessons.

A **medium-level** investigation will involve a more detailed investigation by the relevant supervisor or line manager, the health and safety adviser and employee

Likelihood of recurrence	Potential worst injury consequences of accident/incident			
	Minor	**Serious**	**Major**	**Fatal**
Certain				
Likely				
Possible				
Unlikely				
Rare				

Risk	Minimal	Low	Medium	High
Investigation level	**Minimal Level**	**Low Level**	**Medium Level**	**High Level**

Figure 5.10 Appropriate levels of investigation

representatives and will look for the immediate, underlying and root causes.

A **high-level** investigation will involve a team-based investigation, involving supervisors or line managers, health and safety advisers and employee representatives. It will be carried out under the supervision of senior management or directors and will look for the immediate, underlying and root causes.

5.2.7 Basic incident investigation procedures

Investigations should be led by supervisors, line managers or other people with sufficient status and knowledge to make recommendations that will be respected by the organisation. The person to lead many investigations will be the Department Manager or Supervisor of the person/area involved because they:

▶ know about the situation;
▶ know most about the employees;
▶ have a personal interest in preventing further incidents/accidents affecting 'their' people, equipment, area or materials;
▶ can take immediate action to prevent a similar incident;
▶ can communicate most effectively with the other employees concerned;
▶ can demonstrate practical concern for employees and control over the immediate work situation.

The investigation should be carried out as soon as possible after the incident to allow the maximum amount of information to be obtained. There may be difficulties which should be considered in setting up the investigation quickly – if, for example, the victim is removed from the site of the accident, or if there is a lack of a particular expert. An immediate investigation is advantageous because:

▶ factors are fresh in the minds of witnesses;
▶ witnesses have had less time to talk (there is an almost automatic tendency for people to adjust their story of the events to bring it into line with a consensus view);
▶ physical conditions have had less time to change;
▶ more people are likely to be available, for example delivery drivers, contractors and visitors, who will quickly disperse following an incident, making contact very difficult;
▶ there will probably be the opportunity to take immediate action to prevent a recurrence and to demonstrate management commitment to improvement;
▶ immediate information from the person suffering the accident often proves to be most useful.

Consideration should be given to asking the person to return to site for the accident investigation if they are physically able, rather than wait for them to return to work. A second option, although not as valuable, would

be to visit the injured person at home or even in hospital (with their permission) to discuss the accident.

Although quite a range of individuals may be involved in accident or incident investigation, for most people this will only be a rare event. Training, guidance and help will therefore be required. Training can be provided in accident/incident investigation in courses run on site and also in numerous off-site venues. Computer-based training courses are also available. These are intended to provide refresher training on an individual basis or complete training at office sites, for example, where it may not be feasible to provide practical training.

Initial action

There are a lot of things that have to be done when an incident occurs. The success of an investigation comes in the first few moments. A line manager's initial action varies for every event. The person on the scene must be the judge of what is critical. These steps are guidelines to apply as appropriate.

▶ Take control at the scene – line managers need to take charge, directing and approving everything that is done.
▶ Ensure first-aid is provided and call for emergency services.
▶ Control potential secondary events – these events such as explosions and fire are usually more serious. Positive actions need to be taken quickly after careful thought of the consequences.
▶ Identify sources of evidence at the scene.
▶ Preserve evidence from alteration or removal.
▶ Notify appropriate site management.

Investigation method

There are four basic elements to a sound investigation:

1. Collect facts about what has occurred.
2. Assemble, and analyse, the information obtained.
3. Compare the information with acceptable industry and company standards and legal requirements to draw conclusions.
4. Implement the findings and monitor progress.

Information should be gathered from all available sources, for example, witnesses, supervisors, physical conditions, hazard data sheets, written systems of work, training records. Photographs are invaluable aids to investigation but remember, with digital photography they can easily be altered. It is a good idea to print on the time and date of the picture and be prepared to verify that it is accurate. Printing out the picture as soon as possible captures it for the records. Plans and simple sketches of the incident site are also valuable.

The amount of time spent should not, however, be disproportionate to the risk. The aim of the investigation should be to explore the situation for possible underlying factors, in addition to the immediately obvious causes of the accident. For example, in a machinery accident it would not be sufficient to conclude that an accident occurred because a machine was inadequately guarded. It is necessary to look into the possible underlying system failure that may have occurred.

Investigations have three facets, which are particularly valuable and can be used to check against each other:

▶ direct observation of the scene, premises, workplace, relationship of components, materials and substances being used, possible reconstruction of events and injuries or condition of the person concerned;

▶ documents including written instructions, training records, procedures, safe operating systems, risk assessments, policies, records of inspections or test and examinations carried out;

▶ interviews (including written statements) with persons injured, witnesses, people who have carried out similar functions or examinations and tests on the equipment involved and people with specialist knowledge.

Investigation interview techniques

It must be made clear at the outset and during the course of the interview that the aim is not to apportion blame but to discover the facts and use them to prevent similar accidents or incidents in the future.

A witness should be given the opportunity to explain what happened in their own way without too much interruption and suggestion. Questions should then be asked to elicit more information. These should be of the open type, which do not suggest the answer. Questions starting with the words in Figure 5.11 are useful.

'Why' should not be used at this stage. The facts should be gathered first, with notes being taken at the end of the explanation. The investigator should then

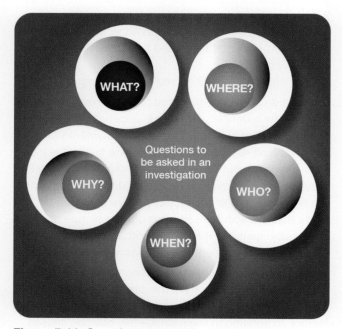

Figure 5.11 Questions to be asked in an investigation

read them or give a summary back to the witness, indicating clearly that they are prepared to alter the notes, if the witness is not content with them.

If possible, indication should be given to the witness about immediate actions that will be taken to prevent a similar occurrence and that there could be further improvements depending on the outcome of the investigation.

Seeing people injured can often be very upsetting for witnesses, which should be borne in mind. This does not mean they will not be prepared to talk about what has happened. They may in fact wish to help, but questions should be sensitive; upsetting the witness further should be avoided.

5.2.8 Incident causes and analysis

Immediate causes – unsafe acts and conditions

An investigation should look at the following factors as they can provide useful information about **immediate causes** that have been manifested in the incident/accident, and result directly from unsafe acts and conditions in the workplace.

Personal factors:

▶ behaviour of the people involved (unsafe acts)
▶ suitability of people doing the work
▶ training and competence.

Task factors:

▶ workplace conditions and precautions or controls (unsafe conditions)
▶ actual method of work adopted at the time
▶ ergonomic factors
▶ normal working practice either written or customary.

Underlying and root causes – management system failures

A thorough investigation should also look at the following factors as they can provide useful information about **underlying and root causes** that have been manifested in the incident/accident.

Underlying causes are the less obvious system or organisational reasons for an accident or incident such as:

▶ pre-start-up machinery checks were not made by supervisors
▶ the hazard had not been considered in the risk assessment
▶ there was no suitable method statement
▶ pressures of production had been more important
▶ the employee was under a lot of personal pressure at the time
▶ have there been previous similar incidents?
▶ was there adequate supervision, control and coordination of the work involved?

Root causes involve an initiating event or failing from which all other causes or failings arise. Root causes are generally management, planning or organisational failings including:

▶ quality of the health and safety policy and procedures;

▶ quality of consultation and cooperation of employees;

▶ the adequacy and quality of communications and information;

▶ deficiencies in risk assessments, plans and control systems;

▶ deficiencies in monitoring and measurement of work activities;

▶ quality and frequency of reviews and audits.

Comparison with relevant standards

There are usually suitable and relevant standards which may come from the ILO, ISO, national authorities, industry or the organisation itself. These should be carefully considered to see if:

▶ suitable standards are available to cover legal standards and the controls required by the risk assessments;

▶ the standards are sufficient and available to the organisation;

▶ the standards were implemented in practice;

▶ the standards were implemented, why there was a failure;

▶ changes should be made to the standards.

Simple root cause analysis – the five whys

This type of simple root cause analysis is ideal for a minimal or low level of investigation. More complex methods like the Tree of Causes is suitable for a higher level of investigation and is not covered here. Now is the time after obtaining all the initial information to ask 'why'.

▶ Basic Question – Keep asking 'What caused or allowed this condition/practice to occur – WHY did it occur?' until you get to root causes.

▶ The 'five whys' is one of the simplest of the root cause analysis methods. It is a question-asking method used to explore the cause/effect relationships underlying a particular problem. Ultimately, the goal of applying the five whys method is to determine a **root cause** of a defect or problem.

The following example demonstrates the basic process:

Car will not start. (the problem)

1. *Why?* – The battery is dead. (first 'why')
2. *Why?* – The alternator is not functioning. (second 'why')
3. *Why?* – The alternator belt has broken. (third 'why')
4. *Why?* – The alternator belt was well beyond its useful service life and has never been replaced. (fourth 'why')

5. *Why?* – The car was not maintained according to the recommended service schedule. (fifth 'why' and the root cause)

Benefits of asking the five whys:

▶ **Simplicity.** It is easy to use and requires no advanced mathematics or tools.

▶ **Effectiveness.** It truly helps to quickly separate symptoms from causes and identify the root case of a problem.

▶ **Comprehensiveness.** It aids in determining the relationships between various problem causes.

▶ **Flexibility.** It works well alone and when combined with other quality improvement and trouble-shooting techniques.

▶ **Engaging.** By its very nature, it fosters and produces teamwork.

▶ **Inexpensive.** It is a guided, team-focused exercise. There are no additional costs.

Often the answer to the one 'why' uncovers another reason and generates another 'why'. It often takes 'five whys' to arrive at the root cause of the problem. Investigators will probably find that they ask more or less than 'five whys' in practice.

5.2.9 Remedial actions

The investigation should have highlighted both immediate causes and underlying causes. Recommendations with priorities, both for immediate action and for longer term improvements, should come out of this. It may be necessary to ensure that the report goes further up the management chain if the improvements recommended require authorisation, which cannot be given by the investigating team.

It is essential that a follow-up is made to check on the implementation of the recommendations. It is also necessary to review the effect of the recommendations to check whether they have achieved the desired result and whether they have had unforeseen 'knock-on' effects, creating additional risks and problems.

Investigation form

Headings which could be used to compile an accident/incident investigation form are given below:

▶ date and location of accident/incident;

▶ circumstances of accident/incident;

▶ immediate cause of accident/incident;

▶ underlying cause of accident/incident;

▶ immediate action taken;

▶ recommendation for further improvement;

▶ report circulation list;

▶ date of investigation;

▶ signature of investigating team leader;

▶ names of investigating team.

157

Follow-up
- ▶ were the recommendations implemented?
- ▶ were the recommendations effective?

Key data for medium level of investigation

The UK's HSE has suggested that the key data included in Box 5.1 should be covered in an investigation report. This level of data is more appropriate to a medium level of investigation.

Box 5.1 Key data for medium level of investigation

Key data to be covered in accident, ill-health and incident reports.

The event

- ▶ Details of any injured person, including age, gender, experience, training, etc.
- ▶ A description of the circumstances, including the place, time of day and conditions.
- ▶ Details of the event, including:
  - ▷ any actions which led directly to the event;
  - ▷ the direct causes of any injuries, ill-health or other loss;
  - ▷ the immediate causes of the event;
  - ▷ the underlying causes – for example failures in workplace precautions, risk control systems or management arrangements.
- ▶ Details of the outcomes, including in particular:
  - ▷ the nature of the outcome – for example injuries or ill-health to employees or members of the public, damage to property, process disruption, emissions to the environment, creation of hazards;
  - ▷ the severity of the harm caused, including injuries, ill-health and losses.
- ▶ The immediate management response to the situation and its adequacy:
  - ▷ Was it dealt with promptly?
  - ▷ Were continuing risks dealt with promptly and adequately?
  - ▷ Was the first-aid response adequate?
  - ▷ Were emergency procedures followed?
  - ▷ Whether the event was preventable and if so how?

The potential consequences

- ▶ What was the worst that could have happened?
- ▶ What prevented the worst from happening?
- ▶ How often could such an event occur (the 'recurrence potential')?

- ▶ What was the worst injury or damage which could have resulted (the 'severity potential')?
- ▶ How many people could the event have affected (the 'population potential')?

Recommendations

- ▶ Prioritised actions with responsibilities and targets for completion.

5.3 Recording and reporting incidents

5.3.1 Internal systems for collecting and analysing incident data

Managers need effective internal systems to know whether the organisation is getting better or worse, to know what is happening and why, and to assess whether objectives are being achieved. Earlier in the chapter, Section 5.1 deals with monitoring generally; but here, the basic requirements of a collection and analysis system for incidents are discussed.

The incident report form is the basic starting point for any internal system. Each organisation needs to lay down what the system involves and who is responsible to do each part of the procedure. This will involve:

- ▶ what type of incidents should be reported;
- ▶ who completes the incident report form – normally the manager responsible for the investigation;
- ▶ how copies should be circulated in the organisation;
- ▶ who is responsible to provide management measurement data;
- ▶ how the incident data should be analysed and at what intervals;
- ▶ the arrangements to ensure that action is taken on the data provided.

The data should seek to answer the following questions:

- ▶ are failure incidents occurring, including injuries, ill-health and other loss incidents?
- ▶ where are they occurring?
- ▶ what is the nature of the failures?
- ▶ how serious are they?
- ▶ what are the potential consequences?
- ▶ what are the reasons for the failures?
- ▶ how much has it cost?
- ▶ what improvements in controls and the management system are required?
- ▶ how do these issues vary with time?
- ▶ is the organisation getting better or worse?

Type of accident/incident

Most organisations will want to collect data on:

- ▶ all injury accidents;
- ▶ cases of ill-health;

- sickness absence;
- dangerous occurrences;
- damage to property, the environment, personal effects and work in progress;
- incidents with the potential to cause serious injury, ill-health or damage (near misses or undesired circumstances).

Not all of these are required by law, but this should not deter the organisation that wishes to control risks effectively.

Analysis

All the information, whether in accident books or report forms, will need to be analysed so that useful management data can be prepared. Many organisations look at the analysis both monthly and annually. However, where there are very few accidents/incidents, quarterly may be sufficient. The health and safety information should be used alongside other business measures and there are several ways in which data can be analysed and presented. Appendices 5.2 and 5.3 show details of some of the classifications used in the ILO Code of Practice. The most common ways are:

- by the nature of the injury, such as cuts, abrasions, asphyxiation and amputations;
- by the part of the body affected, such as hands, arms, feet, lower leg, upper leg, head, eyes, back and so on. Sub-divisions of these categories could be useful if there were sufficient incidents;
- by causation (see Appendix 5.2);
- by time of day;
- by occupation or location of the job;
- by physical agency involved such as machines, means of transport and substances (see Appendix 5.3).

The simplest measure of accident rate is called the incidence rate and is defined in Chapter 3.

The UK HSE's formula for calculating an annual injury incidence rate is:

$$\frac{\text{Number of reportable injuries in financial year}}{\text{Average number employed during the year}} \times 100,000$$

This gives the rate per 100,000 employees. The formula makes no allowances for variations in part-time employment or overtime. It is an annual calculation and the figures need to be adjusted pro-rata if they cover a shorter period. Such shorter-term rates should be compared only with rates for exactly similar periods – not the national annual rates.

While some national authorities like the UK HSE and industry calculate injury **incidence rates** per 100,000 or 1,000 employees, some parts of industry prefer to calculate injury **frequency rates**, usually per million hours worked. This method, by counting hours worked rather than the number of employees, avoids distortions which may be caused in the incidence rate calculations

by part- and full-time employees and by overtime working. Frequency rates can be calculated for any time period.

The calculation is:

$$\frac{\text{Number of injuries in period}}{\text{Total hours worked during the period}} \times 1,000,000$$

There are a number of up-to-date computer recording programs which can be used to manipulate the data if significant numbers are involved. The trends can be shown against monthly, quarterly and annual past performance of, preferably, the same organisation. If indices are calculated, such as Incident Rate, comparisons can usually be made nationally using figures from national authorities, and with other similar organisations or businesses in the same industrial group. Comparisons can also be made internationally with ILO figures. This is really of major value only to larger organisations with significant numbers of events.

There may be some difficulties in comparisons between national figures or zones like Europe where the definitions of accidents or time lost may vary.

Reports should be prepared with simple tables and graphs showing trends and comparisons. Line graphs, bar charts and pie charts are all used quite extensively with good effect. All analysis reports should be made available to employees as well as managers. This can often be done through the Health and Safety Committee and safety representatives, where they exist, or directly to all employees in small organisations. Other routine meetings, team briefings and notice boards can all be used to communicate the message.

It is particularly important to make sure that any actions recommended or highlighted by the reports are taken quickly and employees kept informed.

The report form shown in Chapter 18 uses the UK-based immediate causes which can be used for analysis purposes. The categories can easily be changed to suit local needs or legislation.

5.3.2 Organisational requirements for incident records and reporting

Employers must follow national authorities' specified way(s) in which incidents have to be recorded, kept and what the records should contain. Some authorities simply specify the information which must be recorded while others may well provide a specific form or incident/accident book in which to keep records. Records will normally cover occupational accidents, occupational diseases, commuting accidents (ILO recommendation), dangerous occurrences and incidents. The records should be kept available and readily retrievable at all reasonable times.

The recording and reporting procedures should be the responsibility of a competent person in the organisation.

Workers or their representatives need to be informed of the person responsible and the arrangements for recording.

The ILO Code of Practice states that the following information should be required by national laws or regulations and should include the following:

(a) enterprise, establishment and employer:
  (i) name and address of the employer, and his or her telephone and fax numbers (if available);
  (ii) name and address of the enterprise;
  (iii) name and address of the establishment (if different);
  (iv) economic activity of the establishment; and
  (v) number of workers (size of the establishment);

(b) injured person:
  (i) name, address, gender and date of birth;
  (ii) employment status;
  (iii) occupation;
  (iv) length of service for present employer;

(c) injury:
  (i) fatal accident;
  (ii) non-fatal accident;
  (iii) nature of the injury (e.g. fracture, etc.);
  (iv) location of the injury (e.g. leg, etc.);
  (v) incapacity for work in calendar days;

(d) accident and its sequence:
  (i) geographical location of the place of the accident (usual workplace, another workplace within the establishment or outside the establishment);
  (ii) date and time;
  (iii) shift, start time of work of the injured person and hours worked in the activity in which the accident occurred;
  (iv) work environment (e.g. workshop area, office, road, etc.);
  (v) work process (e.g. welding, maintenance, manual transport, etc.);
  (vi) activity of the injured person at time of the accident (e.g. welding, maintaining press, operating machine, driving, walking, etc.);
  (vii) item or items associated with activity of the injured person (e.g. machine, tool, power press, vehicle, etc.);
  (viii) action leading to injury – type of accident (e.g. fall, etc.);
  (ix) agency related to injury (e.g. ladder, etc.) [see Appendix 5.3].

5.3.3 Reporting to external authorities

The employer or a responsible person is required to notify the authorities according to national laws and regulations.

The ILO Code of Practice paragraph 6.1.4 requires the following to be implemented:

The timing of the notification, which should preferably be made by the employer:
  (i) by the quickest possible means immediately after reporting of an occupational accident causing loss of life;
  (ii) within a prescribed time for other occupational accidents and occupational diseases.

In a number of countries, like the UK, reporting can now be done on the internet using online versions of the official forms. Confirmation is sent back via email to the notifier with a copy of the completed form. Some national notification requirements also involve notifying insurance institutions and the statistics-producing official body.

Typical incidents which need to be reported

The ILO Code of Practice requires that occupational accidents are classified as shown in Box 5.2 in two stages depending on the maturity of the national reporting system. As a national system becomes more established they recommend that statistics are improved as shown in paragraph 9.2.2. of the code (see Box 5.2).

To provide evidence for these statistics organisations will have to report as a minimum:

▶ Occupational accidents resulting in death;
▶ Occupational non-fatal accidents with at least three consecutive days of incapacity excluding the day of the accident;
▶ Commuting accidents;
▶ Occupational diseases as included in Appendix 5.4. National laws or regulations should specify that notification of an occupational disease by an employer is mandatory, at least whenever the employer receives a medical certificate to the effect that one of his or her workers is suffering from an occupational disease;
▶ Dangerous occurrences as defined by national laws (no specimen list is given by ILO).

Box 5.2 ILO Code of Practice – types of incidents which should be recorded for national statistics

9.2.1. Occupational accidents should initially be classified as follows:

(a) total number of victims, divided into:
  (i) accidents resulting in death;
  (ii) non-fatal injuries resulting in incapacity for work of at least three consecutive days, excluding the day of the accident;
(b) total days lost, including the first three days, for non-fatal injuries.

9.2.2. As more detailed information becomes progressively more readily available, the

---

competent authority should as soon as practicable classify accidents as follows:

(a) total number of victims of:

(i) accidents resulting in death, divided into deaths which occurred within 30 days of the accident, and those which occurred between 31 and 365 days of the accident;

(ii) non-fatal accidents, divided into the following categories: no lost time or absence from work (as specified under the national definition); or lost time (excluding the day of the accident) of up to three days and more than three days;

(b) total days lost for non-fatal injuries, divided into the following categories: lost time of up to three days and more than three days.

9.2.6. Statistics on commuting accidents and for self-employed persons should be shown separately.

9.3.1. Statistics of occupational diseases published by the competent authority should give the total number of cases reported for each of the diseases included in the list of occupational diseases prescribed by the competent authorities.

9.3.2. The period covered by the statistics of occupational diseases should not exceed a calendar year.

9.4.1. The competent authority should publish statistics of the numbers and types of dangerous occurrence that have been notified.

5.3.4 Lessons learnt from an incident

After an appropriate investigation there should be an action plan for the implementation of additional risk control measures. The action plan should have SMART objectives, i.e. Specific, Measurable, Achievable, and Realistic, with Timescales. See Chapter 3 for objective settings.

A good knowledge of the organisation and the way it carries out its work is essential to know where improvements are needed. Management, safety professionals, employees and their representatives should all contribute to a constructive discussion on what should be in the action plan in order to make the proposals SMART.

Not every recommendation for further risk controls will be implemented, but the ones accorded the highest priority should be implemented immediately. Organisations need to ask 'What is essential to securing the health and safety of the workforce today? What cannot be left until another day? How high is the risk to employees if this risk control measure is not implemented immediately?'

Despite financial constraints, failing to put in place measures to control serious and imminent risks is totally

unacceptable – either reduce the risks to an acceptable level, or stop the work. Each risk control measure should be assigned a priority, and a timescale with a designated person to carry out the recommendation. It is crucial that the action plan as a whole is properly monitored with a specific person, preferably a director, partner or senior manager, made responsible for its implementation.

Progress on the action plan should be regularly monitored and significant departures should be explained and risk control measures rescheduled, if necessary. There should be regular consultation with employees and their representatives to keep them fully informed of progress with implementation of the action plan.

Relevant safety instructions, safe working procedures and risk assessments should be reviewed after an incident. It is important to ask what the findings of the investigation indicate about risk assessments and procedures in general, to see if they really are suitable and sufficient.

It is also useful to estimate the cost of incidents to fully appreciate the true cost of accidents and ill-health to the organisation. To find out more about the costs of accidents and incidents visit HSE's website cost calculator at: www.hse.gov.uk/costs

The accident or incident investigation should not only be used to generate recommendations but should also be used to generate safety awareness. The investigation report or a summary should therefore be circulated locally to relevant people and, when appropriate, summaries circulated widely in the organisation. The accident or incident does not need to have resulted in a lost time injury for this procedure to be used.

5.3.5 Collection of information, compensation and insurance issues

Accidents/incidents arising out of the organisation's activities resulting in injuries to people and incidents resulting in damage to property can lead to compensation claims. The second objective of an investigation should be to collect and record relevant information for the purposes of dealing with any claim. It must be remembered that, in the longer term, prevention is the best way to reduce claims and must be the first objective in the investigation. An overzealous approach to gathering information concentrating on the compensation aspect can, in fact, prompt a claim from the injured party where there was no particular intention to take this route before the investigation. Nevertheless, relevant information should be collected. Sticking to the collection of facts is usually the best approach.

Box 5.3 shows a checklist of headings, which may assist in the collection of information. It is not expected that all accidents and incidents will be investigated in depth and a dossier with full information prepared.

161

Judgement has to be applied as to which incidents might give rise to a claim and when a full record of information is required. All accident/incident report forms should include the names of all witnesses as a minimum. Where the injury is likely to give rise to lost time, a photograph(s) of the situation should be taken.

Box 5.3 Information for insurance/compensation purposes following accident or incident

Factual information needs to be collected where there is the likelihood of some form of claim either against the organisation or by the organisation (e.g. damage to equipment). This aspect should be considered as a second objective in accident/incident investigation, the first being to learn from the accident/incident to reduce the possibility of accidents/incidents occurring in the future.

Workplace claims

- accident book entry;
- first-aider report;
- surgery record;
- foreman/supervisor accident report;
- safety representatives accident report;
- statutory report to national authorities;
- other communications between defendants and national authorities;
- minutes of Health and Safety Committee meeting(s) where accident/incident considered;
- report to national authorities concerned with social security payments;
- documents listed above relative to any previous accident/incident identified by the claimant and relied upon as proof of negligence;
- earnings information where defendant is employer;
- pre-accident/incident risk assessment;
- post-accident/incident re-assessment;
- accident/incident Investigation Report prepared in implementing the requirements;
- health surveillance records in appropriate cases; information provided to employees;
- documents relating to the employee's health and safety training;
- repair and maintenance records;
- housekeeping records;
- hazard warning signs or notices (traffic routes).

Work equipment claims

- manufacturers' specifications and instructions in respect of relevant work equipment establishing its suitability to comply with legislation;
- maintenance log/maintenance records required; documents providing information and instructions to employees; documents provided to the employee in respect of training for use;
- any notice, sign or document relied upon as a defence against alleged breaches dealing with controls and control systems.

Personal protective equipment claims

- documents relating to the assessment of the PPE;
- documents relating to the maintenance and replacement of PPE;
- record of maintenance procedures for PPE;
- records of tests and examinations of PPE;
- documents providing information, instruction and training in relation to the PPE;
- instructions for use of PPE to include the manufacturers' instructions.

Manual handling claims

- manual handling risk assessment carried out;
- re-assessment carried out post-accident;
- documents showing the information provided to the employee to give general indications related to the load and precise indications on the weight of the load and the heaviest side of the load if the centre of gravity was not positioned centrally;
- documents relating to training in respect of manual handling operations and training records.

Display screen equipment/computer workstation claims

▶ analysis of workstations to assess and reduce risks;

▶ re-assessment of analysis of workstations to assess and reduce risks following development of symptoms by the claimant;

▶ documents detailing the provision of training including training records;

▶ documents providing information to employees.

Control of Substances Hazardous to Health claims

▶ risk assessments and any reviews;

▶ copy labels from containers used for storage handling and disposal of carcinogens;

▶ warning signs identifying designation of areas and installations which may be contaminated by carcinogens;

▶ documents relating to the assessment of the PPE;

▶ documents relating to the maintenance and replacement of PPE;

▶ record of maintenance procedures for PPE;

▶ records of tests and examinations of PPE;

▶ documents providing information, instruction and training in relation to the PPE;

▶ instructions for use of PPE to include the manufacturers' instructions;

▶ air monitoring records for substances assigned a workplace exposure limit;

▶ maintenance examination and test of control measures records;

▶ monitoring records;

▶ health surveillance records;

▶ documents detailing information, instruction and training including training records for employees;

▶ labels and health and safety data sheets supplied to the employers.

5.4 Further information

Guidelines on Occupational Safety and Health Management Systems (ILO-OSH 2001), ISBN 0-580-37805-5 http://www.ilo.org/

Investigating Incidents and Accidents at Work HSG245, HSE Books, 2004, ISBN 978 0 7176 2827 8 http://www.hse.gov.uk/pubns/books/hsg245.htm

Occupational Health and Safety Assessment Series (OHSAS 18000): Occupational Health and Safety

Management Systems – Requirements OHSAS 18001:2007 ISBN 978 0 580 50802 8, OHSAS 18002:2008 ISBN 9780 580 61674 7 http://www.bsigroup.com/

Recording and Notification of Occupational Accidents and Diseases, ILO Code of Practice, Geneva, 1996, ISBN 92-2-109451-0 http://www.ilo.org/

5.5 Practice revision questions

1. (a) **Outline** the importance of monitoring as part of a health and safety management system.

   (b) **Explain** why monitoring reports should be submitted to the Chief Executive or Managing Director of the organisation.

   (c) **Explain** how accident data can be used to improve health and safety performance within an organisation.

2. **Identify FIVE** active (or proactive) and **FIVE** reactive measures that can be used to monitor an organisation's health and safety performance.

3. A large company is planning to introduce a programme of regular inspections of the workplace.

   (a) **Outline** the factors that should be considered when planning such inspections.

   (b) **Outline** the factors that determine the frequency with which health and safety inspections should be undertaken.

4. (a) **Define** the terms 'safety survey', 'safety tour' and 'safety sampling'.

   (b) **Outline** the issues which should be considered when a safety survey of workplace is to be undertaken.

5. (a) **Outline** the strengths **AND** weaknesses of using a checklist to undertake a health and safety inspection of a workplace.

(b) **Identify** the questions that might be included on a checklist to gather information following an accident involving slips, trips and falls.

6. An employee of a company is to be given duties to undertake a safety inspection.

    (a) **Outline** the competencies needed to carry out the duties.

    (b) **Identify** the principal issues that should be included in a safety inspection report so that managers can make decisions on any required remedial actions.

    (c) **Explain** how the report should be structured and presented so as to increase the likelihood of action being taken by managers.

7. (a) **Give FOUR** reasons why an organisation should have a system for the internal reporting of accidents.

    (b) **Identify** the issues that should be included in a typical workplace accident reporting procedure.

    (c) **Outline** factors that may discourage employees from reporting accidents at work.

8. An employee has been seriously injured after being struck by a fork-lift truck in a warehouse.

    (a) **Give FOUR** reasons why the accident should be investigated by the person's employer.

    (b) **Outline** the information that should be included in the investigation report.

    (c) **Outline FOUR** possible immediate causes and **FOUR** possible underlying (root) causes of the accident.

    (d) Giving reasons in **EACH** case, **identify FOUR** people who may be considered useful members of the accident investigation team.

9. **Outline** the issues to be considered to ensure an effective witness interview following a workplace accident.

10. (a) **Outline**, using a workplace example, the meaning of the terms:

    (i) near miss;
    (ii) dangerous occurrence.

    (b) **Explain** the purpose and benefits of collecting 'near miss' incident data.

    (c) **Outline** how an 'accident ratio study' can contribute to an understanding of accident prevention.

11. An employee sustained a serious injury while using an unguarded drilling machine and was admitted to hospital where he remained for several days. The machine had been unguarded for several days before the accident.

    (a) **Outline** the ILO requirement for reporting the accident to the enforcing authority.

    (b) **Identify** the possible immediate **AND** root causes of the accident.

    (c) **Outline** the immediate **AND** longer term actions that should be taken following the accident.

12. **Outline** reasons why employers should keep records of occupational ill-health amongst employees.

13. An employee is claiming compensation for injuries received during a fall down a flight of stairs at the place of work. **Identify** the documented information required when preparing a possible defence against the claim.

APPENDIX 5.1 Workplace inspection exercises

Figures 5.12–5.16 show workplaces with numerous inadequately controlled hazards. They can be used to practise workplace inspections and risk assessments.

To see the safe versions of these scenes with the corrected faults listed, visit the book's companion site at: www.routledge.com/cw/hughes

Figure 5.12 Office

Figure 5.13 Road repair

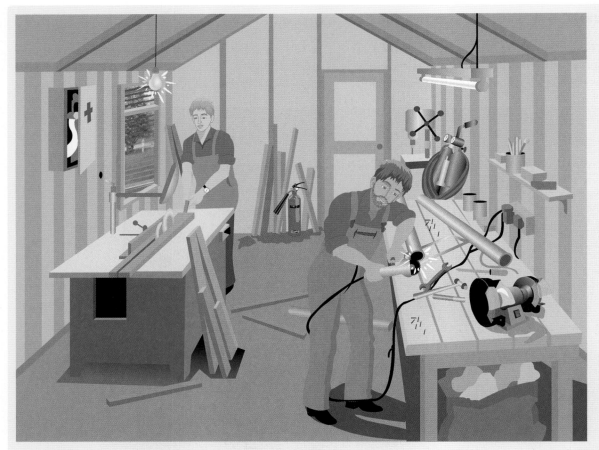

Figure 5.14 Workshop

Figure 5.15 Roof repair/unloading

Figure 5.16 Construction site

APPENDIX 5.2 ILO Code of Practice: Annex H: Classification of industrial accidents according to type of accident

This classification identifies the type of event which directly resulted in the injury, i.e. the manner in which the object or substance causing the injury enters into contact with the injured person.

1. Falls of persons

11 Falls of persons from heights (trees, buildings, scaffolds, ladders, machines, vehicles) and into depths (wells, ditches, excavations, holes in the ground)
12 Falls of persons on the same level

2. Struck by falling objects

21 Slides and cave-ins (earth, rocks, stones, snow)
22 Collapse (buildings, walls, scaffolds, ladders, piles of goods)
23 Struck by falling objects during handling
24 Struck by falling objects, not elsewhere classified

3. Stepping on, striking against or struck by objects excluding falling objects

31 Stepping on objects
32 Striking against stationary objects (except impacts due to a previous fall)
33 Striking against moving objects
34 Struck by moving objects (including flying fragments and particles) excluding falling objects

4. Caught in or between objects

41 Caught in an object
42 Caught between a stationary object and a moving object
43 Caught between moving objects (except flying or falling objects)

5. Overexertion or strenuous movements

51 Overexertion in lifting objects
52 Overexertion in pushing or pulling objects
53 Overexertion in handling or throwing objects
54 Strenuous movements

6. Exposure to or contact with extreme temperatures

61 Exposure to heat (atmosphere or environment)
62 Exposure to cold (atmosphere or environment)
63 Contact with hot substances or objects
64 Contact with very cold substances or objects

7. Exposure to or contact with electric current

8. Exposure to or contact with harmful substances or radiations

81 Contact by inhalation, ingestion or absorption of harmful substances

82 Exposure to ionising radiations
83 Exposure to radiations other than ionising radiations

9. Other types of accident, not elsewhere classified, including accidents not classified for lack of sufficient data

91 Other types of accident, not elsewhere classified
92 Accidents not classified for lack of sufficient data

APPENDIX 5.3 ILO Code of Practice: Annex I: Classification of industrial accidents according to agency

This classification may be used for classifying either the agency related to the injury or the agency related to the accident:

(a) when this classification is used to classify an agency related to the injury, the items selected for coding shall be those which directly inflicted the injury without regard to their influence in initiating the event designated as the accident type (see Annex H) [Appendix 5.2];

(b) when this classification is used to classify an agency related to the accident, the items selected for coding shall be those which because of their hazardous nature or condition precipitated the event designated as the accident type (see Annex H) [Appendix 5.2].

1. Machines

11 Prime-movers, except electrical motors
  111 Steam engines
  112 Internal combustion engines
  113 Others

12 Transmission machinery
  121 Transmission shafts
  122 Transmission belts, cables, pulleys, pinions, chains, gears
  129 Others

13 Metalworking machines
  131 Power presses
  132 Lathes
  133 Milling machines
  134 Abrasive wheels
  135 Mechanical shears
  136 Forging machines
  137 Rolling-mills
  139 Others

14 Wood and assimilated machines
  141 Circular saws
  142 Other saws
  143 Moulding machines
  144 Overhand planes
  149 Others

15 Agricultural machines
  151 Reapers (including combine reapers)
  152 Threshers
  159 Others

16 Mining machinery
  161 Under-cutters
  169 Others

19 Other machines not elsewhere classified
  191 Earth-moving machines, excavating and scraping machines, except means of transport
  192 Spinning, weaving and other textile machines
  193 Machines for the manufacture of foodstuffs and beverages
  194 Machines for the manufacture of paper
  195 Printing machines
  199 Others

2. Means of transport and lifting equipment

21 Lifting machines and appliances
  211 Cranes
  212 Lifts and elevators
  213 Winches
  214 Pulley blocks
  219 Others

22 Means of rail transport
  221 Inter-urban railways
  222 Rail transport in mines, tunnels, quarries, industrial establishments, docks, etc.
  229 Others

23 Other wheeled means of transport, excluding rail transport
  231 Tractors
  232 Lorries
  233 Trucks
  234 Motor vehicles, not elsewhere classified
  235 Animal-drawn vehicles
  236 Hand-drawn vehicles
  239 Others

24 Means of air transport

25 Means of water transport
  251 Motorised means of water transport
  252 Non-motorised means of water transport

26 Other means of transport
  261 Cable-cars
  262 Mechanical conveyors, except cable-cars
  269 Others

3. Other equipment

31 Pressure vessels
   311 Boilers
   312 Pressurised container
   313 Pressurised piping and accessories
   314 Gas cylinders
   315 Caissons, diving equipment
   319 Others

32 Furnaces, ovens, kilns
   321 Blast furnaces
   322 Refining furnaces
   323 Other furnaces
   324 Kilns
   325 Ovens

33 Refrigerating plants

34 Refrigerating installations, including electric motors, but excluding electric hand tools
   341 Rotating machines
   342 Conductors
   343 Transformers
   344 Control apparatus
   349 Others

35 Electric hand tools

36 Tools, implements and appliances, except electric hand tools
   361 Power-driven hand tools, except electric hand tools
   362 Hand tools, not power-driven
   369 Others

37 Ladders, mobile ramps

38 Scaffolding

39 Other equipment, not elsewhere classified

4. Materials, substances and radiations

41 Explosives

42 Dusts, gases, liquids and chemicals, excluding explosives
   421 Dusts
   422 Gases, vapours, fumes
   423 Liquids not elsewhere classified
   424 Chemicals not elsewhere classified
   429 Others

43 Flying fragments

44 Radiations
   441 Ionising radiations
   449 Others

49 Other materials and substances not elsewhere classified

5. Working environment

51 Outdoor
   511 Weather
   512 Traffic and working surfaces
   513 Water
   519 Others

52 Indoor
   521 Floors
   522 Confined quarters
   523 Stairs
   524 Other traffic and working surfaces
   525 Floor openings and wall openings
   526 Environmental factors (lighting, ventilation, temperature, noise, etc.)
   529 Others

53 Underground
   531 Roofs and faces of mine roads and tunnels, etc.
   532 Floors of mine roads and tunnels, etc.
   533 Working faces of mines, tunnels, etc.
   534 Mine shafts
   535 Fire
   536 Water
   539 Others

6. Other agencies, not elsewhere classified

61 Animals
   611 Live animals
   612 Animal products

69 Other agencies, not elsewhere classified

7. Agencies not classified for lack of sufficient data

APPENDIX 5.4 ILO Code of Practice: Annex B: Proposed list of occupational diseases

Summary of occupational diseases proposed by the Informal Consultation on the Revision of the List of Occupational Diseases, appended to the Employment Injury Benefits Convention, 1964 (No. 121), Geneva, 9–12 December 1991. For a full list see ILO: http://www.ilo.org/global/standards/subjects-covered-by-international-labour-standards/occupational-safety-and-health/lang--en/index.htm

1. Diseases caused by agent

1.1. **Diseases caused by chemical agents**, e.g. heavy metals, toxic gases

1.2. **Diseases caused by physical agents**, e.g. noise, vibration, compressed air and radiations

1.3. **Biological agents**, e.g. infectious or parasitic diseases

2. Diseases by target organ systems

2.1. **Occupational respiratory diseases**, e.g. pneumoconioses, silicosis, asbestosis, occupational asthma and recognised sensitising agents

2.2. **Occupational skin diseases**

2.3. **Occupational musculoskeletal disorders**, e.g. repetitive motions, vibration, awkward non-natural postures. Local or environmental cold may potentiate risk

3. Occupational cancer

3.1. **Cancer caused by agents**, e.g. Asbestos, Benzidine, chromium, Vinyl chloride and ionising radiation

4. Others

4.1. **Miners' nystagmus**

CHAPTER 6

Health and safety management systems – Audit and review – ACT

6.1 Health and safety auditing ▶ 172

6.2 Review of health and safety performance ▶ 176

6.3 Further information ▶ 179

6.4 Practice revision questions ▶ 179

This chapter covers the following NEBOSH learning objectives:

1. Explain the purpose of, and procedures for, health and safety auditing
2. Explain the purpose of, and procedures for, regular reviews of health and safety performance

6.1 Health and safety auditing

6.1.1 Audits – definition, scope and purpose

The final ACT steps in the health and safety management control cycle are auditing, performance review and taking action on lessons learned. Organisations need to be able to reinforce, maintain and develop the ability to reduce risks and control hazards in the workplace. This is not a once and for all procedure and should form part of a continual improvement programme.

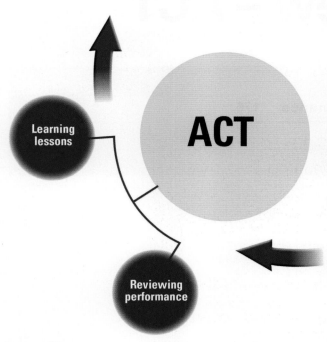

Figure 6.1 ACT part of the health and safety management system

Audit is a business discipline which is frequently used, for example, in finance, environmental matters and quality. It can equally well be applied to health and safety.

The term is often used to mean inspection or other monitoring activity. Here, the following definition is used, which follows the UK HSG65:

'The structured process of collecting independent information on the efficiency, effectiveness and reliability of the total health and safety management system and drawing up plans for corrective action.'

The **Safety Management System (SMS) Audit** is a major exercise, typically carried out every 2–4 years, as a means of assessing the adequacy of the whole organisation's SMS. It addresses all aspects of safety management in a structured manner, using written questions. Answers will be confirmed by a review of records, staff interviews, and observation of workplaces and operations. The SMS Audit may benefit from being undertaken by a person or persons independent of the organisation and should, where practicable, be carried out in real time.

Over time, it is inevitable that control systems will decay and may even become obsolete as things change. Auditing is a way of supporting monitoring by providing managers with information. It will show how effectively plans and the components of health and safety management systems are being implemented. In addition, it will provide a check on the adequacy and effectiveness of the management arrangements and risk control systems (RCS).

Auditing is critical to a health and safety management system, but it is not a substitute for other essential parts of the system like workplace inspections (see Chapter 5). Companies need systems in place to manage cash flow and pay the bills – this cannot be managed through an annual or less frequent audit. In the same way, health and safety needs to be managed on a day-to-day basis and, for this, organisations need to have systems in place. A periodic audit will not achieve this.

The main difference between audits and other forms of monitoring (safety surveys, inspections, tours and sampling,) is primarily the breadth and depth of an **audit**. **Surveys** look at only one aspect of the safety management system, inspections are frequent regular local monitoring normally carried out by line managers, **tours** concentrate on management commitment, **sampling** looks at only one area or subject over a short limited time.

The aims of auditing should be to establish that the three major components of a safety management system are in place and operating effectively. It should show that:

▶ appropriate management arrangements are in place;
▶ adequate risk control systems (RCS) exist, are implemented and are consistent with the hazard profile of the organisation;
▶ appropriate workplace precautions are in place.

Where the organisation is spread over a number of sites, the management arrangements linking the centre with the business units and sites should be covered by the audit.

There are a number of ways in which this can be achieved and some parts of the system do not need auditing as often as others. For example an audit to verify the implementation of RCSs would be made more frequently than a more overall audit of the capability of the organisation or of the management arrangements for health and safety. Critical RCSs, which control the principal hazards of the business, would need to be audited more frequently. Where there are complex workplace precautions, it may be necessary to undertake technical audits. An example would be chemical process plant integrity and control systems.

A well-structured auditing programme will give a comprehensive picture of the effectiveness of the health and safety management system in controlling risks. Such a programme will indicate when and how each component part will be audited. Managers, safety representatives and employees, working as a team, will effectively widen involvement and cooperation needed to put together the programme and implement it.

The process of auditing involves:

▶ gathering information from all levels of an organisation about the health and safety management system;
▶ making informed judgements about its adequacy and performance.

6.1.2 ILO-OSH 2001 requirements for audits

The ILO-OSH 2001 guidelines contain the following requirements for audits:

(a) Arrangements to conduct periodic audits are to be established in order to determine whether the OSH management system and its elements are in place, adequate, and effective in protecting the safety and health of workers and preventing incidents.

(b) An audit policy and programme should be developed, which includes a designation of auditor competency, the audit scope, the frequency of audits, audit methodology and reporting.

(c) The audit includes an evaluation of the organisation's OSH management system elements or a subset of these, as appropriate. The audit should cover:
(i) OSH policy;
(ii) worker participation;
(iii) responsibility and accountability;
(iv) competence and training;
(v) OSH management system documentation;
(vi) communication;

6

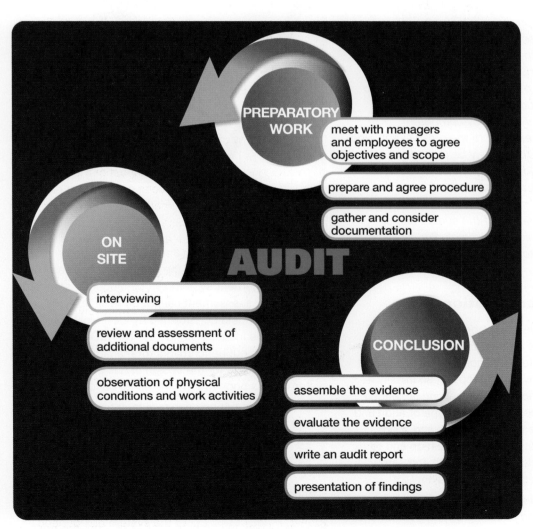

Figure 6.2 The Audit Process

(vii) *system planning, development and implementation;*

(viii) *prevention and control measures;*

(ix) *management of change;*

(x) *emergency prevention, preparedness and response;*

(xi) *procurement;*

(xii) *contracting;*

(xiii) *performance monitoring and measurement;*

(xiv) *investigation of work-related injuries, ill-health, diseases and incidents, and their impact on safety and health performance;*

(xv) *audit;*

(xvi) *management review;*

(xvii) *preventive and corrective action;*

(xviii) *continual improvement; and*

(xix) *any other audit criteria or elements that may be appropriate.*

(d) *The audit conclusion should determine whether the implemented OSH management system elements or a subset of these:*

(i) *are effective in meeting the organisation's OSH policy and objectives;*

(ii) *are effective in promoting full worker participation;*

(iii) *respond to the results of OSH performance evaluation and previous audits;*

(iv) *enable the organisation to achieve compliance with relevant national laws and regulations; and*

(v) *fulfill the goals of continual improvement and best OSH practice.*

(e) *Audits should be conducted by competent persons internal or external to the organisation who are independent of the activity being audited.*

(f) *The audit results and audit conclusions should be communicated to those responsible for corrective action.*

(g) *Consultation on selection of the auditor and all stages of the workplace audit, including analysis of results, are subject to worker participation, as appropriate.*

6.1.3 Pre-audit preparations

Decisions will need to be made about the level and detail of the audit before starting to gather information about the health and safety management of an organisation. Auditing involves sampling; so initially it is necessary to decide how much sampling is needed for the assessment to be reliable. The type of audit and its complexity will relate to its objectives and scope, to the size and complexity of the organisation and to the length of time that the existing health and safety management system has been in operation.

Information sources including interviewing people, looking at documents and checking physical conditions are usually approached in the following order:

(a) Preparatory work
  ▶ meet with relevant managers and employee representatives to discuss and agree the objectives and scope of the audit;
  ▶ prepare and agree the audit procedure with managers;
  ▶ gather and consider documentation.

(b) On site
  ▶ interviewing;
  ▶ review and assessment of additional documents;
  ▶ observation of physical conditions and work activities.

(c) Conclusion
  ▶ assemble the evidence;
  ▶ evaluate the evidence;
  ▶ write an audit report;
  ▶ presentation of findings to management and workforce representatives where appropriate.

Figure 6.3 Using the audit questions for interviews and collecting information

It is essential to start with a relevant standard or benchmark against which the adequacy of a health and safety management system can be judged. If standards are not clear, assessment cannot be reliable. Audit judgements should be informed by local legal standards and applicable industry standards. The ILO-OSH 2001 guidelines set out benchmarks for management arrangements and for the design of risk control systems. This book follows the same concepts.

Auditing should not be seen as a fault-finding activity. It should make a valuable contribution to the health and safety management system and to learning. It should recognise achievement as well as highlight areas where more needs to be done.

Scoring systems can be used in auditing along with judgements and recommendations. This can be seen as a useful way to compare sites or monitor progress over time. However, there is no evidence that quantified results produce a more effective response than the use of qualitative evidence. Indeed, the introduction of a scoring system can have a negative effect, as it may encourage managers to place more emphasis on high-scoring questions which may not be as relevant to the development of an effective health and safety management system.

An organisation can use its own auditing system or one of the proprietary systems on the market or, as it is unlikely that any ready-made system will provide a perfect fit, a combination of both. With any scheme, cost and benefits have to be taken into account. Common problems include:

▶ systems can be too general in their approach. These may need considerable work to make them fit the needs and risks of the organisation;
▶ systems can be too cumbersome for the size and culture of the organisation;
▶ scoring systems may conceal problems in underlying detail;
▶ organisations may design their management system to gain maximum points rather than using one which suits the needs and hazard profile of the business.

Some authorities like the UK's HSE encourage organisations to assess their health and safety management systems using in-house or proprietary schemes but without endorsing any particular one. Other countries have adopted the ILO-OSH 2001 or OHSAS 18001:2007 and incorporated it into their statutory requirements or guidance.

People selected for interviews need reasonable notice and should agree to be interviewed. They should be assured that the audit is not to apportion blame but to improve health and safety management in the organisation.

To achieve the best results, auditors should be competent people who are independent of the area and of the activities being audited. External consultants or staff from other areas of the organisation can be used. It is essential that auditors are experienced in meeting and interviewing people as well as knowledgeable in the subject being audited. In limited audits and smaller companies one person may carry out the audit but in larger sites and organisations it is often necessary to have a team approach with a variety of disciplines. Sufficient time must be allocated to achieve the objectives of the audit and it is essential that local management organise themselves to give adequate time to the audit process including pre-audit preparation, interviews, accompanying auditors on plant inspection and post-audit remedial work.

6.1.4 Responsibility for audits – external v. internal audits

Directors and senior managers have a primary duty to ensure that they establish adequate systems for managing health and safety issues. Equally (as set out in 'Leading health and safety at work', HSE UK INDG417(rev1) and the ILO-OSH 2001) they have a duty to ensure that:

▶ appropriate weight is given to reporting both preventative information (such as progress of training and maintenance programmes) and incident data (such as accident and sickness absence rates);
▶ periodic audits of the effectiveness of management structures and risk controls for health and safety are carried out;
▶ the impact of changes such as the introduction of new procedures, work processes or products, or any major health and safety failure, is reported as soon as possible to the board;
▶ there are procedures to implement new and changed legal requirements and to consider other external developments and events.

The summary of audit results and the fact that they are carried out should be part of directors' annual reporting procedures just as they are required to have financial audited reports. This gives confidence to workers, investors and customers that the company is operating proper health and safety management systems and considers the welfare of all those affected by its business operations. A failure to do this adequately has severely damaged the reputation of many large organisations.

Auditing may be carried out by internal or external consultants or, as is often the case, a combination of the two. Internal auditors:

▶ are familiar with the workplace, systems, processes and the organisation;
▶ are likely to be aware of what is practicable for the industry;
▶ have the ability to see improvements or a deterioration from the last audit;
▶ are familiar with the workforce and an individual's qualities and attitude; and
▶ may be less costly and easier to arrange than an external auditor.

But on the other hand internal auditors:

▶ may miss or gloss over some issues because of their familiarity;
▶ may not get honest views from the work force for fear of the consequences;
▶ may not be in possession of recognised auditing skills;
▶ may not be up to date with legal requirements and may be less likely to be aware of best practice in other organisations;
▶ may be subject to pressure from management and the workforce; and
▶ have time constraints imposed upon them.

6

External auditors:

▶ come with a new perspective, a fresh pair of eyes;
▶ need to ask more questions to understand the systems in operation which can elicit underlying problems;
▶ may have solutions learnt elsewhere that would benefit the organisation;
▶ can be more impartial in their presentation of the audit results;
▶ are more likely to have the necessary auditing skills;
▶ will not be inhibited from criticising members of management or the workforce;
▶ are more likely to be up to date with legal requirements and best practice in other companies.

The disadvantage of external auditors is that they:

▶ are unlikely to be familiar with the workplace, tasks and processes;
▶ will not be familiar with the workforce and their individual attitudes to health and safety and will have difficulty in obtaining their full cooperation;
▶ may be unfamiliar with the industry and seek unrealistic standards; and
▶ may well be more costly than an internal member of staff.

Multi-site organisations often use a hybrid arrangement with cross-site audits where people from other locations inside the organisation will audit a particular site, perhaps with one independent external expert on the team.

6.1.5 Actions taken after the audit

It is essential for the organisation, including and especially the board of directors, to take the appropriate action after the audit report is received. It is often a very good idea for the lead auditor to make a presentation to senior managers on the findings of the audit and the recommendations. It may be necessary to ensure that some or all board members have some health and safety training to promote their understanding of the importance and recommendations from audit and other health and safety monitoring or investigation reports.

An implementation programme will probably be needed to ensure that adequate capital finance is made available. This will need to be closely monitored by a responsible manager allocated to the task. This should be periodically reviewed by senior managers to ensure that implementation is on track. A follow-up independent audit to check progress might be useful.

The audit report and recommendations should be a live document which helps to improve the health and safety management system and produce a new impetus to improve performance and enhance risk control systems. The results should be fed into formal internal reviews covered in Section 6.2, to see whether policy or other changes are required.

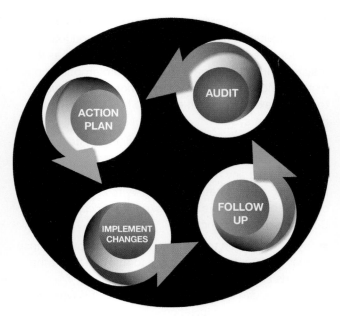

Figure 6.4 The audit report should be reviewed by senior managers with an action plan and follow-up

6.2 Review of health and safety performance

6.2.1 Purpose of reviewing performance

When health and safety performance is reviewed, judgements are made about its adequacy and decisions are taken about how and when to rectify problems. This ACT part in the health and safety management system is needed by organisations so that they can see whether it is working as intended. Feedback on both successes and failures is an essential element to keep employees at all levels motivated to improve health and safety performance. Many successful organisations encourage positive reviews and concentrate on those indicators that demonstrate effective risk control improvements.

The information required for a review of performance comes from audits of RCSs (risk control systems) and workplace precautions, and from the measurement of activities. There may be other influences, both internal and external, such as reorganisation, new legislation or changes in current good practice. These may result in the necessity to re-design or change parts of the health and safety management system or to alter its direction or objectives.

SMART (see Figure 2.6) performance standards or objectives need to be established which will identify the systems requiring change, responsibilities and completion dates. It is essential to feed back the information about success and failure so that employees are motivated to maintain and improve performance.

In a review, the following areas will need to be examined:

▶ the operation and maintenance of the existing system;

▶ how the safety management system is designed, developed and installed to accommodate changing circumstances.

6.2.2 People involved and planned intervals

Reviewing is a continuous process. It should be undertaken at different levels within the organisation. Responses will be needed as follows:

▶ by first-line supervisors or other managers to remedy failures to implement workplace precautions which they observe in the course of routine activities;
▶ to remedy sub-standard performance identified by active and reactive monitoring;
▶ to the assessment of plans at individual, departmental, site, group or organisational level;
▶ to the results of audits.

Senior managers, departmental managers and health and safety professionals will be involved in reviews where these are relevant to their responsibilities. Review plans may include:

Figure 6.5 Review of performance

▶ monthly reviews of individuals, supervisors or sections;
▶ three-monthly reviews of departments;
▶ annual reviews of sites or of the organisation as a whole.

The frequency of review at each level should be decided upon by the organisation and reviewing activities should be devised which will suit the measuring and auditing activities. The review will identify specific remedial actions which establish who is responsible for implementation and set deadlines for completion. Reviews should always be looking for opportunities to improve performance and be prepared to recommend

changes to policies and the organisation to achieve the necessary change. The output from reviews must be consistent with the organisation's policy, performance, resources and objectives. They must be properly documented since they will be the base line from which continual improvement will be measured.

6.2.3 Items to be considered in reviews

Reviews will be wide ranging and may cover one specific subject or a range of subjects for an area of the organisation. They should aim to include:

▶ evaluation of compliance with legal and organisational requirements;
▶ incident data, recommendations and action plans from investigations;
▶ inspections, surveys, tours and sampling;
▶ absences and sickness records and their analysis;
▶ any reports on quality assurance or environmental protection;
▶ audit results and implementation;
▶ monitoring of data, reports and records;
▶ communications from enforcing authorities and insurers;
▶ any developments in legal requirements or best practice within the industry;
▶ changed circumstances or processes;
▶ benchmarking with other similar organisations;
▶ complaints from neighbours, customers and the public;
▶ effectiveness of consultation and internal communications;
▶ whether health and safety objectives have been met;
▶ whether actions from previous reviews have been completed.

6.2.4 Role of directors and senior managers

a) Reporting on performance

The management systems must allow the board and senior managers to receive both specific (e.g. incident-led) and routine reports on the performance of health and safety policy. Much day-to-day health and safety information need be reported only at the time of a formal review. However, only a strong system of monitoring can ensure that the formal review can proceed as planned – and that relevant events in the interim are brought to the board's attention.

The board should ensure that:

▶ appropriate weight is given to reporting both preventative information (results of active monitoring), such as progress of training and maintenance programmes, and incident data (results of reactive monitoring), such as accident and sickness absence rate;

▶ periodic audits of the effectiveness of management structures and risk controls for health and safety are carried out;

▶ the impact of changes such as the introduction of new procedures, work processes or products, or any major health and safety failure, is reported as soon as possible to the board;

▶ there are procedures to implement new and changed legal requirements and to consider other external developments and events.

Good practice involves:

▶ effective monitoring of sickness absence and workplace health. This can alert the board to underlying problems that could seriously damage performance or result in accidents and long-term illness;

▶ the collection of workplace health and safety data. This can allow the board to benchmark the organisation's performance against others in its sector;

▶ appraisals of senior managers which includes an assessment of their contribution to health and safety performance;

▶ boards receiving regular reports on the health and safety performance and actions of contractors;

▶ winning greater support for health and safety by involving workers in monitoring.

b) Reviewing health and safety

Reviewing performance should be supported at the highest level in the organisation and built into the safety management system. A formal boardroom review of health and safety performance is essential. It allows the board to establish whether the essential health and safety principles – strong and active leadership, worker involvement, and assessment and review – have been embedded in the organisation. It tells senior managers whether their system is effective in managing risk and protecting people. Directors will need to know whether they are provided with sufficient information to make sound health and safety judgements about the organisation.

Performance on health and safety and well-being is increasingly being recorded in organisations' annual reports to investors and stakeholders. Board members can make extra 'shop floor' visits to gather information for the formal review. Good health and safety performance can be celebrated at central and local level both inside and outside the organisation (for example at national annual safety awards).

The board should review health and safety performance at least once a year. The review process should:

▶ examine whether the health and safety policy reflects the organisation's current priorities, plans and targets;

▶ examine whether risk management and other health and safety systems have been effectively reported to the board;

▶ report health and safety shortcomings, and the effect of all relevant board and management decisions;

▶ decide actions to address any weaknesses and a system to monitor their implementation;

▶ consider immediate reviews in the light of major shortcomings or events.

Setting up a separate risk management or health and safety committee as a subset of the board, chaired by a senior executive, can make sure the key issues are addressed and guard against time and effort being wasted on trivial risks and unnecessary bureaucracy. This is referred to in Chapter 3 in more detail. The results of these reviews need to be properly recorded and fed into action and development plans for the coming year or so. This committee can ensure ongoing monitoring and review of performance and provide an

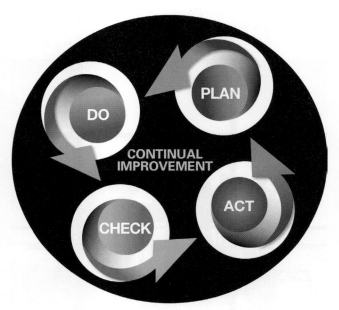

Figure 6.6 Continual improvement part of the health and safety management process

annual report for the formal board review.

6.2.5 Continual improvement

Many organisations have a policy on continual improvement which should be applied to health and safety management in the same way as other management issues. The health and safety commitment on continuous improvement might include statements like: 'The organisation will continuously seek to improve its safety performance.' Continual improvement of safety performance will be achieved through:

▶ active and reactive evaluations of facilities, equipment, documentation and procedures through safety audits and surveys;

▶ active evaluation of each individual's performance to verify the fulfilment of their safety responsibilities; and

▶ a reactive evaluation in order to verify the effectiveness of the system for control and mitigation of risk.

The organisation will also continuously seek to improve its safety management processes.

Measures that could be used to improve safety management include:

▶ more succinct procedures;

▶ improved safety reviews, studies and audits;

▶ improved reporting and analysis tools;

▶ improved hazards identification and risk assessment processes and improved awareness of risks in the organisation;

▶ improved relations with the sub-contractors, suppliers and customers regarding safety;

▶ improved communication processes, including feedback from the personnel.

6.3 Further information

Guidelines on Occupational Safety and Health Management Systems (ILO-OSH 2001), ISBN 0-580-37805-5 http://www.ilo.org/global/publications/ilo-bookstore/order-online/books/WCMS_PUBL_9221116344_EN/lang--en/index.htm

Occupational Health and Safety Assessment Series (OHSAS 18000): Occupational Health and Safety Management Systems– Requirements OHSAS 18001:2007 ISBN 978 0 580 50802 8, OHSAS 18002:2008 ISBN 9780 580 61674 7 http://www.bsigroup.com/

ISO 19011:2011 – Guidelines for auditing management systems

6

6.4 Practice revision questions

1. **Outline** how the following techniques may be used to improve health and safety performance within an organisation and the differences between them.

   (a) Safety inspections.
   (b) Externally led health and safety audits.
   (c) Analysis of accident statistics.

2. (a) **Explain** the meaning of the term health and safety *'audit'*.
   (b) **Outline** the issues that need to be considered at the planning stage of the audit.
   (c) **Identify TWO** methods of gathering information during an audit.

3. (a) **Outline** key areas that may be covered within a health and safety audit.
   (b) **Identify** the documents that are likely to be examined during a health and safety audit.
   (c) **Explain** how the findings of a health and safety audit can be used to improve health and safety performance.

4. **Outline FOUR** advantages **AND FOUR** disadvantages of undertaking a health and safety audit of an organisation's activities by:

   (a) an internal auditor; or
   (b) an external auditor.

5. A health and safety audit of a manufacturing company has identified a lack of compliance with many of its health and safety procedures.

   (a) **Describe** the possible reasons for procedures not being followed.
   (b) **Outline** the practical measures that could be taken to motivate employees to comply with health and safety procedures.

6. (a) **Outline** ways in which an organisation can monitor its health and safety performance.
   (b) **Identify EIGHT** measures that could be used by an organisation in order to monitor its health and safety performance.
   (c) **Outline** the reasons why an organisation should review and monitor its health and safety performance.

7. The Board of Directors of a large company decides to review its health and safety performance over the past year.

   (a) **Outline** the role of the Board in such a review.
   (b) **List** the topics that should be covered in the review.
   (c) **Outline** the possible review periods for the various parts of the company.

CHAPTER 7

Workplace hazards and risk control

7.1 **Health, welfare and work environment requirements** ▶ 182

7.2 **Violence at work** ▶ 187

7.3 **Substance misuse at work** ▶ 191

7.4 **Safe movement of people in the workplace** ▶ 192

7.5 **Working at height** ▶ 197

7.6 **Hazards and control measures for works of a temporary nature** ▶ 210

7.7 **Construction activities** ▶ 212

7.8 **Further information** ▶ 221

7.9 **Practice revision questions** ▶ 221

Appendix 7.1 **Scaffolds and ladders** ▶ 223

Appendix 7.2 **Inspection recording form with timing and frequency chart** ▶ 224

Appendix 7.3 **Checklist of typical scaffolding faults** ▶ 226

Appendix 7.4 **Recommendations for excavation work in the ILO Code of Practice 'Safety and Health in Construction'** ▶ 227

This chapter covers the following NEBOSH learning objectives:

1. Outline common health, welfare and work environment requirements in the workplace
2. Explain the risk factors and appropriate controls for violence at work
3. Explain the effects of substance misuse on health and safety at work and control measures to reduce such risks
4. Explain the hazards and control measures for the safe movement of people in the workplace
5. Explain the hazards and control measures for safe working at height
6. Outline the hazards and controls measures associated with works of a temporary nature

7.1 Health, welfare and work environment requirements

7.1.1 Introduction

People are most often involved in accidents as they walk around the workplace, or when they come into contact with vehicles in or around the workplace. It is therefore important to understand the various common causes of accidents, and the control strategies that can be employed to reduce them. Slips, trips and falls account for the majority of accidents in the workplace. However, the workplace may be a construction site where working at height or excavation work will present risks to health and safety. There are also health hazards and risks present in some workplaces due to violence and substance abuse. Many of the risks associated with these hazards can be significantly reduced by an effective management system.

Over the last 20 years, psychological hazards have been included among the occupational health hazards faced by many workers. This is now the most rapidly expanding area of occupational health and includes topics such as mental health and workplace stress, violence to staff, passive smoking, drugs and alcohol.

The construction industry covers a wide range of activities from large-scale civil engineering projects to very small house extensions. The use of sub-contractors is very common at all levels of the industry. It is most likely that everybody will be aware of or involved with some aspect of the construction industry at their place of work – either in terms of the repair and modification of existing buildings or a major new engineering project. It is, therefore, important that the health and safety practitioner has some basic knowledge of the hazards and health and safety legal requirements associated with construction.

The global number of accidents and diseases in the construction industry is difficult to quantify, as statistical information is not available for many countries. But ILO global estimates for 2003 of work-related fatalities showed that the construction industry recorded some 60,000 fatalities out of a world total of 355,000, nearly 17%.

The construction industry has a disproportionately high rate of recorded accidents. Each year at least 60,000 fatal accidents occur on construction sites around the world – or one fatal accident every ten minutes. These figures include deaths of members of the public, including children playing on construction sites. Most of these fatalities were caused by falls from height. According to ILO estimates:

1. One in six fatal accidents at work occurs on a construction site.
2. In industrialised countries, as many as 25–40% of work-related deaths occur on construction sites, even though the sector employs only 6–10% of the workforce.
3. In some countries, it is estimated that 30% of construction workers suffer from back pains or other musculoskeletal disorders.

The chapter begins with the welfare and work environment requirements to ensure a healthy and safe workplace.

7.1.2 Welfare

Welfare arrangements include the provision of sanitary conveniences and washing facilities, drinking water, accommodation for clothing, facilities for changing clothing and facilities for rest and eating meals. First-aid provision is also a welfare issue, but is covered in Chapter 4. Welfare and work equipment issues are covered by the ILO R102 Welfare Facilities Recommendation 1956.

Sanitary conveniences and washing facilities must be provided together and in proportion to the size of the workforce. Guidance is available on the requisite number of water closets, wash stations and urinals for varying sizes of workforce (approximately one of each for every 25 employees). Special provision should be made for disabled workers and there should normally be separate facilities for men and women. A single convenience would only be acceptable if it were situated in a separate room whose door could be locked from the inside. There should be adequate protection from the weather and only as a last resort should public conveniences be used. A good supply of warm water, soap and towels must be provided as close to the sanitary facilities as possible. The facilities should be well lit and ventilated and their walls and floors easy to clean. It may be necessary to install a shower for certain types of work. Hand dryers are permitted but there are concerns about their effectiveness in drying hands completely and thus

7.1.3 Workplace environment

The issues governing the workplace environment are ventilation, heating and temperature, lighting, workstations and seating.

Ventilation

The ventilation of the workplace should be effective and sufficient and free of any impurity, and air inlets should be sited clear of any potential contaminant (e.g. a chimney flue). Care needs to be taken to ensure that workers are not subject to uncomfortable draughts. The ventilation plant should have an effective visual or audible warning device fitted to indicate any failure of the plant. The plant should be properly maintained and records kept. The supply of fresh air should not normally fall below 5–8 litres per second per occupant.

Every enclosed workplace should be ventilated by a sufficient quantity of fresh or purified air and the fresh air should be free of contaminants such as vehicle exhaust fumes or boiler chimney emissions. If, following a risk assessment, there is a problem of particulate pollution of the incoming air to a workplace, then the air should be filtered.

Heating and temperature

During working hours, the temperature in all workplaces inside buildings shall be reasonable (not uncomfortably high or low). 'Reasonable' is usually defined as at least 16°C, unless much of the work involves severe physical effort in which case the temperature should be at least 13°C. These temperatures refer to readings taken close to the workstation at working height and away from windows. These minimum temperatures cannot be maintained where rooms open to the outside or where food or other products have to be kept cold. A heating or cooling method must not be used in the workplace which produces fumes, or is injurious or offensive to any person. Such equipment needs to be regularly maintained to prevent this problem.

There is no maximum workplace temperature given in the UK Workplace (Health, Safety and Welfare) Regulations, which place a legal obligation on employers to provide a 'reasonable' temperature in the workplace. The Approved Code of Practice (ACOP) suggests a minimum temperature in workrooms should normally be at least 16°C or 13°C if much of the work indoors involves severe physical effort.

The UK Health and Safety Executive (HSE) guidance on workplace temperature states that 'a meaningful figure cannot be given at the upper end of the scale. This is because the factors, other than air temperature which determine thermal comfort (radiant temperature, humidity and air velocity), become more significant and the interplay between them more complex as temperatures rise.' The HSE recommends that

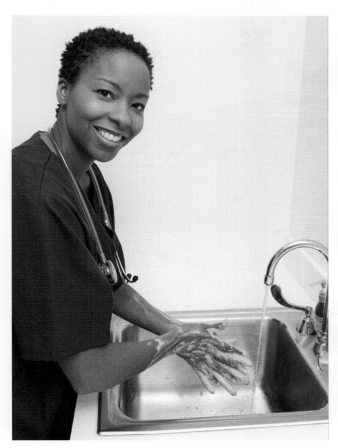

Figure 7.1 Welfare washing facilities: washbasin large enough for people to wash their forearms

removing all bacteria. In the case of temporary or remote worksites, sufficient chemical closets and sufficient washing water in containers must be provided.

All such facilities should be well ventilated and lit, and cleaned regularly.

Drinking water must be readily accessible to the entire workforce. The supply of drinking water must be adequate and wholesome. Normally mains water is provided and should be marked as 'drinking water' if water unfit for drinking is also available. On remote sites, potable water should be provided.

Accommodation for clothing and facilities for changing clothing must be provided. This should be clean, warm, dry, well ventilated and secure. Such accommodation is only necessary when the work activity requires employees to change into specialist clothing. Where workers are required to wear special or protective clothing, arrangements should be such that the workers' own clothing is not contaminated by any hazardous substances.

Facilities for rest and eating meals must be provided so that workers may sit down during break times in areas where they do not need to wear personal protective equipment. Facilities should also be provided for pregnant women and nursing mothers to rest. Arrangements must be in place to ensure that food is not contaminated by hazardous substances.

7

183

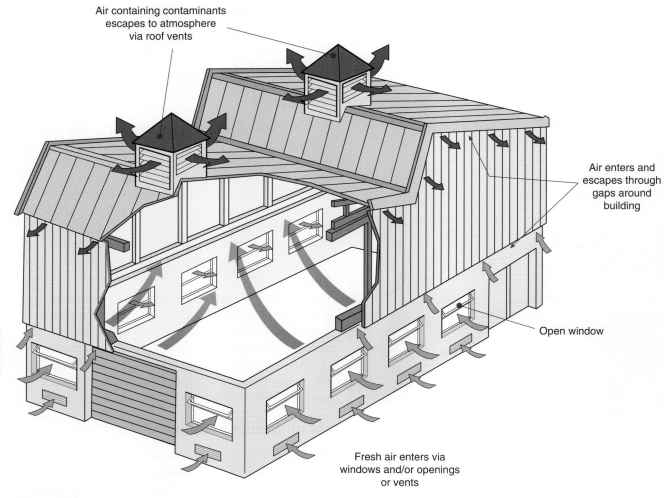

Air containing contaminants escapes to atmosphere via roof vents

Air enters and escapes through gaps around building

Open window

Fresh air enters via windows and/or openings or vents

Figure 7.2 Natural ventilation in a building

employers should consult with employees or their representatives to establish sensible means to cope with high temperatures. Advice is available from professional organisations.

A sufficient number of thermometers should be provided and maintained to enable workers to determine the temperature in any workplace inside a building (but need not be provided in every workroom).

Where, despite the provision of local heating or cooling, the temperatures are still unreasonable, suitable protective clothing and rest facilities should be provided.

Lighting

Every workplace shall have suitable and sufficient lighting and this shall be natural lighting so far as is reasonably practicable. Suitable and sufficient emergency lighting must also be provided and maintained in any room where workers are particularly exposed to danger in the event of a failure of artificial lighting (normally due to a power cut and/or a fire). Windows and skylights should be kept clean and free from obstruction so far as is reasonably practicable unless it would prevent the shading of windows or skylights or prevent excessive heat or glare.

Figure 7.3 A well-lit workplace

When deciding on the suitability of a lighting system, the general lighting requirements will be affected by the following factors:

▶ the availability of natural light;
▶ the specific areas and processes, in particular any colour rendition aspects or concerns over stroboscopic effects (associated with fluorescent lights);
▶ the type of equipment to be used and the need for specific local lighting;

- the lighting characteristics required (type of lighting, its colour, intensity and local adjustability);
- the location of visual display units and any problems of glare;
- structural aspects of the workroom, such as the use of screens in open office layout and the reduction of shadows;
- the presence of atmospheric dust;
- the heating effects of the lighting;
- lamp and window cleaning and repair (and disposal issues);
- the need and required quantity of emergency lighting.

Light levels are measured in illuminance, having units of lux (lx), using a light meter. A general guide to lighting levels in different workplaces is given in Table 7.1.

Table 7.1 Typical workplace lighting levels

Workplace or type of work	Illuminance (lx)
Warehouses and stores	150
General factories or workshops	300
Offices	500
Drawing offices (detailed work)	700
Fine working (ceramics or textiles)	1000
Very fine work (watch repairs or engraving)	1400

Poor lighting levels will increase the risk of accidents such as slips, trips and falls. More information is available on lighting from *Lighting at Work*, HSG38, HSE Books.

Workstations and seating

Workstations should be arranged so that work may be done safely and comfortably. The worker should be at a suitable height relative to the work surface and there should be no need for undue bending and stretching. Workers must not be expected to stand for long periods of time, particularly on solid floors. A suitable seat should be provided when a substantial part of the task can or must be done sitting. The seat should, where possible, provide adequate support for the lower back and a footrest be provided for any worker whose feet cannot be placed flat on the floor. It should be made of materials suitable for the environment, be stable and, possibly, have arm rests.

It is also worth noting that sitting for prolonged periods can present health risks, such as blood circulation and pressure problems, and vertebral and muscular damage. *Seating at Work*, HSG57, HSE Books, provides useful guidance on how to ensure that seating in the workplace is safe and suitable.

Other factors

The condition of floors, stairways and traffic routes should be suitable for the purpose and well maintained and undue space constraints anywhere in the workplace should be avoided. Translucent or transparent doors should be constructed with safety glass and properly marked to warn pedestrians of their presence. Windows and skylights should be designed so that, when they are opened, they do not present an obstruction to passing pedestrians. There must be adequate arrangements in place to ensure the safe cleaning of windows and skylights. Finally, there need to be adequate provisions for the needs of disabled workers.

7.1.4 Extremes of temperature

The human body is very sensitive to relatively small changes in external temperatures. Food not only provides energy and the build-up of fat reserves, but also generates heat, which needs to be dissipated to the surrounding environment. The body also receives heat from its surroundings. The temperature of the body is normally around 37°C, and it will attempt to maintain this temperature irrespective of the temperature of the surroundings. Therefore, if the surroundings are hot, sweating will allow heat loss to take place by evaporation caused by air movement over the skin. On the other hand, if the surroundings are cold, shivering causes internal muscular activity, which generates body heat.

At high temperatures, the body has more and more difficulty in maintaining its natural temperature unless sweating can take place and therefore water must be replaced by drinking. If the surrounding air has high humidity, evaporation of the sweat cannot take place and the body begins to overheat. This leads to heart strain and, in extreme cases, heat stroke. It follows that when working is required at high temperatures, a good supply of drinking water should be available and, further, if the humidity is high, a good supply of ventilation air is also needed. Heat exhaustion is a particular hazard in confined spaces so more breaks and air-moving plant are needed.

At low temperatures, the body will lose heat too rapidly and the extremities of the body will become very cold leading to frostbite and possibly the loss of limbs. Under these conditions, thick, warm (thermal) clothing, the provision of hot drinks and external heating will be required. For those who work in sub-zero temperatures, such as cold store workers, additional precautions will be needed. The store doors must be capable of being unlocked from the inside and an emergency alarm system should be installed. Appropriate equipment selection and a good preventative maintenance system is very important as well as a regular health surveillance programme for the workers, who should be provided with information and training on the hazards associated with working in very low temperatures.

In summary, extremes of temperature require special measures, particularly if accompanied by extremes of

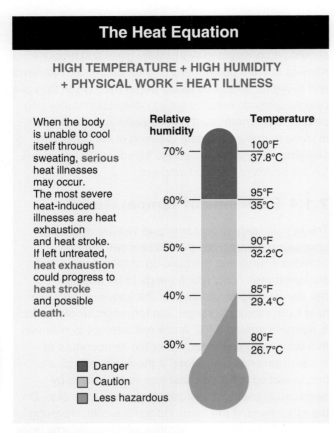

Figure 7.4 The heat equation

humidity. Frequent rest periods will be necessary to allow the body to acclimatise to the conditions. An index called WBGT (Wet Bulb Globe Temperature) is normally used.

ILO recommendations for working in hot and cold environments

The ILO Code of Practice 'Ambient factors in the workplace' applies to any workplace where workers may be occupationally exposed to high or low temperatures and humidity. It applies to conditions in which:

(a) temperatures and/or humidity are unusually high;
(b) workers are exposed to high radiant heat;
(c) high temperature and/or humidity occur in combination with protective clothing or high work rate;
(d) temperatures are unusually low (e.g. in outdoor work during winter or in cold storage work);
(e) high wind speeds (> 5 m/s) prevail with unusually low temperatures;
(f) work with bare hands is carried out for extended periods of time at temperatures below 15°C.

Employers should assess the risks to the health and safety of workers from high and low temperatures, and determine the controls necessary to remove such hazards or to reduce risks to the lowest practicable level. The assessment should also examine the risks, arising from working with hazardous substances in hot or cold environments, caused by:

(a) the use of protective clothing against hazardous substances that may increase the risk from heat stress; and
(b) a hot environment that makes respiratory protection uncomfortable and less likely to be used and necessitates restructuring of jobs in order to reduce the risks, for example by:
   (i) minimising exposure to hazardous substances so that there is less need for protective clothing;
   (ii) changing the tasks so that work rates in hot conditions can be reduced.

The risk assessment should include:

(a) all stages of work cycles and the range of temperature and humidity under which the tasks are performed;
(b) the range of clothing worn during the tasks;
(c) major changes in physical activity level (metabolic heat production);
(d) occasional tasks such as cleaning and maintenance of hot equipment and cold areas, and renewal of hot or cold insulation.

Where assessment shows that the workers may be at risk from heat stress, employers should, if practicable, eliminate the need for work in hot conditions or, if elimination is not practicable, take measures to reduce the thermal load from the environment. For workers who are at risk from exposure to radiant heat by working near hot surfaces, various methods are proposed in the Code of Practice including increasing the distance between the equipment and the exposed workers.

If it is not practicable to reduce the surface temperature, employers are recommended to consider:

(a) the use of radiation barriers between the surface and the workplace (ensuring that they are maintained in a clean state);
(b) water-cooling the hot surfaces, where practicable;
(c) the use of portable reflective shielding;
(d) the remote control of operations.

The Code of Practice requires employers to make water at low salt concentration or dilute flavoured drinks readily available to workers, and to encourage them to drink at least hourly, either by providing a close source or arranging for a regular delivery of drinks. Drinks at temperatures between 15 to 20°C are preferable to iced drinks. Alcohol, caffeine, carbonated drinks or drinks with a high salt or sugar content are unsuitable. Drinking fountains are not recommended because they are too difficult to drink from in sufficient volume.

If a residual risk of heat stress remains even after all the control measures have been taken, workers should be adequately supervised so that they can be withdrawn from the hot conditions if heat stress symptoms occur. First-aid facilities and trained first-aiders should also be readily available.

If the assessment shows that the workers may be at risk from exposure to cold, the employers should, if practicable, eliminate the need for work in cold conditions (for example by rescheduling work to be performed in warmer weather, or by moving the work from outdoors to indoors, or separating the cold parts of a process from the workers, as far as practicable). If elimination of such work is impracticable, employers should introduce other control measures to reduce risk from cold conditions.

If the work is done outdoors, or the temperature at the workplace depends on outdoor temperature, employers should take into account present and forecast weather conditions in scheduling work, and monitor conditions if the work is to last a long time.

The code recommends that if work is carried out at unusually low temperatures:

(a) employers should implement work–rest cycles with warm shelters for recovery when:

  (i) work is likely to last for some time;
  (ii) the temperature and wind speed are likely to vary;
  (iii) workers are experiencing or showing symptoms of discomfort;

(b) work scheduling should allow for the extra time taken by tasks in the cold, and the need for adequate drink and food;

(c) where practicable, work rates should be designed to avoid heavy sweating, but if this does occur, employers should ensure that dry replacement clothing is available with warm changing facilities.

Under cold conditions, suitable protection should be given to the hands and fingers, particularly where dexterity is needed, as well as other exposed parts of the body. Employers should provide:

(a) facilities for warming the hands, for example by warm air;

(b) tools with insulated handles, especially in temperatures below freezing point;

(c) measures to ensure that the bare hand does not touch surfaces below $-7°C$ (e.g. by using workplace design or protective clothing);

(d) measures to ensure that bare skin does not touch liquids below $4°C$;

(e) appropriate measures to be taken in the event of insulating clothing getting wet;

(f) face and eye protection, as appropriate, for outdoor work and working in snow (e.g. safety goggles against glare).

Health surveillance is important for workers employed in extreme temperature environments. Where the control measures include work–rest systems or protective clothing, workers should be examined by qualified occupational health personnel who should determine:

(a) their fitness for the conditions of work;
(b) any limitations that should be applied to their work;

(c) the programme of training and information of workers;

(d) the measures for providing such training and information;

(e) any pre-existing conditions which might affect their tolerance to heat or cold (such as heart disease, overweight or some skin diseases); and

(f) measures to minimise risks among vulnerable groups (such as older workers).

Workers exposed to heat or cold and their supervisors should be trained:

(a) to recognise symptoms which may lead to heat stress or hypothermia, in themselves or others, and the steps to be taken to prevent onset and/or emergencies;

(b) in the use of rescue and first-aid measures;

(c) about action to be taken in the event of increased risks of accidents because of high and low temperatures;

(d) to recognise the importance of drinking sufficient quantities of liquid and the dietary requirements providing intake of salt and potassium and other elements that are depleted due to sweating; and

(e) to be aware of the importance of physical fitness for work in hot or cold environments and the effects of drugs which can reduce their tolerance to thermal extremes.

Workers should be allowed sufficient time to acclimatise to an extremely hot or cold environment, particularly when they have recently moved to a country with a warmer or colder climate.

7.2 Violence at work

Violence at work, particularly from dissatisfied customers, clients, claimants or patients, can cause stress and in some cases injury. This is not only physical violence as people may face verbal and mental abuse, discrimination, harassment and bullying. Fortunately, physical violence is still rare, but violence of all types has risen significantly in recent years. Violence at work is known to cause pain, suffering, anxiety and stress, leading to financial costs due to absenteeism and higher insurance premiums to cover increased civil claims. It can be very costly to ignore the problem. Violence from members of the public is a higher risk with several occupations, e.g. the health and social services, police and fire-fighters, various types of enforcement officers, education, benefit services, various service industries and debt collectors.

A recent survey in the UK found that strangers were the offenders in 60% of cases of workplace violence. Victims of actual or threatened violence at work said that the offender was under the influence of alcohol in 38% of incidents, and that the offender was under the influence of drugs in 26% of incidents. The survey found 51% of assaults at work resulted in injury, with minor bruising or a black eye accounting for the majority of the injuries recorded.

In 1999 the UK Home Office and the HSE published a comprehensive report entitled *Violence at Work: Findings from the British Crime Survey*. This report is updated annually amd the HSE publishes an annual report – *Violence at Work 2013/14: Findings from the Crime Survey for England and Wales*. It shows the extent of violence at work and how it changed during the period 2006–2014 (Table 7.2). Such incidents, whilst they are still high, have halved since 1999. Over this period many of the protections outlined in this chapter and advocated by the HSE have been in place. For example several hospitals now employ their own police officers to police accident and emergency departments and there has been much more use of closed circuit television (CCTV).

Table 7.2 Trend in physical assaults and threats at work, 2006–2014 (based on working adults of working age)

Number of incidents (000s)	2006	2008	2010	2012	2014
All violence	752	554	683	628	583
Assaults	338	202	300	310	269
Threats	414	352	363	318	314
Number of victims (000s)					
All violence	350	302	328	318	257
Assaults	200	110	148	150	125
Threats	150	192	180	168	152

Source: British Crime Survey 2014.

Workers who are most at risk from violence are those that:

▶ handle money;
▶ provide a service to the public (such as shop workers, teachers and nurses);
▶ are lone workers;
▶ represent authority (such as police, traffic wardens and even school crossing patrols).

Many people resort to violence due to frustration. Common causes of such frustration are the following:

▶ dissatisfaction with a product or service, including the cost;
▶ a perception of being unreasonably penalised over an incident such as car parking; and
▶ a general lack of information following a problem, such as aircraft delays or long delays at hospitals.

The report defines violence at work as:

'All assaults or threats which occurred while the victim was working and were perpetrated by members of the public.'

Physical assaults include the offences of common assault, wounding, robbery and snatch theft. Threats include both verbal threats, made to or against the victim, and non-verbal intimidation. These are mainly threats to assault the victim and, in some cases, to damage property.

Excluded from the survey are violent incidents where there was a relationship between the victim and the offender and also where the offender was a work colleague. The latter category was excluded because of the different nature of such incidents.

The British Crime UK HSE Survey reported approximately 314,000 threats of violence and 269,000 physical assaults by members of the public on workers in the UK during the year. Approximately 257,000 workers had experienced at least one incident of violence at work. This resulted in one fatality, 866 major or specified injuries and 4, 069 over-7-day injuries. Those workers particularly affected in 2008/09 were:

▶ social welfare workers (2.6%);
▶ healthcare professionals (3.8%); and
▶ police officers (9%).

In over a third of the incidents, the victim believed the offender was under the influence of alcohol, and in nearly a fifth of incidents under the influence of drugs.

Figure 7.5 Security access and surveillance CCTV camera

It is interesting that almost half of the assaults and a third of the threats happened after 1800 hrs, which suggests that the risks are higher if people work at night or in the late evening. About 16% of the assaults involved offenders under the age of 16 and were mainly against teachers or other education workers.

Violence at work is defined by the HSE as:

'any incident in which an employee is abused, threatened or assaulted in circumstances relating to their work.'

In recognition of this, the HSE has produced a useful guide to employers which includes a four-stage action plan and some advice on precautionary measures ('Violence at Work: a Guide for Employers', INDG69 (rev)). The employer is just as responsible, under health and safety legislation, for protecting employees from violence as they are for any other aspects of their safety.

The UK HSE recommends the following four-point action plan:

1. find out if there is a problem;
2. decide on what action to take;
3. take the appropriate action;
4. check that the action is effective.

7.2.1 Find out if there is a problem

This involves a risk assessment to determine what the real hazards are. It is essential to ask people at the workplace and, in some cases, a short questionnaire may be useful. Record all incidents to get a picture of what is happening over time, making sure that all relevant detail is recorded. The records should include:

▶ a description of what happened;
▶ details of who was attacked, the attacker and any witnesses;
▶ the outcome, including how people were affected and how much time was lost;
▶ information on the location of the event.

Owing to the sensitive nature of some aggressive or violent actions, employees may need to be encouraged to report incidents and be protected from future aggression.

All incidents should be classified so that an analysis of the trends can be examined.

Consider the following:

▶ fatalities;
▶ major injury;
▶ less severe injury or shock which requires first-aid treatment, outpatient treatment, time off work or expert counselling;
▶ threat or feeling of being at risk or in a worried or distressed state.

7.2.2 Decide on what action to take

It is important to evaluate the risks and decide who may be harmed and how this is likely to occur. The threats may be from the public or co-workers at the workplace or may be as a result of visiting the homes of customers. Consultation with employees or other people at risk will improve their commitment to control

measures and will make the precautions much more effective. The level of training and information provided, together with the general working environment and the design of the job, all have a significant influence on the level of risk.

Those people at risk could include those working in:

▶ reception or customer service points;
▶ enforcement and inspection;
▶ lone working situations and community-based activities;
▶ front-line service delivery;
▶ education and welfare;
▶ catering and hospitality;
▶ petrol retail and late-night shopping operations;
▶ leisure facilities, especially if alcohol is sold;
▶ healthcare and voluntary roles;
▶ policing and security;
▶ mental health units or in contact with disturbed people;
▶ cash handling or control of high value goods.

Consider the following issues:

▶ quality of service provided;
▶ design of the operating environment;
▶ type of equipment used;
▶ designing the job.

Some violence may be deterred if measures are taken which suggest that any violence may be recorded. Many public bodies use the following measures:

▶ informing telephone callers that their calls will be recorded;
▶ displaying prominent notices that violent behaviour may lead to the withdrawal of services and prosecution;
▶ using CCTV or security personnel.

Four in ten of all employment enquiries to one UK employment law firm in 2009 were related to bullying. Of those who reported suffering bullying, 80% said it had affected their physical and mental health and a third had taken time off work or left their jobs as a result.

Quality of service provided

The type and quality of service provision has a significant effect on the likelihood of violence occurring in the workplace. Frustrated people whose expectations have not been met and who are treated in an unprofessional way may believe they have the justification to cause trouble.

Sometimes circumstances are beyond the control of the staff member and potentially violent situations need to be defused. The use of correct skills can turn a dissatisfied customer into a confirmed supporter simply by careful response to their concerns. The perceived lack of, or incorrect, information can cause significant frustrations.

Design of the operating environment

Personal safety and service delivery are very closely connected and have been widely researched in recent years. This has resulted in many organisations altering their facilities to reduce customer frustration and enhance sales. It is interesting that most service points experience less violence when they remove barriers or screens, but the transition needs to be carefully planned in consultation with staff, and other measures adopted to reduce the risks and improve their protection.

The layout, ambience, colours, lighting, type of background music, furnishings including their comfort, information, things to do while waiting and even smell all have a major impact. Queue-jumping causes a lot of anger and frustration and needs effective signs and proper queue management, which can help to reduce the potential for conflict.

Wider desks, raised floors and access for special needs, escape arrangements for staff, carefully arranged furniture and screening for staff areas can all be utilised.

Type of security equipment used

There is a large amount of equipment available and expert advice is necessary to ensure that it is suitable and sufficient for the task. Some measures that could be considered include the following:

- **Access control** to protect people and property. There are many variations from staffed and friendly receptions, barriers with swipe-cards and simple coded security locks. The building layout and design may well partly dictate what is chosen. People inside the premises need access passes so they can be identified easily.
- **Closed circuit television** is one of the most effective security arrangements to deter crime and violence. Because of the high cost of the equipment, it is essential to ensure that proper independent advice is obtained on the type and the extent of the system required.
- **Alarms** – there are three main types:
  - ▷ intruder alarms fitted in buildings to protect against unlawful entry, particularly after working hours;
  - ▷ panic alarms used in areas such as receptions and interview rooms covertly located so that they can be operated by the staff member threatened;
  - ▷ personal alarms carried by an individual to attract attention and to temporarily distract the attacker.
- **Radios and pagers** can be a great asset to lone workers in particular, but special training is necessary, as good radio discipline with a special language and codes are required.

- **Mobile phones** are an effective means of communicating and keeping colleagues informed of people's movements and problems such as travel delays. Key numbers should be inserted for rapid use in an emergency.

Job design

Many things can be done to improve the way in which the job is carried out to improve security and avoid violence. These include:

- using cashless payment methods;
- keeping money on the premises to a minimum;
- careful check of customer or client's credentials;
- careful planning of meetings away from the workplace;
- team work where suspected aggressors may be involved;
- regular contact with workers away from their base. There are special services available to provide contact arrangements;
- avoidance of lone working as far as is reasonably practicable;
- thinking about how staff who have to work shifts or late hours will get home. Safe transport and/or parking areas may be required;
- setting up support services to help victims of violence and, if necessary, other staff who could be affected. They may need debriefing, legal assistance, time off work to recover or counselling by experts.

A busy accident and emergency department of a general hospital, for example, has to balance the protection of staff from violent attack with the need to offer patients a calm and open environment. Protection could be given to staff by the installation of wide counters, coded locks on doors, CCTV systems, panic buttons and alarm systems. The employment of security staff and strict security procedures for the storage and issuing of drugs are two further precautions taken by such departments. Awareness training for staff so that they can recognise early signs of aggressive behaviour and an effective counselling service for those who have suffered from violent behaviour should be provided.

7.2.3 Take the appropriate action

The arrangements for dealing with violence should be included in the safety policy and managed like any other aspect of the health and safety procedures. Action plans should be drawn up and followed through using the consultation arrangements as appropriate. The police should also be consulted to ensure that they are happy with the plan and are prepared to play their part in providing back-up and the like.

7.2.4 Check that the action is effective

Ensure that the records are being maintained and any reported incidents are investigated and suitable action taken. The procedures should be regularly audited and changes made if they are not working properly.

Victims should be provided with help and assistance to overcome their distress, through debriefing, counselling, time off to recover, legal advice and support from colleagues.

7.3 Substance misuse at work

Alcohol and drug abuse damages health and causes absenteeism and reduced productivity. The HSE is keen to see employers address the problem and offers advice in two separate booklets.

Alcohol abuse is a considerable problem when vehicle driving is part of the job, especially if driving is required on public roads. Misuse of alcohol can reduce productivity, increase absenteeism, increase accidents at work and, in some cases, endanger the public. The HSE has estimated that between 3% and 5% of all absences from work are due to alcohol and result in approximately 14 million working days lost each year. Employers need to adopt an alcohol policy following employee consultation. The following matters need to be considered:

▶ how the organisation expects employees to restrict their level of drinking;
▶ how drinking problems can be recognised and help offered; and
▶ at what point and under what circumstances will an employee's drinking be treated as a disciplinary rather than a health problem.

Prevention of the problem is better than remedial action after a problem has occurred. During working hours, there should be no drinking, and drinking during break periods should be discouraged. Induction training should stress this policy and managers should set it. Posters can also help to communicate the message. However, it is important to recognise possible symptoms of an alcohol problem, such as lateness and absenteeism, poor work standards, impaired concentration, memory and judgement, mood swings and the slurring of speech and deteriorating relations with colleagues. It is always better to offer counselling rather than dismissal. The policy should be monitored to check on its effectiveness.

Drug and solvent abuse presents similar problems to those found with alcohol abuse – absenteeism, reduced productivity and an increase in the risk of accidents. The drugs that are involved in drug abuse may be legal drugs, such as pain killers of various types, or illegal drugs such as heroin. A study undertaken by Cardiff University found that 'although drug use was lower among workers than the unemployed, one in four workers under the age of 30 years reported having used drugs in the previous year. There are well documented links between drug use and impairments in cognition, perception and motor skills, both at the acute and chronic levels. Associations may therefore exist between drug use and work performance'. The following conclusions were also drawn from the study:

▶ about 13% of working respondents reported drug use in the previous year. The rate varied with age, from 3% of those over 50 years to 29% of those under 30 years;
▶ drug use is strongly linked to smoking and heavy drinking in that order;
▶ there is an association between drug use and minor injuries among those who are also experiencing other minor injury risk factors;
▶ the project has shown that recreational drug use may reduce performance efficiency and safety at work.

A successful drug misuse policy will benefit the organisation and employees by reducing absenteeism, poor productivity and the risk of accidents. There is no simple guide to the detection of drug abuse, but the HSE has suggested the following signals:

▶ sudden mood changes;
▶ unusual irritability or aggression;
▶ a tendency to become confused;
▶ abnormal fluctuations in concentration and energy;
▶ impaired job performance;
▶ poor time-keeping;
▶ increased short-term sickness leave;
▶ a deterioration in relationships with colleagues, customers or management;
▶ dishonesty and theft (arising from the need to maintain an expensive habit).

A policy on drug abuse can be established by:

1. **Investigation of the size of the problem**
   Examination of sicknesses, behavioural and productivity changes and accident and disciplinary records is a good starting point.
2. **Planning actions**
   Develop an awareness programme for all staff and a special training programme for managers and supervisors. Employees with a drug problem should be encouraged to seek help in a confidential setting.
3. **Taking action**
   Produce a written policy that includes everyone in the organisation and names the person responsible for implementing the policy. It should include details of the safeguards to employees and the confidentiality given to anyone with a drug problem. It should also clearly outline the circumstances in which disciplinary and/or reported action will be taken (the refusal of help, gross misconduct and possession of/dealing in drugs).

7

4. **Monitoring the policy**

The policy can be monitored by checking for positive changes in the measures made during the initial investigation (improvements in the rates of sickness and accidents). Drug screening and testing is a sensitive issue and should only be considered with the agreement of the workforce (except in the case of pre-employment testing). Screening will only be acceptable if it is seen as part of the health policy of the organisation and its purpose is to reduce risks to the misusers and others.

It is important to stress that some drugs are prescribed and controlled. The side effects of these can also affect performance and pose risks to colleagues. Employers should encourage employees to inform them of any possible side effects from prescribed medication and be prepared to alter work programmes accordingly.

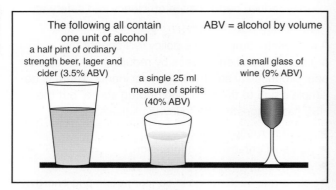

Figure 7.6 It takes a healthy liver about one hour to break down and remove one unit of alcohol. A unit is equivalent to 8 mg or 10 ml (1 cl) of pure alcohol.

7.4 Safe movement of people in the workplace

7.4.1 Hazards in the workplace

The most common hazards to pedestrians at work are slips, trips and falls on the same level, falls from height, collisions with moving vehicles, being struck by moving, falling or flying objects and striking against fixed or stationary objects. Each of these will be considered in turn, including the conditions and environment in which the particular hazard may arise.

Slips, trips and falls on the same level

Slipping and tripping is the single most common cause of injury in the workplace. These are the most common of the hazards faced by pedestrians and account for 38% of all the major accidents every year and 24% of over 3-day injuries reported in the UK to the Health and Safety Executive (HSE). The HSE itself has reported that every 25 minutes in the UK someone breaks or fractures a bone due to slipping, tripping or falling at work.

It has been estimated that the annual cost of these accidents is over £650 million and a direct cost to employers of £512 million. Civil compensation claims are becoming more common and costly to employers. The highest reported injuries are reported in the food and related industries. Older workers, especially women, are the most severely injured group from falls resulting in fractures of the hips and/or femur. Civil compensation claims are now being made by members of the public who have tripped on uneven paving slabs on pavements or in shopping centres. A worker in the UK was awarded compensation after slipping on a wet wheelchair ramp. The accident could have been avoided if the ramp had been covered in an anti-slip material.

Figure 7.7 Tripping hazards

The UK HSE has been so concerned at the large number of such accidents that it has identified slips, trips and falls on the same level as a key risk area. The costs of slips, trips and falls on the same level are high to the injured employee (lost income and pain), the employer (direct and indirect costs including lost production) and to society as a whole in terms of health and social security costs.

Slip hazards are caused by:

▶ wet or dusty floors;
▶ the spillage of wet or dry substances – oil, water, flour dust and plastic pellets used in plastic manufacture;
▶ loose mats on slippery floors;

wet and/or icy weather conditions;

unsuitable footwear or floor coverings or sloping floors.

Trip hazards are caused by:

- loose floorboards or carpets;
- obstructions, low walls, low fixtures on the floor;
- cables or trailing leads across walkways or uneven surfaces; leads to portable electrical hand tools and other electrical appliances (vacuum cleaners and overhead projectors);
- raised telephone and electrical sockets – also a serious trip hazard (this can be a significant problem when the display screen workstations are re-orientated in an office);
- rugs and mats – particularly when worn or placed on a polished surface;
- poor housekeeping – obstacles left on walkways, rubbish not removed regularly;
- poor lighting levels – particularly near steps or other changes in level;
- sloping or uneven floors – particularly where there is poor lighting or no handrails; and
- unsuitable footwear – shoes with a slippery sole or lack of ankle support.

Figure 7.8 Cleaning must be done carefully to prevent slipping

Over half of all trip accidents are caused by poor housekeeping. Many of these housekeeping problems can easily be solved by:

- ensuring that all walkways are suitable for purpose and clear of obstructions (e.g. trailing leads);
- ensuring that workers wear suitable footwear;
- training workers in the maintenance of trip free working areas;
- regular inspections of work areas by supervisors.

The selection of suitable footwear for the workplace, should involve consideration of the types of hazard present, the general environment and ergonomic issues.

The vast majority of major accidents involving slips, trips and falls on the same level result in dislocated or fractured bones.

Falls from work at height

More workplace deaths are triggered by falls from height than any other cause. These are the most common cause of serious injury or death in the construction industry. These accidents are often concerned with falls of greater than about 2 m and often result in fractured bones, serious head injuries, loss of consciousness and death. Twenty-five per cent of all deaths at work and 19% of all major accidents are due to falls from a height. Falls down staircases and stairways, through fragile surfaces, off landings and stepladders and from vehicles, all come into this category. Often injuries, sometimes serious, can also result from falls **below** 2 m, for example using swivel chairs for access to high shelves.

Collisions with moving vehicles

These can occur within the workplace premises or on the access roads around the building. It is a particular problem where there is no separation between pedestrians and vehicles or where vehicles are speeding. Poor lighting, blind corners, and the lack of warning signs and barriers at road crossing points also increase the risk of this type of accident. Eighteen per cent of fatalities at work are caused by collisions between pedestrians and moving vehicles with the greatest number occurring in the service sector (primarily in retail and warehouse activities).

Being struck by moving, falling or flying objects

In the UK, this causes 18% of fatalities at work and is the second highest cause of fatality in the construction industry. It also causes 15% of all major and 14% of over 3-day accidents. Moving objects include articles being moved, moving parts of machinery or conveyor belt systems, and flying objects are often generated by the disintegration of a moving part or a failure of a system under pressure. Falling objects are a major problem in construction (due to careless working at height) and in warehouse work (due to careless stacking of pallets on racking). The head is particularly vulnerable to these hazards. Items falling off high shelves and moving loads are also significant hazards in many sectors of industry.

Striking against fixed or stationary objects

This accounts for over 1,000 major accidents every year in the UK. Injuries are caused to a person either by colliding with a fixed part of the building structure, work in progress, a machine member or a stationary vehicle or by falling against such objects. The head appears to be the most vulnerable part of the body to this particular hazard and this is invariably caused by the misjudgement of the height of an obstacle. Concussion in a mild form is the most common outcome and a medical check-up is normally recommended. It is a very common injury during maintenance operations when there is, perhaps, less familiarity with particular space restrictions around a machine. Effective solutions to all these hazards need not be expensive, time consuming or complicated. Employee awareness and common sense combined with a good housekeeping regime will solve many of the problems.

Maintenance activities

Maintenance-related accidents are a serious cause of concern. For example, analysis of data from recent years indicates that 25–30% of manufacturing industry fatalities in Great Britain were related to maintenance activity. Maintenance work is often more hazardous than other types of work because it is often performed under pressure – possibly due to a plant failure during a production run – or the areas requiring access are difficult to reach.

The majority of recent accidents to workers engaged in maintenance work occurred during the following activities:

▶ roofwork;
▶ machinery and plant maintenance;
▶ building maintenance including painting work;
▶ maintenance of services (gas, electricity, telecommunications and water);
▶ vehicle repair work; and
▶ window and building cleaning.

The main hazards associated with maintenance work are due to:

▶ electricity – electric shock;
▶ mechanical, including entanglement, ejection and unexpected start-up;
▶ hazardous substances – gases, dusts and fume;
▶ physical – noise and vibration;
▶ obstructions of various types;
▶ access – confined spaces and work at height; and
▶ unexpected releases of pressure or other forms of energy.

Inspections and even thorough examinations are not substitutes for properly maintaining equipment. The information gained in the maintenance work, inspections and thorough technical examinations should inform one another. A maintenance log should be kept and be up to date. The whole maintenance system will require proper management systems. The frequency will depend on the equipment, the conditions in which it is used and the manufacturers' instructions.

7.4.2 Control measures for the safe movement of people in the workplace

Slips, trips and falls on the same level

These may be prevented or, at least, reduced by several control strategies. These and all the other pedestrian hazards discussed should be included in the workplace risk assessments by identifying slip or trip hazards, such as poor or uneven floor/pavement surfaces, badly lit stairways and puddles from leaking roofs. All floors should be suitable for the workplace, in good condition and free from obstructions. Traffic routes must be organised to enable people to move around the workplace safely.

The key elements of a health and safety management system are as relevant to these as to any other hazards:

Planning – remove or minimise the risks by using appropriate control measures and defined working practices (e.g. covering all trailing leads).

Organisation – involve employees and supervisors in the planning process by defining responsibility for keeping given areas tidy and free from trip hazards.

Control – record all cleaning and maintenance work. Ensure that anti-slip covers and cappings are placed on stairs, ladders, catwalks, kitchen floors and smooth walkways. Use warning signs when floor surfaces have recently been washed.

Monitoring and review – carry out regular safety audits of cleaning and housekeeping procedures and include trip hazards in safety surveys. Check on accident records to see whether there has been an improvement or if an accident black spot can be identified.

Slip and trip accidents are a major problem for large retail stores for both customers and employees. The provision of non-slip flooring, a good standard of lighting and minimising the need to block aisles during the re-stocking of merchandise are typical measures that many stores use to reduce such accidents. Other measures include the wearing of suitable footwear by employees, adequate handrails on stairways, the highlighting of any floor level changes and procedures to ensure a quick and effective response to any reports of floor damage or spillages. Good housekeeping procedures are essential. The design of the store layout and any associated warehouse can also ensure a reduction in all types of accident. Many of these measures are valid for a range of workplaces.

Cleaning arrangements should be chosen to suit both the type of floor and the users. Floors with surface roughness are not difficult to clean despite the popular belief that they are. Stairs are a particularly hazardous part of the building and become even more so when being cleaned.

The prevention of access during cleaning or drying needs to be effective. Signs that warn people that floors are wet are only partially effective. The HSE make the following suggestions:

▶ Physically exclude people from wet cleaning areas by the use of barriers or locking off an area.
▶ Clean during quiet hours.
▶ Clean in sections so there is always a dry path through the area.

Floor cleaners must be provided with appropriate slip-resistant footwear which can help reduce the risk of slipping.

Stairs can also present significant potential for harm to their users, with falls on stairs often leading to serious injury or even death. Around 20% of all major injuries reported to the HSE each year from slips and trips occurred on stairs.

Falls from work at height

The principal means of preventing falls of people or materials includes the use of fencing, guard rails, toe boards, working platforms, access boards, ladder hoops, safety nets and safety harnesses. Safety harnesses arrest the fall by restricting the fall to a given distance due to the fixing of the harness to a point on an adjacent rigid structure. They should only be used when all other possibilities are not reasonably practicable. The use of banisters on open sides of stairways and handrails fitted on adjacent walls will also help to prevent people from falling. Holes in floors and pits should always be fenced or adequately covered. Precautions should be taken when working on fragile surfaces.

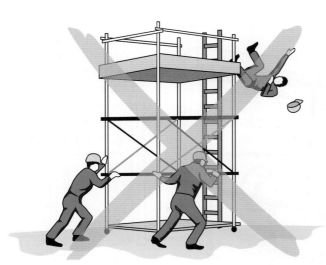

Figure 7.9 Falling from a height – tower scaffold with inadequate handrails (no centre rail) and should never be moved when in use

Permanent staircases are also a source of accidents included within this category of falling from a height and the following design and safety features will help to reduce the risk of such accidents:

▶ adequate width of the stairway, depth of the tread and provision of landings and banisters or handrails and intermediate rails; the treads and risers should always be of uniform size throughout the staircase and designed to meet the national building standards for angle of incline (i.e. steepness of staircase);
▶ provision of non-slip surfaces and reflective edging;
▶ adequate lighting;
▶ adequate maintenance;
▶ special or alternative provision for disabled people (e.g. personnel elevator at the side of the staircase).

Great care should be used when people are loading or unloading vehicles; as far as possible people should avoid climbing on to vehicles or their loads. For example, sheeting of lorries should be carried out in designated places using properly designed access equipment.

Collisions with moving vehicles

These are best prevented by completely separating pedestrians and vehicles, providing well-marked, protected and laid-out pedestrian walkways. People should cross roads by designated and clearly marked pedestrian crossings. Suitable guard rails and barriers should be erected at entrances and exits from buildings and at 'blind' corners at the end of racking in warehouses. Particular care must be taken in areas where lorries are being loaded or unloaded. It is important that separate doorways are provided for pedestrians and vehicles and all such doorways should be provided with a vision panel and an indication of the safe clearance height, if used by vehicles. Finally, the enforcement of a sensible speed limit, coupled, where practicable, with speed governing devices, is another effective control measure.

Being struck by moving, falling or flying objects

These hazards may be prevented by guarding or fencing the moving part (as discussed in Chapter 10) or by adopting the measures outlined later in this chapter (see 7.5.9). Both construction workers and members of the public need to be protected from the hazards associated with falling objects. Both groups should be protected by the use of covered walkways or suitable netting to catch falling debris where this is a significant hazard. Waste material should be brought to ground level by the use of chutes or hoists. Waste should not be thrown from a height and only minimal quantities of building materials should be stored on working platforms. Appropriate personal protective equipment, such as hard hats or safety glasses, should be worn at all times when construction operations are taking place.

Striking against fixed or stationary objects

This hazard can only be effectively controlled by:

▶ having good standards of lighting and housekeeping;

▶ defining walkways and making sure they are used;

▶ the use of awareness measures, such as training and information in the form of signs or distinctive colouring;

▶ the use of appropriate personal protective equipment, such as head protection, as discussed previously.

General preventative measures for pedestrian hazards

Minimising pedestrian hazards and promoting good work practices requires a mixture of sensible planning, good housekeeping and common sense. Few of the required measures are costly or difficult to introduce and, although they are mainly applicable to slips, trips and falls on the same level and collisions with moving vehicles, they can be adapted to all types of pedestrian hazard. Typical measures include the following:

▶ Develop a safe workplace as early as possible and ensure that suitable floor surfaces and lighting are selected and vehicle and pedestrian routes are carefully planned. Lighting should not dazzle approaching vehicles nor should pedestrians be obscured by stored products. Lighting is very important where there are changes of level or stairways. Any physical hazards, such as low beams, vehicular movements or pedestrian crossings, should be clearly marked. Staircases need particular

attention to ensure that they are slip resistant and the edges of the stairs marked to indicate a trip hazard.

▶ Consider pedestrian safety when re-orientating the workplace layout (e.g. the need to reposition lighting and emergency lighting).

▶ Adopt and mark designated walkways.

▶ Apply good housekeeping principles by keeping all areas, particularly walkways, as tidy as possible and ensure that any spillages are quickly removed.

▶ Ensure that all workers are suitably trained in the correct use of any safety devices (such as machine guarding or personal protective equipment) or cleaning equipment provided by the employer.

▶ Only use cleaning materials and substances that are effective and compatible with the surfaces being cleaned, so that additional slip hazards are not created.

▶ Ensure that a suitable system of maintenance, cleaning, fault reporting and repair are in place and working effectively. Areas that are being cleaned must be fenced and warning signs erected. Care must also be taken with trailing electrical leads used with the cleaning equipment. Records of cleaning, repairs and maintenance should be kept.

▶ Ensure that all workers are wearing high visibility clothing and appropriate footwear with the correct type of slip-resistant soles suitable for the type of flooring.

▶ Consider whether there are significant pedestrian hazards present in the area when any workplace risk assessments are being undertaken.

Figure 7.10 Typical pedestrian/vehicle crossing area

Figure 7.11 Walkway and traffic route separated with kerb and partial rails. Traffic movement stopped during factory visit.

It can be seen, therefore, that floors and traffic routes should be of sound construction. If there are frequent, possibly transient, slip hazards, the provision of slip-resistant coating and/or mats should be considered and warning notices posted. Any damaged areas must be cordoned off until repairs are completed. Risk assessments should review past accidents and near misses to enable relevant controls, such as suitable footwear, to be introduced. Employees can often indicate problem areas, so employee consultation is important.

Electrically powered gates

The UK HSE has recently issued a warning on the risks to pedestrians from crushing zones on electrically powered gates, such as those commonly used at car parks. Two young children suffered fatal injuries after being trapped by automatic sliding gates that were between the closing edge of the gate and the gate post at the end of the gates' travel. They were trapped because:

▶ Their presence in the vicinity of the closing edge was not detected; and
▶ The closing force of the gate when they obstructed it was not limited to the values specified in Annex A of BS EN 12453:2001.

The HSE Safety Notice reinforced and updated previous information to organisations and individuals involved in the design, construction, installation and commissioning of electrically powered gates and organisations in control of their use and/or maintenance. It is also relevant to companies carrying out ongoing maintenance of these types of gates. The British Standard (BS EN 12453:2001) recommends a minimum level of safeguarding against the crushing hazard at the closing edge of the gate depending on the type of environment in which the gate is operating. For electrically powered gates that operate in an

area accessible to the public, the advised level of safeguarding is to:

▶ Limit forces according to Annex A of the Standard using force limitation devices or sensitive protective equipment AND provide a means for the detection of the presence of a person or an obstacle standing on the floor at one side of the gate; or
▶ Provide a means for detection of the presence a person, which is designed in a way that in no circumstances can that person be touched by the moving gate leaf.

The HSE recommend the following actions:

▶ All designers and installers of electrically powered gates should ensure that the forces generated by a gate when meeting a person or an obstacle are limited and that they do not exceed the values specified in Annex A of BS EN 12453:2001.
▶ Forces should be periodically re-measured and checked as part of the planned preventative maintenance schedule for the gates.
▶ In addition to force limitation, additional safeguards, such as pressure sensitive strips on the closing edge and photoelectric sensing devices, should be fitted where the risk assessment identifies the gate as high risk, in that it is operating automatically in a public place where children and other members of the public may be present.
▶ Persons or organisations in control of powered gates should periodically review their risk assessments to ensure that they identify any changes to the environment or operating conditions and that they have taken appropriate steps to address them. This is particularly important when the responsibility for management of the gate passes from one person or organisation to another.
▶ All safety devices and features should be checked on a regular basis and in accordance with the manufacturer's instructions to ensure they continue to function as designed to ensure that safety is maintained. This should be specified in a planned preventative maintenance schedule agreed by persons responsible for the gate's management and their appointed maintenance company.

7.5 Working at height

For many countries, including the UK, there is no minimum height requirement for work at height although those countries that comply with the Occupational Safety and Health Administration (OHSA) have defined work at height as work above 2 m. Generally, work at height includes all work activities where there is a need to control a risk of falling a distance liable to cause personal injury. This is regardless of the work equipment being used, the duration of the work at height involved or the height at which the work is performed. It includes access to and egress from a place of work and therefore includes:

▶ working on a scaffold or from a mobile elevating work platform (MEWP);
▶ sheeting a lorry or dipping a road tanker;
▶ working on the top of a container in docks or on a ship or storage area;
▶ tree surgery and other forestry work at height;
▶ using cradles or rope for access to a building or other structure like a ship under repair;
▶ climbing permanent structures like a gantry or telephone pole;
▶ working near an excavation area or cellar opening if a person could fall into it and be injured;
▶ painting or pasting and erecting bill posters at height;
▶ work on staging or trestles, for example for filming or events;
▶ using a ladder/stepladder or kick stool for shelf filling, window cleaning and the like;
▶ using manriding harnesses in ship repair, or offshore or steeple jack work;
▶ working in a mine shaft or chimney;
▶ work carried out at a private house by a person employed for the purpose, for example a painter and decorator (but not if the private individual carries out work on their own home).

However, it would not include:

▶ slips, trips and falls on the same level;
▶ falls on permanent stairs if there is no structural or maintenance work being done;
▶ work in the upper floor of a multi-storey building where there is no risk of falling (except separate activities like using a stepladder).

7.5.1 Hazards and controls associated with working above ground level

The significance of injuries resulting from falls from height, such as fatalities and other major injuries, has been dealt with earlier in the chapter. Also covered were the many hazards involved in working at height, including fragile roofs and the deterioration of materials, unprotected edges and falling materials. Additional hazards include the weather and unstable or poorly maintained access equipment, such as ladders and various types of scaffold.

The employer should apply a three-stage hierarchy to all work which is to be carried out at height. The three steps are the avoidance of work at height, the prevention of workers from falling and the mitigation of the effect on workers of falls should they occur. It follows from this hierarchy that:

▶ work is not carried out at height when it is reasonably practicable to carry the work out safely other than at height (e.g. the assembly of components should be done at ground level);

▶ when work is carried out at height, the employer shall take suitable and sufficient measures to prevent, so far as is reasonably practicable, any person falling a distance liable to cause injury (e.g. the use of guard rails);
▶ the employer shall take suitable and sufficient measures to minimise the distance and consequences of a fall (collective measures, for example airbags or safety nets, must take precedence over individual measures, for example safety harnesses).

A risk assessment will be needed to determine the controls required for any particular work at height. The risk assessment and action required to control risks from using a kick stool to collect books from a shelf should be simple (not overloading, not overstretching, etc.). However, the action required for a complex construction project would involve significantly greater consideration and assessment of risk.

A risk assessment for working at height should first consider whether the work could be avoided. If this is not possible, then the risk assessment should consider the following issues:

▶ the nature and duration of the work;
▶ type of roof: flat, sloping or pitched;
▶ the competence level of all those involved with the work and any additional training requirements;
▶ the required level of supervision;
▶ use of guard rails, toe boards, working platforms and means of access and egress;
▶ provision of foot and hand holds;
▶ required personal protective equipment (PPE), such as helmets and harnesses;
▶ the presence of fall arrest systems, such as netting or soft landing systems;
▶ the health status of the workers;
▶ the possible weather conditions;
▶ compliance with the relevant national legislation;
▶ protection for those at ground level;
▶ adequate barriers and edge protection; and
▶ identification of fragile material.

The principal means of preventing falls of people or materials includes the use of fencing, guard rails, toe boards, working platforms, access boards, ladder hoops, safety nets and safety harnesses. Safety harnesses arrest the fall by restricting the fall to a given distance due to the fixing of the harness to a point on an adjacent rigid structure. They should only be used when all other possibilities are not practical.

7.5.2 Protection against falls from work at height

Those responsible for work at height should ensure that:

▶ all work at height is properly planned and organised;
▶ those involved in work at height are competent;

▶ the risks from work at height are assessed, and appropriate work equipment is selected and used;
▶ the risks of working on or near fragile surfaces are properly managed; and
▶ the equipment used for work at height is properly inspected and maintained.

The following hierarchy of control measures should be used for working at height and to ensure that appropriate access equipment is used:

▶ Eliminate or avoid working at height.
▶ Work from an existing safe workplace provided, such as a properly constructed working platform, complete with toe boards and guard rails.
▶ Ensure that there is sufficient work equipment or other measures to minimise the distance and consequence of a fall by the use of:
  ▷ collective measures rather than individual measures (e.g. a handrail instead of a harness); or where this is not practicable
  ▷ collective fall arrest equipment (airbags or safety nets); or where this is not practicable
  ▷ individual fall restrainers (safety harnesses) should be used.
▶ Provide supervision, training and instruction.
▶ Only when none of the above measures is practicable, should ladders or stepladders be considered.

Safe working at height, in Europe and the UK, requires that guard rails on scaffolds are at a minimum of 950 mm and the maximum unprotected gap between the toe and guard rail of a scaffold is 470 mm (these distances vary from country to country). This implies the use of an intermediate guard rail although other means, such as additional toe boards or screening, may be used. They also specify requirements for personal suspension equipment and means of arresting falls (such as safety nets).

Much of the work which is done at height could often be done, or partly done, at ground level – thus avoiding the hazards of working at height. The partial erection of scaffolding or edge protection at ground level and the use of cranes to lift it into place at height are examples of this. The manufacture of complete window frames in a workshop and then the final installation of the frame into the building is another. By the use of suitable extension equipment high windows can be cleaned from the ground and high walls can be painted from the ground. However, in most construction work at height, the work cannot be done at ground level and suitable control measures to address the hazards of working at height will be required.

7.5.3 Fragile roofs and surfaces

Roof work, particularly work on pitched roofs, is hazardous and requires a specific risk assessment and method statement (see later under the management

of construction activities for a definition) prior to the commencement of work. Particular hazards are fragile roofing materials, including those materials which deteriorate and become more brittle with age and exposure to sunlight, exposed edges, unsafe access equipment and falls from girders, ridges or purlins. There must be suitable means of access such as scaffolding, ladders and crawling boards; suitable barriers, guard rails or covers where people work near to fragile materials and roof lights; and suitable warning signs indicating that a roof is fragile should be on display at ground level.

Where possible, work on a fragile roof should be avoided by doing the following:

▶ work from underneath the roof using a suitable work platform or
▶ where this is not possible, use a mobile elevating work platform that allows people to work from within the basket without having to stand on the roof.

If access on to the fragile roof cannot be avoided, perimeter edge protection should be installed and staging used to spread the load. Unless all the work and access is on staging or platforms that are fitted with guard rails then safety nets should be installed underneath the roof or a harness system used. Where a harness is used, adequate anchorage points will be required.

A roof should always be treated as fragile until a competent person has ruled otherwise. Fragility can be caused by:

▶ general deterioration of the roof through ageing, neglect and lack of maintenance;
▶ corrosion of cladding and fixings;
▶ quality of the original installation and selection of materials;
▶ thermal and impact damage;
▶ deterioration of the supporting structure; and
▶ weather damage.

Asbestos cement sheets and old roof lights should always be treated as fragile.

Figure 7.12 (a) A mobile elevating work platform being used on a fragile roof to replace a roof sheet

Figure 7.12 (b) Workmen wearing harnesses attached to a work positioning line which is fitted to the staging

There are other hazards associated with roof work – overhead services and obstructions, the presence of asbestos or other hazardous substances, the use of equipment such as gas cylinders and bitumen boilers and manual handling hazards.

The following roofing materials are likely to be fragile:

▶ fibre-cement sheets – non-reinforced sheets irrespective of profile type;
▶ liner panels – on built-up sheeted roofs;
▶ metal sheets – where corroded;
▶ glass – including wired glass;
▶ chipboard – or similar material where rotted; and
▶ roof lights.

It is essential that only trained and competent persons are allowed to work on roofs and that they wear footwear having a good grip. It is a good practice to ensure that a person does not work alone on a roof or if slippery conditions exist on the roof. Work should not be undertaken on any roof in poor weather conditions such as rain, ice, frost or strong winds (particularly gusting) in excess of 23mph (Force 5). Strong winds will affect the balance of a roof worker.

Guidance from the UK Advisory Committee for Roof Work states that any person undertaking roof work needs to be 'both mentally and physically fit, competent to do the work, and be fully aware of all the dangers that exist and the actions necessary to overcome those dangers'. The HSE has published the third edition of a guidance booklet on safety during roof work – *HSG33, Health and safety in roof work.*

A roof ladder should only be used if more suitable equipment cannot be used and should only be used for low risk, short duration work. Roof ladders should be of an industrial grade, in good condition and secured to prevent movement. The anchorage at the top of the roof ladder should be by some method which does not depend on the ridge capping, as this is liable to break away from the ridge. The anchorage should

bear on the opposite slope by a properly designed and manufactured ridge hook or be secured by other means.

7.5.4 Protection against falling objects

Both construction workers and members of the public need to be protected from the hazards associated with falling objects. Both groups should be protected by the use of covered walkways or suitable netting to catch falling debris. Waste material should be brought to ground level by the use of chutes or hoists. Waste should not be thrown and only minimal quantities of building materials should be stored on working platforms.

The ILO Code of Practice 'Safety and Health in Construction' recommends that employers should supply head protection (hard hats) to employees whenever there is a risk of head injury from falling objects and ensure that the hard hats are properly maintained and replaced when they are damaged in any way. Self-employed workers should supply and maintain their own head protection. Visitors to construction sites should always be supplied with head protection and head protection signs must be displayed around the site.

7.5.5 Fall arrest equipment

The three most common types of fall arrest equipment are safety harnesses, safety nets and air bags.

Safety harnesses should only be used alone when conventional protection, using guard rails, is no longer practicable. Such conditions occur when it is possible to fall 2 m or more from an open edge. It is important that the following points are considered when safety harnesses are to be used:

1. The length of fall only is reduced by a safety harness. The worker may still be injured due to the shock load applied to him when the fall is arrested. A free fall limit of about 2 m is maintained to reduce this shock loading. Lanyards are often fitted with shock absorbers to reduce the effect of the shock loading.
2. The worker must be attached to a secure anchorage point before they move into an unsafe position. The lanyard should always be attached above the worker, whenever possible.
3. Only specifically trained and competent workers should attach lanyards to anchorage points and work in safety harnesses. Those who wear safety harnesses must be able to undertake safety checks and adjust the harness before it is used.

Three levels of inspection are recommended for safety harnesses as follows:

1. A pre-use check undertaken by the user at the beginning of each shift to check there are no visible or surface defects.

Figure 7.13 Fall arrest harness and device

2. A detailed inspection undertaken at least every six months. However, for frequently used equipment, this should be increased to at least every three months, particularly when the equipment is used in arduous environments (e.g. demolition, steel erection, scaffolding, steel masts/towers with edges). The results of the inspection should be formally recorded.

3. An interim inspection is also an in-depth, recorded inspection and may be needed between detailed inspections because the associated risk assessment has identified a risk that could result in significant deterioration, affecting the safety of the lanyard before the next detailed inspection is due. The need for and frequency of interim inspections will depend on use. Examples of situations where they may be appropriate include risks from arduous working environments involving paints, chemicals or grit blasting.

Any defects discovered by any of these inspections must be reported to the employer as soon as possible.

Safety nets are widely used to arrest falls of people, tools and materials from height but competent installation is essential. The correct tensioning of the net is important and normally specialist companies are available to fit nets. The popularity of nets has grown in the UK since the UK Construction Regulations came into force in 1996 and the subsequent advocacy of their

use by the UK HSE. Nets are used for roofing work and for some refurbishment work. Nets, however, have a limited application since they are not suitable for use in low-level construction where there is insufficient clearance below the net to allow it to deflect the required distance after impact. Nets should be positioned so that workers will not fall more than 2 m, in case they hit the ground or other obstructions.

Air bags are used when it is either not possible or practical to use safety nets. Therefore they are used extensively in domestic house building or when it is difficult to position anchorage points for safety harnesses. When air bags are used, it is important to ensure that the bags are of sufficient strength and the air pressure high enough to ensure that any falling person does not make contact with the ground. Only reputable suppliers should be employed for the provision of air bags. Air bags or bean bags are known as soft landing systems and are used to protect workers from the effects of inward falls. Other possible solutions to this problem of the inward fall are the use of internal scaffolding or lightweight 'crash deck systems'.

Air bags may be linked together to form an inflated crash deck system. Such a system is suitable when a safety net is not viable. It consists of a series of interlinked air mattresses that are positioned beneath the working area and is suitable for working at height inside a building where safety harnesses would not be practicable. The mattresses are interconnected by secure couplings and inflated in position on site using an air pump. The mattresses are made in various sizes so that any floor area configuration can be covered. A similar form of crash deck can be made from bean bags that are clipped together.

After the completion of the construction project, cable-based fall arrest systems are normally suitable for ongoing building maintenance work. They may also be used for the construction of complex roof structures, such as parabolic or dome structures, where a safety net may be more than 2 m below the highest point of the structure.

7.5.6 Emergency procedures (including rescue)

A suitable emergency and rescue procedure or plan needs to be in place for situations that could be expected to occur on the construction site. Such foreseeable situations could include a crane driver trapped in his cab due to a power failure or a serious ill-health problem, or the rescue of a person who has fallen into a safety net. The employer is responsible for such a plan and reliance should not be placed on others, such as the emergency medical services. The method of rescue may be simple and straightforward, such as putting a ladder up to a net and allowing the fallen person to descend or the use of a MEWP to lower the person to the ground.

201

The UK Work at Height Safety Association (WAHSA) recommends that an appropriate rescue plan should consider:

- the safety of the persons undertaking or assisting with the rescue;
- the anchor points to be used for the rescue equipment;
- the suitability of equipment (anchors, harnesses, attachments and connectors) that has already arrested the fall of the person for use during the rescue;
- the method that will be used to attach the person to the rescue system;
- the direction in which the person needs to be moved to get them to the point of safety; and
- the possible needs of the person after the rescue, including first-aid.

The UK HSE guidance states that the key is to release the person from the suspended-harness position to the nearest point of safety, in the shortest possible time, and as soon as is safely possible. Otherwise, the person can suffer from **presyncope** (loss of consciousness). The symptoms for this include light-headedness, nausea, sensations of flushing, tingling or numbness of the arms or legs, anxiety, visual disturbance, or a feeling they are about to faint. Presyncope can occur within one hour (or sooner for some people) and if the rescuer is unable to immediately release the conscious person, elevation of the legs by the person or rescuer may prolong the tolerance of suspension.

Proprietary rescue systems are available and may well be suitable for many more complex rescues. The system selected needs to be proportionate to the risk and there should not be undue reliance on the emergency services. However, arrangements should be in place to inform the emergency services of any serious incident. It is important that rescue teams are available at all times with a designated leader. Instruction and training are essential for the teams, preferably by simulated rescue exercises.

7.5.7 Safe working practices for common types of access equipment

There are many different types of access equipment, but only the following four categories will be considered here:

- ladders
- fixed scaffold
- pre-fabricated mobile scaffold towers
- mobile elevating work platforms (MEWPs).

Ladders

The main cause of accidents involving ladders is ladder movement while in use. This occurs when they have not been secured to a fixed point, particularly at the foot. Other causes include over-reaching by the worker, slipping on a rung, ladder defects and, in the case of metal ladders, contact with electricity. The main types of accident are falls from ladders.

There are three common materials used in the construction of ladders: aluminium, timber and glass fibre. Aluminium ladders have the advantage of being light but should not be used in high winds or near live electricity. Timber ladders need regular inspection for damage and should not be painted, as this could hide cracks. Glass fibre ladders can be used near electrical equipment and in food processing areas.

Every time a ladder is used, a pre-use check should be made. Such a check should be undertaken by the user to check the condition of the feet and rungs at the beginning of the working day or whenever the ladder is dropped or moved from a dirty area to a clean area.

The following factors should be considered when using ladders:

- Undertake as much work as possible from the ground.
- Ensure that the equipment is suitable, stable and strong enough for the job, maintained and checked regularly.
- Ensure that the use of a ladder is the safest means of access given the work to be done and the height to be climbed.
- The location itself needs to be checked. The supporting wall and supporting ground surface should be dry and slip free. Extra care will be needed if the area is busy with pedestrians or vehicles or the ladder is rested against weak upper surfaces, such as glazing or plastic gutters.
- The ladder needs to be stable in use. This means that the inclination should be as near the optimum as possible (1:4 ratio of distance from the wall to distance up the wall).
- Wherever possible, a ladder should be tied to prevent it from slipping. This can either be at the top, the bottom or both, making sure both stiles are tied. Never tie a ladder by its rungs.
- If the ladder cannot be tied, use an 'effective ladder' or one with an 'effective ladder-stability device' that the suppliers or manufacturers can confirm is stable enough to use unsecured in the worst-case scenario.
- If the above two precautions are not possible, then the ladder stiles can be wedged against a wall or other similar heavy object; or, as a last resort, have a second person 'foot' the ladder.
- Weather conditions must be suitable (no high winds or heavy rain).
- The proximity of live electricity should also be considered, particularly when ladders are carried near or under power lines.
- There should be at least 1 m of ladder above the stepping off point.
- The work activity must be considered in some detail. Over-reaching must be eliminated and consideration

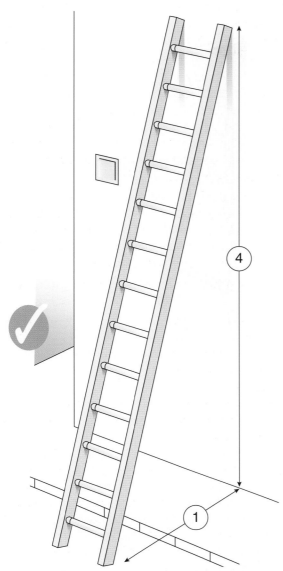

Figure 7.14 Ladder showing correct 1 in 4 angle (means of securing omitted for clarity)

given to the storage of paints or tools which are to be used from the ladder and any loads to be carried up the ladder. The ladder must be matched to work required.

▶ Workers who are to use ladders must be trained in the correct method of use and selection. Such training should include the use of both hands during climbing, clean non-slippery footwear, clean rungs and an undamaged ladder.

▶ Ladders should be inspected (particularly for damaged or missing rungs) and maintained on a regular basis and they should only be repaired by competent persons.

▶ The transportation and storage of ladders is important as much damage can occur at these times. They need to be handled carefully and stored in a dry place.

▶ When a ladder is left secured to a structure during non-working hours, a plank should be tied to

the rungs to prevent unauthorised access to the structure.

▶ Emergency evacuation and rescue procedures should be considered.

Certain work should not be attempted using ladders. This includes work where:

▶ a secure handhold is not available;
▶ the work is at an excessive height;
▶ the ladder cannot be secured or made stable;
▶ the work is of long duration;
▶ the work area is very large;
▶ the equipment or materials to be used are heavy or bulky;
▶ the weather conditions are adverse;
▶ there is no protection from passing vehicles.

Ladders may be used for access when a risk assessment shows that the risk of injury is low and the task is of short duration or there are unalterable features of the work site and that it is not reasonably practicable to use potentially safer alternative means of access such as a MEWP or a mobile tower scaffold. There is no maximum height for using a ladder. However, where a ladder rises 9 m or more above its base, landing areas or rest platforms should be provided at suitable intervals. Ladder guidance from the UK HSE recommends that users should maintain three points of contact when climbing a ladder and wherever possible at the work position. The three points of contact are a hand and two feet.

Stepladders, trestles and staging

Many of the points discussed for ladders apply to stepladders and trestles, where stability and over-reaching are the main hazards.

All equipment must be checked by the supervisor before use to ensure that there are no defects and must be checked at least weekly whilst in use on site. If a defect is noted, or the equipment is damaged, it must be taken out of use immediately. Any repairs must only be carried out by competent persons.

Supervisors must also check that the equipment is being used correctly and not being used where a safer method should be provided.

Where staging, such as a Youngmans staging platform, is being used in roof areas, supervisors must ensure that only experienced operatives are permitted to carry out this work and that all necessary safety harnesses and anchorage points are provided and used.

The main hazards associated with stepladders, trestles and staging are:

▶ unsuitable base (uneven or loose materials);
▶ unsafe and incorrect use of equipment (e.g. the use of staging for barrow ramps);
▶ overloading;
▶ use of equipment where a safer method should be provided;

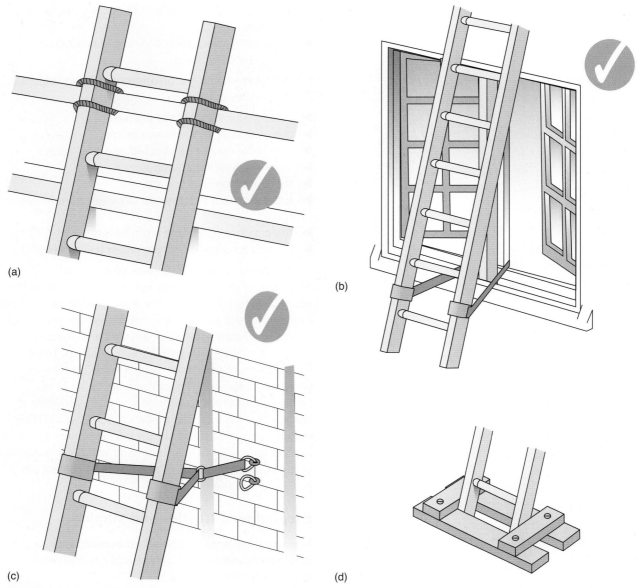

Figure 7.15 (a) Ladder tied at top stiles (correct for working on, not for access); (b) tying part way down; (c) tying at base; (d) securing at the base

▶ overhang of boards or staging at supports ('trap ends');

▶ use of defective equipment.

Stepladders and trestles must be:

▶ manufactured to a recognised industrial specification;

▶ stored and handled with care to prevent damage and deterioration;

▶ subject to a programme of regular inspection (there should be a marking, coding or tagging system to confirm that the inspection has taken place);

▶ checked by the user before use;

▶ taken out of use if damaged, and destroyed or repaired;

▶ used on a secure surface, and with due regard to ensuring stability at all times;

▶ kept away from overhead cables and similar hazards.

The small platform fitted at the top of many stepladders is designed to support tools, paint pots and other working materials. It should not be used as a working place unless the stepladder has been constructed with a suitable handhold above the platform. Stepladders must not be used if they are too short for the work being undertaken, or if there is not enough space to open them out fully. If, when on a stepladder, two hands need to be free for a brief period of light work (e.g. to change a light bulb), keep two feet on the same step and the knees and chest supported by the stepladder to maintain three points of contact.

Platforms based on trestles should be fully boarded, adequately supported (at least one support for each 1.5 m of board for standard scaffold boards) and provided with edge protection when there is a risk of falling a distance liable to cause injury.

✗ Wrong way

Right way ✓

✗ Stepladder too short
✗ Hazard overhead
✗ Over-reaching up and sideways
✗ No grip on ladder
✗ Sideways-on to work
✗ Foot on handrail
✗ Wearing slippers
✗ Loose tools on ladder
✗ Slippery and damaged steps
✗ Uneven soft ground
✗ Damaged stiles
✗ Non-slip rubber foot missing

Steps at right height ✓
No need to over-reach ✓
Good grip on handrail ✓
Working front-on ✓
Wearing good flat shoes ✓
Clean undamaged steps ✓
Firm level base ✓
Undamaged stiles ✓
Rubber non-slip feet all in position ✓
Meets British or European standards ✓

Figure 7.16 Working with stepladders

Fixed scaffolds

It is quicker and easier to use a ladder as a means of access, but it is not always the safest. Jobs, such as painting, gutter repair, demolition work or window replacement, are often easier done using a scaffold. If the work can be completed comfortably using ladders, a scaffold need not be considered. Scaffolds must be capable of supporting building workers, equipment, materials, tools and any accumulated waste. A common cause of scaffold collapse is the 'borrowing' of boards and tubes from the scaffold, thus weakening it. Falls from scaffolds are often caused by badly constructed working platforms, inadequate guard rails or climbing up the outside of a scaffold. Falls also occur during the assembly or dismantling process.

There are two basic types of external scaffold:

▶ **Independent tied** – These are scaffolding structures which are independent of the building but tied to it often using a window or window recess. This is the most common form of scaffolding.
▶ **Putlog** – This form of scaffolding is usually used during the construction of a building. A putlog is a scaffold tube which spans horizontally from the scaffold into the building – the end of the tube is flattened and is usually positioned between two brick courses.

The important components of a scaffold are as follows.

▶ **Standard** – An upright tube or pole used as a vertical support in a scaffold.
▶ **Ledger** – A tube spanning horizontally and tying standards longitudinally.
▶ **Transom** – A tube spanning across ledgers to tie a scaffold transversely. It may also support a working platform.

▶ **Bracing** – Tubes which span diagonally to strengthen and prevent movement of the scaffold.
▶ **Guard rail** – A horizontal tube fitted to standards along working platforms to prevent persons from falling.
▶ **Toe boards** – These are fitted at the base of working platforms to prevent persons, materials or tools falling from the scaffold.
▶ **Base plate** – A square steel plate fitted to the bottom of a standard at ground level.
▶ **Sole board** – Normally a timber plank positioned beneath at least two base plates to provide a more uniform distribution of the scaffold load over the ground.
▶ **Ties** – Used to secure the scaffold by anchoring it to the building. The scaffold in Figure 7.17(a) is tied to the building using a through-tie.
▶ **Working platform** – An important part of the scaffold, as it is the platform on which the building workers operate and where building materials are stored prior to use. These are laid on the transoms in various quantities; usually for general purpose, they should be four boards wide. A working platform can be almost any surface from which work can be undertaken, such as:
  ▷ a roof
  ▷ a floor
  ▷ a platform on a scaffold
  ▷ mobile elevating work platforms (MEWPs)
  ▷ the treads of a stepladder.

Other components of a scaffold include access ladders, brick or block guards and chutes to dispose of waste materials.

The following factors must be addressed if a scaffold is being considered for use for construction purposes:

7

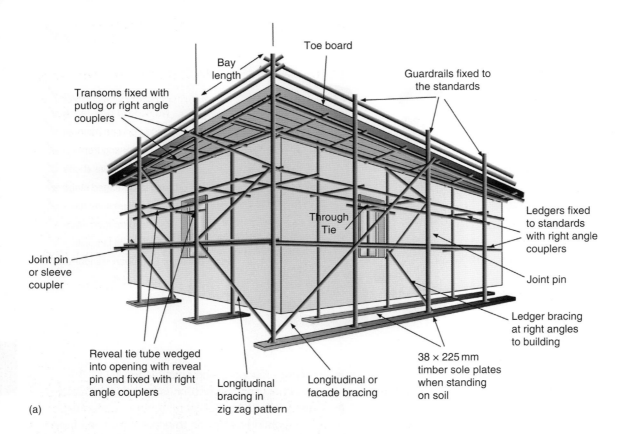

Transoms fixed with putlog or right angle couplers

Bay length

Toe board

Guardrails fixed to the standards

Through Tie

Ledgers fixed to standards with right angle couplers

Joint pin

Joint pin or sleeve coupler

Ledger bracing at right angles to building

Reveal tie tube wedged into opening with reveal pin end fixed with right angle couplers

Longitudinal bracing in zig zag pattern

Longitudinal or facade bracing

38 × 225 mm timber sole plates when standing on soil

(a)

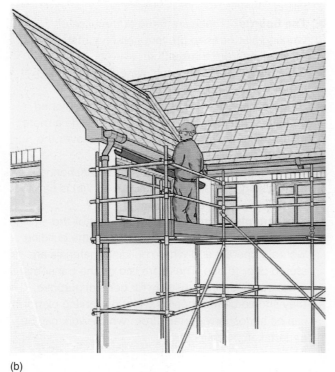

(b)

Figure 7.17 (a) Typical independent tied scaffold; (b) fixed scaffold left in place to fit the gutters

▶ Scaffolding must only be erected by competent people who have attended recognised training courses. Any work carried out on the scaffold must be supervised by a competent person. Any changes to the scaffold must be done by a competent person.

▶ Adequate toe boards, guard rails and intermediate rails must be fitted to prevent people or materials from falling. In Europe and many other countries, the top guard rail should be a minimum of 950 mm above the working platform and any gap between the top rail and the intermediate rail should not exceed 470 mm. The toe boards need to be suitable and sufficient to prevent people or materials from falling.

▶ The scaffold must rest on a stable surface; uprights should have base plates and timber sole plates if necessary.

▶ The scaffold must have safe access and egress.

▶ Work platforms should be fully boarded with no tipping or tripping hazards.

▶ The scaffold should be sited away from or protected from traffic routes so that it is not damaged by vehicles.

▶ Lower level uprights should be prominently marked in red and white stripes.

▶ The scaffold should be properly braced, secured to the building or structure.

▶ Overloading of the scaffold must be avoided.

▶ The public must be protected at all stages of the work.

▶ Regular inspections of the scaffold must be made and recorded: by a competent person; before first use; after a substantial alteration; after an event that could affect stability (e.g. severe wind); at regular intervals – e.g. not exceeding seven days.

A checklist of typical scaffolding faults is given in Appendix 7.3.

Pre-fabricated mobile scaffold towers

Mobile scaffold towers are frequently used throughout industry. It is essential that the workers are trained in their use as recent research has revealed that, in the UK, 75% of lightweight mobile pre-fabricated tower scaffolding is erected, used, moved or dismantled in an unsafe manner.

The following points must be considered when mobile scaffold towers are to be used:

▶ The selection, erection and dismantling of mobile scaffold towers must be undertaken by competent and trained persons with maximum height to base ratios not being exceeded.
▶ Diagonal bracing and stabilisers should always be used.
▶ Access ladders must be fitted to the narrowest side of the tower or inside the tower and persons should not climb up the frame of the tower.
▶ All wheels must be locked while work is in progress and all persons must vacate the tower before it is moved.
▶ The tower working platform must be boarded, fitted with guard rails and toe boards and not overloaded.
▶ Towers must be tied to a rigid structure if exposed to windy weather or to be used for work such as jet blasting.
▶ Persons working from a tower must not over-reach or use ladders from the work platform.
▶ Safe distances must be maintained between the tower and overhead power lines both during working operations and when the tower is moved.
▶ The tower should be inspected on a regular basis and a report made.

Anyone erecting a tower scaffold should be competent to do so and should have received training under an industry recognised training scheme or under a recognised manufacturer or supplier scheme.

Mobile elevating work platforms (MEWPs)

MEWPs are very suitable for high-level work such as changing light bulbs in a warehouse, working on building exteriors or reaching electrical distribution cables. The following factors must be considered when using MEWPs:

▶ The MEWP must only be operated by trained and competent persons.
▶ It must never be moved in the elevated position.
▶ It must be operated on level and stable ground with consideration being given for the stability and loading of floors.
▶ The tyres must be properly inflated and the wheels immobilised.

(a)

(b)

Figure 7.19 Mobile elevating work platforms: (a) scissor lift; (b) hydraulic access platform in Belgium

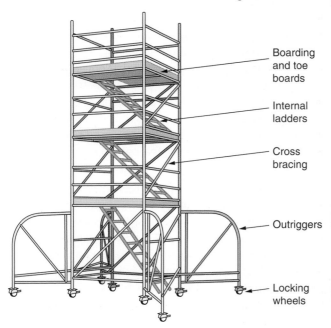

Boarding and toe boards

Internal ladders

Cross bracing

Outriggers

Locking wheels

Figure 7.18 Typical pre-fabricated tower scaffold with outriggers

- Outriggers should be fully extended and locked in position.
- Due care must be exercised with overhead power supplies, obstructions and adverse weather conditions.
- Warning signs should be displayed and barriers erected to avoid collisions.
- It should be maintained regularly and procedures should be in place in the event of machine failure.
- Drivers of MEWPs must be instructed in emergency procedures, particularly to cover instances of power failure.
- All workers on MEWPs should wear safety harnesses.

Workers using mobile elevating work platforms have been injured by falling from the platform due to a lack of handrails or inadvertent movement of the equipment because the brakes had not been applied before raising the platform. Injuries have also resulted from the mechanical failure of the lifting mechanism and by workers becoming trapped in the scissor mechanism.

When working on a MEWP, there is a danger that the operator may become trapped against an overhead or adjacent object, preventing him/her from releasing the controls. There have also been accidents caused when a MEWP is reversed into areas where there is poor pedestrian segregation and the driver has limited visibility. During any manoeuvring operation, a dedicated banksman should be used.

There is evidence of a growing number of serious and fatal accidents from entrapment accidents between guard rails on MEWPs and adjacent obstructions. There have been several serious accidents caused by workers becoming trapped by overhead obstructions while the MEWP was in motion. These accidents often occur during maintenance operations. It is important that risk assessments are undertaken and safe systems of work developed for such work. The Strategic Forum for Construction Plant Safety Group have published guidance on best practice when using MEWPs in confined overhead spaces. The guidance covers hazards, risk assessment, controls and responsibilities of managers.

If **a MEWP is to be used**, the HSE suggest that the following questions should be considered:

- Height – How high is the job from the ground?
- Application – Do you have the appropriate MEWP for the job? (If you're not sure, check with the hirer or manufacturer.)
- Conditions – What are the ground conditions like? Is there a risk of the MEWP becoming unstable or overturning?
- Operators – Are the people using the MEWP trained, competent and fit to do so?
- Obstructions – Could the MEWP be caught on any protruding features or overhead hazards, such as steelwork, tree branches or power lines?
- Traffic – Is there passing traffic and, if so, what do you need to do to prevent collisions?
- Restraint – Do you need to use either work restraint (to prevent people climbing out of the MEWP) or a fall arrest system (which will stop a person hitting the ground if they fall out)? Allowing people to climb out of the basket is not normally recommended – Do you need to do this as part of the job?
- Checks – Has the MEWP been examined, inspected and maintained as required by the manufacturer's instructions and have daily checks been carried out?

7.5.8 Inspection

Equipment for work at height needs regular inspection to ensure that it is fit for use. A marking system is probably required to show when the next inspection is due. Formal inspections should not be a substitute for any pre-use checks or routine maintenance. Inspection does not necessarily cover the checks that are made during maintenance although there may be some common features. Inspections need to be recorded but checks do not.

A weekly inspection of any scaffolding used in construction work should be undertaken by a competent person. A regular scaffold inspection is a requirement of the ILO Convention (Safety and Health) (167) (see Appendix 7.1). The following information should be included in the report:

- the name and address of the person for whom the inspection was carried out;
- the location of the work equipment inspected;
- a description of the work equipment inspected;
- the date and time of the inspection;
- details of any matter identified that could give rise to a risk to the health or safety of any person;
- details of any action taken as a result of any matter identified above;
- details of any further action considered necessary;
- the name and position of the person making the report.

All scaffolding inspection should be carried out by a competent person whose combination of knowledge, training and experience is appropriate for the type and complexity of the particular scaffold to be inspected. Competence may have been assessed by a professional body or an individual may be suitably experienced in scaffolding work and have received additional training under a recognised manufacturer/supplier scheme for the specific configuration being inspected. A non-scaffolder, such as a site manager, who has attended a suitable scaffold inspection course and has the necessary background experience would be considered competent to inspect a basic scaffold.

Appendix 7.2 shows the inspection format that could be used for a scaffold inspection.

7.5.9 Prevention of falling materials through safe stacking and storage

It is estimated that two million working days are lost each year on account of handling injuries and slip and trip incidents. Such incidents can occur in warehouses and storage facilities when palletised goods are stacked higher than two storeys and often weigh several tonnes. Implementing tried and tested methods on safe racking and storage is therefore essential to mitigate the risks of an incident arising.

It is often possible to remove high-level storage from offices and other general workplaces, such as construction sites, and provide storage in warehouses or similar storage facilities. If a storage facility is to be installed for the first time, then the following points should be considered:

▶ The racking must be erected on and fixed securely to a sound, level floor.
▶ The storage system must be installed in accordance with manufacturers' instructions.
▶ If the racking is to be secured to the wall of a building, has this been proved by structural calculations which should be made to ensure that the walls can support the racking and its contents securely.
▶ Beam-connector locks must be fixed at both ends of the beam.
▶ Maximum-load notices must be displayed.
▶ Correct pallets must be used.
▶ Sufficient protective equipment must be used, such as column guards and rack-end protectors.

Management and employees should familiarise themselves with the racking systems used and ensure they understand the difference between general wear and tear and real damage, in order to help them identify potentially dangerous situations as early as possible.

Storage racking is particularly vulnerable and should be strong and stable enough for the loads it has to carry. Damage from vehicles in a warehouse can easily weaken the structure and cause collapse. Uprights need protection, particularly at corners.

The following action can be taken to keep racking serviceable:

▶ Inspect them regularly and encourage workers to report any problems/defects.
▶ Post notices with maximum permissible loads and never exceed the loading.

▶ Use good pallets and safe stacking methods.
▶ Band, box or wrap articles to prevent items falling.
▶ Set limits on the height of stacks and regularly inspect to make sure that limits are being followed.
▶ Provide instruction and training for staff and special procedures for difficult objects.

Regular visual inspections should be carried out and documented, so that any damage can be quickly resolved. In particular, staff should be trained to act if damage occurs to and affects:

▶ the cross-sectional profile of a main load beam;
▶ the straightness of beams, bracing, or uprights; and
▶ the welds and joints, or bolts and clips.

To guard against back strains and other injuries, shelving and storage solutions can be installed that allow access and retrieval of stock at a comfortable, ergonomic height. Products such as vertical storage machines or pallet pull-out units are possible solutions, as they are designed so that stock can be reached without unnecessary straining.

Generally, storage racks should be examined by a qualified inspector at least once or twice a year.

During an inspection, particular attention will be paid to beams, uprights, frame bracing, floor fixings and lock-in clips. The following will also be subject to general observations:

▶ pallet locations on beams;
▶ conditions and types of pallets;
▶ positioning of loads and types of loads stored on pallets;
▶ general fork-lift operatives' use of the racking;
▶ the condition and type of floor on which the racking is fixed;
▶ general housekeeping of the installation; and
▶ possible changes from the original design requirements.

With warehouse managers relying increasingly on temporary and agency workers, who sometimes have little or no prior experience of working in warehouse or storage environments, it is vital that safety training becomes part of the induction process.

All materials used in the construction process on site must be either stacked or stored safely. This will keep the site tidy and reduce slip and trip hazards. It will also help to reduce the risk of fire. Dangerous substances should be kept in a safe place in a separate building or the open air. Only small quantities of hazardous and/or flammable substances should be kept on the site. Any larger amounts, which cannot be kept outside in a safe area, should be kept in a special fire-resisting store which is well ventilated and free of ignition sources. Flammable gas cylinders also need to be stored and used safely. More information on this is given in Chapter 12.

7.6 Hazards and control measures for works of a temporary nature

7.6.1 The impact on workplaces from hazards associated with works of a temporary nature

Works of a temporary nature include such projects as building maintenance, renovation and refurbishment. These works usually take place in a host organisation by contractors and often include building, demolition and excavation work required to supply new underground services. The projects must be properly planned, supervised and monitored, particularly when the project is to take place within or near an occupied building. There are many other possible hazards associated with work of a temporary nature. Some of them are:

► slips, trips and falls are a frequent source of injury to members of the host organisation and the public;
► unfenced excavations;
► inadequate protection of holes, uneven surfaces, poor reinstatement, trailing leads and cables (especially on stairways), spillage of oils and gravel;
► poor lighting and ventilation in and around the temporary works;
► noise and vibration;
► hazardous substances including dust;
► falling or flying objects;
► the accumulation of waste materials;
► falling or moving objects;
► poor storage of materials and equipment and other obstructions in public areas, including inadequate control of waste materials, are common causes of accidents;
► fire and/or explosions;
► lack of project management and supervision; and
► hazards created outside the site perimeter (for example, unloading materials from a delivery lorry outside the perimeter).

Two relatively common problems with building projects concern asbestos and demolition. An up-to-date asbestos survey together with relevant plans and drawings should, therefore, be available before any renovation work is commenced.

Most of the health and safety risks in demolition activities are related to an unplanned or premature collapse of a structure or of a part of it. Various accident investigation reports following such incidents have found:

► the absence of temporary structures to support unstable elements;
► the lack of risk assessment at design stage;
► the lack of any preliminary structural survey or site investigation;
► poor planning of demolition sequences;
► the lack of demolition method statements; and

► the lack of supervision while undertaking demolition activities.

Several of these conclusions are valid for a wide range of large and small building projects, namely the lack of good communications, supervision, planning and risk assessment and the neglect of legislative requirements.

The majority of temporary work projects involve some form of construction work. Details of some of the basic hazards and controls for construction work are given later in this section.

7.6.2 Main control measures relating to the management of works of a temporary nature

The important controls for work of a temporary nature are communication and the cooperation of both contractors and host employees, a comprehensive risk assessment for the project and the management and supervision of the project to ensure that all the agreed controls are implemented.

Communication and cooperation

The organisation of liaison and consultation arrangements is essential to ensure good communication with both the host employees and the contractors. This should provide the cooperation and coordination of all those responsible for the project to ensure the health and safety of everyone in the workplace and enable any concerns to be addressed. Good communication between clients, contractors and residents is an important means of controlling the risks and minimising discomfort. It can also be essential when arranging alternative escape routes during work. Advice should be sought from the local fire authority where it is necessary to block or alter the permanent escape routes.

Risk assessment

A detailed risk assessment should be made to cover the temporary works. This should include the identification of:

► all aspects of the works including the health and safety implications to the host employees and the contractors;
► the people at risk from the work including, where appropriate, members of the public;
► any risks to the contractors from the business of the host organisation; and
► any additional controls and procedures required to protect everyone affected by the project including:
  ▷ any equipment that should or should not be worked on or used;
  ▷ the personal protective equipment that should be used and who will provide it;

▷ all the working procedures, including any permits-to-work and method statements that are to be used;

▷ details of any site access restrictions;

▷ any amendment required to existing emergency procedures and access to the host building; and

▷ details of any safety signs and/or barriers required.

Often work must be undertaken in areas which need to remain occupied. The risk assessment should indicate the nature of the perimeter of the project and how it will be maintained. It might be possible for the work to take place outside normal hours – if not then the work areas must be segregated or access controlled by physical barriers or warning signs.

Management and supervision of the project

The first two important tasks for the management and supervision of the project are the formulation of a detailed specification of the project and the appointment of competent managers and contractors to complete the project (see Chapter 3).

A Temporary Works Coordinator should be appointed for all temporary works to oversee the project and to sign it off on completion. Such an appointment is particularly important for refurbishment projects involving complex activities that require information related to the existing structure and interrelated health and safety issues.

The coordinator would be responsible for:

▶ the appointment of competent contractors to undertake the project;

▶ the provision of information, instruction and training to the employees of the host organisation on any hazards associated with the temporary works;

▶ the provision of information to the temporary works contractors on any hazards from the activities of the host organisation and the controls in place to address them;

▶ the management and supervision of the work of the contractors and agreement on the nature of any temporary controls required before work begins; and

▶ the formulation of a detailed job specification for the work required which should include:

▷ safe access to both the temporary work site and to the rest of the building;

▷ safe method statements to cover all aspects of the work; and

▷ the arrangements for the examination of any temporary access equipment.

The provision of welfare facilities including first-aid arrangements must be discussed and agreed prior to the commencement of the project. It may be that project contractors can have use of the same facilities as the host employees. If this is not possible, then separate arrangements should be included in the project specification.

Other control measures

Where construction work is carried out in, or near, occupied premises, such as houses, hospitals, factories and shops, it may be necessary to evacuate part or all of these premises, either for the full duration of the work, or for a limited period during hazardous operations. Certain organisations such as housing associations and care homes often have an established policy which identifies the circumstances when it is appropriate to evacuate premises.

Other control measures that will be relevant to many projects include the:

▶ use of temporary flooring material, such as plywood or steel plates, to cover uneven ground or potholes;

▶ covering or fixing of any trailing cables which need to cross pedestrian areas;

▶ provision of lighting at night and in dark areas;

▶ prompt removal of all waste and rubbish as it arises;

▶ storage of all hazardous substances in suitable containers or in secure compounds when not in use;

▶ provision of site storage compounds to accommodate all the plant, equipment and materials outside working hours. Strict control over the amount and timing of deliveries will help keep storage to a minimum outside this compound area;

▶ safe storing and stacking of materials and a limit to the height of all material stacks;

▶ use of suitably protected electrical cordless tools or reduced voltage equipment;

▶ fitting of gas cylinders and similar appliances with valves which require special tools to turn on the supply;

▶ isolation of gas cylinders when not in use and locking them in a secure cage out of hours;

▶ protection against hazards that are created within the project site as the work develops, such as around deep excavations which will need to be covered or fenced;

▶ protection of vulnerable groups such as the elderly, children and people with certain disabilities – work in certain premises such as schools and hospitals requires careful planning,

▶ protection of wheelchair users and partially sighted people where construction work affects pedestrian routes;

▶ provision of suitable protection:

▷ from dust, asbestos fibres, noise and vibration (see chapters 13 and 14);

▷ to contractors, host employees and members of the public from being struck by falling or ejected materials; and

▷ from delivery and other moving site vehicles (see Chapter 8). Consideration should be given to ensuring that deliveries are scheduled at times outside of large movements of people such as rush hours or the journeys to and from school;

▶ posting of warning signs around the site and at all entrances to the main building.

Whenever possible, school projects which present high risks to staff and pupils should be carried out during school holidays, at weekends or out of school hours. If this is not possible, the work should be programmed to avoid busy periods such as school start, finish and break times. Deliveries should also be arranged to avoid these times.

Certain premises have their particular hazards that will need to be addressed. Residents in care homes may need to be evacuated when hazardous substances are being used (such as timber treatment or damp course work). Where work takes place within occupied factories, offices, shops and other premises, the host organisation and the contractor will need to manage the risks created by their own work, and cooperate closely to manage those risks created by sharing the workplace.

In healthcare premises, the vulnerability of those who are within the premises must be considered – children and outpatients with restricted mobility or with partial sight. Some patients, such as those who have had major surgery, will be more open to infection than healthy people. Refurbishment works may disturb fungal spores and other organisms which could be a serious hazard to patients. Site vehicles and pedestrians must be separated from the hospital traffic as much as possible.

7.7 Construction activities

The construction industry has a Safety and Health in Construction Convention, 1988 (No. 167) that obliges signatory ILO member states to comply with the construction standards laid out in the Convention – the Convention is a relatively brief statement of those standards. The accompanying Recommendation (No. 175) gives additional information on the Convention statements. The Code of Practice gives more detailed information than the Recommendation. This can best be illustrated by contrasting the coverage of scaffolds and ladders by the three documents shown in Appendix 7.1.

The main causes of fatalities and serious accidents in the construction industry are:

▶ falling through fragile roofs and roof lights;
▶ falling from ladders, scaffolds and other workplaces;
▶ being struck by excavators, lift trucks or dumpers;
▶ overturning vehicles; and
▶ being crushed by collapsing structures.

7.7.1 The scope of construction

The scope of the construction industry is very wide. The most common activity is general building work which is domestic, commercial or industrial in nature. This work may be new building work, such as a building extension or, more commonly, the refurbishment, maintenance or repair of existing buildings. Larger civil engineering projects involving road and bridge building, water supply and sewage schemes and river and canal work all come within the scope of construction.

The work could involve hazardous operations, such as demolition or roof work, or contact with hazardous materials, such as asbestos or lead. Construction also includes the use of woodworking workshops together with woodworking machines and their associated hazards, painting and decorating and the use of heavy machinery. It will often require work to take place in confined spaces, such as excavations and underground chambers.

Finally, at any given time, there are many young people receiving training on site in the various construction trades. These trainees need supervision and structured training programmes.

7.7.2 Construction hazards and controls

There are many hazards likely to be found on a building site, some of which are specific to construction activity and others are more general hazards (manual handling, electricity, noise, vibration, etc.) which are discussed in more detail in later chapters. The principal hazards and controls present in construction work are as follows and are described in the ILO Code of Practice 'Safety and Health in Construction'.

Safe place of work

Safe access to and egress from the site and the individual places of work on the site are fundamental to a good health and safety environment. This clearly requires that all ladders, scaffolds, gangways, stairways and passenger hoists are safe for use. It further requires that all excavations are fenced, the site is tidy and proper arrangements are in place for the storage of materials and the disposal of waste. The site needs to be adequately lit and secured against intruders, particularly children, when it is unoccupied. Such security will include:

▶ secure and locked gates with appropriate notices posted;
▶ a secure and undamaged perimeter fence with appropriate notices posted;
▶ all ladders either stored securely or boarded across their rungs;
▶ all excavations covered;

Figure 7.20 Secure site access gates

▶ all mobile plant immobilised and fuel removed, where practicable, and services isolated;

▶ secure storage of all flammable and hazardous substances;

▶ visits to local schools to explain the dangers present on a construction site. This has been shown to reduce the number of child trespassers; and

▶ if unauthorised entry persists, then security patrols and closed circuit television may need to be considered.

Work at height

Work at height accounts for the majority of fatalities and serious injuries in the construction industry. Safe working procedures for working at height apply to all operations carried out at height, not just construction work, so they are also relevant to, for example, window cleaning, tree surgery, maintenance work at height and the changing of bulbs in street lamps. This topic was covered in more detail earlier in this chapter.

Demolition

Demolition is one of the most hazardous construction operations and is responsible for more deaths and major injuries than any other activity. The principal hazards associated with demolition work are:

▶ falls from height or on the same level;

▶ falling debris;

▶ premature collapse of the structure being demolished;

▶ dust and fumes;

▶ the silting up of drainage systems by dust;

▶ the problems arising from spilt fuel oils;

▶ manual handling;

▶ presence of asbestos and other hazardous substances;

▶ noise and vibration from heavy plant and equipment;

▶ electric shock;

▶ fires and explosions from the use of flammable and explosive substances;

▶ smoke from burning waste timber;

▶ pneumatic drills and power tools;

▶ the existence of services, such as electricity, gas and water;

▶ collision with heavy plant; and

▶ plant and vehicles overturning.

Before any work is started, a full site investigation should be made by a competent person to determine the hazards and associated risks which may affect the demolition workers and members of the public who may pass close to the demolition site. The investigation should cover the following topics:

▶ construction details of the structures or buildings to be demolished and those of neighbouring structures or buildings;

▶ the presence of asbestos, lead or other hazardous substances;

▶ the location of any underground or overhead services (water, electricity, gas, etc.);

▶ the location of any underground cellars, storage tanks or bunkers, particularly if flammable or explosive substances were previously stored; and

▶ the location of any public thoroughfares adjacent to the structure or building.

Figure 7.21 Demolition in progress

A risk assessment should be made by the project designer who will also plan the demolition work. A further risk assessment should then be made by the contractor undertaking the demolition – this risk assessment will be used to draw up a suitable method statement. A written method statement should be produced before demolition takes place. The contents of the method statement should include the following:

▶ details of the method of demolition to be used, including the means of preventing premature collapse or the collapse of adjacent buildings and the safe removal of debris from upper levels;

▶ details of equipment, including access equipment, required and any hazardous substances to be used;

- arrangements for the protection of the public and the construction workforce, particularly if hazardous substances, such as asbestos or other dust, are likely to be released;
- details of the isolation methods of any services which may have been supplied to the site and any temporary services required on the site;
- details of PPE which must be worn;
- first-aid, emergency and accident arrangements;
- training and welfare arrangements;
- arrangements for waste disposal;
- names of site foremen and those with responsibility for health and safety and the monitoring of the work; and
- relevant risk assessments should be appended to the method statement.

There are two forms of demolition:

- **piecemeal** – where the demolition is done using hand and mechanical tools such as pneumatic drills and demolition balls; and
- **deliberate controlled collapse** – where explosives are used to demolish the structure. This technique should be used only by trained, specialist competent persons.

A very important element of demolition is the training required by all construction workers involved in the work. Specialist training courses are available throughout the world for those concerned with the management of the process, from the initial survey to the final demolition. However, induction training, which outlines the hazards and the required control measures, should be given to all workers before the start of the demolition work. The site should be made secure with relevant signs posted to warn members of the public of the dangers.

Excavations

This topic will be covered in more detail later in this chapter (see 7.7.4). Excavations must be constructed so that they are safe environments for construction work to take place. They must also be fenced and suitable notices posted so that neither people nor vehicles fall into them.

Prevention of drowning

Where construction work takes place over water, steps should be taken to prevent people falling into the water and rescue equipment should be available at all times.

Vehicles and traffic routes

All vehicles used on site should be regularly maintained and records kept. Only trained drivers should be allowed to drive vehicles and the training should be relevant to the particular vehicle (fork-lift truck, dumper truck, etc.). Vehicles should be fitted with reversing

Figure 7.22 Barriers around excavation by footpath

warning systems. Many accidents involving dumper trucks have been caused by drivers having little experience and no training. Common forms of accident include driving into excavations, overturning while driving up steep inclines and runaway vehicles which have been left unattended with the engine running. Many vehicles such as mobile cranes require regular inspection and test certificates.

The small dumper truck is widely used on all sizes of construction site. Compact dumper trucks are involved in about 30% of construction transport accidents. The three main causes of such accidents are:

- overturning on slopes and at the edges of excavations;
- poorly maintained braking systems; and
- driver error due to lack of training and/or inexperience.

Some of the hazards associated with these vehicles are: collisions with pedestrians, other vehicles or structures such as scaffolding. They can be struck by falling materials and tools or be overloaded. The person driving the truck can be thrown from the vehicle, come into contact with moving parts on the truck, suffer the effects of whole body vibrations due to driving over potholes in the roadway and suffer from the effects of noise and dust. The precautions that can be taken to address these hazards include the use of authorised, trained, competent and supervised drivers only. As with so many other construction operations, risks should be assessed, safe systems of work followed and drivers forbidden from taking shortcuts. The following site controls should also be in place:

- designated traffic routes and signs;
- speed limits;
- stop blocks used when the vehicle is stationary;
- proper inspection and maintenance procedures;
- procedures for starting, loading and unloading the vehicle;
- provision of rollover protective structures (ROPS) and seat restraints;

▶ provision of falling-object protective structures (FOPS) when there is a risk of being hit by falling materials;
▶ visual and audible warning of approach; and
▶ where necessary, hearing protection.

For other forms of mobile construction equipment, such as fork-lift trucks (covered in Chapter 9), the risk to people from the overturning of the equipment must always be minimised. This can usually be achieved by the avoidance of working on steep slopes, the provision of stabilisers and ensuring that the load carried does not affect the stability of the equipment/vehicle. Chapter 10 described the hazards, safeguards and precautions for several pieces of equipment and machinery used in construction work, such as the cement/concrete mixer and the bench-mounted circular saw.

Traffic routes and loading and storage areas need to be well designed with enforced speed limits, good visibility and the separation of vehicles and pedestrians being considered. The use of one-way systems and separate site access gates for vehicles and pedestrians may be required. Finally, the safety of members of the public must be considered, particularly where vehicles cross public footpaths.

Fire and other emergencies

Emergency procedures relevant to the site should be in place to prevent or reduce injury arising from fire, explosions, flooding or structural collapse. These procedures should include the location of fire points and assembly points, extinguisher provision, site evacuation, contact with the emergency services, accident reporting and investigation, and rescue from excavations and confined spaces. There also needs to be training in these procedures at the induction of new workers and ongoing for all workers.

Welfare facilities

The standard of welfare facilities on many construction sites is very poor. Sanitary and washing facilities (including showers if necessary) with an adequate supply of drinking water should be provided for every person working on the site. Accommodation will be required for the changing and storage of clothes and rest facilities for break times. There should be adequate first-aid provision, an accident book and protective clothing against adverse weather conditions.

Electricity

Electrical hazards are covered in detail later in Chapter 11, and all the control measures mentioned apply on a construction site. However, due to the possibility of wet conditions, it is recommended that only 110V equipment is used on site. Where mains electricity is used (perhaps during the final fitting out of a building), residual current devices should be used with all equipment. Where workers or tall vehicles are working near or under overhead power lines, either the power should be turned off or 'goal posts' or taped markers used to prevent contact with the lines. Similarly, underground supply lines should be located and marked before digging takes place.

Noise

Noisy machinery should be fitted with silencers. When machinery is used in a workshop (such as woodworking machines), a noise survey should be undertaken and, if the noise levels are excessive (above a national exposure limit), then ear defenders should be issued to those workers affected by the noise.

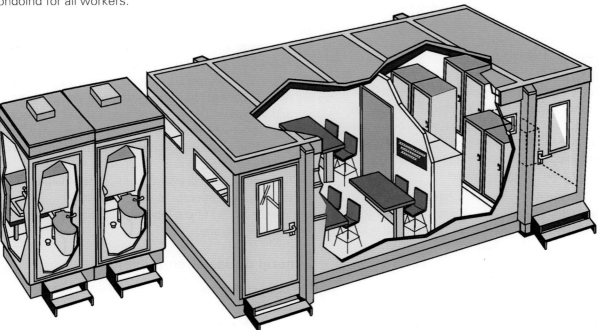

Figure 7.23 Portable toilet, washing, eating and locker facilities for a building site

A cement mixer can be a particularly noisy piece of machinery on a construction site. The levels of noise exposure to workers on the site can be reduced by fixing silencers to diesel-powered mixers, ensuring that the mixer is regularly maintained, minimising the exposure time of the workers by job rotation and providing ear defenders to those working near to the mixer. A better solution would be to use a mixer with lower noise emissions.

Excessive noise hazards may also be present during demolition work and by the use of pneumatic drills, compressors and vehicles used on site.

Health hazards

Health hazards are present on any construction site. These hazards include vibration, dust (including asbestos), cement, solvents and paints and cleaners. A hazardous substance assessment is essential before work starts with regular updates as new substances are introduced. Copies of the assessment and the related safety data sheets should be kept in the Site Office for reference after accidents or fires. They will also be required to check that the correct personal protective equipment (PPE) is available. A manual handling assessment should also be made to ensure that the lifting and handling of heavy objects is kept to a minimum. The findings of all assessments must be communicated to the workforce.

Many of the health hazards (both chemical and biological) are covered in Chapter 13 including silica, which is commonly produced during construction activities. Such activities which can expose workers and members of the public to silica dust include:

▶ cutting building blocks and other stone masonry work;
▶ cutting and/or drilling paving slabs and concrete paths;
▶ demolition work;
▶ sand blasting of buildings; and
▶ tunnelling.

In general, the use of power tools to cut or dress stone and other silica-containing materials will lead to very high exposure levels while the work is occurring. In most cases, exposure levels are in excess of occupational exposure limits (OELs) by factors greater than 2 and in some cases as high as 12.

In addition to silica, there are three hazardous substances that are particularly relevant to construction activities – cement dust and wet cement, wood dust and the biological hazard tetanus.

Waste disposal

The collection and removal of waste from a construction site is normally accomplished using a skip. The skip should be located on firm, level ground away from the main construction work, particularly excavation work. This will allow clear access to the skip for filling and removal from site. On arrival at site, the integrity of the skip should be checked. It should be filled either by chute or by mechanical means unless items can be placed in by hand. Skips should not be overfilled and be netted or sheeted over when full. Any hazardous waste should be segregated as described in Chapter 13, which also describes disposal procedures.

7.7.3 The management of construction activities

Introduction

ILO Code of Practice 'Safety and Health in Construction' provides guidance on the implementation of the provisions of the Safety and Health in Construction Convention, 1988 (No. 167), and the Safety and Health in Construction Recommendation, 1988 (No. 175). The objective of this code is to provide practical guidance on a legal, administrative, technical and educational framework for safety and health in construction with a view to:

(a) preventing accidents and diseases and harmful effects on the health of workers arising from employment in construction;
(b) ensuring appropriate design and implementation of construction projects;
(c) providing means of analysing from the point of view of safety, health and working conditions, construction processes, activities, technologies and operations, and of taking appropriate measures of planning, control and enforcement.

The code applies to construction activities which cover:

(a) building, including excavation and the construction, structural alteration, renovation, repair, maintenance (including cleaning and painting) and demolition of all types of buildings or structures;
(b) civil engineering, including excavation and the construction, structural alteration, repair, maintenance and demolition of, for example, airports, docks, harbours, inland waterways, dams, river and avalanche and sea defence works, roads and highways, railways, bridges, tunnels, viaducts and works related to the provision of services such as communications, drainage, sewerage, water and energy supplies;
(c) the erection and dismantling of pre-fabricated buildings and structures, as well as the manufacturing of pre-fabricated elements on the construction site.

General duties of employers

The Code of Practice places the following duties on construction employers:

1. Employers should provide adequate means and organisation and should establish a suitable programme on the safety and health of workers consistent with national laws and regulations and should comply with the prescribed safety and health measures at the workplace.

2. Employers should provide and maintain workplaces, plant, equipment, tools and machinery and so organise construction work that as far as is reasonably practicable there is no risk of accident or injury to health of workers. In particular, construction work should be so planned, prepared and undertaken that:

    (a) serious or dangerous hazards that are liable to arise at the workplace are prevented as soon as possible;

    (b) excessively or unnecessarily strenuous work positions and movements are avoided;

    (c) organisation of work takes into account the safety and health of workers;

    (d) materials and products are used which are suitable from a safety and health point of view;

    (e) working methods are employed which protect workers against the harmful effects of chemical, physical and biological agents.

3. Employers should establish committees with representatives of workers and management or make other suitable arrangements consistent with national laws and regulations for the participation of workers to ensure safe working conditions.

4. Employers should take all appropriate precautions to protect persons present at, or in the vicinity of, a construction site from all risks which may arise from such site.

5. Employers should arrange for regular safety inspections by competent persons at suitable intervals of all buildings, plant, equipment, tools, machinery, workplaces and systems of work under the control of the employer at construction sites in accordance with national laws, regulations, standards or codes of practice. As appropriate, the competent person should examine and test by type or individually to ascertain the safety of construction machinery and equipment.

6. When acquiring plant, equipment or machinery, employers should ensure that it takes account of ergonomic principles in its design and conforms to relevant national laws, regulations, standards or codes of practice and, if there are none, that it is so designed or protected that it can be operated safely and without risk to health.

7. Employers should provide such supervision as will ensure that workers perform their work with due regard to their safety and health.

8. Employers should assign workers only to employment for which they are suited by their age, physique, state of health and skill.

9. Employers should satisfy themselves that all workers are suitably instructed in the hazards connected with their work and environment and trained in the precautions necessary to avoid accidents and injury to health.

10. Employers should take all practicable steps to ensure that workers are made aware of the relevant national or local laws, regulations, standards, codes of practice, instructions and advice relating to prevention of accidents and injuries to health.

11. Buildings, plant, equipment, tools, machinery or workplaces in which a dangerous defect has been found should not be used until the defect has been remedied.

12. Where there is an imminent danger to the safety of workers, the employer should take immediate steps to stop the operation and evacuate workers as appropriate.

13. On dispersed sites and where small groups of workers operate in isolation, employers should establish a checking system by which it can be ascertained that all the members of a shift, including operators of mobile equipment, have returned to the camp or base at the close of work.

14. Employers should provide appropriate first-aid, training and welfare facilities to workers and, whenever collective measures are not feasible or are insufficient, provide and maintain personal protective equipment and clothing. Employers should also ensure access for workers to occupational health services.

15. Self-employed persons should comply with the prescribed safety and health measures at the workplace according to national laws or regulations.

Cooperation and coordination

The ILO Code of Practice recommends the following duties of cooperation and coordination when several employers operate on a construction site:

1. Whenever two or more employers undertake activities at one construction site, they should cooperate with one another as well as with the client or client's representative and with other persons participating in the construction work being undertaken in the application of the prescribed safety and health measures.

2. Whenever two or more employers undertake activities simultaneously or successively at one construction site, the principal contractor, or other person or body with actual control over or primary responsibility for overall construction site activities, should be responsible for planning and coordinating safety and health measures and, in so far as is compatible with national laws and regulations, for ensuring compliance with such measures.

3. In so far as is compatible with national laws and regulations, where the principal contractor, or other person or body with actual control over or primary

responsibility for overall construction site activities, is not present at the site, they should nominate a competent person or body at the site with the authority and means necessary to ensure on their behalf coordination and compliance with safety and health measures.

4. Employers should remain responsible for the application of the safety and health measures in respect of the workers placed under their authority.

5. Employers and self-employed persons undertaking activities simultaneously at a construction site should cooperate fully in the application of safety and health measures.

6. Employers and designers should liaise effectively on factors affecting safety and health.

General rights and duties of workers

The ILO Code of Practice for Safety and Health in Construction makes several recommendations on the rights and duties of construction workers:

1. Workers should have the right and the duty at any workplace to participate in ensuring safe working conditions on any equipment and methods of work used by them and to express views on working procedures adopted that may affect their safety and health.

2. Workers should have the right to obtain proper information from the employer regarding safety and health risks and safety and health measures related to the work processes. This information should be presented in forms and languages which the workers easily understand.

3. Workers should have the right to remove themselves from danger when they have good reason to believe that there is an imminent and serious danger to their safety or health. They should have the duty to inform their supervisor immediately.

4. In accordance with national legislation, workers should:
   (a) cooperate as closely as possible with their employer in the application of the prescribed safety and health measures;
   (b) take reasonable care for their own safety and health and that of other persons who may be affected by their acts or omissions at work;
   (c) use and take care of personal protective equipment, protective clothing and facilities placed at their disposal and not misuse anything provided for their own protection or the protection of others;
   (d) report forthwith to their immediate supervisor, and to the workers' safety representative where one exists, any situation which they believe could present a risk and which they cannot properly deal with themselves;
   (e) comply with the prescribed safety and health measures;

(f) participate in regular safety and health meetings.

5. Except in an emergency, workers, unless duly authorised, should not interfere with, remove, alter or displace any safety device or other appliance furnished for their protection or the protection of others, or interfere with any method or process adopted with a view to avoiding accidents and injury to health.

6. Workers should not operate or interfere with plant and equipment that they have not been duly authorised to operate, maintain or use.

7. Workers should not sleep or rest in dangerous places such as scaffolds, railway tracks, garages, or in the vicinity of fires, dangerous or toxic substances, running machines or vehicles and heavy equipment.

General duties of designers, engineers, architects and clients

The ILO Code of Practice for Safety and Health in Construction makes several recommendations on the general duties of construction designers, engineers and architects:

1. Those concerned with the design and planning of a construction project should receive training in safety and health and should integrate the safety and health of the construction workers into the design and planning process in accordance with national laws, regulations and practice.

2. Care should be exercised by engineers, architects and other professional persons, not to include anything in the design which would necessitate the use of dangerous structural or other procedures or materials hazardous to health or safety which could be avoided by design modifications or by substitute materials.

3. Those designing buildings, structures or other construction projects should take into account the safety problems associated with subsequent maintenance and upkeep where maintenance and upkeep would involve special hazards.

4. Facilities should be included in the design for such work to be performed with the minimum risk.

The code also covers the general duties of clients of construction projects:

1. Clients should:
   (a) coordinate or nominate a competent person to coordinate all activities relating to safety and health on their construction projects;
   (b) inform all contractors on the project of special risks to health and safety of which the clients are or should be aware;
   (c) require those submitting tenders to make provision for the cost of safety and health measures during the construction process.

2. In estimating the periods for completion of work stages and overall completion of the project,

clients should take account of safety and health requirements during the construction process.

Selection and control of contractors

It is important that health and safety factors are considered as well as technical or professional competence when potential contractors are being shortlisted or employed. The following items will give a guide to health and safety attitudes:

▶ a current health and safety policy;
▶ details of any risk assessments made and control measures introduced;
▶ any method statements required to perform the contract;
▶ details of competence certification, particularly when working with gas or electricity may be involved;
▶ details of insurance arrangements in force at the time of the contract;
▶ details of emergency procedures, including fire precautions, for contractor employees;
▶ details of any previous serious accidents or incidents;
▶ details of accident reporting procedure;
▶ details of previous work undertaken by the contractor;
▶ references from previous employers or main contractors;
▶ details of any health and safety training undertaken by the contractor and his/her employees.

On being selected, contractors should be expected to:

▶ familiarise themselves with the health and safety aspects of the project that may affect their work;
▶ cooperate with the main site contractor.

At the site, sub-contractors should ensure that:

▶ they report to the Site Office on arrival at the site and report to the Site Manager;
▶ they abide by any site rules, e.g. PPE, speed limits, smoking, excluded areas, etc.
▶ the performance of their work does not place others at risk;
▶ they are familiar with the first-aid and accident reporting arrangements on the site;
▶ they are familiar with all emergency procedures on the site;
▶ any materials brought on to the site are safely handled, stored and disposed of;
▶ they adopt adequate fire precaution and prevention measures when using equipment which could cause fires;
▶ they minimise noise and vibration produced by their equipment and activities;
▶ any ladders, scaffolds and other means of access are erected in conformance with good working practice;
▶ any welding or burning equipment brought to the site is in a safe operating condition and used safely with a suitable fire extinguisher to hand;
▶ any lifting equipment brought on to the site complies with any relevant national legislation;

Figure 7.24 Contractors at work in market square, Bruges

▶ all electrical equipment complies with the local statutory requirements;
▶ connections to the electricity supply are from a point specified by the main contractor and are by proper cables and connectors. For outside construction work, only 110V equipment should be used;
▶ any restricted access to areas on the site is observed;
▶ welfare facilities provided on site are treated with respect;
▶ any vehicles brought on to the site observe any speed, condition or parking restrictions.

The control of sub-contractors can be exercised by monitoring them against the criteria listed above and by regular site inspections. On completion of the contract, the work should be checked to ensure that the agreed standard has been reached and that any waste material has been removed from the site.

7.7.4 Hazards associated with excavations

There are about seven deaths each year in the UK due to work in excavations. Many types of soil, such as clays, are self-supporting but others, such as sands and gravel, are not. Many excavations collapse without any warning, resulting in death or serious injury. Many such accidents occur in shallow workings. It is important to note that, although most of these accidents affect workers, members of the public can also be injured. The specific hazards associated with excavations are as follows:

▶ collapse of the sides;
▶ materials falling on workers in the excavation;
▶ falls of people and/or vehicles into the excavation;
▶ workers being struck by plant;
▶ specialist equipment such as pneumatic drills;
▶ hazardous substances, particularly near the site of current or former industrial processes;
▶ influx of ground or surface water and entrapment in silt or mud;
▶ proximity of stored materials, waste materials or plant;

- proximity of adjacent buildings or structures and their stability;
- the presence of toxic decomposition products (methane), when excavating in the vicinity of buildings;
- contact with underground services;
- access and egress to the excavation;
- fumes, lack of oxygen and other health hazards (such as Weil's disease); and
- contaminated ground.

Clearly, alongside these specific hazards, more general hazards, such as manual handling, electricity, noise and vibrations, will also be present.

7.7.5 Precautions and controls required for excavations

The following precautions and controls should be adopted:

- At all stages of the excavation, a competent person must supervise the work and the workers must be given clear instructions on working safely in the excavation.
- The sides of the excavation must be prevented from collapsing either by digging them at a safe angle (between 5° and 45° dependent on soil and dryness) or by shoring them up with timber, sheeting or a proprietary support system. Falls of material into the workings can also be prevented by not storing spoil material near the top of the excavation.
- The workers should wear hard hats.
- If the excavation is more than 2 m deep, a substantial barrier consisting of guard rails and toe boards should be provided around the surface of the workings.
- Vehicles should be kept away as far as possible using warning signs and barriers. Where a vehicle is tipping materials into the excavation, stop blocks should be placed behind its wheels.
- The excavation site **must** be well lit at night.
- All plant and equipment operators must be competent and non-operators should be kept away from moving plant.
- PPE must be worn by operators of noisy plant.
- Nearby structures and buildings may need to be shored up if the excavation may reduce their stability. Scaffolding could also be de-stabilised by adjacent excavation trenches.
- The influx of water can only be controlled by the use of pumps after the water has been channelled into sumps. The risk of flooding can be reduced by the isolation of the mains water supply.
- The presence of hazardous substances or health hazards should become apparent during the original survey work and, when possible, removed or suitable control measures adopted. Any such hazards found after work has started must be reported and noted in the inspection report and

remedial measures taken. Exhaust fumes can be dangerous and petrol or diesel plant should not be sited near the top of the excavation.

- The presence of buried services is one of the biggest hazards and the position of such services must be ascertained using all available service location drawings before work commences. Such plans and maps should be consulted and the location of street lights and manhole covers noted. As these will probably not be accurate, service location equipment should be used by specifically trained people. The area around the excavation should be checked for service boxes. If possible, the supply should be isolated. Only hand tools should be used in the vicinity of underground services. Overhead services may also present risks to cranes and other tall equipment. If the supply cannot be isolated then 'goal posts' beneath the overhead supply together with suitable bunting and signs must be used.
- Safe access by ladders is essential, as are crossing points for pedestrians and vehicles. Whenever possible, the workings should be completely covered outside working hours, particularly if there is a possibility of children entering the site.
- Finally, care is needed during the filling-in process.

Wells and disused mine shafts are found during construction work and must be treated with caution, and in the same way as an excavation. The obvious hazards include falling in and/or drowning and those associated with confined spaces (see Chapter 4) – oxygen deficiency, the presence of toxic gases and the possible collapse of the walls. Controls include fencing off the well and covering it until the situation has been reviewed by specialists. Shallow wells would normally be drained and filled with hard core whereas deeper ones would be capped.

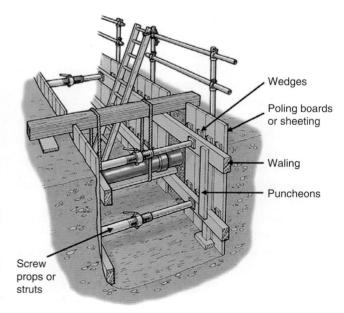

Wedges

Poling boards or sheeting

Waling

Puncheons

Screw props or struts

Figure 7.25 Timbered excavation with ladder access and supported services (guard removed on one side for clarity)

Recommendations for excavation work in the ILO Code of Practice Safety and Health in Construction are shown in Appendix 7.4.

7.7.6 Inspection and reporting requirements

The duty to inspect and prepare a report only applies to excavations which need to be supported to prevent accidental fall of material. Only persons with a recognised and relevant competence should carry out the inspection and write the report. Inspections should take place by a competent person at the following timing and frequency:

▶ after any event likely to affect the strength or stability of the excavation;
▶ before work at the start of every shift;
▶ after an accidental fall of any material.

Although an inspection must be made at the start of every shift, only one report is required of such inspections every seven days. However, reports must be completed following all other inspections. The report should be completed before the end of the relevant working period and a copy given to the manager responsible for the excavation within 24 hours. The report must be kept at the inspection site until the work is completed and then retained for three months at an office of the organisation which carried out the work.

A suitable form is shown in Appendix 7.2.

7.8 Further information

Ambient factors in the workplace, International Labour Organisation (ILO) Code of Practice (CoP),

ISBN 92-2-11628-X http://www.ilo.org/safework/info/standards-and-instruments/WCMS_107729/lang--en/index.htm

Directive 89/656/EEC – use of personal protective equipment https://osha.europa.eu/en/legislation/directives/workplaces-equipment-signs-personal-protective-equipment/osh-directives/4

Hygiene (Commerce and Offices), ILO Convention, 1964 (No 120) – C120 http://www.ilo.org/dyn/normlex/en/f?p=1000:12100:0::NO::P12100_ILO_CODE:C120

Graphical symbols – Registered safety signs, ISO 7010:2011, International Organisation for Standardisation (ISO)

Management of alcohol and drug related issues in the workplace, ILO CoP, 1999, ISBN 92-2-109455-3 http://www.ilo.org/wcmsp5/groups/public/---ed_protect/---protrav/---safework/documents/normativeinstrument/wcms_107799.pdf

Safety and Health in Construction, ILO CoP, ILO Geneva, 1992, ISBN 92-2-107104-9 http://www.ilo.org/safework/info/standards-and-instruments/codes/WCMS_107826/lang--en/index.htm

Safety and Health in Construction Convention, C167, 1988, ILO http://www.ilo.org/dyn/normlex/en/f?p=1000:12100:0::NO::P12100_ILO_CODE:C167

Safety and Health in Construction Recommendation, R175, 1988, ILO http://www.ilo.org/dyn/normlex/en/f?p=1000:12100:0::NO::P12100_ILO_CODE:R175

Welfare Facilities Recommendation, R102, 1956 http://www.ilo.org/dyn/normlex/en/f?p=1000:12100:0::NO::P12100_ILO_CODE:R102

7.9 Practice revision questions

1. A manufacturing company has to relocate its workforce and workshops while a new factory is to be built. There will be 24-hour working at the new premises.
   (a) **Outline** the sanitary and washing facilities, drinking water supply and other welfare facilities that will be required at the new location.
   (b) **Identify** other aspects of the work environment that the company will also need to consider.

2. Extremes of temperature in the workplace may have an adverse effect on the health of workers.
   (a) **Identify** the possible effects on health that may be caused by working in a hot and cold environment.
   (b) **Outline** the controls required when workers are exposed to:
      (i) a hot environment
      (ii) a cold environment.

3. An office building is to be fitted out and equipped by a magazine publishing company.
   (a) **Identify** the issues that should be considered by the company when the lighting requirements are being specified for the building.
   (b) **Outline FOUR** other health and safety issues associated with the physical working environment that should also be considered by the company.

4. (a) **Identify FOUR** types of work where employees may be at an increased risk of

violence when dealing with members of the public.

(b) **Outline** an action plan that can be developed to protect employees from the risk of violence in the workplace.

(c) **Identify FOUR** types of security equipment that could be employed to protect vulnerable workers from the risk of violence when dealing with members of the public.

5. (a) **Outline FIVE** possible indications that an employee may be suffering from alcohol, solvent or drug abuse.

(b) **Identify** the risks to health from the misuse of these substances.

(c) **Outline** the control measures that could reduce these risks to health.

6. (a) **Identify EIGHT** types of hazard that may cause slips or trips at work.

(b) **Outline** the measures that may be needed to reduce the risk of slip and trip accidents in a workplace that contains manufacturing machinery.

7. Permanent staircases are a source of accidents involving a fall from height. **Identify FIVE** design features and/or safe practices intended to reduce the risk of such accidents.

8. Both pedestrians and vehicles use the same traffic routes in a busy warehouse. **Identify** control measures that should be in place to reduce the risks to pedestrians in the warehouse.

9. During the maintenance of a machine that is in a fixed position, a fitter suffers a head injury when he collides with on overhead beam on the machine. **Describe** possible control measures that could have prevented this accident.

10. (a) **Identify FOUR** hazards associated with work at height above ground level.

(b) **Outline** a hierarchy of control measures to be considered when a construction worker is at risk of falling while working at height.

(c) **Outline** the risks facing construction workers required to undertake repair work on a fragile roof.

(d) **Outline** the precautions to be taken to ensure the safety of roof workers.

11. The three most common types of fall arrest equipment are safety harnesses, safety nets and airbags. **Outline** the application, advantages and disadvantages of each of these devices to protect those working at height.

12. **Describe** safe practices required when ladders are to be used in the workplace.

13. (a) **Outline THREE** causes of a scaffold collapse.

(b) **Outline** the precautions that should be taken to reduce the risk of injury to members of the public during the erection and use of the scaffold.

14. Mobile tower scaffolds must be used on stable, level ground.

(a) **Give** reasons why a mobile tower scaffold may become unstable.

(b) **Identify SIX** additional points that should be considered to ensure the safe use of a mobile tower scaffold.

(c) **Identify** measures that should be adopted to protect the risk of people and/or materials falling from the scaffold.

15. A major construction project will involve the demolition of several tall buildings.

(a) **Outline SEVEN** controls that should be introduced to prevent children gaining access to the demolition site.

(b) **Identify** the main hazards that may be present during the demolition of the buildings.

(c) **Outline** the contents of a suitable method statement to be used during the demolition process.

16. Ceiling repairs are to be undertaken in a busy warehouse by construction workers using a mobile elevating work platform (MEWP).

(a) **Identify** the hazards associated with this work.

(b) **Outline** the controls required to ensure the safety of the maintenance workers and others who may be affected by the work.

17. A deep trench needs to be excavated and supported next to a four-storey building.

(a) **Identify** the hazards that may be present in these circumstances.

(b) **Outline** the control measures required so that the excavation workers, members of the public and the adjacent building are adequately protected.

(c) **Identify** when inspections of the supported excavation must be carried out by the competent person.

APPENDIX 7.1 Scaffolds and ladders

1.1 Convention (Safety and Health in Construction) (167)

Article 14

1. Where work cannot safely be done on or from the ground or from part of a building or other permanent structure, a safe and suitable scaffold shall be provided and maintained, or other equally safe and suitable provision shall be made.

2. In the absence of alternative safe means of access to elevated working places, suitable and sound ladders shall be provided. They shall be properly secured against inadvertent movement.

3. All scaffolds and ladders shall be constructed and used in accordance with national laws and regulations.

4. Scaffolds shall be inspected by a competent person in such cases and at such times as shall be prescribed by national laws or regulations.

1.2 Recommendation (Safety and Health in Construction) (175)

Scaffolds

16. Every scaffold and part thereof should be of suitable and sound material and of adequate size and strength for the purpose for which it is used and be maintained in a proper condition.

17. Every scaffold should be properly designed, erected and maintained so as to prevent collapse or accidental displacement when properly used.

18. The working platforms, gangways and stairways of scaffolds should be of such dimensions and so constructed and guarded as to protect persons against falling or being endangered by falling objects.

19. No scaffold should be overloaded or otherwise misused.

20. A scaffold should not be erected, substantially altered or dismantled except by or under the supervision of a competent person.

21. Scaffolds as prescribed by national laws or regulations should be inspected, and the results recorded, by a competent person:
    (i) before being taken into use
    (ii) at periodic intervals thereafter
    (iii) after any alteration, interruption in use, exposure to weather or seismic condition or any other occurrence likely to have affected their strength or stability.

1.3 Code of Practice – Safety and Health in Construction

The Code of Practice covers scaffolds and ladders under the following topics over five pages:

1. general provisions
2. materials
3. design and construction
4. inspection and maintenance
5. lifting appliances on scaffolds
6. pre-fabricated scaffolds
7. use of scaffolds
8. suspended scaffolds.

APPENDIX 7.2 Inspection recording form with timing and frequency chart

See form C1 in Chapter 18 for a sample construction inspection report form.

Timing and frequency chart (reproduced from HSG150).

Place of work or work equipment	Timing and frequency of checks, inspections and examinations equipment								
	Inspect before work at the start of every shift (see note 1)	Inspect after any event likely to have affected its strength or stability	Inspect after accidental fall of rock or other material	Inspect after installation or assembly in any position (see notes 2 and 3)	Inspect at suitable intervals	Inspect after exceptional circumstances which are liable to jeopardise the safety at work equipment	Inspect at intervals not exceeding 7 days (see note 3)	Check on each occasion before use (REPORT NOT REQUIRED)	Thorough Examination
Excavations which are supported to prevent any person being buried or trapped by an accidental collapse or a fall or dislodgement of material	✓	✓	✓						
Cofferdams and caissons	✓	✓							
The surface and every parapet or permanent rail of every existing place of work at height								✓	
Guard rails, toe boards, barriers and similar collective means of fall protection				✓	✓	✓			
Scaffolds and other working platforms (including tower scaffolds and MEWPs) used for construction work and from which a person could fall more than 2 m				✓	✓	✓	✓	✓	✓
All other working platforms				✓	✓	✓	✓	✓	
Collective safeguards for arresting falls (e.g. nets, airbags, soft landing systems)				✓	✓	✓			✓

Timing and frequency of checks, inspections and examinations equipment

Place of work or work equipment	Inspect before work at the start of every shift (see note 1)	Inspect after any event likely to have affected its strength or stability	Inspect after accidental fall of rock or other material	Inspect after installation or assembly in any position (see notes 2 and 3)	Inspect at suitable intervals	Inspect after exceptional circumstances which are liable to jeopardise the safety at work equipment	Inspect at intervals not exceeding 7 days (see note 3)	Check on each occasion before use (REPORT NOT REQUIRED)	Thorough Examination
Personal fall protection systems (including work positioning, rope access, work restraint and fall arrest systems)				✓	✓	✓	✓	✓	✓
Ladders and stepladders					✓	✓	✓	✓	

Notes

1. Although an excavation must be inspected at the start of every shift, only one report is needed in any seven-day period. However, if something happens to affect its strength or stability, and/or an additional inspection is carried out, a report must then be completed. A record of this inspection must be made and retained for three months.

2. 'Installation' means putting into position and 'assembly' means putting together. You are not required to inspect and provide a report every time a ladder, tower scaffold or mobile elevating work platform (MEWP) is moved on site or a personal fall protection system is clipped to a new location.

3. An inspection and a report (see Inspection Report in Chapter 18) is required for a tower scaffold or MEWP (used for construction work and from which a person could fall 2 metres) after installation or assembly and every seven days thereafter, provided the equipment is being used on the same site. A record of this inspection must be made and retained for three months. If a tower scaffold is reassembled rather than simply moved, then an additional, pre-use inspection and report is required. It is acceptable for this inspection to be carried out by the person responsible for erecting the tower scaffold, provided they are trained and competent. A visible tag system, which supplements inspection records as it is updated following each pre-use inspection, is a way of recording and keeping the results until the next inspection.

7

APPENDIX 7.3 Checklist of typical scaffolding faults

Footings	Standards	Ledgers	Bracing	Putlogs and transoms	Couplings	Bridles	Ties	Boarding	Guard rails and toe boards	Ladders
Soft and uneven	Not plumb	Not level	Some missing	Wrongly spaced	Wrong fitting	Wrong spacing	Some missing	Bad boards	Wrong height	Damaged
No base plates	Jointed at same height	Joints in same bay	Loose	Loose	Loose	Wrong couplings	Loose	Trap boards	Loose	Insufficient length
No sole plates	Wrong spacing	Loose	Wrong fittings	Wrongly supported	Damaged	No check couplers	Not enough	Incomplete	Some missing	Not tied
Undermined	Damaged	Damaged			No check couplers			Insufficient supports		

APPENDIX 7.4 Recommendations for excavation work in the ILO Code of Practice 'Safety and Health in Construction'

1. General provisions

1.1. Adequate precautions should be taken in any excavation, shaft, earthworks, underground works or tunnel:

(a) by suitable shoring or otherwise, to guard against danger to workers from a fall or dislodgement of earth, rock or other material;

(b) to guard against dangers arising from the fall of persons, materials or objects or the inrush of water into the excavation, shaft, earthworks, underground works or tunnel;

(c) to secure adequate ventilation at every workplace so as to maintain an atmosphere fit for respiration and to limit any fumes, gases, vapours, dust or other impurities to levels which are not dangerous or injurious to health and are within limits laid down by national laws or regulations;

(d) to enable the workers to reach safety in the event of fire, or an inrush of water or material;

(e) to avoid risk to workers arising from possible underground dangers such as the circulation of fluids or the presence of pockets of gas, by undertaking appropriate investigations to locate them.

1.2. Shoring or other support for any part of an excavation, shaft, earthworks, underground works or tunnel should not be erected, altered or dismantled except under the supervision of a competent person.

1.3. Every part of an excavation, shaft, earthworks, underground works and tunnel where persons are employed should be inspected by a competent person at times and in cases prescribed by national laws or regulations, and the results recorded.

1.4. Work should not commence therein until the inspection by the competent person as prescribed by national laws or regulations has been carried out and the part of the excavation, shaft, earthworks, underground works or tunnel has been found safe for work.

2. Excavations

2.1. Before digging begins on site:

(a) all excavation work should be planned and the method of excavation and the type of support work required decided;

(b) the stability of the ground should be verified by a competent person;

(c) a competent person should check that the excavation will not affect adjoining buildings, structures or roadways;

(d) the employer should verify the position of all the public utilities such as underground sewers, gas pipes, water pipes and electrical conductors that may cause danger during work;

(e) if necessary to prevent danger, the gas, water, electrical and other public utilities should be shut off or disconnected;

(f) if underground pipes, cable conductors, etc., cannot be removed or disconnected, they should be fenced, hung up and adequately marked or otherwise protected;

(g) the position of bridges, temporary roads and spoil heaps should be determined;

(h) if necessary to prevent danger, land should be cleared of trees, boulders and other obstructions;

(i) the employer should see that the land to be excavated is not contaminated by harmful chemicals or gases, or by any hazardous waste material such as asbestos.

2.2. All excavation work should be supervised by a competent person and operatives doing the work should be given clear instructions.

2.3. Sides of excavations should be thoroughly inspected:

(a) daily, prior to each shift and after interruption in work of more than one day;

(b) after every blasting operation;

(c) after an unexpected fall of ground;

(d) after substantial damage to supports;

(e) after a heavy rain, frost or snow;

(f) when boulder formations are encountered.

2.4. No load, plant or equipment should be placed or moved near the edge of any excavation where it is likely to cause its collapse and thereby endanger any person unless precautions such as the provision of shoring or piling are taken to prevent the sides from collapsing.

2.5. Adequately anchored stop blocks and barriers should be provided to prevent vehicles being driven into the excavation. Heavy vehicles should not be allowed near the excavation unless the support work has been specially designed to permit it.

2.6. If an excavation is likely to affect the security of a structure on which persons are working, precautions should be taken to protect the structure from collapse.

2.7. Sides of excavations where workers are exposed to danger from moving ground should be made safe by sloping, shoring, portable shields or other effective means.

227

2.8. All support work should be regularly checked to ensure that the props, wedges, etc., are tight and no undue deflection or distortion is taking place.

2.9. All timber subject to the varying weather conditions should be regularly checked for dryness, shrinkage and rot.

CHAPTER 8

Transport hazards and risk control

8.1 **Safe movement of vehicles in the workplace** ▸ 230

8.2 **Driving at work** ▸ 236

8.3 **Further information** ▸ 240

8.4 **Practice revision questions** ▸ 241

Appendix 8.1 **Advice on driving and the use of taxis** ▸ 242

This chapter covers the following NEBOSH learning objectives:

1. Explain the hazards and control measures for the safe movement of vehicles in the workplace
2. Outline the factors associated with driving at work that increase the risk of an incident and the control measures to reduce work-related driving risks

8.1 Safe movement of vehicles in the workplace

8.1.1 Introduction

The safe movement of vehicles in the workplace is essential if accidents are to be avoided. The more serious accidents between pedestrians and vehicles can often be traced back to excessive speed or other unsafe vehicle practices, such as lack of driver training. More and more workers drive vehicles on public highways as part of their jobs, and some are involved in road accidents that may be classed as work – many of the risks associated with these hazards can be significantly reduced by an effective management system.

As more and more workers spend a considerable amount of time travelling and commuting by road, occupational road safety becomes an important issue. Indeed the term 'commuting accident' has been defined by the ILO in the Protocol of 2002 to the Occupational Safety and Health Convention, 1981 P155, as an accident resulting in death or personal injury occurring on the direct way between the place of work and:

(a) the worker's principal or secondary residence; or
(b) the place where the worker usually takes a meal; or
(c) the place where the worker usually receives his or her remuneration.

This chapter examines vehicular hazards and the controls available to address them.

8.1.2 Hazards in vehicle operations

Many different kinds of vehicle are used in the workplace including dumper trucks, heavy goods vehicles, all-terrain vehicles and, perhaps the most common, the fork-lift truck. Approximately 70 persons are killed annually following vehicle accidents in the workplace in the UK. In 2008/09, 45 workers were killed and over 5,000 injured in workplace transport accidents. The third highest cause of accidental death in the workplace was being struck by a moving vehicle. Approximately a quarter of all workplace vehicle-related deaths occur while a vehicle is reversing. There are also over 1,000 major accidents (involving serious fractures, head injuries and amputations) caused by:

▶ collisions between pedestrians and vehicles;
▶ people falling from vehicles;
▶ people being struck by objects falling from vehicles;
▶ people being struck by an overturning vehicle;
▶ communication problems between vehicle drivers and employees or members of the public.

The term 'workplace transport' covers any vehicle that is used in a work setting, such as fork-lift trucks, compact dumpers, tractors and mobile cranes. It can also include cars, vans and large goods vehicles when these are operating at the works premises. A goods vehicle that is loading or unloading on the public highway outside a works premises is considered as workplace transport.

The main types of accident associated with workplace transport are people being struck by moving vehicles, falling from vehicles (jumping out of cabs or from the tops of high-sided trailers) and being hit by objects falling from vehicles (usually part of a load).

A key cause of these accidents is the lack of competent and documented driver training. The UK HSE investigations, for example, have shown that in over 30% of dumper truck accidents on construction sites, the drivers had little experience and no training. Common forms of these accidents include driving into excavations, overturning while driving up steep inclines and runaway vehicles which have been left unattended with the engine running. Each year, the driving of tractors and other agricultural vehicles, such as all-terrain vehicles, causes several fatalities in farming and other agriculture-related activities.

Risks of injuries to employees and members of the public involving vehicles could arise due to the following occurrences:

▶ collision with pedestrians;
▶ collision with other vehicles;
▶ overloading of vehicles;
▶ overturning of vehicles;
▶ general vehicle movements and parking;
▶ dangerous occurrences or other emergency incidents (including fire);
▶ access and egress from the buildings and the site;
▶ reversing of vehicles, especially inside buildings.

There are several other more general hazardous situations involving pedestrians and vehicles. These include the following:

▶ poor road surfaces and/or poorly drained road surfaces;
▶ roadways too narrow with insufficient safe parking areas;
▶ roadways poorly marked out and inappropriate or unfamiliar signs used;
▶ too few pedestrian crossing points;
▶ the non-separation of pedestrians and vehicles;
▶ lack of barriers along roadways;

- lack of directional and other signs;
- poor environmental factors, such as lighting, dust and noise;
- ill-defined speed limits and/or speed limits which are not enforced;
- poor or no regular maintenance checks;
- vehicles used by untrained and/or unauthorised personnel;
- poor training or lack of refresher training.

Vehicle operations need to be carefully planned so that the possibility of accidents is minimised.

8.1.3 Hazards of mobile work equipment

Mobile work equipment is used extensively throughout industry – in factories, warehouses and construction sites. As mentioned in the previous section, the most common is the fork-lift truck (Figure 8.1).

Figure 8.1 Industrial counterbalanced lift truck

Accidents, possibly causing injuries to people, often arise from one or more of the following events:

- poor maintenance with defective brakes, tyres and steering;
- poor visibility because of dirty mirrors and windows or loads which obstruct the driver's view. Good visibility is essential at all times for mobile plant operators. Operators of mobile construction plant

must ensure they regularly clean their windows so they can safely see all around. This should be combined with constant use of mirrors and a banksman where appropriate. Lights on all vehicles should be cleaned regularly to ensure vehicles are visible at all times, and vehicle depots should be well lit to avoid slip and trip hazards;
- operating on rough ground or steep gradients which causes the mobile equipment to turn on its side 90° plus or roll over 180° or more;
- carrying of passengers without the proper accommodation for them;
- people being flung out as the vehicle overturns and being crushed by it;
- being crushed under wheels as the vehicle moves;
- being struck by a vehicle or an attachment;
- lack of driver training or experience;
- underlying causes of poor management procedures and controls, safe working practices, information, instruction, training and supervision;
- collision with other vehicles;
- overloading of vehicles;
- general vehicle movements and parking;
- dangerous occurrences or other emergency incidents (including fire);
- access and egress from the buildings and the site.

The machines most at risk of rollover according to the UK HSE are:

- compact dumpers frequently used in construction sites;
- agricultural tractors;
- variable reach rough terrain trucks (telehandlers) (Figure 8.2).

Figure 8.2 Telescopic materials handler

8.1.4 Control strategies for safe vehicle and mobile plant operations

Any control strategy involving vehicle operations will involve a risk assessment to ascertain where, on traffic routes, accidents are most likely to happen. It

231

is important that the risk assessment examines both internal and external traffic routes, particularly when goods are loaded and unloaded from lorries. It should also assess whether designated traffic routes are suitable for the purpose and sufficient for the volume of traffic.

The following need to be addressed:

► Traffic routes, loading and storage areas need to be well designed with enforced speed limits, good visibility and the separation of vehicles and pedestrians whenever reasonably practicable.

► Environmental considerations, such as visibility, road surface conditions, road gradients and changes in road level, must also be taken into account.

► The use of one-way systems and separate site access gates for vehicles and pedestrians may be required.

► The safety of members of the public must be considered, particularly where vehicles cross public footpaths.

► All external roadways must be appropriately marked, particularly where there could be doubt on right of way, and suitable direction and speed limit signs erected along the roadways. While there may well be a difference between internal and external speed limits, it is important that all speed limits are observed.

► Induction training for all new employees must include the location and designation of pedestrian walkways and crossings and the location of areas in the factory where pedestrians and fork-lift trucks use the same roadways.

► The identification of recognised and prohibited parking areas around the site should also be given during these training sessions.

Many industries have vehicles designed and used for specific workplace activities. The safe system of work for those activities should include:

► details of the work area (e.g. vehicle routes, provision for pedestrians, signage);

► details of vehicles (e.g. type, safety features and checks, maintenance requirements);

► information and training for employees (e.g. driver training, traffic hazard briefing);

► type of vehicle activities (e.g. loading and unloading, refuelling or recharging, reversing, tipping).

To summarise, there are three aspects to a control strategy for safe vehicle and mobile plant operations:

1. **The design of the site (safe site)** – involves managers in:

► planning routes to separate pedestrians from vehicles whenever possible;

► reducing the need to reverse by using one-way systems;

► avoiding steep gradients and overhead cables and provide traffic routes on firm ground, minimising sharp and blind corners;

► marking out parking areas for vehicles;

► providing speed limit signs and traffic warning signs;

► ensuring a well-lit environment; and

► maintaining good housekeeping and a tidy site.

2. **Vehicle selection and maintenance (safe vehicle)** requires the provision of:

► a vehicle with suitable and effective headlights, brakes, bumpers and horns, sufficient mirrors to reduce blind spots and seat belts for drivers and passengers;

► some additional vehicle features such as rear lens or radar sensors to provide extra safety when reversing and speed governors; and

► a regular and documented inspection and maintenance regime.

3. **Systems of work for system operatives (safe driver)** to include:

► effective supervision of everyone who is in areas where vehicles operate, including the provision of banksmen, when required;

► adequate training and refresher training for all drivers;

► relevant information for all drivers, including speed limits and parking areas;

► regular health checks on the suitability of employees for driving roles;

► the provision of high visibility clothing, appropriate protective clothing (such as steel toe capped boots and hard hats);

► the control of vehicle movements at times of day when there are more people moving around. Access to vehicle areas should be restricted to those that need to be there.

8.1.5 Mobile work equipment safeguards

The main purpose of the mobile work equipment safeguards is to protect workers who could come into contact with such equipment while it is travelling from one location to another or where it does work while moving. Mobile equipment normally moves on wheels, tracks, rollers or skids. It may be self-propelled, towed or remote controlled and may incorporate attachments. No employee may be carried on mobile work equipment:

► unless it is suitable for carrying persons;

► unless it incorporates features to reduce risks as low as is reasonably practicable, including risks from wheels and tracks.

Where there is significant risk of falling materials, falling-object protective structures (FOPS) should be fitted.

Where there is a risk of overturning it must be minimised by:

► stabilising the equipment;

► fitting a structure so that it only falls on its side;

▶ fitting a structure which gives sufficient clearance for anyone being carried if it turns over further – rollover protective structure (ROPS);

▶ a device giving comparable protection;

▶ fitting a suitable restraining system for people if there is a risk of being crushed by rolling over.

Where self-propelled work equipment may involve risks while in motion it shall have:

▶ facilities to prevent unauthorised starting;

▶ facilities to minimise the consequences of collision (with multiple rail-mounted equipment);

▶ a device for braking and stopping;

▶ emergency facilities for braking and stopping, in the event of failure of the main facility, which have readily accessible or automatic controls (where safety constraints so require);

▶ devices fitted to improve vision (where the driver's vision is inadequate);

▶ appropriate lighting fitted or otherwise it shall be made sufficiently safe for its use (if used at night or in dark places);

▶ if there is anything carried or towed that constitutes a fire hazard liable to endanger employees (particularly if escape is difficult such as from a tower crane), appropriate fire-fighting equipment carried, unless it is sufficiently close by.

Figure 8.3 Various construction plant with driver protection

Rollover and falling-object protection (ROPS and FOPS)

Rollover protective structures are now becoming much more affordable and available for most types of mobile equipment where there is a high risk of turning over. Their use is spreading across most developed countries and even some less well-developed areas. A ROPS is a cab or frame that provides a safe zone for the vehicle operator in the event of a rollover (see Figure 8.3).

The ROPS frame must pass a series of static and dynamic crush tests. These tests examine the ability of the ROPS to withstand various loads to see if the protective zone around the operator remains intact in an overturn.

A home-made bar attached to a tractor axle or simple shelter from the sun or rain cannot protect the operator if the equipment overturns.

The ROPS must meet International Standards such as ISO 3471:1994. All mobile equipment safeguards should comply with the essential health and safety requirements of the national health and safety law.

ROPS must also be correctly installed strictly following the manufacturers' instructions and using the correct strength bolts and fixings. They should never be modified by drilling, cutting, welding or other means as this may seriously weaken the structure.

ROPS provide some safety during overturning but only when operatives are confined to the protective zone of the ROPS. So where ROPS are fitted, a suitable restraining system must be provided for all seats. The use of seat restraints could avoid accidents where drivers are thrown from machines, thrown through windows or doors or thrown around inside the cab. In agriculture and forestry, 50% of overturning accidents occur on slopes of less than 10° and 25% on slopes of 5° or less. This means that seat restraints should be used most of the time that the vehicle is being operated.

Falling-object protective structures (FOPS) are required where there is a significant risk of objects falling on the equipment operator or other authorised person using the mobile equipment. Canopies that protect against falling objects (FOPS) must be properly designed and certified for that purpose. Front loaders work in woods or construction sites near scaffolding or buildings under construction and high bay storage areas, these all being locations where there is a risk of falling objects. Purchasers of equipment should check that any canopies fitted are FOPS. ROPS should never be modified by the user to fit a canopy without consultation with the manufacturers.

8.1.6 ILO recommendations for transport, earth-moving and materials-handling equipment

The ILO outlines a series of recommendations on the use of transport, earth-moving and materials handling equipment in its Code of Practice 'Safety and Health in Construction'.

All vehicles and earth-moving or materials-handling equipment should:

(a) be of good design and construction, taking into account as far as possible ergonomic principles particularly with reference to the seat;

(b) be maintained in good working order;

(c) be properly used with due regard to health and safety;

(d) be operated by workers who have received appropriate training in accordance with national laws and regulations.

Figure 8.4 Excavator digging out a swimming pool in France

The drivers and operators of vehicles and earth-moving or materials-handling equipment should be medically fit, trained and tested and of a prescribed minimum age as required by national laws and regulations.

Adequate signalling or other control arrangements or devices should be provided to guard against danger from the movement of vehicles and earth-moving or materials-handling equipment. Special safety precautions should be taken for vehicles and equipment when manoeuvring backwards. The assistance of a trained and authorised signaller should be available when the view of the driver or operator is restricted. The signalling code should be understood by all involved.

When cranes and shovels are being moved or are out of service, the boom should be in the direction of travel and the scoop or bucket should be raised and without load, except when travelling downhill. Vehicles and earth-moving or materials-handling equipment should not be left on a slope with the engine running. Vehicles and earth-moving or materials-handling equipment should not travel on bridges, viaducts or embankments,

unless it has been established that it is safe to do so. Earth-moving and materials-handling equipment should be fitted with safety structures, such as those designed to protect the operator from being crushed either from the overturn of the machine or from falling material.

All vehicles and earth-moving or materials-handling equipment should be provided with a plate indicating:

(a) the gross laden weight;

(b) the maximum axle weight or, in the case of caterpillar equipment, ground pressure;

(c) the tare weight.

All vehicles and earth-moving or materials-handling equipment should be equipped with:

(a) an electrically operated acoustic signalling device;

(b) searchlights for forward and backward movement;

(c) power and hand brakes;

(d) tail lights;

(e) silencers;

(f) a reversing alarm.

Operators of vehicles and earth-moving or materials-handling equipment should be adequately protected against the weather or accidents due to impact, crushing or contact with a moving load by a cab:

(a) which is designed and constructed in accordance with ergonomic principles and provides full protection from adverse weather conditions;

(b) which is fully enclosed where dusty conditions are likely to be encountered;

(c) which provides the driver with a clear and unrestricted view of the area of operation;

(d) which is equipped with a direction indicator and a rear-view mirror on both sides.

Deck plates and steps of vehicles and equipment should be kept free from oil, grease, mud or other slippery substances. The motors, brakes, steering gear, chassis, blades, blade-holders, tracks, wire ropes, sheaves, hydraulic mechanisms, transmissions, bolts and other parts on earth-moving and materials-handling equipment on which safety depends should be inspected daily.

On all construction sites on which vehicles, earth-moving or materials-handling equipment are used:

(a) safe and suitable access ways should be provided for them;

(b) traffic should be so organised and controlled as to secure their safe operation;

(c) when earth-moving or materials-handling equipment is required to operate in dangerous proximity to live electrical conductors, adequate precautions should be taken, such as isolating the electrical supply or erecting overhead barriers of a safe height;

(d) preventative measures should be taken to avoid the fall of vehicles and earth-moving or materials-handling equipment into excavations or into water.

The cab of vehicles and earth-moving or materials-handling equipment should be kept at least 1 m from a face being excavated. Bucket excavators should not be used at the top or bottom of earth walls with a slope exceeding 60°. Dredge-type excavators should not be used on earth walls more than 1 m higher than the reach of the excavator if they are installed at the bottom of the wall.

8.1.7 Safe driving of mobile equipment

Drivers have an important role to play in the safe use of mobile equipment. They should include the following in their safe working practice checklist:

▶ Make sure they understand fully the operating procedures and controls on the equipment being used.
▶ Only operate equipment for which they are trained and authorised.
▶ Never drive if abilities are impaired by, for example, alcohol, poor vision or hearing, ill-health or drugs whether prescribed or not.
▶ Use the seat restraints where provided.
▶ Know the site rules and signals.
▶ Know the safe operating limits relating to the terrain and loads being carried.
▶ Keep vehicles in a suitably clean and tidy condition with particular attention to mirrors and windows or loose items which could interfere with the controls.
▶ Drive at suitable speeds and follow site rules and routes at all times.
▶ Allow passengers only when there are safe seats provided on the equipment.
▶ Park vehicles on suitable flat ground with the engine switched off and the parking brakes applied; use wheel chocks if necessary.
▶ Make use of visibility aids or a signaller when vision is restricted.
▶ Get off the vehicle during loading operations unless adequate protection is provided.
▶ Ensure that the load is safe to move.
▶ Do not get off the vehicle until it is stationary, engine stopped and parking brake applied.
▶ Where practicable, remove the operating key when getting off the vehicle.
▶ Take the correct precautions such as not smoking and switching off the engine when refuelling.
▶ Report any defects immediately.

8.1.8 The management of vehicle movements

The movement of vehicles should be properly managed, as should vehicle maintenance and driver training. The development of an agreed code of practice for drivers, to which all drivers should sign up, and the enforcement of site rules covering all vehicular movements are essential for effective vehicle management.

All vehicles should be subject to appropriate regular preventative maintenance programmes with appropriate records kept and all vehicle maintenance procedures properly documented. Many vehicles, such as mobile cranes, require regular inspection by a competent person and test certificates.

Certain vehicle movements, such as reversing, are more hazardous than others and particular safe systems should be set up. The reversing of lorries, for example, must be kept to a minimum (and then restricted to particular areas). Vehicles should be fitted with reversing warning systems as well as being able to give warning of approach. Refuges, where pedestrians can stand to avoid reversing vehicles, are a useful safety measure. Banksmen, who direct reversing vehicles, should also be aware of the possibility of pedestrians crossing in the path of the vehicle. Where there are many vehicle movements, consideration should be given to the provision of high visibility clothing. Pedestrians must keep to designated walkways and crossing points, observe safety signs and use doors that are separate to those used by vehicles. Visitors who are unfamiliar with the site and access points should be escorted through the workplace.

Fire is often a hazard which is associated with many vehicular activities, such as battery charging and the storage of warehouse pallets. All batteries should be recharged in a separate well-ventilated area.

As mentioned earlier, driver training, given by competent people, is essential. Only trained drivers should be allowed to drive vehicles and the training should be relevant to the particular vehicle (fork-lift truck, dumper truck, lorry, etc.). All drivers must receive specific training and instruction before they are permitted to drive vehicles. They must also be given refresher training and medical examinations at regular intervals. This involves a management system for ensuring driver competence, which must include detailed records of all drivers with appropriate training dates and certification in the form of a driving licence or authorisation. Competence and its definition was discussed in Chapter 3.

Where large vehicles are routinely stopping to load or unload at loading bays, a certain amount of reversing is probably inevitable. When large vehicles need to reverse in the workplace, the following precautions should be taken:

▶ Undertake a risk assessment and develop a written safe system of work.
▶ Restrict reversing to places where it can be carried out safely.
▶ Keep people on foot or in wheelchairs away from the area.
▶ Provide suitable high visibility clothing for those people who are permitted in the area.
▶ Fit reversing alarms to alert or a detection device to warn the driver of an obstruction, or automatically apply the brakes.

▶ Employ banksmen to supervise the safe movement of vehicles.

The design features that may need to be considered to minimise risks associated with movement of vehicles in the workplace include:

▶ providing traffic routes with smooth and stable surfaces and with the right width and headroom for the types of vehicles that will use them;
▶ eliminating sharp bends, blind corners and steep gradients, and siting convex mirrors on those corners that are unavoidably blind;
▶ installing a one-way system, to minimise the need for reversing;
▶ providing passing places for vehicles;
▶ introducing speed limits and providing speed retarders;
▶ providing a good standard of lighting for the traffic routes, and particularly at the transition areas between the inside and outside of buildings;
▶ segregating vehicles and pedestrians, including separate access and egress, and providing clearly marked crossing places (zebra crossings).

The procedural arrangements that should accompany these design features are:

▶ selecting and training competent drivers;
▶ implementing a regular health screening programme for all drivers;
▶ providing information on site rules for visitors, such as delivery drivers;
▶ procedures for the regular maintenance of the traffic routes and the in-house vehicles, including a system for the reporting of defects and near-miss accidents;
▶ rigorously enforcing speed limits, with the possibility of a points system on drivers' licences or permits.

The UK HSE publications *Workplace Transport Safety. Guidance for Employers* HSG136, and *Managing Vehicle Safety at the Workplace* INDG199 (revised) provide useful checklists of relevant safety requirements that should be in place when vehicles are used in a workplace.

8.2 Driving at work

8.2.1 Introduction

The first fatal car accident in the UK occurred in 1896 when Mrs Bridget Driscoll was run over by a Roger-Benz which had a maximum speed of 8mph. Since then, the Royal Society for the Prevention of Accidents has estimated that more than 550,000 people have been killed on UK roads.

It has been estimated that up to a third of all road traffic accidents involve somebody who is at work at the time – accounting in the UK for over 20 fatalities and 250 serious injuries every week. Based on recent annual statistics, this means around 800–1,060 deaths a year on the road, compared with 241 fatal injuries to workers in the 'traditional workplace'. Accident rates are 30–40% higher for business drivers than for private drivers. Some employers believe, incorrectly, that if they comply with certain road traffic law requirements, so that company vehicles have a valid roadworthy test certificate, and drivers hold a valid licence, this is enough to ensure the safety of their employees, and others, when they are on the road. However, health and safety law applies to on-the-road work activities as it does to all work activities, and the risks should be managed effectively within a health and safety management system.

Just under half of all those who drive while at work in the UK suffer from a road-rage incident at least once a year, and 11% are assaulted. Furthermore, pain in the lower back, neck and shoulder is linked to prolonged driving, as is poor mental health.

Figure 8.5 Professional driver at work on vehicle recovery

8.2.2 Benefits of managing work-related road safety

The true costs of accidents to organisations are nearly always higher than just the costs of repairs and insurance claims. The benefits of managing work-related road safety can be considerable, no matter what the size of the organisation. There will be benefits in the area of:

▶ control: costs, such as wear and tear and fuel, insurance premiums and claims can be better controlled;
▶ driver training and vehicle purchase: better informed decisions can be made;
▶ lost time: fewer days will be lost due to injury, ill-health and work rescheduling;
▶ vehicles: fewer will need to be off the road for repair;
▶ orders: fewer orders will be missed;
▶ key employees: there is likely to be a reduction in driving bans.

8.2.3 Managing occupational road risks

Where work-related road safety is integrated into the arrangements for managing health and safety at work, it can be managed effectively. The main areas to be addressed are policy, responsibility, organisation, systems and monitoring. Employees should be encouraged to report all work-related road incidents and be assured that punitive action will not be taken against them.

The risk assessment should:

▶ consider the use, for example, of air or rail transport as a partial alternative to driving;

▶ consider the factors that might increase the risk of becoming involved in a road traffic incident – distance, driving hours, work schedules, stress due to traffic and road conditions and weather conditions;

▶ attempt to avoid situations where employees feel under pressure;

▶ make sure that maintenance work is organised to reduce the risk of vehicle failure. This is particularly important when pool cars are used because pool car users often assume another user is checking on maintenance and legality. The safety critical systems that need to be properly maintained are the brakes, steering and tyres. Similarly, if the car is leased and serviced by the leasing company, a system should be in place to confirm that servicing is being done to a reasonable standard;

▶ insist that drivers and passengers are adequately protected in the event of an incident. Crash helmets and protective clothing for those who ride motorcycles and other two-wheeled vehicles should be of the appropriate colour and standard;

▶ ensure that company policy covers the important aspects of the (UK) Highway Code.

Figure 8.6 Occupational road risk – animals can be very unpredictable on busy roads with major potential consequences when elephants are involved

8.2.4 Evaluating the risks

The European Parliament and Council have introduced the Driving CPC (Certificate of Professional Competence) – under EU Directive 2003/59. There are two types of qualification – one for goods vehicles and one for drivers of passenger vehicles – and the Driver CPC will only apply to operators of vehicles over three-and-a-half tonnes or with more than eight passenger seats.

The following actions should be considered by employers to reduce the risks to employees who drive as part of their work:

▶ Journeys should be planned to reduce driving time.

▶ Rest breaks should be included in journey times.

▶ Encourage drivers to remain physically fit to reduce chronic fatigue.

▶ Ensure that drivers have hands-free mobile equipment.

▶ Monitor related working procedures to ensure safety.

▶ Ensure that drivers have necessary communication equipment (mobile phones, GPS, personal alarms).

▶ Keep a record of the location of driver destinations.

▶ Encourage drivers to regularly report back to colleagues in the base office.

The following considerations can be used to check on work-related road safety management.

The driver

Competency

▶ Is the driver competent, experienced and capable of doing the work safely?

▶ Is his or her licence valid for the type of vehicle to be driven?

▶ Is the vehicle suitable for the task or is it restricted by the driver's licence?

▶ Does the recruitment procedure include appropriate pre-appointment checks?

▶ Is the driving licence checked for validity on recruitment and periodically thereafter?

▶ When the driver is at work, is he or she aware of company policy on work-related road safety?

▶ Are written instructions and guidance available?

▶ Has the company specified and monitored the standards of skill and expertise required for the circumstances for the job?

Training

▶ Are drivers properly trained?

▶ Do drivers need additional training to carry out their duties safely?

▶ Does the company provide induction training for drivers?

▶ Are those drivers whose work exposes them to the highest risk given priority in training?

- ▶ Do drivers need to know how to carry out routine safety checks such as those on lights, tyres and wheel fixings?
- ▶ Do drivers know how to adjust safety equipment correctly, for example seat belts and head restraints?
- ▶ Is the headrest 3.8 cm (1.5 inches) behind the driver's head?
- ▶ Is the front of the seat higher than the back and are the legs 45° to the floor?
- ▶ Is the steering wheel adjustable and set low to avoid shoulder stress?
- ▶ Are drivers able to use anti-lock brakes (ABS) properly?
- ▶ Do drivers have the expertise to ensure safe load distribution?
- ▶ If the vehicle breaks down, do drivers know what to do to ensure their own safety?
- ▶ Is there a handbook for drivers?
- ▶ Are drivers aware of the dangers of fatigue?
- ▶ Do drivers know the height of their vehicle, both laden and empty?

Figure 8.7 Specialised training is required for heavy goods vehicles and other similar categories

Fitness and health

- ▶ The driver's level of health and fitness should be sufficient for safe driving.
- ▶ Drivers of Heavy Goods Vehicles (HGVs) must have the appropriate medical certificate.
- ▶ Drivers who are most at risk should also undergo regular medicals. Staff should not drive, or undertake other duties, while taking a course of medicine that might impair their judgement.
- ▶ All drivers should have regular (every two years) eyesight tests. Research has shown that one in four motorists have a level of eyesight below the legal standard for driving, which is to be able to read a car number plate from a distance of 20.5 m.
- ▶ Drivers should rest their eyes by taking a break of at least 15 minutes every two hours.

In the UK new offences under the Road Safety Act allow courts to imprison drivers who cause deaths by not paying due care to the road or to other road users. Avoidable distractions which courts will consider when sentencing motorists who have killed include:

- ▶ using a mobile phone (for either calling or sending text messages);
- ▶ drinking and eating;
- ▶ applying make-up;
- ▶ anything else which takes their attention away from the road and which a court judges to have been an avoidable distraction.

Every year in the UK, over 87,000 motorists are disqualified for drink-driving or driving while under the influence of drugs and up to 20% of drink-drivers are caught the morning after drinking. The Department for Transport have calculated that 5% of drivers who failed a breath test after a crash were driving for work purposes at the time.

The vehicle

Suitability

All vehicles should be fit for the purpose for which they are used. When purchasing new or replacement vehicles, the employer should select those that are most suitable for both driving and the health and safety of the public. The fleet should be suitable for the job in hand. Where privately owned vehicles are used for work, they should be insured for business use and have an appropriate roadworthy certificate test (e.g. MOT test for vehicles over three years old, in the UK).

Condition and safety equipment

Are vehicles maintained in a safe and fit condition? There will need to be:

- ▶ maintenance arrangements to acceptable standards;
- ▶ basic safety checks for drivers;
- ▶ a method of ensuring that the vehicle does not exceed its maximum load weight;

Figure 8.8 Road accidents are a significant risk when vehicles are overloaded – hay bales in transit in Morocco

- reliable methods to secure goods and equipment in transit;
- checks to make sure that safety equipment is in good working order;
- checks on seat belts and head restraints. (Are they fitted correctly and functioning properly?);
- a defect reporting system for drivers to use if they consider their vehicle is unsafe.

Ergonomic considerations

The health of the drivers, and possibly also their safety, may be put at risk from an inappropriate seating position or driving posture. Ergonomic considerations should therefore be considered before purchasing or leasing new vehicles. Information may need to be provided to drivers about good posture and, where appropriate, on how to set their seat correctly.

The load

For any lorry driving, most of the topics covered in this section are relevant. However, the load being carried is an additional issue. If the load is hazardous, emergency procedures (and possibly equipment) must be in place and the driver trained in those procedures. The load should be stacked safely in the lorry so that it cannot move during the journey. There must also be satisfactory arrangements for handling the load at either end of the journey.

The journey

Routes

Route planning is crucial. Safe routes should be chosen which are appropriate for the type of vehicle undertaking the journey wherever practicable. Motorways are the safest roads. Minor roads are suitable for cars, but they are less safe and could present difficulties for larger vehicles. Overhead restrictions, for example bridges, tunnels and other hazards such as level crossings, may present dangers for long and/or high vehicles, so route planning should take particular account of these.

Scheduling

There are danger periods during the day and night when people are most likely, on average, to feel sleepy. These are between 2 a.m. and 6 a.m. and between 2 p.m. and 4 p.m. Schedules need to take sufficient account of these periods. Where tachographs are carried, they should be checked regularly to make sure that drivers are not putting themselves and others at risk by driving for long periods without a break. Periods of peak traffic flow should be avoided if possible and new drivers should be given extra support while training.

Time

Has enough time been allowed to complete the driving job safely? A realistic schedule would take into account the type and condition of the road and allow

the driver rest breaks. A non-vocational driver should not be expected to drive and work for longer than a professional driver. The recommendation of the UK Highway Code is for a 15-minute break every two hours.

- Are drivers put under pressure by the policy of the company? Are they encouraged to take unnecessary risks, for example exceeding safe speeds because of agreed arrival times?
- Is it possible for the driver to make an overnight stay? This may be preferable to having to complete a long road journey at the end of the working day.
- Are staff aware that working irregular hours can add to the dangers of driving? They need to be advised of the dangers of driving home from work when they are excessively tired. In such circumstances they may wish to consider an alternative, such as a taxi.

Distance

Managers need to satisfy themselves that drivers will not be put at risk from fatigue caused by driving excessive distances without appropriate breaks. Combining driving with other methods of transport may make it possible for long road journeys to be eliminated or reduced. Employees should not be asked to work an exceptionally long day.

Weather conditions

When planning journeys, sufficient consideration will need to be given to adverse weather conditions, such as snow, ice, heavy rain and high winds. Routes should be rescheduled and journey times adapted to take adverse weather conditions into consideration. Where poor weather conditions are likely to be encountered, vehicles should be properly equipped to operate, with, for example, ABS (anti-lock braking system).

Where there are ways of reducing risk, for example when driving a high-sided vehicle in strong winds with a light load, drivers should have the expertise to deal with the situation. In addition, they should not feel pressured to complete journeys where weather conditions are exceptionally difficult and this should be made clear by management.

Appendix 8.1 gives advice on driving and the use of taxis for those involved with international travel.

8.2.5 Typical health and safety rules for drivers of cars on company business

At least 25% of all road accidents are work-related accidents involving people who are using the vehicle on company business. The following are typical rules that have been produced to reduce accidents by car drivers at work. Any breach of these rules will be treated as a disciplinary offence.

- All drivers must have a current and valid driving licence.

▶ All vehicles must carry comprehensive insurance for use at work.

▶ Plan the journey in advance to avoid, where possible, dangerous roads or traffic delays.

▶ Use headlights in poor weather conditions and fog lights in foggy conditions (visibility, 100 m).

▶ Use hazard warning lights if an accident or severe traffic congestion is approached (particularly on motorways).

▶ All speed limits must be observed but speeds should always be safe for the conditions encountered.

▶ Drivers must not drive continuously for more than two hours without a break of at least 15 minutes.

▶ Mobile phones, including hands-free equipment, must not be used whilst driving. They must be turned off during the journey and only used during the rest periods or when the vehicle is safely parked and the handbrake on.

▶ No alcohol must be consumed during the day of the journey until the journey is completed. Only minimal amounts of alcohol should be consumed on the day before a journey is to be made.

▶ No recreational drugs should be taken on the day of a journey. Some prescribed and over-the-counter drugs and medicines can also affect driver awareness and speed of reaction. Always check with a doctor or pharmacist to ensure that it is safe to drive.

8.2.6 ILO recommendations for road transport drivers

The ILO covers some aspects of driver health and safety in the Hours of Work and Rest Periods (Road Transport) Convention 1979 (No. 153) and its accompanying Hours of Work and Rest Periods (Road Transport) Recommendation 1979 (No. 153).

The Convention states that no driver shall be allowed to drive continuously for more than four hours without a break, although this may be exceeded by no more than one hour under certain circumstances. The maximum total driving time, including overtime, must not exceed nine hours per day or 48 hours per week. The length of the break and, as appropriate, the way in which the break may be split shall be determined by the competent authority or body in each country.

The Convention also states that the daily rest of drivers shall be at least ten consecutive hours during any 24-hour period starting from the beginning of the working day. The daily rest may be calculated as an average over periods to be determined by the competent authority or body in each country. During the daily rest the driver shall not be required to remain in or near the vehicle if he/she has taken the necessary precautions to ensure the safety of the vehicle and its load.

The Recommendation adds the following detail:

▶ The normal hours of work should not exceed eight per day as an average.

▶ The length of the break after the four-hour driving period and, as appropriate, the way in which the break may be split should be determined by the competent authority or body in each country.

▶ The maximum total driving time, including overtime, should exceed neither nine hours per day nor 48 hours per week (averaged, if necessary, over a maximum period of four weeks).

▶ The daily rest of drivers should be at least 11 consecutive hours during any 24-hour period starting from the beginning of the working day.

▶ The daily rest may be calculated as an average over periods to be determined by the competent authority or body in each country; provided that the daily rest should in no case be less than eight hours (exceptions to the recommended duration of the daily rest periods and the manner of taking such rest periods may be provided in the cases of vehicles having a crew of two drivers and of vehicles using a ferry-boat or a train).

▶ The minimum duration of the weekly rest should be 24 consecutive hours, preceded or followed by the daily rest.

▶ In long-distance transport, it should be possible to accumulate weekly rest over two consecutive weeks. In appropriate cases, the competent authority or body in each country may approve the accumulation of this rest over a longer time.

8.3 Further information

Graphical symbols – Safety signs – registered safety signs – ISO 7010:2011, International Organisation for Standardisation (ISO)

C153 Hours of Work and Rest Periods (Road Transport) Convention 1979 (No. 153) http://www.ilo.org/dyn/normlex/en/f?p=1000:12100:0::NO::P12100_ILO_CODE:C153 and its accompanying R161 Hours of Work and Rest Periods (Road Transport) Recommendation 1979 (No. 161): http://www.ilo.org/dyn/normlex/en/f?p=NORMLEXPUB:12100:0::NO:12100:P12100_INSTRUMENT_ID:312499:NO

Safety and Health in Construction, ILO CoP, ILO Geneva, 1992, ISBN 92-2-107104-9 http://www.ilo.org/safework/info/standards-and-instruments/codes/WCMS_107826/lang--en/index.htm

8.4 Practice revision questions

1. (a) **Define** the term 'workplace transport' and give **THREE** examples of such transport.
   (b) **Outline** the hazards associated with workplace transport operations.
   (c) **Outline** the factors to be considered when planning traffic routes within a busy warehouse.
   (d) **Describe** the main aspects of a control strategy for safe workplace transport operations.

2. (a) **Identify SIX** ways in which people may be injured by vehicles that are operating in a workplace.
   (b) **Identify** the important contents of a training programme for vehicle drivers so that the risk of accidents to themselves and other employees may be reduced.
   (c) **Outline** the actions required by both drivers and pedestrians to improve the safety of pedestrians in vehicle manoeuvring areas.
   (d) **Outline** the hazards and controls associated with reversing vehicles in a warehouse.

3. The warehouse of a plastic products manufacturer is to be enlarged to enable the use of internal transport to transfer the goods between racking and lorry loading bays.
   (a) **Outline** the design features of the traffic routes that should be addressed in order to minimise the risk of fork-lift truck related accidents.

(b) **Describe** *additional* measures that need to be taken to protect pedestrians from the risk of being struck by a fork-lift truck in the warehouse.

4. **Describe** the rollover (ROPS) and falling-object (FOPS) protection techniques that are used to protect drivers of mobile work equipment. **Outline** the types of unsafe incidents from which drivers may be protected by these techniques.

5. Following an accident between a vehicle and a worker on a construction site, it was decided to examine the vehicle routes on the site. **Outline** suitable control measures that will need to be introduced to reduce the risk of further accidents.

6. Managing work-related road risk is an important part of a health and safety management system. **Outline** the possible risks and controls associated with:
   (a) The driver
   (b) The vehicle
   (c) The journey.

7. (a) **Identify SIX** health and safety rules for drivers on company business.
   (b) **Identify** the particular risks associated with long-distance lorry driving.

8

APPENDIX 8.1 Advice on driving and the use of taxis

The UK Institution of Occupational Safety and Health (IOSH) have produced advice for international travellers when driving or using taxis.

Driving

▶ Carry an up-to-date driving licence and insurance documentation.

▶ Understand local driving practices and ask about bad driving habits, such as for giving way and overtaking. Check on local police methods and carry money for fines.

▶ Carry a local map, be aware of 'no-go' areas, and plan the route thoroughly.

▶ Learn some useful local phrases in case you break down or have an accident.

▶ Ask to inspect and try out a hired vehicle before accepting it – ask for a demonstration. Remember to check tyres, brakes, oil and water levels.

▶ Make sure there's enough fuel for your journey and check ahead for petrol stations on long journeys.

▶ Drive unobtrusively and be observant, particularly of following vehicles. Note familiar landmarks.

▶ Lock the vehicle even if you're leaving it for only a few minutes, such as when refuelling. Keep the passenger doors locked while driving. Leave nothing valuable inside.

▶ Carry emergency equipment (e.g. fire extinguisher, first-aid kit, tool kit, spare bulbs and warning triangle) in the vehicle. In many countries, this is a legal requirement.

▶ Don't get out of the vehicle if you're unsure of your surroundings, or if you're involved in an accident that appears in any way contrived.

▶ Be wary of locals pointing out 'problems' with the car. Carry on to the next busy public place to inspect the vehicle.

▶ The police in some countries aren't always sympathetic to travellers. If possible, tell your office that you're going to the police station before you go. Don't give the police your passport unless you have to – try to use some other form of identification such as an ID card or driving licence.

▶ Make sure that you take lots of water with you if you're driving in a hot climate. It may be impossible to walk very far to get help if you break down.

▶ Always leave enough room between you and the car in front to drive out if you're approached by potential hijackers.

▶ Don't wind your car window down fully when speaking to strangers.

▶ Don't drink and drive. Some countries have lower limits than your home country.

Taxis and drivers

▶ If you're not confident about driving or there's a high risk of carjacking or kidnap, hire a reliable driver.

▶ If possible, book taxis through your hotel or a reliable local contact.

▶ Make a note of the taxi company and the driver's name, car registration, make and colour, and the approximate fare when you book, and check them again before you get into the taxi.

▶ Travel in a licensed taxi with a meter, and make sure the driver uses it.

▶ Don't get into a cab if there's another passenger already there.

▶ Taxi drivers could take criminal advantage if they see a passenger as a newcomer – act naturally and don't ask too many questions.

▶ Always ask drivers who are to meet you at the airport to use your organisation's logo on the meeting card. (This makes it harder for other people to copy your name and try and get your attention before your official driver.) Before getting into the car, make sure they know your name and either put your luggage in the boot yourself or watch as the driver does it.

CHAPTER 9

Musculoskeletal hazards and risk control

9.1 Work-related upper limb disorders ▶ 244

9.2 Manual handling hazards and control measures ▶ 248

9.3 Manually operated load handling equipment ▶ 254

9.4 Powered load handling equipment ▶ 256

9.5 Further information ▶ 266

9.6 Practice revision questions ▶ 267

Appendix 9.1 **A typical UK risk assessment for the use of lifting equipment** ▶ 269

Appendix 9.2 **Examples of manually operated load handling equipment** ▶ 270

This chapter covers the following NEBOSH learning objectives:

1. Explain work processes and practices that may give rise to work-related upper limb disorders and appropriate control measures
2. Explain the hazards and control measures which should be considered when assessing risks from manual handling activities
3. Explain the hazards and controls to reduce the risk in the use of lifting and moving equipment with specific reference to manually operated load moving equipment
4. Explain the hazards and the precautions and procedures to reduce the risk in the use of lifting and moving equipment with specific reference to powered load moving equipment

9.1 Work-related upper limb disorders

9.1.1 Introduction

Musculoskeletal diseases are often caused by poor workstation design, particularly with display screen equipment (DSE), and the manual handling of loads. Until a few years ago, accidents caused by the manual handling of loads were the largest single cause of over-3-day accidents reported to the UK Health and Safety Executive (HSE). The Manual Handling Operations Regulations recognised this fact and helped to reduce the number of these accidents. However, accidents due to poor manual handling technique still account for over 25% of all reported accidents and in some occupational sectors, such as the UK National Health Service, the figure rises above 50%. An understanding of the factors causing some of these accidents is essential if they are to be further reduced. Mechanical handling methods should always be used whenever possible, but they are not without their hazards. Much mechanical handling involves the use of lifting equipment, such as cranes and lifts, which present specific hazards to both the users and persons/workers in the vicinity. The risks from these hazards are reduced by thorough examinations and inspections. It is interesting to note that the ILO recognised the health problems arising from manual handling in 1967 when it introduced the Maximum Weight Convention (No. 127) and its accompanying Maximum Weight Recommendation (No. 128). More information is given on these two documents later in this chapter (see 9.2.6).

9.1.2 The principles and scope of ergonomics

Ergonomics is the study of the interaction between workers and their work in the broadest sense, in that it encompasses the whole system surrounding the work process. It is, therefore, as concerned with the work organisation, process and design of the workplace and work methods as it is with work equipment. The common definitions of ergonomics, the 'man–machine interface' or 'fitting the man to the machine rather than vice versa' are far too narrow.

It is concerned about the physical and mental capabilities of an individual as well as their understanding of the job under consideration. Ergonomics includes the limitations of the worker in terms of skill level, perception and other personal factors in the overall design of the job and the system supporting and surrounding it. It is the study of the relationship between the worker, the machine and the environment in which it operates and attempts to optimise the whole work system, including the job, to the capabilities of the worker so that maximum output is achieved for minimum effort and discomfort by the worker. Cars, buses and lorries are all ergonomically designed so that all the important controls, such as the steering wheel, brakes, gear stick and instrument panel, are easily accessed by most drivers within a wide range of sizes. Ergonomics is sometimes described as human engineering and as working practices become more and more automated, the need for good ergonomic design becomes essential.

The scope of ergonomics and an ergonomic assessment is very wide incorporating the following areas of study:

▶ personal factors of the worker, in particular physical, mental and intellectual abilities, body dimensions and competence in the task required;
▶ the machine and associated equipment under examination;

Figure 9.1 Handling roofing slates onto a roof using a teleporter lift truck

▶ the interface between the worker and the machine – controls, instrument panel or gauges and any aids including seating arrangements and hand tools;

▶ environmental issues affecting the work process such as lighting, temperature, humidity, noise and atmospheric pollutants;

▶ the interaction between the worker and the task, such as the production rate, posture and system of working;

▶ the task or job itself – the design of a safe system of work, checking that the job is not too strenuous or repetitive and the development of suitable training packages;

▶ the organisation of the work, such as shift work, breaks and supervision.

The reduction of the possibility of human error is one of the major aims of ergonomics and an ergonomic assessment. An important part of an ergonomic study is to design the workstation or equipment to fit the worker. For this to be successful, the physical measurement of the human body and an understanding of the variations in these measurements between people are essential. Such a study is known as **anthropometry,** which is defined as the scientific measurement of the human body and its movement. Since there are considerable variations in, for example, the heights of people, it is common for some part of the workstation to be variable (e.g. an adjustable seat).

9.1.3 The ill-health effects of poor ergonomics

Ergonomic hazards are those hazards to health resulting from poor ergonomic design. They generally fall within the physical hazard category and include the manual handling and lifting of loads, pulling and pushing loads, prolonged periods of repetitive activities and work with vibrating tools. The condition of the working environment, such as low lighting levels, can present health hazards to the eyes. It is also possible for psychological conditions, such as occupational stress, to result from ergonomic hazards.

The common ill-health effects of ergonomic hazards are musculoskeletal disorders (MSDs), work-related upper limb disorders (WRULDs) including repetitive strain injury (RSI), and deteriorating eyesight.

9.1.4 Work-related upper limb disorders

Work-related upper limb disorders (WRULDs) describe a group of conditions which can affect the neck, shoulders, arms, elbows, wrists, hands and fingers. **Tenosynovitis** (affecting the tendons), **carpal tunnel syndrome** (affecting the tendons which pass through the carpal bone in the hand) and **frozen shoulder** are all examples of WRULDs which differ in the manifestation and site of the illness. The term **RSI** is commonly used to describe WRULDs. *Repetitive Strain Injury*

WRULDs are caused by repetitive movements of the fingers, hands or arms which involve pulling, pushing, reaching, twisting, lifting, squeezing or hammering. These disorders can occur to workers in offices as well as in factories or on construction sites. Typical occupational groups at risk include painters and decorators, riveters and pneumatic drill operators and desktop computer users.

The main symptoms of WRULDs are aching pain to the back, neck and shoulders, swollen joints and muscle

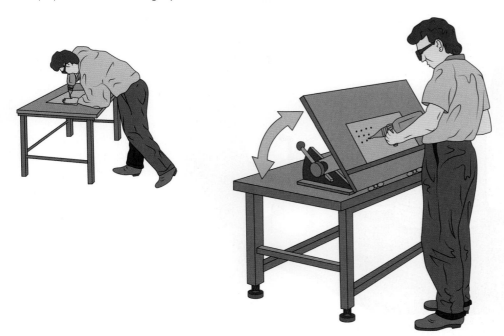

Figure 9.2 A tilted worktable. The distance between the operator and the work can be reduced by putting the table at a more vertical angle. The table is adjustable in height and angle to suit the particular job

fatigue accompanied by tingling, soft tissue swelling, similar to bruising, and a restriction in joint movement. The sense of touch and movement of fingers may be affected. The condition is normally a chronic one in that it gets worse with time and may lead eventually to permanent damage. The injury occurs to muscle, tendons and/or nerves. If the injury is allowed to heal before being exposed to the repetitive work again, no long-term damage should result. However, if the work is repeated again and again, healing cannot take place and permanent damage can result leading to a restricted blood flow to the arms, hands and fingers.

The risk factors, which can lead to the onset of WRULDs, are repetitive actions of lengthy duration, the application of significant force and unnatural postures, possibly involving twisting and over-reaching, and the use of vibrating tools. Cold working environments, work organisation and worker perception of the work organisation have all been shown in studies to be risk factors, as is the involvement of vulnerable workers such as those with pre-existing ill-health conditions and pregnant women.

Ergonomic improvements have been implemented in several small-scale industries in developing countries in Africa, Asia and Latin America. Most improvements have been made in areas directly related to work processes such as materials handling, workstation design (Figure 9.3), isolating hazards, lighting, welfare facilities and work organisation. All these changes have been low cost and have often resulted from training courses employing the Work Improvement in Small Enterprises (WISE) methodology developed by the ILO. Table 9.1 gives some examples reported by the ILO.

Table 9.1 Examples of low-cost ergonomic and other improvements resulting from an ILO initiative

Topic	Examples
Premises	Good layout, heat insulation, shades, natural ventilation, smooth floors
Workstation design	Easy reach, work height, fixtures, easy-to-read displays, good chairs
Work organisation	Combining tasks, buffer stocks, group work, rotation, breaks
Welfare facilities	Drinking water, clean toilets, rest corners, eating place, first-aid kits
Lighting	Skylights, re-positioning lights, light-coloured walls, avoiding glare
Materials handling	Marking passage borders, multi-shelves, mobile racks, carts, lifts
Isolating hazards	Covers, guards, machine-feeding devices, isolating hazard sources

9.1.5 Display screen equipment

Display screen equipment (DSE), which includes visual display units, is a good example of a common work activity which relies on an understanding of ergonomics and the ill-health conditions which can be associated with poor ergonomic design. A recent survey of safety representatives by the UK Trades Union Congress found that injuries/illnesses caused by the poor use of DSE and repetitive strain injuries together with stress or overwork were among their major concerns.

The basic recommendations for safeguarding the health and safety of DSE workers are:

▶ a suitable and sufficient risk assessment of the workstation, including the software in use, trip and electrical hazards from trailing cables and the surrounding environment;

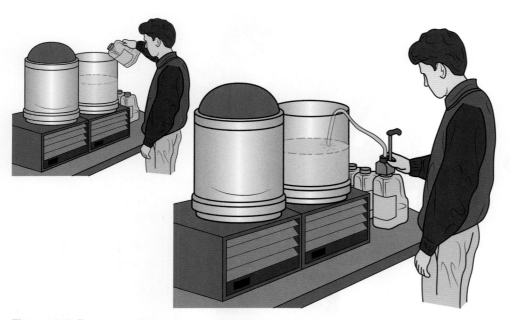

Figure 9.3 Feed material into a machine from a bulk bin or drum by suction or pumping to save workers having to climb steps carrying loads. This reduces the risk of trips and falls as well as avoiding manual handling and improves the efficiency of your process. Consider this solution wherever you have to transfer powders, granules and liquids

▶ workstation compliance with the minimum specifications laid down in the national or international standards;

▶ a plan of the work programme to ensure that there are adequate breaks in the work pattern of workers;

▶ the provision of eye sight tests and, if required, spectacles to users of DSE;

▶ a suitable programme of training and sufficient information given to all DSE users.

The risk assessment of a DSE workstation needs to consider the following factors, many of which are shown in Figure 9.4:

▶ the height and adjustability of the monitor;

▶ the adjustability of the keyboard, the suitability of the mouse and the provision of wrist support;

▶ the stability and adjustability of the DSE user's chair;

▶ the provision of ample foot room and suitable foot support;

▶ the effect of any lighting and window glare at the workstation;

▶ the storage of materials around the workstation;

▶ the safety of trailing cables, plugs and sockets;

▶ environmental issues: noise, heating, humidity and draughts.

Chapter 18 gives an example of a checklist that can be used for a DSE workstation assessment by the user. There are three basic ill-health hazards associated with DSE. These are:

▶ musculoskeletal problems;
▶ visual problems;
▶ psychological problems.

A fourth hazard, of radiation, has been shown from several studies to be very small and is now no longer normally considered in the risk assessment.

Similarly, in the past, there have been suggestions that DSE could cause epilepsy and there were concerns about adverse health effects on pregnant women and their unborn children. All these risks have been shown in various studies to be very low.

The provision of DSE training and risk assessments online has become more common. The risk assessment, however, still has to be managed, made appropriate to the particular workplace setting and reviewed from time to time. Various studies have shown that users of any e-learning package lose concentration after 30–40 minutes.

Musculoskeletal problems

Tenosynovitis is the most common and well-known problem which affects the wrist of the user. The symptoms and effects of this condition have already been covered. Suffice it to say that if the condition is ignored, then the tendon and tendon sheath around the wrist will become permanently injured.

Tenosynovitis is caused by the continual use of a keyboard and can be relieved by the use of wrist

Seating and posture for typical office tasks

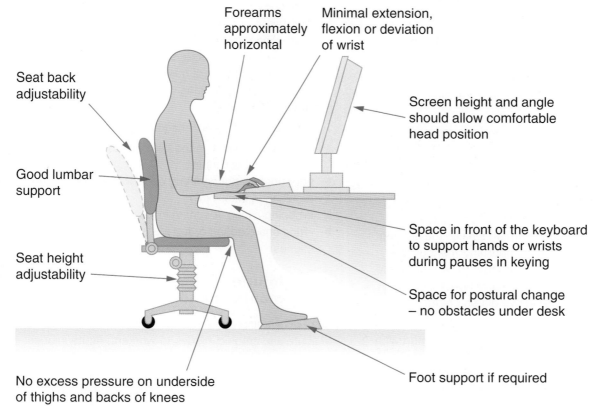

Forearms approximately horizontal

Minimal extension, flexion or deviation of wrist

Seat back adjustability

Screen height and angle should allow comfortable head position

Good lumbar support

Seat height adjustability

Space in front of the keyboard to support hands or wrists during pauses in keying

Space for postural change – no obstacles under desk

No excess pressure on underside of thighs and backs of knees

Foot support if required

Figure 9.4 Workstation design

supports. Other WRULDs are caused by poor posture and can produce pains in the back, shoulders, neck or arms. Less commonly, pain may also be experienced in the thighs, calves and ankles. These problems can be mitigated by the application of ergonomic principles in the selection of working desks, chairs, footrests and document holders. It is also important to ensure that the desk is at the correct height and the computer screen is tilted at the correct angle to avoid putting too much strain on the neck. (Ideally the user's eyes should be at the same height as the top of the screen.)

The keyboard should be detachable so that it can be positioned anywhere on the desktop and a correct posture adopted while working at the keyboard. The chair should be adjustable in height, stable and have an adjustable backrest. If the knees of the user are lower than the hips when seated, then a footrest should be provided. The surface of the desk should be non-reflecting and uncluttered but ancillary equipment (e.g. telephone and printer) should be easily accessible.

Visual problems

There does not appear to be much medical evidence that DSE causes deterioration in eye sight, but users may suffer from visual fatigue which results in eye strain, sore eyes and headaches. Less common ailments are skin rashes and nausea.

The use of DSE may indicate that reading spectacles are needed and it is possible that any prescribed lenses may only be suitable for DSE work as they will be designed to give optimum clarity at the normal distance at which screens are viewed (50–60 cm).

Eye strain is a particular problem for people who spend a large proportion of their working day using DSE. A survey has indicated that up to 90% of DSE users complain of eye fatigue. Eye strain can be reduced by the following steps additional to those already identified in this section:

▶ train staff in the correct use of the equipment;
▶ ensure that a font size of at least 12 is used on the screen;
▶ ensure that users take regular breaks away from the screen (up to 10 minutes every hour).

The screen should be adjustable in tilt angle and screen brightness and contrast. Finally, the lighting around the workstation is important. It should be bright enough to allow documents to be read easily but not too bright such that either headaches are caused or there are reflective glares on the computer screen.

Sore eyes are a common complaint associated with low humidity. Various professional building research organisations have agreed that an appropriate range of relative humidity in an office environment is 40 to 60%. Air conditioning systems should be properly maintained to ensure that they maintain a reasonable temperature

and humidity level in the workplace. Sore eyes are often a product of dry eyes which may be controlled by:

▶ restraining from rubbing the eyes;
▶ focusing on an object in the near distance (preferably green in colour);
▶ blinking can encourage the moistening of the eye;
▶ drinking water regularly;
▶ looking away from the screen periodically to rest the eyes.

Psychological problems

These are generally stress-related problems. They may have environmental causes, such as noise, heat, humidity or poor lighting, but they are usually due to high-speed working, lack of breaks, poor training and poor workstation design. One of the most common problems is the lack of understanding of all or some of the software packages being used.

There are several other processes and activities where ergonomic considerations are important. These include the assembly of small components (microelectronics assembly lines) and continually moving assembly lines (car assembly plants).

9.2 Manual handling hazards and control measures

The term 'manual handling' is defined as the movement, transporting or supporting of a load by human effort alone. This effort may be applied directly or indirectly using a rope or a lever. Manual handling may involve the transportation of the load or the direct support of the load including pushing, pulling, carrying, moving using bodily force and, of course, straightforward lifting. Back injury due to the lifting of heavy loads is very common and several million working days are lost each year as a result of such injuries.

Typical hazards of manual handling include:

▶ lifting a load which is too heavy or too cumbersome, resulting in back injury;
▶ poor posture during lifting or poor lifting technique, resulting in back injury;
▶ dropping a load, resulting in foot injury;
▶ lifting sharp-edged or hot loads resulting in hand injuries.

9.2.1 Injuries caused by manual handling

Manual handling operations can cause a wide range of acute and chronic injuries to workers. Acute injuries normally lead to sickness leave from work and a period of rest during which time the damage heals. Chronic injuries build up over a long period of time and are usually irreversible, producing illnesses such as arthritic and spinal disorders. There is considerable

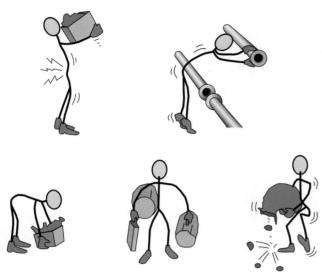

Figure 9.5 Manual handling: there are many potential hazards

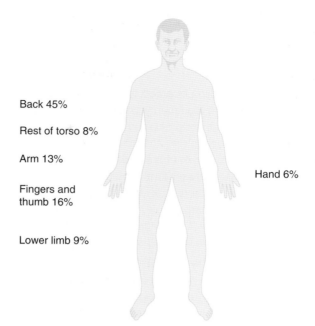

Back 45%

Rest of torso 8%

Arm 13%

Fingers and thumb 16%

Lower limb 9%

Hand 6%

Figure 9.6 Main injury sites caused by manual handling accidents

evidence to suggest that modern lifestyles, such as a lack of exercise and regular physical effort, have contributed to the long-term serious effects of these injuries.

The most common injuries associated with poor manual handling techniques are all musculoskeletal in nature and are:

▶ muscular sprains and strains – caused when a muscular tissue (or ligament or tendon) is stretched beyond its normal capability leading to a weakening, bruising and painful inflammation of the area affected. Such injuries normally occur in the back or in the arms and wrists;

▶ back injuries – include injuries to the discs situated between the spinal vertebrae (i.e. bones) and can lead to a very painful prolapsed disc lesion (commonly known as a slipped disc). This type of injury can lead to other conditions known as lumbago and sciatica (where pain travels down the leg);

▶ trapped nerve – usually occurring in the back as a result of another injury but aggravated by manual handling;

▶ hernia – this is a rupture of the body cavity wall in the lower abdomen, causing a protrusion of part of the intestine. This condition eventually requires surgery to repair the damage;

▶ cuts, bruising and abrasions – caused by handling loads with unprotected sharp corners or edges;

▶ fractures – normally of the feet due to the dropping of a load. Fractures of the hand also occur but are less common;

▶ work-related upper limb disorders (WRULDs) – cover a wide range of musculoskeletal disorders, which are discussed later in this chapter;

▶ rheumatism – this is a chronic disorder involving severe pain in the joints. It has many causes, one of which is believed to be the muscular strains induced by poor manual handling lifting technique.

The sites on the body of injuries caused by manual handling accidents are shown in Figure 9.6.

Musculoskeletal problems are the most common cause of absence, followed by viral infections and stress-related illnesses. These findings are based on an analysis of sickness management records for 11,000 individual employees across a range of private sector organisations.

A recent study has found that musculoskeletal disorders (MSDs) account for nearly half (49%) of all absences from work and 60% of permanent work incapacity in the European Union. It is estimated that this costs the UK economy £7 billion every year and costs Europe £240 billion each year. The study, conducted across 25 European countries, found that 100 million Europeans suffer from chronic musculoskeletal pain – over 40 million of whom are workers – with up to 40% having to give up work due to their condition. In the UK alone, 9.5 million working days were lost in one year due to musculoskeletal problems.

In general, pulling a load is much easier for the body than pushing one. If a load can only be pushed, then pushing backwards using the back is less stressful on body muscles. Lifting a load from a surface at waist level is easier than lifting from floor level and most injuries during lifting are caused by lifting and twisting at the same time. If a load has to be carried, it is easier to carry it at waist level and close to the body trunk. A firm grip is essential when moving any type of load.

9.2.2 Hierarchy of measures for manual handling operations

The emphasis during the assessment of lifting operations has changed from a simple reliance on safe lifting techniques to an analysis, using risk assessment, of the need for manual handling. A clear hierarchy of measures to be taken when an employer is confronted with a manual handling operation has been established as follows:

1. Avoid manual handling operations that involve a risk of injury so far as is reasonably practicable by either redesigning the task to avoid moving the load or by automating or mechanising the operations.
2. If manual handling cannot be avoided, then a suitable and sufficient risk assessment should be made.
3. Reduce the risk of injury from those operations so far as is reasonably practicable, either by the use of mechanical handling or making improvements to the task, the load and the working environment.

The guidance given in the UK HSE Manual Handling (Guidance) (L23) (the full reference is given at the end of this chapter) is a very useful document. It gives very helpful advice on manual handling assessments and manual handling training. The advice is applicable to all occupational sectors. Chapter 18 gives an example of manual handling assessment forms.

9.2.3 Manual handling assessments

There are four main factors which must be taken into account when undertaking a manual handling assessment. These are the task, the load, the working environment and the capability of the individual who is expected to do the lifting.

The **task** should be analysed in detail so that all aspects of manual handling are covered including the use of mechanical assistance. The number of people involved and the cost of the task should also be considered. Some or all of the following questions are relevant to most manual handling tasks:

▶ Is the load held or manipulated at a distance from the trunk? The further from the trunk, the more difficult it is to control the load and the stress imposed on the back is greater.
▶ Is a satisfactory body posture being adopted? Feet should be firmly on the ground and slightly apart and there should be no stooping or twisting of the trunk. It should not be necessary to reach upwards, as this will place additional stresses on the arms, back and shoulders. The effect of these risk factors is significantly increased if several are present while the task is being performed.
▶ Are there excessive distances to carry or lift the load? Over distances greater than 10 m, the physical demands of carrying the load will dominate the operation. The frequency of lifting and the vertical and horizontal distances the load needs to be carried (particularly if it has to be lifted from the ground and/or placed on a high shelf) are very important considerations.
▶ Is there excessive pulling and pushing of the load? The state of floor surfaces and the footwear of the

individual should be noted so that slips and trips may be avoided.
▶ Is there a risk of a sudden movement of the load? The load may be restricted or jammed in someway.
▶ Is frequent or prolonged physical effort required? Frequent and prolonged tasks can lead to fatigue and a greater risk of injury.
▶ Are there sufficient rest or recovery periods? Breaks and/or the changing of tasks enables the body to recover more easily from strenuous activity.
▶ Is there an imposed rate of work on the task? This is a particular problem with some automated production lines and can be addressed by spells on other operations away from the line.
▶ Are the loads being handled while the individual is seated? In these cases, the legs are not used during the lifting processes and stress is placed on the arms and back.
▶ Does the handling involve two or more people? The handling capability of an individual reduces when he/she becomes a member of a team (e.g. for a three-person team, the capability is half the sum of the individual capabilities). Visibility, obstructions and the roughness of the ground must all be considered when team handling takes place.

The **load** must be carefully considered during the assessment and the following questions asked:

▶ Is the load too heavy? The maximum load that an individual can lift will depend on the capability of the individual and the position of the load relative to the body. There is therefore no safe load. Figure 9.7 is reproduced from the UK HSE guidance, which does give some advice on loading levels. It recommends that loads in excess of 25 kg should not be lifted or carried by a man (and this is only permissible when the load is at the level of and adjacent to the thighs). For women, the guideline figures should be reduced by about one-third.
▶ Is the load too bulky or unwieldy? In general, if any dimension of the load exceeds 0.75 m (approx. 2 ft.), its handling is likely to pose a risk of injury. Visibility around the load is important. It may hit obstructions or become unstable in windy conditions. The position of the centre of gravity is very important for stable lifting – it should be as close to the body as possible.
▶ Is the load difficult to grasp? Grip difficulties will be caused by slippery surfaces, rounded corners or a lack of foot room.
▶ Are the contents of the load likely to shift? This is a particular problem when the load is a container full of smaller items, such as a sack full of nuts and bolts. The movements of people (in a nursing home) or animals (in a veterinary surgery) are loads which fall into this category.
▶ Is the load sharp, hot or cold? Personal protective equipment may be required.

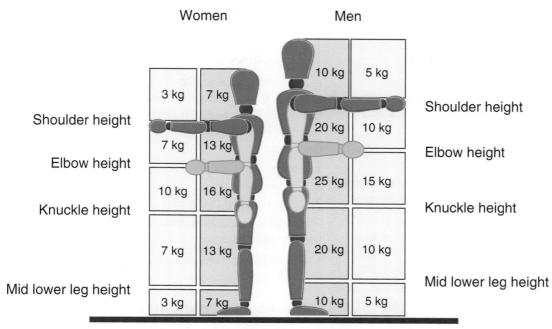

Figure 9.7 UK HSE guidance for manual lifting – recommended weights

The **working environment** in which the manual handling operation is to take place must be considered during the assessment. The following areas will need to be assessed:

▶ any space constraints which might inhibit good posture. Such constraints include lack of headroom, narrow walkways and items of furniture;

▶ slippery, uneven or unstable floors;

▶ variations in levels of floors or work surfaces, possibly requiring the use of ladders;

▶ extremes of temperature and humidity. These effects were discussed in detail in Chapter 7;

▶ ventilation problems or gusts of wind;

▶ poor lighting conditions.

Finally, the capability of the **individual** to lift or carry the load must be assessed. The following questions will need to be asked:

▶ Does the task require unusual characteristics of the individual (e.g. strength or height)? It is important to remember that strength and general manual handling ability depends on age, gender, state of health and fitness.

▶ Are employees who might reasonably be considered to be pregnant or to have a health problem put at risk by the task? Particular care should be taken to protect pregnant women or those who have recently given birth from handling loads. Allowance should also be given to any employee who has a health problem which could be exacerbated by manual handling.

The assessment must be reviewed if there is reason to suspect that it is no longer valid or there has been a significant change to the manual handling operations to which it relates.

9.2.4 Reducing the risk of injury

This involves the introduction of control measures resulting from the manual handling risk assessment. The UK HSE guidance to Manual Handling (L23) and the HSE publication *Manual Handling – Solutions You Can Handle* (HSG115) contain many ideas to reduce the risk of injury from manual handling operations. An ergonomic approach is generally required to design and develop the manual handling operation as a whole. The control measures can be grouped under five headings. However, the first consideration, when it is reasonably practicable, is mechanical assistance.

Mechanical assistance involves the use of mechanical aids to assist the manual handling operation such as wheelbarrows, hand-powered hydraulic hoists, specially adapted trolleys, hoists for lifting patients, roller conveyors and automated systems using robots. This topic is discussed later in this chapter (see 9.3).

The **task** can be improved by changing the layout of the workstation by, for example, storing frequently used loads at waist level. The removal of obstacles and the use of a better lifting technique that relies on the leg rather than back muscles should be encouraged. When pushing, the hands should be positioned correctly. The work routine should also be examined to see whether job rotation is being used as effectively as it could be. Special attention should be paid to seated manual handlers to ensure that loads are not lifted from the floor while they are seated. Employees should be encouraged to seek help if a difficult load is to be moved so that a team of people can move the load. Adequate and suitable personal protective equipment should be provided where there is a risk of loss of grip or injury. Care must be taken to ensure that the clothing does not become a hazard in itself (e.g. the snagging of fasteners and pockets).

The **load** should be examined to see whether it could be made lighter, smaller or easier to grasp or manage. This could be achieved by splitting the load, the positioning of handholds or a sling, or ensuring that the centre of gravity is brought closer to the handler's body. Attempts should be made to make the load more stable and any surface hazards, such as slippery deposits or sharp edges, should be removed. It is very important to ensure that any improvements do not, inadvertently, lead to the creation of additional hazards.

The **working environment** can be improved in many ways. Space constraints should be removed or reduced. Floors should be regularly cleaned and repaired when damaged. Adequate lighting is essential and working at more than one level should be minimised so that hazardous ladder work is avoided. Attention should be given to the need for suitable temperatures and ventilation in the working area.

The **capability of the individual** is the fifth area where control measures can be applied to reduce the risk of injury. The state of health of the employee and his/her medical record will provide the first indication as to whether the individual is capable of undertaking the task. A period of sick leave or a change of job can make an individual vulnerable to manual handling injury. The ILO requires that the employee be given information and training. The information includes the provision, where it is reasonably practicable to do so, of precise information on the weight of each load and the heaviest side of any load whose centre of gravity is not centrally positioned. In a more detailed risk assessment, other factors will need to be considered such as the effect of personal protective equipment and psychosocial factors in the work organisation. The following points may need to be assessed:

1. Does protective clothing hinder movement or posture?
2. Is the correct personal protective equipment being worn?
3. Is proper consideration given to the planning and scheduling of rest breaks?
4. Is there good communication between managers and employees during risk assessment or workstation design?
5. Is there a mechanism in place to deal with sudden changes in the volume of workload?
6. Have employees been given sufficient training and information?
7. Does the worker have any learning disabilities and, if so, has this been taken into account in the assessment?
8. Is the worker physically unsuited to carry out the tasks in question because he or she:

    (a) is wearing unsuitable clothing, footwear or other personal effects?

    (b) does not have adequate or appropriate knowledge or training?

The UK HSE has developed a Manual Handling Assessment Chart (MAC) to help with the assessment of common risks associated with lifting, carrying and handling. It incorporates a numerical and a colour coding score system to highlight high-risk manual handling tasks. There are three types of assessment that can be carried out with the MAC:

1. lifting operations;
2. carrying operations;
3. team handling operations.

The MAC is available on the HSE website: http://www.hse.gov.uk/msd/mac/

The training requirements are given in the following section.

9.2.5 Manual handling training

Training alone will not reduce manual handling injuries – there still needs to be safe systems of work in place and the full implementation of the control measures highlighted in the manual handling assessment. The following topics should be addressed in a manual handling training session:

▶ types of injuries associated with manual handling activities;
▶ the findings of the manual handling assessment;
▶ the recognition of potentially hazardous manual handling operations;
▶ the correct use of mechanical handling aids;
▶ the correct use of personal protective equipment;
▶ features of the working environment which aid safety in manual handling operations;
▶ good housekeeping issues;
▶ factors which affect the capability of the individual;
▶ good lifting or manual handling technique as shown in Figure 9.8.

The 'good handling technique' given in the UK HSE Guidance on the Manual Handling Operations Regulations (L23) advises:

▶ think before you lift;
▶ keep the load close to your waist;
▶ adopt a stable position;
▶ ensure a good hold on the load;
▶ at the start of the lift, moderate flexion (slight bending) of the back, hips and knees is preferable to fully flexing the back (stooping) or the hips and knees (squatting);
▶ don't flex your spine any further as you lift;
▶ avoid twisting the trunk or leaning sideways, especially while the back is bent;
▶ keep your head up when handling;
▶ move smoothly;
▶ don't lift more than you can easily manage; and
▶ put down then adjust.

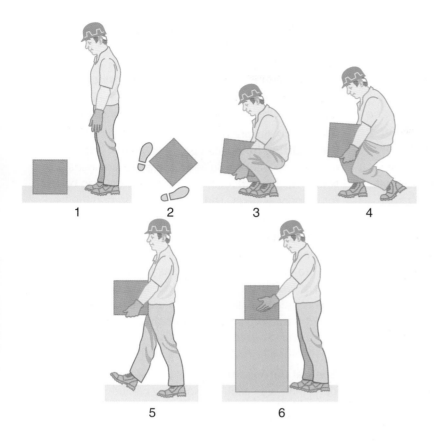

1. Check suitable clothing and assess load. Heaviest side to body.

2. Place feet apart – bend knees.

3. Firm grip – close to body Slight bending of back, hips and knees at start.

4. Lift smoothly to knee level and then waist level. No further bending of back.

5. With clear visibility move forward without twisting. Keep load close to the waist. Turn by moving feet. Keep head up. Do not look at load.

6. Set load down at waist level or to knee level and then on the floor.

Figure 9.8 The main elements of a good lifting technique

Finally, it needs to be stressed that if injuries involving manual handling operations are to be avoided, planning, control and effective supervision are essential.

9.2.6 ILO Recommendations on manual handling

Manual handling of loads is covered by the ILO in the Maximum Weight Convention, 1967 (No. 127) and its accompanying Maximum Weight Recommendation, 1967 (No. 128).

In the Convention, the ILO requires that no worker shall be required, or permitted to, engage in the manual transport of a load which, by reason of its weight, is likely to threaten his/her health or safety. Also each member state must take appropriate steps to ensure that any worker assigned to the manual transport of loads other than light loads receives adequate training or instruction in working techniques, with a view to safeguarding health and preventing accidents. In order to limit or to facilitate the manual transport of loads, suitable technical devices shall be used as much as possible.

The Convention restricts the assignment of women and young workers to manually transport loads other than light loads – which shall be limited. Where women and young workers are engaged in the manual transport of loads, the maximum weight of such loads shall be substantially less than that permitted for adult male workers.

The Recommendation (R128) states any worker assigned to regular manual transport of loads should,

prior to such assignment, receive adequate training or instruction in manual handling techniques. Such training or instruction should include methods of lifting, carrying, putting down, unloading and stacking of different types of loads, and should be given by suitably qualified persons or institutions. It should be followed up by supervision on the job to ensure that the correct methods are used. Any worker occasionally assigned to manual transport of loads should be given appropriate instructions on the manner in which such operations may be safely carried out. A medical examination for fitness for employment should, as far as practicable and appropriate, be required before assignment to regular manual transport of loads.

The Recommendation continues to discuss suitable technical devices and packaging to facilitate the manual transport of loads. The packaging of loads which may be transported manually should be compact and of suitable material and should, as far as possible and appropriate, be equipped with devices for holding and designed so as not to create risk of injury (e.g. it should not have sharp edges, projections or rough surfaces). Member states should take account of:

(a) physiological characteristics, environmental conditions and the nature of the work to be done;

(b) any other conditions which may influence the health and safety of the worker.

Where the maximum permissible weight which may be transported manually by one adult male worker is more than 25 kg, measures should be taken as speedily

253

as possible to reduce it to that level. Where adult women workers are engaged in the manual transport of loads, the maximum weight of such loads should be substantially less than that permitted for adult male workers.

As far as possible, adult women workers should not be assigned to regular manual transport of loads. Where adult women workers are assigned to regular manual transport of loads, provision should be made:

▶ as appropriate, to reduce the time spent on actual lifting, carrying and putting down of loads by such workers;

▶ to prohibit the assignment of such workers to certain specified jobs, comprised in manual transport of loads, which are especially arduous.

No woman should be assigned to manual transport of loads during a pregnancy which has been medically determined or during the ten weeks following confinement if in the opinion of a qualified physician such work is likely to impair her health or that of her child.

Where young workers are engaged in the manual transport of loads, the maximum weight of such loads should be substantially less than that permitted for adult workers of the same sex.

As far as possible, young workers should not be assigned to regular manual transport of loads. Where the minimum age for assignment to manual transport of loads is less than 16 years, measures should be taken as speedily as possible to raise it to that level. The minimum age for assignment to regular manual transport of loads should be raised, with a view to attaining a minimum age of 18 years. Where young workers are assigned to regular manual transport of loads, provision should be made:

(a) as appropriate, to reduce the time spent on actual lifting, carrying and putting down of loads by such workers;

(b) to prohibit the assignment of such workers to certain specified jobs, comprised in manual transport of loads, which are especially arduous.

9.3 Manually operated load handling equipment

9.3.1 Types of manually operated load handling equipment

As mentioned earlier, the first consideration to reduce the risk of injury due to manual handling is, when it is reasonably practicable, to use mechanical assistance in the form of:

▶ simple tools;
▶ wheelbarrows;
▶ trucks and trolleys;
▶ roller tracks and chutes;
▶ pallet trucks;

▶ conveyors; and
▶ various types of hoists that can be used to lift people as well as other loads.

The aids vary from simple, manually operated tools to power-assisted trucks and lifting devices. All of them 'lighten the load' and reduce the risk of injury.

All kinds of **simple tools** can aid the manual handling of loads. **Lifting hooks** can be used to lift sheets of steel or glass, timber boards and large awkward loads. **Log tongs** will help lift logs, and other devices can be used for cylindrical loads. These tools help to grip the load and reduce the need for bending.

Trucks and trolleys allow one person to transport loads between different locations. They can be inexpensive and are available in various sizes to suit the load and type of workplace. **Sack trucks,** like **wheelbarrows,** move loads by balancing them on the truck axle. Some trucks have lifting mechanisms that allow loads to be raised and lowered and others are fitted with special wheels for climbing and descending stairs.

General purpose trucks can be used to support larger loads than with sack trucks. They can be flat-topped or fitted with a variety of sides and wheels to suit different uses. The wheels can be fitted with swivels to improve manoeuvrability. Some **platform trucks** are fitted with detachable tug units. This helps to reduce congestion and obstacles in busy production areas. The platform can be designed to be raised or lowered – this further reduces the manual handling because it reduces the need for bending when loading and unloading. Trucks can have big wheels for use on rough ground. When fitted with removable sides they can be used for bulk loads like sand and gravel. **Balance trucks** have a central axle with swivel wheels at each end so they can spin around their centres. This makes them highly manoeuvrable so they can be used in restricted spaces.

There is no clear distinction between general purpose trucks and trolleys. **Trolleys** tend to be of lighter weight construction and designed for more specific applications. Container trolleys allow mixed loads to be carried. A supermarket trolley is a simple example of this. Shelf trolleys can have fixed or removable shelves. Drum trolleys are useful for transporting drums. Brakes can be fitted to trucks and trolleys to keep them stationary and may be necessary where the ground slopes. Garment rails can be used in clothes factories, shops and theatres.

Roller tracks and chutes allow heavy and bulky loads to be moved manually or by gravity under their own weight. They can be portable or set into the floor. Carefully designed work areas with appropriate tracks and chutes can reduce the number of manual handling tasks. When using tracks, the potential for creating a tripping hazard must be kept as low as possible. Using gentle gradients can ensure that the load is moved with very little effort. Chutes are normally used instead of tracks when there are significant changes of level, for

Figure 9.9 A pallet truck

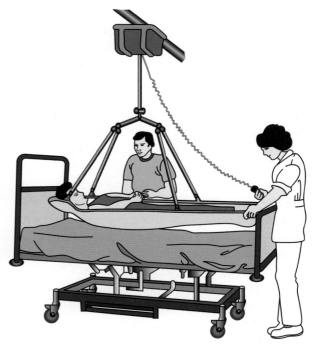

Figure 9.10 Mechanical aids to lift patients in hospital

example movement between floors. Spiral chutes are typically used for sacks but almost any kind of load can be handled on a straight chute.

Pallet trucks are moved by pedestrians. Manual effort is required to transfer the load but hydraulic power is normally used to raise and lower the load. They can be used in fairly congested and confined areas and are designed to move different types of loads.

Portable conveyors are used to transport loads between places at the same level or different heights. Different types can transport a wide variety of loads including bulk materials like sand and grain. Using portable conveyors can significantly reduce (and often avoid) manual handling. The height of the conveyor at loading and unloading points is important. Recommended heights to reduce manual handling efforts are:

▶ around 0.9 m above floor level when light loads which can normally be lifted with one hand are to be lifted;

▶ around 0.75 m above floor level when heavier loads, normally lifted with two hands, are loaded; and

▶ around floor level when heavy loads, like drums, are loaded onto conveyors, either by rolling or by lifting devices or lift trucks.

Portable conveyors for occasional or semi-permanent occasions can be used in different work areas. Some conveyors are fitted with wheels to increase their mobility. They can be used to move loads at the same level or at different levels. Some types can be inclined by hydraulic rams, to adjust the height of transfer to suit a particular elevation. Non-powered roller conveyors on a slight incline allow loads to move under gravity. Powered conveyers and elevators are covered in 9.4.2.

Mechanical assistance, therefore, involves the use of mechanical aids to assist the manual handling operation such as wheelbarrows, hand-powered hydraulic hoists, specially adapted trolleys, hoists for lifting patients, roller conveyors and automated systems using robots.

Some examples of manually operated load handling equipment are shown in Appendix 9.2.

9.3.2 Hazards associated with manually operated load handling equipment

The principal hazards with the use of load handling equipment are due to its incorrect use, such as overloading or attempting to carry unstable loads. Pushing or pulling a truck or trolley is still a manual handling operation and can create different kinds of risks. Injuries could also be suffered by pedestrians using the same walkway.

Lack of maintenance can also create additional hazards. This is particularly important for certain lifting machines and tackle where there may be a legal requirement for a regular examination by a competent person (see 9.4.3).

9.3.3 Precautions with the use of manually operated load handling equipment

When handling aids are being selected, the subsequent user should be consulted whenever it is possible. Moving and handling tasks are often made easier by good design. For example, in the health and social care sector, workers are often required to care for service users in bed and the provision of adjustable height beds can prevent the risks of back injuries. Raising the height of laundry equipment, such as washing machines or driers, by placing it on a platform can also reduce bending and stooping. Further measures to reduce the amount of handling by care staff are:

- raising the height of beds and chairs using wooden blocks;
- handrails at strategic heights adjacent to the bath and toilet;
- the use of bath hoists and overhead hoists;
- walk-in showers with seats; and
- cutting slots in bath panels to allow for mobile hoist wheels to fit underneath a bath fixed against a wall.

All operators of mechanical handling equipment must be properly trained in its use and supervised while they are using it. These precautions are particularly important when people are being lifted or aided using a manual hoist. Other precautions include:

- a safe system of work in place;
- sufficient room available to easily manoeuvre the equipment;
- adequate visibility and lighting available;
- the floors in a stable condition;
- regular safety checks to identify any faults with the equipment; and
- a regular maintenance schedule for the equipment.

Other important considerations are that:

- the proposed use will be within the safe working load of the equipment;
- it is suitable for the area in which it will operate (e.g. is there enough room to manoeuvre and enough headroom?);
- it suits the terrain in terms of stability and ground surface;
- the lifting equipment is CE-marked; and
- advice is sought from the suppliers/hirers on its suitability for the proposed task and any maintenance requirements.

A useful precaution is to request the equipment on a trial basis, if possible, to check that it is suitable for the required task and to involve the employees who will be expected to use it in this appraisal exercise. Consideration should also be given to other risks associated with the introduction of the lifting aid, such as site safety and the lack of communication with fork-lift truck drivers.

9.4 Powered load handling equipment

9.4.1 Safety in the use of lifting and moving equipment

Positioning and installing lifting equipment

Lifting equipment must be positioned and installed so as to reduce the risks, so far as is reasonably practicable, from:

- equipment or a load striking a person;
- a load drifting, falling freely or being released unintentionally.

Lifting equipment should be positioned and installed to minimise the need to lift loads over people and to prevent crushing in extreme positions. It should be designed to stop safely in the event of a power failure and not release its load. Lifting equipment, which follows a fixed path, should be enclosed with suitable and substantial interlocked gates and any necessary protection in the event of power failure.

The organisation of lifting operations

Every lifting operation, that is lifting or lowering of a load, shall be:

- properly planned by a competent person;
- appropriately supervised;
- carried out in a safe manner.

The person planning the operation should have adequate practical and theoretical knowledge and experience of planning lifting operations. The plan needs to address the risks identified by the risk assessment and identify the resources, the procedures and the responsibilities required so that any lifting operation is carried out safely. For routine simple lifts, a plan will normally be the responsibility of the people using the lifting equipment. For complex lifting operations, for example where two cranes are used to lift one load, a written plan may need to be produced each time.

The planning should include the need to avoid suspending loads over occupied areas, visibility, the attaching/detaching and securing of loads, the environment, the location, the possibility of overturning, the proximity to other objects, any lifting of people and the pre-use checks required for the equipment.

Summary of the requirements for lifting operations

There are four general requirements for all lifting operations:

- use strong, stable and suitable lifting equipment;
- the equipment should be positioned and installed correctly;
- the equipment should be visibly marked with the safe working load (SWL);
- lifting operations must be planned, supervised and performed in a safe manner by competent people.

9.4.2 Types of mechanical handling and lifting equipment

There are four elements to mechanical handling, each of which can present hazards. These are handling equipment, the load, the workplace and the employees involved.

The **mechanical handling equipment** must be capable of lifting and/or moving the load. It must be fault-free, well-maintained and inspected on a regular basis. The hazards related to such equipment include collisions between people and the equipment and personal injury from being trapped in moving parts of the equipment (such as belt and screw conveyors).

The **load** should be prepared for transportation in such a way as to minimise the possibility of accidents. The hazards will be related to the nature of the load (e.g. substances which are flammable or hazardous to health) or the security and stability of the load (e.g. collapse of bales or incorrectly stacked pallets).

The **workplace** should be designed so that, whenever possible, workers and the load are kept apart. If, for example, an overhead crane is to be used, then people should be segregated away or barred from the path of the load.

The **employees** and any other people who are to use the equipment must be properly trained and competent in its safe use.

Conveyors and elevators

Conveyors transport loads along a given level which may not be completely horizontal, whereas elevators move loads from one level or floor to another. Conveyors are shown in Figure 9.11.

There are three common forms of **conveyor**: belt, roller and screw conveyors. The most common hazards and preventative measures are:

▶ the in-running nip, where a hand is trapped between the rotating rollers and the belt. Protection from this hazard can be provided by nip guards and trip devices;

▶ entanglement with the power drive requiring the fitting of fixed guards and the restriction of

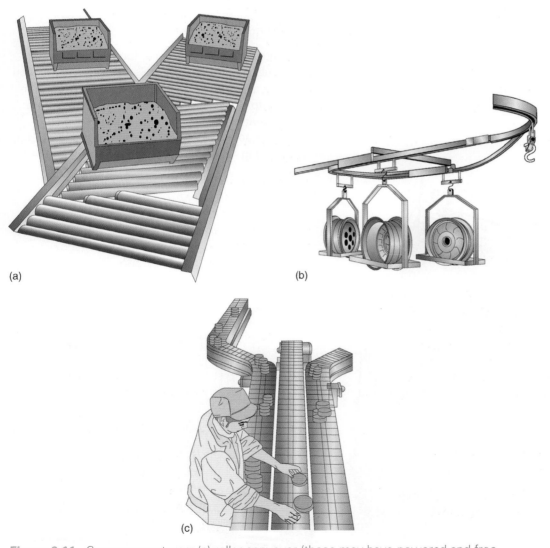

(a)

(b)

(c)

Figure 9.11 Conveyor systems: (a) roller conveyor (these may have powered and free running rollers); (b) an overhead conveyor moving wheels. Other designs of overhead conveyor are useful for transferring components and garments between workstations in, for example, manufacture of machines or clothing; (c) a slat conveyor in use in a food factory

loose clothing which could become caught in the drive;

▶ loads falling from the conveyor. This can be avoided by edge guards and barriers;

▶ impact against overhead systems. Protection against this hazard may be given by the use of bump caps, warning signs and restricted access;

▶ contact hazards prevented by the removal of sharp edges, conveyor edge protection and restricted access;

▶ manual handling hazards;

▶ noise and vibration hazards.

Screw conveyors, often used to move very viscous substances, must be provided with either fixed guards or covers to prevent accidental access. People should be prohibited from riding on belt conveyors, and emergency trip wires or stop buttons must be fitted and be operational at all times.

Elevators are used to transport goods between floors, such as the transportation of building bricks to upper storeys during the construction of a building or the transportation of grain sacks into the loft of a barn. Guards should be fitted at either end of the elevator and around the power drive. The most common hazard is injury due to loads falling from elevators. There are also potential manual handling problems at both the feed and discharge ends of the elevator.

Fork-lift trucks

The most common form of mobile handling equipment is the fork-lift truck. It comes from the group of vehicles, known as lift trucks, and can be used in factories, on construction sites and on farms. The term fork-lift truck is normally applied to the counterbalanced lift truck, where the load on the forks is counterbalanced by the weight of the vehicle over the rear wheels. The reach truck is designed to operate in narrower aisles in warehouses and enables the load to be retracted within the wheelbase. The very narrow aisle (VNA) truck does not turn within the aisle to deposit or retrieve a load. It is often guided by guides or rails on the floor. Other forms of lift truck include the pallet truck and the pallet stacker truck, both of which may be pedestrian or rider controlled. Around 400 workers are seriously injured in fork-lift truck accidents every year in the UK and some of these accidents cause about 10 fatalities annually. Accidents frequently occur when fork-lift truck drivers or pedestrians in warehouses are distracted, or simply assume that they have been seen by the other party.

There are many hazards associated with the use of fork-lift trucks. These include:

▶ overturning – manoeuvring at too high a speed (particularly cornering); wheels hitting an obstruction such as a kerb; sudden braking; poor tyre condition leading to skidding; driving forwards down a ramp; movement of the load; insecure, excessive or uneven loading; incorrect tilt or driving along a ramp;

▶ overloading – exceeding the rated capacity of the machine;

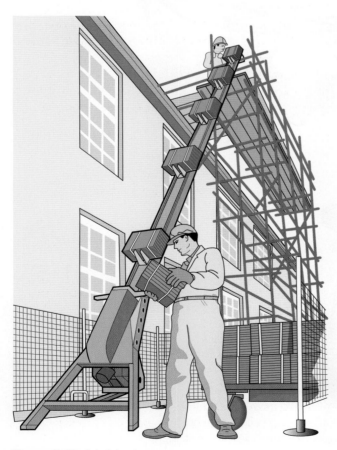

Figure 9.12 A brick elevator

Figure 9.13 Reach truck – designed so that the load retracts within the wheel base to save space

▷ collisions – particularly with warehouse racking which can lead to a collapse of the whole racking system;

▷ silent operation of the electrically powered fork-lift truck – can make pedestrians unaware of its presence;

▷ uneven road surface – can cause the vehicle to overturn and/or cause musculoskeletal problems for the driver;

▷ overhead obstructions – a particular problem for inexperienced drivers;

▷ loss of load – shrink wrapping or sheeting will reduce this hazard;

▷ inadequate maintenance leading to mechanical failure;

▷ use as a work platform;

▷ speeding – strict enforcement of speed limits is essential;

▷ poor vision around the load;

▷ pedestrians – particularly when pedestrians and vehicles use the same roadways. Warning signs, indicating the presence of fork-lift trucks, should be posted at regular intervals;

▷ dangerous stacking or de-stacking technique – this can destabilise a complete racking column;

▷ carrying passengers – this should be a disciplinary offence;

▷ battery charging – presents an explosion and fire risk;

▷ fire – often caused by poor maintenance resulting in fuel leakages or engine/motor burnout, or through using an unsuitable fork-lift truck in areas where flammable liquids or gases are used and stored;

▷ lack of driver training.

If fork-lift trucks are to be used outside, visibility and lighting, weather conditions and the movement of other vehicles become additional hazards.

There are also the following physical hazards:

▷ noise – caused by poor silencing of the power unit;

▷ exhaust fumes – should only be a problem when the maintenance regime is poor;

▷ vibrations – often caused by a rough road surface or wide expansion joints. Badly inflated tyres will exacerbate this problem;

▷ manual handling – resulting from manoeuvring the load by hand or lifting batteries or gas cylinders;

▷ ergonomic – musculoskeletal injuries caused by soft tyres and/or undulating road surface or holes or cracks in the road surface (e.g. expansion joints).

Regular and documented maintenance by competent mechanics is essential. However, the driver should undertake the following checks at the beginning of each shift:

▷ condition of tyres and correct tyre pressures;

▷ effectiveness of all brakes;

▷ audible reversing horn and light working properly;

▷ lights, if fitted, working correctly;

▷ mirrors, if fitted, in good working order and properly set;

▷ secure and properly adjusted seat;

▷ correct fluid levels, when appropriate;

▷ fully charged batteries, when appropriate;

▷ correct working of all lifting and tilting systems.

A more detailed inspection should be undertaken by a competent person within the organisation on a weekly basis to include the mast and the steering gear. Driver training is essential and should be given by a competent trainer. The training session must include the site rules covering items such as the fork-lift truck driver code of practice for the organisation, speed limits, stacking procedures and reversing rules. Refresher training should be provided at regular intervals and a detailed record kept of all training received. Table 9.2 illustrates some key requirements of fork-lift truck drivers and the points listed should be included in most codes of practice.

Finally, care must be taken with the selection of drivers, including relevant health checks and previous experience. The UK HSE recommend that drivers should be at least 18 years of age and their fitness to drive should be reassessed regularly (every five years after the age of 40 and every year after 65) (*HSG6 – Safety in working with lift trucks*).

9

Table 9.2 Safe driving of lift trucks

Drivers must:
▷ drive at a suitable speed to suit road conditions and visibility
▷ use the horn when necessary (at blind corners and doorways)
▷ always be aware of pedestrians and other vehicles
▷ take special care when reversing (do not rely on mirrors)
▷ take special care when handling loads which restrict visibility
▷ travel with the forks (or other equipment fitted to the mast) lowered
▷ use the prescribed lanes
▷ obey the speed limits
▷ take special care on wet and uneven surfaces
▷ use the handbrake, tilt and other controls correctly
▷ take special care on ramps
▷ always leave the truck in a state which is safe and discourages unauthorised use (brakes on, motor off, forks down, key out).

Drivers must not:
▷ operate in conditions in which it is not possible to drive and handle loads safely (e.g. partially blocked aisles)
▷ travel with the forks raised
▷ use the forks to raise or lower persons unless a purpose-built working cage is used
▷ carry passengers
▷ park in an unsafe place (e.g. obstructing emergency exits)
▷ turn round on ramps
▷ drive into areas where the truck would cause a hazard (flammable substance store)
▷ allow unauthorised use.

Other forms of lifting equipment

The other types of lifting equipment to be considered are cranes (mobile overhead and jib), lifts and hoists and lifting tackle. A sample risk assessment for the use of lifting equipment is given in Appendix 9.1.

The lifting operation should be properly prepared and planned. This involves the selection of a suitable crane having up-to-date test certificates and examination reports that have been checked. A risk assessment of the task will be needed which would ascertain the weight, size and shape of the load and its final resting place. A written plan for completing the lift should be drafted and a competent person appointed to supervise the operation.

Lift plans

A lift plan identifies the ways in which the risks involved in a lifting operation can be eliminated or controlled. The degree of planning for a lifting operation should be proportional to the risk and will vary considerably depending upon the complexity of the lifting operation. The complexity will depend on the load to be lifted, the equipment to be used and the environment in which the operation is to be undertaken.

Well-planned lifting operations are a combination of two parts:

1. initial planning to ensure that the lifting equipment provided is suitable for the range of tasks that it will have to undertake; and
2. the planning of individual lifting operations so that they can be performed safely with the lifting equipment provided.

The balance between the two parts of the planning process will vary depending upon the lifting equipment and the particular lifting operation.

The person planning the operation should have adequate practical and theoretical knowledge and experience of planning lifting operations. The plan will need to address the risks identified by the risk assessment and identify the resources, the procedures and the responsibilities required so that any lifting operation is carried out safely. For routine simple lifts a plan will normally be left to the people using the lifting equipment. For complex lifting operations, for example where two cranes are used to lift one load, a written plan may need to be produced each time.

The planning should take account of avoiding suspending loads over occupied areas, visibility, attaching/detaching and securing loads, the environment, location, overturning, proximity to other objects, lifting of people and pre-use checks of the equipment.

Cranes

Cranes may be either a **jib crane** or an **overhead gantry travelling crane**. The safety requirements are similar for each type. All cranes need to be properly designed, constructed, installed and maintained. They must also be operated in accordance with a safe system of work. They should only be driven by authorised persons who are fit and trained. Each crane is issued with a certificate by its manufacturer giving details of the **safe working load** (SWL). The SWL must **never be exceeded** and should be marked on the crane structure. If the SWL is variable, as with a jib crane (the SWL decreases as the operating radius increases), an SWL indicator should be fitted. Care should be taken to avoid sudden shock loading, as this will impose very high stresses on the crane structure. It is also very important that the load is properly shackled and all eyebolts tightened. Safe slinging should be included in any training programme. All controls should be clearly marked and be of the 'hold-to-run' type.

Large cranes, which incorporate a driving cab, often work in conjunction with a banksman, who will direct the lifting operation from the ground. Banksmen are operatives trained to direct vehicle movement on or around site. They are often called traffic marshals. Banksmen should only be used in circumstances where other control measures are not possible. It is important that banksmen are trained so that they understand recognised crane signals.

Signallers are operatives who are trained to direct crane drivers during lifting operations. It is a hazardous job and measures will help to keep the signaller safe:

▶ Ensure that they are trained and competent to direct lifting operations.
▶ Provide a protected position from which they can work in safety and where they can be seen at all times by the crane driver.
▶ Provide distinctive 'hi viz' clothing for identification.
▶ Tell drivers that if they cannot see the signaller they should stop immediately.
▶ Agree on the use of standard signals.

The lifting operation should be properly prepared and planned. This involves the selection of a suitable crane

Figure 9.14 Manoeuvring a yacht using a large overhead travelling gantry and slings in a Turkish marina

having up-to-date test certificates and examination reports that have been checked. A risk assessment of the task will be needed which would ascertain the weight, size and shape of the load and its final resting place. A written plan for completing the lift should be drafted and a competent person appointed to supervise the operation.

During the planning of a lifting operation close to a public highway, there should be liaison with the police and local authority over any road closures and the lift should be timed to take place when there will be a minimum number of people about. Arrangements may be needed to give advanced warning to motorists and other road users and appropriate signs and traffic controls used.

The basic principles for the safe operation of cranes are as follows. **For all cranes**, the driver must:

▶ undertake a brief inspection of the crane and associated lifting tackle each time before it is used;
▶ check that all lifting accessory statutory inspections are in place and up to date;
▶ check that tyre pressures, where appropriate, are correct;
▶ ensure that loads are not left suspended when the crane is not in use;
▶ before a lift is made, ensure that nobody can be struck by the crane or the load;
▶ ensure that loads are never carried over people;
▶ ensure good visibility and communications;
▶ lift loads vertically – cranes must not be used to drag a load;
▶ travel with the load as close to the ground as possible;
▶ switch off power to the crane when it is left unattended and lock the cab to prevent unauthorised access.

For **mobile jib cranes,** the following points should be considered:

▶ Each lift must be properly planned, with the maximum load and radius of operation known.
▶ Overhead obstructions or hazards must be identified; it may be necessary to protect the crane from overhead power lines by using goal posts and bunting to mark the safe headroom.
▶ The ground on which the crane is to stand should be assessed for its load-bearing capacity.
▶ If fitted, outriggers should be used with appropriate outrigger mats where ground conditions dictate.

The principal reasons for crane failure, including loss of load, are:

▶ overloading;
▶ poor slinging of load;
▶ insecure or unbalanced load;
▶ loss of load;
▶ overturning;
▶ collision with another structure or overhead power lines;

▶ foundation failure;
▶ structural failure of the crane;
▶ operator error;
▶ lack of maintenance and/or regular inspections;
▶ no signaller used when driver's view is obscured;
▶ incorrect signals given or a misunderstanding of a signal.

Tower cranes

Typical causes of recent serious incidents with tower cranes include:

▶ mechanical failure of the brake or lifting ram;
▶ overturn of the crane;
▶ jib collapse;
▶ a load or dropped load striking a worker;
▶ sling failure.

The reasons for some of these incidents were:

▶ poor site induction training – not dealing with site-specific risks and lasting too long (20–30 minutes maximum is sufficient time);
▶ problems with crane maintenance and thorough examinations;
▶ operators working long hours without a break;
▶ poor operator cabin design and too high a climbing distance;
▶ operator health problems;
▶ problems in communicating health and safety issues by crane operators on site.

During lifting operations using cranes, it must be ensured that:

▶ the driver has good visibility;
▶ there are no pedestrians below the load by using barriers, if necessary;
▶ there is adequate space for the installation, manoeuvring and operation of the crane;
▶ all pedestrians are re-routed during the lifting operation;
▶ everyone within the lifting area is wearing a safety helmet; and
▶ an audible warning is given prior to the lifting operation.

If lifting takes place in windy conditions, tag lines may need to be attached to the load to control its movement.

Tower cranes are being used more and more often on construction sites globally. Over recent years, there have been many serious accidents, including fatalities, involving tower cranes. Site-specific induction training should be given to everybody involved with the tower crane erection, operation and dismantling. A specific risk assessment and a regular inspection of the crane should be undertaken and only competent and trained operators should be involved with tower crane activities.

The key aspects that apply to all cranes including tower cranes are:

1. The planning of safe lifting operations by a competent person following a detailed risk assessment;
2. Safe systems of work for the installation, operation and dismantling stages. The main elements of the safe system or method statement are:

    ▶ planning – including site preparation, crane erection and dismantling;

    ▶ selection, provision and use of a suitable crane and work equipment, including safe slinging and signalling arrangements;

    ▶ maintenance and examination of the crane and equipment;

    ▶ provision of properly trained and competent personnel;

    ▶ supervision of operations by personnel having the necessary authority;

    ▶ thorough examinations, reports and other documents;

    ▶ preventing unauthorised movement or use of the crane; and

    ▶ measures to secure the safety of persons not involved in the lifting.

3. Supervision of all lifting operations; and
4. The thorough examination of the crane.

A **lift or hoist** incorporates a platform or cage and is restricted in its movement by guides. Hoists are generally used in industrial settings (e.g. construction sites and garages), whereas lifts are normally used inside buildings. Lifts and hoists may be designed to carry passengers and/or goods alone. They should be of sound mechanical construction and have interlocking doors or gates, which must be completely closed before the lift or hoist moves. The hoistway should be properly enclosed so that the moving parts of the hoist are guarded. The loads should be secured on the hoist platform so that they cannot fall and the operator should have a clear view of the landing levels. There should be no unauthorised use of the hoist by untrained personnel.

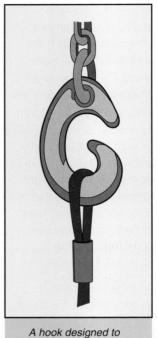 

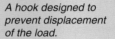
A hook designed to prevent displacement of the load.

A hook with a safety catch.

Figure 9.16 Specially designed safety hooks

Other items of lifting tackle, usually used with cranes, include chain slings and hooks, wire and fibre rope slings, eyebolts and shackles (Figure 9.16). Special care should be taken, when slings are used, to ensure that the load is properly secured and balanced. Lifting hooks should be checked for signs of wear and any distortion of the hook. Shackles and eyebolts must be correctly tightened. Slings should always be checked for any damage before they are used and only competent people should use them. Training and instruction in the use of lifting tackle is essential and should include regular inspections of the tackle, in addition to the mandatory thorough examinations. Finally, care should be taken when these items are being stored between use.

Passenger lifts

Passenger-carrying lifts must be fitted with an automatic braking system to prevent overrunning, at least two suspension ropes, each capable alone of supporting the maximum working load, and a safety device which could support the lift in the event of suspension rope failure. Maintenance procedures must be rigorous, recorded and only undertaken by competent persons. It is very important that a safe system of work is employed during maintenance operations to protect others, such as members of the public, from falling down the lift shaft and other hazards.

The following additional issues are relevant to the safe operation of passenger lifts:

▶ An emergency mushroom stop button should be fitted in the lift.

Figure 9.15 Hoists for lifting cars in a high quality French repair garage

▷ A direct telephone line is required if the internal alarm sounder/siren cannot be heard by rescue staff in the building.

▷ A risk assessment for possible injuries to the lift passengers should be undertaken on the design, sensitivity and performance of the door edges. The lift inspector should have the relevant standard to help with this.

▷ A lift stop switch should be available in the pit as well as the motor room at the top of the lift.

▷ The lift beam should be inspected by the lift inspector – the safe working load should be stated on it to prevent overload.

▷ If there are any moving or revolving mechanisms accessible by lift technicians performing their maintenance tasks, then they should be protected from inadvertent contact by suitable guards.

▷ It is advisable to locate a fire extinguisher inside the access door of the motor room.

▷ Emergency door release devices must not be routinely operated during non-emergency situations. The emergency keys are intended to allow emergency access to the lifting platform in the event of people becoming trapped and should be under strict control.

▷ Emergency unlocking should be undertaken only in exceptional circumstances and by suitably trained and authorised people.

▷ Safe working procedures and arrangements should be in place setting out what to do in the event of an emergency or failure. For example, how to deal with trapped people and the arrangements for repairing faults.

▷ All lifts and lifting platforms must be inspected, serviced and maintained. Where the lift is used for work purposes, it must be thoroughly examined by a competent person.

9.4.3 The examination of lifting equipment

Several countries have statutory requirements for the examination of lifting equipment particularly when the equipment is used to lift people. The ILO recommendations are given later in this chapter. Two terms are often used when defining those requirements and recommendations – an **inspection** and a **thorough examination**.

Lifting equipment includes any equipment used at work to lift or lower loads including any anchoring, fixing or supporting attachments.

An **inspection** is used to identify whether the equipment can be operated, adjusted and maintained safely so that any defect, damage or wear can be detected before it results in unacceptable risks. It is normally performed by a competent person appointed by the employer (often an employee).

Figure 9.17 Lifting equipment in the forest

A **thorough examination** is a detailed examination, which may involve a visual check, a disassembly and testing of components and/or an equipment test under operating conditions. Such an examination must normally be carried out by a competent person who is independent of the employer. The examination is usually carried out according to a written scheme and a written report is submitted to the employer.

The UK HSE Code of Practice for the Safe Use of Lifting Equipment (L113) defines a thorough examination as a visual examination carried out by a competent person carefully and critically and, where appropriate, supplemented by other means, such as measurement

Figure 9.18 Truck-mounted lifting equipment

and testing, in order to check whether the equipment is safe to use.

A thorough examination of lifting equipment should be undertaken at the following times:

▶ before the equipment is used for the first time;
▶ after it has been assembled at a new location;
▶ at least every 6 months for equipment used for lifting persons or a lifting accessory;
▶ at least every 12 months for all other lifting equipment including the lifting of loads over people;
▶ in accordance with a particular examination scheme drawn up by an independent competent person;
▶ each time that exceptional circumstances, which are likely to jeopardise the safety of the lifting equipment, have occurred (such as severe weather).

The person making the thorough examination of lifting equipment should:

▶ notify the employer forthwith of any defect which, in their opinion, is or could become dangerous;
▶ as soon as practicable (normally within 28 days) write an authenticated report to the employer and any person who leased or hired the equipment.

The initial report should be kept for as long as the lifting equipment is used (except for a lifting accessory which need only be kept for two years). For all other examinations, a copy of the report should be kept until the next thorough examination is made or for two years (whichever is the longer). If the report shows that a defect exists that could lead to an existing or imminent risk of serious personal injury, a copy of the report should be sent, by the person making the thorough examination, to the appropriate enforcing authority.

The equipment should be inspected at suitable intervals between thorough examinations. The frequency and the extent of the inspection are determined by the level of risk presented by the lifting equipment. A report or record should be made of the inspection which should be kept until the next inspection. Unless stated otherwise, lifts and hoists should be inspected every week.

It is important to stress that thorough examinations must be accompanied by meticulous in-service inspection that will detect any damage every time the equipment is used. For most workplaces, it is better to store every item of portable lifting equipment in a central store where these inspections can take place and records of the equipment be kept. Users must also be encouraged to report any defects in equipment that they have used.

9.4.4 ILO recommendations on the use of lifting equipment

The ILO gives a series of recommendations on the selection, installation, examination, testing, maintenance, operation and dismantling of lifting equipment in its Code of Practice 'Safety and Health in Construction'.

It recommends that employers should have a well-planned safety programme so that all the lifting appliances and lifting gear are selected, installed, examined, tested, maintained, operated and dismantled:

1. with a view to preventing the occurrence of any accident;
2. in accordance with the requirements laid down in the national laws, regulations and standards.

Every lifting appliance including its constituent elements, attachments, anchorages and supports should be of good design and construction, sound material and adequate strength for the purpose for which it is used.

Every lifting appliance and every item of lifting gear should be accompanied at the time of purchase with instructions for use and with a test certificate from a competent person or a guarantee of conformity with national laws and regulations concerning:

(a) the maximum safe working load;
(b) safe working loads at different radii if the lifting appliance has a variable radius;
(c) the conditions of use under which the maximum or variable safe working loads can be lifted or lowered.

Every lifting appliance and every item of lifting gear having a single safe working load should be clearly marked at a conspicuous place with the maximum safe working load in accordance with national laws and regulations.

Every lifting appliance having a variable safe working load should be fitted with a load indicator or other effective means to indicate clearly to the driver each maximum safe working load and the conditions under which it is applicable.

All lifting appliances should be adequately and securely supported; the weight-bearing characteristics of the ground on which the lifting appliance is to operate should be surveyed in advance of use.

Installation

Fixed lifting appliances should be installed:

(a) by competent persons;
(b) so that they cannot be displaced by the load, vibration or other influences;
(c) so that the operator is not exposed to danger from loads, ropes or drums;
(d) so that the operator can either see over the zone of operations or communicate with all loading and unloading points by telephone, signals or other adequate means.

A clearance of at least 60 cm or more should be provided between moving parts or loads of lifting appliances and:

(a) fixed objects in the surrounding environment such as walls and posts; or

(b) electrical conductors.

The clearance from electrical conductors should be more for high voltage transmission lines. The strength and stability of lifting appliances should take into account the effect of any wind forces to which they may be exposed. No structural alterations or repairs should be made to any part of a lifting appliance which may affect the safety of the appliance without the permission and supervision of the competent person.

Examinations and tests

Lifting appliances and items of lifting gear, as prescribed by national laws or regulations, should be examined and tested by a competent person:

(a) before being taken into use for the first time;
(b) after erection on a site;
(c) subsequently at intervals prescribed by national laws and regulations;
(d) after any substantial alteration or repair.

The manner in which the examinations and tests are to be carried out by the competent person and the test loads to be applied for different types of lifting appliances and lifting gear should be in accordance with national laws and regulations. The results of the examinations and tests on lifting appliances and lifting gear should be recorded in prescribed forms and, in conformity with national laws and regulations, made available to the competent authority and to employers and workers or their representatives.

Controls, control devices and cabins

Controls of lifting appliances should be:

(a) designed and constructed as far as possible in accordance with ergonomic principles;
(b) conveniently situated with ample room for operation and an unrestricted view for the operator;
(c) provided, where necessary, with a suitable locking device to prevent accidental movement or displacement;
(d) in a position free from danger from the passage of the load;
(e) clearly marked to show their purpose and method of operation.

Lifting appliances should be equipped with devices that would prevent the load from overrunning and prevent the load from moving if power fails.

The operator of every lifting appliance used outdoors except those used for short periods should be provided with:

(a) a safe cabin with full protection from weather and adverse climatic conditions, and designed and constructed in accordance with ergonomic principles;

(b) a clear and unrestricted view of the area of operation;
(c) safe access to and egress from the cabin, including situations where the operator is taken ill.

Operation

No lifting appliance should be operated by a worker who:

(a) is below 18 years of age;
(b) is not medically fit;
(c) has not received appropriate training in accordance with national laws and regulations or is not properly qualified.

A lifting appliance or item of lifting gear should not be loaded beyond its safe working load or loads, except for testing purposes as specified by and under the direction of a competent person. Where necessary to guard against danger, no lifting appliance should be used without the provision of suitable signalling arrangements or devices. No person should be raised, lowered or carried by a lifting appliance unless it is constructed, installed and used for that purpose in accordance with national laws and regulations, except in an emergency situation:

(a) in which serious personal injury or fatality may occur;
(b) for which the lifting appliance can safely be used.

Every part of a load in the course of being hoisted or lowered should be adequately suspended or supported so as to prevent danger.

Every platform or receptacle used for hoisting bricks, tiles, slates or other loose material should be so enclosed as to prevent the fall of any of the material. Loaded wheelbarrows placed directly on a platform for raising or lowering should be taped or secured so that they cannot move and the platform should be enclosed as necessary to prevent the fall of the contents. In hoisting a barrow, the wheel should not be used as a means of lifting unless efficient steps are taken to prevent the axle from slipping out of the bearings. Landings should be so designed and arranged that workers are not obliged to lean out into empty space for loading and unloading.

To avoid danger, long objects such as girders should be guided with a tag line while being raised or lowered. The hoisting of loads at points where there is a regular flow of traffic should be carried out in an enclosed space, or if this is impracticable (e.g. in the case of bulky objects), measures should be taken to hold up or divert the traffic for the time necessary.

Tower cranes

Where tower cranes have cabs at high level, persons should only be employed as crane operators who are capable and trained to work at heights. The characteristics of the various machines available should be considered against the operating requirements and the surroundings in which the crane will operate before a particular type of crane is selected. Care should be taken in the assessment of wind loads both during operations and out of service. Account should also be taken of the effects of high structures on wind forces in the vicinity of the crane.

The ground on which the tower crane stands should have adequate bearing capacity. Account should be taken of seasonal variations in ground conditions. Bases for tower cranes and tracks for rail-mounted tower cranes should be firm and level. Tower cranes should only operate on gradients within limits specified by the manufacturer. Tower cranes should only be erected at a safe distance from excavations and ditches.

Tower cranes should be sited where there is clear space available for erection, operation and dismantling. As far as possible, cranes should be sited so that loads do not have to be handled over occupied premises, over public thoroughfares, other construction works and railways or near power cables. Where two or more tower cranes are sited in positions where their jibs could touch any part of the other crane, there should be direct means of communication between them and a distinct warning system operated from the cab so that one driver may alert the other to impending danger.

The manufacturer's instructions on the methods and sequence of erection and dismantling should be followed. The crane should be tested in accordance with national laws or regulations before being taken into use. The operation of climbing tower cranes should be carried out in accordance with the manufacturer's instructions and national laws or regulations. The free-standing height of the tower crane should not extend beyond what is safe and is permissible in the manufacturer's instructions.

When the tower crane is left unattended, loads should be removed from the hook, the hook raised, the power switched off and the boom brought to the horizontal. For longer periods or at times when adverse weather conditions are expected, out-of-service procedures should be followed. The main jib should be slewed to the side of the tower away from the wind, put into free slew and the crane immobilised.

A wind speed measuring device should be provided at an elevated position on the tower crane with the indicator fitted in the driver's cab. Devices should be provided to prevent loads being moved to a point where the corresponding safe working load of the crane would be exceeded. Name boards or other items liable to catch the wind should not be mounted on a tower crane other than in accordance with the manufacturer's instructions.

Tower cranes should not be used for magnet or demolition ball service, piling operations or other duties which could impose excessive loadings on the crane structure.

Lifting ropes

Only ropes with a known and adequate safe working capacity should be used as lifting ropes.

Lifting ropes should be installed, maintained and inspected in accordance with the manufacturer's instructions and national laws or regulations. Repaired steel ropes should not be used on hoists. Where multiple independent ropes are used, for the purpose of stability, to lift a work platform, each rope should be capable of carrying the load independently.

9.5 Further information

Ambient factors in the workplace, International Labour Organisation (ILO) Code of Practice (CoP), ISBN 92-2-11628-X http://www.ilo.org/safework/info/standards-and-instruments/WCMS_107729/lang--en/index.htm

Directive 2009/104/EC – use of work equipment https://osha.europa.eu/en/legislation/directives/workplaces-equipment-signs-personal-protective-equipment/osh-directives/3

Ergonomic Checkpoints: Practical and easy-to-implement solutions for improving safety, health and working conditions, second edition, ILO Geneva, 2010, ISBN 978-92-2-122666-6 http://www.ilo.org/wcmsp5/groups/public/---dgreports/---dcomm/---publ/documents/publication/wcms_120133.pdf

Safety and health in the use of machinery, ILO CoP http://www.ilo.org/wcmsp5/groups/public/---ed_protect/---protrav/---safework/documents/normativeinstrument/wcms_164653.pdf

Safety of machinery – General principles for design – Risk assessment and risk reduction, ISO 12100:2010, ISBN 978-0-580-74262-0

Safety of machinery, basic concepts, general principles for design, basic terminology, methodology, ISO 12100-1:2003+A1:2009, ISBN 978-0-580-68672-6

Safety of machinery, basic concepts, general principles for design, technical principles, ISO 12100-2:2003+A1:2009, ISBN 978-0-580-68673-3

9.6 Practice revision questions

1. (a) **Define** the term 'ergonomics'.
   (b) **Identify SIX** factors that need to be considered during an ergonomic assessment.

2. Work-related upper limb disorders (WRULDs) are responsible for many cases of work-related ill-health:
   (a) **Identify TWO** examples of WRULDs.
   (b) **Identify THREE** work activities that may cause WRULDs and the typical symptoms that might be experienced by affected individuals.
   (c) **Describe** the measures that should be taken to minimise the risk to these individuals.

3. Employees who work for a publishing company are required to use display screen equipment (computers) for eight hours a day.
   (a) **Identify** the risks to the health and safety of these employees.
   (b) **Outline** the factors to consider when making an assessment of a display screen equipment (DSE) workstation.
   (c) **Outline** the control measures that may be taken to minimise the risks to these employees.
   (d) **Identify** the features of a suitable seat for use at a DSE workstation.

4. (a) **Identify FOUR** types of injury that may be caused by the incorrect manual handling of loads.
   (b) **Outline** a good lifting technique that should be adopted by a person required to lift a load from the ground to a work bench.
   (c) **Give TWO** examples of how a manual handling task might be avoided.

5. **Outline** the factors that may affect the risk from manual handling activities in relation to:
   (a) The task
   (b) The individual
   (c) The load
   (d) The environment.

6. Metal boxes, each weighing 5 kg, are lifted by workers from the ground to a conveyor belt for onward transmission to a paint shop.
   (a) **Outline** the issues that should be considered when undertaking a manual handling assessment of this task.
   (b) **Identify FOUR** factors associated with the worker that may affect the risk of injury when lifting the boxes.

   (c) **Outline** the measures that may be needed in order to reduce the risk of injury to the workers undertaking this task.

7. In order to minimise the risk of injury when undertaking a manual handling operation:
   (a) **Identify FOUR** types of manually operated load handling equipment that can be used to assist the manual handling operation;
   (b) **Outline FOUR** hazards and the corresponding precautions to be taken when using conveyor systems for moving materials within a workplace.

8. Fork-lift trucks are used extensively in busy warehouses.
   (a) **Identify EIGHT** general and **FOUR** physical hazards associated with their use.
   (b) **Outline** the checks that the driver should undertake at the beginning of a shift.
   (c) **Identify FOUR** rules that the driver should follow when a fork-lift truck is left unattended.
   (d) **Identify EIGHT** ways in which a fork-lift truck may become unstable whilst in operation.

9. A pedestrian has been injured after a collision with a fork-lift truck in a warehouse. **Outline** the possible **immediate** causes of the accident associated with:
   (a) the pedestrian
   (b) the way in which the vehicle was driven
   (c) the workplace
   (d) the vehicle.

10. (a) **Identify THREE** types of crane used for lifting operations.
    (b) **Outline** the key issues for the safe management of all crane operations.
    (c) **Identify EIGHT** reasons for accidents during the operation of cranes.
    (d) **Identify EIGHT** requirements of crane drivers to ensure the safe operation of the crane.

11. A mobile crane is to be used on a construction site.
    (a) **Outline** the principal requirements for lifting operations.
    (b) **Outline** the precautions that should be taken when using the crane.
    (c) **Outline** issues that could be included in a lifting plan.

12. An overhead gantry crane is used in a plastic manufacturing factory to transport heavy equipment between machines.

9

(a) **Identify FOUR** reasons why loads may fall from the crane.

(b) **Outline** the precautions to be taken so that accidents to those working at ground level are prevented when gantry cranes are in use.

13. A mechanical hoist with specially designed lifting tackle is used to remove engines from vehicles in a motor repair shop.

(a) **Outline** the precautions to be taken to reduce the risk of injury to employees and others during the lifting operation.

(b) **Identify** issues concerning the condition of lifting accessories that could form part of a pre-use checklist.

(c) **Identify** the two types of safety inspection required for this equipment.

14. (a) **Identify** the occasions when a thorough examination of a fork-lift truck should be undertaken.

(b) **Outline** the differences between an inspection and a thorough examination of lifting equipment.

(c) **Identify** the duties of the person making a thorough examination of lifting equipment.

APPENDIX 9.1 A typical UK risk assessment for the use of lifting equipment

INITIAL RISK ASSESSMENT	Use of lifting equipment		
SIGNIFICANT HAZARDS	Low	Medium	High
1. Unintentional release of load			✓
2. Unplanned movement of load	✓		
3. Damage to equipment	✓		
4. Crush injuries to personnel		✓	

ACTION ALREADY TAKEN TO REDUCE THE RISKS:

Compliance with:

Lifting Operations and Lifting Equipment Regulations (LOLER)

Safety Signs and Signals Regulations (SSSR). Provision and Use of Work Equipment Regulations (PUWER)

British Standard – Specification for flat woven webbing slings

BS – Guide to selection and use of lifting slings for multi-purposes

Planning:

Copies of statutory thorough examinations of lifting equipment will be kept on site. Before selection of lifting equipment, the above standards will be considered as well as the weight, size, shape and centre of gravity of the load. Lifting equipment is subject to the planned maintenance programme.

Physical:

All items of lifting equipment will be identified individually and stored so as to prevent physical damage or deterioration. Safe working loads of lifting equipment will be established before use. Packing will be used to protect slings from sharp edges on the load. All items of lifting equipment will be visually examined for signs of damage before use. Ensuring the eyes of strops are directly below the appliance hook and that tail ropes are fitted to larger loads will check swinging of the load. Banksmen will be used where the lifting equipment operator's vision is obstructed. Approved hand signals will be used.

Managerial/Supervisory:

Only lifting equipment that is in date for statutory examination will be used. Manufacturer's instructions will be checked to ensure that methods of sling attachment and slinging arrangements generally are correct.

Training:

Personnel involved in the slinging of loads and use of lifting equipment will be required to be trained to CITB or equivalent standard. Supervisors will be trained in the supervision of lifting operations.

Date of Assessment..........................Assessment made by.......................

Risk Re-Assessment Date...................Site Manager's Comments:

9

APPENDIX 9.2 Examples of manually operated load handling equipment

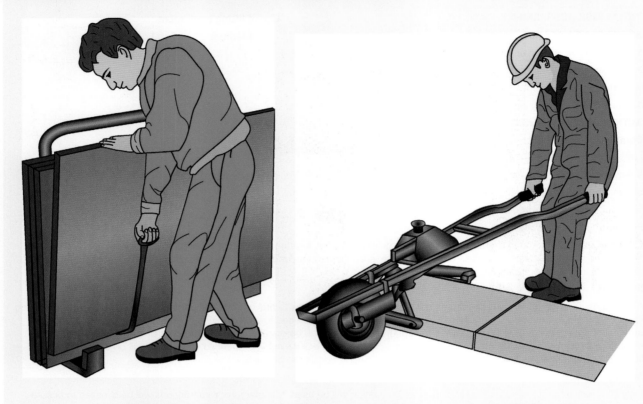

Lifting hooks

Paving slab and general purpose handler

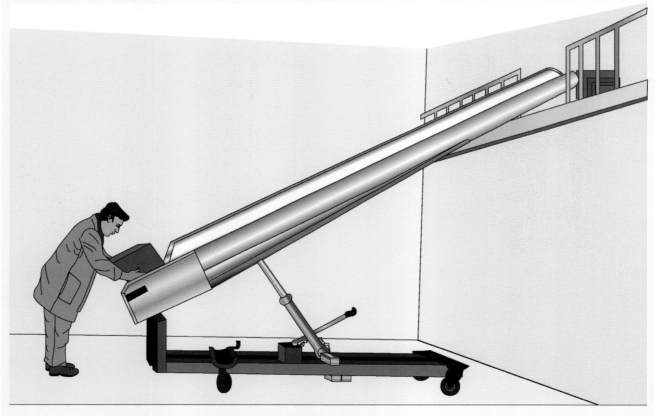

A hydraulically-lifted mobile belt conveyor

A chute being used to transfer sacks from a high to low level

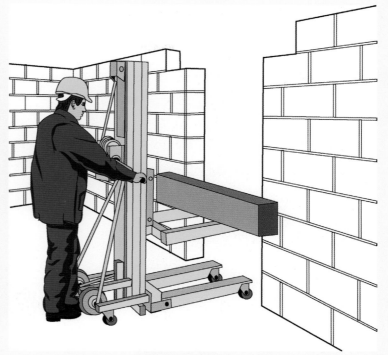

Use of hand-operated lift truck to place lintels

A sack truck with star truck wheels for climbing and descending stairs

Platform truck

9

CHAPTER 10

Work equipment hazards and risk control

10.1 General requirements for work equipment ▶ 274

10.2 Hazards and controls for hand-held tools ▶ 284

10.3 Mechanical and non-mechanical hazards of machinery ▶ 290

10.4 Control measures for reducing risks from machinery hazards ▶ 295

10.5 Further information ▶ 307

10.6 Practice revision questions ▶ 308

This chapter covers the following NEBOSH learning objectives:

1. Outline general requirements for work equipment
2. Explain the hazards and controls for hand-held tools
3. Describe the main mechanical and non-mechanical hazards of machinery
4. Explain the main control measures for reducing risk from machinery hazards

---

10.1 General requirements for work equipment

10.1.1 Introduction and types of equipment

This chapter covers the scope and main safety and health requirements for the selection, use and maintenance of work equipment as covered by the UK's HSE Safe Use of Work Equipment L22, 'Safety and health in the use of machinery' ILO Code of Practice and ISO 12100:2010 safety of machinery, General principles for design. It also covers relevant issues in the ILO Code of Practice 'Ambient factors in the workplace 2001'. The safe use of hand tools, hand-held power tools and the proper safeguarding of a small range of machinery used in industry and commerce are included.

Any equipment used by an employee at work is generally covered by the term 'work equipment'. The scope is extremely wide and includes hand tools, power tools, ladders (Chapter 6), photocopiers, laboratory apparatus, lifting equipment (Chapter 9), fork-lift trucks (Chapter 9) and motor vehicles (which are not privately owned – Chapter 8). Virtually anything used to do a job of work, including employees' own equipment, is covered. The uses covered include starting or stopping the equipment, repairing, modifying, maintaining, servicing, cleaning and transporting.

Employers and sometimes the self-employed must ensure that work equipment is suitable, maintained, inspected if necessary, provided with adequate information and instruction and only used by people who have received sufficient training.

Many serious accidents at work involve machinery. Hair or clothing can become entangled in moving parts, people can be struck by moving parts of machinery, parts of the body can be drawn into or trapped in machinery, and/or parts of the machinery or work tool can be ejected.

Many circumstances can increase the risks, including:

- not using the right equipment for the task, e.g. ladders instead of access towers for an extended task at high level;
- not fitting adequate controls on machines, or fitting the wrong type of controls, so that equipment cannot be stopped quickly and safely, or it starts accidentally;
- not guarding machines properly, leading to accidents caused by entanglement, shearing, crushing, trapping or cutting;

- not properly maintaining guards and other safety devices;
- not providing the right information, instruction and training;
- not fitting rollover protective structures (ROPS) and seat belts on mobile work equipment where there is a risk of rollover (*excluding* quad bikes);
- not maintaining work equipment or doing the regular inspections and thorough examinations;
- not providing (free) adequate personal protective equipment to use.

When identifying the risks, think about:

- the work being done during normal use of the equipment and also during setting up, maintenance, cleaning and clearing blockages;
- which workers will use the equipment, including those who are inexperienced, have changed jobs or those who may have particular difficulties, e.g. those with language problems or impaired hearing;
- people who may act stupidly or carelessly or make mistakes;
- guards or safety devices that may be badly designed and difficult to use or are easy to defeat;
- other features of the equipment which could cause risks like vibration, electricity, wet or cold conditions.

Consider the following:

- Is the equipment suitable for the task?
- Are all the necessary safety devices fitted and in working order?
- Are there proper instructions for the equipment?
- Is the area around the machine safe and level with no obstructions?
- Has suitable lighting been provided?
- Has extraction ventilation been provided where required, e.g. on grinding and woodworking machinery?
- Has a risk assessment been done to establish a person's competence or training requirements to control particular machinery – this is very important for everyone;
- Are machine operators trained and do they have enough information, instruction, training?
- Are people adequately supervised?
- Are safety instructions and procedures being used and followed?
- Are machine operators using appropriate work clothing without loose sleeves, open jackets, dangling jewellery or sandals?

▶ Has the employer supplied all necessary special personal protective equipment (PPE)?

▶ Are safety guards or devices being used properly?

▶ Is maintenance carried out correctly and in a safe way?

▶ Are hand tools being used correctly and properly maintained, and only used by people who have received sufficient training?

ISO 12100:2010 requires that protective measures to achieve the necessary risk reduction with work equipment are undertaken in the following three-step sequence:

1. Inherently safe design measures which eliminate hazards or reduce risks for people using the machinery;

2. The provision of safeguarding and/or complementary protective measures which take into account the use and reasonably foreseeable misuse of the equipment;

3. The provision of information for use which covers operating procedures, recommended safe working practices, warning of residual risks and other information for the different phases of the life of the equipment; the description of any PPE required.

10.1.2 Suitability of work equipment and CE markings

(a) Standards and requirements

When work equipment is provided it has to conform to standards which cover its supply as a new or second-hand piece of equipment and its use in the workplace. This involves:

▶ its initial integrity;
▶ the place where it will be used;
▶ the purpose for which it will be used.

In many national legal systems there are two groups of law that deal with the provision of work equipment:

▶ the first deals with what manufacturers and suppliers have to do. This can be called the 'supply' law (for example in the UK one set of regulations is the Supply of Machinery (Safety) Regulations 2008 which complies with the EU Directive 2006/42/EC – Machinery Directive). Such legislation requires manufacturers and suppliers to ensure that machinery is safe when supplied and is properly marked with the appropriate standards.

▶ The other deals with what the users of work equipment have to do. This can be called the 'user' law and applies to most pieces of work equipment. Its **primary** purpose is to protect people at work (for example the UK Provision and Use of Work Equipment Regulations).

Under 'user' law employers have to provide safe equipment of the correct type, ensure that it is correctly used and maintain it in a safe condition.

When buying new equipment, the 'user' has to check that the equipment complies with all the 'supply' law that is relevant. Figure 10.1(b) shows the division of responsibility between designers, manufacturers and suppliers, and those of employers. However, whatever hazards have or have not been controlled effectively **the user must check that the machine is safe before it is used**.

Most new work equipment, including machinery in particular, needs to comply with the relevant standards that are applicable to that equipment and the territory into which the machine is being supplied. Throughout the world there are many standards bodies such as CEN (Comité Européen de Normalisation), CENELEC (Comité Européen de Normalisation Électrotechnique) in Europe, ANSI (American National Standards Institute), ASME (American Society of Mechanical Engineers), CSA (Canadian Standards Association) and ISO (International Organisation for Standardisation). The ILO have also produced a Code of Practice on Safety and health in the use of machinery which was adopted in March 2012.

The objective of this code is to protect workers from the hazards of machinery and to prevent accidents, incidents and ill-health resulting from the use of machinery at work by providing guidelines for:

(a) ensuring that all machinery for use at work is designed and manufactured to eliminate or minimise the hazards associated with its use;

(b) ensuring that employers are provided with a mechanism for obtaining from their suppliers necessary and sufficient safety information about machinery to enable them to implement effective protective measures for workers; and

(c) ensuring that proper workplace safety and health measures are implemented to identify, eliminate, prevent and control risks arising from the use of machinery.

The code requires manufacturers to monitor and study any reports of malfunctions, dangerous occurrences, accidents and diseases involving the actual machinery in question or similar machinery. When designing machinery the manufacturer is required to carry out a process of risk assessment and risk reduction as part of the design process. This should be repeated regularly throughout the design and manufacturing stages to ensure that no new hazards and risks are introduced. Through the risk assessment the code requires the manufacturer to:

(a) determine the full range of uses to which the machinery may be put, which should include both the intended use and any reasonably foreseeable misuse;

(b) with reference to (a), identify the hazards or hazardous situations which the use or misuse of such machinery may present;

(c) eliminate any hazards as far as is reasonably practicable;

10

275

(d) estimate the risks, taking into account the severity of a possible injury or damage to health and the probability of its occurrence;

(e) evaluate whether the risk level is adequately controlled with a view to determining whether risk reduction is required; and

(f) reduce the risks identified in (e) by the application of protective measures.

The ILO Code of Practice on Ambient factors in the workplace also requires the following which relates to the hazardous ambient factors at work:

Measures should be taken, in accordance with national law and practice, to ensure that those who design, manufacture, import, provide or transfer machinery, equipment or substances for occupational use:

(a) satisfy themselves, as far as is reasonable and practicable, that the machinery, equipment or substance does not entail dangers for the safety and health of those using it correctly;

(b) make available:

(i) information concerning the correct installation and use of machinery and equipment and the correct use of substances;

(ii) information concerning hazards of machinery and equipment; dangerous properties of hazardous substances; and physical agents or products;

(iii) instructions on how known hazards are to be avoided.

Suppliers of equipment, processes and hazardous substances, whether manufacturers, importers or distributors, should ensure so far as is practicable that the design is such as to eliminate or control the hazards and risks to safety and health from hazardous ambient factors at work. Where suppliers become aware of new information concerning the hazards and risks presented by equipment, processes and hazardous substances, they should provide, as appropriate, updated information and instructions.

Designers should ensure, as far as is practicable, that the levels of hazardous ambient factors emitted from plant and processes are minimised and that they conform to internationally recognised plant and equipment standards.

Within the EU machinery should have 'CE' marking when purchased (Figure 10.1).

Before buying new equipment the buyer will need to think about:

▶ where and how it will be used;
▶ what it will be used for;
▶ who will use it (skilled employees, trainees);
▶ what risks to health and safety might result;
▶ how well health and safety risks are controlled by different manufacturers.

This can help in deciding which equipment may be suitable, particularly if buying a standard piece of equipment 'off the shelf'.

If buying a more complex or custom-built machine the buyer should discuss their requirements with potential suppliers. For a custom-built piece of equipment, there is the opportunity to work with the supplier to design out the causes of injury and ill-health. Time spent now on agreeing the necessary safeguards, to control health and safety risks, could save time and money later.

Note: Sometimes equipment is supplied via another organisation, for example an importer, rather than direct from the manufacturer, so this other organisation is referred to as the manufacturer. It is important to realise that the supplier may not be the manufacturer.

When the equipment has been supplied the buyer should look for the relevant Standards marking (e.g. CE, ANSI, ISO), check for a copy of any Declaration of Conformity (Figure 10.2) and that there is a set of instructions in English (or local language) on how the

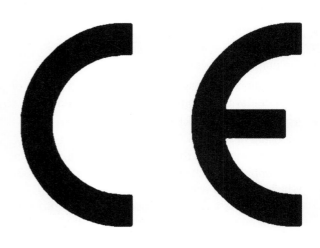

Figure 10.1 (a) CE mark

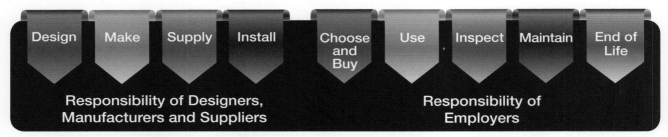

Figure 10.1 (b) Division of responsibility for the safety of machinery

machine should be used and, most important of all, **check to see if they think that it is safe**.

(b) EU CE marking requirements

'CE' marking is a claim by the manufacturer that the equipment is safe and that they have met relevant supply law. If this is done properly manufacturers or suppliers within the EU will have to do the following:

- ▶ find out about the health and safety hazards (trapping, noise, crushing, electrical shock, dust, vibration, etc.) that are likely to be present when the machine is used;
- ▶ assess the likely risks;
- ▶ design out the hazards that result in risks; or, if that is not possible
- ▶ provide safeguards (e.g. guarding dangerous parts of the machine, providing noise enclosures for noisy parts); or, if that is not possible
- ▶ use warning signs on the machine to warn of hazards that cannot be designed out or safeguarded (e.g. 'noisy machine' signs).

Manufacturers also have to:

- ▶ keep information, explaining what they have done and why, in a technical file;
- ▶ fix CE marking to the machine where necessary, to show that they have complied with all the relevant supply laws;
- ▶ issue a 'Declaration of Conformity' for the machine (see Figure 10.2). This is a statement that the machine complies with the relevant essential health and safety requirements or with the example that underwent type-examination or complies with a quality assurance system (2008). A Declaration of Conformity must:
  - ▷ state the name and address of the manufacturer or importer into the EU;
  - ▷ state the name and address of the person authorised to compile a technical file (a new requirement);
  - ▷ contain a description of the machine, and its make, type and serial number;
  - ▷ indicate all relevant European Directives with which the machinery complies;
  - ▷ state details of any notified body that has been involved;
  - ▷ specify which standards have been used in the manufacture (if any);
  - ▷ be signed by a person with authority to do so.

(c) Limitations of CE marking

CE marking is not a guarantee that the machine is safe. It is a claim by the manufacturer that the machinery complies with the law. CE marking has many advantages if done properly, for example:

- ▶ it allows a common standard across Europe;

EC-Declaration of Conformity

MACHINERY DIRECTIVE

DC720, DC721, DC722, DC725, DC727, DC730, DC731, DC732, DC735, DC737, DC742, DC743, DC745

DeWALT declares that these products described under "technical data" are in compliance with: 98/37/EC (until Dec. 28, 2009), 2006/42/EC (from Dec. 29, 2009), EN 60745-1, EN 60745-2-1, EN 60745-2-2.

These products also comply with Directive 2004/108/EC. For more information, please contact DeWALT at the following address or refer to the back of the manual.

Horst Grossmann
Vice President Engineering and Product Development
DeWALT, Richard-Klinger-Strase 11,
D-65510, Idstein, Germany
30.07.2009

Figure 10.2 Typical Declaration of Conformity

- ▶ it provides a means of selling to all European Union member states without barriers to trade;
- ▶ it ensures that instructions and safety information are supplied in a fairly standard way in most languages in the EU;
- ▶ it has encouraged the use of diagrams and pictorials which are common to all languages;
- ▶ it allows for independent type-examination for some machinery like woodworking machinery which has not been made to an EU-harmonised standard – identified by an EN marking before the standard number (e.g. BS EN …).

Clearly there are disadvantages as well, for example:

- ▶ instruction manuals have become very long. Sometimes two volumes are provided, because of the number of languages required;
- ▶ translations can be very poor and disguise the proper meaning of the instruction;
- ▶ manufacturers can fraudulently put on the CE marking;
- ▶ manufacturers might make mistakes in claiming conformity with safety laws.

10.1.3 Prevention of access to dangerous parts of machinery

Article 6 of the ILO C119 – Guarding of Machinery Convention, 1963 states that:

1. *The use of machinery any dangerous part of which, including the point of operation, is without appropriate guards shall be prohibited by national laws or regulations or prevented by other equally effective measures: Provided that where this prohibition cannot fully apply without preventing the use of the machinery it shall apply to the extent that the use of the machinery permits.*

2. *Machinery shall be so guarded as to ensure that national regulations and standards of occupational safety and hygiene are not infringed.*

Under R118 Guarding of machinery Recommendation, 1963, the following is stated under paragraphs 8 and 9:

The obligation to ensure compliance with the provisions of paragraph 7 (Article 6 of C119) should rest on the employer.

1. *The provisions of paragraph 7 (Article 6) do not apply to machinery or parts thereof which, by virtue of their construction, installation or position, are as safe as if they were guarded by appropriate safety devices.*

2. *The provisions of paragraph 7 (Article 6) and paragraph 12 (see below) do not prevent the maintenance, lubrication, setting-up or adjustment of machinery or parts thereof carried out in conformity with accepted standards of safety.*

Paragraph 12 of R118:

1. *No worker should use any machinery without the guards provided being in position, nor should any worker be required to use any machinery without the guards provided being in position.*

2. *No worker using machinery should make inoperative the guards provided, nor should such guards be made inoperative on any machinery to be used by any worker.*

The term '**dangerous part**', in practice, means that if a piece of work equipment or machinery could cause injury and if it is being used in a foreseeable way, it can be considered a dangerous part. The risk assessment carried out by the manufacturer at the design stage (see 10.1.2) and by the employer at the user stage should identify all the hazards presented by machinery. The risk assessment should evaluate the nature of the injury, its severity and likelihood of occurrence for each hazard identified. This will enable employers to decide whether the level of risk is acceptable or if risk reduction measures are needed. In most cases the objective of risk reduction measures is to prevent contact of part of the body or clothing with any dangerous part of the machine, for example by providing guards.

The measures taken to prevent access to dangerous parts of machinery are ranked in the order they should be implemented in order to achieve an adequate level of protection. The levels of protection are:

1. *First, hazards should be eliminated by technical means such as substitution of hazardous materials.*

2. *Where this is not possible, the employer should ensure that safety and health issues are managed through technical measures such as engineering controls, layout design, barriers, upgraded guards and protective devices (such as pressure mats and trip devices), protection appliances (such as jigs, push sticks and holders), ventilation, noise enclosure and ergonomic solutions.*

3. *If that is not possible, the safety of workers should be ensured, where appropriate, through training and safe systems of work and supervision and,*

4. *where residual hazards cannot be controlled by these measures, through the use of PPE, backed up by appropriate safety information and signs.*

Note the Code of Practice states that:

*Guards designed to protect persons against the hazards generated by moving **transmission parts** should be either:*

(a) *fixed guards; or*

(b) *interlocking movable guards should be used where frequent access is envisaged.*

When a process requires access to a danger zone and a fixed guard is impracticable, an interlocking guard should be considered. Guards or protective devices designed to protect persons against the hazards generated by moving parts involved in the process should be:

(a) *fixed guards;*

(b) *interlocking movable guards;*

(c) *protective devices; or*

(d) *a combination of the above.*

When selecting measures employers should consider each level of protection from the first level of the scale listed above, and they should use measures from that level so far as it is practicable to do so, provided that they contribute to the reduction of risk. It may be necessary to select a combination of measures. The selection process should continue down the scale until the combined measures are effective in reducing the risks to an acceptable level. In selecting the appropriate combination employers will need to take account of:

▶ the requirements of the work;
▶ evaluation of the risks; and
▶ the technical features of possible safeguarding solutions.

Most machinery will present more than one mechanical and non-mechanical hazard, and employers will need to deal with the risks associated with all of these. For example, at belt conveyors there is a risk of entanglement with the rotating shafts and of being trapped by the intake between drum and moving belt – so appropriate safety measures should be adopted.

Any risk assessment carried out should not just deal with the machine when it is operating normally, but must also cover activities such as setting, maintenance, cleaning or repair. The assessment may indicate that these activities require a different combination of protective measures from those appropriate to the machine doing its normal work. In particular, parts of machinery that are not dangerous in normal use because they are not then accessible may become accessible and therefore dangerous while this type of work is being carried out.

Certain setting or adjustment operations which may have to be done with the machine running may require a greater reliance on the provision of information, instruction, training and supervision than for normal use.

10.1.4 Use and maintenance of equipment with specific risks

Some pieces of work equipment involve specific risks to health and safety where it is not possible to control adequately the hazards by physical measures alone, for example the use of a bench-mounted circular saw or an abrasive wheel (Figure 10.3). In all cases the hierarchy of controls should be adopted to reduce the risks by:

▶ eliminating the risks; or, if this is not possible
▶ taking physical measures to control the risks such as guards, but if the risks cannot be adequately controlled
▶ taking appropriate software measures, such as a safe system of work.

The use of such equipment should be restricted to the persons designated to use it. These people need to have received sufficient information, instruction and training so that they can carry out the work using the equipment safely. Use is normally restricted to persons over 18 years old but this may vary across different national jurisdictions.

Figure 10.3 A typical bench-mounted abrasive wheel

Repairs, modifications, maintenance or servicing is also restricted to designated persons. A designated person may be the operator if he/she has the necessary skills and has received specific instruction and training. Another person specifically trained to carry out a particular maintenance task, for example dressing an abrasive wheel, may not be the operator but may be designated to do this type of servicing task on a range of machines.

10.1.5 Information, instruction and training for specific risks

People using and maintaining work equipment, where there are residual risks that cannot be sufficiently reduced by physical means, require enough information, instruction and training to operate safely. Managers and supervisors will also need to be provided with sufficient information about a piece of equipment to enable them to fulfil their responsibilities towards people using and maintaining equipment.

The information and instructions are likely to come from the manufacturer in the form of operating and maintenance manuals. It is up to the employer to ensure that what is provided is easily understood, and set out logically with illustrations and standard symbols where appropriate. The information should normally be in the language or languages of the country or market in which the machinery is put into service.

The extent of the information and instructions will depend on the complexity of the equipment and the specific risks associated with its use. They should cover:

▶ all safety and health aspects;
▶ any limitations on the use of the equipment;
▶ any foreseeable problems that could occur;
▶ safe methods to deal with the problems;
▶ any relevant experience with the equipment that would reduce the risks or help others to work more safely, being recorded and circulated to everyone concerned.

Everyone who uses and maintains work equipment needs to be adequately trained. For general training requirements see section 10.4.6. Examples of specific requirements for woodworking and abrasive wheels are given below.

Employers should ensure that information and instructions provided on the use of woodworking machinery includes, where relevant:

(a) the speed, range, type and dimensions of tools suitable for the machine;
(b) any limitation on the cutting speeds of the machine, particular operations or size and material of any work-piece;
(c) procedures relating to the repair or replacement of any guard or protection device;

(d) the availability, suitability and use of any additional protection device or protection appliance;

(e) the correct procedures to be followed for setting and adjusting operations;

(f) safe methods of handling tools;

(g) correct procedures for start-up and shutdown, isolation and how to discharge any residual energy;

(h) procedures for cleaning saw blades by hand (which should be carried out with the machine isolated and the blade stopped);

(i) procedures for adjusting any guard, tool, clamp or other part of a machine (which should not be carried out while any part of the machine is in motion, unless they can be done safely).

Training of someone to use a grinding machine should cover the proper methods of mounting an abrasive wheel. They are as follows:

Summary of mounting procedures for an abrasive wheel:

1. Wheel mounting should be carried out only by an appropriately trained person. A wheel should be mounted only on the machine for which it was intended. Before mounting, all wheels should be closely inspected to ensure that they have not been damaged in storage or transit.

2. The speed marked on the machine should not exceed the speed marked on the wheel, blotter or identification label.

3. The bush, if any, should not project beyond the sides of the wheel and blotters. The wheel should fit freely but not loosely on the spindle.

4. Flanges should not be smaller than their specified minimum diameter, and their bearing surfaces should be true and free from burrs.

5. With the exception of the single flange used with threaded-hole wheels, all flanges should be properly recessed or undercut.

6. Flanges should be of equal diameter and have equal bearing surfaces. Protection flanges should have the same degree of taper as the wheel. Blotters, slightly larger than the flanges, should be used with all abrasive wheels except those listed below in a-g. Wrinkles in blotters should be avoided.

Blotters should not be used with the following types of wheels:

(a) mounted wheels and points;

(b) abrasive discs (inserted nut discs and cylinders);

(c) plate-mounted wheels;

(d) cylinder wheels mounted in chucks;

(e) rubber-bonded cutting-off wheels 0.5 mm or less in thickness;

(f) taper-sided wheels;

(g) wheels with threaded inserts.

7 Wheels, blotters and flanges should be free from foreign matter. Clamping nuts should be tightened only sufficiently to hold the wheel firmly. When the flanges are clamped by a series of screws they should be tightened uniformly in a criss-cross sequence. Screws for inserted nut mounting of discs, cylinders and cones should be long enough to engage a sufficient length of thread, but not so long that they contact the abrasive.

8. When mounting the wheels and points, the overhang appropriate to the speed, diameter of the mandrel and size of the wheel should not be exceeded, and there should be sufficient length of mandrel in the collet or chuck.

The training and supervision of young persons is particularly important because of their relative immaturity, unfamiliarity with a working environment and lack of awareness of existing or potential risks. Only young persons with sufficient maturity and competence who have finished their training may use the equipment unsupervised.

10.1.6 Maintenance and inspection

(a) Maintenance

Work equipment needs to be correctly maintained so that it continues to operate safely and in the way it was designed to perform. The amount of maintenance will be stipulated in the manufacturer's instructions and will depend on the amount of use, the working environment and the type of equipment. High-speed, high-risk machines, which are heavily used in an adverse environment like salt water, may require very frequent maintenance, whereas a simple hand tool, like a shovel, may require very little.

Maintenance management schemes can be based around a number of techniques designed to focus on those parts which deteriorate and need to be maintained to prevent health and safety risks. These techniques include the following:

▶ **Preventative planned maintenance** – this involves replacing parts and consumables or making necessary adjustments at preset intervals normally set by the manufacturer, so that there are no hazards created by component deterioration or failure. Vehicles are normally maintained on this basis.

▶ **Condition-based maintenance** – this involves monitoring the condition of critical parts and carrying out maintenance whenever necessary to avoid hazards which could otherwise occur.

▶ **Breakdown-based maintenance** – here maintenance is only carried out when faults or failures have occurred. This is only acceptable if the failure does not present an immediate hazard and can be corrected before the risk is increased. If, for example, a bearing overheating can be detected by a monitoring device, it is acceptable to wait for the overheating to occur as long as the equipment can be stopped and repairs carried out before the fault becomes dangerous to persons employed.

In the context of health and safety, maintenance is not concerned with operational efficiency but only with avoiding risks to people. It is essential to ensure that maintenance work can be carried out safely. This will involve the following:

▶ competent, well-trained maintenance people;
▶ the equipment being made safe for the maintenance work to be carried out. In many cases, the normal safeguards for operating the equipment may not be sufficient as maintenance sometimes involves going inside guards to observe and subsequently adjust, lubricate or repair the equipment. Careful design that allows adjustments, lubrication and observation from outside the guards, for example, can often eliminate the hazard. Making equipment safe will usually involve disconnecting the power supply and then preventing anything moving, falling or starting during the work. It may also involve waiting for equipment to cool down or warm up to room temperature. Where machines cannot easily be disconnected the power supply should be locked off to prevent accidental starting;
▶ a safe system of work being used to carry out the necessary procedures to make and keep the equipment safe and perform the maintenance tasks. This can often involve a formal 'permit-to-work' scheme to ensure that the correct sequence of safety critical tasks, including isolation, has been performed and all necessary precautions taken;
▶ correct tools and safety equipment being available to perform the maintenance work without risks to people. For example special lighting or ventilation may be required.

(b) Periodic examination and testing of pressure systems

A wide range of pressure vessels and systems require thorough examination by a competent person to an agreed specifically written scheme. This includes steam boilers, pressurised hot water plants and air receivers (see Figure 10.4).

Figure 10.4 Typical electrically powered compressor with air receiver tank attached

National laws normally place duties on designers and manufacturers as well as the users of the equipment. This section is only concerned with the duty on users to have the vessels examined and tested as required. An employer who operates a steam boiler and/or a pressurised hot water plant and/or an air receiver must ensure:

▶ that it is supplied with the correct written information and markings;
▶ that the equipment is properly installed by a competent person;
▶ that it is used within its operating limits;
▶ there is a written scheme for periodic examination of the equipment certified by a competent person (in the case of standard steam boilers and air receivers the scheme is likely to be provided by the manufacturer) or the examinations are carried out within the specific requirements of national laws;
▶ that the equipment is examined in accordance with the written scheme by a competent person within the specified period or as required by national laws;
▶ that a report of the periodic examination is reported as required by national laws and held on file giving the required particulars;
▶ that the actions required by the report are carried out;
▶ that any other safety critical maintenance work is carried out, whether or not covered by the report.

In these cases the competent person is usually a specialist inspector from an external inspection organisation. Many of these organisations are linked to the insurance companies that cover the financial risks of use of the pressure vessel/system.

10.1.7 Operation and working environment

To operate work equipment safely it must be fitted with easily reached and operated controls, kept stable, properly lit, kept clear and provided with adequate markings and warning signs. These are covered by ISO 12100:2010, which applies to all types of machinery. The ILO Code of Practice 'Safety and health in the use of machinery', also covers these requirements.

Controls

Equipment should be provided with efficient means of:

▶ starting or making a significant change in operating conditions;

▶ stopping in normal circumstances;

▶ emergency stopping as necessary to prevent danger.

All controls should be well positioned, clearly visible and identifiable, so that it is easy for the operator to know what each control does. Markings should be clearly visible and remain so under the conditions met at the workplace.

Design features of equipment controls should:

▶ be easily reached from the operating positions;

▶ not permit accidental starting of equipment;

▶ move in the same direction as the motion being controlled;

▶ vary in mode, shape and direction of movement to prevent inadvertent operation of the wrong control;

▶ incorporate adequate red emergency stop buttons of the mushroom-headed type with lock-off;

▶ have shrouded or sunken green start buttons to prevent accidental starting of the equipment;

▶ be clearly marked to show what they do.

(a) Start controls

It should only be possible to start the work equipment by using the designed start control. Equipment may well have a start sequence which is electronically controlled to meet certain conditions before starting can be achieved, for example preheating a diesel engine, or purge cycle for gas-fed equipment. Restarting after a stoppage will require the same sequence to be performed.

Stoppage may have been deliberate or as a result of opening an interlocked guard or tripping a switch accidentally. In most cases it should not be possible to restart the equipment simply by shutting the guard or resetting the trip. Operation of the start control should be required.

Any other change to the operating conditions such as speed, pressure or temperature should only be done by using a control designed for the purpose.

(b) Stop controls

The action of normal stopping controls should bring the equipment to a safe condition in a safe manner. In some cases immediate stopping may cause other risks to occur. The stop controls do not have to be instantaneous and can bring the equipment to rest in a

safe sequence or at the end of an operating cycle. It is only the parts necessary for safety, that is, accessible dangerous parts, that have to be stopped. So, for example, suitably guarded cooling fans may need to run continuously and be left on.

In some cases where there is, for example, stored energy in hydraulic systems, it may be necessary to insert physical scotches to prevent movement and/or to exhaust residual hydraulic pressure. These should be incorporated into the stopping cycle, which should be designed to dissipate or isolate all stored energy to prevent danger.

It should not be possible to reach dangerous parts of the equipment until it has come to a safe condition, e.g. stopped, cooled, electrically safe.

(c) Emergency stop controls

Emergency stop must be provided where the other safeguards in place are not sufficient to prevent danger to operatives and any other persons who may be affected. Where appropriate, there should be an emergency stop at each control point and at other locations around the equipment so that action can be taken quickly. Emergency stops should bring the equipment to a halt rapidly but this should be controlled where necessary so as not to create any additional hazards. Crash shutdowns of complex systems have to be carefully designed to optimise safety without causing additional risks.

Emergency stops are not a substitute for effective guarding of dangerous parts of equipment and should not be used for normal stopping of the equipment.

Emergency stop buttons should be easily identified, reached and operated. Common types are mushroom-headed buttons, bars, levers, kick-plates or pressure-sensitive cables. They are normally red and should need to be reset after use. With stop buttons this is either by twisting or a security key (see Figure 10.5).

Mobile work equipment is normally provided with effective means of stopping the engine or power source. In some cases large equipment may need

Figure 10.5 Emergency stop button

emergency stop controls away from the operator's position.

(d) Isolation of equipment

Equipment should be provided with efficient means of isolating it from all sources of energy. The purpose is to make the equipment safe under particular conditions, for example when maintenance is to be carried out or where adverse weather conditions may make it unsafe to use. On static equipment isolation will usually be for mains electrical energy; however, in some cases there may be additional or alternative sources of energy.

The isolation should cover all sources of energy such as diesel and petrol engines, LPG, steam, compressed air, hydraulics, batteries and heat. In some cases special consideration is necessary where, for example, hydraulic pumps are switched off, so as not to allow heavy pieces of equipment to fall due to gravity. An example would be the loading shovel on an excavator.

Isolation should be secure so that accidental starting is not possible by, for example, locking off electrical isolators, removing keys from vehicles, disconnecting starting mechanisms, or locking off valves. Good practice for maintenance is to adopt a 'Lock, tag and try' procedure, where the isolator is locked off, fitted with a tag and then the operator tries to start the machine to test that it is dead.

(e) Stability and lighting

Stability is important and is normally achieved by bolting equipment in place or, if this is not possible, by using clamps. Some equipment can be tied down, counterbalanced or weighted, so that it remains stable under all operating conditions. If portable equipment is weighted or counterbalanced, it should be reappraised when the equipment is moved to another position. If outriggers are needed for stability in certain conditions, for example to stabilise mobile access towers, they should be employed whenever conditions warrant the additional support. In severe weather conditions it may be necessary to stop using the equipment or reappraise the situation to ensure stability is maintained.

The quality of general and local lighting will need to be considered to ensure the safe operation of the equipment. The level of lighting and its position relative to the working area are often critical to the safe use of work equipment. Poor levels of lighting, glare and shadows can be dangerous when operating equipment. Some types of lighting, for example sodium lights, can change the colour of equipment, which may increase the level of risk. This is particularly important if the colour coding of pipework or cables is essential for safety.

(f) Markings and warnings

Markings on equipment must be clearly visible and durable. They should follow international conventions for some hazards like radiation and lasers and, as far as possible, conform to the ISO 7010:2003, Graphical symbols: Safety colours and safety signs – Safety signs used in workplaces and public areas (see Chapter 4). The contents, or the hazards of the contents, as well as controls, will need to be marked on some equipment.

There are many circumstances in which marking of equipment is appropriate for health or safety reasons. Stop and start controls for equipment need to be identified. The maximum rotation speed of an abrasive wheel should be marked upon it. The maximum safe working load (rated capacity) should be marked on lifting equipment. Gas cylinders should indicate (normally by colour) the gas in them. Storage and feed vessels containing hazardous substances should be marked to show their contents, and any hazard associated with them. Pipework for water and compressed air and other mains services should be colour-coded to indicate contents.

Any other marking that might be appropriate for the user's own purposes should be considered, for example numbering machines to aid identification, particularly if the controls or isolators for the machines are not directly attached to them and there could otherwise be confusion.

Warnings or warning devices may be appropriate where risks to health or safety remain after other hardware measures have been taken. They may be incorporated into systems of work (including permit-to-work systems), and can enforce measures of information, instruction and training. A warning is normally in the form of a notice or similar. Examples are positive instructions ('hard hats must be worn'), prohibitions ('not to be operated by people under 18 years'), restrictions ('do not heat above 60°C'). A warning device is an active unit giving a signal; the signal may typically be visible or audible, and is often connected into equipment so that it is active only when a hazard exists.

Warning devices can be:

(a) audible, for example reversing alarms on construction vehicles;

(b) visible, for example a light on a control panel that a fan on a microbiological cabinet has broken down or a blockage has occurred on a particular machine;

(c) an indication of imminent danger, for example machine about to start, or development of a fault condition (i.e. pump failure or conveyor blockage indicator on a control panel); or

(d) the continued presence of a potential hazard (for example, hotplate or laser on).

(g) Work space and operating stations

The controls at a machine should be so designed and positioned that an operative has adequate vision for control of the process being carried out. The operator should have adequate clear space in their working position and have all controls within comfortable reach.

10

Where operatives have to stand or sit to operate, machinery platforms or seats should be provided that are situated so that they are protected from dangerous parts of the machine.

Where work platforms are used, they should be designed to prevent slipping and should be large enough and provide a level standing space. Suitable guard rails and access steps/bridges should be provided where needed.

Machinery shall be so designed as to enable operation and all routine tasks relating to setting and/or maintenance to be carried out, as far as possible, by a person remaining at ground level.

Where this is not possible, machines shall have built-in platforms, stairs or other facilities to provide safe access for those tasks, but care should be taken to ensure that such platforms or stairs do not give access to danger zones of machinery.

ILO Code of Practice at Section 7.4 requires:

Marking of machinery

All machinery should be marked visibly, legibly and indelibly with the following minimum particulars:

(a) the business name and full address of the manufacturer;

(b) designation of the machinery;

(c) designation of series or type;

(d) serial number, if any; and

(e) the year of construction, that is the year in which the manufacturing process is completed.

Machinery designed and constructed for use in a potentially explosive atmosphere should be marked accordingly.

Machinery should also bear full information relevant to its type and essential for safe use, such as the maximum permissible speed of certain rotating parts, the maximum diameter of tools to be fitted, and weight.

Where a machinery part needs to be handled during use and transportation with lifting equipment, its weight should be indicated legibly, indelibly and unambiguously.

Signs and pictograms should only be used if they are understood in the culture in which the machinery is to be used.

10.1.8 User responsibilities

Under the ILO Code of Practice workers have the duty, in accordance with their training, the instructions and the means given by their employers:

(a) to comply with prescribed safety and health measures;

(b) to take all reasonable steps to eliminate or control

hazards or risks to themselves and to others from hazardous ambient factors at work, including proper care and use of protective clothing, facilities and equipment placed at their disposal for this purpose;

(c) to report forthwith to their immediate supervisor any situation which they believe could present a hazard or risk to their safety and health or that of other persons arising from hazardous ambient factors at work, and which they cannot properly deal with themselves;

(d) to cooperate with the employer and other workers to permit compliance with the duties and responsibilities placed on the employer and workers pursuant to national laws and regulations.

It is not always possible to eliminate every hazard or design safeguards that protect people against all machinery hazards, particularly during commissioning, setting, adjustment, cleaning and maintenance. Safe working practices, in some cases even permits to work, need to be adopted and followed by machinery operators or maintenance staff. These should be considered at the machine design and installation stages as the production of special jigs, fixtures, controls and isolation facilities may be needed.

Where mechanical hazards cannot be avoided there are many precautions which should be observed by properly trained and supervised people, these include:

▶ keeping the area around the machine clear and free of obstructions,

▶ wearing suitable clothing and footwear that does not have loose ends which could become entangled;

▶ avoiding the use of neckties, rings, necklaces and other jewellery;

▶ wearing appropriate eye protection where there is a risk of particles being ejected;

▶ using precautions against kickback of work-pieces such as riving knives on circular saws, work rests on abrasive wheels, cutting speeds are correct for the task and cutting tool;

▶ taking precautions against bursting of abrasive wheels by ensuring they are not damaged and are running within their design speed;

▶ limiting closeness of approach to machines like overhead cranes, or removing work from the rear of circular saws;

▶ using manual handling devices such as tongs for hot steel handling or push sticks for pushing timber through a circular saw;

▶ using jigs or holders for work-pieces.

10.2 Hazards and controls for hand-held tools

10.2.1 Introduction

Work equipment includes hand tools and hand-held power tools. This section deals with hand tools. These tools need to be correct for the task, well maintained and properly used by trained people.

Five basic safety rules can help prevent hazards associated with the use of hand-held tools:

► keep all tools in good condition with regular maintenance;
► use the right tool for the job;
► examine each tool for damage before use and do not use damaged tools;
► use tools according to the manufacturer's instructions;
► provide and use properly the right personal protective equipment (PPE).

10.2.2 Hazards of non-powered hand tools

Hazards from the misuse or poor maintenance of non-powered hand tools (Figure 10.6) include:

► broken handles on files/chisels/screwdrivers/hammers which can cause cut hands or hammer heads to fly off;
► incorrect use of knives, saws and chisels with hands getting injured in the path of the cutting edge;
► tools that slip causing stab wounds;
► poor-quality uncomfortable handles that damage hands;
► splayed spanners that slip and damage hands or faces;
► chipped or loose hammer heads that fly off or slip;
► incorrectly sharpened or blunt chisels or scissors that slip and cut hands. Dull tools can cause more injuries than sharp ones. Cracked saw blades must be removed from service;
► flying particles that damage eyes from breaking up stone or concrete;

Figure 10.6 Range of good non-powered hand tools

► electrocution, electric shock, or burns by using incorrect or damaged tools for electrical work;
► use of poorly insulated tools for hot work in the catering or food industry;
► use of pipes or similar equipment as extension handles for a spanner which is likely to slip causing hand or face injury;
► mushroom-headed chisels or drifts which can damage hands or cause hammers (not suitable for chisels) and mallets to slip;
► use of spark-producing or percussion tools in flammable atmospheres;
► painful wrists and arms (upper limb disorders) from the frequent twisting from using screwdrivers;
► when using saw blades, knives or other tools, they should be directed away from aisle areas and away from other people working in close proximity.

10.2.3 Non-powered hand tools safety considerations

Use of non-powered hand tools should be properly controlled including those tools owned by employees. The following controls are important.

(a) Suitability

All tools should be suitable for the purpose and location in which they are to be used. Using the correct tool for the job is the first step in safe hand tool use. Tools are designed for specific needs. That is why screwdrivers have various lengths and tip styles and pliers have different head shapes. Using any tool inappropriately is a step in the wrong direction. To avoid personal injury and tool damage, select the proper tool to do the job well and safely.

High-quality professional hand tools will last many years if they are taken care of and treated with respect. Manufacturers design tools for specific applications. Use tools only for their intended purpose.

Suitability will include:

► specially protected and insulated tools for electricians;
► non-sparking tools for flammable atmospheres and the use of non-percussion tools and cold cutting methods;
► tools made of suitable quality materials which will not chip or splay in normal use;
► the correct tools for the job, for example using the right-sized spanner and the use of mallets not hammers on chisel heads. The wooden handles of tools must not be splintered;
► safety knives with enclosed blades for regular cutting operations;
► impact tools such as drift pins, wedges and cold-chisels being kept free of mushroomed heads;
► spanners not being used when jaws are sprung to the point that slippage occurs.

(b) Inspection

All tools should be maintained in a safe and proper condition. This can be achieved through:

- the regular inspection of hand tools;
- discarding or prompt repair of defective tools;
- taking time to keep tools in the proper condition and ready for use;
- proper storage to prevent damage and corrosion;
- locking tools away when not in use to prevent them being used by unauthorised people.

(c) Training

All users of hand tools should be properly trained in their use. This may well have been done through apprenticeships and similar training. This will be particularly important with specialist working conditions or work involving young people.

Always wear approved eye protection when using hand tools, particularly when percussion tools are being used. Metal and wood particles will fly out when the material is cut or planed, so other workers in the vicinity should wear eye protection as well.

(d) Use well-designed, high-quality tools

Finally, investing in high-quality tools makes the professional's job safer and easier:

- if extra leverage is needed, use high-leverage pliers, which give more cutting and gripping power than standard pliers. This helps, in particular, when making repetitive cuts or twisting numerous wire pairs;
- serrated jaws provide sure gripping action when pulling or twisting wires;
- some side-cutting and diagonal-cutting pliers are designed for heavy-duty cutting. When cutting screws, nails and hardened wire, only use pliers that are recommended for that use;
- pliers with hot riveting at the joint ensure smooth movement across the full action range of the pliers, which reduces handle wobble, resulting in a positive cut. The knives align perfectly every time;
- induction hardening on the cutting knives adds to long life, so the pliers cut cleanly day after day;
- sharp cutting knives and tempered handles also contribute to cutting ease;
- some pliers are designed to perform special functions. For example some high-leverage pliers have features that allow crimping connectors and pulling fish tapes;
- tool handles with dual-moulded material allow for a softer, more comfortable grip on the outer surface and a harder, more durable grip on the inner surface and handle ends;
- well-designed tools often include a contoured thumb area for a firmer grip or colour-coded handles for easy tool identification;

- insulated tools reduce the chance of injury where the tool may make contact with an energised source.

Well-designed tools are a pleasure to use. They save time, give professional results and help to do the job more safely.

10.2.4 Hand-held portable power tools

(a) Introduction

The electrical hazards of portable hand-held tools and portable appliance testing (PAT) are covered in more detail in Chapter 11. This section deals mainly with other physical hazards and safeguards relating to hand-held power tools.

The section covers, in particular, electric drills and sanders which are commonly used in the workplace. Chainsaws are covered later.

(b) General hazards of hand-held power tools

The general hazards involve:

- mechanical entanglement in rotating spindles or sanding discs;
- waste material flying out of the cutting area;
- coming into contact with the cutting blades or drill bits;
- risk of hitting electrical, gas or water services when drilling into building surfaces;
- electrocution/electric shock from poorly maintained equipment and cables or cutting the electrical cable;
- manual handling problem with a risk of injury if the tool is heavy or very powerful;
- hand–arm vibration, especially with pneumatic drills and chainsaws, disc cutters and petrol-driven units;
- tripping hazard from trailing cables, hoses or power supplies;
- eye hazard from flying particles;
- injury from poorly secured or clamped work-pieces;
- fire and explosion hazard with petrol-driven tools or when used near flammable liquids, explosive dusts or gases;
- high noise levels with pneumatic chisels, planes and saws in particular (see Chapter 14);

Figure 10.7 Range of hand-held power tools

► levels of dust and fumes given off during the use of the tools (but see Chapter 13).

(c) Typical safety controls and instructions

Guarding

The exposed moving parts of power tools need to be safeguarded. Belts, gears, shafts, pulleys, sprockets, spindles, drums, flywheels, chains or other reciprocating, rotating or moving parts of equipment must be guarded. Machine guards, as appropriate, must be provided to protect the operator and others from the following:

► point of operation;
► drawing running nip points;
► rotating parts;
► flying chips and sparks.

Safety guards must never be removed when a tool is being used. Portable circular saws, for example, must be equipped at all times with guards. An upper guard must cover the entire blade of the saw. A retractable lower guard must cover the teeth of the saw, except where it makes contact with the work material. The lower guard must automatically return to the covering position when the tool is withdrawn from the work material.

Operating controls and switches

Most hand-held power tools should be equipped with a constant-pressure switch or control that shuts off the power when pressure is released. On/off switches should be easily accessible without removing hands from the equipment.

Handles should be designed to protect operators from excessive vibration and keep their hands away from danger areas. In some cases handles are also designed to activate a brake of the cutting chain or blade, for example in a chainsaw. Equipment should be designed to reduce lifting and manual handling problems, with special harnesses being used as necessary, for example when using large strimmers.

Means of starting engines and holding equipment should be designed to minimise any musculoskeletal problems.

Safe operations/instructions

When using power tools, the following basic safety measures should be observed to protect against electrical shock, personal injury, ill-health and risk of fire. See also more detailed electrical precautions in Chapter 11. Operators should read these instructions before using the equipment and ensure that they are followed; that is, they must:

► maintain a clean and tidy working area that is well lit and clear of obstructions;
► never expose power tools to rain. Do not use power tools in damp or wet surroundings;

► not use power tools in the vicinity of combustible fluids, dusts or gases unless they are specially protected and certified for use in these areas;
► protect against electric shock (if tools are electrically powered) by avoiding body contact with grounded objects such as pipes, scaffolds and metal ladders (see Chapter 11 for more electrical safety precautions);
► keep children away;
► not let other persons handle the tool or the cable. Keep them away from the working area;
► store tools in a safe place when not in use where they are in a dry, locked area which is inaccessible to children;
► ensure tools are not overloaded as they operate better and more safely in the performance range for which they were intended;
► use the right tool. Do not use small tools or attachments for heavy work. Do not use tools for purposes and tasks for which they were not intended; for example, do not use a hand-held circular saw to cut down trees or cut off branches;
► wear suitable work clothes. Do not wear loose-fitting clothing or jewellery. They can get entangled in moving parts. For outdoor work, rubber gloves and non-skid footwear are recommended. Long hair should be protected with a hair net;
► use safety glasses;
► also use a filtering respirator mask for work that generates dust;
► not abuse the power cable;
► not carry the tool by the power cable and do not use the cable to pull the plug out of the power socket. Protect the cable from heat, oil and sharp edges;
► secure the work-piece. Use clamps or a vice to hold the work-piece. It is safer than using hands and it frees both hands for operating the tool;
► not over-reach the work area. Avoid abnormal body postures. Maintain a safe stance and maintain a proper balance at all times;
► maintain tools with care. Keep your tools clean and sharp for efficient and safe work. Follow the maintenance regulations and instructions for the changing of tools. Check the plug and cable regularly and, in the case of damage, have them repaired by a qualified service engineer. Also inspect extension cables regularly and replace if damaged;
► keep the handle dry and free of oil or grease;
► disconnect the power plug when not in use, before servicing and when changing the tool parts, that is blade, bits, cutter, sanding disc, etc.;
► not forget to remove the key. Check before switching on that the key and any tools for adjustment are removed;
► avoid unintentional switch-on. Do not carry tools that are connected to power with your finger on the power switch. Check that the switch is turned off before connecting the power cable;

10

▶ when working outdoors, use extension cables, and only those which are intended for such use and marked accordingly;

▶ stay alert, keep eyes on the work. Use common sense. Do not operate tools when there are significant distractions;

▶ check the equipment for damage. Before further use of a tool, check carefully the protection devices or lightly damaged parts for proper operation and performance of their intended functions. Check movable parts for proper function, for whether there is binding or for damaged parts. All parts must be correctly mounted and meet all conditions necessary to ensure proper operation of the equipment;

▶ ensure damaged protection devices and parts are repaired or replaced by a competent service centre unless otherwise stated in the operating instructions. Damaged switches must be replaced by a competent service centre. Do not use any tool which cannot be turned on and off with the switch;

▶ only use accessories and attachments that are described in the operating instructions or are provided or recommended by the tool manufacturer. The use of tools other than those described in the operating instructions or in the catalogue of recommended tool inserts or accessories can result in a risk of personal injury;

▶ use engine-driven power tools in well-ventilated areas. Store petrol in a safe place in approved storage cans. Stop and let engines cool before refuelling.

(d) Specific hazards and control measures for specified hand-held power tools

The following hand-held power tools have been put in the International General Certificate syllabus: electric drills and sanders, both of which are commonly used in various forms. The hazards and safety control measures in addition to the general ones covered earlier are set out for each type of equipment.

Electric drills

Hazards (Figure 10.8) are:

▶ entanglement, particularly of loose clothing or long hair in rotating drill bits;

▶ high noise levels from the drill or attachment;

▶ eye injury from flying particles and chips, particularly from chisels;

▶ injury from poorly secured work-pieces;

▶ electrocution/electric shock from poorly maintained equipment;

▶ electric shock from drilling into a live hidden cable;

▶ hand–arm vibration hazard in hammer mode;

▶ dust given off from material being worked on;

▶ tripping hazard from trailing cables;

▶ upper limb disorder from powerful machines with a strong torque, particularly if they jam and kick back;

▶ foot injury hazards from dropping heavy units onto unprotected feet;

▶ manual handling hazards, particularly with heavy machines and intensive use or using at awkward heights and/or reaches;

▶ fire and explosion hazard when used near flammable liquids, explosive dusts or gases;

▶ using equipment in poor weather conditions with wet, slippery surfaces, poor visibility and cold conditions.

Specific control measures include the following:

Figure 10.8 Electric drill with percussion hammer action to drill holes in masonry

▶ use double-insulated tools or earthed reduced voltage tools with a residual current device (see Chapter 11 for electrical safeguards in detail);

▶ use a pilot hole or punch to start holes whenever possible;

▶ select the correct drill bit for the material being drilled;

▶ secure small pieces to be drilled to prevent spinning;

▶ protect against damage or injury on the far side if the bit is long enough to pass through the material or there are buried services in, say, plaster walls;

▶ take care to prevent loose sleeves or long hair from being wound around the drill bit; for example wear short or close-fitting sleeves;

▶ wear suitable eye protection.

Sanders

There is a large range of hand-held sanders on the market, from rotating discs (Figure 10.9), random orbital, rectangular orbital (Figure 10.10(b)), belt sanders and heavy-duty floor sanders of both the rotating drum (Figure 10.10(a)) and, recently, the orbital floor sander. The high-speed rotating discs and drum types are the most hazardous but they all need using with care.

Hazards include:

▶ high noise levels from the sander in operation;

▶ injury from poorly secured work-pieces;

▶ electrocution/electric shock from poorly maintained electrical equipment;

▶ potential of entanglement with rotating disc and drum sanders;

▶ sanding attachments can become loose in the chuck and fling off;

▶ injury from contact with abrasive surfaces, particularly with course abrasives and high-speed rotating sanding discs and drums;

▶ hand–arm vibration hazard, particularly from reciprocating equipment;

▶ health hazards from extensive dust given off from material being worked on;

▶ fire and health hazards from overheating of abraded surfaces, particularly if plastics are being sanded;

▶ tripping hazard from trailing cables;

▶ large powerful sanders suddenly gripping the surface and pulling the operative off their feet;

▶ foot injury hazards from dropping heavy units onto unprotected feet;

▶ manual handling hazards, particularly with heavy machines and intensive use or using at awkward heights and/or reaches;

▶ fire and explosion hazard when used near flammable liquids, explosive dusts or gases;

▶ using equipment in poor weather conditions with wet, slippery surfaces, poor visibility and cold conditions.

Specific control measures include the following:

▶ work-pieces must be securely clamped or held in position during sanding. In some cases a jig will be necessary. The direction of spin of disc sanders (normally anti-clockwise) is important to ensure that a small work-piece is pushed towards a stop or fence which is normally at the left side of the work-piece, particularly when clamping is impossible;

▶ abrasive sanding belts, discs and sheets should be properly and firmly attached to the machine without any torn parts or debris underneath. The

Figure 10.10 (a) Rotary drum floor sander; (b) orbital finishing sander

manufacturer's instructions and fixing accessories should be used to ensure correct attachment of the abrasive. Operators should be trained, competent and registered to fit abrasive discs;

▶ old nails and fixings should be sunk below the surface or removed to prevent the sander snagging;

▶ always hold the equipment by the proper handles and particularly on large disc and floor sanders always use both hands. Excessive pressure should not be used as the surface will be rutted and the machine may malfunction;

▶ ensure that the dust extraction is working properly

Figure 10.9 Disc sander with dual hand holds

and has been emptied when about one-third full. Some extraction systems draw dust and air through the sanding sheet, which must have correctly pre-punched holes to allow the passage of air;

▶ operators should wear suitable dust respirators, eye protection and, where necessary, hearing protection;

▶ operators should wear suitable clothing avoiding loose garments, long hair and jewellery, which could catch in the equipment. Protective gloves and footwear are recommended.

10.3 Mechanical and non-mechanical hazards of machinery

10.3.1 Hazard identification

Most machinery has the potential to cause injury to people, and machinery accidents figure prominently in official accident statistics. These injuries may range in severity from a minor cut or bruise, through various degrees of wounding and disabling mutilation, to crushing, decapitation or other fatal injury. It is not solely powered machinery that is hazardous, for many manually-operated machines (e.g. hand-operated guillotines and fly presses) can still cause injury if not properly safeguarded.

In the risk assessment of machinery (see ISO 12100:2010) it is essential to consider both permanent hazards and those which can appear unexpectedly. Hazardous situations and events during all phases of the machine life cycle need to be covered. These include: transport, assembly and installation; commissioning; use; maintenance; dismantling, disabling and scrapping.

The human interaction with the machine involving numerous tasks throughout the life cycle of the machine should be considered. They include: setting; testing; teaching/programming; process/tool changeover; start-up; all modes of operation; feeding the machine; removal of product from the machine; stopping the machine; stopping the machine in case of emergency; recovery of operation from jam or blockage; restart after unscheduled stop; fault-finding/trouble-shooting (operator intervention); cleaning and housekeeping; preventative maintenance; corrective maintenance.

The possible states of the machine should be taken into account, which are:

1. the machine performs the intended function (the machine operates normally);
2. the machine does not perform the intended function (i.e. it malfunctions) due to a variety of reasons, including:

   ▶ variation of a property or of a dimension of the processed material or of the work-piece;

   ▶ failure of one or more of its component parts or services;

   ▶ external disturbances (for example, shocks, vibration, electromagnetic interference);

▶ design error or deficiency (for example, software errors);

▶ disturbance of its power supply; and

▶ surrounding conditions (for example, damaged floor surfaces).

Finally the unintended behaviour of the operator or reasonably foreseeable misuse of the machine must be considered. Examples include:

▶ loss of control of the machine by the operator (especially for hand-held or mobile machines);

▶ reflex behaviour of a person in case of malfunction, incident or failure during the use of the machine;

▶ behaviour resulting from lack of concentration or carelessness;

▶ behaviour resulting from taking the 'line of least resistance' in carrying out a task;

▶ behaviour resulting from pressures to keep the machine running in all circumstances; and

▶ behaviour of certain persons (for example, children, people with a disability).

10.3.2 Mechanical hazards

The hazards of machinery are set out in ISO 12100:2010, which covers the classification of machinery hazards and how harm may occur. The following mechanical machinery hazards follow this standard (Figure 10.11 shows a number of these hazards).

A person may be injured at machinery as a result of:

▶ **acceleration, deceleration** – potential for impact, being thrown, run-over, slipping, tripping or falling, for example from a piece of mobile plant or an overhead travelling crane;

▶ **angular parts** – potential consequences are impact, crushing or shearing when a person hits a stationary part of the machine or a moving part impacts on an operator particularly in mobile plant;

▶ **approach of a moving element to a fixed part** – potential consequences are crushing or impact through being trapped between a moving part of a machine and a fixed structure, such as a wall or any material in a machine;

▶ **cutting parts** – potential consequences are cutting or severing through contact with a cutting edge, such as a band saw or rotating cutting disc;

▶ **elastic elements** – consequences are crushing or impact when an elastic part is deformed and suddenly straightens or returns to its original dimensions or explodes when an airbag used for lifting fails;

▶ **falling objects** – potential consequences are crushing or impact, for example if a load fell from an overhead hoist or rail system;

- **gravity** – potential consequences are crushing or trapping, for example if a vehicle lift failed while a person was underneath;
- **height from the ground** – potential consequences are crushing, impact, slipping, tripping or falling,

for example getting out of a truck or large piece of mobile plant or gaining access to a tower crane;
- **high pressure** – potential consequences are injection, stabbing or puncture, impact, for example from a high pressure hydraulic system leak;

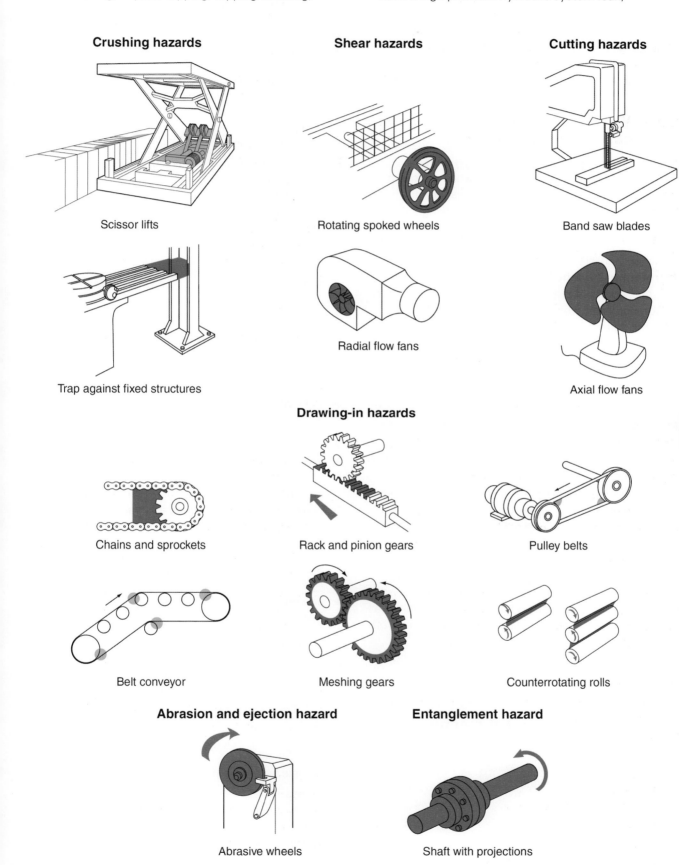

Crushing hazards

Scissor lifts

Trap against fixed structures

Shear hazards

Rotating spoked wheels

Radial flow fans

Cutting hazards

Band saw blades

Axial flow fans

Drawing-in hazards

Chains and sprockets

Rack and pinion gears

Pulley belts

Belt conveyor

Meshing gears

Counterrotating rolls

Abrasion and ejection hazard

Abrasive wheels

Entanglement hazard

Shaft with projections

Figure 10.11 Range of mechanical hazards

▶ **instability** – potential for being thrown, crushing or impact when a machine breaks its holding down bolts and falls over or a tower crane is not left free to slew to the wide direction and is blown over;

▶ **kinetic energy** – potential for impact, puncture or severing when, for example, a kinetic recovery rope is used to tow a vehicle or ship and it snaps the whiplash can be very hazardous;

▶ **machinery mobility** – potential consequences are being run over, impact, crushing when, for example, a large machine is being manoeuvred in a small space;

▶ **moving elements** – potential consequences are crushing, friction, abrasion, impact, shearing, severing, drawing-in, for example on grinding wheels or sanding machines. Or when a moving part directly strikes a person, such as with the accidental movement of a robot's working arm when maintenance is taking place. A further example is when persons are drawn-in between in-running gear wheels or rollers or between belts and pulley drives;

▶ **rotating elements** – potential consequences are severing or entanglement, for example an exposed rotating shaft which grips loose clothing, hair or working material, such as emery paper. The smaller the diameter of the revolving part the easier it is to get a wrap or entanglement. A further example is a shearing which occurs when part of the body, typically a hand or fingers, is trapped between rotating and fixed parts of the machine;

▶ **rough, slippery surface** – potential consequences are friction, abrasion, slipping, tripping and falling, impact when, for example, the operating area round a machine is very slippery from product spillages and the operator falls over and/or into the machinery hazard zone;

▶ **sharp edges** – potential consequences are severing, impact, shearing, stabbing or puncture;

▶ **stored energy** – potential consequences are crushing, impact, puncture, suffocation when the stored energy is released suddenly, for example when the valve on a welding gas bottle (other hazards such as explosion or fire may also be present) is broken off or a high pressure air hose breaks;

▶ **vacuum** – potential consequences are crushing, drawing-in, suffocation, for example the sudden failure of a large vacuum vessel.

10.3.3 Non-mechanical machinery hazards

Non-mechanical machinery hazards are also set out in ISO 12100:2010 and include:

Electrical hazards from:

▶ arc;
▶ electromagnetic phenomena;
▶ electrostatic phenomena;
▶ live parts;

▶ not enough distance to live parts under high voltage;
▶ overload;
▶ parts which have become live under fault conditions;
▶ short-circuit;
▶ thermal radiation.

Thermal hazards from:

▶ explosion;
▶ flame;
▶ objects or materials with a high or low temperature;
▶ radiation from heat sources.

Noise hazards from:

▶ cavitation phenomena;
▶ exhausting system;
▶ gas leaking at high speed;
▶ manufacturing process;
▶ moving parts;
▶ scraping surfaces;
▶ unbalanced rotating parts;
▶ whistling pneumatics;
▶ worn parts.

Vibrations hazards from:

▶ cavitation phenomena;
▶ misalignment of moving parts;
▶ mobile equipment;
▶ scraping surfaces;
▶ unbalanced rotating parts;
▶ vibrating equipment;
▶ worn parts.

Radiation hazards from:

▶ ionising radiation sources;
▶ low frequency electromagnetic radiation;
▶ optical radiation (infrared, visible and ultraviolet), including laser;
▶ radio frequency electromagnetic radiation.

Material/substance hazard from:

▶ aerosol;
▶ biological and microbiological (viral and bacteria) agent;
▶ combustible;
▶ dust;
▶ explosive;
▶ fibre;
▶ flammable;
▶ fluid;
▶ fume;
▶ gas;
▶ mist;
▶ oxidiser.

Ergonomic hazard from:

▶ access;
▶ design or location of indicators and visual display units;
▶ design, location or identification of control devices;
▶ effort;
▶ flicker, dazzling, shadow, stroboscopic effect;

- local lighting;
- mental overload/underload;
- posture;
- repetitive activity;
- visibility.

Hazards associated with the environment in which the machine is used from:

- dust and fog;
- electromagnetic disturbance;
- lightning;
- moisture;
- pollution;
- snow;
- temperature;
- water;
- wind;
- lack of oxygen.

Combination of hazards from:

- for example, repetitive activity + effort + high environmental temperature (or repetitive hard work in a hot climate).

In many cases it will be practicable to install safeguards which protect the operator from both mechanical and non-mechanical hazards.

For example a guard may prevent access to hot or electrically live parts as well as to moving ones. The use of guards which reduce noise levels at the same time is also common.

As a matter of policy, machinery hazards should be dealt with in this integrated way instead of dealing with each hazard in isolation.

See Chapter 18 for a form which could be used to assess machinery risk.

10.3.4 Examples of machinery hazards

The following examples are given to demonstrate a small range of machines found in industry and commerce, which are included in the International General Certificate syllabus. The typical photographs of these machines are shown in section 10.4 under the application of safeguards.

(a) Office – photocopier

The hazards are:

- contact (cutting or abrasion) with moving parts when clearing a jam;
- electrical – when clearing a jam, maintaining the machine or through poorly maintained plug and wiring;
- heat through contact with hot parts when clearing a jam;
- health hazard from ozone or lack of ventilation in the area or very rarely an allergy to paper dust;
- inhalation of dust from emptying toner;
- getting dust from toners in the eyes.

(b) Office – document shredder

The hazards are:

- drawing-in between the rotating cutters when feeding paper into the shredder;
- contact (cutting or severing) with the rotating cutters when emptying the waste container or clearing a jam;
- electrical through faulty plug and wiring or during maintenance;
- possible noise from the cutting action of the machine;
- possible dust from the cutting action.

(c) Manufacturing and maintenance – bench-top grinding machine

The hazards are:

- contact (friction or abrasion) with the rotating wheel causing abrasion;
- drawing-in between the rotating wheel and a badly adjusted tool rest;
- bursting of the wheel, ejecting fragments which puncture the operator;
- electrical through faulty wiring and/or earth bonding or during maintenance;
- fragments given off during the grinding process causing eye injury;
- hot fragments given off which could cause a fire or burns;
- noise produced during the grinding process;
- possible health hazard from dust/particles/fumes given off during grinding.

(d) Manufacturing and maintenance – pedestal drill

The hazards are:

- entanglement around the rotating spindle and chuck;
- contact (cutting, puncturing, severing) with the cutting drill or work-piece;
- being struck by the work-piece if it rotates;
- being cut or punctured by fragments ejected from the rotating spindle and cutting device;
- drawing-in to the rotating drive belt and pulley;
- contact or entanglement with the rotating motor;
- electrical from faulty wiring and/or earth bonding or during maintenance;
- possible health hazard from cutting fluids or dust given off during the process.

(e) Agricultural/horticultural – cylinder mower

The hazards are:

- trapping (cutting or severing), typically hands or fingers, in the shear caused by the rotating cutters;
- contact (impact, cutting) and entanglement with moving parts of the drive motor;

10

293

▶ drawing-in between chain and sprocket drives;
▶ impact and cutting injuries from the machine starting accidentally;
▶ burns from hot parts of the engine;
▶ fire from the use of highly flammable petrol as a fuel;
▶ possible noise hazard from the drive motor;
▶ electrical if electrically powered, but this is unlikely;
▶ possible sensitisation health hazard from cutting grass, for example, hay fever;
▶ possible health hazard from exhaust fumes.

(f) Agricultural/horticultural – brush cutter/strimmer

The hazards are:

▶ entanglement with rotating parts of motor and shaft;
▶ cutting or severing from contact with cutting head/line;
▶ electric shock, if electrically powered, but this is unlikely;
▶ burns from hot parts of the engine;
▶ fire from the use of highly flammable petrol as a fuel;
▶ possible noise hazard from the drive motor and cutting action;
▶ eye and face puncture wounds from ejected particles;
▶ health hazard from hand–arm vibration causing white finger and other problems;
▶ back strain from carrying the machine while operating;
▶ health hazards from animal faeces.

(g) Chainsaw

The hazards are:

▶ very serious cutting or severing by contact with the high-speed cutting chain;
▶ kickback (impact, cutting, severing) due to being caught on the wood being cut or contact with the top front corner of the chain in motion with the saw chain being kicked upwards towards the face in particular;
▶ pull in (drawing-in) when the chain is caught and the saw is pulled forward;
▶ push-back (impact) when the chain on the top of the saw bar is suddenly pinched and the saw is driven straight back towards the operator;
▶ burns from hot parts of the engine;
▶ high noise levels;
▶ hand–arm vibration causing white finger and other problems;
▶ fire from the use of highly flammable petrol as a fuel;
▶ eye and face puncture wounds from ejected particles;
▶ back strain, WRULDs, due to supporting the weight of the chainsaw while operating;
▶ electric shock, if electrically powered;
▶ falls from height if using the chainsaw in trees and the like;

▶ lone working and the risk of serious injury;
▶ contact with overhead power lines if felling trees;
▶ being hit by falling branches or whole trees while felling;
▶ possible health hazards from cutting due to wood dust, partically if wood has been seasoned;
▶ using chainsaw in poor weather conditions with slippery surfaces, poor visibility and cold conditions;
▶ health hazards from the engine fumes (carbon dioxide and carbon monoxide), particularly if used inside a shed or other building.

(h) Retail – compactor

The hazards are:

▶ crushing hazard between the ram and the machine sides;
▶ trapping (severing, crushing) between the ram and machine frame (shear action);
▶ crushing when the waste unit is being changed if removed by truck;
▶ entanglement with moving parts of pump motor;
▶ electrical from faulty wiring and/or earth bonding or during maintenance;
▶ failure of hydraulic hoses with liquid released under pressure causing puncture to eyes or other parts of body;
▶ falling of vertical ram under gravity if the hydraulic system fails;
▶ handling hazards during loading and unloading.

(i) Retail – checkout conveyor system

The hazards are:

▶ entanglement with belt fasteners if fitted;
▶ drawing-in between belt and rollers if under tension;
▶ drawing-in between drive belt and pulley;
▶ contact (impact, drawing-in, entanglement) with motor drive;
▶ electrical from faulty wiring and/or earth bonding or during maintenance.

(j) Construction – cement/concrete mixer

The hazards are:

▶ contact (impact, drawing-in, entanglement) with moving parts of the drive motor;
▶ crushing between loading hopper (if fitted) and drum;
▶ drawing-in between chain and sprocket drives;
▶ electric shock, if electrically powered;
▶ burns from hot parts of the engine;
▶ fire if highly flammable liquids used as fuel;
▶ possible noise hazards from the motor and dry mixing of aggregates;
▶ eye injury from splashing cement slurry;
▶ possible health hazard from cement dust and cement slurry while handling.

(k) Construction – bench-mounted circular saw

The hazards are:

▶ contact (cutting, severing) with the cutting blade above and below the bench;
▶ ejection of the work-piece or timber as it closes after passing the cutting blade;
▶ drawing-in between chain and sprocket or V belt drives;
▶ contact (impact, drawing-in, entanglement) with moving parts of the drive motor;
▶ likely noise hazards from the cutting action and motor;
▶ health hazards from wood dust given off during cutting;
▶ electric shock from faulty wiring and/or earth bonding or during maintenance;
▶ climatic conditions such as wet/extreme heat or cold;
▶ possible fire and/or explosion from wood dust caused by overheated blades from excessive friction or an electrical fault.

10.4 Control measures for reducing risks from machinery hazards

10.4.1 Introduction

As discussed in 10.1.1 and 10.1.3, ISO 12100:2010 requires that access to dangerous parts of machinery should be prevented in a preferred order or hierarchy of control measures. The standard required is a 'practicable' one, so that the only acceptable reason for non-compliance is that there is no technical solution. Cost is not a factor. (See Chapter 1 for more details on standards of compliance.)

As the mechanical hazard of machinery arises principally from someone coming into contact or entanglement with dangerous components, risk reduction is based on preventing this contact occurring.

This may be by means of:

▶ a physical barrier between the individual and the component (e.g. a fixed enclosing guard);
▶ a device which allows access only when the component is in a safe state (e.g. an interlocked guard which prevents the machine starting unless a guard is closed and acts to stop the machine if the guard is opened);
▶ a device which detects that the individual is entering a risk area and then stops the machine (e.g. certain photoelectric guards and pressure-sensitive mats).

The best method should, ideally, be chosen by the designer as early in the life of the machine as possible. It is often found that safeguards which are 'bolted on' instead of 'built in' are not only less effective in reducing risk, but are also more likely to inhibit the normal

operation of the machine. In addition, they may in themselves create hazards and are likely to be difficult and hence expensive to maintain.

10.4.2 Fixed guards

Fixed guards have the advantage of being simple, always in position, difficult to remove and almost maintenance-free. Their disadvantage is that they do not always properly prevent access, they may be left off by maintenance staff and they can create difficulties for the operation of the machine.

A fixed guard has no moving parts and should, by its design, prevent access to the dangerous parts of the machinery. It must be of robust construction and sufficient to withstand the stresses of the process and environmental conditions. If visibility or free air flow (e.g. for cooling) is necessary, this must be allowed for in the design and construction of the guard. If the guard can be opened or removed, this must only be possible with the aid of a tool.

An alternative fixed guard is the distance fixed guard, which does not completely enclose a hazard, but which reduces access by virtue of its dimensions and its distance from the hazard. Where perimeter fence guards are used, the guard must follow the contours of the machinery as far as possible, thus minimising space between the guard and the machinery. With this type of guard, it is important that the safety devices and operating systems prevent the machinery being operated with the guards closed and someone inside the guard, that is in the danger area. Figure 10.12 shows a range of fixed guards for some of the examples shown in Figure 10.11.

10.4.3 Adjustable guards

(a) User-adjusted guard

These are fixed or movable guards, which are adjustable for a particular operation during which they remain fixed. They are particularly used with machine tools where some access to the dangerous part is required (e.g. drills, circular saws, milling machines) and where the clearance required will vary (e.g. with the size of the cutter in use on a horizontal milling machine or with the size of the timber being sawn on a circular saw bench).

Adjustable guards may be the only option with cutting tools, which are otherwise very difficult to guard, but they have the disadvantage of requiring frequent readjustment. By the nature of the machines on which they are most frequently used, there will still be some access to the dangerous parts, so these machines must only be used by suitably trained operators (see Figure 10.13). Jigs, push sticks and false tables must be used wherever possible to minimise hazards during the feeding of the work-piece. The working area should be well lit and kept free of anything which might cause the operator to slip or trip.

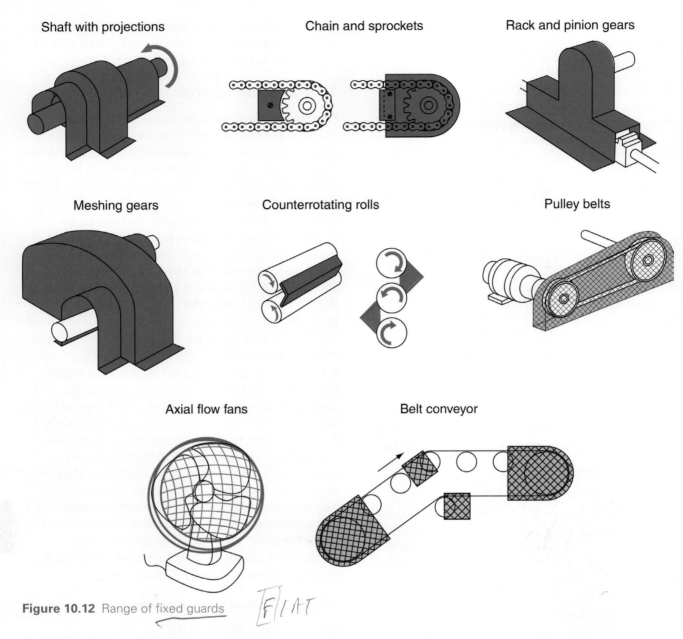

Shaft with projections

Chain and sprockets

Rack and pinion gears

Meshing gears

Counterrotating rolls

Pulley belts

Axial flow fans

Belt conveyor

Figure 10.12 Range of fixed guards

(b) Self-adjusting guard

A self-adjusting guard is one which adjusts itself to accommodate, for example, the passage of material. A good example is the spring-loaded guard fitted to many portable circular saws.

As with adjustable guards they only provide a partial solution in that they may well still allow access to the dangerous part of the machinery. They require careful maintenance to ensure they work to the best advantage (see Figure 10.14).

10.4.4 Interlocking guard

The advantages of interlocked guards are that they allow safe access to operate and maintain the machine without dismantling the safety devices. Their disadvantage stems from the constant need to ensure that they are operating correctly and designed to be fail-safe. Maintenance and inspection procedures must be very strict.

This is a guard which is movable (or which has a movable part) whose movement is connected with the power or control system of the machine.

An interlocking guard must be connected to the machine controls such that:

▶ until the guard is closed the interlock prevents the machinery from operating by interrupting the power medium;
▶ either the guard remains locked until the risk of injury from the hazard has passed or opening the guard causes the hazard to be eliminated before access is possible.

A passenger lift or hoist is a good illustration of these principles: the lift will not move unless the doors are closed, and the doors remain closed and locked until the lift is stationary and in such a position that it is safe for the doors to open.

Special care is needed with systems which have stored energy. This might be the momentum of a heavy

296

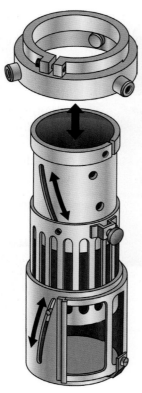

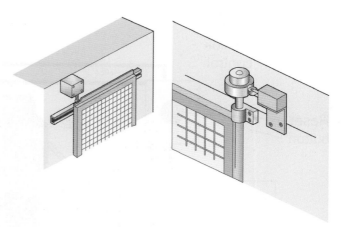

Figure 10.15 Typical interlocked guards: (a) sliding and (b) hinged

10.4.5 Other protective devices

(a) Trip devices

A trip device does not physically keep people away but detects when a person approaches close to a danger point. It should be designed to stop the machine before injury occurs. A trip device depends on the ability of the machine to stop quickly and in some cases a brake may need to be fitted. Trip devices can be:

▶ mechanical in the form of a bar or barrier;
▶ electrical in the form of a trip switch on an actuator rod, wire or other mechanism;
▶ photoelectric or other type of presence-sensing device;
▶ a pressure-sensitive mat.

They should be designed to be self-resetting so that the machine must be restarted using the normal procedure (Figure 10.16).

(b) Two-handed control devices

These are devices which require the operator to have both hands in a safe place (the location of the controls) before the machine can be operated. They are an option on machinery that is otherwise very difficult to guard but they have the drawback that they only protect the operator's hands. It is therefore essential that the design does not allow any other part of the operator's body to enter the danger zone during operation. More significantly, they give no protection to anyone other than the operator.

Where two-handed controls are used, the following principles must be followed:

▶ the controls should be so placed, separated and protected as to prevent spanning with one hand only, being operated with one hand and another part of the body, or being readily bridged;
▶ it should not be possible to set the dangerous parts in motion unless the controls are operated within approximately 0.5 seconds of each other. Having set the dangerous parts in motion, it should not be

Figure 10.13 Adjustable guard for a rotating shaft, such as a pedestal drill

Figure 10.14 Self-adjusting guard on a wood saw

moving part, stored pressure in a hydraulic or pneumatic system, or even the simple fact of a part being able to move under gravity even though the power is disconnected. In these situations, dangerous movement may continue or be possible with the guard open, and these factors need to be considered in the overall design. Braking devices (to arrest movement when the guard is opened) or delay devices (to prevent the guard opening until the machinery is safe) may be needed. All interlocking systems must be designed to minimise the risk of failure-to-danger and should not be easy to defeat (Figure 10.15).

10

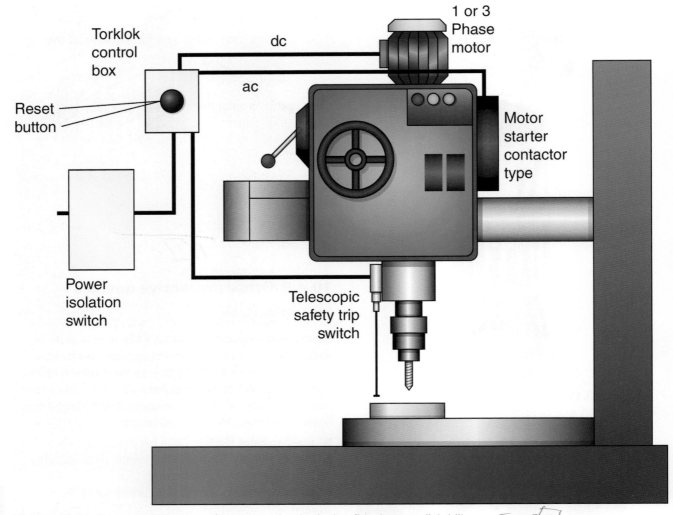

Figure 10.16 Schematic diagram of a telescopic trip device fitted to a radial drill

possible to do so again until both controls have been returned to their off position;

▶ movement of the dangerous parts should be arrested immediately or, where appropriate, arrested and reversed if one or both controls are released while there is still danger from movement of the parts;

▶ the hand controls should be situated at such a distance from the danger point that, on releasing the controls, it is not possible for the operator to reach the danger point before the motion of the dangerous parts has been arrested or, where appropriate, arrested and reversed (Figure 10.17).

(c) Hold-to-run control

This is a control which allows movement of the machinery only as long as the control is held in a set position. The control must return automatically to the stop position when released. Where the machinery runs at crawl speed, this speed should be kept as low as practicable.

Hold-to-run controls give even less protection to the operator than two-handed controls and have the same main drawback in that they give no protection to anyone other than the operator.

However, along with limited movement devices (systems which permit only a limited amount of machine movement on each occasion that the control is operated and are often called 'inching devices'), they are extremely relevant to operations such as setting, where access may well be necessary and safeguarding by any other means is difficult to achieve.

(d) Jigs, holders and push sticks

There are a number of devices to limit the approach of people to mechanical hazards while using machinery. These include manual handling devices such as push sticks for circular saws and spindle moulding machines, or push blocks for wood planing machines. With certain machines jigs and other holding devices may be necessary to hold the work-piece near to the cutter of the machine without exposing the operator's hands to the dangerous part of the machine.

There is a logical progression from push sticks through jigs and holders, through more sophisticated clamping devices and moving work tables for circular saws, to power driven feeds and automatic feeding arrangements. Emergency stops must always be immediately available.

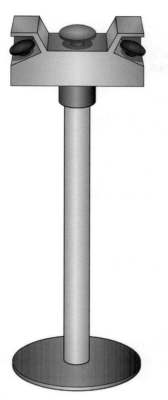

Figure 10.17 Pedestal-mounted free-standing two-handed control device

10.4.6 Information, instruction, training and supervision

(a) Information and instruction

The ILO Code requires manufacturers to provide instructions for the installation and use of machinery, including the information needed by the employer, operator and users of the machinery for its safe operation and maintenance.

This should cover:

▶ the conditions in which the work equipment can be used;

▶ the way in which the work equipment can be used;

▶ any foreseeable difficulties that could arise, and instructions on how to deal with them;

▶ any conclusions drawn from experience using the work equipment, which should be either recorded or steps taken to make sure that all appropriate members of the workforce are aware of them.

Written instructions include all the information provided by manufacturers or suppliers such as manuals, instruction sheets, instruction placards, warning labels and training manuals. In addition there may be in-house instructions and information from training sessions. Manufacturers have a duty to provide sufficient information, including drawings, to enable the correct installation, safe operation and maintenance of the equipment. These are often available to download from the manufacturer's website.

The ILO Code requires that instructions cover:

(a) the business name and full address of the manufacturer;

(b) the make and model of the machinery;

(c) the drawings, diagrams, descriptions and explanations necessary for the safe use, maintenance and repair of the machinery, as well as for checking that it is functioning correctly;

(d) a description of the intended use of the machinery;

(e) warnings indicating foreseeable ways in which the machinery should not be used;

(f) assembly, installation and connection instructions;

(g) instructions for reducing noise or vibration;

(h) instructions for servicing and putting the machinery into use and, if necessary, instructions for training operators;

(i) information about any residual risks that remain, despite inherently safe design measures, safeguarding and complementary protective measures adopted;

(j) instructions for users on protective measures to be adopted, including, where appropriate, PPE to be provided;

(k) the essential characteristics of tools which may be fitted to the machinery;

(l) the conditions in which the machinery meets the requirement of stability during use, transportation, assembly, dismantling when out of service, testing or foreseeable breakdowns;

(m) instructions to ensure that transportation, handling and storage can be effected safely;

(n) operating instructions that are to be followed in the event of an accident or breakdown or, if a blockage is likely to occur, details of the procedure for safely unblocking the equipment;

(o) the description of the setting up, adjustment and maintenance operations that should be carried out by the user and the preventative maintenance measures that should be applied;

(p) instructions designed to enable adjustment and maintenance to be carried out safely, including the protective measures that should be taken during these operations;

(q) the specifications of the spare parts to be used, when these may affect the safety and health of workers or others in the vicinity;

(r) data on exposures generated by the machinery with regard to noise, vibration, radiation, gases, vapours and dust, where

> these may affect the safety and health of
> workers or others in the vicinity; and
>
> **(s)** explanations of any information and
> warnings provided on the machinery in the
> form of symbols or pictograms.

Employers must ensure that written instructions are available to everyone using the equipment. This includes temporary workers and others like maintenance staff who will need information like maintenance instructions and manuals.

Supervisors and managers should also have access to the information and written instructions. The amount of detailed health and safety information they will need to have immediately available for day-to-day operations will vary, but it is important that they know what information is available and where it can be found.

Information should be provided in writing or verbally where that is considered sufficient. Managers should make a decision based on the unusual nature or complexity of the equipment or task. They should consider:

▶ the skill levels of the workers involved;
▶ their experience and existing training;
▶ the level of supervision required;
▶ the complexity and length of the task.

Information and instructions whether verbal or written should be easy to understand and must be provided in an appropriate language for the user/purchaser. They should be logical, illustrated where appropriate and use standard symbols. Particular attention should be paid where people's first language differs from the local official language, or they have language problems or a disability which makes receiving or understanding information or instructions difficult.

(b) Training and supervision

The ILO Code requires that employers should ensure that workers have received the necessary training, information and instructions to perform the work competently and safely. Taking into account information provided by the manufacturer and supplier, the training, information and instructions should include information on:

(a) risks which the use of the machinery may entail;
(b) risk avoidance and foreseeable abnormal situations;
(c) safe working procedures; and
(d) the use of PPE.

Employers should:

▶ consider the existing levels of competence and skills of all users of work equipment including temporary and agency workers;
▶ evaluate the level of management or supervision needed;

▶ carry out a gap analysis to determine the training required to enhance workers' level of competence taking into consideration, for example, whether they work alone or under close supervision;
▶ carry out the training particularly on recruitment but also when:
  ▷ risks change;
  ▷ tasks change;
  ▷ new technology or equipment is introduced;
  ▷ the system of work changes.

Refresher training needs to be provided as necessary as skills will decline particularly if not used regularly, such as after lengthy absence. Supervisors and managers also require adequate training to carry out their function, particularly if they only supervise a particular task occasionally.

The training and supervision of young persons (under 18) is particularly important because of their relative immaturity, unfamiliarity with a working environment and lack of awareness of existing or potential risks. Some codes of practice, for example on the UK's 'Safe Use of Woodworking Machinery', restrict the use of high-risk machinery, so that only young persons with sufficient maturity and competence who have finished their training may use the equipment unsupervised. ISO 12100:2010 states that the unintended behaviour of the operator or foreseeable misuse of the machinery should be taken into consideration when assessing the risks.

10.4.7 Personal Protective Equipment

This is covered in detail in Chapter 13 and also for chainsaws in section 10.4.8. The use of PPE for users of work equipment is the last issue to be considered in the hierarchy of protection (see 10.1) after inherently safe design, safeguards, instructions and training. However, there are many tasks and pieces of work equipment where PPE is necessary for general protection such as the use of suitable gloves which fit properly and do not restrict the operation.

The ILO Code requires that where PPE is necessary to protect the safety and health of workers, it should be fit-for-purpose, suited to the individual and provided at no cost to workers. The employer should implement measures to ensure that it is available, used and stored and maintained safely and in good working order. Workers should be consulted in its selection and trained in its use.

Some pieces of equipment such as chainsaws are so inherently dangerous that despite all the safeguards and training specialised PPE is absolutely essential. These include Kevlar-based footwear and overtrousers to prevent chain cuts on the feet and legs; special face screen and helmets to prevent flying chips and saw kickback damaging the face, head and eyes. Harnesses are necessary when working at height in trees.

Figure 10.18 Typical multifunction printer/photocopier

It is important that the risk assessments identify the equipment which should be used to protect against specific risks as well as general standards required for the use of work equipment. It is becoming increasingly common for all people in process plants to wear eye protection, gloves, high-visibility jackets or overalls and hard hats. These should not be seen as a substitute for well-designed and properly safeguarded equipment.

10.4.8 Application of safeguards to the range of work equipment

The application of the safeguards to the Certificate range of work equipment is as follows.

(a) Office – photocopier

Application of safeguards (Figure 10.18):

▶ the machines are provided with an all-enclosing case which prevents access to the internal moving, hot or electrical parts;
▶ the access doors are interlocked so that the machine is automatically switched off when gaining access to clear jams or maintain the machine. It is good practice to switch off when opening the machine;
▶ internal electrics are insulated and protected to prevent contact;
▶ regular inspection and maintenance should be carried out;
▶ the machine should be on the PAT schedule;
▶ good ventilation in the machine room should be maintained;
▶ wear a suitable dust mask and eye protection when changing toner powder if the cartridge is to be refilled.

Figure 10.19 Typical office paper shredder

(b) Office – document shredder

Application of safeguards (Figure 10.19):

▶ enclosed fixed guards surround the cutters with restricted access for paper only, which prevents fingers reaching the dangerous parts;
▶ interlocks are fitted to the cutter head so that the machine is switched off when the waste bin is emptied;
▶ a trip device is used to start the machine automatically when paper is fed in;
▶ the machine should be on PAT schedule and regularly checked;
▶ general ventilation will cover most dust problems except for very large machines where dust extraction may be necessary;
▶ noise levels should be checked and the equipment perhaps placed on a rubber mat if standing on a hard reflective floor.

(c) Manufacturing and maintenance – bench-top grinder

Application of safeguards (Figure 10.20):

▶ the wheel should be enclosed as much as possible in a strong casing capable of containing a burst wheel;
▶ grinder should be bolted down to prevent movement;
▶ an adjustable tool rest should be adjusted as close as possible to the wheel;

301

Figure 10.20 Typical bench-mounted grinder

Figure 10.21 Typical pedestal drill. Guard should be adjusted close to the work-piece to cover the rotating drill

▶ an adjustable screen should be fitted over the wheel to protect the eyes of the operator. Goggles should also be worn;

▶ only properly trained competent and registered people should mount an abrasive wheel. See 10.1.5 for a suitable mounting procedure;

▶ the maximum speed should be marked on the machine so that the abrasive wheel can be matched to the machine speed to ensure that the wheel permitted speed exceeds or equals the machine max speed;

▶ noise levels should be checked and attenuating screens used if necessary;

▶ the machine should be on the PAT schedule and regularly checked;

▶ if necessary, extract ventilation should be fitted to the wheel encasing to remove dust at source.

(d) Manufacturing and maintenance – pedestal drill

Application of safeguards (Figure 10.21):

▶ motor and drive should be fitted with fixed guard;

▶ the machine should be bolted down to prevent movement;

▶ the spindle should be guarded by an adjustable guard, which is fixed in position during the work;

▶ a clamp should be available on the pedestal base to secure work-pieces;

▶ the machine should be on the PAT schedule and regularly checked;

▶ cutting fluid, if used, should be contained and not allowed to get onto clothing or skin. A splash guard may be required but is unlikely;

▶ goggles should be worn by the operator.

(e) Agricultural/horticultural – cylinder mower

Application of safeguards (Figure 10.22):

▶ the machine should be designed to operate with the grass collection box in position to restrict access to the bottom blade trap. A warning sign should be fitted on the machine;

▶ on pedestrian-controlled machines the control handle should automatically stop the blade rotation when the operator's hands are removed. It should take two separate actions to restart;

▶ ride-on machines should be fitted with a device to automatically stop the blades when the operator leaves the operating position. This is normally a switch under the seat and it should be tested to ensure that it is functioning correctly and is not defective;

▶ drives and motor should be completely encased with a fixed guard;

▶ the machine should only be refuelled in the open air with a cool engine, using the correct container for highly flammable fuels with a pourer to restrict spillage. No smoking should be allowed;

▶ hot surfaces like the exhaust should be covered;

Figure 10.22 (a) Typical large cylinder mower

Figure 10.22 (b) Tractor-mounted cylinder mower for hedge cutting

▶ the engine must only be run in the open air to prevent a build up of fumes;
▶ noise levels should be checked and if necessary an improved silencer fitted to the engine and where required hearing protection used;
▶ hay-fever-like problems from grass cutting are difficult to control. A suitable dust mask may be required to protect the user.

(f) Agricultural/horticultural – brush cutter/strimmer

Application of safeguards (Figure 10.23):

▶ moving engine parts should be enclosed;
▶ rotating shafts should be encased in a fixed drive shaft cover;
▶ rotating cutting head should have a fixed top guard, which extends out on the user side of the machine;
▶ line changes must only be done either automatically or with the engine switched off;
▶ the engine should only be run in the open air;
▶ refuelling should only be done in the open air using the correct container for highly flammable fuel with pouring spout;

Figure 10.23 Typical brush cutter

▶ boots with steel toe cap and good grip, stout trousers and non-snagging upper garments should be worn in addition to a hard hat fitted with full face screen and safety glasses;
▶ if the noise levels are sufficiently high (normally they are with petrol-driven units) suitable hearing protection should be worn;
▶ low-vibration characteristics should be balanced with engine power and speed of work to achieve the minimum overall vibration exposure. Handles should be of an anti-vibration type. Engines should be mounted on flexible mountings. Work periods should be limited to allow recovery;
▶ washing arrangements and warm impervious gloves should be provided to guard against health risks;
▶ a properly constructed harness should be worn which comfortably balances the weight of the machine.

(g) Agricultural/horticulture – chainsaw

Application of safeguards (Figure 10.24):

▶ may only be operated by fully trained, fit and competent people;
▶ avoid working alone with a chainsaw. Where this is not possible, establish procedures to raise the alarm if something goes wrong;
▶ moving engine parts should be enclosed;
▶ electrical units should be double-insulated and cables fitted with residual current devices;
▶ the saw must be fitted with a top handle and effective brake mechanism;
▶ chainsaws expose operators to high levels of noise and hand–arm vibration which can lead to hearing loss and conditions such as vibration white finger. These risks can be controlled by good management practice including:
  ▷ purchasing policies for low-noise/low-vibration chainsaws (e.g. with anti-vibration mounts and heated handles);
  ▷ providing suitable hearing protection;
  ▷ proper maintenance schedules for chainsaws and protective equipment;

10

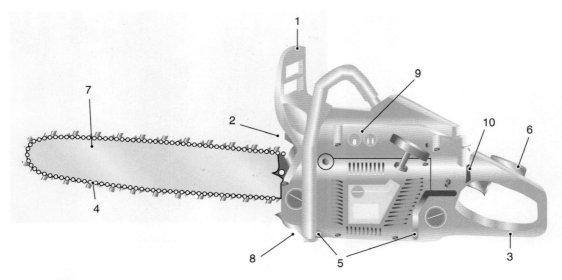

Figure 10.24 Typical chainsaw with rear handle. The rear handle projects from the back of the saw. It is designed to always be gripped with both hands, with the right hand on the rear handle. It may be necessary to have a range of saws with different guide bar lengths available. As a general rule, choose a chainsaw with the shortest guide bar suitable for the work. 1 – hand guard with integral chain brake; 2 – exhaust outlet directed to the right-hand side away from the operator; 3 – chain breakage guard at bottom of rear handle; 4 – chain designed to have low-kickback tendency; 5 – rubber anti-vibration mountings; 6 – lockout for the throttle trigger; 7 – guide bar (should be protected when transporting chainsaw); 8 – bottom chain catcher; 9 – PPE hand/eye/ear defender signs; 10 – on/off switch

▷ giving information and training to operators on the health risks associated with chainsaws and use of PPE, etc.;

▷ operators need to be trained in the correct chain-sharpening techniques and chain and guide bar maintenance to keep the saw in safe working condition;

▷ make sure petrol containers are in good condition and clearly labelled, with securely fitting caps. Use containers which are specially designed for chainsaw fuelling and lubrication;

▷ do not allow operators to use discarded engine oil as a chain lubricant – it is a very poor lubricant and may cause cancer if it is in regular contact with an operator's skin;

▷ when starting the saw, operators should maintain a safe working distance from other people and ensure the saw chain is clear of obstructions;

▷ kickback is the sudden uncontrolled upward and backward movement of the chain and guide bar towards the operator. This can happen when the saw chain at the nose of the guide bar hits an object. A properly maintained chain brake and use of low-kickback chains (safety chains) reduce the effect, but cannot entirely prevent it;

▷ suitable PPE should always be worn, no matter how small the job. European standards for chainsaw PPE are published as part of EN 381 *Protective Clothing for users of Hand-Held Chainsaws* (Figure 10.25). Protective clothing complying with this standard should provide

Figure 10.25 Kevlar gloves, overtrousers and overshoes providing protection against chainsaw cuts; helmet, ear and face shield protect the head. Apprentice under training – first felling

a consistent level of resistance to chainsaw cut-through. Other clothing worn with the PPE should be close-fitting and non-snagging.NB: **No protective equipment can ensure 100% protection against cutting by a hard-held chainsaw;**

Figure 10.26 Typical retail compactor

Figure 10.27 Typical retail checkout conveyor

▷ if conditions are dusty, suitable filtering face masks should be worn.

(h) Retail – compactor

Application of safeguards (Figure 10.26):

▶ access doors to the loading area, which gives access to the ram, should be positively interlocked with electrical and/or hydraulic mechanisms;
▶ the ram pressure should be dumped if it is hydraulic;
▶ drives of motors should be properly guarded;
▶ if the waste unit is removed by truck, the ram mechanism should be interlocked with the unit so that it cannot operate when the unit is changed for an empty one;
▶ the machine should be regularly inspected and tested by a competent person;
▶ if the hydraulic ram can fall under gravity, mechanical restraints should automatically move into position when the doors are opened;
▶ emergency stop buttons should be fitted on each side.

(i) Retail – checkout conveyor

Application of safeguards (Figure 10.27):

▶ all traps between belt and rollers are provided with either fixed guards or interlocked guards;
▶ motor and drive unit should be provided with a fixed guard and access to the underside of the conveyor is prevented by enclosure;
▶ adequate emergency stop buttons must be provided to the checkout operator;
▶ the machine should be on the PAT electrical inspection schedule and regularly checked;
▶ an auto stop system should be fitted so that the conveyor does not push products into the operator's working zone.

(j) Construction – cement mixer

Application of safeguards (Figures 10.28 and 10.29):

Figure 10.28 Small concrete/cement mixer

Figure 10.29 Diesel concrete mixer with hopper

▶ operating position for the hopper hoist (Figure 10.29) should be designed so that anyone in the trapping area is visible to the operator. The use of the machine should be restricted to designated operators only. As far as possible, the trapping point should be designed out. The hoist operating location should be fenced off just allowing access for barrows, etc., to the unloading area;

▶ drives and rotating parts of the engine should be enclosed;

▶ the drum gearing should be enclosed and persons kept away from the rotating drum, which is normally fairly high on large machines;

▶ no one should be allowed to stand on the machine while it is in motion;

▶ goggles should be worn to prevent cement splashes;

▶ if petrol-driven, care is required with flammable liquids and refuelling;

▶ engines must only be run in the open air;

▶ electric machines should be regularly checked and be on the PAT schedule;

▶ noise levels should be checked and noise attenuation, for example silencers and damping, fitted if necessary.

(k) Construction – bench-mounted circular saw

Application of safeguards (Figure 10.30):

▶ a fixed guard should be fitted to the blade below the bench;

▶ fixed guards should be fitted to the motor and drives;

▶ an adjustable top guard should be fitted to the blade above the bench which encloses as much of the blade as possible. An adjustable front section should also be fitted;

▶ a riving knife should be fitted behind the blade to keep the cut timber apart and prevent ejection;

▶ a push stick should be used on short work-pieces (under 300 mm) or for the last 300 mm of longer cuts;

▶ blades should be kept properly sharpened and set with the diameter of the smallest blade marked on the machine;

▶ noise attenuation should be applied to the machine, for example damping, special quiet saw blades, and, if necessary, fitting in an enclosure. Hearing protection may have to be used;

▶ protection against wet weather should be provided;

▶ the electrical parts should be regularly checked in addition to all the mechanical guards;

▶ extraction ventilation will be required for the wood dust and shavings;

▶ suitable dust masks should be worn;

▶ suitable warm or cool clothing will be needed when used in hot or cold locations;

▶ space around the machine should be kept clear.

10.4.9 Basic requirements for guards and safety devices

The design and construction of guards must be appropriate to the risks identified and the mode of operation of the machinery in question. ISO 12100:2010 gives the following general requirements for guards and protective devices which should protect against danger, including risks from moving parts. They should:

▶ be of robust construction;

▶ be securely held in place;

▶ not give rise to any additional hazard;

▶ not be easy to bypass or render non-operational, or be easily defeated;

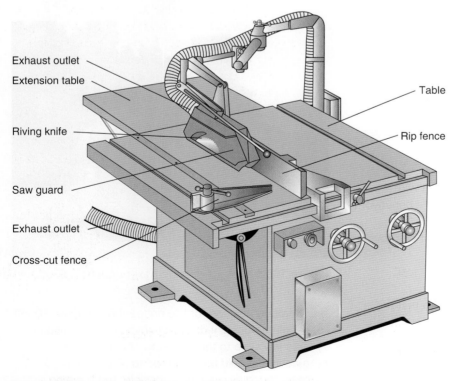

Figure 10.30 Typical bench-mounted circular saw

- be located at an adequate distance from the danger zone;
- cause minimum obstruction of the view of the production process; and
- enable essential work to be carried out on the installation and replacement of tools and for maintenance purposes by restricting access exclusively to the area where the work has to be done, if possible without the guard having to be removed or the protective device having to be disabled.

In addition, guards should protect against the ejection or falling of materials or objects and against emissions generated by the machinery.

These general requirements involve consideration of the following factors:

- strength – guards and protection devices must be of good construction, sound material and adequate strength. They must be capable of doing the job they are intended to do. Several factors can be considered:
  - ▷ material of construction (metal, plastic, laminated glass, etc.);
  - ▷ form of the material (sheet, open mesh, bars, etc.);
  - ▷ method of fixing.

They must be adequate for the purpose, able to resist the forces and vibration involved and able to withstand impact (where applicable). Good construction involves design and layout as well as the mechanical nature and quality of the construction. Foreseeable use and misuse should be taken into account. Guards and protection devices must be designed and installed so that they cannot be easily bypassed or disabled. This refers to accidental or deliberate action that removes the protection offered;

- weight and size – in relation to the need to remove and replace the guard during maintenance;
- compatibility with materials being processed and lubricants, etc.;
- hygiene and the need to comply with food safety regulations;
- visibility – may be necessary to see through the guard for both operational and safety reasons. Guards and protection devices must not unduly restrict the view of the operating cycle of the machinery, where such a view is necessary. It is not usually necessary to be able to see all the machine – the part that needs to be seen is normally that which is acting directly on material or a work-piece. Operations for which it is necessary to provide a view include those where the operator controls and feeds a loose work-piece at a machine. Examples include manually fed woodworking machines and food slicers. Many of these operations involve the use of protection appliances. Where an operation

protected by guards needs to be seen, the guard should be provided with viewing slits or properly constructed panels, perhaps backed up by internal lighting, enabling the operator to see the operation;

- noise attenuation – guards are often utilised to reduce the noise levels produced by a machine. Conversely, the resonance of large undamped panels may exacerbate the noise problem;
- enabling a free flow of air – where necessary (e.g. for ventilation);
- avoidance of additional hazards – for example free of sharp edges. Guards must be constructed so that they are not themselves dangerous parts. If a guard is power operated or assisted, the closing or opening action might create a potentially dangerous trap which needs secondary protection, for example a leading-edge trip bar or pressure-sensitive strip;
- ease of maintenance and cleanliness – it should be necessary to use a tool such as a spanner or screwdriver to remove a fixed guard. Guards should be interlocked with the main drive if they need to be removed frequently for maintenance or cleaning. Movable panels in guards giving access to dangerous parts or movable guards themselves will often need to be fitted with an interlocking device. This device must be designed and installed so that it is difficult or impossible to bypass or defeat. Guards and protection devices must be maintained in an efficient state, in efficient working order and in good repair. This is an important requirement as many accidents have occurred when guards have not been maintained. It is a general requirement under most local legislation to maintain equipment. Compliance can be achieved by the use of an effective check procedure for guards and protection devices, together with any necessary follow-up action. In the case of protection devices or interlocks, some form of functional check or test is desirable.
- openings – the size of openings and their distance from the dangerous parts should not allow anyone to be able to reach into a danger zone. These values can be determined by experiment or by reference to standard tables. If doing so by experiment, it is essential that the machine is first stopped and made safe (e.g. by isolation). The detailed information on openings is contained in EN 294:1992, EN 349:1993 and EN 811:1997. Safeguarding is normally attached to the machine, but the regulation does not preclude the use of free-standing guards or protection devices. In such cases, the guards or protection devices must be fixed in an appropriate position relative to the machine.

10.5 Further information

Ambient factors in the workplace, ILO CoP, ISBN 92-2-11628-X http://www.ilo.org/safework/info/standards-and-instruments/WCMS_107729/lang--en/index.htm

10

Directive 2006/42/EC – machinery directive
http://eur-lex.europa.eu/LexUriServ/LexUriServ.do?uri=
OJ:L:2006:157:0024:0086:EN:PDF

Directive 2009/104/EC – use of work equipment https://
osha.europa.eu/en/legislation/directives/workplaces-
equipment-signs-personal-protective-equipment/
osh-directives/3

Graphical symbols – Safety colours and safety signs
used in workplaces and public areas – ISO 7010:2003,
International Organisation for Standardisation (ISO)

Safety and health in the use of machinery, ILO CoP
http://www.ilo.org/wcmsp5/groups/public/---ed_protect/
---protrav/---safework/documents/normativeinstrument/
wcms_164653.pdf

Safety of machinery – General principles for design –
Risk assessment and risk reduction, ISO 12100:2010,
ISBN 978-0-580-74262-0

Safe Use of Work Equipment Provision and Use of
Work Equipment Regulations 1998 Approved Code of
Practice and guidance L22 (4th edition), HSE Books,
2014, ISBN 978 0 7176 6619 5 http://www.hse.gov.uk/
pubns/books/l22.htm

10.6 Practice revision questions

1. A manufacturing company has decided to purchase a new machine.
   (a) **Identify** the questions that must be answered before a purchase is made.
   (b) **Outline** the issues that will need to be considered before the machine is put into use.
   (c) **Identify** the main issues when assessing the suitability of controls (including emergency controls) of the new machine.

2. (a) **Describe** the contents of a 'Declaration of Conformity' that should be issued with a new machine.
   (b) **Outline** the advantages and limitations of CE marking on a new machine.

3. **Outline** a hierarchy of control measures that may be used to prevent contact with dangerous parts of machinery. Give an example of the level of protection required at each stage of the hierarchy.

4. An operator suffers a serious injury after coming into contact with a dangerous part of a machine, **describe**:
   (a) Possible immediate causes.
   (b) Possible root (underlying) causes.

5. Employers are required to provide adequate information, instruction and training to ensure the safe use of work equipment.
   (a) **Identify THREE** categories of employees that should receive information, instruction and training on the safe use of work equipment.
   (b) **Outline** the issues that could be included in such information, instruction and training.

6. (a) **Describe THREE** preventative maintenance schemes that may be used on machinery to prevent health and safety risks.
   (b) **Outline** the measures to be taken to reduce the risk of accidents associated with the routine maintenance of machinery.
   (c) **Identify FOUR** non-mechanical hazards that could lead to injury and/or ill-health when undertaking maintenance work on machinery in the workplace.

7. A bricklayer uses a hammer and chisel to split a building block.
   (a) **Identify FOUR** unsafe conditions, associated with the tools, which could affect the safety of the bricklayer.
   (b) **Outline** suitable control measures for minimising the risk of injury to the bricklayer when using the tools.

8. A carpenter regularly uses a hand-held electric sander for the preparation of wooden doors before painting them.
   (a) **Outline** the checks that should be made to ensure the electrical safety of the sander.
   (b) Other than electricity, **identify EIGHT** hazards associated with the use of the sander.

9. (a) **Identify FOUR** non-mechanical hazards that may be encountered on woodworking machines and **outline** the possible health and safety effects from exposure in **EACH** case.
   (b) **Identify EIGHT** mechanical hazards associated with moving parts of machinery.

10. **Identify THREE** mechanical hazards, **TWO** non-mechanical hazards and the associated control measures with the use of the following items of equipment:
    (a) a bench-top grinder
    (b) a document shredder
    (c) a photocopier
    (d) a bench-mounted circular saw
    (e) a strimmer.

11. A new pedestal (pillar) drill is to be used in a maintenance workshop.

    (a) **Identify** the issues that should be addressed before it is first used so that the risk of injury to operators of the machine is reduced.

    (b) **Identify FOUR** mechanical hazards presented by the machine and **outline** in each case how injury may occur and the appropriate control measures.

12. A bench-mounted circular saw is used in a carpentry workshop on a regular basis.

    (a) **Identify FOUR** risks to the health **AND FIVE** risks to the safety of the circular saw operators.

    (b) **Outline** the control measures that can be implemented to minimise the health and safety risks to these operators.

13. A petrol-driven chainsaw is to be used to remove several branches from a large tree.

    (a) **Identify EIGHT** hazards associated with the use of the chainsaw for this work.

    (b) **Outline EIGHT** safeguards that should be in place to ensure the safe use of chainsaws.

    (c) **Identify** the items of personal protective equipment that should be used by the chainsaw operative.

14. A local authority uses a rider-operated petrol-powered motor mower to cut the grass on roadside verges.

    (a) **Identify FIVE** mechanical and **FOUR** non-mechanical hazards that the operator could be exposed to while using the mower.

    (b) **Outline** the precautions that should be taken to address the hazards identified in part (a).

15. **Outline** the principles of operation, advantages and limitations of the following guards and safeguarding devices:

    (a) Fixed guards
    (b) Interlocked guards
    (c) Adjustable guards
    (d) Trip devices
    (e) Two-handed control devices.

16. Other than contact with dangerous parts, **identify FOUR** types of hazard against which fixed guards on machines may provide protection.

10

CHAPTER 11

Electrical safety

11.1 **Principles, hazards and risks associated with the use of electricity at work** ▶ 312

11.2 **Control measures when working with electrical systems or using electrical equipment in all workplace conditions** ▶ 320

11.3 **Further information** ▶ 329

11.4 **Practice revision questions** ▶ 330

> **This chapter covers the following NEBOSH learning objectives:**
> 1. Outline the principles, hazards and risks associated with the use of electricity in the workplace
> 2. Outline the control measures that should be taken when working with electrical systems or using electrical equipment in all workplace conditions

11.1 Principles, hazards and risks associated with the use of electricity at work

11.1.1 Introduction

Electricity is a widely used, efficient and convenient, but potentially hazardous, method of transmitting and using energy. It is in use in every factory, workshop, laboratory and office in the country. Any use of electricity has the potential to be very hazardous with possible fatal results. Legislation has been in place for many years to control and regulate the use of electrical energy and the activities associated with its use. Such legislation provides a framework for the standards required in the design, installation, maintenance and use of electrical equipment and systems and the supervision of these activities to minimise the risk of injury. Electrical work from the largest installation to the smallest job must be carried out by people known to be competent to undertake such work. New installations always require expert advice at all appropriate levels to cover both design aspects of the system and its associated equipment. Electrical systems and equipment must be properly selected, installed, used and maintained.

In the UK, approximately 8% of all fatalities at work are caused by electric shock. Over the last few years, there have been 1,000 electrical accidents each year and 25 people die of their injuries. The majority of the fatalities occur in the agriculture, extractive and utility supply and service industries, whereas the majority of the major accidents happen in the manufacturing, construction and service industries.

Only voltages up to and including **mains voltage** (220/240V) and the three principal electrical hazards – electric shock, electric burns and electrical fires and explosions – are considered in detail in this chapter.

11.1.2 Basic principles of electricity

In simple terms, electricity is the flow or movement of electrons through a substance which allows the transfer of electrical energy from one position to another. The substance through which the electricity flows is called a **conductor**. This flow or movement of electrons is known as the **electric current**. There are two forms of electric current – direct and alternating. **Direct current (dc)** involves the flow of electrons along a conductor from one end to the other. This type of current is mainly restricted to batteries, dynamos and similar devices. **Alternating current (ac)** is produced by a rotating

alternator and causes an oscillation of the electrons rather than a flow of electrons so that energy is passed from one electron to the adjacent one and so on through the length of the conductor.

It is sometimes easier to understand the basic principles of electricity by comparing its movement with that of water in a pipe flowing downhill. The flow of water through the pipe (measured in litres per second) is similar to the current flowing through the conductor which is measured in amperes, normally abbreviated to **amps (A)**. Sometimes very small currents are used and these are measured in milliamps (mA).

The higher the pressure drop is along the pipeline, the greater will be the flow rate of water and, in a similar way, the higher the electrical 'pressure difference' along the conductor, the higher the current will be. This electrical 'pressure difference' or potential difference is measured in **volts (V)**.

The flow rate through the pipe will also vary for a fixed pressure drop as the roughness on the inside surface of the pipe varies – the rougher the surface, the slower the flow and the higher the resistance to flow becomes. Similarly, for electricity, the poorer the conductor, the higher the resistance is to electrical current and the lower the current becomes. Electrical resistance is measured in **ohms**.

The voltage (V), the current (I) and the resistance (R) are related by the following formula, known as Ohm's law:

$$V = I \times R \text{ (volts)}$$

and, electrical power (P) is given by:

$$P = V \times I \text{ (watts)}$$

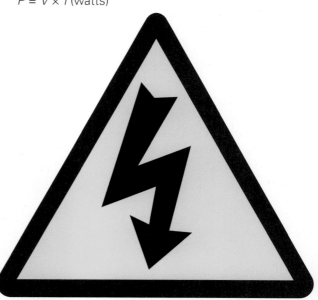

Figure 11.1 Electrical hazard warning sign

These basic formulae enable simple calculations to be made so that, for example, the correct size of fuse may be ascertained for a particular piece of electrical equipment.

Conductors and insulators

Conductors are nearly always metals, copper being a particularly good conductor, and are usually in wire form but they can be gases or liquids, water being a particularly good conductor of electricity. Superconductors is a term given to certain metals which have a very low resistance to electricity at low temperatures.

Very poor conductors are known as **insulators** and include materials such as rubber, timber and plastics. Insulating material is used to protect people from some of the hazards associated with electricity.

Short circuit

Electrical equipment components and an electrical power supply (normally the mains or a battery) are joined together by a conductor to form a **circuit.** If the circuit is broken in some way so that the current flows directly to earth rather than to a piece of equipment, a **short circuit** is made. As the resistance is greatly reduced but the voltage remains the same, a rapid increase in current occurs which could cause significant problems if suitable protection were not available.

Earthing

The principle of **earthing** is that the Earth is basically a huge conductor which accepts any electrical current that is conducted to it. Any metal in an electrical device with which a person might come into contact (for example the metal casing of a food mixer) should be connected to earth so that if, due to a fault, the casing becomes live, the electricity will flow away to earth – an easier path than through the body. In the USA or where OSHA standards are adopted, this process is called grounding.

The electricity supply company has one of its conductors solidly connected to earth and every circuit supplied by the company must have one of its conductors connected to earth. This means that if there is a fault, such as a break in the circuit, the current, known as the fault current, will return directly to earth, which forms the circuit of least resistance, thus maintaining the supply circuit. This process is known as **earthing.** Other devices, such as fuses and residual current devices (RCDs), which will be described later, will also be needed within the circuit to interrupt the current flow to earth so as to protect people from electric shock and equipment from overheating. Good and effective earthing is absolutely essential and must be connected and checked by a competent person. Where a direct contact with earth is not possible, for

example in a motor car, a common voltage reference point is used, such as the vehicle chassis. Hence, earthing protects people from an electric shock by providing a path for a fault current to flow to earth. It also causes any protective device, such as a fuse or residual current device (RCD), to switch off the electricity supply to the circuit that has the fault. If a fault within an electrical device connects a live electrical supply to an exposed conductive surface, anyone touching it will complete a circuit back to earth and receive an electric shock.

Where other potential metallic conductors exist near to electrical conductors in a building, they must be connected to the main earth terminal to ensure **equipotential bonding** of all conductors to earth. This applies to gas, water and central heating pipes and other devices such as lightning protection systems. **Supplementary bonding** is required in bathrooms and kitchens where, for example, metal sinks and other metallic equipment surfaces are present. This involves the connection of a conductor from the sink to a water supply pipe which has been earthed by equipotential bonding. There have been several fatalities due to electric shocks from 'live' service pipes or kitchen sinks.

Some definitions

Certain terms are frequently used with reference to electricity and the more common ones are defined here.

▶ **Low voltage** – This is a voltage normally not exceeding 1000V dc or 600V ac between conductors and earth or 1000V ac between phases. Mains voltage falls into this category.
▶ **High voltage** – This is defined in national and international standards as a voltage exceeding 1000V dc or 600V ac between conductors and earth or 1000V ac between phases.
▶ **Mains voltage** – The common voltage available in domestic premises and many workplaces and normally taken from three-pin socket points. Most of the world, including Europe, uses a 220V/50 Hertz system. However, a small group of other countries (41 out of 260), including the United States, Saudi Arabia and Japan, use a 110V/60 Hertz electrical system.
▶ **Maintenance** – A combination of any actions carried out to retain an item of electrical equipment in, or restore it to, an acceptable and safe condition.
▶ **Testing** – A measurement carried out to monitor the conditions of an item of electrical equipment without physically altering the construction of the item or the electrical system to which it is connected.
▶ **Inspection** – A maintenance action involving the careful scrutiny of an item of electrical equipment, using, if necessary, all the senses to detect any failure to meet an acceptable and safe condition. An inspection does not include any dismantling of the item of equipment.

▶ **Examination** – An inspection together with the possible partial dismantling of an item of electrical equipment, including measurement and non-destructive testing as required, in order to arrive at a reliable conclusion as to its condition and safety.

▶ **Isolation** – Involves cutting off the electrical supply from all or a discrete section of the installation by separating the installation or section from every source of electrical energy. This is the normal practice so as to ensure the safety of persons working on or in the vicinity of electrical components which are normally live and where there is a risk of direct contact with live electricity.

▶ **Competent electrical person** – A person possessing sufficient electrical knowledge and experience to avoid the risks to health and safety associated with electrical equipment and electricity in general.

11.1.3 Hazards, risks and danger of electricity

Electricity is a safe, clean and quiet method of transmitting energy. However, this apparently benign source of energy, when accidentally brought into contact with conducting material, such as people, animals or metals, permits dangerous releases of energy which may result in serious damage or loss of life. Constant awareness is necessary to avoid and prevent danger from accidental releases of electrical energy.

The principal hazards associated with electricity are:

▶ electric shock;
▶ electric burns;
▶ electrical fires and explosions;
▶ arcing;
▶ secondary hazards.

The use of portable electrical equipment can lead to a higher likelihood of these hazards occurring.

Electric shock and burns

There are 1,000 workplace accidents and 30 fatalities involving electric shock and burns reported in the UK to the HSE each year.

Electric shock is the convulsive reaction by the human body to the flow of electric current through it. This sense of shock is accompanied by pain and, in more severe cases, by burning. The shock can be produced by low voltages, high voltages or lightning. Most incidents of electric shock occur when the person becomes the route to earth for a live conductor. The effect of electric shock and the resultant severity of injury depend upon the size of the electric current passing through the body which, in turn, depends on the voltage and the electrical resistance of the skin and body. If a person comes into contact with a voltage above about 50 volts, they can receive a range of injuries, including those directly resulting from electrical shock (such as problems with breathing and heart function), and indirect effects resulting from loss of control (such as falling from height or coming into contact with moving machinery). The chance of being injured by an electric shock increases where it is damp or where there is a lot of metalwork.

If the skin is wet, a shock from mains voltage (220/240V) could well be fatal. The effect of shock is very dependent on conditions at the time but it is always dangerous and must be avoided. Electric burns are usually more severe than those caused by heat, since they can penetrate deep into the tissues of the body.

The effect of electric current on the human body depends on its pathway through the body (e.g. hand to hand or hand to foot), the frequency of the current, the length of time of the shock and the size of the current. Current size is dependent on the duration of contact and the electrical resistance of body tissue. The electrical resistance of the body is greatest in the skin and is approximately 100,000 ohm; however, this may be reduced by a factor of 100 when the skin is wet. The body beneath the skin offers very little resistance to electricity due to its very high water content and, while the overall body resistance varies considerably between people and during the lifetime of each person, it averages at 1,000 ohms. Skin that is wounded, bruised or damaged will considerably reduce human electrical resistance and work should not be undertaken on electrical equipment if damaged skin is unprotected.

An electric current of 1 mA is detectable by touch and one of 10 mA will cause muscle contraction which may prevent the person from being able to release the conductor, and if the chest is in the current path, respiratory movement may be prevented, causing asphyxia. Current passing through the chest may also cause fibrillation of the heart (vibration of the heart muscle) and disrupt the normal rhythm of the heart, though this is likely only within a particular range of currents. The shock can also cause the heart to stop completely (cardiac arrest) and this will lead to the cessation of breathing. Current passing through the respiratory centre of the brain may cause respiratory arrest that does not quickly respond to the breaking of the electrical contact. These effects on the heart and respiratory system can be caused by currents as low as 25 mA. It is not possible to be precise on the threshold current because it is dependent on the environmental conditions at the time, as well as the age, sex, body weight and health of the person.

Alternating current (AC) and direct current (DC) have slightly different effects on the human body, but both are dangerous above a certain voltage. The risk of injury changes according to the frequency of the AC, and it is common for DC to have an AC component (called ripple). The effect on a particular person is very difficult to predict as it depends upon a large number of factors, as mentioned above.

The effect of electricity on the human body may be summarised in the following way:

▶ 2 mA, onset of sensation
▶ 8 mA, mild sensation
▶ 20 mA, painful contraction of muscles prevents victim releasing the live connection
▶ 40 mA, immediate resuscitation necessary to cope with muscular paralysis
▶ 80 mA, extreme distress and breathing difficulties

▶ above 100 mA, onset of ventricular fibrillation
▶ above 200 mA, breathing ceases, severe burns.

These figures are only approximations and there may be considerable variations between people.

Burns of the skin occur at the point of electrical contact due to the high resistance of skin. These burns may be deep, slow to heal and often leave permanent scars. Burns may also occur inside the body along the path of the electric current, causing damage to muscle tissue and blood cells. Burns associated with radiation and microwaves are dealt with in Chapter 14.

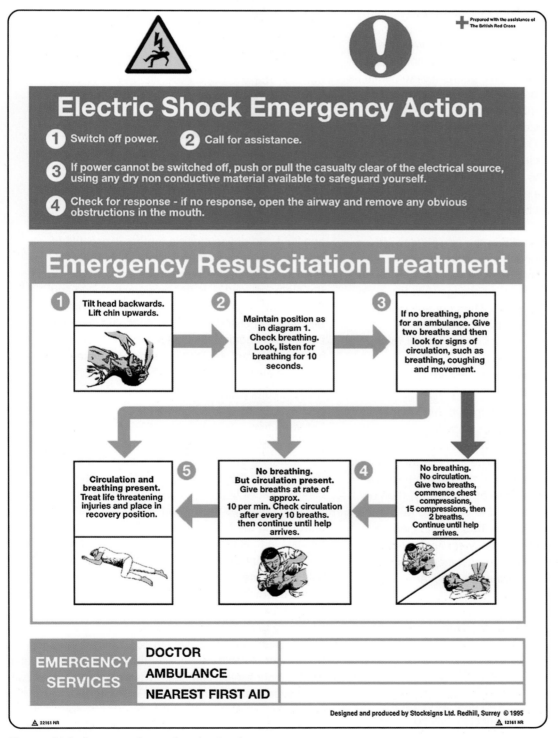

Figure 11.2 A poster about electric shock

Treatment of electric shock and burns

There are many excellent posters available which illustrate first-aid procedures for treating electric shock and such posters should be positioned close to electrical junction boxes or isolation switches (Figure 11.2). The recommended procedure for treating an unconscious person who has received a **low-voltage** electric shock is as follows:

1. On finding a person suffering from electric shock, raise the alarm by calling for help from colleagues (including a trained first-aider).
2. Switch off the power if it is possible and/or the position of the emergency isolation switch is known.
3. Call for an ambulance.
4. If it is not possible to switch off the power, then push or pull the person away from the conductor using an object made from a good insulator, such as a wooden chair or broom. Remember to stand on dry insulating material, for example, a wooden pallet, rubber mat or wooden box. If these precautions are not taken, then the rescuer will also be electrocuted.
5. If the person is breathing, place him/her in the recovery position so that an open airway is maintained and the mouth can drain if necessary.
6. If the person is not breathing, apply mouth-to-mouth resuscitation and, in the absence of a pulse, chest compressions. When the person is breathing normally place them in the recovery position.
7. Treat any burns by placing a sterile dressing over the burn and secure with a bandage. Any loose skin or blisters should not be touched nor any lotions or ointments applied to the burn wound.
8. If the person regains consciousness, treat for normal shock.
9. Remain with the person until they are taken to a hospital or local surgery.

It is important to note that electrocution by high-voltage electricity is normally instantly fatal. On discovering a person who has been electrocuted by high-voltage electricity, the police and electricity supply company should be informed. If the person remains in contact with or within 18 m of the supply, then he/she should not be approached to within 18 m by others until the supply has been switched off and clearance has been given by the emergency services. High-voltage electricity can 'arc' over distances less than 18 m, thus electrocuting the would-be rescuer (Figure 11.3).

Electrical fires and explosions

Over 25% of all fires have a cause linked to a malfunction of either a piece of electrical equipment or wiring or both. Electrical fires are often caused by a lack of reasonable care in the maintenance and use of electrical installations and equipment. The electricity that provides heat

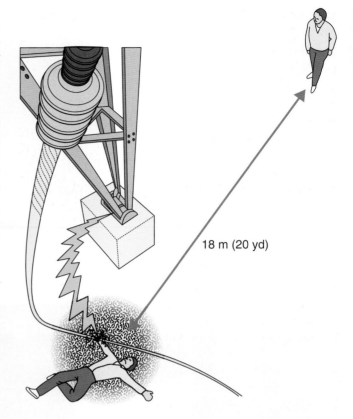

Figure 11.3 Keep 18 m clear of high-voltage lines

and light and drives electric motors is capable of igniting insulating or other combustible material if the equipment is misused, is not adequate to carry the electrical load or is not properly installed and maintained. The most common causes of fire in electrical installations are short circuits, overheating of cables and equipment, the ignition of flammable gases and vapours and the ignition of combustible substances by static electrical discharges.

Short circuits happen, as mentioned earlier, if insulation becomes faulty, and an unintended flow of current between two conductors or between one conductor and earth occurs. The amount of the current depends upon, among other things, the voltage, the condition of the insulating material and the distance between the conductors. At first the current flow will be low, but as the fault develops the current will increase and the area surrounding the fault will heat up. In time, if the fault persists, a total breakdown of insulation will result and excessive current will flow through the fault. If the fuse fails to operate or is in excess of the recommended fuse rating, overheating will occur and a fire will result. A fire can also be caused if combustible material is in close proximity to the heated wire or hot sparks are ejected. Short circuits are most likely to occur where electrical equipment or cables are susceptible to damage by water leaks or mechanical damage. Twisted or bent cables can also cause breakdowns in insulation materials.

Inspection covers and cable boxes are particular problem areas. Effective steps should be taken to prevent the entry of moisture as this will reduce or

Figure 11.4 Over 25% of fires are caused by electrical malfunction

Figure 11.5 Modern European multi-plug – France

eliminate the risk. Covers can themselves be a problem especially in dusty areas where the dust can accumulate on flat insulating surfaces resulting in tracking between conductors at different voltages and a subsequent insulation failure. The interior of inspection panels should be kept clean and dust-free by using a suitable vacuum cleaner.

Overheating of cables and equipment will occur if they become overloaded. Electrical equipment and circuits are normally rated to carry a given safe current which will keep the temperature rise of the conductors in the circuit or appliance within permissible limits and avoid the possibility of fire. These safe currents define the maximum size of the fuse (the fuse rating) required for the appliance. A common cause of circuit overloading is the use of equipment and cables which are too small for the imposed electrical load. This is often caused by the addition of more and more equipment to the circuit, thus taking it beyond its original design specification. In offices, the overuse of multi-socket incorrectly fused outlet adaptors can create overload problems (sometimes known as the Christmas tree effect). The more modern multi-plugs are much safer as they lead to one fused plug and cannot be easily overloaded (see Figure 11.5). Another cause of overloading is mechanical breakdown or wear of an electric motor and the driven machinery. Motors must be maintained in good condition with particular attention paid to bearing surfaces. Fuses do not always provide total protection against the overloading of motors and, in some cases, severe heating may occur without the fuses being activated.

Loose cable connections are one of the most common causes of overheating and may be readily detected (as well as overloaded cables) by a thermal imaging survey (a technique which indicates the presence of hot spots). The bunching of cables together can also cause excessive heat to be developed within the inner cable leading to a fire risk. This can happen with cable extension reels, which have only been partially

unwound, used for high-energy appliances like an electric heater.

Ventilation is necessary to maintain safe temperatures in most electrical equipment and overheating is liable to occur if ventilation is in any way obstructed or reduced. All electrical equipment must be kept free of any obstructions that restrict the free supply of air to the equipment and, in particular, to the ventilation apertures.

Most electrical equipment either sparks in normal operation or is liable to spark under fault conditions. Some electrical appliances, such as electric heaters, are specifically designed to produce high temperatures. These circumstances create fire and explosion hazards, which demand very careful assessment in locations where processes capable of producing flammable concentrations of gas or vapour are used, or where flammable liquids are stored.

It is likely that many fires are caused by static electrical discharges. Static electricity can, in general, be eliminated by the careful design and selection of materials used in equipment and plant, and the materials used in products being manufactured. When it is impractical to avoid the generation of static electricity, a means of control must be devised. Where flammable materials are present, especially if they are gases or dusts, then there is a great danger of fire and explosion, even if there is only a small discharge of static electricity. The control and prevention of static electricity is considered in more detail later in the chapter.

The use of electrical equipment in potentially flammable atmospheres should be avoided as far as possible. However, there will be many cases where electrical equipment must be used and, in these cases, the construction of the equipment should comply with the national standards.

Before electrical equipment is installed in any location where flammable vapours or gases may be present, the area must be zoned in accordance with national

11

standards and records of the zoned areas must be marked on building drawings and revised when any zoned area is changed. The installation and maintenance of electrical equipment in potentially flammable atmospheres is a specialised task. It must only be undertaken by specially trained electricians or instrument mechanics.

In the case of a fire involving electrical equipment, the first action must be the isolation of the power supply so that the circuit is no longer live. This is achieved by switching off the power supply at the mains isolation switch or at another appropriate point in the system. Where it is not possible to switch off the current, the fire must be attacked in a way which will not cause additional danger. The use of a non-conducting extinguishing medium, such as carbon dioxide or powder, is necessary. After extinguishing such a fire, careful watch should be kept for renewed outbreaks until the fault has been rectified. Re-ignition is a particular problem when carbon dioxide extinguishers are used, although less equipment may be damaged than is the case when powder is used.

Finally, the chances of electrical fires occurring are considerably reduced if the original installation was undertaken by competent electricians working to recognised standards, such as the UK Institution of Electrical Engineers' Code of Practice. It is also important to have a system of regular testing and inspection in place so that any remedial maintenance can take place.

Electric arcing

A person who is standing on earth too close to a high-voltage conductor may suffer flash burns as a result of arc formation. Such burns may be extensive and lower the resistance of the skin so that electric shock may add to the ill-effects. Electric arc faults can cause temporary blindness by burning the retina of the eye and this may lead to additional secondary hazards. The quantity of electrical energy is as important as the size of the voltage since the voltage will determine the distance over which the arc will travel. The risk of arcing can be reduced by the insulation of live conductors.

Strong electromagnetic fields induce surface charges on people. If these charges accumulate, skin sensation is affected and spark discharges to earth may cause localised pain or bruising. Whether prolonged exposure to strong fields has any other significant effects on health has not been proved. However, the action of an implanted cardiac pacemaker may be disturbed by the close proximity of its wearer to a powerful electromagnetic field. The health effects of arcing and other non-ionising radiation are covered in Chapter 14.

Static electricity

Static electricity is produced by the build-up of electrons on weak electrical conductors or insulating materials. These materials may be gaseous, liquid or solid and may include flammable liquids, powders, plastic films and granules. Plastics have a high resistance that enables them to retain static charges for long periods of time. The generation of static may be caused by the rapid separation of highly insulated materials by friction or by transfer from one highly charged material to another in an electric field by induction (see Figure 11.6).

There have been many very serious accidents in which an electrostatic discharge has caused massive damage and loss of life; for example, huge concrete silos were completely destroyed and collapsed into working areas following the ignition of flour by an electrostatic discharge. The electrostatic charge had built up by the friction of surfaces and materials rubbing against each other. A static electric shock, perhaps caused by closing a door with a metallic handle, can produce a voltage greater than 10,000V. Since the current flows for a very short period of time, there is seldom any serious harm to an individual. However, discharges of static electricity may be sufficient to cause serious electric shock and are always a potential source of ignition when flammable liquids, dusts or powders are present. This is a particular problem in the parts of the printing industry where solvent-based inks are used on high-speed web presses.

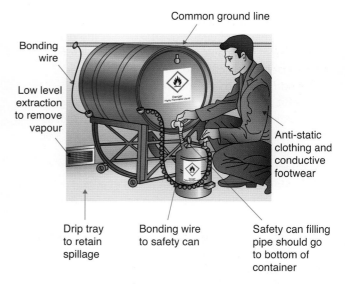

Dispensing Flammable Liquids

Figure 11.6 Earth bonding of containers to prevent a static discharge – anti-static outer clothing and conductive footwear should be used

Static electricity may build up on both materials and people. When a charged person approaches flammable gases or vapours, and a spark ignites the substance, the resulting explosion or fire often causes serious injury. In these situations, effective static control systems must be used.

Lightning strikes are a natural form of static electricity and result in large amounts of electrical energy being dissipated in a short time in a limited space with a varying degree of damage. The current produced in the vast majority of strikes exceeds 3,000 A over a short period of time. Before a strike, the electrical potential between the cloud and earth might be about 100 million volts and the energy released at its peak might be about 100 million watts per metre of strike.

The need to provide lightning protection depends on a number of factors, which include:

▶ the risk of a strike occurring;
▶ the number of people likely to be affected;
▶ the location of the structure and the nearness of other tall structures in the vicinity;
▶ the type of construction, including the materials used;
▶ the contents of the structure or building (including any flammable substances);
▶ the value of the building and its contents.

Expert advice will be required from a specialist company in lightning protection, especially when flammable substances are involved. Lightning strikes can also cause complete destruction and/or significant disruption of electronic equipment.

Thus, the key to the control of static electricity is the provision of a conducting path so that the charge will not continue to build up but will flow away to earth without causing harm.

Portable electrical equipment

Portable and transportable electrical equipment is defined by the UK Health and Safety Executive as 'not part of a fixed installation but may be connected to a fixed installation by means of a flexible cable and either a socket and plug or a spur box or similar means'. It may be hand-held or hand-operated while connected to the supply, or is intended or likely to be moved while connected to the supply. The auxiliary equipment, such as extension leads, plugs and sockets, used with portable tools, is also classified as portable

equipment. The term 'portable' means both portable and transportable.

Almost 25% of all reportable electrical accidents involve portable electrical equipment (known as portable appliances). While most of these accidents were caused by electric shock, over 2,000 fires each year are started by faulty cables used by portable appliances, caused by a lack of effective maintenance. Portable electrical tools often present a high risk of injury, which is frequently caused by the conditions under which they are used. These conditions include the use of defective or unsuitable equipment and, indeed, the misuse of equipment. There must be a system to record the inspection, maintenance and repair of these tools.

Where plugs and sockets are used for portable tools, sufficient sockets must be provided for all the equipment and adaptors should not be used. Many accidents are caused by faulty flexible cables, extension leads, plugs and sockets, particularly when these items become damp or worn. Accidents often occur when contact is made with some part of the tool which has become live (probably at mains voltage), while the user is standing on, or in contact with, an earthed conducting surface. If the electricity supply is at more than 50V ac, then the electric shock that a person may receive from such defective equipment is potentially lethal. In adverse environmental conditions, such as humid or damp atmospheres, even lower voltages can be dangerous. Portable electrical equipment should not be used in flammable atmospheres if it can be avoided and it must also comply with any standard relevant to the particular environment. Air-operated equipment should also be used as an alternative whenever it is practicable.

Some portable equipment requires substantial power to operate and may require voltages higher than those usually used for portable tools, so that the current is kept down to reasonable levels. In these cases, power leads with a separate earth conductor and earth screen must be used. Earth leakage relays and earth monitoring equipment must also be used, together with substantial plugs and sockets designed for this type of system.

Electrical equipment is safe when properly selected, used and maintained. It is important, however, that the environmental conditions are always carefully considered. The hazards associated with portable appliances increase with the frequency of use and the harshness of the environment (construction sites are often particularly hazardous in this respect). These factors must be considered when inspection, testing and maintenance procedures are being developed.

Secondary hazards

It is important to note that there are other hazards associated with portable electrical appliances, such as abrasion and impact, noise and vibration. Trailing leads

Figure 11.7 Portable hand-held electric power tools

used for portable equipment and raised socket points offer serious trip hazards and both should be used with great care near pedestrian walkways. Power drives from electric motors should always be guarded against entanglement hazards.

Secondary hazards are those additional hazards which present themselves as a result of an electrical hazard. It is very important that these hazards are considered during a risk assessment. An electric shock could lead to a fall from height if the shock occurred on a scaffold or it could lead to a collision with a vehicle if the victim collapsed on to a roadway.

Similarly, an electrical fire could lead to all the associated fire hazards outlined in Chapter 12 (e.g. suffocation, burns and structural collapse) and electrical burns can easily lead to infections.

High risks associated with the use of electricity

There are several sources of high risks associated with the use of electricity. These include:

▶ working with poorly maintained electrical equipment;
▶ using electrical equipment in adverse or hazardous environments such as wet, flammable or explosive atmospheres;
▶ working on mains electricity supplies;
▶ contact with underground cables during excavation work; and
▶ contact with live overhead power lines.

11.2 Control measures when working with electrical systems or using electrical equipment in all workplace conditions

11.2.1 Safe systems of work and competence

The principal control measures for electrical hazards are applicable to all electrical equipment and systems found at the workplace and impose duties on employers, employees and the self-employed. The following issues need to be addressed:

▶ the design, construction and maintenance of electrical systems, work activities and protective equipment;
▶ the strength and capability of electrical equipment;
▶ the protection of equipment against adverse and hazardous environments;
▶ the insulation, protection and placing of electrical conductors;
▶ the earthing of conductors and other suitable precautions;
▶ the integrity of referenced conductors;
▶ the suitability of joints and connections used in electrical systems;

▶ means for protection from excess current;
▶ means for cutting off the supply and for isolation;
▶ the precautions to be taken for work on equipment made dead;
▶ working on or near live conductors;
▶ adequate working space, access and lighting;
▶ the competence requirements for persons working on electrical equipment to prevent danger and injury.

Detailed safety standards for designers and installers of electrical systems and equipment are given a code of practice, published by the UK Institution of Electrical Engineers (now the Institution of Engineering and Technology), known as the IEE Wiring Regulations. While these Regulations are not legally binding, they are recognised as a code of good practice and widely used as an industry standard.

BS 7671: Requirements for Electrical Installation – more commonly known as the IEE Wiring Regulations – is the national standard to which all domestic and industrial wiring has to conform. The 17th edition includes substantial changes to harmonise with EU requirements. Among the changes to these Regulations is the inclusion of four new regulations for the protection of people and livestock against voltage disturbances and electromagnetic influences. There is also a specific requirement for appropriate documentation for all installations. Seven new special locations are covered to address the risk associated with certain environments or facilities, including exhibition areas, mobile units and temporary installations.

The standard wiring colours in the UK, shown in Table 11.1, are the same as elsewhere in Europe, Australia, and New Zealand and follow the international standard IEC 60445 which defines basic safety principles for identifying electrical conductors by colours in electricity distribution wiring. The United States, Canada and Japan are mentioned in a note in the standard for using different colours:

▶ White or grey for the neutral conductor instead of blue.
▶ Green for the protective earth instead of green-and-yellow.

Table 11.1 Standard wiring colours

	Colour
Protective earth (PE)	Green-and-yellow
Neutral (N)	Blue
Single phase: Line (L) Three phase: L1	Brown
Three phase: L2	Black
Three phase: L3	Grey

The 17th edition of the Regulations requires that inspection and testing must be carried out by a 'competent person' to check that the electrical work meets required standards. It defines a competent person as someone 'who possesses sufficient technical knowledge and experience for the nature of the electrical work undertaken and is able at all times to prevent danger, and where appropriate, injury to themselves and others'. After the initial testing, the Regulations recommend that every electrical installation is subject to periodic inspection and testing by a competent person.

The risk of injury and damage inherent in the use of electricity can only be controlled effectively by the introduction of employee training, safe operating procedures (safe systems of work) and guidance to cover specific tasks.

Training is required at all levels of the organisation ranging from simple on-the-job instruction to apprenticeship for electrical technicians and supervisory courses for experienced electrical engineers. First-aid training related to the need for cardiovascular resuscitation and treatment of electric burns should be available to all people working on electrical equipment and their supervisors.

A **management system** should be in place to ensure that the electrical systems are installed, operated and maintained in a safe manner. All managers should be responsible for the provision of adequate resources of people, material and advice to ensure that the safety of electrical systems under their control is satisfactory and that **safe systems of work** are in place for all electrical equipment. For certain types of electrical work, such as working with live electricity, a permit to work will be required. The following items should be included in electrical permits to work:

▶ the permit issue number;
▶ the details of the work and its location;
▶ the significant hazards and risks involved in the work and the precautions to be taken including any personal protective equipment to be worn;
▶ details of the electrical items to be isolated and the points at which such isolations are to be made;
▶ the test procedures to be followed to confirm that circuits are dead;
▶ details of any special work tools required;
▶ details of the safety warning signs required;
▶ the emergency procedures required;
▶ the date and time of issue of the permit and the length of its duration and the cross referencing with other permits issued;
▶ acceptance of the permit by the person carrying out the work; and
▶ on completion of the work, a signature of the person who authorised the work to ensure cancellation of the permit.

A permit should not be issued on electrical equipment that is live. It is never absolutely safe to work on live electrical equipment. There are, however, a few circumstances where it is necessary to work live, but

this must only be done after it has been determined that it is unreasonable for the work to be done dead. Even if working live can be justified, many precautions are needed to make sure that the risk is reduced 'so far as is reasonably practicable'. (Chapter 4 gave more information on both safe systems of work and permits to work.)

For small factories and office or shop premises where the system voltages are normally at mains voltage, it may be necessary for an external competent person to be available to offer the necessary advice. Managers must set up a high-voltage permit-to-work system for all work at and above 600V. The system should be appropriate to the extent of the electrical system involved. Consideration should also be given to the introduction of a permit system for voltages under 600V when appropriate and for all work on live conductors.

The additional control measures that should be taken when working with electricity or using electrical equipment are summarised by the following topics:

▶ the selection of suitable equipment;
▶ the use of protective systems;
▶ inspection and maintenance strategies.

These three groups of measures will be discussed in detail.

11.2.2 The selection and suitability of equipment

Many factors which affect the selection of suitable electrical equipment, such as flammable, explosive and damp atmospheres and adverse weather conditions, have already been considered. Other issues include high or low temperatures, dirty or corrosive processes or problems associated with vegetation or animals (e.g. tree roots touching and displacing underground power cables, farm animals urinating near power supply lines and rats gnawing through cables). Temperature extremes will affect, for example, the lubrication of motor bearings and corrosive atmospheres can lead to the breakdown of insulating materials. The equipment selected must be suitable for the task demanded or either it will become overloaded or running costs will be too high.

The equipment should be installed to a recognised standard and be capable of being isolated in the event of an emergency. It is also important that the equipment is effectively and safely earthed. Electric supply failures may affect process plant and equipment. These are certain to happen at some time and the design of the installation should be such that a safe shutdown can be achieved in the event of a total mains failure. This may require the use of a battery-backed shutdown system or emergency standby electric generators (assuming that this is cost-effective).

Suppliers of hired electrical equipment should formally inspect and test the equipment before each hire to

11

ensure that it is safe to use. The person hiring the equipment should also take appropriate steps to ensure it remains safe to use throughout the hire period.

Finally, it is important to stress that electrical equipment must only be used within the rating performance given by the manufacturer and any accompanying instructions from the manufacturer or supplier must be carefully followed.

11.2.3 The advantages and limitations of protective systems

There are several different types of protective systems and techniques that may be used to protect people, plant and premises from electrical hazards, some of which, for example earthing, have already been considered earlier in this chapter. However, only the more common types of protection will be considered here.

Fuse

A fuse will provide protection against faults and continuous marginal current overloads. It is basically a thin strip of conducting wire which will melt when an excess of the rated current passes through it, thus breaking the circuit.

A fuse rated at 13 A will melt when a current in excess of 13 A passes through the fuse thus stopping the flow of current. A **circuit breaker** throws a switch off when excess current passes and is similar in action to a fuse. Protection against overload is provided by fuses which detect a continuous marginal excess flow of current and energy. This overcurrent protection is arranged to operate before damage occurs, either to the supply system or to the load which may, for example, be a motor or heater. When providing protection

against overload, consideration needs to be made as to whether tripping the circuit could give rise to an even more dangerous situation, such as with fire-fighting equipment.

The prime objective of a fuse is to protect equipment or an installation from overheating and damage and becoming a fire hazard. It is not an effective protection against electric shock due to the time that it takes to cut the current flow.

The examination of fuses is a vital part of an inspection programme to ensure that the correct size or rating is fitted at all times.

Insulation

Insulation is used to protect people from electric shock, the short circuiting of live conductors and the dangers associated with fire and explosions. Insulation is achieved by covering the conductor with an insulating material. Insulation is often accompanied by the enclosure of the live conductors so that they are out of reach of people. A breakdown in insulation can cause electric shock, fire, explosion or instrument damage.

Isolation

The isolation of an electrical circuit involves more than 'switching off' the current in that the circuit is made dead and cannot be accidentally re-energised. It, therefore, creates an air gap between the equipment and the electrical supply which only an authorised person should be able to remove. When it is intended to carry out work, such as mechanical maintenance or a cleaning operation on plant or machinery, isolation of electrical equipment will be required to ensure safety during the work process. Isolators should always be locked off when work is to be done on electrical equipment.

Before working on an isolated circuit, checks must be made to ensure that the circuit is dead and that the

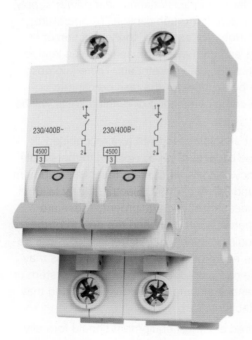

Figure 11.8 Typical 240 volt mini circuit breaker

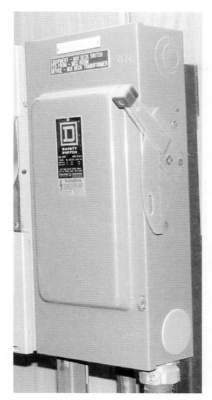

Figure 11.9 A variety of electrical equipment: (a) flush mounted exterior socket with RCD; (b) lockable electrical isolator

isolation switch is 'locked off' and clearly labelled (see Figure 11.9 (b)).

Reduced low-voltage systems

When the working conditions are relatively severe, either due to wet conditions or heavy and frequent usage of equipment, reduced voltage systems should be used.

All portable tools used on construction sites, in vehicle washing stations or near swimming pools should operate on 110V or less, preferably with a centre tapped to earth transformer at 55V. This means that while the full 110V are available to power the tool, only 55V are available should the worker suffer an electric shock. At this level of voltage, the effect of any electric shock should not be severe.

Safety Extra Low Voltage (SELV) – a voltage less than 50 volts – is used in low power tools, hand lights or soldering irons. Another way to reduce the voltage is to use **battery (cordless)-operated hand tools.**

Residual current devices

If electrical equipment must operate at mains voltage, the best form of protection against electric shock is the Residual Current Device (RCD). RCDs, also known as earth leakage circuit breakers, monitor and compare the current flowing in the live and neutral conductors supplying the protected equipment. Such devices are very sensitive to differences of current between the

live and neutral power lines and will cut the supply to the equipment in a very short period of time when a difference of only a few milliamperes occurs. It is the speed of the reaction which offers the protection against electric shock. For protection against electric shock, the RCD must have a rated residual current of 30 mA or less and an operating time of 40 milliseconds or less at a residual current of 250 mA. RCDs rated above 30 mA provide very limited protection against harm from an electric shock.

The protected equipment must be properly protected by insulation and enclosure in addition to the RCD. The RCD will not prevent shock or limit the current resulting from an accidental contact, but it will ensure that the duration of the shock is limited to the time taken for the RCD to operate. Electrical circuits must be securely isolated before any work is done on them. The RCD has a test button which should be tested frequently to ensure that it is working properly (see Figure 11.9(a)). RCDs will also protect installations against fire, as they will interrupt the electrical supply before sufficient energy to start a fire has accumulated. However, this is not their prime

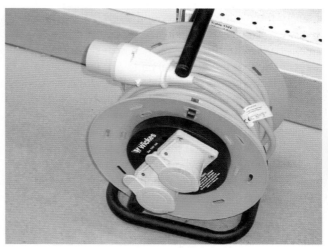

Figure 11.10 Multi-plug extention lead and special plugs and sockets: (a) reduced voltage usually 110 volt; (b) 240 volt mains extension lead with cut out

purpose. Therefore the advantages of a RCD over a fuse are:

▶ a fuse is slower to act than a RCD and protects equipment rather than people;

▶ a blown fuse may be replaced by a fuse having an incorrect value; and

▶ a RCD, with its built-in test button, is easier to test and re-set.

It is advisable to use a residual current device (RCD) whenever possible but particularly in wet or damp locations, such as outdoors. It is best practice to use an RCD that is incorporated into the switchboard of the electrical installation so that all circuits fed from that RCD are protected by it. A RCD that is incorporated into an ordinary mains socket, or plugged into it, will protect anything attached to that socket, but it is possible that equipment may have been accidentally plugged into another unprotected socket.

Double insulation

To remove the need for earthing on some portable power tools, double insulation is used. Double insulation employs two independent layers of insulation over the live conductors, each layer alone being adequate to insulate the electrical equipment safely. An electrical appliance which is double insulated does not have an earth wire fitted. The appliance is designed in such a way that the electrical parts can never come into contact with the outer casing of the device. An appliance which is double insulated has the whole of the inside contained in plastic beneath an outer casing. If an electrical fault occurs within the appliance, no live conductor can touch the outer casing because of the insulating plastic layer. Such equipment is double insulated – the 'single' insulation is the insulation of the wires and the 'double' is the plastic body of the equipment.

Double insulation is used on Class II appliances such as hand-held portable appliances and non-hand held portable appliances. Common double insulated

appliances are hair dryers, radios and electric drills. These devices are regularly handled and the cases must never become live. For a hair dryer this is especially important since these are sometimes used in a bathroom or with wet hands. As such tools are not protected by an earth, they must be inspected and maintained regularly and must be discarded if damaged.

Figure 11.11 shows the symbol which is marked on double-insulated portable power tools.

11.2.4 Protection against contact with live overhead or buried power lines

Where possible all work likely to lead to contact with overhead power lines should be done in an area well clear of the line itself. It may be possible to alter the work and eliminate or reduce the risk. As a general rule, vehicles, plant or equipment should be brought no closer than:

▶ 15 m of overhead lines suspended from steel towers;

▶ 9 m of overhead lines supported on wooden poles.

Where a closer approach is likely either the lines should be made dead or barriers erected to prevent an approach. Permits to work are likely to be required if work close to the lines is necessary.

Where work is necessary directly beneath the lines, or blasting or other unusual activity has to be done adjacent to them, they may need to be made dead and a permit-to-work system operated. In certain situations induced AC voltages and even arcing can be created in fences and pipelines which run parallel to overhead lines. When there are concerns about nearby structures the advice of the safety adviser and/or the electrical supply company must be sought. Risk assessments should also be made and suitably recorded, the information being made available to all workers who are likely to be working near the power lines.

On many construction sites, vehicles will need to move beneath power lines. In these cases, the roadway should be covered by goalposts covered with warning tape. Bunting should be suspended, level with the top of the goalposts and just above ground level (often using empty oil drums), between poles across the site along the length of the power line. This will ensure that vehicles can only pass under the power lines by passing through the goalposts. Suitable warning signs should be placed on either side of the roadway on each side of the power line to warn of the overhead power line, and metal equipment, such as ladders and scaffold materials, should be excluded from the vicinity.

The presence of buried power lines is a common problem. Often official knowledge of the exact location of such services is less than accurate, particularly if the power lines have been in place for many years.

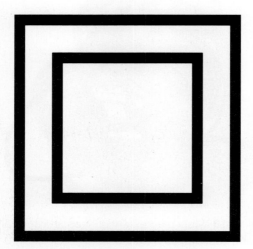

Figure 11.11 Double insulation sign

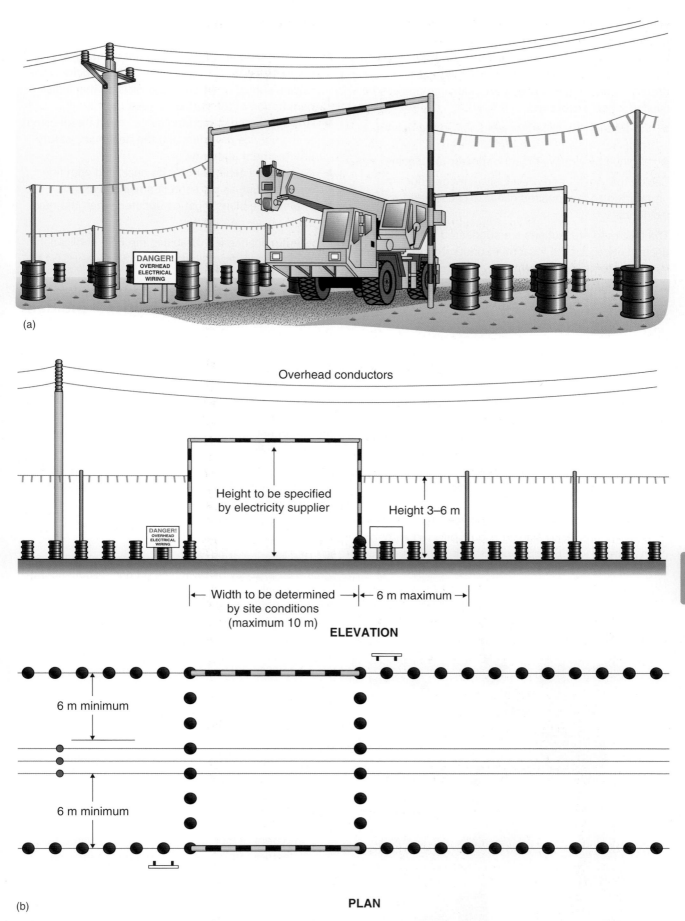

(a)

Overhead conductors

Height to be specified
by electricity supplier

Height 3–6 m

DANGER!
OVERHEAD
ELECTRICAL
WIRING

Width to be determined
by site conditions
(maximum 10 m)

6 m maximum

ELEVATION

6 m minimum

6 m minimum

(b)

PLAN

Figure 11.12 Precautions for overhead lines: (a) 'goalpost' crossing points beneath lines to avoid contact by plant;
(b) diagram showing normal dimensions for 'goalpost' crossing points and barriers

11

When underground cables are damaged, people can be injured or killed by electric shock or electrical arcs. Electric arcs can cause explosions and flames resulting in severe burns to hands, face and body, even if protective clothing is being worn. Such damage can be caused when a cable is:

▶ cut through by a sharp object such as the point of a tool; or

▶ crushed by a heavy object or powerful machine.

Cables that have been previously damaged but left unreported and unrepaired can also cause serious incidents.

The following precautions are suggested when there is uncertainty about the location of underground services in an area to be excavated:

▶ Check for any obvious signs of underground services, for example valve covers or patching of the road surface.

▶ Ensure that the excavation supervisor has the necessary service plans and is competent to use them to locate underground services.

▶ Ensure that all excavation workers are trained in safe digging practices and emergency procedures.

▶ Use locators to trace any services and mark the ground accordingly. A series of trial holes should be dug by hand to confirm the position of the pipes or cables. This is particularly important in the case of plastic pipes which cannot be detected by normal locating equipment.

▶ In areas where underground services may be present, only hand digging should be used with insulated tools. Spades and shovels should be used rather than picks and forks, which are more likely to pierce cables.

Figure 11.13 Using a cable detector

▶ Assume that all cables are 'live' unless it is known otherwise.

▶ Hand-held power tools should not be used within 0.5 m of the marked position of an electricity cable. Collars should be fitted to the tools so that initial penetration of the surface is restricted.

▶ Any suspected damage to cables must be reported to the service providers and the health and safety enforcement authority.

▶ All exposed cables should be backfilled with fine material such as dry sand or small gravel.

▶ The service plans must be updated when the new cables have been laid.

Excellent guidance is available in the HSE publication *Avoiding Danger from Underground Services* HSG47.

11.2.5 Emergency procedures following an electrical incident

General emergency procedures were covered in Chapter 4. However, even with the best of safety procedures in place, electrical incidents will occur so that specific and effective emergency procedures are essential. If someone is found unconscious as a result of electric shock or suffering from an electric burn, then the treatment procedure listed earlier in this chapter should be followed.

The emergency procedures should ensure that, for a serious electrical accident, the emergency services and key personnel in the organisation are notified and the emergency plan is activated. Such a plan should include the following points:

▶ The isolation of the electrical device that has caused the emergency by switching off and disconnecting the supply.

▶ If electric shock is the main emergency, the victim must not be touched until there is no longer a possibility of contact with the electrical current.

▶ If there is a fire as a result of electrical equipment malfunction, the fire procedures should also be activated.

▶ If the emergency could affect hazardous equipment or processes, essential actions, such as emergency plant shutdown, isolation or making processes safe, should be put into action. It is important that important items such as shut-off valves and electrical isolators are easily identifiable.

▶ Work must not resume after an emergency until a competent person has checked and declared that all electrical equipment affected by the emergency is safe and the electrical supply has been switched back on. If there are any doubts, then advice should be sought from the emergency services.

▶ Following the incident, an investigation should take place, a report produced and, if appropriate, reported to the national health and safety authority.

11.2.6 Inspection and maintenance strategies

Inspection strategies

Regular inspection of electrical equipment is an essential component of any preventative maintenance programme. Any strategy for the inspection of electrical equipment, particularly portable appliances, should involve the following considerations:

▶ a means of identifying the equipment to be tested;
▶ the number and type of appliances to be tested;
▶ the competence of those who will undertake the testing (whether in-house or hired for the task);
▶ the legal requirements for portable appliance testing (PAT) and other electrical equipment testing and the guidance available;
▶ organisational duties of those with responsibilities for PAT and other electrical equipment testing;
▶ test equipment selection and re-calibration;
▶ the development of a recording, monitoring and review system; and
▶ the development of any training requirements resulting from the test programme.

Maintenance strategies

Regular maintenance is required to ensure that a serious risk of injury or fire does not result from installed electrical equipment. Maintenance standards should be set as high as possible so that a more reliable and safe electrical system will result.

Inspection and maintenance periods should be determined by referring to the recommendations of the manufacturer, and considering the operating conditions and the environment in which equipment is located. The importance of equipment within the plant, from the plant safety and operational viewpoint, will also have a bearing on inspection and maintenance periods. The mechanical safety of driven machinery is vital and the electrical maintenance and isolation of the electrically powered drives is an essential part of that safety.

The particular areas of interest for inspection and maintenance are:

▶ the cleanliness of insulator and conductor surfaces;
▶ the mechanical and electrical integrity of all joints and connections;
▶ the integrity of mechanical mechanisms, such as switches and relays;
▶ the calibration, condition and operation of all protection equipment, such as circuit breakers, RCDs and switches.

Safe operating procedures for the isolation of plant and machinery during both electrical and mechanical maintenance must be prepared and followed. All electrical isolators must, wherever possible, be fitted with mechanisms which can be locked in the 'open/off' position and there must be a procedure to allow fuse withdrawal wherever isolators are not fitted.

Working on live equipment with voltages in excess of 110V must not be permitted except where fault-finding or testing measurements cannot be done in any other way. Reasons such as the inconvenience of halting production are not acceptable.

Part of the maintenance process should include an appropriate system of visual inspection. By concentrating on a simple, inexpensive system of looking for visible signs of damage or faults, many of the electrical risks can be controlled, although more systematic testing may be necessary at a later stage.

All fixed electrical installations should be inspected and tested periodically by a competent person.

Portable electrical appliances testing

Portable appliances should be subject to three levels of inspection – a user check, a formal visual inspection and a combined inspection and test.

User checks

When any portable electrical hand tool, appliance, extension lead or similar item of equipment is taken into use, at least once each week or, in the case of heavy work, before each shift, the following visual check and associated questions should be asked:

▶ Is there a recent PAT label attached to the equipment?
▶ Are any bare wires visible?
▶ Is the cable covering undamaged and free from cuts and abrasions (apart from light scuffing)?
▶ Is the cable too long or too short? (Does it present a trip hazard?)
▶ Is the plug in good condition (for example the casing is not cracked and the pins are not bent)?
▶ Are there no taped or other non-standard joints in the cable?
▶ Is the outer covering (sheath) of the cable gripped where it enters the plug or the equipment? (The coloured insulation of the internal wires should not be visible.)
▶ Is the outer case of the equipment undamaged or loose and are all screws in place?
▶ Are there any overheating or burn marks on the plug, cable, sockets or the equipment?
▶ Are the trip devices (RCDs) working effectively (by pressing the 'test' button)? (see Figure 11.9 (a)).

Formal visual inspections and tests

There should be a **formal visual inspection** routinely carried out on all portable electrical appliances. Faulty equipment should be taken out of service as soon as the damage is noticed. At this inspection the plug cover (if not moulded) should be removed to check that the correct fuse is included, but the equipment itself should not be taken apart. This work can normally be carried out by a trained person who has sufficient information and knowledge.

Some faults, such as the loss of earth continuity due to wires breaking or loosening within the equipment, the breakdown of insulation and internal contamination (e.g. dust containing metal particles may cause short circuiting if it gets inside the tool), will not be found by visual inspections. To identify these problems, a programme of testing and inspection will be necessary.

This formal **combined testing and inspection** should be carried out by a competent person when there is reason to suspect the equipment may be faulty, damaged or contaminated, but this cannot be confirmed by visual inspection or after any repair, modification or similar work to the equipment, which could have affected its electrical safety. The competent person could be a person who has been specifically trained to carry out the testing of portable appliances using a simple 'pass/fail' type of tester. When more sophisticated tests are required, a competent person with the necessary technical electrical knowledge and experience would be needed.

The inspection and testing should normally include the following checks:

▶ that the polarity is correct;
▶ that the correct fuses are being used;
▶ that all cables and cores are effectively terminated;
▶ that the equipment is suitable for its environment.

Testing need not be expensive in many low-risk premises like shops and offices, if an employee is trained to perform the tests and appropriate equipment is purchased.

Frequency of inspection and testing

Electrical equipment should be visually checked to spot early signs of damage or deterioration. The frequency of inspection and testing should be based on a risk assessment which is related to the usage, type and operational environment of the equipment. The harsher the working environment is, the more frequent the period of inspection. Thus tools used on a construction site should be tested much more frequently than a visual display unit which is never moved from a desk. Manufacturers or suppliers may recommend a suitable testing period. Table 11.2 lists the suggested intervals for inspection and testing

Table 11.2 Suggested intervals for portable appliance inspection and testing

Type of business/equipment	User checks	Formal visual inspection	Combined inspection and electrical tests
Equipment hire	Yes	Before issue and after return	Before issue
Construction	Yes	Before initial use and then every month	3 months
Industrial	Yes	Before initial use and then every 3 months	6–12 months
Hotels and offices, low-risk environments			
Battery operated (less than 20V)	No	No	No
Extra low voltage (less than 50V ac), for example telephone equipment, low-voltage desk lights	No	No	No
Desk-top computers, VDU screens	No	Yes 2–4 years	No if double insulated, otherwise up to 5 years
Photocopiers/fax machines/printers Not hand-held; rarely moved	No	Yes 2–4 years	No if double insulated, otherwise up to 5 years
Double-insulated equipment: not hand-held. Moved occasionally, for example fans, table lamps, slide projectors	No	Yes 2–4 years	No
Double-insulated equipment: hand-held, for example some floor cleaners	Yes	Yes 6 months–1 year	No
Earthed equipment (class 1): for example electric kettles, some floor cleaners, portable electric heaters, some kitchen equipment and irons	Yes	Yes 6 months–1 year	Yes 1–2 years
Cables (leads) and plug connected to the above. Extension leads (mains voltage)	Yes	Yes 6 months–4 years depending on the type of equipment it is connected to	Yes 1–5 years depending on the type of equipment it is connected to

Source: Derived from HSE.

Note: Operational experience may demonstrate that the above intervals can be reviewed.

derived from the UK HSE publications *Maintaining Portable and Transportable Electrical Equipment* (HSG107 and INDG236 and 237).

It is very important to stress that there is no 'correct' interval for testing – it depends on the frequency of usage, type of equipment, and how and where it is used. A few years ago, a young trainee was badly scalded by a boiling kettle of water which exploded while in use. On investigation, an inspection report indicated that the kettle had been checked by a competent person and passed just a few weeks before the accident. Further investigation showed that this kettle was the only method of boiling water on the premises and was in use continuously for 24 hours each day. It was therefore unsuitable for the purpose and a plumbed-in continuous-use hot water heater would have been far more suitable.

Suppliers, who loan equipment, should formally inspect and test the equipment before each hire to ensure that it is safe to use. The person hiring the equipment should also take appropriate steps to ensure it remains safe to use throughout the hire period.

Records of inspection and testing

Schedules which give details of the inspection and maintenance periods and the respective programmes must be kept together with records of the inspection findings and the work done during maintenance. Records must include both individual items of equipment and a description of the complete system or section of the system. They should always be kept up-to-date and with an audit procedure in place to monitor the records and any required actions. The records do not have to be paper-based but could be stored electronically on a computer. It is good practice to label the piece of equipment with the date of the last combined test and inspection.

The effectiveness of the equipment maintenance programme may be monitored and reviewed if a record of tests is kept. It can also be used as an inventory of portable appliances and help to regulate the use of unauthorised appliances. The record will enable any adverse trends to be monitored and to check that suitable equipment has been selected. It may also give an indication as to whether the equipment is being used correctly.

Advantages and limitations of Portable Appliance Testing (PAT)

The advantages of PAT include:

- an earlier recognition of potentially serious equipment faults, such as poor earthing, frayed and damaged cables and cracked plugs;
- discovery of incorrect or inappropriate electrical supply and/or equipment;
- discovery of incorrect fuses being used;
- a reduction in the number of electrical accidents;
- monitoring the misuse of portable appliances;
- equipment selection procedures checkable;
- an increased awareness of the hazards associated with electricity;
- a more regular maintenance regime should result.

The limitations of PAT include:

- some fixed equipment is tested too often leading to excessive costs;
- some unauthorised portable equipment, such as personal kettles, are never tested as there is no record of them;
- equipment may be misused or overused between tests due to a lack of understanding of the meaning of the test results;
- all faults, including trivial ones, are included on the action list, so the list becomes very long and the more significant faults are forgotten or overlooked;
- the level of competence of the tester can be too low;
- the testing equipment has not been properly calibrated and/or checked before testing takes place.

Most of the limitations may be addressed and the reduction in electrical accidents and injuries enables the advantages of PAT to greatly outweigh the limitations.

11.3 Further information

Directive 2006/95/EC – electrical equipment http://ec.europa.eu/enterprise/sectors/electrical/documents/lvd/legislation/

Directive 2009/104/EC – use of work equipment https://osha.europa.eu/en/legislation/directives/workplaces-equipment-signs-personal-protective-equipment/osh-directives/3

11

11.4 Practice revision questions

1. (a) **Explain** the relationship between voltage, current and resistance referring to a basic electric circuit.
   (b) **Outline** the main hazards associated with the use of electricity.
   (c) **Identify** the control measures that will reduce the risk of fire from electrical equipment.

2. (a) **Outline** the effects on the human body from a severe electric shock.
   (b) **Outline** the emergency action to take if a person suffers a severe electric shock.

3. **Explain** the following terms used in electrical work:
   (a) 'isolation'
   (b) 'earthing'
   (c) 'conductors and insulators'
   (d) 'overcurrent protection'
   (e) 'electric arcing'
   (f) 'static electricity'.

4. **Describe** how the following protective measures reduce the risk of electric shock and, in **EACH** case, **give** an example of their application.
   (a) fuse
   (b) reduced low voltage
   (c) residual current devices
   (d) double insulation.

5. Electrical plugs and cables can cause accidents in the workplace.
   (a) **Identify FOUR** examples of faults and bad practices that could lead to such accidents.
   (b) **Outline** the corresponding control measures that should be taken for **EACH** of the examples identified in (a) above.

6. A construction worker is working on a scaffold on the outside of a building. While using a portable 240V electrical drill to drill into the wall of the building, he makes direct contact with the electrical supply cable installed in the wall.
   (a) **Identify FOUR** possible outcomes arising from this incident.
   (b) **Identify FOUR** protective devices and procedures that could have reduced the risk of injury to this worker.
   (c) **Describe** the types of inspection and/or test that should have been made on the drill.

7. (a) **Outline** the **THREE** levels of inspection that should be included in a maintenance and inspection strategy for portable electrical appliances.
   (b) **Identify** the reasons for keeping records of the results of portable appliance testing within an organisation.
   (c) **Outline** the issues to be considered when determining the frequency for the inspection and testing of a portable electrical tool.

8. (a) **Identify** the items that should be included on a checklist for the routine visual inspection of portable electrical appliances.
   (b) **Identify EIGHT** examples of faults and bad practices that could contribute to accidents following the use of portable electrical appliances.
   (c) **Identify** the advantages and limitations of portable appliance testing (PAT).

9. **Outline** the precautions to protect against contact with live electrical conductors when:
   (a) excavating near underground cables
   (b) working in the vicinity of overhead power lines.

CHAPTER 12

Fire hazards and risk control

12.1 Fire initiation, classification, spread and some legal standards ▶ 332

12.2 Fire risk assessment ▶ 340

12.3 Fire prevention and prevention of fire spread ▶ 344

12.4 Fire alarm system and fire-fighting arrangements ▶ 354

12.5 Evacuation of a workplace ▶ 361

12.6 Further information ▶ 367

12.7 Practice revision questions ▶ 367

Appendix 12.1 Fire risk assessment checklist as recommended in Fire Safety Guides published by the UK Department for Communities and Local Government in 2006 ▶ 369

Appendix 12.2 Typical fire notice ▶ 371

This chapter covers the following NEBOSH learning objectives:

1. Describe the principles of fire initiation, classification and spread
2. Outline the principles of fire risk assessment
3. Describe the basic principles of fire prevention and the prevention of fire spread in buildings
4. Identify the appropriate fire alarm systems and fire-fighting equipment for a simple workplace
5. Outline the factors which should be considered when implementing a successful evacuation of a workplace in the event of a fire
6. In addition to the NEBOSH learning outcomes have a basic understanding of some legal and ILO standards for fire prevention

---

12.1 Fire initiation, classification, spread and some legal standards

12.1.1 Introduction

This chapter covers fire prevention in the workplace and how to ensure that people are properly protected if fire does occur. Fire is still a major risk in many workplaces (see Figure 12.1).

> At the beginning of the 21st century, the population of the **Earth** was 6,300,000,000 people, who annually experience a reported 7,000,000–8,000,000 fires with 70,000–80,000 fire deaths and 500,000–800,000 fire injuries.
>
> At the beginning of the 21st century, the population of **Europe** was 700,000,000 people, who annually experience a reported 2,000,000–2,500,000 fires with 20,000–25,000 fire deaths and 250,000–500,000 fire injuries.

The CTIF (Center of Fire Statistics) based in Moscow, analyses the fire statistics of 30–50 countries every year. The combined population of these countries is 1–2,000,000,000 people and their fire services received 25–33,000,000 emergency calls of which 3–4,000,000 were fires (10% of all emergency calls). These fires led to the deaths of 25–35,000 people every year. The

Figure 12.1 Fire is still a significant risk in many workplaces

average death rates are about three fire deaths per 100,000 people and one fire death per 100 fires (see Figure 12.2). However, this average conceals a more than 100-fold variation in death rates from country to country. A better indication of typical fire risk is the median fire death rate per 100,000 people by country, which was 1.0 in 2003 and 0.9 in 2004.

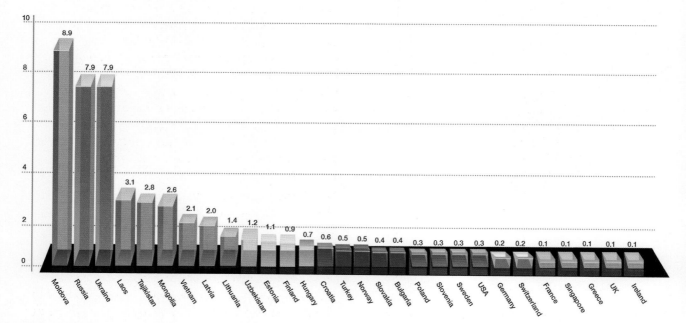

Figure 12.2 Average number of fire deaths per 100 fires

Each year CTIF world fire statistics show that:

▶ direct losses to fire average 0.16% of GDP;
▶ indirect losses average 0.013% (or 1/12 the direct losses);
▶ cost of fire service average 0.15%;
▶ fire protection of buildings average 0.27%;
▶ fire insurance net after payment of claims average 0.06% of GDP.

Figure 12.3 shows how these costs are apportioned between these various cost areas.

On average across the 14 countries, the total 'cost' of fires was 0.65% of GDP. Important components that are included in this calculation include monetary equivalents for human loss (deaths and injuries), monetary equivalents for volunteer fire-fighters and other donated time, and the cost of achieving better fire performance in products, whether mandated by regulation or otherwise.

The financial costs associated with serious fires are very high including, in many cases (believed to be over 40%), the failure to start up business again. Never underestimate the potential of any fire. What may appear to be a small fire in a waste bin, if not dealt with, can quickly spread through a building or a structure. The UK Bradford City Football ground in 1985 or King's Cross Underground station in 1987 are examples of where small fires quickly became raging infernos, resulting in many deaths and serious injuries. The Deep Water Horizon oil platform fire in 2010 may well cost in the order of $40 billion which will go down as the most costly peacetime fire in recent history.

Figure 12.3 Economic-statistical evaluation of 'costs' of fire for 14 countries

12.1.2 Fire legislation and standards

Generally, fire safety legislation exists in nearly every nation. Some are extensive and complete while others are extremely basic, if not primitive. Internationally, there are as many legislative bodies as there are countries. The origins of nearly all of this legislation are as varied as the number of countries where it is applied. The national fire authorities are often involved in formation of the fire legislation concerned with general fire protection, means of escape in case of fire, fire-fighting, fire alarms and emergency evacuation. Safety and health authorities are normally concerned with process fire risks concerned with the use and storage of flammable chemicals and other flammable materials. Large insurance companies have also been active in improving fire prevention standards although they have concentrated more on building protection rather than life safety.

In the United States over the past hundred or so years, the federal government has enacted substantial legislation on fire safety and fire prevention. Each state, county and municipality creates and applies legislation tailored to its particular wants and needs. Standards and codes do not originate on Capitol Hill in Washington, but principally in the American National Standards Institute (ANSI), which coordinates the creation and diffusion of codes and standards related to nearly every activity in the country, including fire safety, created by more than 80 entities in the United States and other countries.

The prime source of fire safety standards in the USA is the National Fire Protection Association (NFPA), which publishes and constantly updates the majority of codes that form the basis for national, state and local legislation. Although the NFPA standards and codes are not legislation but rather documents of recommended good practice, they form the foundation on which nearly all United States fire safety legislation is based, including building codes, municipal ordinances, and so forth.

The 48 countries that comprise Europe have each developed specific regulations for their own territories. The 28 countries of the European Union (EU) have consolidated much of their individual legislation and codes, often sacrificing particular national interests to a pan-European effort toward standardisation, for example making a particular standard on portable fire extinguisher classifications or fire detection system and component specifications commonly applicable throughout all the member countries.

This European standardisation applies to hundreds of fire protection-related subjects such as extinguishing agents, smoke/flame detectors, sprinkler systems, fire-resistance characteristics of materials, testing procedures, equipment for potentially explosive atmospheres and more. The European situation is similar to that of the USA, although a number of these countries have retained many specific national laws and regulations. In these countries, the standards and codes

12

apply to the entire nation, such as the British Standards (BS), the German Deutsches Institut für Normung (DIN) and the Spanish AENOR. On this international level, the International Standards Organisation (ISO) is the worldwide standards-producing body comprising the national entities of 163 countries, providing information, products and services related to property and liability risks. The ISO standards meet and often exceed individual national ones. Many ISO standards have profound influence on national legislation and standards around the world.

In Central and South America, a great many of the independent nations, republics, island states, and protectorates have based their fire protection legislation and standards on those of the USA, specifically NFPA. Some others have created their own legislation based on their particular characteristics, and still others have looked to Europe for guidelines. Some countries such as Mexico and Peru have extensive national and regional regulations, providing ample information on materials or systems specifications, whereas a few countries make direct reference to specific NFPA codes.

Australia and New Zealand both have extensive legislation covering building design and construction as well as standards for equipment, systems and installations. Australia has a very high ratio of research and testing facilities in relation to its population, performing some of the world's most advanced investigation and research projects in fire protection, such as smoke control in various types and sizes of buildings.

In Asia, Japan is probably the leader in fire safety regulations, in part because of the particular characteristics of most of the nation's residential and small to medium-sized business premises construction. The Philippines and China have recently made enormous strides in improving regulations on building characteristics and fire safety, principally because of public pressure in response to numerous multi-fatality fires.

In the continent of Africa, the Republic of South Africa has numerous fire prevention and protection codes in effect, followed by certain other countries in North Africa, such as Algeria, Egypt, Tunisia and Morocco, although these are far behind South Africa in regard to extensive or exacting legislation. However, they are still far ahead of other countries such as Rwanda, Congo, Chad, Namibia and Mali, where fire prevention and protection is given little or no consideration.

12.1.3 ILO standards

(a) ILO-OSH 2001 – Emergency preparedness and response

The ILO-OSH 2001 requirements include:

Emergency prevention, preparedness and response arrangements should be established and maintained. These arrangements should identify the potential for accidents and emergency situations, and address the prevention of OSH risks associated with them. The arrangements should be made according to the size and nature of activity of the organisation. They should:

(a) ensure that the necessary information, internal communication and coordination are provided to protect all people in the event of an emergency at the worksite;

(b) provide information to, and communication with, the relevant competent authorities, and the neighbourhood and emergency response services;

(c) address first-aid and medical assistance, fire-fighting and evacuation of all people at the worksite; and

(d) provide relevant information and training to all members of the organisation, at all levels, including regular exercises in emergency prevention, preparedness and response procedures.

Emergency prevention, preparedness and response arrangements should be established in cooperation with external emergency services and other bodies where applicable.

(b) ILO Convention 155 control measures for chemicals and emergencies

The ILO Convention 155 covers the following in Article 16 and 18:

Article 16–2. Employers shall be required to ensure that, so far as is reasonably practicable, the chemical, physical and biological substances and agents under their control are without risk to their health.

Article 18. Employers shall be required to provide, where necessary, for measures to deal with emergencies and accidents, including adequate first-aid arrangements without risk to health when the appropriate measures of protection are taken.

12.1.4 Basic principles of fire

(a) Fire triangle

Fire cannot take place unless three things are present. These are shown in Figure 12.4.

The absence of any one of these elements will prevent a fire from starting. Prevention depends on avoiding these three coming together. Fire extinguishing depends on removing one of the elements from an existing fire, and is particularly difficult if an oxidising substance is present.

Fuel
Flammable gases,
liquids, solids

Ignition source
Hot surfaces
Electrical equipment
Static electricity
Smoking materials
Naked flame

Oxygen
From the air
Oxidising substances

Figure 12.4 Fire triangle

Once a fire starts, it can spread very quickly from fuel to fuel as the heat increases.

(b) Sources of ignition

Workplaces have numerous sources of ignition, some of which are obvious but others may be hidden inside machinery. Most of the sources may cause an accidental fire from sources inside but, in the case of arson (about 13% of industrial fires), the source of ignition may be brought from outside the workplace and will be deliberately used. The following are potential sources of ignition in the typical workplace:

▶ *Naked flames* – from smoking materials, cooking appliances, heating appliances and process equipment.
▶ *External sparks* – from grinding metals, welding, impact tools, electrical switch gear.
▶ *Internal sparking* – from electrical equipment (faulty and normal), machinery, lighting.
▶ *Hot surfaces* – from lighting, cooking, heating appliances, process equipment, poorly ventilated equipment, faulty and/or badly lubricated equipment, hot bearings and drive belts.
▶ *Static electricity* – causing significant high-voltage sparks from the separation of materials such as unwinding plastic, pouring highly flammable liquids, walking across insulated floors or removing synthetic overalls.

(c) Sources of fuel

If something will burn, it can be fuel for a fire. The things which will burn easily are the most likely to be the initial fuel, which then burns quickly and spreads the fire to other fuels. The most common things that will burn in a typical workplace are listed below. Note the pictograms given are both the international signs for transporting hazardous goods and those in the new UN Global Harmonised System for packaging, being introduced worldwide. See the annex to the UN GHS for comparisons and allocation from existing signs.

▶ *Solids* – these include: wood, paper, cardboard, wrapping materials, plastics, rubber, foam (e.g. polystyrene tiles and furniture upholstery), textiles (e.g. furnishings and clothing), wallpaper, hardboard and chipboard used as building materials, waste materials (e.g. wood shavings, dust, paper), hair (see Figure 12.5 (a) and (b)).

FLAMMABLE SOLID

Danger
Flammable solid

Figure 12.5 (a) Transport of flammable solid sign; (b) GHS packaging sign

▶ *Liquids* – these include: paint, varnish, thinners, adhesives, petrol, white spirit, methylated spirits, paraffin, toluene, acetone and other chemicals. Most flammable liquids give off vapours which are heavier than air so they will fall to the lowest levels. A flash flame or an explosion can occur if the vapour catches fire in the correct concentrations of vapour and air (see Figure 12.6 (a) and (b)).
▶ *Gases* – flammable gases include: LPG (liquefied petroleum gas in cylinders, usually butane or propane), acetylene (used for welding) and hydrogen. An explosion can occur if the air/gas mixture is within the explosive range (see Figure 12.7 (a) and (b)).

(d) Oxygen

Oxygen is of course provided by the air all around but this can be enhanced by wind, or by natural or powered

12

ventilation systems which will provide additional oxygen to continue burning.

Cylinders providing oxygen for medical purposes or welding can also provide an additional very rich source of oxygen. In addition, some chemicals such as nitrates, chlorates, chromates and peroxides can release oxygen as they burn and therefore need no external source of air (see Figure 12.8 (a) and (b)).

Danger
Highly Flammable Liquid

Figure 12.6 (a) Transport of flammable liquid sign; (b) GHS packaging sign

Danger
May cause or intensify fire
Oxidiser

Figure 12.8 (a) Transport of oxidising agent sign; (b) GHS packaging sign

12.1.5 Classification of fire

Fires are classified in accordance with EN 2:1992 Classification of Fires and ISO 3941: Classification of Fires. There are five main classes of fire – A, B, C, D and F – plus fires involving electrical equipment. The categories based on fuel and the means of extinguishing are as follows:

▶ **Class A** – Fires which involve solid materials such as wood, paper, cardboard, textiles, furniture and plastics where there are normally glowing embers during combustion. Such fires are extinguished by cooling, which is achieved using water.

▶ **Class B** – Fires which involve liquids or liquefied solids such as paints, oils or fats. These can be further subdivided into:

Danger
Extremely Flammable Gas

Figure 12.7 (a) Transport of flammable gas sign; (b) GHS packaging sign

▷ **Class B1** – fires which involve liquids that are soluble in water such as methanol. They can be extinguished by carbon dioxide, dry powder, water spray, light water and vaporising liquids;

▷ **Class B2** – fires which involve liquids not soluble in water, such as petrol and oil. They can be extinguished by using foam, carbon dioxide, dry powder, light water and vaporising liquid.

▶ **Class C** – Fires which involve gases such as natural gas, or liquefied gases such as butane or propane. They can be extinguished using foam or dry powder in conjunction with water to cool any containers involved or nearby.

▶ **Class D** – Fires which involve metals such as aluminium or magnesium. Special dry powder extinguishers are required to extinguish these fires, which may contain powdered graphite or talc.

▶ **Class F** – Fires which involve high-temperature cooking oils or fats in large catering establishments or restaurants.

▶ **Electrical fires** – Fires involving electrical equipment or circuitry do not constitute a fire class on their own, as electricity is a source of ignition that will feed a fire until switched off or isolated. But there are some pieces of equipment that can store, within capacitors, lethal voltages even when isolated. Extinguishers specifically designed for electrical use like carbon dioxide or dry powder units should always be used for this type of fire hazard.

Fire extinguishers are usually designed to tackle one or more classes of fire. This is discussed in Section 12.4.

12.1.6 Principles of heat transmission and fire spread

Fire transmits heat in several ways, which needs to be understood in order to prevent, plan escape from, and fight, fires. Heat can be transmitted by convection, conduction, radiation and direct burning (Figure 12.9).

(a) Convection

Hot air becomes less dense and rises, drawing in cold new air to fuel the fire with more oxygen. The heat is transmitted upwards at sufficient intensity to ignite combustible materials in the path of the very hot products of combustion and flames. This is particularly important inside buildings or other structures where the shape may effectively form a chimney for the fire.

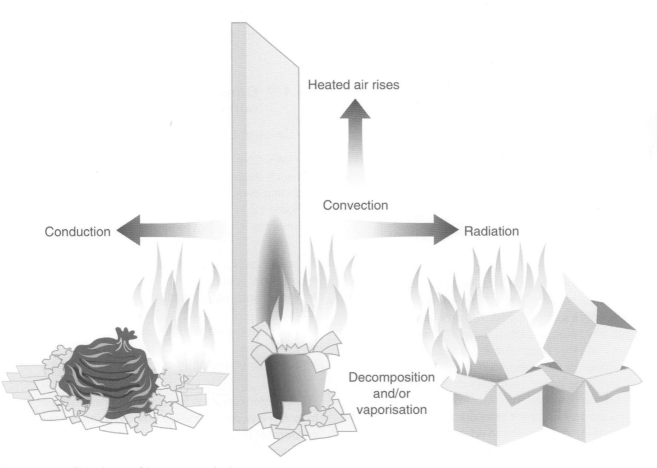

Heated air rises

Conduction

Convection

Radiation

Decomposition and/or vaporisation

Figure 12.9 Principles of heat transmission

12

(b) Conduction

This is the transmission of heat through a material with sufficient intensity to melt or destroy the material and ignite combustible materials which come into contact or close to a hot section. Metals like copper, steel and aluminium are very effective or good conductors of heat. Other materials like concrete, brickwork and insulation materials are very ineffective or poor conductors of heat.

Poor conductors or good insulators are used in fire protection arrangements. When a poor conductor is also incombustible, it is ideal for fire protection. Care is necessary to ensure that there are no other issues, such as health risks, with these materials. Asbestos is a very poor conductor of heat and is incombustible. However, there are very severe health effects which now outweigh its value as a fire protection material and it is banned in the United Kingdom. Unfortunately, asbestos is still found in many buildings where it was used extensively for fire protection; it now has to be managed under legislation in many countries.

(c) Radiation

Often in a fire, the direct transmission of heat through the emission of heat waves from a surface can be so intense that adjacent materials are heated sufficiently to ignite. A metal surface glowing red-hot would be typical of a severe radiation hazard in a fire.

(d) Direct burning

This is the effect of combustible materials catching fire through direct contact with flames which causes fire to spread, in the same way that lighting an open fire, with a range of readily combustible fuels, results in its spread within a grate.

Fire and smoke spread in buildings

Where fire is not contained and people can move away to a safe location, there is little immediate risk to those people. However, where fire is confined inside buildings, the fire behaves differently (Figure 12.10).

The smoke rising from the fire gets trapped inside the space by the immediate ceiling, then spreads horizontally across the space, deepening all the time until the entire space is filled. The smoke will also pass through any holes or gaps in the walls, ceiling or floor and get into other parts of the building. It moves rapidly up staircases or lift wells and into any areas that are left open, or rooms which have open doors connecting to the staircase corridors. The heat from the building gets trapped inside, raising the temperature very rapidly. The toxic smoke and gases are an added danger to people inside the building, who must be able to escape quickly to a safe location.

Figure 12.10 Fire and smoke spread in buildings

12.1.7 Common causes of fire and consequences

(a) Causes of fires

The UK's Communities and Local Government fire statistics show that the causes of fires in buildings, excluding dwellings, in recent years was as shown in Figure 12.11. The total shows a general trend downwards in the last ten years. The majority of fires occurred in:

► private garages and sheds (22%) – 6,700 fires;
► retail distribution (14%) – 4,200 fires;
► restaurants, cafes, public houses, etc. (9%) – 2,600 fires;
► industrial premises (other than construction) (8%) – 2,400 fires;
► recreational and other cultural services (6%) – 1,700 fires.

Further information on fire statistics is available at: **http://www.communities.gov.uk/publications/ corporate/statistics/monitorq1q420091** and Fire Statistics Scotland, see: www.scotland.gov.uk/Topics/ Statistics/Search/ Forthcoming

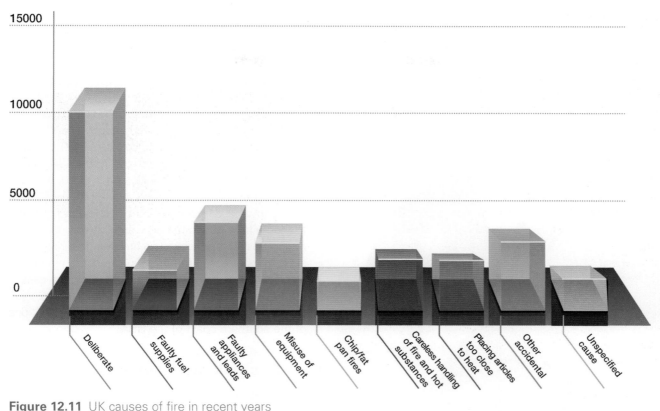

Figure 12.11 UK causes of fire in recent years

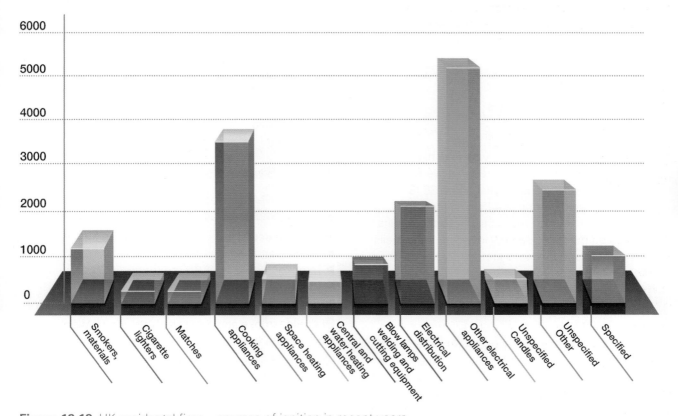

Figure 12.12 UK accidental fires – sources of ignition in recent years

The sources of ignition in the UK are shown in Figure 12.12. Out of approximately 19,000 accidental fires each year, it shows that cooking appliances and electrical equipment account for over 60% of the total.

Figure 12.13 shows the distribution of international fires by place of occurrence as follows:

▶ the structure fire share is about 35%;
▶ other shares by place of occurrence were 2.0–2.5% for chimneys (note that some

339

countries do not provide separate figures for
chimney fires);

▶ 5.2–6.3% for places outside buildings;
▶ 13.8–15.7% for vehicles;
▶ 0.5–2.4% for forests;
▶ 16.4–17.9% for grass and brush;
▶ 8.1–13.4% for outdoor rubbish; and
▶ 11.2–13.1% for all other places.

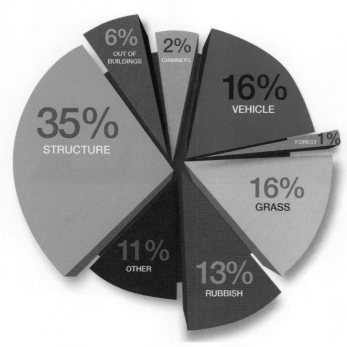

Figure 12.13 Distribution of fires by fire origin

(b) Consequences of fires

The main consequences of fire are:

▶ Death – although this is a very serious risk, relatively
few people died in building fires that were not in
residential dwellings. Each year about 36 (8%)
people die out of a total of 443 in all fires. In
countries with similar rescue services the results are
likely to be much the same. However, if the facility is
very remote and rescue is ineffectual the deaths are
likely to be much higher.

The main causes of these deaths in the UK were:
▷ overcome by gas or smoke – 46%;
▷ burns – 27%;
▷ burns and overcome by gas or smoke – 20%;
▷ other – 7%.

Clearly gas and smoke are the main risks in these
fires.

▶ Personal injury – some 1,282 people were injured,
10% of total injuries in all fires;
▶ Building damage – can be very significant,
particularly if the building materials have poor
resistance to fire and there is little or no built-in fire
protection;
▶ Flora and fauna damage – can be significant,
particularly in a hot draught or forest fire;

▶ Loss of business and jobs – it is estimated that
about 40% of businesses do not start up again after
a significant fire. Many are under- or not insured and
small companies often cannot afford the time and
expense of setting up again when they probably still
have old debts to service;
▶ Transport disruption – rail routes, roads and even
airports are sometimes closed because of a serious
fire. The worst cases were of course 11 September
2001 with the Twin Towers attack in the USA, and
the Icelandic ash cloud over Europe in 2010, when
airports around the world were disrupted;
▶ Environmental damage from the fire and/or fighting
the fire – fire-fighting run-off water, the products
of combustion and exploding building materials,
such as asbestos cement roofs, can contaminate
significant areas around the fire site. Or in the case
of the Deep Water Horizon drilling platform explosion
and fire, the disruption and pollution of large sections
of the southern coast of the USA.

12.2 Fire risk assessment

12.2.1 General

A fire risk assessment will indicate what fire precautions
are needed. There are numerous ways of carrying out a
fire risk assessment: the one described below is based
on the method contained within Fire Safety Guides
published by the UK Department of Communities
and Local Government (see Appendix 12.1). A
systematic approach, considered in five simple stages,
is generally the best practical method. See also the
European Guideline, Introduction to qualitative fire risk
assessment.

12.2.2 Stage 1 – identify fire hazards

There are five main hazards produced by fire that should
be considered when assessing the level of risk:

▶ oxygen depletion;
▶ flames and heat;
▶ smoke;
▶ gaseous combustion products;
▶ structural failure of buildings.

Of these, smoke and other gaseous combustion
products are the most common cause of death in fires.

For a fire to occur, it needs sources of heat and fuel. If
these hazards can be kept apart, removed or reduced,
then the risks to people and businesses are minimised.
Identifying fire hazards in the workplace is the first
stage as follows.

(a) Identify any combustibles

Most workplaces contain combustible materials.
Usually, the presence of normal stock in trade should
not cause concern, provided the materials are used

safely and stored away from sources of ignition. Good standards of housekeeping are essential to minimise the risk of a fire starting or spreading quickly.

The amount of combustible material in a workplace should be kept as low as is reasonably practicable. Materials should not be stored in gangways, corridors or stairways or where they may obstruct exit doors and routes. Fires often start and are assisted to spread by combustible waste in the workplace. Such waste should be collected frequently and removed from the workplace, particularly where processes create large quantities of it.

Some combustible materials, such as flammable liquids, gases or plastic foams, ignite more readily than others and quickly produce large quantities of heat and/or dense toxic smoke. Ideally, such materials should be stored away from the workplace or in fire-resisting stores. The quantity of these materials kept or used in the workplace should be as small as possible, normally no more than half a day's supply.

(b) Identify any sources of heat

All workplaces will contain heat/ignition sources; some will be obvious such as cooking sources, heaters, boilers, engines, smoking materials or heat from processes, whether in normal use or through carelessness or accidental failure. Others may be less obvious such as heat from chemical processes or electrical circuits and equipment.

Where possible, sources of ignition should be removed from the workplace or replaced with safer forms. Where this cannot be done, the ignition source should be kept well away from combustible materials or made the subject of management controls.

Particular care should be taken in areas where portable heaters are used or where smoking is permitted (now banned inside many EU commercial premises). Where heat is used as part of a process, it should be used carefully to reduce the chance of a fire as much as possible. Good security both inside and outside the workplace will help to combat the risk of arson.

Under smoke-free legislation in Europe, smoking is not permitted in enclosed or significantly enclosed areas. Outside designated safe areas should be provided for those who still require to smoke. The smoking rules should be rigorously enforced.

Demolition work can involve a high risk of fire and explosion. In particular:

▶ Dismantling tank structures can cause the ignition of flammable residues. This is especially dangerous if hot methods are used to dismantle tanks before residues are thoroughly cleaned out. The work should only be done by specialists.

▶ Disruption and ignition of buried gas and electrical services is a common problem. It should always be

assumed that buried services are present unless it is positively confirmed that the area is clear. A survey using service detection equipment must be carried out by a competent person to identify any services. The services should then be marked, competently purged or made dead, before any further work is done. A permit to excavate or dig is the normal formal procedure to cover buried services.

(c) Identify any unsafe acts

Persons undertaking unsafe acts such as smoking next to combustible materials, etc.

(d) Identify any unsafe conditions

These are hazards that may assist a fire to spread in the workplace, for example if there are large areas of hardboard or polystyrene tiles etc., or open stairs that can enable a fire to spread quickly, trapping people and engulfing the whole building.

An ideal method of identifying and recording these hazards is by means of a simple single-line plan, an example of which is illustrated in Figure 12.14. Checklists may also be used. See Chapter 18.

12.2.3 Stage 2 – identify persons who are at significant risk

Consider the risk to any people who may be present. In many instances, and particularly for most small workplaces, the risk(s) identified will not be significant, and specific measures for persons in this category will not be required. There will, however, be some occasions when certain people may be especially at risk from fire, because of their specific role, disability, sleeping, location or the workplace activity (see Section 12.5.3 for more information). Special consideration is needed if:

▶ sleeping accommodation is provided;
▶ large numbers of the public may be present;
▶ people may be unfamiliar with the layout of the building and the location of the exit routes;
▶ staff are working in areas where there is a specific risk, such as paint spraying;
▶ people may have lengthy or tortuous escape routes;
▶ contractors are working up ladders or on scaffolds;
▶ people are physically, visually or mentally challenged;
▶ people are unable to react quickly;
▶ people are isolated.

People, such as visitors, the public or other workers, may come into the workplace from outside. The assessor must decide whether the current arrangements are satisfactory or if changes are needed.

Because fire is a dynamic event, which, if unchecked, will spread throughout the workplace, all people present will eventually be at risk if fire occurs. Where people are at risk, adequate means of escape from fire should

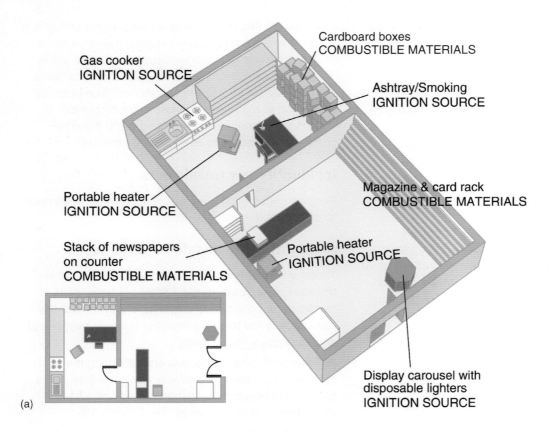

Gas cooker
IGNITION SOURCE

Cardboard boxes
COMBUSTIBLE MATERIALS

Ashtray/Smoking
IGNITION SOURCE

Portable heater
IGNITION SOURCE

Magazine & card rack
COMBUSTIBLE MATERIALS

Stack of newspapers
on counter
COMBUSTIBLE MATERIALS

Portable heater
IGNITION SOURCE

Display carousel with
disposable lighters
IGNITION SOURCE

(a)

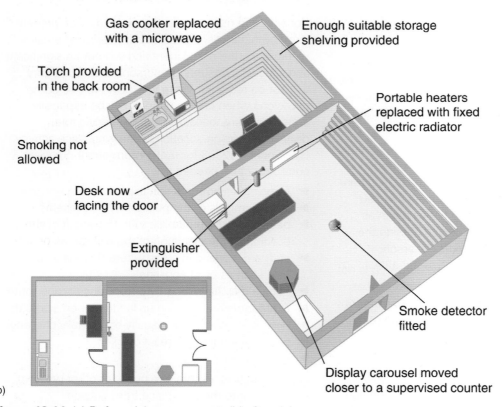

Gas cooker replaced
with a microwave

Enough suitable storage
shelving provided

Torch provided
in the back room

Portable heaters
replaced with fixed
electric radiator

Smoking not
allowed

Desk now
facing the door

Extinguisher
provided

Smoke detector
fitted

Display carousel moved
closer to a supervised counter

(b)

Figure 12.14 (a) Before risk assessment; (b) after risk assessment

be provided together with arrangements for detecting and giving warning of fire. Fire-fighting equipment suitable for the hazards in the workplace should be provided.

Some people may be at significant risk because they work in areas where fire is more likely or where rapid fire growth can be anticipated. Where possible, the hazards creating the high level of risk should be reduced. Specific steps should be taken to ensure that people affected are made aware of the danger and the action they should take to ensure their safety and the safety of others.

12.2.4 Stage 3 – evaluate and reduce the risks

If the building has been built and maintained in accordance with Building Regulations and is being put to its designed use, it is likely that the means of escape provisions will either be adequate, or it will be easy to decide what is required in relation to the risk. Having identified the hazards and the persons at risk, the next stage is to reduce the chance of a fire occurring and spreading, thereby minimising the chance of harm to persons in the workplace. The principles of prevention are based on EC Directive requirements (see Section 4.3 for further details).

Evaluate the risks

Attempt to classify each area as 'high', 'normal' or 'low risk'. If 'high risk', it may be necessary to reconsider the principles of prevention, otherwise additional compensatory measures will be required.

▶ *Low risk* – Areas where there is minimal risk to persons' lives; where the risk of fire occurring is low; or the potential for fire, heat and smoke spreading is negligible and people would have plenty of time to react to an alert of fire.
▶ *Normal risk* – Such areas will account for nearly all parts of most workplaces; where an outbreak of fire is likely to remain confined or spread slowly, with an effective fire warning allowing persons to escape to a place of safety.
▶ *High risk* – Areas where the available time needed to evacuate the area is reduced by the speed of development of a fire, for example highly flammable or explosive materials stored or used (other than small quantities under controlled conditions); also where the reaction time to the fire alarm is slower because of the type of person present or the activity in the workplace, for example the infirm and elderly or persons sleeping on the premises.

Determine if the existing arrangements are adequate, or need improvement. Matters that will have to be considered are:

▶ Means for detecting and giving warning in case of fire – can it be heard by all occupants? (see Section 12.4.)
▶ Means of escape – are they adequate in size, number, location, well lit, unobstructed, safe to use, etc.? (see Section 12.5.)
▶ Signs – for exits, fire routines, etc. (see Section 12.5 and Section 4.2 in Chapter 4.)
▶ Fire-fighting equipment – wall-mounted or in a cradle on fire exit routes, suitable types for hazards present and sufficient in number? (see Section 12.4.)

12.2.5 Stage 4 – the findings (always recommended, see Stage 5 – review)

The findings of the assessment and the actions (including maintenance) arising from it should be recorded. It is suggested that if five or more people are employed a formal record of the significant findings and any measures proposed to deal with them must be recorded as this is the European standard and is reasonable. The record should indicate:

▶ the date the assessment was made;
▶ the hazards identified;
▶ any staff and other people especially at risk;
▶ what action needs to be taken, and by when (the *action plan*);
▶ the conclusions arising.

The above guidelines are to be used with caution. Each part of the workplace must be looked at and a decision made on how quickly persons would react to an alert of fire in each area. Adequate safety measures will be required if persons are identified as being at risk. Where maximum travel distances (see Table 12.5 in Section 12.5) cannot be achieved, extra fire safety precautions will be needed.

Where persons are at risk or an unacceptable hazard still exists, additional fire safety precautions will be required to compensate for this, or alternatively repeat previous stages to manage risk to an acceptable level.

12.2.6 Stage 5 – monitor and review on a regular basis

The fire risk assessment is not a one-off procedure. It should be continually monitored to ensure that the existing fire safety arrangements and fire risk assessment remain realistic. The assessment should be reviewed if there is a significant change in the occupancy, work activity, the materials used or stored when building works are proposed, or when it is no longer thought to be valid. Use the fire safety maintenance checklist in Chapter 18 to check on fire safety standards.

12.2.7 Structural features

The workplace may contain features that could promote the rapid spread of fire, heat or smoke and affect escape routes. These features may include ducts or flues, openings in floors or walls, or combustible wall or ceiling linings. Where people are put at risk from these features, appropriate steps should be taken to reduce the potential for rapid fire spread by, for example, non-combustible automatic dampers fitted in ducts or to provide an early warning of fire so that people can leave the workplace before their escape routes become unusable.

12

Combustible wall or ceiling linings should not be used on escape routes and large areas should be removed wherever they are found. Other holes in fire-resisting floors, walls or ceilings should be filled in with fire-resisting material to prevent the passage of smoke, heat and flames.

12.2.8 Temporary workplaces, maintenance and refurbishment

Temporary workplaces such as construction sites, temporary buildings, festivals and fêtes all have requirements for fire precautions and means of escape in case of fire. The scale of the temporary workplace will dictate the requirements which will depend on:

▶ the number of persons working or visiting the site at any one time;
▶ the nature of the materials being used to construct the workplace or being used in the workplace. Are they flammable or highly flammable?
▶ height above or below the ground floor and how far it is to a place of safety;
▶ the location, whether in a remote area or close to water supplies and/or fire and rescue services;
▶ the size of the premises and whether audible warnings can be heard.

Risk assessments will be needed to determine the level of precautions necessary. Guidance has been given in Fire Safety in Construction 2nd Edition HSG 168, HSE Books, downloadable at: http://www.hse.gov.uk/pubns/books/hsg168.htm and European Guideline No. 4 Introduction to Qualitative fire risk assessment.

Sources of heat or combustible materials may be introduced into the workplace during periods of maintenance or refurbishment. Where the work involves the introduction of heat, such as welding, this should be carefully controlled by a safe system of work, for example Hot Work Permit (see Chapter 4 for details). All materials brought into the workplace in connection with the work being carried out should be stored away from sources of heat and not obstruct exit routes.

12.2.9 Fire evacuation plans

Fire evacuation plans should be produced and attached to the fire risk assessment. A copy should be posted in the workplace. A single-line plan of the area or floor should be produced or an existing plan should be used which needs to show:

▶ escape routes, number of exits, number of stairs, fire-resisting doors, fire-resisting walls and partitions, places of safety, and the like;
▶ fire safety signs and notices including pictorial fire exit signs and fire action notices;
▶ the location of fire warning call points and sounders or rotary gongs;

▶ the location of emergency lights;
▶ the location and type of fire-fighting equipment.

See Figure 12.28 and Section 12.5 for more details.

12.3 Fire prevention and prevention of fire spread

This section provides further information on controlling the risk of a fire and its prevention in premises. The following should be considered:

▶ good housekeeping;
▶ control of flammable and combustible materials;
▶ control of ignition sources;
▶ systems of work.

12.3.1 Control measures

(a) Housekeeping

Good housekeeping will lower the chances of a fire starting, so the accumulation of combustible materials in all premises should be monitored carefully. Good housekeeping is essential to reduce the chances of escape routes and fire doors being blocked or obstructed.

Waste material should be kept in suitable containers before it is removed from the premises. If bins, particularly wheeled bins, are used outside, secure them in a compound to prevent them being moved to a position next to the building and set on fire. Never place skips against a building – they should normally be a minimum of 6 m away from any part of the premises.

If considerable quantities of combustible waste material are generated then a formal plan to manage this effectively needs to be developed.

In higher risk areas special arrangements should be in place for close down, e.g. checking all appliances are turned off and combustible waste has been removed.

(b) Storage

Many of the materials found in premises will be combustible which will increase the fire risk. Combustible materials are not just those generally regarded as highly combustible, such as polystyrene, but all materials that will readily catch fire. Even non-combustible materials may present a fire hazard when packed in combustible materials.

The absence of adequate storage arrangements results in congestion on the factory floor, warehouse or shop. In offices, the retention of large quantities of paper records, especially if not filed away in proprietary cabinets, can increase the fire hazard. Such readily available flammable material makes the potential effect of arson more serious. Many shops or hotels will take great care to present an efficient and attractive image in the retail or public areas, while other areas are neglected and allowed to become over-stocked or

Figure 12.15 Partly blocked fire exit door

dumping areas for unsold material, old furniture and the like (see Figure 12.15).

This may lead to a concealed fire and restriction of access to the fire, fire extinguishers, alarm points and escape routes. Discarded packaging materials, e.g. polystyrene and cardboard, and even piles of wooden pallets, can introduce severe fire hazards. Poorly managed storage areas often become over-stocked or dumping areas for unwanted material. Do not pile combustible material against electrical equipment or heaters, even if turned off for the summer.

To reduce the risk, store excess materials and stock in a dedicated storage area, storeroom or cupboard. Do not store excess stock in areas where the public would normally have access.

Consider how stock is displayed in shops and evaluate any additional risk of fire that it generates. For example, rugs stacked on the floor on top of each other would not present a high fire risk, but rolls of carpet stored vertically up against a wall or hung on displays present a vertical surface for fire to spread rapidly upwards. The display of large quantities of clothing on vertical hangers is also likely to increase the risk of rapid fire development.

The fire risk assessment should also consider any additional risk generated by seasonal products such as fireworks and Christmas decorations.

Consider the following to reduce these risks:

▶ ensure storage and display areas are adequately controlled and monitored;
▶ use fire-resistant display materials wherever possible (suppliers should be able to provide evidence of this); and
▶ ensure electrical lighting used as part of the display does not become a potential source of ignition.

Voids (including roof voids) should not be used for the storage of combustible material. Such voids should be sealed off or kept entirely open to allow for easy access for inspection and the removal of combustible materials.

(c) Equipment and machinery

Common causes of fire in equipment are:

▶ allowing ventilation points to become clogged or blocked, causing overheating;
▶ inadequate cleaning of heat-shrink packaging equipment, such as that used in in-store bakeries;
▶ allowing extraction equipment in catering environments to build up excessive grease deposits;
▶ misuse or lack of maintenance of cooking equipment and appliances; and
▶ disabling or interfering with automatic or manual safety features and cut-outs.

All machinery, apparatus and office equipment should be properly maintained by a competent person. Appropriate signs and instructions on safe use may be necessary.

(d) Fork-lift trucks and other vehicles

There are hazards associated with industrial vehicles, particularly during refuelling and maintenance operations; also when stored or in use. Battery charging of fork-lift trucks can give rise to sparks and hydrogen (a gas that is highly flammable, explosive and is lighter than air). Sparks can occur when connecting and disconnecting power supplies.

Fork-lift truck charging points should be carefully sited in a well-ventilated area (ideally direct to open air), clear of ignition sources and preferably in a separate dedicated non-combustible structure. However, if sited in the building, the charging point should be against a fire-resisting wall (e.g. 30-minute fire resistance).

(e) Heating

Individual heating appliances require particular care if they are to be used safely, particularly those which are kept for emergency use during a power cut or as supplementary heating during severe weather. The greatest risks arise from lack of maintenance and staff unfamiliarity with them. Heaters should preferably be secured in position when in use and fitted with a fire guard if appropriate.

As a general rule, convector or fan heaters should be preferred to radiant heaters because they present a lower risk of fire and injury.

The following rules should be observed:

▶ all heaters should be kept well clear of combustible materials and where they do not cause an obstruction;
▶ heaters which burn a fuel should be sited away from draughts;
▶ they should only be used in places where there is an adequate supply of fresh air and vents are not covered up;
▶ portable fuel burning heaters (including bottled gas (LPG)) should only be used in exceptional

circumstances and if shown to be acceptable in your risk assessment.

All gas heating appliances should be used only in accordance with manufacturer's instructions and should be serviced annually by a competent person.

In general, staff should be discouraged from bringing in their own portable heaters and other electrical equipment (e.g. kettles) into the premises.

(f) Cooking processes

Typical installations used in cooking processes include deep fat fryers, ovens, grills, surface cookers, ductwork, flues, filters, hoods, extract and ventilation ducts and dampers. These cooking processes can operate at high temperatures, involving large quantities of oil and combustible food stuffs. Heat sources used for cooking processes include: gas, electric and microwave. The main cause of fire are ignition of cooking oil, combustion of crumbs and sediment deposits, and ductwork fires from a build-up of fats and grease. The siting of cooking processes close to insulated wall panels with combustible insulation can lead to the likely ignition of the panels and consequent rapid fire spread to other parts of the building. This practice should therefore be avoided.

The following should be considered to reduce the risk from cooking processes:

► regular cleaning to prevent build-up of crumbs and other combustible material;
► fire-resisting containers for waste products;
► a fire suppression system capable of controlling an outbreak of fire;
► monitored heat/oil levels, even after the cooking process is complete, and installation of temperature control/cut-off/shut-off devices as appropriate;
► duct, joints and supports able to withstand high cooking temperatures;
► separation from wall and ceiling panels (with combustible insulation), e.g. 2.5 m for walls, 4 m for ceilings;
► insulation of ducts to prevent heating/ignition of nearby combustible wall and ceiling materials;
► a regular programme for inspection and cleaning;
► a programme of electrical and mechanical maintenance; and
► annual service of all gas heating appliances by a competent person.

(g) Smoking

Carelessly discarded cigarettes and other smoking materials are still a major cause of fire. A cigarette can smoulder for several hours, especially when surrounded by combustible material.

Many fires are started several hours after the smoking materials have been emptied into waste bags and left for future disposal.

The smoke-free legislation in many EU countries, concerning enclosed commercial premises, has greatly reduced the risk of fire inside buildings. However, the risks can still exist in outside areas and designated smoking shelters. Display the smoke-free or no-smoking signs throughout the premises.

In those areas where smoking is permitted, provide deep and substantial metal ashtrays to help prevent unsuitable containers being used. Empty all ashtrays daily into a metal waste bin and keep it outside. It is dangerous to empty ashtrays into plastic waste sacks which are then left inside for disposal later.

(h) Electrical safety

Electrical equipment can be a significant cause of accidental fires in premises. The main causes are:

► overheating cables and equipment, e.g. due to overloading circuits, bunched or coiled cables or impaired cooling fans;
► incorrect installation or use of equipment;
► little or no maintenance and testing of equipment;
► incorrect fuse ratings;
► damaged or inadequate insulation on cables or wiring;
► combustible materials being placed too close to electrical equipment which may give off heat even when operating normally or may become hot due to a fault;
► arcing or sparking by electrical equipment; and
► embrittlement and cracking of cable sheathing in cold environments.

All electrical equipment should be installed and maintained in a safe manner by a competent person. If portable electrical equipment is used, including items brought into a workplace by staff, then the fire risk assessment should ensure that it is visually inspected and undergoes portable appliance testing (PAT) at intervals suitable for the type of equipment and its frequency of use (see Chapter 11). If there is any doubt about the safety of the electrical installation then a competent electrician should be consulted.

Issues to consider include:

► overloading of equipment;
► correct fuse ratings;
► PAT testing and testing of fixed installations;
► protection against overloading of installation;
► protection against short circuit;
► insulation, earthing (or grounding) and electrical isolation requirements;
► frequency of electrical inspection and test;
► temperature rating and mechanical strength of flexible cables;
► portable electrical equipment;
► physical environment in which the equipment is used (e.g. wet or dusty atmospheres); and

▶ suitable use and maintenance of personal protective equipment.

See Chapter 11 for more information.

(i) Systems of work

Safe systems of work have an important part to play in preventing fires. This is particularly important in high hazard areas where dangerous substances are being used. Systems of work may involve a hot work permit where, for example, welding, cutting or burning are taking place and/or a method statement from a contractor stating how hot roof work using heated bitumen will take place. See Chapter 4 for more details.

12.3.2 Safe storage and use of flammable liquids and gases

(a) Introduction

The ILO Code of Practice on the safety in the use of chemicals at work covers all properties of a chemical including fire, explosion and toxic hazards. However, flammable liquids and gases are often called Dangerous Substances rather than Hazardous Substances which involves a health hazard. Dangerous substances include any substance or preparation which, because of its properties or the way it is used, could be harmful because of fires and explosions. The list includes petrol, LPG (Liquefied Petroleum Gas), paints, varnishes, solvents and some dusts. These are dusts which, when mixed with air, can cause an explosive atmosphere. Dusts from milling and sanding operations are examples

of this. Most workplaces contain a certain amount of dangerous substances. The NEBOSH International Syllabus does not include flammable gases but they have been included here as LPG is very commonly used throughout the world.

An explosive atmosphere is an accumulation of gas, mist, dust or vapour, mixed with air, which has the potential to catch fire or explode. Although an explosive atmosphere does not always result in an explosion (detonation), if it catches fire, flames can quickly travel through the workplace. In a confined space (e.g. in plant or equipment) the rapid spread of the flame front or rise in pressure can itself cause an explosion and rupture of the plant and/or building.

The ILO Code of Practice 'Safety in the Use of Chemicals at Work' addresses the issue of control measures for flammable, dangerously reactive or explosive chemicals at paragraph 6.6 as in Box 12.1.

It is necessary to carry out the following to comply with this code and good practice standards:

▶ carry out a risk assessment of any work activities involving dangerous substances;
▶ provide a way of eliminating or reducing risks as far as is reasonably practicable;
▶ provide procedures and equipment to deal with accidents and emergencies;
▶ provide training and information for employees;
▶ classify places where explosive atmospheres may occur into zones and mark the zones where necessary.

Box 12.1 Extract from ILO Code of Practice 'Safety in the Use of Chemicals at Work'

6.6.1. Workers should be protected against risks of injury resulting from the use of *flammable, unstable or explosive chemicals.* A combination of the following measures should be used to reduce the risk of a fire or explosion.

(a) Good design and installation practice:

In addition to the fundamental principles in paragraph 6.5.2 (a) (good design) which should be applied to eliminating flammable vapours, fumes or dusts liable to be given off, the following practices should also be observed where appropriate:

(i) elimination or control of sources of ignition;

(ii) separation of processes that use flammable chemicals from:
  ▷ other processes;
  ▷ bulk storage of the flammable chemicals or bulk storage which may cause a hazard in the event of fire;
  ▷ the boundary and premises off site, which are not under the control of the employer; and
  ▷ fixed sources of ignition;

(iii) provision of an inert atmosphere for totally enclosed processes and handling systems;

(iv) provision of means of fire detection and alarm which, as far as is practicable, should include automatic means of extinguishing incipient fires;

(v) installation of means for detecting increases in pressure and the automatic operation of a gas suppressor to prevent an explosion, e.g. for dust explosions;

(b) Safe work systems and practices:

(i) use and proper maintenance of the engineering control measures provided;

(ii) minimisation of the quantities of chemicals kept in the workplace;

12

> (iii) minimisation of the quantities of chemicals handled and used in buildings;
> (iv) separation of arrangements for storing chemicals from normal process activities;
> (v) separation of incompatible chemicals;
> (vi) reduction of the numbers of workers exposed and exclusion of non-essential access;
> (vii) arrangements for spillages to be cleared up immediately;
> (viii) arrangements for the safe disposal of chemicals;
> (ix) ensuring that appropriate equipment is provided, e.g. non-sparking tools for low-incendive materials in specified situations;
> (x) use of appropriate signs and notices;
>
> (c) Personal protection:
> (i) ensuring that where personal protective equipment and general work clothing are provided, they are not liable to increase the possibility of serious burns. Certain synthetic materials may melt in a fire and thereby cause more serious burns;
> (ii) making adequate preparations for an emergency.
>
> **6.6.2.** The adequacy of the means of escape, fire-fighting arrangements, the fire alarm system and provisions for the evacuation of the premises should be considered, following the assessment of chemicals that may be flammable, unstable or explosive.
>
> **6.7.** Control measures for the storage of hazardous chemicals.
>
> **6.7.1.** Hazardous chemicals should be stored under conditions specific to their inherent properties and characteristics to ensure safety and in accordance with established criteria. Chemicals with typical properties and characteristics that are relevant include:
> (a) flammable liquids;
> (b) flammable gases;
> (c) toxic chemicals;
> (d) corrosive chemicals;
> (e) chemicals that emit highly toxic fumes in the event of a fire;
> (f) chemicals which, in contact with water, give off flammable gas;
> (g) oxidising chemicals;
> (h) explosives;
> (i) unstable chemicals;
> (j) flammable solids;
> (k) compressed gases.
>
> Section 7 of the code deals with the design and installation of plant and equipment to minimise the risk of chemicals (including flammable chemicals) at work.

(b) Risk assessment

This is the process of identifying and carefully examining the dangerous substances present or likely to be present in the workplace, the work activities involving them and how they might fail and cause fire, explosion and similar events that could harm employees and the public. The purpose of a risk assessment is to enable the employer to decide what needs to be done to eliminate or reduce the safety risks from dangerous substances as far as is reasonably practicable. It should take into account the following:

▶ what dangerous properties the substances have;
▶ the way they are used and stored;
▶ the possibility of dangerous explosive atmospheres occurring;
▶ any potential ignition sources.

Regardless of the quantity of dangerous substance present, the employer should carry out a risk assessment. This will enable them to decide whether existing measures are sufficient or whether they need to make any additional controls or precautions. Non-routine activities need to be assessed as well as the normal activities within the workplace. For example in maintenance work there is often a higher potential for fire and explosion incidents to occur.

Employers should ensure that the safety risks from dangerous substances are eliminated or, when this is not reasonably practicable, to take measures to control risks *and* to reduce the harmful effects of any fire, explosion or similar events, so far as is reasonably practicable.

(c) Substitution

Substitution is the best solution. It is much better to replace a dangerous substance with a substance or process that totally eliminates the risk. In practice this is difficult to achieve; so it is more likely that the dangerous substance will be replaced with one that is less hazardous (e.g. by replacing a low-flashpoint solvent with a high-flashpoint one).

Designing the process so that it is less dangerous is an alternative solution. For example, a change could be made from a batch production to a continuous production process, or the manner or sequence in which the dangerous substance is added could be altered. However, care must be taken when carrying out these steps to make sure that no other new safety or health risks are created or increased, as this would outweigh the improvements implemented as a result of this assessment.

The fact is that where a dangerous substance is handled or stored for use as a fuel, there is often no scope to eliminate it and very little chance to reduce the quantities handled.

Where risk cannot be entirely eliminated, control and mitigation measures should be applied. This should reduce risk as follows.

(d) Order of priority for control measures

Control measures should be applied in the following order of priority:

▶ reduce the amount of dangerous substances to a minimum;
▶ avoid or minimise releases;
▶ control releases at source;
▶ prevent the formation of an explosive atmosphere;
▶ use a method such as ventilation to collect, contain and remove any releases to a safe place;
▶ avoid ignition sources;
▶ avoid adverse conditions (e.g. exceeding the limits of temperature or other control settings) that could lead to danger;
▶ keep incompatible substances apart.

(e) Control measures

Choose control measures which are consistent with the risk assessment and appropriate to the nature of the activity or operation. These can include:

▶ preventing fires and explosions from spreading to other plant and equipment or to other parts of the workplace;
▶ making sure that a minimum number of employees is exposed;
▶ in the case of a process plant, providing plant and equipment that can safely contain or suppress an explosion, or vent it to a safe place.

In workplaces where explosive atmospheres may occur, ensure that:

▶ areas where hazardous explosive atmospheres may occur are classified into zones based on their likelihood and persistence;
▶ areas classified into zones are protected from sources of ignition by selecting suitable special equipment and protective systems;

▶ where necessary, areas classified into zones are marked with a specified 'EX' sign at their points of entry;
▶ employees working in zoned areas must be provided with appropriate clothing that does not create a risk of an electrostatic discharge igniting the explosive atmosphere;
▶ before coming into operation for the first time, areas where hazardous explosive atmosphere may be present are confirmed as being safe (verified) by a person (or organisation) competent in the field of explosion protection. The person carrying out the verification must be competent to consider the particular risks at the workplace and the adequacy of control and other measures put in place.

(f) Storage

Dangerous substances should be kept in a safe place in a separate building or the open air. Only small quantities of dangerous substances should be kept in a workroom or area as follows:

▶ For flammable liquids that have a flashpoint above the maximum ambient temperature (normally taken as 32°C), the small quantity that may be stored in the workroom is considered to be an amount up to 250 litres.
▶ For extremely and highly flammable liquids and those flammable liquids with a flashpoint below the maximum ambient temperature, the small quantity is considered to be up to 50 litres and this should be held in a special metal cupboard or container (Figure 12.16).
▶ Any larger amounts should be kept in a special fire-resisting store, which should be:
  ▷ properly ventilated;
  ▷ provided with spillage retaining arrangements such as sills;
  ▷ free of sources of ignition, such as unprotected electrical equipment, sources of static electrical sparks, naked flames or smoking materials;
  ▷ arranged so that incompatible chemicals do not become mixed together either in normal use or in a fire situation;
  ▷ of fire-resisting construction;
  ▷ used for empty as well as full containers – all containers must be kept closed;
  ▷ kept clear of combustible materials such as cardboard or foam plastic packaging materials.

(g) Flammable gases

Flammable gas cylinders also need to be stored and used safely. The following guidance should be adopted:

▶ both full and empty cylinders should be stored outside. They should be kept in a separate secure compound at ground level with sufficient ventilation. Open mesh is preferable;

12

Figure 12.16 (a)–(c) Various storage arrangements for highly flammable liquids

▶ valves should be uppermost during storage to retain them in the vapour phase of the LPG;
▶ cylinders must be protected from mechanical damage. Unstable cylinders should be together, for example, and cylinders must be protected from the heat of the summer sun;
▶ the correct fittings must be used. These include hoses, couplers, clamps and regulators;

▶ gas valves must be turned off after use at the end of the shift;
▶ precautions must be taken to avoid welding flame 'flash back' into the hoses or cylinders. People need training in the proper lighting up and safe systems of work procedures; non-return valves and flame arrestors also need to be fitted;
▶ cylinders must be changed in a well-ventilated area remote from any sources of ignition;
▶ joints should be tested for gas leaks using soapy/detergent water – never use a flame;
▶ flammable material must be removed or protected before welding or similar work;
▶ cylinders should be positioned outside buildings with gas piped through in fixed metal piping;
▶ both high and low ventilation must be maintained where LPG applications are being used;
▶ flame failure devices are necessary to shut off the gas supply in the event of flame failure.

(h) Aerosols

Some aerosols can contain flammable products stored at pressure and they can present a high level of hazard. When ignited they can explode, produce fireballs and rocket to distances of 40 m. Their presence in premises can make it unsafe for fire-fighters to enter a building and they have the potential for starting multiple fires.

The following should be considered to reduce these risks:

▶ All staff involved in the movement, storage and display of aerosol cans should be adequately instructed, trained and supervised.
▶ Damaged and leaking aerosol cans should be removed immediately to a safe, secure, well-ventilated place prior to disposal.
▶ Powered vehicles should not be used to move damaged stock, unless specially adapted for use in flammable atmospheres.
▶ Arrangements should be made for disposal at a licensed or recognised waste management site.

12.3.3 Principles of fire protection in buildings

(a) General

The design of all new buildings and the design of extensions or modifications to existing buildings often need to be approved by local building control authorities.

Design data for new and modified buildings should be retained throughout the life of the structure.

Building legal standards are concerned mainly with safety of life. Therefore, it is necessary to consider the early stage of a fire and how it affects the means of escape, and, also, aim to prevent eventual spread to other buildings.

Asset or building protection measures require extra precautions that will have an effect at both early and later stages of the fire growth by controlling fire spread through and between buildings and preventing structural collapse. However, this extra fire protection will also improve life safety not only for those escaping at the early stages of the fire but also for fire-fighters who will subsequently enter.

If a building is carefully designed and suitable materials are used to build it and maintain it, then the risk of injury or damage from fire can be substantially reduced. Three objectives must be met:

▶ it must be possible for everyone to leave the building quickly and safely;
▶ the building must remain standing for as long as possible;
▶ the spread of fire and smoke must be reduced.

These objectives can be met through the proper selection of materials and good design of buildings.

(b) Fire loading

The fire load of a building is used to classify types of building use. It may be calculated simply by multiplying the weight of all combustible materials by their energy values and dividing by the floor area under consideration. The higher the fire load, the more the effort needed to offset this by building to higher standards of fire resistance.

(c) Surface spread of fire

Combustible materials, when present in a building as large continuous areas, such as for lining walls and ceilings, readily ignite and contribute to spread of fire over their surfaces. This can represent a risk to life in buildings, particularly where walls of fire-escape routes and stairways are lined with materials of this nature.

Materials are tested by insurance bodies and fire research establishments. The purpose of the test is to classify materials according to the tendency for flame to spread over their surfaces. As with all standardised test methods, care must be taken when applying test results to real applications.

In the UK, a material is classified as having a surface in one of the following categories:

▶ Class 1 – Surface of very low flame spread;
▶ Class 2 – Surface of low flame spread;
▶ Class 3 – Surface of medium flame spread;
▶ Class 4 – Surface of rapid flame spread.

The test shows how a material would behave in the initial stages of a fire.

As all materials tested are combustible, in a serious fire they would burn or be consumed. Therefore, there is an additional Class 0 of materials which are non-combustible throughout or, under specified conditions, non-combustible on one face and combustible on the

other. The spread of flame rating of the combined Class 0 product must not be worse than that for Class 1.

Internal partitions of walls and ceilings should be Class 0 materials wherever possible and must not exceed Class 1.

The UK Building Regulations use three of these classes in their Approved Document for lining materials as follows:

Class 0: Materials suitable for circulation spaces and escape routes
Such materials include brickwork, blockwork, concrete, ceramic tiles, plaster finishes (including rendering on wood or metal lathes), wood-wool cement slabs and mineral fibre tiles or sheets with cement or resin binding.

Note: Additional finishes to these surfaces may be detrimental to the fire performance of the surface and if there is any doubt about this then consult the manufacturer of the finish.

Class 1: Materials suitable for use in all rooms but not on escape routes
Such materials include all the Class 0 materials referred to above. Additionally, timber, hardboard, blockboard, particle board, heavy flock wallpapers and thermosetting plastics will be suitable if flame-retardant treated to achieve a Class 1 standard.

Class 3: Materials suitable for use in rooms of less than 30 m²
Such materials include all those referred to in Class 1, including those that have not been flame-retardant treated, and certain dense timber or plywood and standard glass-reinforced polyesters.

The equivalent European classification standard would also be acceptable. Appropriate UK testing procedures are detailed in BS 476-7 and where appropriate BS EN 13501-1.

(d) Fire resistance of structural elements

If structural elements such as walls, floors, beams, columns and doors are to provide effective barriers to fire spread and to contribute to the stability of a building, they should be of a required standard of fire resistance.

In the UK, tests for fire resistance are made on elements of structure, full size if possible, or on a representative portion having minimum dimensions of 3 m long for columns and beams and 1 m² for walls and floors. All elements are exposed to the same standard fire provided by furnaces in which the temperature increases with time at a set rate. The conditions of exposure are appropriate to the element tested. Free-standing columns are subjected to heat all round, and walls and floors are exposed to heat on one side only.

Elements of structure are graded by the length of time they continue to meet three criteria:

▶ the element must not collapse;
▶ the element must not develop cracks through which flames or hot gases can pass;
▶ the element must have enough resistance to the passage of heat so the temperature of the unexposed face does not rise by more than a prescribed amount.

The term fire resistance has a precise meaning. It should not be applied to such properties of materials as resistance to ignition or resistance to flame propagation. For example steel has a high resistance to ignition and flame propagation but will distort quickly in a fire and allow the structure to collapse – it therefore has poor 'fire resistance' (Figure 12.17). It must be insulated to

Figure 12.17 Steel structures can collapse in the heat of a fire

provide good fire protection. This is normally done by encasing steel frames in concrete.

In the past, asbestos has been made into a paste and plastered onto steel frames, giving excellent fire protection, but it has caused major health problems and its use in new work is banned.

Building materials with high fire resistance are, for example, brick, stone, concrete, very heavy timbers (the outside chars and insulates the inside of the timber), and some specially made composite materials used for fire doors.

(e) Insulating materials

Building materials used for thermal or sound insulation could contribute to the spread of fire. Only approved fire-resisting materials should be used. Many buildings have insulated core panels as exterior cladding or for internal structures and partitions. The food industry, in particular, uses insulated core panels because they are easy to clean and facilitate consistent temperature control within the premises. The simple construction of these panels enables alterations and for additional internal partitions to be erected with minimum disruption to business.

They normally consist of a central insulated core, sandwiched between an inner and outer metal skin. There is no air gap. The external surface is then normally coated with a PVC covering to improve weather resistance or the aesthetic appeal of the panel. The central core can be made of various insulating materials, ranging from virtually non-combustible through to highly combustible. Differing fire hazards are associated with common types of insulation, when the panels are subjected to certain temperatures. Typical examples are:

▶ Mineral rock/modified phenolic will produce surface char and little smoke or gaseous combustion products, at temperatures above 230°C.
▶ Polyisocyanurate (PIR)/polyurethane (PUR) will char and will generate smoke and gaseous combustion products, at temperatures above 430°C PIR and 300°C PUR.

Figure 12.18 Insulated core panels

▶ Expanded polystyrene (EPS) will melt and will generate smoke and gaseous combustion products, at temperatures above 430°C.

Insulation charring can lead to panel delamination/collapse, and the gaseous combustion products can fill areas with the toxic gases carbon monoxide and styrene. A number of fires in buildings where insulated core panels have been used extensively in the fabric of the building have highlighted the particular dangers that may be associated with this form of construction, i.e. where the fabric of the building can contribute to the fire hazard. For more information see: http://www.communities.gov.uk/fire/firesafety/firesafetylaw/aboutguides/

(f) Fire compartmentation

A compartment is a part of a building that is separated from all other parts by walls and floors, and is designed to contain a fire for a specified time, for example 30 minutes or 1 hour.

The principal objective is to limit the effect of both direct fire damage and consequential business interruption caused not only by fire spread but also smoke and water damage in the same floor and other storeys.

Buildings are classified into purpose groups, according to their size. To control the spread of fire, any building whose size exceeds that specified for its purpose group must be divided into compartments that do not exceed the prescribed limits of volume and floor area. Otherwise, they must be provided with special fire protection. In the UK, the normal limit for the size of a compartment is 7,000 m³. Compartments must be separated by walls and floors of sufficient fire resistance. Any openings needed in these walls or floors must be protected by fire-resisting doors to ensure fire-tight separation.

Ventilation and heating ducts must be fitted with fire dampers where they pass through compartment walls and floors. Firebreak walls must extend completely across a building from inside wall to outside wall. They must be stable; they must be able to stand even when the part of the building on one side or the other is destroyed.

No portion of the wall should be supported on unprotected steelwork nor should it have the ends of unprotected steel members embedded in it. The wall must extend up to the underside of a non-combustible roof surface, and sometimes above it. Any openings must be protected to the required minimum grade of fire resistance. If an external wall joins a firebreak wall and has an opening near the join, the firebreak wall may need to extend beyond the external wall.

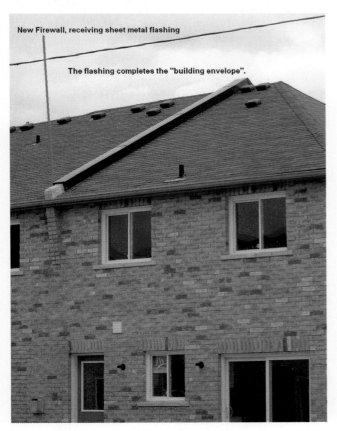

Figure 12.19 Firebreak wall between dwellings

An important function of external walls is to contain a fire within a building, or to prevent fire spreading from outside.

The fire resistance of external walls should be related to the:

▶ purpose for which the building is used;
▶ height, floor area and volume of the building;
▶ distance of the building from relevant boundaries and other buildings;
▶ extent of doors, windows and other openings in the wall.

A wall which separates properties from each other should have no doors or other openings in it.

12.3.4 Electrical and other equipment in potentially flammable atmospheres

Most electrical equipment either sparks in normal operation or is liable to spark under fault conditions. Some electrical appliances such as electric heaters are specifically designed to produce high temperatures. These circumstances create fire and explosion hazards, which demand very careful assessment in locations where processes capable of producing flammable concentrations of gas or vapour are used, or where flammable liquids are stored.

It is likely that many fires are caused by static electrical discharges. Static electricity can, in general, be eliminated by the careful design and selection of materials used in equipment and plant, and the materials used in products being manufactured. When it is impractical to avoid the generation of static electricity, a means of control must be devised. Where flammable materials are present, especially if they are gases or dusts, then there is a great danger of fire and explosion, even if there is only a small discharge of static electricity. The control and prevention of static electricity is considered in more detail later.

The use of electrical equipment in potentially flammable atmospheres should be avoided as far as possible. However, there will be many cases where electrical equipment must be used and, in these cases, the standards for the construction of the equipment should comply with the Equipment and Protective Systems Intended for Use in Potentially Explosive Atmospheres, known as ATEX Directive. This applies throughout the EU and also to manufacturers wishing to import products into the EU. The aim of directive 2014/34/EU is to allow the free trade of 'ATEX' equipment and protective systems within the EU by removing the need for separate testing and documentation for each member state.

Employers must classify areas where hazardous explosive atmospheres may occur into zones. The classification given to a particular zone, and its size and location, depends on the likelihood of an

12

explosive atmosphere occurring and its persistence if it does.

Areas classified into zones (0, 1, 2 for gas-vapour-mist and 20, 21, 22 for dust) must be protected from effective sources of ignition. Equipment and protective systems intended to be used in zoned areas must meet the requirements of the directive. Zone 0 and 20 require Category 1 marked equipment, zone 1 and 21 required Category 2 marked equipment and zone 2 and 22 required Category 3 marked equipment. Zone 0 and 20 are the zones with the highest risk of an explosive atmosphere being present.

Before electrical equipment is installed in any location where flammable vapours or gases may be present, the area must be zoned in accordance with ATEX and records of the zoned areas should be marked on building drawings and revised when any zoned area is changed. The installation and maintenance of electrical equipment in potentially flammable atmospheres is a specialised task. It must only be undertaken by electricians or instrument mechanics who are trained to ATEX standards.

Static electricity is produced by the build-up of electrons on weak electrical conductors or insulating materials. These materials may be gaseous, liquid or solid and may include flammable liquids, powders, plastic films and granules. Plastics have a high resistance that enables them to retain static charges for long periods of time. The generation of static may be caused by the rapid separation of highly insulated materials by friction or by transfer from one highly charged material to another in an electric field by induction (see Figure 12.20).

Dispensing Flammable Liquids

Figure 12.20 Earth bonding of containers to prevent a static discharge – anti-static outer clothing and conductive footwear should be used

A static electric shock, perhaps caused by closing a door with a metallic handle, can produce a voltage greater than 10,000V. Since the current flows for a very short

period of time, there is seldom any serious harm to an individual. However, discharges of static electricity may be sufficient to cause serious electric shock and are always a potential source of ignition when flammable liquid, dusts or powders are present. This is a particular problem in the parts of the printing industry where solvent-based inks are used on high-speed web presses. Flour dust in a mill has also been ignited by static electricity.

Static electricity may build up on both materials and people. When a charged person approaches flammable gases or vapours and a spark ignites the substance, the resulting explosion or fire often causes serious injury. In these situations, effective static control systems must be used which includes:

▶ bonding or earthing continuity between pieces of equipment, particularly portable equipment such as containers for carrying highly flammable substances, piping, filling funnels, drip trays and the like;
▶ not wearing outer clothing which generates static charges. In practice this usually means avoiding man-made fibres and using cotton only;
▶ using conductive footwear to leak static charges to ground (these must not be used by electricians as they conduct electricity and will not protect them against electric shock);
▶ avoiding the free fall of highly flammable liquids from one container to another unless anti-static additives have been put into the liquid or its natural properties will not hold static charges. Filling funnels should reach down to near the bottom of the container being filled.

For more information see Chapter 11 on electrical hazards.

12.4 Fire alarm system and fire-fighting arrangements

12.4.1 Fire detection and alarm systems

In the event of fire, it is vital that everyone in the workplace is alerted as soon as possible. The earlier the fire is discovered, the more likely it is that people will be able to escape before the fire takes hold and before it blocks escape routes or makes escape difficult.

Every workplace should have detection and warning arrangements. Usually the people who work there will detect the fire and in many small workplaces nothing further will be needed.

In small workplaces where occupancy is low, a shouted warning should be all that is needed, so long as the warning can be heard and understood everywhere on the premises. Where a simple shout of 'fire' or the operation of a manual device such as a gong, whistle or air horn is not adequate to warn everyone when operated from any single point within the building, it is likely that an electrical fire warning system may be required. This will typically include the following:

▶ manual call points (break-glass call points) at storey exit and final exit doors;
▶ electronic sirens or bells;
▶ a simple control and indicator panel; and
▶ in some cases a heat or smoke detector will also be fitted in the system.

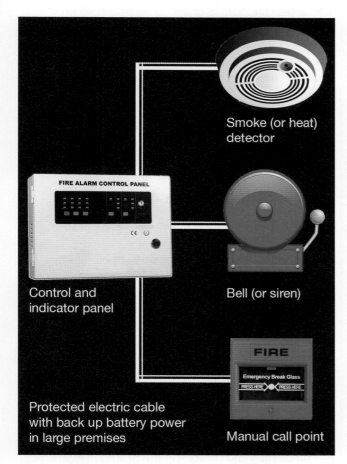

Figure 12.21 Simple electrical fire alarm system components

Some premises (e.g. a community centre) may only require a few interconnected manual call points and sounders. Other premises (e.g. a multilevel pub, or factory) may require a more complex warning system involving staff action. If a building has areas where a fire could develop undetected or where people work alone and might not see a fire (e.g. storerooms or warehouses), then it may be necessary to upgrade the fire-warning system to incorporate some form of automatic fire detection (e.g. in some circumstances, this could be a series of interconnected smoke alarms, if shown to be acceptable in the risk assessment).

If the system fails, people in your premises still need to be warned and escape safely. A temporary arrangement, such as gongs, whistles or air horns, combined with suitable training, may be acceptable for a short period pending system repairs. All systems should be properly installed to recognised standards with an auxiliary power source, if temporary manual alarms would be unacceptable. Cables need protecting from fire by using mineral-insulated, copper-sheathed, silicone-rubber insulated or other approved fire-resistant cabling.

The fire warning sound levels should be loud enough to alert everyone, taking into account background noise; any other sound systems should be muted (automatically or manually) when the fire alarm sounds. In areas of high background noise, or where people may be wearing hearing protectors, the audible warning should be supplemented, e.g. with visual or even vibratory alarms.

People with hearing difficulties

Where people have hearing difficulties, particularly those who are profoundly deaf, then simply hearing the fire warning is likely to be the major difficulty. If these persons are never alone while on the premises then this may not be a serious problem, as it would be reasonable for other occupants to let them know that the building should be evacuated. If a person with hearing difficulties is likely to be alone, then consider other means of raising the alarm. Among the most popular are visual beacons and vibrating devices or pagers that are linked to the existing fire alarm.

Voice alarms

Research has shown that some people and, in particular, members of the public, do not always react quickly to a conventional fire alarm. Voice alarms are therefore becoming increasingly popular and can also incorporate a public address facility. The message or messages sent must be carefully considered. It is therefore essential to ensure that voice-alarm systems are designed and installed by a person with specialist knowledge of these systems.

Schematic plan

In order to quickly determine where a fire has been detected, you should consider displaying a schematic plan showing fire alarm zones in a multi-zoned system adjacent to the control panel.

Manual call points

Manual call points, often known as 'break-glass' call points, enable a person who discovers a fire to immediately raise the alarm and warn other people in the premises of the danger.

People leaving a building because of a fire will normally leave by the way they entered. Consequently, manual call points are normally positioned at exits and storey exits that people may reasonably be expected to use in case of fire, not just those designated as fire exits.

12

However, it is not necessary in every case to provide call points at every exit.

Manual call points should normally be positioned so that, after all fixtures and fittings, machinery and stock are in place, no one should have to travel more than 45 m to the nearest alarm point. This distance may be less if the premises cater for people of limited mobility or there are particularly hazardous areas. They should be conspicuous (red), fitted at a height of about 1.4 m (or less for premises with a significant number of wheelchair users), and not in an area likely to be obstructed.

Automatic fire detection

It is important to consider how long a fire is likely to burn before it is discovered. Fires are likely to be discovered quickly if they occur in places that are frequently visited by employees, or in occupied areas of a building. For example, employees are likely to smell burning or smoke. If it is thought that there might be some delay in fire being detected, automatic fire detection should be considered, linked into an electrical fire alarm system.

Automatic fire detection may be needed for a number of reasons. These can include:

▶ if there are areas where people are isolated or remote and could become trapped by a fire because they are unaware of its development, such as lone workers;

▶ sleeping accommodation is provided;

▶ if there are areas where a fire can develop unobserved (e.g. storerooms or warehouse);

▶ as a compensating feature, e.g. for inadequate structural fire protection, in dead-ends or where there are extended travel distances; and

▶ where smoke control and ventilation systems are controlled by the automatic fire-detection system.

An automatic fire detection system should:

▶ be designed to accommodate the emergency evacuation procedure;

▶ give an automatic indication of the fire warning and its location. If the indicator panel is located in a part of the premises other than the control point (for example, an office) there should ideally be a repeater panel sited in the control point;

▶ be maintained and tested by a competent person; and

▶ communicate with a central control room (if there is one).

New automatic fire detection systems should be designed and installed by a competent person. Further guidance is given in BS 5839-116 or a more recent standard where applicable.

Where the public address system is part of the fire warning system it should be connected to an auxiliary power source to ensure the continued use of the system in the event of fire or other emergency.

Whichever warning or detection systems are in place, the fire and rescue service should always be called immediately if a fire occurs. When properly installed and maintained, these systems can be a significant factor in reducing the risk to life and limiting damage to your property in the event of a fire. These systems can help emergency services to respond very quickly to genuine emergencies.

Unfortunately, these systems can also produce 'unwanted' fire alarm signals; for example, when the signals are triggered by mistake and not because of a fire or a test. The subsequent evacuation can cause unnecessary business disruption and they also mean that emergency services send fire appliances unnecessarily.

'Unwanted' fire alarm signals are a widespread problem and all fire and rescue services are trying to reduce the number of false alarms caused by automatic fire detection systems.

12.4.2 Extinguishing media

There are four main methods of extinguishing fires, which are explained as follows:

▶ **Cooling** – reducing the ignition temperature by taking the heat out of the fire – using water to limit or reduce the temperature.

▶ **Smothering** – limiting the oxygen available by smothering and preventing the mixture of oxygen and flammable vapour – by the use of foam or a fire blanket.

▶ **Starving** – limiting the fuel supply – by removing the source of fuel by switching off electrical power, isolating the flow of flammable liquids or removing wood and textiles, etc.

▶ **Chemical reaction** – by interrupting the chain of combustion and combining the hydrogen atoms with chlorine atoms in the hydrocarbon chain, for example with halon extinguishers. (Halons have generally been withdrawn because of their detrimental effect on the environment, as ozone-depleting agents.)

Advantages and limitations of the main extinguishing media

In these standards fire extinguishers are classified by the type of extinguishing medium they contain. The following gives the advantages and disadvantages of the various types of extinguishing medium. Banding colours follow EN3:7.

Water extinguishers (red band)

This type of extinguisher can only be used on Class A fires. They allow the user to direct water onto a fire from a considerable distance.

A nine-litre water extinguisher can be quite heavy and some water extinguishers with additives can achieve the same rating, although they are smaller and therefore considerably lighter. This type of extinguisher is not suitable for use on live electrical equipment, liquid or metal fires .

Water extinguishers with additives (red band)

This type of extinguisher is suitable for Class A fires. They can also be suitable for use on Class B fires and, where appropriate, this will be indicated on the extinguisher. They are generally more efficient than conventional water extinguishers.

Foam extinguishers (cream band)

This type of extinguisher can be used on Class A or B fires and is particularly suited to extinguishing liquid fires such as petrol and diesel. They should not be used on free-flowing liquid fires unless the operator has been specially trained, as these have the potential to rapidly spread the fire to adjacent material. This type of extinguisher is not suitable for deep-fat fryers or chip pans. They should not be used on electrical or metal fires.

Powder extinguishers (blue band)

This type of extinguisher can be used on most classes of fire and achieve a good 'knock down' of the fire. They can be used on fires involving electrical equipment but will almost certainly render that equipment useless.

Because they do not cool the fire appreciably, it can re-ignite. Powder extinguishers can create a loss of visibility and may affect people who have breathing problems and are not generally suitable for confined spaces. They should not be used on metal fires.

Carbon dioxide extinguishers (black band)

This type of extinguisher is particularly suitable for fires involving electrical equipment as they will extinguish a fire without causing any further damage (except in the case of some electronic equipment, for example computers). As with all fires involving electrical equipment, the power should be disconnected if possible. These extinguishers should not be used on metal fires.

Wet chemical – class 'F' extinguishers

This type of extinguisher is particularly suitable for commercial catering establishments with deep-fat fryers. The intense heat in the fluid generated by fat fires means that when standard foam or carbon dioxide extinguishers stop discharging, re-ignition tends to occur.

Wet chemical extinguishers starve the fire of oxygen by sealing the burning fluid, which prevents flammable vapour reaching the atmosphere.

12.4.3 Portable fire-fighting equipment

Employers or those who have control of non-domestic premises usually have a statutory duty under fire safety and/or health and safety legislation to ensure that there are appropriate means of fighting fires. The employer or controller of non-domestic premises may be known as the 'responsible person'. The responsible person also has to provide suitably trained people to operate non-automatic fire-fighting equipment.

If fire breaks out in the workplace and trained staff can safely extinguish it using suitable fire-fighting equipment, the risk to others will be removed. Therefore, all workplaces where people are at risk from fire should be provided with suitable fire-fighting equipment.

The most useful form of fire-fighting equipment for general fire risks is the water-type extinguisher or suitable alternative. One such extinguisher should be provided for around each 200 m² of floor space with a minimum of one per floor. If each floor has a hose reel, which is known to be in working order and of sufficient length for the floor it serves, there may be no need for water-type extinguishers to be provided.

Areas of special risks involving the use of oil, fats or electrical equipment may need carbon dioxide, dry powder or other types of extinguisher.

Fire extinguishers should be sited on exit routes, preferably near to exit doors, or where they are provided for specific risks, near to the hazards they protect. Notices indicating the location of FFE should be displayed where the location of the equipment is not obvious or in areas of high fire risk where the notice will assist in reducing the risk to people in the workplace.

All halon fire extinguishers should have been decommissioned and disposed of safely as halon affects the ozone layer in the Earth's atmosphere.

Employees carrying out hot work should have appropriate fire extinguishers with them and know how to use them.

The primary purpose of fire extinguishers is to tackle fires at a very early stage to enable people to make their escape. Putting out larger fires is the role of the fire and rescue services.

Case study

It is very important to understand the limitations of portable fire-fighting equipment and ensure the safety of people as a first priority. At one cotton mill, in the UK, the author was involved with the investigation of a fire where they used **28 water extinguishers** fetched from all over the mill on various floors **before** sounding the alarm, evacuating the building and calling the fire and rescue services. Fortunately no one was seriously injured but the outcome could so easily have been very tragic indeed. Such equipment should only be used, if safe to do so, by trained people while the building is evacuated and the professionals are called.

12

Extinguishers should conform to a recognised standard such as ISO 7165:2009 Fire-fighting-Portable fire extinguishers – Performance and construction, or in Europe, EN3:7 Portable Fire Extinguishers. Characteristics, performance requirements and test methods.

Under ISO 7165 the recommended colour for extinguisher bodies is red with no secondary colour coding noted. Under EN3:7 fire extinguishers are all red with 5% of the cylindrical area taken up with a secondary colour code. The colour code (band) denotes on which class of fire the extinguisher can be used. There is no single universal colour coding used: Table 12.1 shows the Australian code, Table 12.2 shows the UK code which is also followed in the EU, Table 12.3 shows the US code.

Table 12.1 Classification for fire extinguishers used in Australia

Type	Pre-1997		Current	SUITABLE FOR USE ON FIRE CLASSES (BRACKETS DENOTE SOMETIMES APPLICABLE)				
Water	Solid red		Solid red	A				
Foam	Solid blue		Red with a blue band	A	B			
Dry chemical (powder)	Red with a white band		Red with a white band	A	B	C	E	
Carbon dioxide	Red with a black band		Red with a black band	(A)	B	C	E	F
Vaporising liquid (not halon)	Red with a yellow band		Red with a yellow band	A	B	C	E	
Halon	Solid green		No longer produced	A	B		E	
Wet chemical	Solid oatmeal		Red with an oatmeal band	A			F	

Table 12.2 Classification for fire extinguishers used in the UK under EN3:7

Type	Old Code		BS EN 3 Colour Code	SUITABLE FOR USE ON FIRE CLASSES (BRACKETS DENOTE SOMETIMES APPLICABLE)				
Water	Signal Red		Signal Red	A				
Foam	Cream		Red with a Cream panel above the operating instructions	A	B			
Dry powder	Blue		Red with a Blue panel above the operating instructions	(A)	B	C	E	
Carbon dioxide	Black		Red with a Black panel above the operating instructions		B		E	
Wet chemical	N/A		Red with a Canary Yellow panel above the operating instructions	A	(B)			F
Class D powder	Blue		Red with a Blue panel above the operating instructions				D	
Halon gas	Green		Now prohibited except under certain situations.					

Table 12.3 Classification for fire extinguishers used in the USA

Fire Class	Geometric Symbol		Pictogram	Intended Use
A	Green Triangle		Garbage can and wood pile burning	Ordinary solid combustibles
B	Red Square		Fuel container and burning puddle	Flammable liquids and gases
C	Blue Circle		Electric plug and burning outlet	Energised electrical equipment
D	Yellow Decagon (star)		Burning gear and bearing	Combustible metals
K	Black Hexagon		Pan burning	Cooking oils and fats

ISO 7165 gives a marking layout template as shown in Figure 12.22 for extinguishers using the use-code symbols shown in Figure 12.23.

12.4.4 Fixed fire-fighting equipment – sprinkler installations

Sprinklers should be considered as only one component part of a total fire safety strategy, which is tailored to the existing and projected needs of a building. They have significant benefits to offer in suppressing fires until those best trained to deal with major incidents are on the scene to extinguish them.

Sprinklers, however, are an emotive topic. In some buildings, they have been used for a long time as the most significant element of a fire safety system. This situation is probably most prolific in warehouses and

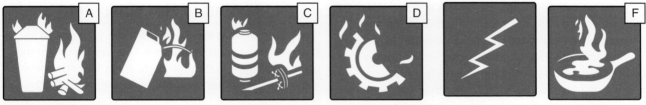

2 kg CARBON-DIOXIDE FIRE EXTINGUISHER	INSTRUCTIONS	APPROVAL MARK

2 kg CARBON-DIOXIDE FIRE EXTINGUISHER

INSPECTION: INSPECT MONTHLY. CHECK THAT EXTINGUISHER IS CHARGED, UNDAMAGED AND SEAL IS INTACT. MAKE SURE HORN IS UNOBSTRUCTED.

MAINTENANCE: EXAMINE CAREFULLY EVERY 12 MONTHS TO ENSURE EXTINGUISHER IS OPERABLE. RECHARGE IF MASS LOSS EXCEEDS 0.2 kg. REPLACE ANY DAMAGED PARTS. CHECK HORN FOR OBSTRUCTIONS. HYDROSTATIC RETEST TO DOT/TC REQUIREMENTS EVERY 5 YEARS.

USE: AFTER ANY USE RECHARGE IMMEDIATELY.

RECHARGE: CO_2 CHARGE IS 2 kg. FULL MASS STAMPED ON VALUE BODY INCLUDES HORN ASSEMBLY.

RECORD: RECORD MAINTENANCE AND RECHARGE DATES ON ATTACHED TAG.

FOR INDUSTRIAL USE.

INSTRUCTIONS

① HOLD UPRIGHT PULL RINGS PIN
② START BACK 3 m AIM AT BASE OF FIRE
③ SQUEEZE LEVER SWEEP SIDE TO SIDE

APPROVAL MARK

CARBON-DIOXIDE FIRE EXTINGUISHER
CLASSIFICATION 21-B
SERIAL NO. XX-XXXXX
MEETS ISO XXXXX STANDARD
2 kg CARBON-DIOXIDE FIRE EXTINGUISHER
SUITABLE FOR USE AT TEMPERATURES FROM −40°C TO 49°C (−40°F TO 120°F)
PRESSURE TESTED TO 20 MPa

MODEL **322** | 1999

MFG. NAME
MFG. ADDRESS

Figure 12.22 Marking layout for an extinguisher under ISO 7165:2009

Figure 12.23 Use-code symbols for extinguishers under ISO 7165:2009. Class A: Ordinary solid material fires. Class B: Flammable liquid fires. Class C: Gas and vapour fires. Class D: Combustible metal fires. Fires involving energised electrical conductors. Class F: Cooking oil fires

Figure 12.24 Various items of fire-fighting equipment: (a) fire blanket; (b) water hose reel; (c) ISO labelled fire extinguisher

359

retail premises. They have unfortunately been resisted in most other buildings because of the initial capital cost and the perceived inherent risk of accidental water discharge.

Sprinkler systems can be very effective in controlling fires. They can be designed to protect life and/or property and may be regarded as a cost-effective solution for reducing the risks created by fire. If a building has a sprinkler installation, it may have been installed as a result of a business decision, for example for the protection of business assets, or it may have been installed as a requirement, for example imposed under local legislation, byelaw, or an integral part of the building design.

Sprinkler systems should normally extend to the entire building. In a well-designed system, only those heads in the immediate vicinity of the fire will actually operate. Sprinkler installations typically comprise a water supply (preferably a stored water supply incorporating tanks), pumps, pipe work and sprinkler heads. There are different types of sprinkler design; sprinklers can be operated to discharge water at roof or ceiling level or within storage racks. Other design types such as ESFR (early suppression fast response) and dry pipe may also be appropriate. In all cases, a competent person/contractor should be used to provide guidance.

The installation should be designed for the fire hazard, taking into account the building occupancy, the fire load and its burning characteristics, and the sprinkler control characteristics. For each hazard the sprinkler installation design should take account of specific matters such as storage height, storage layout, ceiling clearance and sprinkler type (e.g. sprinkler orifice, sprinkler sensitivity).

There are some areas where sprinklers should not be fitted, such as over salt baths and metal melt pans, because water will possibly cause an explosive reaction.

If any significant changes are being made to the premises, for example changing storage arrangements or material stored, the sprinkler installation should be checked to see that it is still appropriate and expert advice sought as necessary.

Sprinkler protection could give additional benefits, such as a reduction in the amount of portable fire-fighting equipment necessary, and the relaxation of restrictions in the design of buildings.

In the UK guidance on the design and installation of new sprinkler systems and the maintenance of all systems is given in the Loss Prevention Council (LPC) Rules, BS EN 12845 or BS 5306 and should be carried out only by a competent person.

Routine maintenance by on-site personnel may include checking of pressure gauges, alarm systems, water supplies, any anti-freezing devices and automatic booster pump(s). For example diesel fire pumps should be given a test run for 30 minutes each week.

Figure 12.25 Various sprinkler heads designed to fit into a high-level water pipe system and spray water at different angles onto a fire below

12.4.5 Inspection, maintenance and testing of fire equipment

It is important that equipment is fit for its purpose and is properly maintained and tested. One way in which this can be achieved is through companies that specialise in the test and maintenance of fire protection and fighting equipment.

All equipment provided to assist escape from the premises, such as fire detection and warning systems

Table 12.4 Maintenance and testing of fire equipment

Equipment	Period	Action
Fire-detection and fire-warning systems including self-contained smoke alarms and manually operated devices	Weekly	• Check all systems for state of repair and operation • Repair or replace defective units • Test operation of systems, self-contained alarms and manually operated devices
	Annually	• Full check and test of system by competent service engineer • Clean self-contained smoke alarms and change batteries
Emergency lighting including self-contained units and torches	Weekly	• Operate torches and replace batteries as required • Repair or replace any defective unit
	Monthly	• Check all systems, units and torches for state of repair and apparent function
	Annually	• Full check and test of systems and units by competent service engineer • Replace batteries in torches
Fire-fighting equipment installation including hose reels	Weekly	• Check all extinguishers including hose reels for correct apparent function
	Annually	• Full check and test by competent service engineer

and emergency lighting, and all equipment provided to assist with fighting fire, should be regularly checked and maintained by a suitably competent person in accordance with the manufacturer's recommendations. Table 12.4 gives guidance on the frequency of test and maintenance and provides a simple guide to good practice.

Inspection instructions on each extinguisher must include the following information under ISO 7165:2009.

The inspection instructions state that the extinguisher shall be checked to ensure that:

▶ the seals and tamper indicators are not broken or missing;

▶ it is full (by weighing or lifting);

▶ it is not obviously damaged, corroded or leaking and does not have a clogged nozzle;

▶ its pressure gauge reading or indicator is in the operable range or position.

See also Chapter 18 for a maintenance checklist.

12.5 Evacuation of a workplace

12.5.1 Means of escape in case of fire

(a) General

It is essential to ensure that people can escape quickly from a workplace if there is a fire. Normally the entrances and exits to the workplace will provide escape routes, particularly if staff have been trained in what to do in case of fire and if it is certain that an early warning will be given.

It is likely that the means of escape will be adequate in modern buildings which have had local Building Regulation approval or have been built to acceptable building construction standards. Where there have not been significant changes to the building or where the workplace has recently been inspected by the fire authorities or other fire expert and found to be satisfactory, no change is likely to be needed. It may occasionally be necessary to improve the fire protection on existing escape routes, or to provide additional exits. In making a decision about the adequacy of means of escape, the following points should be considered:

▶ people need to be able to turn away from a fire as they escape or be able to pass a fire when it is very small;

▶ if a single-direction escape route is in a corridor, the corridor may need to be protected from fire by fire-resisting partitions and self-closing fire doors;

▶ stair openings can act as natural chimneys in fires. This makes escape from the upper parts of some workplaces difficult. Most stairways, therefore, need to be separated from the workplace by fire-resistant partitions and self-closing fire doors. Where stairways serve no more than two open areas, in shops for example, which people may need to use

Figure 12.26 External fire escape from a multi-storey building. Extreme caution would be needed in winter in a cold climate, and enclosure of the staircase may be essential

as escape routes, there may be no need to use this type of protection.

(b) Doors

Some doors may need to open in the direction of travel, such as:

▶ doors from a high-risk area, such as a paint spraying room or large kitchen;

▶ doors that may be used by more than 50 persons;

▶ doors at the foot of stairways where there may be a danger of people being crushed;

▶ some sliding doors may be suitable for escape purposes provided that they do not put people using them at additional risk, slide easily and are marked with the direction of opening;

▶ doors which only revolve and do not have hinged segments are not suitable as escape doors.

(c) Escape routes

Escape routes should meet the following criteria:

▶ where two or more escape routes are needed they should lead in different directions to places of safety;

▶ escape routes need to be short and to lead people directly to a place of safety, such as the open air or an area of the workplace where there is no immediate danger;

▶ it should be possible for people to reach the open air without returning to the area of the fire. They should then be able to move well away from the building;

▶ escape routes should be wide enough for the volume of people using them. A 750 mm door will allow up to 40 people to escape in 1 minute, so most doors and corridors will be wide enough. If the routes are likely to be used by people in wheelchairs, the minimum width will need to be 800 mm.

While the workplace is in use, it must be possible to open all doors easily and immediately from the inside, without using a key or similar device. Doors must be readily opened in the direction of escape. Fire doors should be self-closing (fire doors to cupboards or lockers can be simply latched or locked).

Make sure that there are no obstructions on escape routes, especially on corridors and stairways where people who are escaping could dislodge stored items or be caused to trip. Any fire hazards must be removed from exit routes as a fire on an exit route could have very serious consequences.

Escape routes need regular checks to make sure that they are not obstructed and that exit doors are not locked. Self-closing fire-resisting doors should be checked to ensure doors close fully, including those fitted with automatic release mechanisms.

The maximum advisable travel distances from any area in a workplace to a fire exit door leading out to a relative place of safety should be in accordance with Table 12.5.

Table 12.5 Maximum travel distances

Fire hazard			
	Lower	*Normal*	*Higher*
Enclosed structures:			
Alternative	60 m	45 m	25 m
Dead-end	18 m	18 m	12 m
Semi-open structures:			
Alternative	200 m	100 m	60 m
Dead-end	25 m	18 m	12 m

Source: UK HSE.

(d) Lighting

Escape routes must be well lit. If the route has only artificial lighting or if it is used during the hours of darkness, alternative sources of lighting should be considered in case the power fails during a fire. Check the routes when it is dark as, for example, there may be street lighting outside that provides sufficient illumination. In small workplaces it may be enough to provide the staff with torches that they can use if the power fails. However, it may be necessary to provide

battery-operated emergency lights so that if the mains lighting fails the lights will operate automatically. Candles, matches and cigarette lighters are not adequate forms of emergency lighting.

(e) Signs

Exit signs on doors or indicating exit routes should be provided where they will help people to find a safe escape route. Signs on exit routes should have directional arrows, 'up' for straight on and 'left', 'right' or 'down' according to the route to be taken. Signs should be in accordance with ISO 7010:2003 Graphical symbols – safety colours and safety signs – Safety signs used in workplaces and public areas. Advice on the use of all signs including exit signs can be found in Chapter 4.

Figure 12.27 International Fire Escape pictorial

(f) Escape times

Everyone in the building should be able to get to the nearest place of safety in between 2 and 3 minutes. This means that escape routes should be kept short. Where there is only one means of escape, or where the risk of fire is high, people should be able to reach a place of safety, or a place where there is more than one route available, in 1 minute.

The way to check this is to pace out the routes, walking slowly and noting the time. Start from where people work and walk to the nearest place of safety. Remember that the more people there are using the route, the longer they will take. People take longer to negotiate stairs and they are also likely to take longer if they have a disability.

Where fire drills are held, check how long it takes to evacuate each floor in the workplace. This can be used as a basis for assessment. If escape times are too long, it may be worth rearranging the workplace so that people are closer to the nearest place of safety, rather than undertake expensive alterations to provide additional escape routes.

Reaction time needs to be considered. This is the amount of time people will need for preparation before they escape. It may involve, for example, closing down machinery, issues of security or helping visitors or members of the public out of the premises. Reaction time needs to be as short as possible to reduce risk to staff. Assessment of escape routes should include this. If reaction times are too long, additional routes may need to be provided. It is important that people know what to do in case of fire as this can lessen the time needed to evacuate the premises.

12.5.2 Evacuation procedure and emergency plans

(a) Introduction

Each workplace should have an emergency plan. The plan should include the action to be taken by staff in the event of fire, the evacuation procedure and the arrangements for calling the local Fire Authority.

For small workplaces, this could take the form of a simple fire action notice posted in positions where staff can read it and become familiar with it.

High fire-risk or larger workplaces will need more detailed plans, which take account of the findings of the risk assessment, for example the staff significantly at risk and their location. For large workplaces, notices giving clear and concise instructions of the routine to be followed in the event of fire should be prominently displayed. The notice should include the method of raising an alarm in the event of fire and the location of assembly points to which staff escaping from the workplace should report.

(b) Fire routines

Site managers must make sure that all employees are familiar with the means of escape in case of fire and their use, and with the routine to be followed in the event of fire.

To achieve this, routine procedures must be set up and made known to all employees, generally outlining the action to be taken in the event of fire and specifically laying down the duties of certain nominated persons. Notices should be posted throughout the premises.

While the need in individual premises may vary, there are a number of basic components which should be considered when designing any fire routine procedures:

▶ the action to be taken on discovering a fire;
▶ the method of operating the fire alarm;
▶ the arrangements for calling the fire brigade;
▶ the stopping of machinery and plant;
▶ first-stage fire-fighting by employees;
▶ evacuation of the premises;
▶ assembly of staff, customers and visitors; and
▶ carrying out a roll-call to account for everyone on the premises.

The procedures must take account of those people who may have difficulty in escaping quickly from

a building because of their location or a disability. Insurance companies and other responsible people may need to be consulted where special procedures are necessary to protect buildings and plant during or after people have been evacuated. For example, there may be some special procedures necessary to ensure that a sprinkler system is operating in the event of fire.

(c) Supervisory duties/fire marshals

A member of the staff should be nominated to supervise all fire and emergency arrangements. This person should be in a senior position or at least have direct access to a senior manager. Senior members of the staff should be appointed as departmental fire wardens, with deputies for every occasion of absence, however brief. In the event of fire or other emergency, their duties would be, while it remains safe to do so, to ensure that:

▶ the alarm has been raised;
▶ the whole department, including toilets and small rooms, has been evacuated;
▶ the fire and rescue service has been called;
▶ fire doors are closed to prevent fire spread to adjoining compartments and to protect escape routes;
▶ plant and machinery are shut down wherever possible and any other actions required to safeguard the premises are taken where they do not expose people to undue risks;
▶ a roll-call is carried out at the assembly point and the result reported to whoever is in control of the evacuation.

Under normal conditions, fire wardens should check that good standards of housekeeping and preventative maintenance exist in their department, that exits and escape routes are kept free from obstruction, that all fire-fighting appliances are available for use and fire points are not obstructed, that smoking is rigidly controlled, and that all members of staff under their control are familiar with the emergency procedure and know how to use the fire alarm and fire-fighting equipment.

(d) Assembly and roll-call

Assembly points should be established for use in the event of evacuation. They should be in positions, preferably under cover, which are unlikely to be affected at the time of fire. In some cases, it may be necessary to make mutual arrangements with the occupiers of nearby premises.

In the case of small premises, a complete list of the names of all staff should be maintained so that a roll-call can be made if evacuation becomes necessary.

In those premises where the number of staff would make a single roll-call difficult, each departmental fire

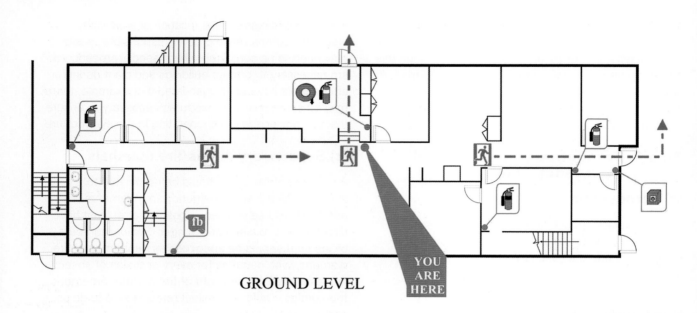

GROUND LEVEL

YOU
ARE
HERE

LEGEND

🧯 CO2 Fire Extinguisher

🧯 Dry Powder Fire Extinguisher

⭕ Fire Hose Reel

▣ Manual Call Point

fb Fire Blanket

◹ Fire Door

— Sliding Fire Door

🏃 Egress Route

St Paul's Terrace

EMERGENCY
ASSEMBLY
AREA A

Misterton St

Baxter Street

Site Plan

Figure 12.28 (a) Fire evacuation diagram in the UK

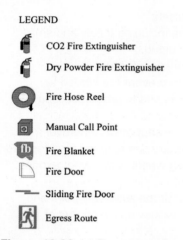

Figure 12.28 (b) Fire evacuation plan in four languages in Bruges, Belgium

warden should maintain a list of the names of staff in their area. Roll-call lists must be updated regularly.

(e) Fire notices

Printed instructions for the action to be taken in the event of fire should be displayed throughout the premises. The information contained in the instructions should be stated briefly and clearly. The staff and their deputies to whom specific duties are allocated should be identified.

Instruction for the immediate calling of the fire brigade in case of fire should be displayed at telephone switchboards, exchange telephone instruments and security lodges.

A typical fire notice is given in Appendix 12.2 and a typical fire evacuation diagram is shown in Figure 12.28.

(f) Fire drills

Once a fire routine has been established, it must be tested at regular intervals in order to ensure that all staff are familiar with the action to be taken in an emergency.

The most effective way of achieving this is by carrying out fire drills at prescribed intervals. Drills should be held at least twice a year other than in areas dealing with hazardous processes, where they should be more frequent. A programme of fire drills should be planned to ensure that all employees, including shift workers and part-time employees, are covered.

12.5.3 People with special needs

(a) General

Of all the people who may be especially at risk, employers will need to pay particular attention to people who have special needs, including those with a disability.

In the UK, the Equality and Human Rights Commission estimates that 11 million people have some form of disability, which may mean that they find it more difficult to leave a building if there is a fire. Under the UK Equalities Act 2010 if disabled people could realistically expect to use premises, then employers or those in charge of premises must anticipate any reasonable adjustments that would make it easier for that right to be exercised. Similar legislation may well be in place in most EU countries and a number of countries elsewhere.

The UK Equalities Act 2010 includes the concept of 'reasonable adjustments' and this can be carried over into fire safety law. It can mean different things in different circumstances. For a small business, it may be considered reasonable to provide contrasting colours on a handrail to help people with vision impairment to follow an escape route more easily. However, it might be unreasonable to expect that same business to install an expensive voice alarm system. Appropriate 'reasonable adjustments' for a large business or organisation may be much more significant.

Where people with special needs use or work on the premises, their needs should, so far as is practicable, be discussed with them. These will often be modest and may require only changes or modifications to existing procedures. There may be a need to develop individual 'personal emergency evacuation plans' (PEEPs) for disabled persons who frequently use a building. They will need to be confident of any plan/PEEP that is put in place after consultation with them. As part of an employer's consultation exercise they will need to consider the matter of personal dignity.

If members of the public use the building, then those in control may need to develop a range of standard PEEPs which can be provided on request to a disabled person or others with special needs.

Guidance on removing barriers to the everyday needs of disabled people is in BS 8300. Much of this advice will also help disabled people during an evacuation.

Advice on the needs of people with a disability, including sensory impairment, is available from the organisations which represent various groups.

Many businesses have discovered the problem of reconciling their duties under disability discrimination and general fire legislation.

Employers need to consider the needs of all their staff and users of their building when fire evacuation strategies are being considered. This can include people with a surprisingly diverse range of access needs, not just those with perhaps more obvious disabilities such as wheelchair users. Access and exit for each of these groups can sometimes seem hard to reconcile with tightly controlled fire safety. For example local Building Regulations may require internal doors held open by electromagnetic devices to self-close when activated by smoke detectors, but a disabled person may not be able to open a heavy fire door once it has closed.

So what is more important, fulfilling fire safety requirements or fulfilling the requirements of the Equalities Act?

Essentially, fire safety concerns life-threatening incidents, the Equalities Act is about dignity and equal treatment. Basically, health and safety (including fire) overrides the Equalities Act if there is a conflict. However, there really should not be a problem achieving both. In respect of fire, it is never acceptable to refuse someone entry to an upper level on the basis of there not being an evacuation lift available. However, it is acceptable to pre-plan and decide to hold public meetings on the ground floor for this reason.

In other words, it should be possible to comply with fire safety legislation and comply with the Equalities Act, as long as there are management procedures in place to make sure both are adhered to successfully. In many cases, the answer lies in planning ahead, and implementing procedures which may involve other staff in personal evacuation plans, for example.

(b) Special needs, fire emergencies and precautions

If disabled people are going to be in premises, then employers must also provide a safe means for them to leave if there is a fire. Staff should be aware that disabled people may not react, or can react differently, to a fire warning or a fire incident. Employers should give similar consideration to others with special needs such as parents with young children or the elderly.

In premises with a simple layout, a common-sense approach, such as offering to help lead a blind person or helping an elderly person down steps may be enough. In more complex premises, more elaborate plans and procedures will be needed, with trained staff assigned to specified duties.

Consider the needs of people with mental disabilities or spatial recognition problems. The range of disabilities encountered can be considerable, extending from mild epilepsy to complete disorientation in an emergency situation. Many of these can be addressed by properly trained staff, discreet and empathetic use of the 'buddy system' or by careful planning of colour and texture to identify escape routes.

People with special needs (including members of the public) need extra consideration when planning for

12

emergencies. But the problems this raises are seldom great. Employers should:

▶ identify everyone who may need special help to get out;
▶ allocate responsibility to specific staff to help people with a disability in emergency situations;
▶ consider possible escape routes;
▶ enable the safe use of lifts;
▶ enable people with a disability to summon help in emergencies;
▶ train staff to be able to help their colleagues;
▶ consider safe havens.

People with impaired vision must be encouraged to familiarise themselves with escape routes, particularly those not in regular use. A 'buddy' system would be helpful. But, to take account of absences, more than one employee working near anyone with impaired vision should be taught how to help them.

Where people have hearing difficulties, particularly those who are profoundly deaf, then simply hearing the fire warning is likely to be the major difficulty. If these persons are never alone while on the premises then this may not be a serious problem, as it would be reasonable for other occupants to let them know that the building should be evacuated. If a person with hearing difficulties is likely to be alone, then consider other means of raising the alarm. Among the most popular are visual beacons and vibrating devices or pagers that are linked to the existing fire alarm.

People with impaired hearing may not hear alarms in the same way as those with normal hearing but may still be able to recognise the sound. This may be tested during the weekly alarm audibility test. There are alternative means of signalling, such as lights or other visual signs, vibrating devices or specially selected sound signals. Ask the fire brigade before installing alternative signals.

Wheelchair users or others with impaired mobility may need help to negotiate stairs, etc. Anyone selected to provide this help should be trained in the correct methods. Advice on the lifting and carrying

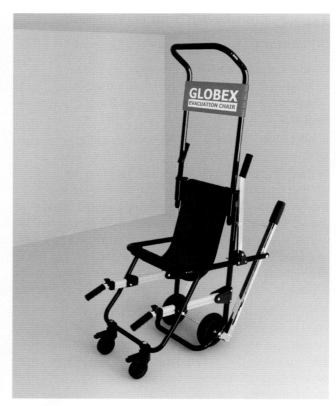

Figure 12.30 Stair evacuation chair for people with a disability

of people can be obtained from the Fire Service, Ambulance Service, Red Cross Society, Red Crescent Society, St John Ambulance Brigade or local disability organisations.

Lifts should not be used as a means of escape in the event of a fire. If the power fails, the lift could stop between floors, trapping occupants in what may become a chimney of fire and smoke. In the UK, BS 5588: Part 8 provides advice on specially designed lifts for use by people with a disability in the event of a fire.

Employees with learning difficulties may also require special provision. Management should ensure that the colleagues of any employee with a learning difficulty know how to reassure them and lead them to safety.

12.5.4 Building plans and specifications

Plans and specifications can be used to assist understanding of a fire risk assessment or emergency plan. Even where not needed for this purpose they can help you and your staff keep your fire risk assessment and emergency plan under review and help the fire and rescue service in the event of fire. Any symbols used should be shown on a key. Plans and specifications could include the following:

▶ essential structural features such as the layout of function rooms, escape doors, wall partitions, corridors, stairways, etc. (including any fire-resisting

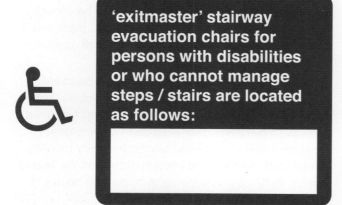

'exitmaster' stairway evacuation chairs for persons with disabilities or who cannot manage steps / stairs are located as follows:

Figure 12.29 Special addition to the fire notice for people with a disability

structure and self-closing fire doors provided to protect the escape routes);

▶ location of refuges and lifts that have been designated suitable for use by people with a disability and others who may need assistance to escape in case of a fire;

▶ methods for fighting fire (details of the number, type and location of the fire-fighting equipment);

▶ location of manually operated fire alarm call points and control equipment for fire alarms;

▶ location of any control rooms and fire staff posts;

▶ location of any emergency lighting equipment and the exit route signs;

▶ location of any high-risk areas, equipment or process that must be immediately shut down by staff on hearing the fire alarm;

▶ location of any automatic fire-fighting systems, risers and sprinkler control valves;

▶ location of the main electrical supply switch, the main water shut-off valve and, where appropriate, the main gas or oil shut-off valves; and

▶ plans and specifications relating to all recent constructions.

This information should be passed on to any later users or owners of the premises.

12.6 Further Information

ATEX Guidelines: 3rd Edition, June 2009 (updated May 2011) http://ec.europa.eu/enterprise/sectors/mechanical/files/atex/guide/atexguidelines_june2009_en.pdf

Emergency plan, CFPA-European Guideline, CFPA-E No 25:2010 F http://www.cfpa-e.eu/files/PDF/Guidelines/Ratified%20Guidelines/CFPA_E_Guideline_No_25_2010_F.pdf

Guidelines on Occupational Safety and Health Management Systems (ILO-OSH 2001), ISBN-0-580-37805-5

Introduction to qualitative fire risk assessment, CFPA-European Guideline, CFPA-E No 4:2010 F http://www.cfpa-e.eu/files/PDF/Guidelines/Ratified%20Guidelines/CFPA_E_Guideline_No_4_2010_F.pdf

Occupational safety and Health Convention (C155), ILO http://www.ilo.org/ilolex/cgi-lex/convde.pl?C155

Safety in the Use of Chemicals at Work, International Labour Office (ILO) Code of Practice (CoP), ILO, 1993, ISBN 92-2-108006-4

Section 6: Operational control measures and

Section 7: Design and installation

12.7 Practice revision questions

1. (a) **Describe** the three sides of a 'fire triangle'.
   (b) **Outline THREE** methods of extinguishing fires.
   (c) **Identify FOUR** different types of ignition source that may cause a fire to occur, and give a typical workplace example of **EACH** type.
   (d) **Identify FIVE** classes of fire, the associated fuel sources and the appropriate fire extinguisher to use for each class.

2. **Outline THREE** methods of heat transfer (other than direct burning) and **explain** how each can cause the spread of fire in buildings.

3. (a) **Identify SIX** causes of fire in a workplace.
   (b) **Identify** the possible consequences on people of fires in a workplace.
   (c) **Outline** the main control measures to minimise the risk of fire in a workplace.

4. (a) **Identify** the five stages of a fire risk assessment in a workplace.
   (b) **Outline** the particular issues that should be considered in the fire risk assessment of temporary workplaces.
   (c) **Identify** additional factors to be taken into account during building refurbishment work.

5. **Outline** the contents of a typical fire plan that should be attached to a fire risk assessment.

6. (a) **Outline** effective control measures for the safe storage and handling of small containers containing flammable solvents.
   (b) **Identify EIGHT** types of unsatisfactory working practices using flammable gas cylinders that could increase the risk of a fire or explosion.

7. Fire protection is an important feature in the design of a new building.
   (a) **Identify SIX** structural measures that can help to prevent the spread of fire and smoke.
   (b) **Outline** the issues involved in the fire compartmentation of a building.

8. The failure of electrical equipment is one of the main causes of workplace fires.
   (a) **Outline** how fires could be caused by the failure of electrical cables and equipment.
   (b) **Identify TWO** types of extinguisher that can be used safely on fires involving electrical equipment and **identify** those extinguishers that should not be used in such fires.
   (c) **Outline** suitable control measures that should be taken to minimise the risk of fire from electrical equipment when used in a flammable atmosphere.

12

9. (a) **Identify TWO** methods of raising an alarm in the event of a fire.

   (b) **Identify FOUR** types of portable fire extinguisher **AND**, in **EACH** case, **identify** the class of fire on which they should be used.

10. (a) **Identify TWO** types of emergency warning systems that can be installed in the building to ensure that all occupants can be made aware of the need to evacuate the building.

    (b) **Outline** the main factors to be considered in the siting of fire extinguishers.

    (c) **Outline** suitable arrangements for the inspection and maintenance of fire extinguishers in the workplace.

11. (a) **Outline ONE** advantage and **ONE** limitation of each of the main types of fire extinguishers.

    (b) **Outline TWO** advantages and **TWO** disadvantages of using hose reels as a means of extinguishing fires.

    (c) **Identify** the issues associated with the access for fire and rescue services to a workplace.

12. (a) **Outline** the principal requirements for a safe means of escape from a building in the event of a fire.

    (b) **Outline** the role of fire marshals in the event of a fire emergency.

    (c) **Outline** reasons that may delay the safe evacuation of occupants from a workplace during a fire.

13. (a) **Identify** the issues to consider for the location of an assembly point for use in a workplace evacuation.

    (b) **Outline** the reasons for undertaking regular fire drills in the workplace.

    (c) **Outline** the contents of a training programme for employees on the emergency action to take in the event of fire.

APPENDIX 12.1 Fire risk assessment checklist

As recommended in Fire Safety Guides published by the UK Department for Communities and Local Government in 2006

Checklist

Follow the five key steps – fill in the checklist – assess your fire risk and plan fire safety.

1. Fire safety risk assessment

Fire starts when heat (source of ignition) comes into contact with fuel (anything that burns) and oxygen (air). You need to keep sources of ignition and fuel apart.

How could a fire start?

Think about heaters, lighting, naked flames, electrical equipment, hot processes such as welding or grinding, cigarettes, matches and anything else that gets very hot or causes sparks.

What could burn?

Packaging, rubbish and furniture could all burn just like the more obvious fuels such as petrol, paint, varnish and white spirit. Also think about wood, paper, plastic, rubber and foam. Do the walls or ceiling have hardboard, chipboard or polystyrene? Check outside, too.

Check:

▶ Have you found anything that could start a fire? Make a note of it.
▶ Have you found anything that could burn? Make a note of it.

2. People at risk

People at risk

Everyone is at risk if there is a fire. Think whether the risk is greater for some because of when or where they work, such as night staff, or because they're not familiar with the premises, such as visitors or customers.

Children, the elderly or disabled people are especially vulnerable.

Check:

Have you identified:
▶ Who could be at risk?
▶ Who could be especially at risk?
Make a note of what you have found.

3. Evaluate and act

Evaluate and act

First, think about what you have found in Steps 1 and 2: what are the risks of a fire starting, and what are the risks to people in the building and nearby?

Remove and reduce risk

How can you avoid accidental fires? Could a source of heat or sparks fall, be knocked or pushed into something that would burn? Could that happen the other way round?

Protect

Take action to protect your premises and people from fire.

Check:

▶ Have you assessed the risks of fire in your workplace?
▶ Have you assessed the risk to staff and visitors?
▶ Have you kept any source of fuel and heat/sparks apart?
▶ If someone wanted to start a fire deliberately, is there anything around they could use?
▶ Have you removed or secured any fuel an arsonist could use?
▶ Have you protected your premises from accidental fire or arson?
▶ **How can you make sure everyone is safe in case of fire?**
  ▷ Will you know there is a fire?
  ▷ Do you have a plan to warn others?
  ▷ Who will make sure everyone gets out?
  ▷ Who will call the fire service?
  ▷ Could you put out a small fire quickly and stop it spreading?
▶ **How will everyone escape?**
  ▷ Have you planned escape routes?
  ▷ Have you made sure people will be able to safely find their way out, even at night if necessary?
  ▷ Does all your safety equipment work?
  ▷ Will people know what to do and how to use equipment?

Make a note of what you have found.

4. Record, plan and train

Record

Keep a record of any fire hazards and what you have done to reduce or remove them. If your premises are small, a record is a good idea. If you have five or more staff or have a licence, then you must keep a record of what you have found and what you have done.

Plan

You must have a clear plan of how to prevent fire and how you will keep people safe in case of fire. If you share a building with others, you need to coordinate your plan with them.

Train

You need to make sure your staff know what to do in case of fire, and if necessary, are trained for their roles.

Check:

- ▶ Have you made a record of what you have found, and action you have taken?
- ▶ Have you planned what everyone will do if there is a fire?
- ▶ Have you discussed the plan with all staff?
- ▶ **Have you:**
  - ▷ informed and trained people (practised a fire drill and recorded how it went)?
  - ▷ nominated staff to put in place your fire prevention measures, and trained them?
  - ▷ made sure everyone can fulfil their role?

  - ▷ informed temporary staff?
  - ▷ consulted others who share a building with you, and included them in your plan?

5. Review

Keep your risk assessment under regular review. Over time, the risks may change.

If you identify significant changes in risk or make any significant changes to your plan, you must tell others who share the premises and, where appropriate, retrain staff.

Have you:

- ▶ made any changes to the building inside or out?
- ▶ had a fire or near miss?
- ▶ changed work practices?
- ▶ begun to store chemicals or dangerous substances?
- ▶ significantly changed your stock or stock levels?
- ▶ planned your next fire drill?

The checklist above can help you with the fire risk assessment but you may need additional information, especially if you have large or complex premises.

Source: Department for Community and Local government.

APPENDIX 12.2 Typical fire notice

Chemical and biological health hazards and risk control

13.1	Forms of, classification of, and health risks from hazardous substances ▶ 374
13.2	Assessment of health risks ▶ 379
13.3	Occupational exposure limits ▶ 387
13.4	Control measures ▶ 389
13.5	Specific agents ▶ 400
13.6	Safe handling and storage of waste ▶ 411
13.7	Further information ▶ 415
13.8	Practice revision questions ▶ 415
Appendix 13.1	GHS Hazard (H) Statements (Health only) ▶ 418
Appendix 13.2	Hazardous properties of waste ▶ 419
Appendix 13.3	Different types of protective gloves ▶ 420

> **This chapter covers the following NEBOSH learning objectives:**
>
> 1. Outline the forms of, the classification of, and health risks from exposure to hazardous substances
> 2. Explain the factors to be considered when undertaking an assessment of the health risks from substances commonly encountered in the workplace
> 3. Explain the use and limitations of occupational exposure limits including the purpose of long-term and short-term exposure limits
> 4. Outline control measures that should be used to reduce the risk of ill-health from exposure to hazardous substances
> 5. Outline the hazards, risks and controls associated with specific agents
> 6. Outline the basic requirements related to the safe handling and storage of waste

13.1 Forms of, classification of, and health risks from hazardous substances

13.1.1 Introduction

Occupational health is as important as occupational safety but generally receives less attention from managers. Every year twice as many people suffer ill-health caused or exacerbated by the workplace than suffer workplace injury. Although these illnesses do not usually kill people, they can lead to many years of discomfort and pain. Such illnesses include respiratory disease, hearing problems, asthmatic conditions and back pain. Furthermore, it has been estimated in the UK that 30% of all cancers probably have an occupational link – that linkage is known for certain in 8% of cancer cases. Occupational exposures to fumes, dusts and chemicals in various forms account for around 4,000 deaths a year and some 38,000 individuals suffer breathing or lung problems possibly caused by their work.

Work in the field of occupational health has been taking place for the last four centuries and possibly longer. The main reason for the relatively low profile for occupational health over the years has been the difficulty in linking the ill-health effect with the workplace cause. Many illnesses, such as asthma or back pain, can have a workplace cause but can also have other causes. Many of the advances in occupational health have been as a result of statistical and epidemiological studies (one well-known such study linked the incidence of lung cancer to cigarette smoking). While such studies are invaluable in the assessment of health risk, there is always an element of doubt when trying to link cause and effect. The measurement of gas and dust concentrations is also subject to doubt when a correlation is made between a measured sample and the workplace environment from which it was taken. Occupational health, unlike occupational safety, is generally more concerned with probabilities than certainties.

The hazards of working in unfamiliar countries and/or climates are important health and safety issues – snake bites, diseases (such as malaria and yellow fever) and sunstroke. The course provider will relate these and any other particular hazards to the country for which the course is being provided.

In this chapter, chemical and biological health hazards will be considered – other forms of health hazard will be covered in Chapter 14.

Risk assessment is not only concerned with injuries in the workplace but also needs to consider the possibility of occupational ill-health. Health risks fall into the following four categories:

▶ chemical (e.g. paint solvents, exhaust fumes);
▶ biological (e.g. bacteria, pathogens);
▶ physical (e.g. noise, vibrations);
▶ psychological (e.g. occupational stress).

13.1.2 Forms of chemical agent

Chemicals can be transported by a variety of agents and in a variety of forms. They are normally defined in the following ways.

Dusts are solid particles slightly heavier than air but often suspended in it for a period of time. The size of the particles ranges from about 0.4 μm (fine) to 10 μm (coarse). Dusts are created either by mechanical processes (e.g. grinding or pulverising) or construction processes (e.g. concrete laying, demolition or sanding), or by specific tasks (e.g. furnace ash removal). The fine dust is much more hazardous because it penetrates deep into the lungs and remains there – known as **respirable dust**. In rare cases, respirable dust enters the bloodstream, directly causing damage to other organs. Examples of such fine dust are cement, granulated plastic materials and silica dust produced from stone or concrete dust. Repeated exposure may lead to permanent lung disease. Any dusts, which are capable of entering the nose and mouth during breathing, are known as **inhalable dusts.**

Fibres are threads or filaments that can occur naturally (e.g. asbestos) or be man-made such as glass-fibre, nylon and polyester. Man-made fibres are commonly used in insulation boards, blankets for the purpose of heat treatment, electrical insulation and in the reinforcement of plastic and cement. Fibres have a very high length to width ratio of at least 100 and many fibres are in the respirable range causing concern about the effects of exposure to many fibres – fibrosis of the lung and various cancers. Fibres with diameters in excess of 4 µm can cause irritation of the skin and eyes. Exposure to large concentration of such fibres may also cause irritation in the upper respiratory tract. Although the hazards and risks of asbestos and glass-fibre have been well documented, there are many other types of fibre in use. These include synthetic and semi-synthetic fibres (e.g. cellulosic fibres), a range of other non-organic fibres and whiskers and specialist technical fibres including various types of carbon fibres and nanofibres.

Gases are any substances at a temperature above their boiling point. Steam is the gaseous form of water. Common gases include carbon monoxide, carbon dioxide, nitrogen and oxygen. Gases are absorbed into the bloodstream where they may be beneficial (oxygen) or harmful (carbon monoxide).

Vapours are substances which are at or very close to their boiling temperatures. They are gaseous in form. Many solvents, such as cleaning fluids, fall into this category. The vapours, if inhaled, enter the bloodstream and some can cause short-term effects (dizziness) and long-term effects (brain damage).

Liquids are substances which normally exist at a temperature between freezing (solid) and boiling (vapours and gases). They are sometimes referred to as fluids in health and safety legislation.

Mists are similar to vapours in that they exist at or near their boiling temperature but are closer to the liquid phase. This means that there are very small liquid droplets, suspended and present in the vapour. A mist is produced during a spraying process (such as paint spraying). Many industrially produced mists can be very damaging if inhaled, producing similar effects to vapours. It is possible for some mists to enter the body through the skin or by ingestion with food.

Fume is a collection of very small metallic particles (less than 1 µm) which have condensed from the gaseous state. They are most commonly generated by the welding process. The particles tend to be within the respirable range (approximately 0.4–1.0 µm) and can lead to long-term permanent lung damage. The exact nature of any harm depends on the metals used in the welding process and the duration of the exposure.

13.1.3 Forms of biological agent

As with chemicals, biological hazards may be transported by any of the following forms of agent.

Fungi are very small organisms, sometimes consisting of a single cell, and can appear plant like (e.g. mushrooms and yeast). Unlike plants, they cannot produce their own food but either live on dead organic matter or on living animals or plants as parasites. Fungi reproduce by producing spores, which can cause

allergic reactions when inhaled. The infections produced by fungi in humans may be mild, such as athlete's foot, or severe, such as ringworm. Many fungal infections can be treated with antibiotics.

Moulds are a particular group of very small fungi which, under damp conditions, will grow on surfaces such as walls, bread, cheese, leather and canvas. They can be beneficial (penicillin) or cause allergic reactions (asthma). Asthma attacks, athlete's foot and farmer's lung are all examples of fungal infections.

Bacteria are very small single-celled organisms which are much smaller than cells within the human body. They can live outside the body and be controlled and destroyed by antibiotic drugs. There is evidence that some bacteria are becoming resistant to antibiotics. This has been caused by the widespread misuse of antibiotics. It is important to note that not all bacteria are harmful to humans. Bacteria aid the digestion of food, and babies would not survive without their aid to break down the milk in their digestive system. Legionellosis, tuberculosis and tetanus are all bacterial diseases.

Viruses are minute non-cellular organisms which can only reproduce within a host cell. They are very much smaller than bacteria and cannot be controlled by antibiotics. They appear in various shapes and are continually developing new strains. They are usually only defeated by the defence and healing mechanisms of the body. Drugs can be used to relieve the symptoms of a viral attack but cannot cure it. The common cold is a viral infection as are hepatitis, AIDS (HIV) and influenza.

A number of diseases including bovine spongiform encephalopathy (BSE) in cattle and Creutzfeldt–Jakob disease (CJD) in humans are caused by another biological agent known as a prion. A **prion** is an infectious agent that is composed primarily of protein. Such agents induce existing substances, called polypeptides, in the host organism to take on a rogue form. All known prion diseases affect the structure of the brain or other neural tissue, are currently untreatable and are always fatal.

13.1.4 Classification of hazardous substances and their associated health risks

A hazardous substance is one which can cause ill-health to people at work. Such substances may include those used directly in the work processes (glues and paints), those produced by work activities (welding fumes) or those which occur naturally (dust). Hazardous substances are classified according to the severity and type of hazard which they may present to people who may come into contact with them. The contact may occur while working or transporting the substances or might occur during a fire or accidental spillage.

Until 2015, the supply and hazard classification of substances was regulated in the UK by the Chemicals

13

(Hazard Information and Packaging for Supply) 2009 Regulations (CHIP 4). These regulations required suppliers to identify the hazards of the chemicals that they supplied, provide information regarding those hazards and package their products safely. From June 2015 CHIP was revoked and fully superseded by the Classification, Labelling and Packaging of Substances and Mixtures Regulation (EC) (known as CLP).

A hazard pictogram is an image on a label that includes a warning symbol and specific colours intended to provide information about the damage a particular substance or mixture can cause to health or the environment (see Figure 13.1). The CLP Regulation has introduced a new classification and labelling system for hazardous chemicals in the European Union. The pictograms have also changed and are in line with the United Nations Globally Harmonised System (GHS). The new pictograms are in the shape of a red diamond with a white background, and replace the old orange square symbols which applied under CHIP 4. Since 1 December 2010, some substances and mixtures have already been labelled according to the new legislation, but the old pictogram can still be on the market until 1 June 2017.

There are several classifications but here only the most common under each classification will be described.

CHIP 4 classifications

Irritant is a non-corrosive substance which can cause skin (dermatitic) or lung (bronchial) inflammation after repeated contact. People who react in this way to a particular substance are **sensitised** or **allergic** to that substance. In most cases, it is likely that the concentration of the irritant may be more significant than the exposure time. Many household substances, such as wood preservatives, bleaches and glues, are irritants. Many chemicals used as solvents are also irritants (white spirit, toluene and acetone). Formaldehyde and ozone are other examples of irritants.

Corrosive substances are ones that may destroy living tissue on contact – usually by burning the skin. Usually strong acids or alkalis, examples include sulphuric acid and caustic soda. Many tough cleaning substances, such as kitchen oven cleaners, are corrosive as are many dishwasher crystals.

Harmful is the most commonly used classification and describes a substance which, if swallowed, inhaled or penetrates the skin, *may* pose limited health risks. These risks can usually be minimised or removed by following the instructions provided with the substance (e.g. by using personal protective equipment). There are many household substances which fall into this category including bitumen-based paints and paintbrush restorers. Many chemical cleansers are categorised as harmful. It is very common for substances labelled harmful also to be categorised as irritant.

Toxic substances are poisonous and impede or prevent the function of one or more organs within the body, such as the kidney, liver and heart. A toxic substance is, therefore, a poisonous one. Lead, mercury, pesticides and the gas carbon monoxide are toxic substances. The effect on the health of a person exposed to a toxic substance depends on the concentration and toxicity of the substance, the frequency of the exposure and the effectiveness of the control measures in place. The state of health and age of the person and the route of entry into the body have influence on the effect of the toxic substance.

Very toxic substances are very hazardous to health when inhaled, swallowed or when they come in contact with the skin and could be fatal. Examples of such substances include prussic acid, nicotine and carbon monoxide.

GHS classifications

The **health hazard** classification combines the CHIP 4 'harmful' and 'irritant' classifications. Such a substance will be harmful if swallowed, inhaled or contacts the skin. It may cause respiratory irritation, drowsiness or dizziness, an allergic skin reaction, a serious eye irritation or skin irritation. It could also be harmful to the environment by destroying ozone in the upper atmosphere. Typical substances include washing detergents, toilet cleaner and coolant fluid.

A **serious health hazard** substance may be fatal if swallowed and/or enters the airways and cause damage to organs. It may damage human fertility or the unborn child and may cause genetic defects. It may also cause allergy or asthma symptoms or breathing difficulties if inhaled or may cause cancer. Turpentine and petrol carry a serious health hazard warning. It is similar to the toxic CHIP 4 classification.

The **corrosive** classification is very similar to the CHIP 4 corrosive classification. A corrosive substance will probably be corrosive to metals and can cause both severe skin burns and eye damage. Strong drain cleaners, acetic acid and hydrochloric acid are all examples of corrosive substances.

Acute toxicity is similar to the very toxic CHIP 4 classification. Such a substance will always be toxic but can also be fatal if swallowed, inhaled or comes into contact with the skin. Biocides, methanol and many pesticides can be acutely toxic.

Other types of hazardous substances

Carcinogenic substances are ones which are known for, or suspected of, promoting abnormal development of body cells to become cancers. Asbestos, hardwood dust, creosote and some mineral oils are carcinogenic. It is very important that the health and safety rules accompanying the substance are strictly followed.

Mutagenic substances are those which damage genetic material within cells, causing abnormal changes that can be passed from one generation to another. A **reproductive toxin** is a chemical which affects adversely the reproductive process including mutations and other effects on the unborn child. One group of such toxins are teratogens that can cause congenital abnormalities, which are also called birth defects.

Figure 13.1 (a) Use of GHS symbols

Each of the classifications may be identified by a symbol and a symbolic letter – the most common of these are shown in Figure 13.1(b), which shows the relationship between the CHIP 4 and GHS classifications but also note that these are changing to the new Global Harmonisation Scheme (GHS).

The effects on health of hazardous substances may be either acute or chronic.

Acute effects are of short duration and appear fairly rapidly, usually during or after a single or short-term exposure to a hazardous substance. Such effects may be severe and require hospital treatment but are usually reversible. Examples include asthma-type attacks, nausea and fainting.

Chronic effects develop over a period of time which may extend to many years. The word 'chronic' means 'with time' and should not be confused with 'severe' as its use in everyday speech often implies. Chronic health effects are produced from prolonged or repeated exposures to hazardous substances resulting in a gradual, latent and often irreversible illness, which may remain undiagnosed for many years. Many cancers and mental diseases fall into the chronic category. During the development stage of a chronic disease, the individual may experience no symptoms.

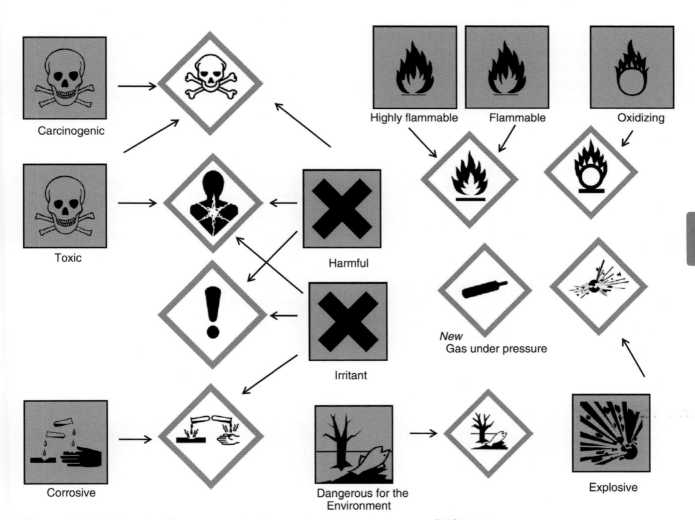

Figure 13.1 (b) How the European packaging symbols relate to the new GHS labels

13.1.5 Hazard warnings and precautionary statements

The European Regulation on classification, labelling and packaging of substances and mixtures became law across the EU in 2009 and is known as either the 'CLP Regulation' or simply as 'CLP'. It adopted the United Nations Globally Harmonised System on the classification and labelling of chemicals (GHS) across all EU states.

Worldwide, there are many different laws to identify hazardous chemicals (classification) and to communicate this information to users. This is often confusing as the same chemical can have different hazard descriptions in different countries. The UN brought together experts from different countries to create the GHS.

The aim of the GHS is to have, throughout the world, the same:

▶ criteria for classifying chemicals according to their health, environmental and physical hazards;
▶ hazard communication requirements for labelling and safety data sheets.

The GHS provides a basis for globally uniform physical, environmental, health and safety information on hazardous chemicals through the harmonisation of the criteria for their classification and labelling. It was developed at UN level with the aim of overcoming differing labelling information requirements on physical, health and environmental hazards for the same chemicals around the world. Moreover, it also aims to lower barriers to trade caused by the fact that every time a product was exported, it mostly had to be classified and labelled differently because of differing criteria.

Outside the EU many other countries have subscribed to implementing the GHS into domestic law, including the USA, Canada, New Zealand, Brazil, China, the Philippines, Russia, Japan, Mexico, South Africa and various other African countries. The stage of implementation ranges from those countries which already have or are about to have in place their own GHS implementing scheme (e.g. the EU, Japan, New Zealand, South Korea) to those countries where focused activities on and development of a GHS implementing scheme are ongoing (e.g. the USA) and to further countries which have just started their discussions with the view to implement the GHS. Each individual country employs its specific domestic legal instruments, e.g. sector-specific acts or national standards, to implement the GHS.

Since the GHS is a voluntary agreement, it has to be adopted through a suitable national or regional legal mechanism to ensure it becomes legally binding. That is achieved by the CLP Regulation. As the GHS was heavily influenced by the old EU system, the CLP Regulation is very similar in many ways. The duties on suppliers are broadly the same: classification, labelling and packaging. The rules for the classification of hazardous substances have changed and a new set of hazard pictograms (quite similar to the old ones) are used (see Figure 13.1).

The EU legislation on classification, labelling and packaging consists of three acts: The Dangerous Substances Directive (Directive 67/548/EEC, DSD), the Dangerous Preparations Directive (Directive 1999/45/EC, DPD) and the Regulation on Classification, Labelling and Packaging of Substances and Mixtures, Regulation (EC) No 1272/2008 (CLP Regulation or CLP) which applies directly in all member states.

Further information on the stage of implementation of the UN GHS in different countries is available on the UN ECE website, see: http://unece.org/trans/danger/publi/ghs/implementation_e.html

Under the CLP Regulations there are:

▶ new scientific criteria to assess hazardous properties of chemicals;
▶ two new harmonised hazard warning symbols for labels (known as 'pictograms');
▶ a new design for existing symbols;
▶ new harmonised **hazard warning (H)** and **precautionary statements (P)** for labels; and
▶ two new signal words – 'Danger' and 'Warning'.

A **hazard statement** is a phrase that describes the nature of the hazard in the substance or mixture. A hazard statement will be determined by the application of the classification criteria. It replaces the 'risk or R-phrase' used previously. A list of some of these is given in Appendix 13.1.

A **precautionary statement** is a phrase that describes recommended measure(s) to minimise or prevent adverse effects resulting from exposure to a hazardous substance or mixture due to its use or disposal. Suppliers determine the appropriate precautionary statements (usually no more than six) based on the required hazard statements. It replaces the 'safety or S-phrase' used previously.

As mentioned above, the CLP Regulation also introduces two new **signal words**: 'Danger' and 'Warning'. If the chemical has a more severe hazard, the label includes the signal word 'Danger'; in case of less severe hazards, the signal word is 'Warning'.

Examples of the new hazard statements and the new precautionary statements for labels are shown in Table 13.1.

Table 13.1 Examples of the new hazard warning (H) and precautionary statements (P)

Hazard warning statement (H)	Precautionary statement (P)
H240 – Heating may cause an explosion	P102 – Keep out of reach of children
H320 – Causes eye irritation	P271 – Use only outdoors or in well-ventilated area
H401 – Toxic to aquatic life	P410 – Protect from sunlight

13.2 Assessment of health risks

13.2.1 Types of health risk

The principles of control for health risks are the same as those for safety. However, the nature of health risks can make the link between work activities and employee ill-health less obvious than in the case of injury from an accident.

Unlike safety risks, which can lead to immediate injury, the result of daily exposure to health risks may not manifest itself for months, years and, in some cases, decades. Irreversible health damage may occur before any symptoms are apparent. It is, therefore, essential to develop a preventative strategy to identify and control risks before anyone is exposed to them.

Risks to health from work activities include:

▶ skin contact with irritant substances, leading to dermatitis, etc.;

▶ inhalation of respiratory sensitisers, triggering immune responses such as asthma;

▶ badly designed workstations requiring awkward body postures or repetitive movements, resulting in upper limb disorders, repetitive strain injury and other musculoskeletal conditions;

▶ noise levels which are too high, causing deafness and conditions such as tinnitus;

▶ too much vibration, for example from hand-held tools, leading to hand–arm vibration syndrome and circulatory problems;

▶ exposure to ionising and non-ionising radiation including ultraviolet in the sun's rays, causing burns, sickness and skin cancer;

▶ infections ranging from minor sickness to life-threatening conditions, caused by inhaling or being contaminated with microbiological organisms;

▶ stress causing mental and physical disorders.

Some illnesses or conditions, such as asthma and back pain, have both occupational and non-occupational causes and it may be difficult to establish a definite causal link with a person's work activity or their exposure to particular agents or substances. But, if there is evidence that shows the illness or condition is prevalent among the type of workers to which the person belongs or among workers exposed to similar agents or substances, it is likely that their work and exposure has contributed in some way.

13.2.2 Routes of entry to the human body

There are three principal routes of entry of hazardous substances into the human body:

▶ **inhalation** – breathing in the substance with normal air intake. This is the main route of contaminants into the body. These contaminants may be chemical (e.g. solvents or welding fume) or biological (e.g. bacteria or fungi) and become airborne by a variety of modes, such as sweeping, spraying, grinding and bagging. They enter the lungs where they have access to the bloodstream and many other organs;

▶ **absorption through the skin** – the substance comes into contact with the skin and enters through either the pores or a wound. Tetanus can enter in this way as can toluene, benzene and various phenols;

▶ **ingestion** – through the mouth and swallowed into the stomach and the digestive system. This is not a significant route of entry to the body. The most common occurrences are due to airborne dust or poor personal hygiene (not washing hands before eating food, drinking or smoking) (see Figure 13.3).

Another rare entry route is by **injection** or **skin puncture**. The abuse of compressed air lines by shooting high pressure air at the skin can lead to air bubbles entering the bloodstream. Accidents involving hypodermic syringes in a health or veterinary service setting are rare but illustrate this form of entry route.

Body cells have a natural defence system against illness caused by harmful invasive foreign particles

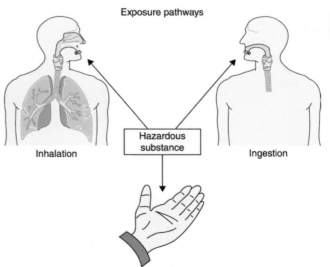

Figure 13.3 Hazardous substances – principal routes of entry into the human body

Figure 13.2 Paint spraying – risk of sensitising, particularly if isocyarate based paint used and inadequate local exhaust ventilation

13

and bacteria. White blood cells, known as phagocytes, give protection by destroying these harmful foreign particles, bacteria and dead cells. This protection is basically a scavenging action at the level of cells. Other defence mechanisms include the secretion of defensive substances, the prevention of excessive blood loss and the repair of damaged tissue.

The most effective control measures which can reduce the risk of infection from biological organisms are disinfection, proper disposal of clinical waste (including syringes), good personal hygiene and, where appropriate, personal protective equipment. Other measures include vermin control, water treatment and immunisation.

There are five major functional systems within the human body – respiratory, nervous, cardiovascular (blood), urinary and the skin.

The respiratory system

This comprises the lungs and associated organs (e.g. the nose). Air is breathed in through the nose, and passes through the trachea (windpipe) and the bronchi into the two lungs. Within the lungs, the air enters many smaller passageways (bronchioles) and thence to one of 300,000 terminal sacs called alveoli. The alveoli are approximately 0.1 mm across, although the entrance is much smaller. On arrival in the alveoli, there is a diffusion of oxygen into the bloodstream through blood capillaries and an effusion of carbon dioxide from the bloodstream. While soluble dust which enters the alveoli will be absorbed into the bloodstream, insoluble dust (respirable dust) will remain permanently, leading to possible chronic illness (Figure 13.4).

The whole of the bronchial system is lined with hairs, known as cilia. The cilia offer some protection against insoluble dusts. These hairs will arrest all non-respirable dust (above 5 μm) and, with the aid of mucus, pass the dust from one hair to a higher one and thus bring the dust back to the throat (this is known as the ciliary escalator). It has been shown that smoking damages this action. The nasal hairs in the nose will normally trap large particles (greater than 20 μm) before they enter the trachea. There are over 40 conditions that can affect the lungs and/or airways and impinge on the ability of a person to breathe normally.

Respirable dust tends to be long thin particles with sharp edges which puncture the alveoli walls. The puncture heals producing scar tissues which are less flexible than the original walls – this can lead to fibrosis. Such dusts include asbestos, coal, silica, some plastics and talc. The possible indicators of a dust problem in the workplace are fine deposits on surfaces, people and products or blocked filters on extraction equipment. Ill-health reports or complaints from the workforce could also indicate a dust problem.

Acute effects on the respiratory system include bronchitis and asthma and chronic effects include fibrosis and cancer. Hardwood dust, for example, can produce asthma attacks and nasal cancer.

Finally, asphyxiation, due to a lack of oxygen, is a problem in confined spaces particularly when MIG (metal inert gas) welding is taking place.

The nervous system

The nervous system consists primarily of the brain, the spinal cord and nerves extending throughout the body. Any muscle movement or sensation (e.g. hot and cold) is controlled or sensed by the brain through small electrical impulses transmitted through the spinal cord and nervous system.

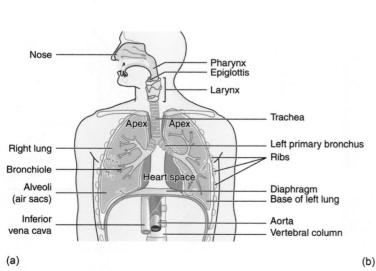

(a)

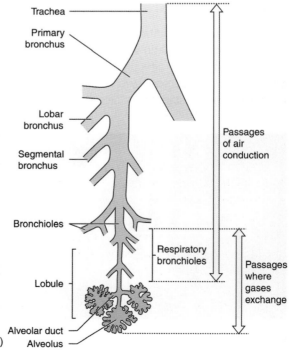

(b)

Figure 13.4 The (a) upper and (b) lower respiratory system

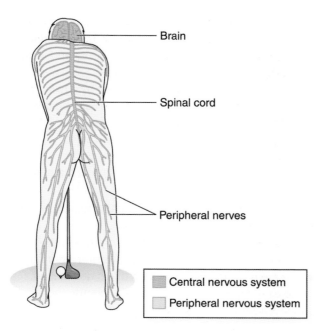

Figure 13.5 The nervous system

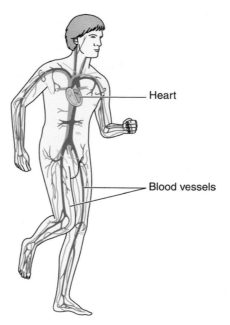

Figure 13.6 The cardiovascular system

The effectiveness of the nervous system can be reduced by **neurotoxins** and lead to changes in mental ability (loss of memory and anxiety), epilepsy and narcosis (dizziness and loss of consciousness). Organic solvents (trichloroethylene) and heavy metals (mercury) are well-known neurotoxins. The expression 'mad hatters' originated from the mental deterioration of top hat polishers in the 19th century who used mercury to produce a shiny finish on the top hats.

The cardiovascular system

The blood system uses the heart to pump blood around the body through arteries, veins and capillaries.

Blood is produced in the bone marrow and consists of a plasma within which are red cells, white cells and platelets. The system has three basic objectives:

▶ to transport oxygen to vital organs, tissues and the brain and carbon dioxide back to the lungs (red cell function);
▶ to attack foreign organisms and build up a defence system (white cell function);
▶ to aid the healing of damaged tissue and prevent excessive bleeding by clotting (platelets).

There are several ways in which hazardous substances can interfere with the cardiovascular system. Benzene can affect the bone marrow by reducing the number of blood cells produced. Carbon monoxide prevents the red cells from absorbing sufficient oxygen and the effects depend on its concentration. Symptoms begin with headaches and end with unconsciousness and possibly death.

The urinary system

The urinary system extracts waste and other products from the blood. The two most important organs are the liver (normally considered part of the digestive system) and kidney, both of which can be affected by hazardous substances within the bloodstream.

The liver removes toxins from the blood, maintains the levels of blood sugars and produces protein for the blood plasma. Hazardous substances can cause

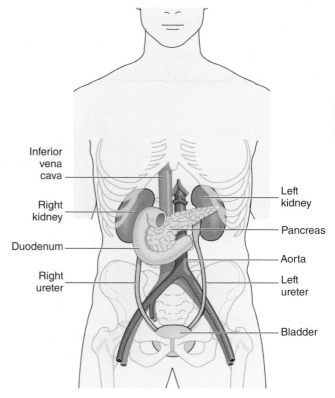

Figure 13.7 Parts of the urinary system

the liver to be too active or inactive (e.g. xylene), lead to liver enlargement (e.g. cirrhosis caused by alcohol) or liver cancer (e.g. vinyl chloride).

The kidneys filter waste products from the blood as urine, regulate blood pressure and liquid volume in the body and produce hormones for making red blood cells. Heavy metals (e.g. cadmium and lead) and organic solvents (e.g. glycol ethers used in screen printing) can restrict the operation of the kidneys possibly leading to failure.

The skin

The skin holds the body together and is the first line of defence against infection. It regulates body temperature, is a sensing mechanism, provides an emergency food store (in the form of fat) and helps to conserve water. There are two layers – an outer layer called the epidermis (0.2 mm) and an inner layer called the dermis (4 mm). The epidermis is a tough protective layer and the dermis contains the sweat glands, nerve endings and hairs.

The most common industrial disease of the skin is **dermatitis** (non-infective dermatitis). It begins with a mild irritation on the skin and develops into blisters which can peel and weep and may become septic. It can be caused by various chemicals, mineral oils and solvents. There are two types:

▶ **irritant contact dermatitis** – occurs soon after contact with the irritant substance and the condition reverses after contact ceases (detergents and weak acids);

▶ **allergic contact dermatitis** – caused by a skin sensitiser such as turpentine, epoxy resin, solder flux and formaldehyde that exerts its effects via the immune system. Once sensitised to a substance, a severe dermatitis may occur following a small exposure to the same substance at a later date.

Dermatitis is on the rise and, in the UK, is costing business more than £20 million a year, even though the cost of control measures to prevent the disease is minimal. Workers in the hotel and catering industry are particularly vulnerable to this debilitating disease as are print workers who have contracted dermatitis after coming into contact with UV curable lacquers used to produce a glossy finish to magazine covers.

For many years, dermatitis was seen as a 'nervous' disease which was psychological in nature. Nowadays, it is recognised as an industrial disease which can be controlled by good personal hygiene, personal protective equipment, use of barrier creams and health screening of employees. Dermatitis can appear anywhere on the body but it is normally found on the hands. Therefore, gloves should always be worn when there is a risk of dermatitis.

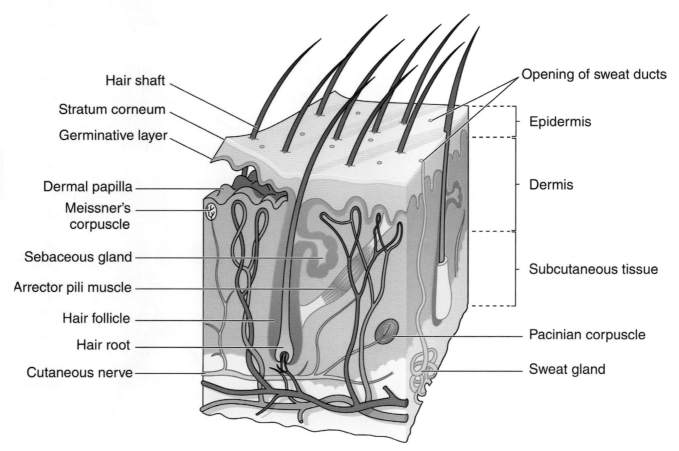

Figure 13.8 The skin – main structures in the dermis

The risk of dermatitis occurring increases with the presence of cuts or abrasions on the skin, which allow chemicals to be more easily absorbed. The risk also depends on the type, sensitivity and existing condition of the skin.

13.2.3 ILO requirements to protect workers from chemical agents

Hazardous substances or chemical agents are found in a wide range of workplaces. Therefore it is important to determine whether hazardous substances are present. Risks to the health and safety of workers must be assessed, and risk management measures established with the aim of reducing exposure to those risks.

The ILO has produced two detailed Codes of Practice 'Safety in the Use of Chemicals' and 'Ambient factors in the workplace' that give very detailed information on working with hazardous substances. It is derived from policy on occupational safety and health and the working environment required by the Occupational Safety and Health Convention, 1981 (No.155).

In the Code of Practice 'Safety in the Use of Chemicals' the ILO expects the competent authority (normally an arm of the national government) to ensure that criteria are established on measures which provide for safety of workers, in particular:

(a) in the production and handling of hazardous chemicals;

(b) in the storage of hazardous chemicals;

(c) in the transport of hazardous chemicals, consistent with national or international transport regulations;

(d) in the disposal and treatment of hazardous chemicals and hazardous waste products, consistent with national or international regulations.

Employers are expected to:

(a) produce a written policy on safety in the use of chemicals, as part of their general policy and arrangements in the field of occupational health and safety;

(b) ensure that all chemicals used at work are labelled or marked in accordance with the provisions of the ILO Code of Practice, and that chemical safety data sheets have been provided in respect of all hazardous chemicals used at work. They should also ensure that the chemical safety data sheets provided by the supplier, or similar relevant information where such data sheets have not been provided, are made available to workers and their representatives;

(c) not use chemicals which have not been provided with chemical safety data sheets until the relevant information has been obtained from the supplier or from other reasonably available sources, and has been made available to workers and their representatives;

(d) use, wherever practical, non-hazardous or low-hazard chemicals;

(e) maintain a record of hazardous chemicals used at the workplace, cross-referenced to the appropriate chemical safety data sheets. The record should be accessible to all workers in the workplace who might be affected by the use of the chemicals, and to their representatives;

(f) make an assessment of the risks arising from the use of chemicals at work, taking into account the information provided by the supplier or, where this is not available, obtained from reasonably available sources, and should protect the workers by appropriate preventative measures;

(g) take appropriate measures to protect workers against the risks identified by the assessment of risks. Where the risks cannot be eliminated or adequately controlled, employers should provide and maintain personal protective equipment, including clothing, as appropriate, at no cost to the worker, and should implement measures to ensure its use;

(h) comply with appropriate standards, codes and guidelines formulated, approved or recognised by the competent authority concerning safety in the use of chemicals;

(i) ensure adequate and competent supervision of work and work practices, and the application and use of the control measures provided;

(j) make adequate arrangements to deal with incidents and accidents involving chemicals;

(k) provide their workers with necessary, appropriate and periodic instructions and training, taking account of the functions and capacities of different categories of workers and, where appropriate, workers' representatives.

Where an employer is also a national or multinational enterprise with more than one establishment, the employer should provide safety and health measures relating to the prevention and control of, and protection against, risks from hazardous chemicals, without discrimination, to all workers who may be affected regardless of the place or country in which they are situated. In all countries in which they operate, multinational and multi-site enterprises should make available to:

(a) the workers concerned;

(b) workers' representatives;

(c) the competent authority;

(d) employers' and workers' organisations;

Information on the standards and procedures related to the use of hazardous chemicals relevant to their local operations, which they observe in other countries.

The ILO Codes of Practice also expect workers to:

(a) take all reasonable steps to eliminate or minimise risk to themselves and to others from the use of chemicals or other hazardous substances at work;

13

(b) take care of their own health and safety and that of other persons who may be affected by their acts or omissions at work, as far as possible and in accordance with their training and with instructions given by their employer;

(c) make proper use of all devices provided for their protection or the protection of others;

(d) report forthwith to their supervisor any situation which they believe could present a risk, and which they cannot properly deal with themselves.

The ILO Code of Practice 'Ambient factors in the workplace' should also be considered as the basis for eliminating or controlling exposure to hazardous airborne chemicals.

The European Union in its Directive (98/24/E(C) lays down minimum requirements which apply to all hazardous substances that are or may be present in EU workplaces. These minimum requirements are very similar to those required by the ILO and comprise:

▶ indicative and binding Occupational Exposure Limit values (OELs) and biological limit values (employers and workers are to be kept informed of these);

▶ determination and risk assessment of hazardous chemical agents;

▶ general principles for prevention of risks associated with hazardous chemical agents;

▶ specific protection and prevention measures;

▶ arrangements to deal with accidents, incidents and emergencies;

▶ information and training for workers;

▶ prohibition of the production, manufacture or use of certain chemicals to prevent workers' exposure;

▶ health surveillance;

▶ consultation and participation of workers;

▶ preparation and adoption of technical guidance.

The European Commission, Directorate General for Employment, Social Affairs and Equal Opportunities, has published practical, non-binding guidelines on this.

13.2.4 Details of a hazardous substance assessment

Not all hazardous substances are covered by the legislation. If there is no warning symbol on the substance container or it is a biological agent which is not directly used in the workplace (such as an influenza virus), then no hazardous substance assessment is required. A hazardous substance assessment should be made for the following substances:

▶ substances having documented occupational exposure limits such as those listed in the HSE publication EH40 (Occupational Exposure Limits) and in the ILO Code of Practice 'Ambient factors in the workplace' (including its associated references);

▶ biological agents connected with the workplace;

▶ substantial quantities of airborne dust (more than 10 mg/m^3 of total inhalable dust or 4 mg/m^3 of respirable dust, both 8-hour time-weighted average (TWA), when there is no indication of a lower value);

▶ any substance creating a comparable hazard which for technical reasons may not be documented.

The following factors should be considered when assessing health risks:

▶ the health risk assessment should be suitable and sufficient (see Chapter 4);

▶ the exposure of employees to hazardous substances is adequately controlled;

▶ the control measures provided are properly used;

▶ the control measures and any associated equipment is properly maintained;

▶ monitor employees exposed to hazardous substances;

▶ there is adequate health surveillance; and

▶ information, instruction and training.

Assessment requirements

A hazardous substance assessment is very similar to a risk assessment but is applied specifically to hazardous substances. The UK HSE has suggested five steps for such assessments but within these steps there are a number of sub-sections. The steps are as follows.

Step 1

Gather information about the substances, the work and working practices. Assessors should:

▶ identify the hazardous substances present or likely to be present in the workplace;

▶ monitor the categories and numbers of persons (e.g. employees and visitors) likely to be affected;

▶ gather information about the hazardous substances including the quantity of the substances used;

▶ identify the hazards from these substances by reviewing labels, material safety data sheets, HSE guidance and published literature;

▶ decide who could be affected by the hazardous substances and the possible routes of entry to people exposed (i.e. inhalation, ingestion or absorption). There is a need to look at both the substances and the activities where people could be exposed to hazardous substances.

Step 2

Evaluate the risks to health either individually or collectively. Assessors should:

▶ evaluate the risks to health including the duration and frequency of the exposure of those persons to the substances;

▶ evaluate the level of exposure, for example the concentration and length of exposure to any airborne dusts, gases, fumes or vapours;

- consider any Workplace Exposure Limits (WELs) (see 13.3);
- decide if existing and potential exposure present any insignificant risks to health or they pose a significant risk to health.

Step 3

Decide what needs to be done to control the exposure to hazardous substances. Assessors should:

- evaluate the existing control measures, including any PPE and RPE, for their effectiveness (using any available records of environmental monitoring) and compliance with relevant legislation;
- decide on additional control measures, if any are required (see 13.4);
- decide what maintenance and supervision of the use of the control measures are needed;
- plan what to do in an emergency;
- set out how exposure should be monitored;
- decide what, if any, health surveillance is necessary;
- decide what information, instruction and training is required.

Step 4

Record the assessment. Assessors should:

- decide if a record is required (five or more employed and significant findings);
- decide on the format of the record;
- decide on storage and how to make records available to employees, safety representatives, etc.

Step 5

Review the assessment. Assessors should:

- decide when a review is necessary (e.g. changes in substances used, processes or people exposed);
- decide what needs to be reviewed.

An example of a typical form that can be used for a hazardous substance assessment is given in Chapter 18.

It is important that the assessment is conducted by somebody who is competent to undertake it. Such competence will require some training, the extent of which will depend on the complexities of the workplace. For large organisations with many high-risk operations, a team of competent assessors will be needed. If the assessment is simple and easily repeated, a written record is not necessary. In other cases, a concise and dated record of the assessment together with recommended control measures should be made available to all those likely to be affected by the hazardous substances. The assessment should be reviewed on a regular basis, particularly when there are changes in work processes or substances or when adverse ill-health is reported.

13.2.5 Assessing exposure and health surveillance

Some aspects of health exposure will need input from specialist or professional advisers, such as occupational health hygienists, nurses and doctors. However, considerable progress can be made by taking straightforward measures such as:

- consulting the workforce on the design of workplaces;
- talking to manufacturers and suppliers of substances and work equipment about minimising exposure;
- enclosing machinery to cut down dust, fumes and noise;
- researching the use of less hazardous substances;
- ensuring that employees are given appropriate information and are trained in the safe handling of all the substances and materials to which they may be exposed.

To assess health risks and to make sure that control measures are working properly, it may be necessary, for example, to measure the concentration of substances in the air to make sure that exposures remain within the relevant workplace exposure levels. Sometimes health surveillance of workers who may be exposed will be needed. This will enable data to be collected to check control measures and for early detection of any adverse changes to health. Health surveillance procedures available include biological monitoring for bodily uptake of substances, examination for symptoms and medical surveillance – which may entail clinical examinations and physiological or psychological measurements by occupationally qualified registered medical practitioners. The procedure chosen should be suitable for the case concerned. Sometimes a method of surveillance is specified for a particular substance in national legislation or guidance. Whenever surveillance is undertaken, a health record has to be kept for the person concerned.

Health surveillance should be supervised by a registered/qualified medical practitioner or, where appropriate, it should be done by a suitably qualified person (e.g. an occupational nurse). In the case of inspections for easily detectable symptoms like chrome ulceration or early signs of dermatitis, health surveillance should be done by a suitably trained responsible person. There is more on health surveillance later in this chapter.

13.2.6 Sources of information

There are other important sources of information available for a hazardous substance assessment.

Product labels include details of the hazards associated with the substances contained in the product and any precautions recommended. They may also bear one or more of the hazard classification symbols and/or risk phrases.

13

Material safety data sheets are another very useful source of information for hazard identification and associated advice. Manufacturers of hazardous substances are obliged to supply such sheets to users, giving details of the name, chemical composition and properties of the substance. Information on the nature of the health hazards and any relevant exposure standard (OEL) should also be given together with recommended exposure control measures and personal protective equipment. The sheets contain useful additional information on first-aid and fire-fighting measures and handling, storage, transport and disposal information. The data sheets should be stored in a readily accessible and known place for use in the event of an emergency, such as an accidental release.

Product J

(abc chemical)

Danger

Fatal if swallowed
Causes skin irritation

Precautions:
Wear protective gloves.
Take off contaminated clothing and wash before reuse.
Wash hands thoroughly after handling.
Do not eat, drink or smoke when using this product.

Store locked up.
Dispose of contents/containers in accordance with local regulations.

IF ON SKIN: Rinse skin with water/shower.
IF IN EYES: Rinse cautiously with water.
IF SWALLOWED: Immediately call a Poison Center or doctor/physician. Do not induce vomiting.

ABC Chemical Co., 123 Anywhere St., (123) 456-7890
See the SDS for more information

Figure 13.9 (a) Typical symbols and (b) product label on containers

Other sources of information include trade association publications, industrial codes of practice and specialist reference manuals.

13.2.7 Survey techniques for health risks

An essential part of the hazardous substance assessment is the measurement of the quantity of the substance in the atmosphere surrounding the workplace. This is known as air sampling. There are four common types of air-sampling technique used for the measurement of air quality:

1. **Stain tube detectors** use direct reading glass indicator tubes filled with chemical crystals which change colour when a particular hazardous substance passes through them. The method of operation is very similar to the breathalyser used by the police to check alcohol levels in motorists. The glass tube is opened at each end and fitted into a pumping device (either hand or electrically operated). A specific quantity of contaminated air containing the hazardous substance is drawn through the tube and the crystals in the tube change colour in the direction of the air flow. The tube is calibrated such that the extent of the colour change along the tube indicates the concentration of the hazardous substance within the air sample. This method can only be effective if there are no leakages within the instrument and the correct volume of sampled air is used. The instrument should be held within 30 cm of the nose of the person whose atmosphere is being tested. A large range of different tubes is available. This technique of sampling is known as **grab** or **spot sampling** as it is taken at one point.

The advantages of the technique are that it is quick, relatively simple to use, portable and inexpensive. There are, however, several disadvantages:

▶ The instrument cannot be used to measure concentrations of dust or fumes.
▶ The substance to be detected has to be known prior to sampling.

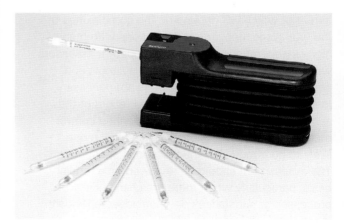

Figure 13.10 Hand pump and stain detectors

- The accuracy of the reading is approximately 25%; it will yield a false reading if other contaminants present react with the crystals.
- The instrument can only give an instantaneous reading, not an average reading over the working period (TWA).
- The tubes are very fragile with a limited shelf life.
- Some disposal issues.

2. **Passive sampling** is measured over a full working period by the worker wearing a badge containing absorbent material. The material will absorb the contaminant substance, and, at the end of the measuring period, the sample is sent to a laboratory for analysis. The advantages of this method over the stain tube are that there is less possibility of instrument errors and it gives a TWA reading.

3. **Sampling pumps and heads** can be used to measure gases and dusts. The worker, whose breathing zone is being monitored, wears a collection head as a badge and a battery-operated pump on his/her back at waist level. The pump draws air continuously through a filter, fitted in the head, which will either absorb the contaminant gas or trap hazardous dust particles. If this filter is used for dust measurement, it is sometimes called a static dust sampler. Before sampling takes place, the filter is weighed and the pump flow rate preset. After the designated testing period, it is sent to a specialist laboratory for analysis. The quantity of dust present would be determined either by measuring the weight change of the filter or by chemical analysis (e.g. for lead) or by using microscopes to count the number of fibres (e.g. for asbestos). This system is more accurate than stain tubes and gives a TWA result but can be uncomfortable to wear over long periods. Such equipment can only be used by trained personnel.

4. **Direct reading instruments** are available in the form of sophisticated analysers which can only be used by trained and experienced operatives. Infrared gas analysers are the most common but other types of analysers are also available. They are very accurate and give continuous or TWA readings. They tend to be very expensive and are normally hired or used by specialist consultants.

Other common monitoring instruments include **vane anemometers**, used for measuring air flow speeds, and hygrometers, which are used for measuring air humidity.

Qualitative monitoring techniques include smoke tubes and the dust observation lamp. **Smoke tubes** generate a white smoke which may be used to indicate the direction of flow of air – this is particularly useful when the air speed is very low or when testing the effectiveness of ventilation ducting. A **dust observation lamp** enables dust particles which are normally invisible to the human eye to be observed in the light beam. This dust is usually in the respirable range and, although the lamp does not enable any measurements of the dust to be made, it will illustrate the operation of a ventilation system and the presence of such dust.

Records of air monitoring should be kept for five years unless an employee is identifiable in the records, in which case they should be kept for 40 years.

13.3 Occupational exposure limits

An Occupational Exposure Limit (OEL) is the acceptable upper limit on the concentration of a hazardous substance in the air within a workplace for a particular or a group of hazardous substances. OELs are normally established by competent national authorities and enforced by appropriate legislation.

One of the main purposes of a hazardous substance assessment is to adequately control the exposure of employees and others to hazardous substances. This means that such substances should be reduced to levels which do not pose a health threat to those exposed to them day after day at work. Recommended or mandatory occupational exposure limits (OELs) have been developed in many countries for airborne exposure to gases, vapours and particulates. The levels of the OEL values for some hazardous substances vary between countries. The reasons for this variation include differences in assessment methods and the assessments on the actual risks of the hazardous substances themselves. It is often not possible to compare exposure limits between countries because of the differing approaches.

The most widely used limits, called threshold limit values (TLVs), are those issued in the USA by the American Conference of Governmental Industrial Hygienists (ACGIH).

For airborne exposures, there are four types of limits in common use:

- the time-weighted average (TWA) exposure limit – the maximum average concentration of a chemical in air for a normal 8-hour working day and 40-hour week;
- the short-term exposure limit (STEL) – the maximum average concentration to which workers can be exposed for a short period (usually 15 minutes);
- the ceiling value – the concentration that should not be exceeded at any time.
- maximum permissible concentrations or threshold level values (TLVs).

The variations between countries are shown by the following examples.

In the **USA,** a permissible exposure limit (PEL) is also used and is the maximum amount or concentration of a chemical that a worker may be exposed to under

13

OSHA Regulations. PELs can be defined in two different ways:

▶ Ceiling values (denoted by C) – at no time should this exposure limit be exceeded.
▶ 8-hour time-weighted averages (TWA) – are an average value of exposure over the course of an 8-hour work shift. TWA levels are usually lower than ceiling values.

Recommended exposure limits are developed and periodically revised by the National Institute for Occupational Safety and Health (NIOSH). These recommendations are then published and transmitted to the Occupational Safety and Health Administration (OSHA) for use in promulgating legal standards. Permissible exposure limits are published in the US Code of Federal Regulations, Occupational Safety and Health Standards on Toxic and Hazardous Substances.

Some states also have their own exposure limits, for example: in California, permissible exposure limits for chemical contaminants are listed in the California Code of Regulations, Control of Hazardous Substances Order; in Michigan, permissible exposure limits for air contaminants are listed in the Occupational Health Standards published by the Department of Consumer and Industry Services; in Minnesota, the Minnesota Department of Labor and Industry publishes permissible exposure limits for air contaminants; in Washington, permissible exposure limits for air contaminants are published in the Safety and Health Rules of the Washington State Department of Labor and Industries.

In **Australia**, exposure standards are available on the Hazardous Substances Information System (HSIS) database of the Australian Safety and Compensation Council (formerly the Australian National Occupational Health and Safety Commission, NOHSC). The HSIS provides access to two data sets, one for hazardous substance information and the other for exposure standard information.

In **Canada**, occupational exposure limits are regulated within each Province. In Alberta, exposure limits are listed in the Chemical Hazards Regulation. In British Columbia, exposure limits are generally determined with reference to the Threshold Limit Values (TLVs) adopted by the American Conference of Governmental Industrial Hygienists (ACGIH).

In **Germany**, rules for limiting exposure to hazardous substances in the workplace and a list of occupational exposure limits are published by the Federal Institute for Occupational Safety and Health.

In **New Zealand**, exposure limits are available in Workplace Exposure Standards published by the Occupational Safety and Health Service of the Department of Labour.

In **South Africa**, occupational exposure limits for airborne pollutants are issued by both the Department of Labour and the Department of Minerals and Energy.

Values are listed on the site of the South Africa Institute of Occupational Hygiene.

In the **European Union**, the legal basis for the preparation of occupational exposure limits and biological limits is contained in Directive 98/24/EC on chemical agents and Directive 2004/37/EC on carcinogens and mutagens. Indicative Occupational Exposure Limit Values (IOELVs) are adopted through Commission Directives while Binding Occupational Exposure Limit Values (BOELVs) are adopted through Council and European Parliament Directives. In 2006, the Commission adopted Directive 2006/15/CE establishing a second list of Indicative Occupational Exposure Limit Values (IOELVs). Figure 13.11 shows an example of chemical storage in France.

In the **United Kingdom**, under the 2004 amendments to the COSHH Regulations 2002, the HSE has assigned WELs to a large number of hazardous substances and publishes any updates in a publication called 'Occupational Exposure Limits' EH40. The WEL is related to the concentration of airborne hazardous substances that people breathe over a specified period of time – known as 'time-weighted average'. Before the introduction of WELs, there were two types of exposure limit published: the maximum exposure limit (MEL) and the occupational exposure standard (OES).

The COSHH (Amendment) Regulations 2004 replaced the OES/MEL system with a single WEL. This removed the concern of the HSE that the OES was seen as a 'safe' limit rather than a 'likely safe' limit. Hence, the WEL must not be exceeded. Hazardous substances which have been assigned a WEL fall into two groups.

1. Substances which are carcinogenic or mutagenic (having a risk phase R45, R46 or R49) or could cause occupational asthma (or listed in section C of the HSE publication *'Asthmagen? Critical assessment for the agents implicated in occupational asthma'* as updated from time to time) or are listed in

Figure 13.11 Chemical storage in France which needs to comply with European standards

Schedule 1 of the COSHH Regulations. These are substances which were assigned a MEL before 2005. The level of exposure to these substances should be reduced as far as is reasonably practicable.

2. All other hazardous substances which have been assigned a WEL. Exposure to these substances by inhalation must be controlled adequately to ensure that the WEL is not exceeded. These substances were previously assigned an OES before 2005. For these substances, employers should achieve adequate control of exposure by inhalation by applying the principles of good practice outlined in the Approved Code of Practice and listed in 13.4.1. The implication of these principles is discussed later in this chapter.

The WELs are subject to time-weighted averaging. There are two such time-weighted averages (TWA): the long-term exposure limit (LTEL) or 8-hour reference period and the short-term exposure limit (STEL) or 15-minute reference period. The 8-hour TWA is the maximum exposure allowed over an 8-hour period so that if the exposure period was less than 8 hours the WEL is increased accordingly with the proviso that exposure above the LTEL value continues for no longer than 1 hour. Table 13.2 shows some typical WELs for various hazardous substances.

Table 13.2 Examples of workplace exposure limits (WELs)

Group 1 WELs	LTEL (8 h TWA)	STEL (15 min)
All isocyanates	0.02 mg/m^3	0.07 mg/m^3
Styrene	430 mg/m^3	1080 mg/m^3
Group 2 WELs		
Ammonia	18 mg/m^3	25 mg/m^3
Toluene	191 mg/m^3	574 mg/m^3

For example, if a person was exposed to a hazardous substance with a WEL of 100 mg/m^3 (8-hour TWA) for 4 hours, no action would be required until an exposure level of 200 mg/m^3 was reached (exposure at levels between 100 and 200 mg/m^3 should be restricted to 1 hour).

If, however, the substance has a STEL of 150 mg/m^3, then action would be required when the exposure level rose above 150 mg/m^3 for more than 15 minutes.

The STEL always takes precedence over the LTEL. When a STEL is not given, it should be assumed that it is three times the LTEL value.

The publication EH40 is a valuable document for the health and safety professional as it contains much additional advice on hazardous substances for use during the assessment of health risks, particularly where new medical information has been made public. The HSE is constantly revising WELs and introducing new ones and it is important to refer to the latest publication of EH40.

It is important to stress that if a WEL from Group 1 is exceeded, the process and use of the substance should cease immediately and employees should be removed from the immediate area until it can be made safe. In the longer term the process and the control and monitoring measures should be reviewed and health surveillance of the affected employees considered.

The overriding requirement for any hazardous substance which has a WEL from Group 1, is to reduce exposure to as low as is reasonably practicable.

Finally, there are certain limitations on the use of the published WELs:

▶ They are specifically quoted for an 8-hour period (with an additional STEL for many hazardous substances). Adjustments must be made when exposure occurs over a continuous period longer than 8 hours.

▶ They can only be used for exposure in a workplace and not to evaluate or control non-occupational exposure (e.g. to evaluate exposure levels in a neighbourhood close to the workplace, such as a playground).

▶ WELs are only approved where the atmospheric pressure varies from 900 to 1100 millibars. This could exclude their use in mining and tunnelling operations.

▶ They should also not be used when there is a rapid build-up of a hazardous substance due to a serious accident or other emergency. Emergency arrangements should cover these eventualities.

The fact that a substance has not been allocated a WEL does not mean that it is safe. The exposure to these substances should be controlled to a level which nearly all of the working population could experience all the time without any adverse effects to their health.

13.4 Control measures

13.4.1 The principles of good practice for the control of exposure to substances hazardous to health

To prevent ill-health due to the exposure to hazardous substances, employers are expected to develop suitable and sufficient control measures by:

1. identifying hazards and potentially significant risks;
2. taking action to reduce and control risks;
3. keeping control measures under regular review.

In order to assist employers with these duties, the UK HSE has produced the following eight principles of good practice:

(a) Design and operate processes and activities to minimise the emission, release and spread of substances hazardous to health.

(b) Take into account all relevant routes of exposure – inhalation, skin absorption and ingestion – when developing control measures.

(c) Control exposure by measures that are proportionate to the health risk.

(d) Choose the most effective and reliable control options which minimise the escape and spread of substances hazardous to health.

(e) When adequate control of exposure cannot be achieved by other means, provide, in combination with other control measures, suitable personal protective equipment.

(f) Check and review regularly all elements of control measures for their continuing effectiveness.

(g) Inform and train all employees on the hazards and risks from the substances with which they work and the use of control measures developed to minimise the risks.

(h) Ensure that the introduction of control measures does not increase the overall risk to health and safety.

All these principles are embodied in the following sections on the control measures for hazardous substances.

The frequency and type of future monitoring of exposure levels will depend on the exposure found in relation to recognised exposure limits. If the exposure level is very much lower than the limit and there is no change in process or other reason, then repeat measurement may only be needed occasionally.

If the exposure level is relatively high, then measurement may be needed several times between assessment reviews, to ensure that these levels have not been altered by some unidentified factor.

13.4.2 Hierarchy of control measures

The prevention or adequate control of exposure to hazardous substances by measures other than personal protective equipment, taking into account the degree of exposure and current knowledge of the health risks and associated technical remedies, can be listed in a hierarchy of control measures as follows:

► elimination;
► substitution;
► provision of engineering controls;
► provision of supervisory (people) controls;
► provision of personal protective equipment.

Examples where engineering controls are not reasonably practicable include emergency and maintenance work, short-term and infrequent exposure and where such controls are not technically feasible.

Measures for preventing or controlling exposure to hazardous substances include one or a combination of the following:

► elimination of the substance;
► substitution of the substance (or the reduction in the quantity used);
► total or partial enclosure of the process;
► local exhaust ventilation;

► dilution or general ventilation;
► reduction of the number of employees exposed to a strict minimum;
► reduced time exposure by task rotation and the provision of adequate breaks;
► good housekeeping;
► training and information on the risks involved;
► effective supervision to ensure that the control measures are being followed;
► personal protective equipment (such as clothing, gloves and masks);
► welfare (including first-aid);
► medical records;
► health surveillance.

13.4.3 ILO recommendations for the prevention and control of risks from hazardous substances

The ILO Code of Practice 'Ambient factors in the workplace' makes a series of recommendations for the prevention and control of risks from hazardous substances. It recommends that where the assessment of hazards or risks shows that control measures are inadequate or likely to become inadequate, risks should be:

(a) eliminated by ceasing to use such hazardous substances or replacing them with less hazardous substances or modified processes;

(b) minimised by designing and implementing a programme of action;

(c) reduced by minimising the use of toxic substances, where feasible.

Recommended control measures for implementing such a programme could include any combination of the following:

(a) good design and installation practice:
  (i) totally enclosed process and handling systems;
  (ii) segregation of the hazardous process from the operators or from other processes;
  (iii) plants, processes or work systems which minimise generation of, or suppress or contain hazardous dusts, fumes and gases and which limit the area of contamination in the event of spills and leaks;
  (iv) partial enclosure, with local exhaust ventilation;
  (v) local exhaust ventilation;
  (vi) sufficient general ventilation;

(b) work systems and practices:
  (i) reduction of the numbers of workers exposed and exclusion of non-essential access;
  (ii) reduction in the period of exposure of workers;
  (iii) regular cleaning of contaminated walls, surfaces, etc.;
  (iv) use and proper maintenance of engineering control measures;
  (v) provision of means for safe storage and disposal of substances hazardous to health;

(c) personal protection:
  (i) where the above measures do not suffice, suitable personal protective equipment should be provided until such time as the risk is eliminated or minimised to a level that would not pose a threat to health;
  (ii) prohibition of eating, chewing, drinking and smoking in contaminated areas;
  (iii) provision of adequate facilities for washing and changing and for storage of clothing (everyday clothing separated from work clothing), including arrangements for laundering contaminated clothing;
  (iv) use of signs and notices;
  (v) adequate arrangements in the event of an emergency.

Exposure to the following types of hazardous substances may require appropriate health surveillance:

(a) substances (dusts, fibres, solids, liquids, fumes, gases) that have a recognised systemic toxicity;
(b) substances known to cause chronic effects (e.g. occupational asthma);
(c) substances known to be sensitisers, irritants or allergens;
(d) substances that are known or suspected carcinogens, teratogens (affects embryos) or mutagens;
(e) other substances likely to have adverse health effects under particular work conditions or in case of fluctuations in ambient conditions.

The code also requires that employers should ensure that workers have sufficient, specific and systematic training and information on:

(a) the nature and degree of hazards and risks from hazardous substances which may occur, particularly in the case of an emergency;
(b) the protection of their safety and health and that of others from hazardous substances which may be present, in particular by using correct and prescribed methods for the handling, storage and transport of hazardous substances, and waste disposal;
(c) the correct and effective use of control and protection measures and of personal protective equipment.

This information should also be transmitted, where appropriate, to sub-contractors and their workers.

13.4.4 Preventative control measures

Prevention is the safest and most effective of the control measures and is achieved either by changing the process completely or by substituting for a less hazardous substance (the change from oil-based to water-based paints is an example of this). It may be possible to use a substance in a safer form, such as a brush paint rather than a spray.

The EU has introduced chemical safety regime REACH (Registration, Evaluation, Authorisation and restriction of Chemicals) Regulations, which restrict the use of high-risk substances or substances of very high concern and require that safer substitutes must be used. Manufacturers and importers of chemicals are responsible for understanding and managing the risks associated with their products. Authorisation seeks to ensure that risks from these substances of very high concern (SVHCs) are properly controlled. These are substances that are carcinogenic, mutagenic or toxic for reproduction or substances that are persistent, bio-accumulative and toxic. The objective is that all such substances should be replaced where possible with less dangerous alternatives.

The Regulations apply to many common items, such as glues, paints, solvents, detergents, plastics, additives, polishes, pens and computers. The three main types of REACH duty-holder are:

▶ **Manufacturers/importers** – businesses that manufacture or import (from outside the EU) 1 tonne or more of any given substance each year are responsible for registering a dossier of information about that substance with the European Chemicals Agency. If substances are not registered, then the data on them will not be available and it will no longer be legal to manufacture or supply them within the EU. Suppliers will be obliged to carry out an inventory and identify where in the supply chain the chemicals come from. Under the REACH system, industry will also have to prepare risk assessments and provide control measures for safe use of the substance by downstream users and get community-wide authorisations for the use of any substances considered to be of high concern.

▶ **Downstream users** – downstream users include any businesses using chemicals, which probably includes most businesses in some way. Companies that use chemicals have a duty to use them in a safe way, and according to the information on risk management measures that should be passed down the supply chain.

▶ **Other users in the supply chain** – in order for suppliers to be able to assess these risks they need information from the downstream users about how they are used. REACH provides a framework in which information can be passed both up and down supply chains by using the safety data sheet. This should accompany materials down through the supply chain so that users are provided with the information that they need to ensure chemicals are safely managed. It is envisaged that, in the future, the safety data sheets will include information on safe handling and use. Some users may also be importers and have a duty to register.

An important aspect of chemical safety is the need for clear information about any hazardous chemical

13

properties. The classification of different chemicals according to their characteristics (for example, those that are corrosive, or toxic to fish) currently follows an established system, which is reflected in REACH.

13.4.5 Engineering controls

The simplest and most efficient engineering control is the segregation of people from the process; a chemical fume cupboard is an example of this as is the handling of toxic substances in a glove box. Modification of the process is another effective control to reduce human contact with hazardous substances.

More common methods, however, involve the use of forced ventilation – local exhaust ventilation and dilution ventilation.

Local exhaust ventilation

Local exhaust ventilation (LEV) removes the hazardous gas, vapour or fume at its source before it can contaminate the surrounding atmosphere and harm people working in the vicinity. Such systems are commonly used for the extraction of welding fumes and dust from woodworking machines. All exhaust ventilation systems have the following five basic components (Figure 13.12).

1. **A collection hood and intake** – sometimes this is a nozzle-shaped point which is nearest to the work-piece, while at other times it is simply a hood placed over the workstation. The speed of the air entering the intake nozzle is important; if it is too low then hazardous fumes may not be removed (air speeds of up to 1 m/s are normally required). A suitable airflow indicator should make it easy to see whether airflow is adequate. New LEV systems are likely to be fitted with airflow indicators as standard.
   LEV hood design is very important and poor design is the main reason that LEV fails to control exposure. It should be as close as possible to the work-piece and ergonomic principles used in the designing of LEV hoods. LEV suppliers tend to concentrate on

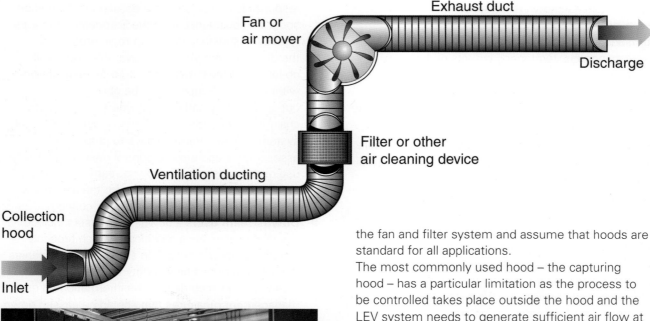

Figure 13.12 (a) A LEV system; (b) welding with an adjustable LEV system to remove dust and fumes

the fan and filter system and assume that hoods are standard for all applications.
   The most commonly used hood – the capturing hood – has a particular limitation as the process to be controlled takes place outside the hood and the LEV system needs to generate sufficient air flow at and around the process to 'capture' and draw in the airborne contaminant cloud. One company replaced a capturing hood with an enclosing hood fitted with a glass panel so that soldering fumes were separated from the operative's breathing zone.

2. **Ventilation ducting** – this normally acts as a conduit for the contaminated air and transports it to a filter and settling section. It is very important that this section is inspected regularly and any dust deposits removed. It has been known for ventilation ducting attached to a workshop ceiling to collapse under the added weight of metal dust deposits. It has also been known for them to catch fire.

3. **Filter or other air cleaning device** – normally located between the hood and the fan, the filter removes the contaminant from the air stream.

The filter requires regular attention to remove the contaminant and to ensure that it continues to work effectively.

4. **Fan** – this moves the air through the system. It is crucial that the correct type and size of fan is fitted to a given system and it should only be selected by a competent person. It should also be positioned so that it can easily be maintained but does not create a noise hazard to nearby workers.

5. **Exhaust duct** – this exhausts the air to the outside of the building. The duct should be positioned on the outside wall of the building such that the air is not discharged into a public area or close to an air inlet for an air conditioning system of any building. It should be checked regularly to ensure that the correct volume of air is leaving the system and that there are no leakages. The exhaust duct should also be checked to ensure that there is no corrosion due to adverse weather conditions.

Controlling Airborne Contaminants at Work, HSG258, HSE Books, is a very useful document on ventilation systems.

Such ventilation systems should be inspected at least every 14 months by a competent person to ensure that they are still operating effectively.

The effectiveness of a ventilation system will be reduced by damaged ducting, blocked or defective filters and poor fan performance. More common problems include the unauthorised extension of the system, poor initial design, poor maintenance, incorrect adjustments and a lack of inspection or testing.

Routine maintenance should include repair of any damaged ducting, checking filters, examination of the fan blades to ensure that there has been no dust accumulation, tightening all drive belts and a general lubrication of moving parts. LEV systems will degrade if they are not regularly checked and maintained. Dirty ventilation systems can seriously affect staff well-being and working environments, and both bacteria, MRSA and Clostridium difficile, have been found in duct systems. A 'responsible person' should be appointed who follows the instructions in the user manual, arranges monitoring and maintenance, and keeps records. The employer will also need to check that operators are using the LEV correctly.

The local exhaust ventilation system will have an effect on the outside environment in the form of noise and odour. Both these problems can be reduced by regular routine maintenance of the fan and filter. The waste material from the filter may be hazardous and require the special disposal arrangements described later in this chapter.

Dilution (or general) ventilation

Dilution (or general) ventilation uses either natural ventilation (doors and windows) or a fan-assisted forced ventilation system to ventilate the whole working room by inducing a flow of clean air, using extraction fans fitted into the walls and the roof, sometimes assisted by inlet fans. It operates by either removing the contaminant or reducing its concentration to an acceptable level. It is used when airborne contaminants are of low toxicity, low concentration and low vapour density or contamination occurs uniformly across the workroom.

Paint-spraying operations often use this form of ventilation as does the glass-reinforced plastics (GRP) boat-building industry – these being instances where there are no discrete points of release of the hazardous substances. It is also widely used in kitchens and bathrooms. It is not suitable for dust extraction or where it is reasonably practicable to reduce levels by other means.

There are limitations to the use of dilution ventilation. Certain areas of the workroom (e.g. corners and beside cupboards) will not receive the ventilated air and a build-up of hazardous substances occurs. These areas are known as 'dead areas'. The flow patterns are also significantly affected by doors and windows being opened or the rearrangement of furniture or equipment.

13.4.6 Supervisory or people controls

Many of the supervisory controls required for hazardous substance assessments are part of a good safety culture and were discussed in detail in Chapter 3. These include items such as systems of work, arrangements and procedures, effective communications and training. Additional controls when hazardous substances are involved are as follows:

▶ Reduced time exposure – thus ensuring that workers have breaks in their exposure periods. The use of this method of control depends very much on the nature of the hazardous substance and its STEL.
▶ Reduced number of workers exposed – only persons essential to the process should be allowed in the vicinity of the hazardous substance. Walkways and other traffic routes should avoid any area where hazardous substances are in use.
▶ Eating, drinking and smoking must be prohibited in areas where hazardous substances are in use.
▶ Any special rules, such as the use of personal protective equipment, must be strictly enforced.

13.4.7 Personal protective equipment

Personal protective equipment (PPE) is to be used as a control measure only as a last resort. It does not eliminate the hazard and will present the wearer with the maximum health risk if the equipment fails. Successful use of PPE relies on good user training, the availability of the correct

13

Figure 13.13 (a) Local exhaust ventilation systems can quickly corrode and need regular maintenance of fans and filters

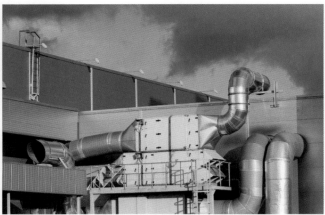

Figure 13.13 (b) Large roof level system with access platform for maintenance of the filters and fans

Figure 13.13 (c) Paint-spraying inside a special room with general ventilation to remove fumes

equipment at all times and good supervision and enforcement.

The 'last resort' rule applies in particular to RPE within the context of hazardous substances. There are some working conditions when RPE may be necessary:

▶ during maintenance operations;
▶ as a result of a new assessment, perhaps following the introduction of a new substance;
▶ during emergency situations, such as fire or plant breakdown;
▶ where alternatives are not technically feasible.

The principal requirements of PPE are as follows:

▶ PPE which is suitable for the wearer and the task;
▶ compatibility and effectiveness of the use of multiple personal protective equipment;
▶ a risk assessment to determine the need and suitability of proposed PPE;
▶ a suitable maintenance programme for the PPE;
▶ suitable accommodation for the storage of the PPE when not in use;
▶ information, instruction and training for the user of PPE including a demonstration on how to use the equipment properly;
▶ the supervision of the use of PPE by employees and a reporting system for defects.

Figure 13.14 Personal protective equipment at work

Types of personal protective equipment

There are several types of PPE such as footwear, hearing protectors and hard hats which are not primarily concerned with protection from hazardous substances; those which are used for such protection include:

▶ respiratory protection PPE;
▶ hand and skin protection PPE;
▶ eye protection PPE;
▶ protective clothing.

Table 13.3 shows the types of PPE recommended by the UK HSE for various parts of the body.

Table 13.3 The hazards and types of PPE for various parts of the body

	Hazards	PPE
Eyes	chemical or metal splash, dust, projectiles, gas and vapour, radiation.	safety spectacles, goggles, face shields, visors.
Head	impact from falling or flying objects, risk of head bumping, hair.	a range of helmets and bump caps.
Respiratory system	dust, vapour, gas, oxygen-deficient atmospheres.	disposable filtering face-piece or respirator, half- or full-face respirators, air-fed helmets, breathing apparatus.
Hand and arms	abrasion, temperature extremes, cuts and punctures, impact, chemicals, electric shock, skin infection, disease or contamination.	gloves, gauntlets, mitts, wrist-cuffs, armlets.
Feet and legs	wet, electrostatic build-up, slipping, cuts and punctures, falling objects, metal and chemical splash, abrasion.	safety boots and shoes with protective toe caps and penetration-resistant mid-sole, gaiters, leggings, spats.
Whole body	temperature extremes, adverse weather, chemical or metal splash, spray from pressure leaks or spray guns, impact or penetration, contaminated dust, excessive wear or entanglement of own clothing.	conventional or disposable overalls, boiler suits, specialist protective clothing, e.g. chain-mail aprons, high visibility clothing.

Source: HSE.

For all types of PPE, there are some basic standards that should be reached. The PPE should fit well, be comfortable to wear and not interfere with other equipment being worn or present the user with additional hazards (e.g. impaired vision due to scratched eye goggles). Training in the use of particular PPE is essential, so that it is not only used correctly, but the user knows when to change an air filter or to change a type of glove. Supervision is essential, with disciplinary procedures invoked for non-compliance with PPE rules.

It is also essential that everyone who enters the proscribed area, particularly senior managers, wear the specified PPE.

Respiratory protective equipment

Respiratory protective equipment (RPE) can be used to reduce the risk of harm from dusts, gases, vapours, mists, fume and micro-organisms. RPE can be subdivided into two categories: respirators (or face masks) that filter and clean the air, and breathing apparatus that supplies breathable air.

Respirators should not be worn in air which is dangerous to health, including oxygen deficient atmospheres. They are available in several different forms but the common ones are:

▶ a **filtering half-mask** often called disposable respirator – made of the filtering material. It covers the nose and mouth and removes respirable size dust particles. It is normally replaced after 8–10 hours of use. It offers protection against some vapours and gases;
▶ a **half-mask respirator** – made of rubber or plastic and covering the nose and mouth. Air is drawn through a replaceable filter cartridge. It can be used for vapours, gases or dusts but it is very important that the correct filter is used (a dust filter will not filter vapours);
▶ a **full-face mask respirator** – similar to the half-mask type but covers the eyes with a visor;
▶ a **powered respirator** – a battery-operated fan delivers air through a filter to the face mask, hood, helmet or visor.

Breathing apparatus is used in one of three forms (Figure 13.15):

▶ **self-contained breathing apparatus** – where air is supplied from compressed air in a cylinder and forms a completely sealed system;
▶ **fresh air hose apparatus** – fresh air is delivered through a hose to a sealed face mask from an uncontaminated source. The air may be delivered by the wearer, by natural breathing or mechanically by a fan;
▶ **compressed air-line apparatus** – air is delivered through a hose from a compressed air line. This can be either continuous flow or on demand. The air must be properly filtered to remove oil, excess water and other contaminants and the air pressure must be reduced. Special compressors are normally used.

The selection of appropriate RPE and correct filters for particular hazardous substances is best done by a competent specialist person.

There are several important technical standards which must be considered during the selection process. The following information will be needed before a selection can be made:

▶ details of the hazardous substance, in particular whether it is a gas, vapour or dust or a combination of all three;

13

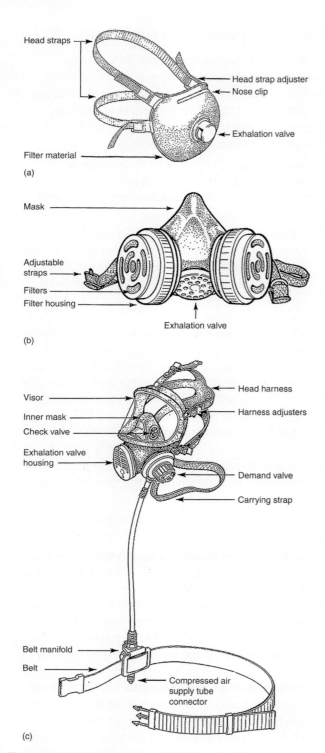

Figure 13.15 Types of respiratory protective equipment: (a) filtering half-mask; (b) half-mask reusable with filters; (c) compressed air-line breathing apparatus with full-face mask fitted with demand valve

▶ presence of a beard or other facial hair which could prevent a good leak-free fit (a simple test to see whether the fit is tight or not is to close off the air supply, breathe in and hold the breath. The respirator should collapse onto the face. It should then be possible to check to see if there is a leak);

▶ the size and shape of the face of the wearer and physical fitness;

▶ compatibility with other personal protective equipment, such as ear defenders;

▶ the nature of the work and agility and mobility required.

Filters and masks should be replaced at the intervals recommended by the supplier or when taste or smell is detected by the wearer.

The performance of RPE with a tight-fitting face piece depends on a good contact between the wearer's skin and the face seal of the mask. However, research by the British Safety Industry Federation indicated that up to 50% of all RPE used does not offer the wearer the level of protection assumed, usually because it was not fitted correctly.

The UK HSE Approved Code of Practice states that 'Employers should ensure that the selected face piece is of the right size and can correctly fit each wearer. For a tight-fitting face piece, the initial selection should include fit testing to ensure the wearer has the correct device. Also, employers must ensure that whoever carries out the fit testing is competent to do so.'

The British Safety Industry Federation (working with the HSE) has developed 'Fit2Fit', an accreditation scheme for people performing face piece fit testing. An HSE document 'Fit testing of respiratory protective equipment face pieces' provides more detail on fit testing methods. The HSE uses the term 'face piece' and defines this to mean a full-face mask, a half-face mask, or a filtering face piece/disposable mask. Fit2Fit standards are based on this HSE document.

Fit testing is needed when RPE is used as a control measure not when it is worn by choice for comfort. Fit testing should be repeated if the shape of the face of the wearer changes for any reason, such as weight loss or gain. Re-testing is recommended to check that the face piece remains suitable and that the wearer is taking care to wear and remove the mask correctly. A two-year cycle has been suggested, and a one-year retest is suggested for work with asbestos.

Fit testing can be qualitative or quantitative. Qualitative testing involves a simple pass or fail based on whether the user can taste or smell a harmless aerosol or odour through the RPE. Qualitative testing is only appropriate for disposable or half-face masks.

A quantitative fit test may use a laboratory test chamber or a portable fit test device, and produces a numerical 'fit factor' measure. The British Safety Industry Federation 'Fit2Fit' website provides details of accredited Fit2Fit training providers. The results of fit tests should be recorded. Fit test records must be made available to the HSE on request, and to the employees who wear RPE. RPE should be checked every time that it is used to make sure it has not been damaged and is being worn correctly.

A filtering face piece (FFP3) device is a mask which is certified to the PPE Directive. It provides a high level of filtering capability and face fit. It can be supplied with an exhale valve so that it can be worn comfortably over a fairly long period of time. It will provide an effective barrier to both droplets and fine aerosols and is the type recommended particularly for people in the healthcare sector dealing with symptomatic patients undergoing treatment where aerosols are likely to be generated.

All RPE should be examined at least once a month except for disposable respirators. A record of the inspection should be kept for at least five years. There should be a routine cleaning system in place and proper storage arrangements.

Respiratory Protective Equipment – A Practical Guide for Users, HSG53, HSE Books, contains comprehensive advice and guidance on RPE selection, use, storage, maintenance and training, and should be consulted for more information.

Hand and skin protection

Hand and skin protection is mainly provided by gloves (arm shields are also available). A wide range of safety gloves is available for protection from chemicals, sharp objects, rough working and temperature extremes. Many health and safety catalogues give helpful guidance for the selection of gloves. For protection from chemicals, including paints and solvents, impervious gloves are recommended. These may be made of PVC, nitrile or neoprene. For sharp objects, such as trimming knives, a Kevlar-based glove is the most effective. Gloves should be regularly inspected for tears or holes since this will obviously allow skin contact to take place. See Appendix 13.3 for types of gloves.

Another effective form of skin protection is the use of barrier creams and these come in two forms – pre-work and after-work. Pre-work creams are designed to provide a barrier between the hazardous substance and the skin. After-work creams are general purpose moisteners which replace the natural skin oils removed either by solvents or by washing.

Eye protection

Eye protection comes in three forms – spectacles (safety glasses), goggles and face visors. Eyes may be damaged by chemical and solvent splashes or vapours, flying particles, molten metals or plastics, non-ionising radiation (arc welding and lasers) and dust. Spectacles are suitable for low-risk hazards (low-speed particles such as machine swarf). Some protection against scratching of the lenses can be provided but this is the most common reason for replacement. Prescription lenses are also available for people who normally wear spectacles.

Goggles are best to protect the eyes from dust or solvent vapours because they fit tightly around the

Figure 13.16 Variety of eye protection goggles

eyes. Visors offer protection to the face as well as the eyes and do not steam up so readily in hot and humid environments. For protection against very bright lights, special light filtering lenses are used (e.g. in arc welding). Maintenance and regular cleaning are essential for the efficient operation of eye protection.

When selecting eye protection, several factors need to be considered. These include the nature of the hazard (the severity of the hazard and its associated risk will determine the quality of protection required), comfort and user acceptability, compatibility with other personal protective equipment, training and maintenance requirements and costs.

Protective clothing

Protective clothing includes aprons, boots and headgear (hard hats and bump caps). Aprons are normally made of PVC and protect against spillages but can become uncomfortable to wear in hot environments. Other lighter fabrics are available for use in these circumstances. Safety footwear protects against falling objects, collision with hard or sharp objects, hot or molten materials, slippery surfaces and chemical spills. It has metal toe caps and comes in the form of shoes, ankle boots or knee-length boots and is made of a variety of materials dependent on the particular hazard (e.g. thermally insulated against cold environments). It must be used with care near live, unprotected-against electricity. Specialist advice is needed for use with flammable liquids.

Appropriate selection of safety footwear involves the matching of the workplace hazards to the performance requirements of the footwear. The key issues are:

▶ the type of hazard (e.g. physical, chemical or thermal);
▶ the type of environment (e.g. indoors or out of doors);
▶ the ergonomics of the job (e.g. standing up, constant movement).

The footwear must have the correct grip for the environment, a hard-wearing sole unit and, possibly, a good shock-absorbing capability.

13

It is important to note that appropriate personal protective equipment should be made available to work-related visitors and other members of the public visiting workplaces where hazardous substances are being used. It is also important to stress that managers and supervisors must lead by example, particularly if there is a legal requirement to wear particular personal protective equipment. Refusals by employees to wear mandatory personal protective equipment must lead to some form of disciplinary action.

13.4.8 Health surveillance and personal hygiene

The need for health surveillance enables the identification of those employees most at risk from occupational ill-health. It should not be confused with health monitoring procedures such as pre-employment health checks or drugs and alcohol testing, but it covers a wide range of situations, from a responsible person looking for skin damage on hands to medical surveillance by a medical doctor. Health surveillance allows for early identification of ill-health and helps identify any corrective action needed. It may be required by law if employees are exposed to noise or vibration, solvents, isocyanates or other respiratory sensitisers, fumes, dusts, biological agents and asbestos, lead, work in compressed air or with ionising radiation. Health surveillance is a system of ongoing health checks and is important for:

▶ detecting ill-health effects at an early stage, so employers can introduce better controls to prevent them getting worse;
▶ providing data to help employers evaluate health risks;
▶ enabling employees to raise concerns about how work affects their health;
▶ highlighting lapses in workplace control measures, therefore providing invaluable feedback to the risk assessment;
▶ providing an opportunity to reinforce training and education of employees (e.g. on the impact of health effects and the use of protective equipment).

When initiating a health surveillance programme, it is important to avoid blanket coverage for all employees as this can produce misleading results and waste money.

The risk assessment for the organisation should indicate when and where health surveillance is required. In its simplest form, health surveillance could involve employees checking themselves for signs or symptoms of ill-health following a training session – for example, soreness, redness and itching on their hands and arms if they work with substances that can irritate or damage the skin. A responsible person, who has been trained by a competent medical practitioner, can also make routine basic checks, such as skin inspections or signs of rashes.

For more complicated assessments, an occupational health nurse or an occupational health doctor can ask about symptoms or carry out periodic examinations. Statutory medical surveillance involves a medical examination and possibly tests by a doctor with appropriate training and experience.

A health record must be kept for all employees under health surveillance. Health records, or a copy, should be kept in a suitable form for at least 40 years from the date of last entry because often there is a long period between exposure and onset of ill-health. Recorded details of each health surveillance check should include:

▶ the date they were carried out and by whom;
▶ the outcome of the test/check;
▶ the decision made by the occupational health professional in terms of fitness for task and any restrictions required. This should be factual and only relate to the employee's functional ability and fitness for specific work, with any advised restrictions.

Health records are different to medical records in that they should not contain confidential medical information.

Personal hygiene has already been covered under supervisory controls. It is very important for workers exposed to hazardous substances to wash their hands thoroughly before eating, drinking or smoking. Protection against biological hazards can be increased significantly by vaccination (e.g. tetanus). Finally, contaminated clothing and overalls need to be removed and cleaned on a regular basis.

13.4.9 Further controls required to prevent exposure to a carcinogen, mutagen or substances that cause asthma

Under the ILO Code of Practice 'Ambient factors in the workplace', the employer should assess the risk of exposure (exposure means taking in chemicals by breathing in, by skin contact or by swallowing). If that risk is significant, then the employer should consider the substitution of harmful substances with less harmful ones. Where it is not reasonably practicable to prevent exposure to a carcinogen or mutagen, the employer should apply the following additional measures:

(a) totally enclosing the process and handling systems, unless this is not reasonably practicable;
(b) the prohibition of eating, drinking and smoking in areas that may be contaminated by carcinogens;
(c) cleaning floors, walls and other surfaces at regular intervals and whenever necessary;
(d) designating those areas and installations which may be contaminated by carcinogens and using suitable and sufficient warning signs; and
(e) storing, handling and disposing of carcinogens safely, including using closed and clearly labelled containers.

All these controls should be kept in good working order, including the following recommendations for asthma sufferers:

- mechanical controls, e.g. local exhaust ventilation (LEV), protective gloves;
- administrative controls, e.g. supervision;
- operator controls, e.g. following instructions.

Health surveillance is essential when workers are dealing with any of these particular substances. This means regularly looking for early signs of work-related ill-health and putting procedures in place to achieve this. The purpose of health surveillance is to monitor and protect the health of individual employees. The collection of simple information may lead to early detection of ill-health caused by work and identify the need for improved control measures.

All employees exposed or likely to be exposed to an asthmagen should receive suitable health surveillance. This might involve examinations by a doctor or trained nurse.

Control measures to reduce the risk of exposure to biological agents include thorough cleaning, sterilisation and disinfection; the use of personal protective equipment to reduce the risk of exposure to biological agents include thorough cleaning, sterilisation and disinfection; the use of personal protective equipment such as gloves, respiratory and eye protection and an overall; containment of the agent in a microbiological safety cabinet; the use of spill trays; the prohibition of smoking and eating and drinking in the work area; a good standard of personal hygiene.

13.4.10 Maintenance and emergency controls

Engineering control measures will only remain effective if there is a programme of preventative maintenance available. Maintenance will involve the cleaning, testing and, possibly, the dismantling of equipment. It could involve the changing of filters in extraction plant or entering confined spaces. It will almost certainly require hazardous substances to be handled and waste material to be safely disposed of. It may also require a permit-to-work procedure to be in place since the control equipment will be inoperative during the maintenance operations. Records of maintenance should be kept for at least five years.

Emergencies can range from fairly trivial spillages to major fires involving serious air pollution incidents. The following points should be considered when emergency procedures are being developed:

- the possible results of a loss of control (e.g. lack of ventilation);
- dealing with spillages and leakages (availability of effective absorbent materials);
- raising the alarm for more serious emergencies;

- evacuation procedures, including the alerting of neighbours;
- fire-fighting procedures and organisation;
- availability of respiratory protective equipment;
- information and training.

The Emergency Services should be informed of the final emergency procedures and, in the case of the Fire and Rescue Service, consulted for advice during the planning of the procedures. (See Chapter 4 for more details on emergency procedures.)

13.4.11 The transport of hazardous substances by road

Although this topic is not in the NEBOSH International General Certificate syllabus, a brief mention of the main precautions required to safeguard the health and safety of those directly involved in the transport of hazardous substances and of general members of the public is important.

Data sheets from the manufacturer of the hazardous substance should indicate the safest method of handling it and will give information on emergency procedures (e.g. for spillages and fire). These sheets should be available to all concerned with the transportation of the substance, in particular those responsible for loading/unloading, as well as the driver. The hazardous substance should be loaded correctly on the vehicle in suitable containers and segregated from incompatible materials. There must be adequate emergency information with the substance containers and attached to the vehicle. Drivers of the vehicles must receive special training which covers issues such as emergency procedures and route planning. There should also be emergency provisions for first-aid and personal protective equipment on the vehicle.

13.4.12 An illustrative example using hazardous substance controls

Organic solvents are widely used throughout industry and commerce in paints, inks, glues and adhesives. The control hierarchy discussed earlier should be applied to minimise the health risks from the use of these solvents. The top of this hierarchy is to eliminate or substitute the use of the organic solvent by using a less volatile or water-based alternative. If this is not possible, then some form of engineering control should be applied, such as dilution or local exhaust ventilation. Alternatively, the workplace, where the solvents are being used, could be enclosed or isolated from the main work activities. Other engineering-type controls could include the use of properly labelled anti-spill containers, the use of covered disposal units for any used cloths and the transfer of large quantities using pumping/pipe arrangements rather than simply pouring the solvent.

Supervisory controls include the reduction in the length of time any employee is exposed to the solvent and

13

the provision of good housekeeping, such as ensuring that containers are kept closed when not in use and any spills are quickly removed. The provision of barrier creams and personal protective equipment (eye protection, gloves and aprons) and RPE may also be required. Welfare issues will include first-aid provision, washing facilities and the encouragement of high levels of personal hygiene. Smoking and the consumption of food and drink should be prohibited where there might be contamination from organic solvents. All these supervisory items should be reinforced in training sessions and employees given appropriate information on the risks associated with the solvents. Finally, some form of health surveillance will be needed so that employees who show allergies to the solvents can be treated and, possibly, assigned other duties.

13.5 Specific agents

13.5.1 Health risks and controls associated with asbestos

Approximately 125 million people in the world are exposed to asbestos in the workplace and, according to estimates by the World Health Organisation (WHO), more than 107,000 deaths each year are attributable to occupational exposure to asbestos. The WHO, ILO and the United Nations Environment Programme has called on states to stop using all types of asbestos in an attempt to eliminate all asbestos-related diseases throughout the world. The use of asbestos has declined by 55% from its peak in 1980 but more than 2 million tonnes are still used worldwide.

Younger people, if routinely exposed to asbestos fibres over time, are at greater risk of developing asbestos-related disease than older workers. This is due to the time it takes for the body to develop symptoms after exposure to asbestos (latency).

Exposure to asbestos can cause four main diseases:

▶ Mesothelioma (a cancer of the lining of the lungs; it is always fatal and is almost exclusively caused by exposure to asbestos).
▶ Asbestos-related lung cancer (which is almost always fatal).
▶ Asbestosis (a scarring of the lungs which is not always fatal but can be a very debilitating disease, greatly affecting quality of life).
▶ Diffuse pleural thickening (a thickening of the membrane surrounding the lungs which can restrict lung expansion leading to breathlessness).

In 2005, occupational exposure to asbestos was estimated to cause 43,000 mesothelioma deaths globally of which 7,000 were attributed to Europe. Most mesothelioma deaths occurring now are a legacy of past occupational exposures to asbestos when it was widely used in the building industry.

Asbestos appears in three main forms: crocidolite (blue), amosite (brown) and chrysotile (white). The blue and brown asbestos are considered to be the most dangerous and may be found in older buildings where they were used as heat insulators around boilers and hot water pipes and as fire protection of structures. White asbestos has been used in asbestos cement products and brake linings. It is difficult to identify an asbestos product by its colour alone – laboratory identification is usually required. Many asbestos-containing materials (ACMs) are difficult to distinguish from other materials. It is easy to drill or cut ACMs unwittingly and release large quantities of airborne fibres that could cause long-term health problems to the operator. Asbestos produces a fine fibrous dust of respirable dust size which can become lodged in the lungs. The fibres can be very sharp and hard, causing damage to the lining of the lungs over a period of many years.

The condition of the ACM will need to be monitored regularly and a record kept. The time between these monitoring checks should not exceed 12 months and may need to be more frequent. Monitoring involves visual inspection to see whether there has been any deterioration at the surface of the ACM. This may be remedied by resealing the surface or removing a section of the ACM.

If asbestos is discovered during the performance of a contract, work should cease immediately and the employer informed. Typical sites of asbestos include ceiling tiles, asbestos cement roof and wall sheets, sprayed asbestos coatings on structural members, loft insulation and asbestos gaskets. In most countries, asbestos has its own legislation. This normally covers the need for a risk assessment, a method statement covering removal and disposal, air-monitoring procedures and the control measures (including personal protective equipment and training) to be used.

Training is required for the majority of workers involved in maintenance, refurbishment and demolition. The UK Health and Safety Executive (HSE) has estimated that approximately 50% of buildings in the UK still contain some form of asbestos and about 1.5 million workers require asbestos training.

Additional training requirement for asbestos awareness should include:

▶ the health risks caused by exposure to asbestos;
▶ the materials that are likely to contain asbestos and where they are likely to be found;
▶ the methods to reduce asbestos risks during work;
▶ the action to take in an emergency, such as an uncontrolled release of asbestos dust.

Asbestos is the single biggest workplace killer. According to HSE statistics, there are 15 times as many deaths from asbestos as there are deaths from workplace accidents. Asbestos is responsible for at least 4,000 deaths in the UK each year, and the HSE felt that there was a need to increase awareness amongst the workforce of the risks associated with this material.

It is important to stress that hazards associated with asbestos occur throughout the world. The WHO, the ILO and other intergovernmental organisations are working with countries with the objective of eliminating asbestos-related diseases in the following ways:

▶ by recognising that the most efficient way to eliminate asbestos-related diseases is to stop the use of all types of asbestos;

▶ by providing information about solutions for replacing asbestos with safer substitutes and developing economic and technological mechanisms to stimulate its replacement;

▶ by taking measures to prevent exposure to asbestos in place and during asbestos removal;

▶ by improving early diagnosis, treatment and rehabilitation services for asbestos-related diseases;

▶ by establishing registries of people with past and/ or current exposures to asbestos and organising medical surveillance of exposed workers;

▶ by providing information on the hazards associated with asbestos-containing materials and products; and

▶ by raising awareness that waste containing asbestos should be treated as hazardous waste.

ILO requirements to protect workers from asbestos

Asbestos may be found in a wide range of workplaces. Therefore it is important to determine whether asbestos is present. Risks to the health and safety of workers must be assessed, and risk management measures established with the aim of reducing exposure to those risks.

The ILO has produced a detailed Code of Practice 'Safety in the Use of Asbestos' that gives detailed advice on dealing with asbestos. It is derived from the ILO policy on asbestos required by the Asbestos Convention, 1986 (No. 162), and Recommendation, 1986 (No. 172).

This ILO Code of Practice has the following objectives:

(a) to prevent the risk of exposure to asbestos dust at work;

(b) to prevent harmful effects on the health of workers arising from exposure to asbestos dust;

(c) to provide reasonably practicable control procedures and practices for minimising occupational exposure to asbestos dust.

In the code, the ILO expects the competent authority (normally an arm of the national government) to ensure that criteria are established on measures which provide for safety of workers; in particular, it should:

1. In consultation with the most representative organisations of employers and workers, issue or approve and periodically update regulations or other suitable provisions for the protection of workers' health against hazards due to occupational exposure to asbestos dust. These regulations on the prevention of airborne asbestos in the working environment should clearly define the agencies and persons responsible for carrying them out.

2. Have the expertise to supervise the enforcement of such regulations or provisions and to supply relevant advice and information.

3. Establish those procedures which it considers necessary for the notification by the employer of working operations and workplaces where asbestos or materials containing asbestos are or will be produced, handled, processed, stored or used and any other work situations where asbestos or materials containing asbestos are being used or handled in such a manner as to cause dust. This notification may include the following information:
   (a) nature and place of work;
   (b) type and quantity of asbestos or materials containing asbestos;
   (c) total number of workers exposed;
   (d) duration or anticipated duration of the work period;
   (e) protective and preventative measures to be taken.

4. Determine those dangerous working operations or techniques which should be prohibited or made subject to specific authorisation and which will require compliance with particular measures of prevention and protection.

5. Establish procedures for:
   (a) the setting up of airborne asbestos exposure limits in the working environment;
   (b) the standardisation of methods for monitoring airborne asbestos in the working environment;
   (c) the approval of personal protective equipment.

6. Endeavour to promote close cooperation between employers and workers on matters of prevention and assist in providing safety and health committees, employers and workers with information on the hazards and on safety in the use of asbestos.

The Code of Practice places the following duties on employers to:

1. Control and prevent the exposure of airborne asbestos in the working environment and limit the exposure of workers as far as is reasonably practicable to at least within the asbestos exposure limits imposed by the national government.

2. Notify the competent authority of those working operations and workplaces where asbestos or materials containing asbestos are present according to the terms of any authorisation procedures.

3. Take appropriate measures to prevent, as far as is reasonably practicable, the presence of asbestos dust in the working environment when buildings and installations are being designed or altered or equipment is purchased.

13

4. Investigate asbestos health hazards before any production or use so as to identify the preventative measures required appropriate to the hazard; without such measures, asbestos should not be produced, used or handled.

5. Establish and implement a general control programme to reduce the workers' exposure to asbestos dust.

6. Provide the necessary equipment and services for monitoring the working environment. All such equipment should be maintained and calibrated properly.

7. Provide the surveillance necessary to enable workers to perform their tasks in the best possible conditions with respect to occupational safety and health; in particular, provision should be made for the regular inspection and maintenance of installations, machinery and equipment to prevent the contamination of the working environment with asbestos dust.

8. Inform all workers of the asbestos hazards associated with the tasks assigned to them and the measures to be taken to prevent damage to their health. This information should also be transmitted to sub-contractors and their workers. In particular, provisions may be needed for newly recruited workers, foreign workers who may encounter language difficulties and for all other workers who may have difficulties in understanding written instructions.

9. Ensure that all members of the managerial staff are steadily trying to improve prevention, that they are fully aware of their duties with regard to occupational hazards from asbestos dust exposure and, in particular, that they are appropriately trained and are constantly updating their knowledge in this field so that they may thoroughly instruct the workers regarding the precautions to be taken in their jobs and in the event of unexpected circumstances giving rise to asbestos dust.

10. Provide and maintain without cost to the workers such personal protective equipment and clothing as are necessary when airborne asbestos hazards cannot be otherwise prevented or controlled.

11. Inform sub-contractors of the appropriate regulations and safety precautions and should ensure, so far as is reasonably practicable, that persons present at the workplace who are not under his/her direct control follow such regulations and take the necessary safety precautions.

The code also places the following general duties on workers:

1. Within the limits of their responsibilities, they should do everything in their power to prevent the presence of airborne asbestos in the working environment.

2. They should abide by any instructions given to them in connection with the prevention of airborne asbestos in the working environment; submit themselves to medical surveillance according to national practice and wear personal samplers when necessary to measure personal exposure to asbestos dust.

3. They should wear the personal protective equipment and clothing provided when either other methods for the control of asbestos dust cannot be applied or it is necessary to wear personal protective equipment and clothing in addition to other methods of control.

4. They should draw to the attention of management any change of circumstance in the work process which might give rise to asbestos dust exposure.

The Code of Practice requires records to be kept by the employer on any asbestos dust exposure and such records should be clearly marked by date, work area and plant location. All relevant data from measurements of airborne asbestos in the working environment should be systematically recorded and the workers, or their representatives, should have access to these records.

Besides the numerical results of measurements, the monitoring data should include, for example:

(a) the composition and trade names of materials containing asbestos;

(b) the location, nature, dimensions and other distinctive features of the workplace where static measurements were made; the exact location at which personal monitoring measurements were made and the names and job titles of the workers involved;

(c) the source or sources of airborne asbestos emission, their location and the type of work and operations being performed during sampling;

(d) relevant information on the functioning of the process, engineering controls, ventilation and weather conditions with respect to emission of asbestos dust;

(e) the sampling instrument used, its accessories and the method of analysis;

(f) the date and exact time of sampling;

(g) the duration of the workers' exposure, the use or non-use of respiratory protection and other comments relating to the exposure evaluation; and

(h) the names of the persons responsible for the sampling and for the analytical determinations.

Finally, the code deals with the actions that should be taken in construction, demolition and alteration work where ACMs are likely to be present.

13.5.2 Managing asbestos in buildings

A study published by the *British Medical Journal* has found that there are over 1,800 mesothelioma deaths each year in Britain. Since this disease can take between 15 and 60 years to develop, the peak of the epidemic has still to be reached. In the construction industry, those at risk are asbestos removal workers and those, such as electricians, plumbers and carpenters, who are involved in refurbishment, maintenance or repair of buildings.

The policy of the ILO is to ensure that those involved in the repair, removal or disturbance of asbestos-containing materials (ACMs), such as insulation, coatings or insulation boards, are competent and working to the requirements of the Asbestos Convention, 1986 (No. 162), the Recommendation, 1986 (No. 172) and the Code of Practice 'Safety in the Use of Asbestos'. This requires the identification of ACMs and the planning of any subsequent work. This should prevent inadvertent exposure to asbestos and minimise the risks to those who have to work with it.

The Code of Practice requires training for anyone liable to be exposed to asbestos fibres at work, including maintenance workers and others who may come into contact with or who may disturb asbestos (e.g. cable installers) as well as those involved in asbestos removal work.

When work with asbestos or which may disturb asbestos is being carried out, the code requires employers and the self-employed to prevent exposure to asbestos fibres. Where this is not reasonably practicable, they must make sure that exposure is kept as low as reasonably practicable by measures other than the use of respiratory protective equipment. The spread of asbestos must be prevented.

Worker exposure must be below the airborne exposure limit (the Control Limit), in the UK and many other countries, of 0.1 fibres per cm^3 for all types of asbestos. The Control Limit is the maximum concentration of asbestos fibres in the air (averaged over any continuous 4-hour period) and must not be exceeded. Short-term exposures must be strictly controlled and worker exposure should not exceed 0.6 fibres per cm^3 of air averaged over any continuous 10-minute period using respiratory protective equipment if exposure cannot be reduced sufficiently using other means. Respiratory protective equipment is an important part of the control regime but it must not be the sole measure used to reduce exposure and should only be used to supplement other measures.

Most asbestos removal work must be undertaken by a licensed contractor but any decision on whether particular work needs to be licensed is based on the level of risk.

Those in control of premises and the duty-holders should:

▶ take reasonable steps to determine the location and condition of materials likely to contain asbestos;
▶ presume materials contain asbestos unless there is strong evidence that they do not;

▶ make and keep an up-to-date record of the location and condition of the ACMs or presumed ACMs in the premises;
▶ assess the risk of the likelihood of anyone being exposed to fibres from these materials;
▶ prepare a plan setting out how the risks from the materials are to be managed;
▶ take the necessary steps to put the plan into action;
▶ review and monitor the plan periodically;
▶ provide such information and asbestos awareness training to anyone who is liable to work on these materials or otherwise disturb them.

The UK HSE has produced a document – *HSG 264 Asbestos: The survey guide* – that gives useful guidance on asbestos surveys. There are **two** types of survey, known in the UK as the **management survey** and the **refurbishment and demolition survey.**

A **management survey** is the standard survey. Its purpose is to locate, as far as reasonably practicable, the presence and extent of any suspect ACMs in the building which could be damaged or disturbed during normal occupancy, including foreseeable maintenance and installation, and to assess their condition. A management survey is required during the normal use of the building to ensure continued management of the ACMs. It should include an assessment of the condition of the various ACMs and their ability to release fibres into the air should they be disturbed in some way. It will often involve minor intrusive work and some disturbance.

Management surveys can involve a combination of sampling to confirm asbestos is present or presuming asbestos to be present. All ACMs should be identified as far as is reasonably practicable. The areas inspected should include under floor coverings, above false ceilings (ceiling voids), cladding and partitions, under carpets, tiles and floors, lofts, inside risers, service ducts and lift shafts and basements, cellars or underground rooms. In these situations, controls should be put in place to prevent the spread of debris, which may include asbestos.

A **refurbishment or demolition survey** is used to locate and describe, as far as reasonably practicable, all ACMs in the area where the refurbishment work will take place or in the whole building if demolition is planned. It is necessary when the building (or part of it) is to be upgraded, refurbished or demolished. It is required for all work which disturbs the fabric of the building in areas particularly where the management survey has not been intrusive. A refurbishment and demolition survey is needed before any refurbishment or demolition work is carried out.

Refurbishment and demolition surveys should only be conducted in unoccupied areas to minimise risks to the public or employees on the premises. Ideally, the building should not be in service and all furnishings removed.

13

For minor refurbishment, this would only apply to the room involved or even part of the room where the work is small and the room large. In these situations, there should be effective isolation of the survey area (e.g. full floor to ceiling partition), and furnishings should be removed as far as possible or protected using sheeting.

The person who undertakes any of these surveys must be suitably trained and experienced in such work.

Some types of work, of an intermittent and low intensity nature, will not have to be done by a licensed contractor – artex work is an example of this.

There are many issues which need to be addressed when asbestos is possibly present in a workplace. Several publications are available from HSE Books which cover all these issues in some detail and the reader should refer to them for more information. Here a brief summary of the principal issues will be given.

Identification of the presence of asbestos is the first action. Asbestos is commonly found as boiler and pipe lagging, insulation panels around pillars and ducting for fire protection and heat insulation, ceiling tiles and asbestos cement products, including asbestos cement sheets. The main duty-holder should ensure that a written plan is prepared that shows where the ACM is located and how it will be managed to prevent exposure to asbestos, including to contractors and other workers who may undertake work on the fabric of the building that could disturb the ACM. This plan then needs to be put into action and communicated to those affected. The duty-holder should ensure that the plan is reviewed regularly and updated as circumstances change, in consultation with all those who may be affected. The plan, including drawings, should be available on site for the entire life of the premises and should be kept up to date.

Initial investigations will involve the examination of building plans, the determination of the age of the building and a thorough examination of the building. Advice is available from a number of reputable specialist consultants and details may be obtained from the competent authority, who often offers such a service. If the specialist is in any doubt, a sample of the suspect material will be sent to a specialist laboratory for analysis. It is important for a specialist to take the sample because the operation is likely to expose loose fibres.

The **asbestos risk register** is a key component of the asbestos management. The management plan must contain current information about the presence and condition of any asbestos in the building. The asbestos risk register will therefore need to be updated on a regular basis (at least once a year). This will involve:

▶ regular inspections to check the current condition of asbestos materials;

▶ deletions to the register when any asbestos is removed;

Figure 13.17 Damaged asbestos lagging on pipework

▶ additions to the register when new areas are surveyed and asbestos is located;

▶ changes to the register (at any time asbestos-containing materials are found to have deteriorated).

The register can be kept as a paper or electronic record and it is very important that this is kept up to date and easily accessible should any future maintenance be necessary. Paper copies may be easier to pass on to visiting maintenance workers, who will need them to know the location and condition of any asbestos before they start work. Electronic copies are easier to update and are probably better suited for people responsible for large numbers of properties or bigger premises.

An example of an asbestos register is available on the UK HSE website.

Assessment is an evaluation or asbestos survey to determine whether the location or the condition could lead to the asbestos being disturbed. If it is in good condition, undamaged and not likely to be disturbed, then it is usually safer to leave it in place and monitor it. However, if it is in a poor condition, it may need to be repaired, sealed, enclosed or removed. If there is doubt, then specialist advice should be sought. The condition of the ACM will need to be monitored regularly and a record kept. The time between these monitoring checks should not exceed 12 months and may need to be more frequent. Monitoring involves visual inspection to see whether there has been any deterioration at the surface of the ACM. This may be remedied by resealing the surface or removing a section of the ACM.

The asbestos survey should provide enough information so that an asbestos register, a risk assessment and a management plan can then be prepared. The survey will usually involve sampling and analysis to determine the presence of asbestos so asbestos surveys should only be carried out by competent surveyors who can clearly demonstrate they have the necessary skills, experience and qualifications.

Removal must only be done by a licensed contractor. A detailed plan of work is essential before work begins. The plan should give details of any equipment to be used for the protection and decontamination of employees and others. This process will also require an assessment to be made to ensure that people within the building and neighbours are properly protected. At the planning stage, generally, the competent authority must be notified of the intention to remove asbestos and again when the work begins. This is particularly important when the exposure to asbestos is likely to exceed the action level. The assessment should include details of the type and location of the asbestos, the number of people who could be affected, the controls to be used to prevent or control exposure, the nature of the work, the removal methods, the procedures for the provision of personal and respiratory protective equipment and details of emergency procedures. If asbestos cement sheeting is to be removed the following procedure is recommended:

▶ Where reasonably practicable, remove the asbestos sheets before any other operation, such as demolition.
▶ Avoid any breaking of the sheets.
▶ Dampen the sheets while working on them.
▶ Lower the sheets on to a clean hard surface.
▶ Remove all waste and debris from the site as soon as possible to prevent its spread around the site.

Figure 13.18 High hazard vacuum cleaner to clear up asbestos material

▶ Do not bulldoze broken asbestos cement or sheets into piles.
▶ Do not dry sweep asbestos cement debris.
▶ Dispose of the waste and debris safely, separate from general waste as hazardous waste.

Control measures during the removal of asbestos include the provision of personal and RPE including overalls, good ventilation arrangements in and the segregation or sealing of the working area, suitable method statements and air-monitoring procedures. A decontamination unit should also be provided. The sealed area will need to be tested for leaks. Good supervision and induction of the workforce are also essential. A high level of personal hygiene must be expected for all workers and the provision of welfare amenity arrangements, particularly washing and catering facilities and the separation of working and personal clothing. Suitable warning signs must be displayed and extra controls provided if the work is taking place at height. After the work is completed, the area must be thoroughly cleaned and a clean air certificate provided after a successful air test.

Work with high-risk asbestos products, such as asbestos insulating board, sprayed fire or insulation coatings or lagging, must only be carried out by a suitably qualified and competent asbestos specialist.

Medical surveillance in the form of a regular medical examination by an appointed doctor should be given to any employee who has been exposed to asbestos at levels above the action level. The first medical examination should take place within two years of exposure and further examinations at intervals of not more than two years. A health record of such surveillance should be kept for a period of at least 40 years after the last entry.

Awareness training is an important feature of the ILO Code of Practice and requires that adequate information, instruction and training are given to those employees:

(a) who are or who are liable to be exposed to asbestos, or who supervise such employees; and/or
(b) who carry out work in connection with the employer's duties under these Regulations, so that they can carry out that work effectively.

Such training should cover the following topics:

▶ properties of asbestos and its effects on health, including its interaction with smoking;
▶ the types of products or materials likely to contain asbestos;
▶ the uses and location of ACMs in buildings and plant;
▶ the operations which could result in asbestos exposure and the importance of preventative controls to minimise exposure;
▶ the presence of other hazards such as working at height;

13

- ▶ the requirements of the ILO Code of Practice;
- ▶ safe work practices, control measures, and protective equipment;
- ▶ the purpose, choice, limitations, proper use and maintenance of RPE;
- ▶ emergency procedures;
- ▶ hygiene requirements;
- ▶ decontamination procedures;
- ▶ waste handling procedures;
- ▶ medical examination requirements; and
- ▶ the control limit and the need for air monitoring.

Refresher training should be given at regular intervals (at least annually) and adapted to take account of significant changes in the type of work carried out or methods of work used by the employer. It should be provided in a manner appropriate to the nature and degree of exposure identified by the risk assessment, and so that the employees are aware of:

- ▶ the significant findings of the risk assessment; and
- ▶ the results of any air monitoring carried out with an explanation of the findings.

Disposal of asbestos waste should be to an authorised asbestos waste site only. Asbestos waste describes any asbestos products or materials that are ready to be disposed. This includes any contaminated building materials, dust, rubble, used tools that cannot be decontaminated, disposable PPE and damp rags that have been used for cleaning. Asbestos waste must be placed in suitable packaging to prevent any fibres being released. This should be double wrapped and appropriately labelled. Standard practice is to use a red inner bag marked up with asbestos warning labels and a clear outer bag with appropriate hazard markings. Intact asbestos cement sheets and textured coatings that are firmly attached to a board should not be broken up into smaller pieces. These should instead be carefully double wrapped in suitable polythene sheeting (1000 gauge) and labelled.

The waste container must be strong enough to securely contain the waste and not become punctured; it must be easily decontaminated, kept securely on the site until required and properly labelled. The waste must only be carried by a licensed carrier.

Accidental exposure to ACMs can occur even when all reasonable precautions have been taken. If an ACM has been worked on by a worker who did not realise that it was an ACM or has accidentally damaged an ACM, then all work must stop immediately and nobody should be allowed to enter the area in question. The work supervisor must be informed and a sample of the material sent for analysis. If the result is positive, then a specialist licensed contractor should be employed unless the work is exempt from the need for a licence. Any contaminated clothing must be removed and placed in a plastic bag. The affected worker should shower as soon as possible or wash thoroughly.

13.5.3 Health hazards of specific agents

The health hazards associated with hazardous substances can vary from very mild (momentary dizziness or a skin irritation) to very serious, such as a cancer.

Cancer is a serious body cell disorder in which the cells develop into tumours. There are two types of tumour: benign and malignant. Benign tumours do not spread but remain localised within the body and grow slowly. Malignant tumours are called cancers and often grow rapidly, spreading to other organs using the bloodstream and lymphatic system. Survival rates have improved dramatically in recent years as detection methods have improved and the tumours can be found in their early stages of development.

Workplace exposures leading to occupational illness and work-related cancers are a leading cause of death worldwide, with 12.7 million new cases and 7.6 million deaths in 2008. Currently, 63% of all cancer deaths are reported from low- and middle-income countries and this figure is predicted to increase. Globally, 19% of all cancers are attributable to the environment, including work setting, resulting in 1.3 million deaths each year. Five of the key causes of occupational cancer registrations and deaths are:

- ▶ Diesel engine exhaust emissions – related deaths in the UK are estimated at 650 a year to lung or bladder cancer.
- ▶ Solar radiation – at over 1,500 a year.
- ▶ Asbestos – as covered earlier.
- ▶ Shift work – causes an estimated 2,000 cases of breast cancer a year.
- ▶ Silica – over 600 cancer registrations.

A minority of cancers are believed to be occupational in origin.

Occupational asthma has approximately 4 million sufferers in the UK and it is estimated that 13 million working days are lost each year as a result of it. It is mainly caused by breathing in respiratory sensitisers, such as wood dusts, organic solvents, solder flux fumes or animal hair. The symptoms are coughing, wheezing, tightness of the chest and breathlessness due to a constriction of the airways. It can be a mild attack or a serious one that requires hospitalisation. There is some evidence that stress can trigger an attack.

The following common agents of health hazards will be described together with the circumstances in which they may be found (those marked * are **not** in the NEBOSH International General Certificate syllabus):

Ammonia* is a colourless gas with a distinctive odour which, even in small concentrations, causes the eyes to smart and run and a tightening of the chest. It is a corrosive substance which can burn the skin, burn and seriously damage the eye, cause soreness and ulceration of the throat and severe bronchitis and oedema (excess of fluid) of the lungs. Good eye and respiratory protective equipment (RPE) is essential when maintaining equipment containing ammonia. Any such equipment should be tested regularly for leaks and repaired promptly if required. Ammonia is also used in the production of fertilisers and synthetic fibres. Most work on ammonia plant should require a permit-to-work procedure.

Chlorine* is a greenish, toxic gas with a pungent smell which is highly irritant to the respiratory system, producing severe bronchitis and oedema of the lungs and may also cause abdominal pain, nausea and vomiting. It is used as a disinfectant for drinking water and swimming pool water and in the manufacture of chemicals.

Organic solvents* are used widely in industry as cleansing and degreasing agents. There are two main groups: the hydrocarbons (includes the aromatic and aliphatic hydrocarbons, such as toluene and white spirit) and the non-hydrocarbons (such as trichloroethylene and carbon tetrachloride). All organic solvents are heavier than air and most are sensitisers and irritants. Some are narcotics, while others can cause dermatitis and, after long periods of exposure, liver and kidney failure. It is very important that the safety data sheet accompanying the particular solvent is read and the recommended control measures adhered to and the correct personal protective equipment is worn at all times.

Solvents are used extensively in a wide variety of industries as varnishes, paints, adhesives, glue strippers, printing inks and thinners. They are highest risk when used as sprays. One of the most hazardous is dichloromethane (DCM), also known as methylene chloride. It is used as a paint stripper, normally as a gel. It can produce narcotic effects and has been classified as a Category 3 carcinogen in the European Union. The minimum personal protective equipment requirements are impermeable overalls, apron, footwear, long gloves and gauntlet and chemically resistant goggles or visor. Respiratory protective equipment is also required if it cannot be demonstrated that exposure is below the appropriate occupational exposure limit.

Carbon dioxide* is a colourless and odourless gas which is heavier than air. It represses the respiratory system, eventually causing death by asphyxiation. At low concentrations it will cause headaches and sweating followed by a loss of consciousness. The greatest hazard occurs in confined spaces, particularly where the gas is produced as a by-product.

Carbon monoxide is a colourless, tasteless and odourless gas which makes it impossible to detect in the blood without special measuring equipment. As explained earlier, carbon monoxide enters the blood (red cells) more readily than oxygen and restricts the supply of oxygen to vital organs. At low concentrations in the blood (less than 5%), headaches and breathlessness will occur, while at higher concentrations unconsciousness and death will result. The most common occurrence of carbon monoxide is as an exhaust gas either from a vehicle or a heating system. In either case, it results from inefficient combustion and, possibly, poor maintenance.

Carbon monoxide exposure may be prevented by ensuring that any work carried out in relation to gas appliances in domestic or commercial premises is undertaken by an engineer, competent in that area of work, and that there is enough fresh air in the room containing a gas appliance. In addition the use of audible carbon monoxide (CO) alarms is recommended as a useful back-up precaution.

Employers and landlords should ensure their appliances are serviced regularly by an approved and competent engineer.

Diesel-engine exhaust emissions* – Particulates from diesel engines emissions have been classified as carcinogenic, by experts from the World Health Organisation (WHO). The International Agency for Research on Cancer (IARC) found sufficient evidence that exposure to diesel exhausts is associated with an increased risk of lung cancer. They also noted limited evidence that exposure could lead to an increased risk of bladder cancer. The classification follows the findings in a US study of occupational exposure to such emissions in underground miners, which showed an increased risk of death from lung cancer in exposed workers. Exposure to these particulates may also aggravate other respiratory diseases such as asthma.

Isocyanates* are volatile organic compounds widely used in industry for products such as printing inks, adhesives, and two-pack paints (particularly in vehicle body shops) and in the manufacture of plastics (polyurethane products). They are irritants and sensitisers. Inflammation of the nasal passages, the throat and bronchitis are typical reactions to many isocyanates. When a person becomes sensitised to an isocyanate, very small amounts of the substance often provoke a serious reaction similar to an extreme asthma attack. Isocyanates also present a health hazard to fire-fighters. They are subject to an occupational exposure limit and respiratory protective equipment should be worn.

Lead* is a heavy, soft and easily worked metal. It is used in many industries but is most commonly associated with plumbing and roofing work. Lead enters the body normally by inhalation but can also enter by ingestion and skin contact. It is a cumulative poison that is stored in, and then released from, the bones. Excretion is via the faeces and urine.

The toxicity of lead affects nerve conduction and red blood cell production. Symptoms include abdominal pains, muscular weakness and tiredness. Inhalation is the most common means for the entry of lead into the body. From the lungs, lead enters the bloodstream and is distributed throughout the body. Regular health surveillance of lead workers by a medical practitioner is required.

The main targets for lead are the spinal cord and the brain, the blood and blood production. The effects are normally chronic and develop as the quantity of lead builds up. Headaches and nausea are the early symptoms followed by anaemia, muscle weakening and (eventually) coma. Regular blood tests are a legal and sensible requirement as are good ventilation and the use of appropriate personal protective equipment. High personal hygiene standards and adequate welfare (washing) facilities are essential and must be used before smoking or food is consumed. The reduction in the use of leaded petrol was an acknowledgement of the health hazard represented by lead in the air. If lead is to be used in the workplace, risk assessments should be undertaken and engineering controls put in place. Lead can be transferred to an unborn child through the placenta and, therefore, additional protection must be offered to women of reproductive capacity. Medical surveillance in the form of a blood test of all employees who come into contact with lead operations is recommended and such tests should take place at least once a year.

Silica is the main component of most rocks and is a crystalline substance made of silicon and oxygen. It occurs in quartz (found in granite), sand and flint. Harm is caused by the inhalation of silica dust, which can lead to silicosis (acute and chronic), fibrosis and pneumoconiosis. The dust which causes the most harm is respirable dust which becomes trapped in the alveoli. This type of dust is sharp and very hard and, probably, causes wounding and scarring of lung tissue. As silicosis develops, breathing becomes more and more difficult and eventually, as it reaches its advanced stage, lung and heart failure occur. It has also been noted that silicosis can result in the development of tuberculosis as a further complication. Hard rock miners, quarry workers, stone and pottery workers are most at risk. Health surveillance is recommended for workers in these occupations at initial employment and at subsequent regular intervals. Prevention is best achieved by the use of good dust extraction systems and respiratory protective equipment.

Silica is commonly produced during construction activities. Such activities which can expose workers and members of the public to silica dust include:

▶ cutting building blocks and other stone masonry work;
▶ cutting and/or drilling paving slabs and concrete paths;
▶ demolition work;
▶ sand blasting of buildings;
▶ tunnelling.

In general, the use of power tools to cut or dress stone and other silica-containing materials will lead to very high exposure levels while the work is occurring. In most cases, exposure levels are in excess of workplace exposure limits by factors greater than 2 and in some cases as high as 13.

The use of compressed air to remove dust is not recommended. There is a risk of compressed air entering the operator's bloodstream, which can result in death. Eye injury including blindness can also occur if dust particles bounce back at the operator. The inhalation of very fine silica dust can lead to the development of silicosis.

Cement dust and wet cement is an important construction material and is also a hazardous substance. Contact with wet cement can cause serious burns or ulcers which will take several months to heal and may need a skin graft. Dermatitis, both irritant and allergic, can be caused by skin contact with either wet cement or cement powder. Allergic dermatitis is caused by an allergic reaction to hexavalent chromium (chromate) which is present in cement. Cement powder can also cause inflammation and irritation of the eye, irritation of the nose and throat, and, possibly, chronic lung problems. Research has shown that between 5% and 10% of construction workers are probably allergic to cement. And plasterers, concreters and bricklayers or masons are particularly at risk. A plasterer, who knelt in wet cement for 5 hours while working, required skin grafts to his legs.

Manual handling of wet cement or cement bags can lead to musculoskeletal health problems and cement bags weighing more than 25 kg should not be carried by a single worker. PPE in the form of gloves, overalls with long sleeves and full-length trousers and waterproof boots must be worn on all occasions. If the atmosphere is dusty, goggles and respiratory protection equipment must be worn. An important factor in the possibility of dermatitis occurring is the sensitivity of the worker to the chromate in the cement and the existing condition of the skin including cuts and abrasions. Finally, adequate welfare facilities are essential so that workers

Figure 13.19 Example of dermatitis due to wet cement

can wash their hands at the end of the job and before eating, drinking or using the toilet. If cement is left on the skin for long periods without being washed off, the risk of an allergic reaction to hexavalent chromium will increase.

In the UK, it is prohibited to supply cement which has a concentration of more than 2 parts per million of chromium VI. This measure is designed to prevent allergic contact dermatitis when wet cement comes into contact with the skin. However, since the strong alkalinity of cement will remain, there is still the potential for skin burns. Workers exposed to cement need to be provided with adequate washing, changing, eating and drinking facilities.

Wood dust can be hazardous, particularly when it is hard wood dust which is known, in rare cases, to lead to nasal cancer. Composite boards, such as **medium-density fibreboard (MDF)**, are hazardous due to the resin bonding material used, which can also be carcinogenic. There are three types of wood-based boards available: laminated board, particle board and fibreboard. The resins used to bond the fibreboard together contain formaldehyde (usually urea formaldehyde). It is generally recognised that formaldehyde is 'probably carcinogenic to humans' and is subject to an OEL. At low exposure levels, it can cause irritation to the eyes, nose and throat and can lead to dermatitis, asthma and rhinitis. The main problems are most likely to occur when the MDF is being machined and dust is produced. A suitable risk assessment should be made and gloves and appropriate masks should always be worn when machining MDF. However, it is important to stress that safer materials are available which do not contain formaldehyde and these should be considered for use in the first instance.

Wood dust is produced whenever wood materials are machined, particularly sawed, sanded, bagged as dust from dust extraction units or during cleaning operations, especially if compressed air is used. The main hazards associated with all wood dusts are skin disorders, nasal problems, such as rhinitis, and asthma. There is also a hazard from fire and explosion. A hazardous substance assessment is essential to show whether the particular wood dust is hazardous. When the wood dust is created inside a woodworking shop, a well-designed extraction system is essential. PPE in the form of gloves, suitable respiratory protective equipment, overalls and eye protection may also be necessary as a result of the assessment. Finally, good washing and welfare facilities are also essential.

Tetanus is a serious, sometimes fatal, disease caused by a bacterium that lives in the soil. It usually enters the human body through a wound from an infected object, such as a nail, wood splinter or thorn. On entering the wound, it produces a powerful toxin which attacks the nerves that supply muscle tissue. It is commonly known as lockjaw because after an incubation period of approximately a week, stiffness around the jaw area occurs. Later the disease spreads to other muscles including the breathing system and this can be fatal. The disease has been well controlled with anti-tetanus immunisation and it is important that all construction workers are so immunised. Booster shots should be obtained every few years. Any flesh wound should be thoroughly cleaned immediately, an anti-septic cream applied and the wound covered.

Leptospirosis (Weil's disease) is caused by contact with Leptospira bacteria found in the urine of rats. Weil's disease starts with flu-like symptoms, severe headache and vomiting and muscle pains followed by jaundice, liver and kidney failure and, in up to 20% of cases, it can be fatal. It enters the body either through the skin or by ingestion. The most common source is contaminated water in a river, sewer or ditch and workers such as canal or sewer workers are most at risk. Leptospirosis is always a risk where rats are present, particularly if the associated environment is damp. Good, impervious protective clothing, particularly wellington boots, and the covering of any skin wounds is essential in these situations. For workers who are frequently in high-risk environments (sewer workers), immunisation with a vaccine may be the best protection. Weil's disease is, strictly, a severe form of leptospirosis. The symptoms of leptospirosis are similar to influenza but those for Weil's disease are anaemia, nose bleeds and jaundice. While the most common source of infection is from the urine of rats, Leptospirosis (Hardjo) has been found in other animals, such as cattle; therefore, farm and veterinary workers may also be at risk.

Legionella is an airborne bacterium and is found in a variety of water sources. It produces a form of pneumonia caused by the bacteria penetrating to the alveoli in the lungs. This disease is known as Legionnaires' disease, named after the first documented outbreak at a State Convention of the American Legion held at Pennsylvania in 1976. During this outbreak, 200 men were affected, of whom 29 died. That outbreak and many subsequent ones were attributed to air-conditioning systems. Legionnaires' disease is a potentially fatal form of pneumonia and everyone is susceptible to infection. However, some people are at higher risk, including:

▶ people over 45 years of age;
▶ smokers and heavy drinkers;
▶ people suffering from chronic respiratory or kidney disease;
▶ anyone with an impaired immune system.

The symptoms are a high temperature, fever and chills, cough, muscle pains and headache. In a severe case, there may also be pneumonia, and occasionally diarrhoea, as well as signs of mental confusion. Legionnaires' disease is not known to spread from person to person.

13

409

The legionella bacterium cannot survive at temperatures above 60°C but grows between 20°C and 45°C, being most virulent at 37°C. It also requires food in the form of algae and other bacteria. Control of the bacteria involves the avoidance of water temperatures between 20°C and 45°C, avoidance of water stagnation and the build-up of algae, sludge, sediments and organic materials and the use of suitable water-treatment chemicals. This work is often done by a specialist contractor.

The most common systems at risk from the bacterium are:

▶ water systems incorporating a cooling tower;
▶ water systems incorporating an evaporative condenser;
▶ hot and cold water systems and other plant where the water temperature may exceed 20°C.

An Approved Code of Practice (*Legionnaires' disease – The control of legionella bacteria in water systems* – L8) is available from the UK HSE. The latest edition of the ACoP and guidance contains revisions to simplify and clarify the text. The main changes are removing Part 2, the technical guidance, which is published separately in HSG274, and giving the following issues ACoP status:

▶ risk assessment;
▶ the specific role of an appointed competent person, known as the 'responsible person';
▶ the control scheme;
▶ review of control measures;
▶ duties and responsibilities of those involved in the supply of water systems.

Where plant at risk of the development of legionella exists, the following is recommended:

▶ a written 'suitable and sufficient' risk assessment;
▶ the preparation and implementation of a written control scheme involving the treatment, cleaning and maintenance of the system;
▶ temperature control between 20°C and 60°C;
▶ appointment of a named person with responsibility for the management of the control scheme;
▶ the monitoring of the system by a competent person;
▶ record keeping and the review of procedures developed within the control scheme.

The Code of Practice also covers the design and construction of hot and cold water systems and cleaning and disinfection guidance. There have been several cases of members of the public becoming infected from a contaminated cooling tower situated on the roof of a building. It is required that all cooling towers are registered with the local authority. People are more susceptible to the disease if they are older or weakened by some other illness. It is, therefore, important that residential and nursing homes and hospitals are particularly vigilant. The most common source of isolated outbreaks of legionella is showerheads, particularly when they remain unused for

a period of time. Showerheads should be cleaned and descaled at least every three months.

All warm water systems should be checked annually by a competent person. However, there are some simple checks that can be made as follows:

▶ each month, check that the temperature of water from hot water taps reaches 50°C after the water has run for 1 minute;
▶ each month, check that the temperature of water from cold water taps is below 20°C after the water has run for 2 minutes;
▶ each quarter, clean and descale shower heads and hoses;
▶ each week, purge little-used water outlets.

Hepatitis is a disease of the liver and can cause high temperatures, nausea and jaundice. It can be caused by hazardous substances (some organic solvents) or by a virus. The virus can be transmitted from infected faeces (hepatitis A) or by infected blood (hepatitis B and C). The normal precautions include good personal hygiene, particularly when handling food and in the use of blood products. Hospital workers who come into contact with blood products are at risk of hepatitis as are drug addicts who share needles. It is also important that workers at risk regularly wash their hands and wear protective disposable gloves.

Hepatitis is an example of a **blood-borne virus**. Another example is human immunodeficiency virus (HIV) which causes acquired immune deficiency syndrome (AIDS), affecting the immune system of the body. Some people carry a blood-borne virus in their blood which may cause disease in some people and few or no symptoms in others. The virus can spread to another person, whether the carrier of the virus is ill or not.

In the workplace, direct exposure can happen through accidental contamination by a sharp instrument or through contamination of open wounds, skin abrasions or through splashes to the eyes, nose or mouth. In occupations where there is a risk of exposure to blood-borne viruses, the following measures to prevent or control risks are recommended by the UK HSE:

▶ prohibit eating, drinking, smoking and the application of cosmetics in working areas where there is a risk of contamination;
▶ prevent puncture wounds, cuts and abrasions, especially in the presence of blood and body fluids;
▶ when possible avoid use of, or exposure to, sharps such as needles, glass or metal or if unavoidable take care in handling and disposal;
▶ consider the use of devices incorporating safety features, such as safer needle devices and blunt-ended scissors;
▶ cover all breaks in exposed skin by using waterproof dressings and suitable gloves;

▶ protect the eyes and mouth by using a visor/goggles/safety spectacles and a mask, where splashing is possible;

▶ avoid contamination by using water-resistant protective clothing;

▶ wear rubber boots or plastic disposable overshoes when the floor or ground is likely to be contaminated;

▶ use good basic hygiene practices, such as hand washing;

▶ control contamination of surfaces by containment and using appropriate decontamination procedures; and

▶ dispose of contaminated waste safely.

The risk of first-aiders being infected with a blood-borne virus while carrying out their duties is small. There has been no recorded case of HIV or hepatitis B being passed on during mouth-to-mouth resuscitation. The following precautions are recommended to reduce the risk of infection:

▶ cover any cuts or grazes on your skin with a waterproof dressing;

▶ wear suitable disposable gloves when dealing with blood or any other body fluids;

▶ use suitable eye protection and a disposable plastic apron where splashing is possible;

▶ use devices such as face shields when giving mouth-to-mouth resuscitation, but only if training in the use of them has been received; and

▶ wash hands after each procedure.

13.6 Safe handling and storage of waste

13.6.1 Environmental considerations

The current NEBOSH International General Certificate no longer includes general environmental considerations other than waste disposal (13.6.2). However, many health and safety advisers also have an 'environmental' responsibility. Organisations must also be concerned with aspects of the environment. A brief summary is given here of some of the more important issues that may be the responsibility of the health and safety manager.

There will be an interaction between the health and safety policy and the environmental policy which many organisations are now developing. Many of these interactions will be concerned with good practice, the reputation of the organisation within the wider community and the establishment of a good health and safety culture. The Environmental Management System should cover all aspects of an organisation's activities, products and services that could interact with the environment. Such aspects or elements include:

▶ the raw materials used and their source;

▶ by-products from processes;

▶ waste material produced;

▶ the energy and water used;

▶ emissions to air and water;

▶ lifespan of products and end-of-life disposal;

▶ packaging and transport issues; and

▶ environmental hazards associated with any materials used.

There are three environmental issues that are directly related to the health and safety function. These are:

▶ air pollution;

▶ water pollution;

▶ waste disposal.

Pollution is a term that covers more than the effect on the environment of atmospheric emissions, effluent discharges and solid waste disposal from industrial processes. It also includes the effect of noise, vibration, heat and light on the environment. The EU Solvents Emissions Directive (SED) has produced some tougher rules on the use of solvents in industry to protect the environment. The control of pesticides is an important function of regulatory authorities throughout the world.

Air pollution

The most common airborne pollutants are carbon monoxide, benzene, 1,3-butadiene, sulphur dioxide, nitrogen dioxide and lead. Air pollution is monitored by environmental agencies throughout the world. The causes, principles of formation and consequences of four common air pollution effects are as follows:

▶ **The greenhouse effect** is caused by the emission of, and increase in, atmospheric levels of specific pollutants such as carbon dioxide, methane and refrigerant gases. Greenhouse gases trap long wave thermal radiation causing a warming effect upon the atmosphere. As concentrations increase, this warming effect increases to the point where there is an imbalance between incoming short wave radiation and outgoing long wave radiation leading to net global warming of the atmosphere. The consequences of this warming include climate change, possible sea level rises and effects on agricultural and natural ecosystems.

▶ **Stratospheric ozone depletion** is caused by emissions of volatile organic compounds containing halogens which have a long half-life in the atmosphere – such as refrigerants, carbon tetrachloride, halons and other chlorinated solvents. The Earth is protected by a natural layer of ozone that absorbs incoming UV radiation. Natural ozone formation is by the process of photo-dissociation of oxygen molecules and reformation as ozone. Chlorine (and other halogen) atoms catalyse the destruction of ozone back to oxygen molecules and are not consumed in the reaction. Therefore, one atom of chlorine can destroy many ozone molecules. The main consequence is an increased level of

13

cell and genetic damage, ultimately leading to skin cancer and crop damage.

▶ **Acid rain** is caused by the emission of gases that form acidic compounds in the presence of water or water vapour, such as sulphur dioxide, hydrogen chloride and nitrogen dioxide. These acid-forming gases dissolve in water in the atmosphere to produce acids such as sulphurous or sulphuric acid. When the acidic water forms raindrops or mist and reaches the ground, it affects the acidity (pH) balance in upland ecosystems, leading to the release of toxic metals into surface run-off with subsequent toxic effects on vegetation, aquatic invertebrates and fish. Acid rain can also damage materials such as limestone in buildings, and cause corrosion to metals such as galvanised steel.

▶ **Photochemical smog** is associated with the emission of volatile organic compounds such as solvents, petrol vapour and other compounds into a warm atmosphere in the presence of sunlight. The emissions react with each other in the presence of ultraviolet radiation and reactive gases such as nitrogen dioxide to create complex mixtures of secondary pollutants and ozone, many of which cause irritation to respiratory systems and degrade materials such as rubber. On a warm day, levels of photochemical air pollutants accumulate in the atmosphere around larger cities to create a brown photochemical haze.

Water pollution

An ILO report states that 'Water is an integral part of most development activities, from health and sanitation, to the location of human settlements, agricultural production, nutrition, and the maintenance of ecological balance. It may be difficult to accept the idea in many countries where water has been considered as a "free commodity" for generations that it can be an economic good with distribution costs that have to be paid if the service is to remain sustainable'. While the global demand for water has doubled, the amount of usable water available has reduced by 40%.

The problem is particularly serious in many developing economies and it has been predicted that by 2025, approximately two-thirds of the world's population might live in countries with moderate to severe water shortages.

The pollution of rivers and other water courses can produce very serious effects on the health of plants and animals which rely on that water supply. National governments throughout the world are responsible for coastal waters, inland freshwater and groundwaters. The European Union Groundwater Directive seeks to

(a)

(b)

Figure 13.21 (a) Water pollution from an oil spillage; (b) water pollution from plastic and other solid waste

Figure 13.20 Example of heavy industrial air pollution

protect groundwater from pollution since this is a source of drinking water. Such sources can become polluted by leakage from industrial soakaways. Discharges to a sewer are also controlled by environmental agencies and these agencies define those substances which are prohibited from discharge (e.g. petroleum spirit). If hazardous substances are being used by the organisation, safety data sheets give advice on the safe disposal of any residues that remain after the particular process has been completed.

In some countries, the local water companies and environmental agencies have a legal right to sample discharges into its sewers because it is required to keep a public trade effluent register. These organisations often publish lists of proscribed substances which can only be discharged into a public sewer with permission from the agencies.

Finally, if oil is stored on the premises, a retaining bund wall should surround the oil store. This will not only ensure that any oil leakage is contained but will also stop the contamination of groundwater by fire-fighting foam in the event of a fire. One major environmental problem is that of contaminated land. Contaminated land is produced by leakage, accidental spillage and uncontrolled waste disposal. The contaminator of the land has the primary responsibility and liability to clean up the contamination.

13.6.2 Waste disposal

Basic principles of waste disposal

The principal requirements for safe waste disposal are as follows:

▶ to handle waste so as to prevent any unauthorised escape into the environment;
▶ to pass waste only to an officially authorised person; and
▶ to ensure that a written description accompanies all waste. Producers of waste should complete a document that gives full details of the type and quantity of waste for collection and disposal. Copies of the note should be kept for at least two years.

There are several different categories of waste. The important ones are as follows:

▶ Controlled waste comprises household, clinical, industrial or commercial waste. Controlled waste should only be removed by those holding a government licence so that the waste is not disposed in a manner likely to cause environmental pollution or harm to human health.
▶ Hazardous waste which can only be disposed of using special arrangements. These are sometimes substances which are life threatening (toxic, corrosive or carcinogenic) or highly flammable. Clinical waste falls within this category. A

consignment note system should accompany this waste at all the stages to its final destination. Before hazardous waste is removed from the originating premises, a contract should be in place with a licensed carrier and it should be stored securely prior to collection to ensure that the environment is protected.

Figure 13.22 Commercial waste collection

Hazardous waste also covers computer monitors, fluorescent tubes, end-of-life vehicles and television sets. The hazardous properties are listed in Appendix 13.2. It is important to ensure that hazardous waste is safely managed and its movement is documented. The following points also are important for construction sites:

▶ Sites that produce more than 200 kg of hazardous waste each year for removal, treatment or disposal may need to register with a government agency.
▶ Different types of hazardous waste must not be mixed.
▶ Producers must maintain registers of their hazardous wastes.

Some form of training may be required to ensure that employees segregate hazardous and non-hazardous wastes on site and fully understand the risks and necessary safety precautions which must be taken. Personal protective equipment, including overalls, gloves and eye protection, must be provided and used. The storage site should be protected against trespassers, fire and adverse weather conditions. If flammable or combustible wastes are being stored, adequate fire protection systems must be in place. There may also be manual handling issues to be considered. Finally, in the case of liquid wastes, any drains must be protected and bunds used to restrict spreading of the substance as a result of spills.

A hierarchy for the management of waste streams has been recommended by the UK Environment Agency.

1. **Prevention** – by changing the process so that the waste is not produced (e.g. substitution of a particular material).

13

413

2. **Reduction** – by improving the efficiency of the process (e.g. better machine maintenance).
3. **Reuse** – by recycling the waste back into the process (e.g. using reground waste plastic products as a feed for new products).
4. **Recovery** – by releasing energy through the combustion, recycling or composting of waste (e.g. the incineration of combustible waste to heat a building).
5. **Responsible disposal**– by disposal in accordance with regulatory requirements.

The control measures for hazards associated with waste disposal:

▶ All workers should wear suitable PPE, such as high-visibility jackets, gloves and suitable footwear.
▶ Training so that employees segregate hazardous and non-hazardous wastes on site and fully understand the risks and necessary safety precautions to be taken.
▶ The storage site should be protected against trespassers, fire and adverse weather conditions.
▶ If flammable or combustible wastes are being stored, adequate fire protection systems must be in place.
▶ Manual handling issues may need to be considered.
▶ All lifting equipment, including chains and shackles, must be subject to a periodic statutory examination.
▶ Separate storage required for incompatible waste streams.
▶ Finally, for liquid wastes, drains must be protected and bunds used to restrict spreading of the substance as a result of spills.

Waste disposal practice

The collection and removal of waste from a workplace, such as a construction site, is normally accomplished using a skip. Every year, activities involving the movement of skips and containers cause death and serious injury. The hazards include:

▶ being struck by vehicles;
▶ falling and slipping;
▶ failures of lifting equipment;
▶ striking overhead cables/obstructions;
▶ vehicle overturns;
▶ runaway vehicles.

The skip should be located on firm, level ground away from the main construction work, particularly excavation work. This will allow clear access to the skip for filling and removal from site. On arrival at the site, the integrity of the skip should be checked. It should be filled either by chute or by mechanical means unless items can be placed in by hand. Skips should not be overfilled and be netted or sheeted over when they are full. Any hazardous waste should be segregated.

Waste skip selection should be made during the site planning process. The selected skip must be suitable for the particular job. The following points should be considered:

▶ sufficient strength to cope with its load;
▶ stability while being filled;
▶ a reasonable uniform load distribution within the skip at all times;
▶ the immediate removal of any damaged skip from service and the skip inspected after repair before it is used again;
▶ sufficient space around the skip to work safely at all times;
▶ the skip should be resting on firm level ground;
▶ the skip should never be overloaded or overfilled;
▶ there must be sufficient headroom for the safe removal of the skip when it is filled; and
▶ there should be suitable security arrangements such as ensuring that the skip is situated away from boundary fences to reduce the risk of arson.

There are hazards present during the movement of a loaded skip from the ground to the back of a skip loader vehicle. Entanglement with the vehicle lifting mechanisms, such as the hydraulic arms and lifting chains, is a major hazard. Other hazards include contact by the skip with overhead obstructions, movement of the skip contents and skip overload leading to mechanical or structural failure. Slip hazards may be present due to spillages from the skip and the skip contents could be contaminated with biological material, asbestos or syringes. Passing traffic during the loading operation may also present a hazard.

The control measures for these hazards include the use of outriggers to increase the stability of the loader vehicle and the provision of steps for the driver to alight from the cab or the vehicle flatbed. The contents of the skip should be secured using netting or tarpaulin. Adhering to the safe working loads of the skip and lifting equipment and the use of a banksman during the lifting process are additional controls. The area around the vehicle may need to be cordoned off to protect passing pedestrians and road traffic. All workers concerned with

Figure 13.23 A designated waste collection area with two types of skip commonly used for waste collection. Heavy materials would be transported in the smaller skip. Sizes of skip range from about 4 cu metres (small skip shown) to about 35 cu metres (large skip shown)

the operation should wear suitable PPE, such as high-visibility jackets, gloves and suitable footwear. Finally, all lifting equipment, including chains and shackles, must be subject to a periodic examination.

The European Union has introduced the Waste Electrical and Electronic Equipment (WEEE) Directive, the Restrictions of the use of certain Hazardous Substances in electrical and electronic equipment (RoHS) and the End of Life Vehicle (ELV) Directive.

The aim of the WEEE Directive is to minimise the environmental impact of electrical and electronic equipment both during their lifetime and when they are discarded. All electrical and electronic equipment must be returned to the retailer from its end user and reused or reprocessed by the manufacturer. Manufacturers must register with the Environment Agency, who will advise on these obligations. The WEEE Regulations have been amended so that producer-compliance schemes report their activities in a more precise way (by providing evidence of recycling in kilograms instead of tonnes). It is hoped that this will reduce the delays in recycling and lead to the recycling of smaller amounts at more frequent intervals. The European Commission has proposed changes to both WEEE and RoHS. On WEEE, a major proposal is to increase the amounts of electrical and electronic waste that are separately collected and recycled, while the proposals on RoHS aim for a higher

Figure 13.24 Electronic waste under EU WEEE

level of environmental protection by revising the scope of the restrictions and the applicable substances.

13.7 Further information

Ambient factors in the workplace, International Labour Organisation (ILO) Code of Practice (CoP), ISBN 92-2-11628-X http://www.ilo.org/safework/info/standards-and-instruments/WCMS_107729/lang--en/index.htm

Asbestos, Convention and Recommendation, C162 and R172, 1986 http://www.ilo.org/dyn/normlex/en/f?p=1000:12100:0::NO::P12100_ILO_CODE:C162 http://www.ilo.org/dyn/normlex/en/f?p=1000:12100:0::NO::P12100_ILO_CODE:R172

Chemicals, Convention and Recommendation, C170 and R177, 1990 http://www.ilo.org/dyn/normlex/en/f?p=1000:12100:0::NO::P12100_ILO_CODE:C170 http://www.ilo.org/dyn/normlex/en/f?p=1000:12100:0::NO::P12100_ILO_CODE:R177

Classification, Labelling and Packaging of Substances and Mixtures Regulations EC No 1272/2008 – http://ec.europa.eu/enterprise/sectors/chemicals/documents/classification/

Occupational Cancer, Convention and Recommendation, C139 and R147, 1974 http://www.ilo.org/dyn/normlex/en/f?p=1000:12100:0::NO::P12100_ILO_CODE:R147

Occupational Exposure to Airborne Substances Harmful to Health, ILO CoP, ILO Geneva, ISBN 92-2-102442-3 http://www.ilo.org/wcmsp5/groups/public/---ed_protect/--protrav/---safework/documents/normativeinstrument/wcms_107851.pdf

Occupational health services, Convention and Recommendation, C161 and R171, 1985 http://www.ilo.org/dyn/normlex/en/f?p=1000:12100:0::NO::P12100_ILO_CODE:R171

Registration, Evaluation, Authorisation and Restriction of Chemicals (REACH) http://www.hse.gov.uk/reach/

Safety in the Use of Chemicals at Work, International Labour Office (ILO) Code of Practice (CoP), ILO, 1993, ISBN 92-2-108006-4 http://www.ilocarib.org.tt/images/stories/contenido/pdf/OccupationSafetyandHealth/codes/safety-in-use-chemicals-1993.pdf

13

13.8 Practice revision questions

1. Health hazards may be present in the workplace through various chemical and biological agents.

   (a) **Describe FOUR** chemical and **THREE** biological hazardous agents and **give** an example of each agent.

   (b) **Explain** the term 'respirable dust' and **outline** its health effect on the human body.

2. For each of the following types of hazardous substance, **give** a typical example and **outline** its primary effect on the human body:

   (a) irritant
   (b) corrosive
   (c) toxic
   (d) carcinogenic
   (e) mutagenic.

3. (a) **Describe** the differences between acute and chronic health effects giving an example of each effect.
   (b) **Identify TWO** types of cellular defence mechanisms that the body has as a natural defence system.
   (c) **Identify** the factors that could affect the level of harm experienced by a worker who has been exposed to a toxic substance.

4. (a) **Identify FOUR** possible routes of entry of a hazardous substance into the body **AND, in EACH** case, **give** an example of how an employee might be at risk of such exposure.
   (b) **Define** the term 'target organ' within the context of occupational health.
   (c) **Outline** the personal hygiene practices that should be followed to reduce the risk of ingestion of a hazardous substance.

5. (a) **Identify** possible routes of entry of biological organisms into the body.
   (b) **Outline** control measures that could be used to reduce the risk of infection from biological organisms.

6. (a) **Identify FOUR** respiratory diseases that could be caused by exposure to dust at work.
   (b) **Describe** the respiratory defence mechanisms of the body against atmospheric dust.

7. (a) **Describe** the typical symptoms of occupational dermatitis.
   (b) **Identify TWO** common causative agents.
   (c) **Outline** control measures which may be taken to prevent the occurrence of occupational dermatitis.

8. (a) **Outline** the issues to be considered when undertaking an assessment of health risks from hazardous substances that may be used in a workplace.
   (b) **Identify** the information that should be included on a manufacturer's safety data sheet supplied with a hazardous substance.
   (c) **Outline FOUR** sources of information that might be consulted when assessing the risk of a hazardous substance in a workplace.

9. (a) **Identify** the main advantages and limitations of chemical indicator (stain detector) tubes.
   (b) **Explain** how a dust lamp can give an indication of airborne dust levels in a workplace.
   (c) **Outline** how a personal dust sampler is used to measure the levels of airborne dust in a workplace.

10. (a) **Outline** the main purposes of occupational exposure limits (OELs).
    (b) **Explain** the meaning of the following terms:
        (i) the short-term exposure limit (STEL)
        (ii) the long-term exposure limit (LTEL)
        (iii) the maximum permissible concentration.
    (c) **Outline** the limitations of occupational exposure limits.

11. (a) **Explain** the meaning of the term 'occupational exposure limit' (OEL).
    (b) **Identify** the two classifications of hazardous substances that require exposure to be reduced to as low a level as is reasonably practicable below the OEL.
    (c) **Outline FOUR** actions that could be taken when a WEL has been exceeded.

12. With respect to workplace exposure to airborne hazardous substances:
    (a) **Outline TWO** purposes of exposure limits.
    (b) **Give** the meaning of the following terms:
        (i) maximum allowable concentration
        (ii) threshold limit value
        (iii) short-term exposure limit
        (iv) time-weighted average (TWA) exposure limit.

13. (a) **Outline** the Principles of Good Practice that will help to prevent ill-health due to exposure of workers to hazardous substances.
    (b) **Outline** a hierarchy of control measures that could be used to implement the Principles of Good Practice for the control of exposure to hazardous substances.

14. A local exhaust ventilation (LEV) system is to be installed to extract welding fume from an engineering workshop.
    (a) **Identify** the five basic components of the LEV system.
    (b) **Outline** the issues that might reduce the effectiveness of the LEV system.
    (c) **Outline** the routine maintenance that should be carried out between periodic examinations so that the ventilation system continues to operate effectively.

15. (a) **Explain** the meaning of the term 'dilution ventilation'.
    (b) **Identify THREE** instances when dilution ventilation may be used in a workplace.
    (c) **Identify** the limitations to the use of dilute ventilation.

16. **Identify** the possible health effects on workers and **identify FOUR** control measures to minimise those health effects for the following hazardous substances:

(a) solvent-based adhesives and paints
(b) wood dust
(c) glass fibres
(d) isocyanates.

17. (a) **Outline** the issues that should be considered when selecting and using personal protective equipment.
(b) **Identify** the hazards that the following items of personal protective equipment should offer protection from and **outline** the practical limitations of each item:
(i) safety gloves
(ii) safety footwear
(iii) safety goggles
(iv) safety spectacles.

18. (a) **Outline** the health and safety hazards associated with welding operations.
(b) **Outline** the factors to be considered in the selection of respiratory protective equipment for persons carrying out welding activities.
(c) **Identify** the difference between breathing apparatus and respirators.

19. (a) **Outline** the purpose of health surveillance of workers who may have come into contact with hazardous substances.
(b) **Give THREE** examples of health surveillance in the workplace.
(c) **Outline** the importance of personal hygiene in the workplace.

20. (a) **Identify TWO** respiratory diseases that may be caused by exposure to asbestos.
(b) **Identify** the **THREE** types of asbestos commonly found in buildings and **identify** the common sources of asbestos in buildings.
(c) **Outline** the control measures required before and during the removal of asbestos from a building.

21. For **EACH** of the following agents, **outline** the principal health effects **AND identify** a typical workplace situation in which a person might be exposed:
(a) carbon monoxide
(b) legionella bacteria

(c) leptospira
(d) blood-borne viruses.

22. **Identify** the possible health hazards to which construction workers may be exposed from the following substances. In each case **give** an example of a likely source and suitable control measures:
(a) silica dust
(b) wet cement.

23. A toxic substance has been transported by lorry to a warehouse in a number of drums. At the warehouse, one of the drums is ruptured during unloading and some of the contents are spilt on the floor.
(a) **Outline** the main controls that should be in place to ensure the safe transport of the hazardous substance by road.
(b) **Outline** a procedure that would deal safely with the spillage.
(c) **Identify THREE** ways in which persons working near the spillage might be harmed.

24. A company produces a range of solid and liquid wastes, both hazardous and non-hazardous.
(a) **Identify** the main hazards associated with the storage of wastes.
(b) **Outline** the required arrangements to ensure the safe storage of the wastes prior to their collection and disposal.
(c) **Outline** the issues that should be addressed by the company when developing a system for the safe collection and disposal of waste.

25. Skips are used to store solid waste materials prior to collection and disposal.
(a) **Identify** the hazards a skip collector could be exposed to when moving a full skip from the ground onto the back of a skip loader vehicle.
(b) **Identify EIGHT** safe practices to be followed when using a skip for the collection and removal of waste from a construction site.

13

APPENDIX 13.1 GHS Hazard (H) Statements (Health only)

H-statement	Phrase
300	Fatal if swallowed
301	Toxic if swallowed
302	Harmful if swallowed
303	May be harmful if swallowed
304	May be fatal if swallowed and enters airways
305	May be harmful if swallowed and enters airways
310	Fatal in contact with skin
311	Toxic in contact with skin
312	Harmful in contact with skin
313	May be harmful in contact with skin
314	Causes severe burns and eye damage
315	Causes skin irritation
316	Causes mild skin irritation
317	May cause an allergic skin reaction
318	Causes serious eye damage
319	Causes serious eye irritation
320	Causes eye irritation
330	Fatal if inhaled
331	Toxic if inhaled
332	Harmful if inhaled
333	May be harmful if inhaled
334	May cause allergy or asthma symptoms or breathing difficulties if inhaled
335	May cause respiratory irritation
336	May cause dizziness or drowsiness
340	May cause genetic defects *(route if relevant)*
341	Suspected of causing genetic defects *(route if relevant)*
350	May cause cancer *(route if relevant)*
351	Suspected of causing cancer *(route if relevant)*
360	May damage fertility or the unborn child *(effect if known, route if relevant)*
361	Suspected of damaging fertility or the unborn child *(effect if known, route if relevant)*
362	May cause harm to breast-fed children
370	Causes damage to organs *(organ if known, route if relevant)*
371	May cause damage to organs *(organ if known, route if relevant)*
372	Causes damage to organs through prolonged or repeated exposure *(organ if known, route if relevant)*
373	May cause damage to organs through prolonged or repeated exposure *(organ if known, route if relevant)*
EU66	Repeated exposure may cause skin dryness or cracking
EU70	Toxic by eye contact
EU71	Corrosive to the respiratory tract

APPENDIX 13.2 Hazardous properties of waste

	Hazard	Description
H1	Explosive	Substances and preparations which may explode under the effect of flame or which are more sensitive to shocks or friction than dinitrobenzene.
H2	Oxidising	Substances and preparations which exhibit highly exothermic reactions when in contact with other substances, particularly flammable substances.
H3-A	Highly flammable	▶ liquid substances and preparations having a flash point below 21°C (including extremely flammable liquids), or ▶ substances and preparations which may become hot and finally catch fire in contact with air at ambient temperature without any application of energy, or ▶ solid substances and preparations which may readily catch fire after brief contact with a source of ignition and which continue to burn or to be consumed after removal of the source of ignition, or ▶ gaseous substances and preparations which are flammable in air at normal pressure, or ▶ substances and preparations which, in contact with water or damp air, evolve highly flammable gases in dangerous quantities.
H3-B	Flammable	Liquid substances and preparations having a flash point equal to or greater than 21°C and less than or equal to 55°C.
H4	Irritant	Non-corrosive substances and preparations which, through immediate, prolonged or repeated contact with the skin or mucous membrane, can cause inflammation.
H5	Harmful	Substances and preparations which, if they are inhaled or ingested or if they penetrate the skin, may involve limited health risks.
H6	Toxic	Substances and preparations (including very toxic substances and preparations) which, if they are inhaled or ingested or if they penetrate the skin, may involve serious, acute or chronic health risks and even death.
H7	Carcinogenic	Substances and preparations which, if they are inhaled or ingested or if they penetrate the skin, may induce cancer or increase its incidence.
H8	Corrosive	Substances and preparations which may destroy living tissue on contact.
H9	Infectious	Substances containing viable micro-organisms or their toxins which are known or reliably believed to cause disease in humans or other living organisms.
H10	Teratogenic	Substances and preparations which, if they are inhaled or ingested or if they penetrate the skin, may induce non-hereditary congenital malformations or increase their incidence.
H11	Mutagenic	Substances and preparations which, if they are inhaled or ingested or if they penetrate the skin, may induce hereditary genetic defects or increase their incidence.
H12		Substances and preparations which release toxic or very toxic gases in contact with water, air or an acid.
H13		Substances and preparations capable by any means, after disposal, of yielding another substance, e.g. a leachate, which possesses any of the characteristics listed above.
H14	Ecotoxic	Substances and preparations which present or may present immediate or delayed risks for one or more sectors of the environment.

13

APPENDIX 13.3 Different types of protective gloves

Remember: Select your gloves according to the Hazard and Risk Assessment. Keep your gloves in good condition. Seek advice if you are in any doubt.

GENERAL USE

TYPE 1 Heavy work, General Protection (Rigger's type replacement)

- ✓ Some Abrasion & Good Tear Resistance
- ✗ Low Cut & Puncture Protection No Resistance to Fluids
- 2142

TYPE 2 Fine, Dextrous Work

- ✓ Abrasion & Tear Resistant
- ✗ Low Cut & Puncture Protection No Resistance to Fluids
- 3131

TYPE 3 Work involving Water & Oil

- ✓ Water & Oil Resistant, Good Abrasion Protection
- ✗ Low Cut Resistance
- 3121

TYPE 4 Handling Acid and other Chemicals

- ✓ Chemical Resistant, Long to Protect Arm
- ✗ Minimal Cut & Abrasion Protection
- N/A

TYPE 5 Skin Protection from Oils and other Low Hazard Substances
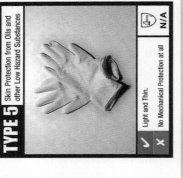
- ✓ Light and Thin.
- ✗ No Mechanical Protection at all
- N/A

SPECIALIST USE

TYPE 6 Welders Leather Gauntlets For protection against heat and burns
- ✓ Very Durable for Heat Protection
- ✗ Poor Chemical & Cut Resistance
- N/A

TYPE 7 High Pressure Water Jetting Slip Resistant Grip
- ✓ Waterproof with Good Wet Grip Abrasion & Tear Resistant
- ✗ Low Cut & Puncture Resistance
- 4131

TYPE 8 Kevlar, with High Cut Resistance Use with Sharps, e.g. cert screens, glass

- ✓ Strong & Dextrous High Cut Resistance
- ✗ No Resistance to Fluids
- 3444

TYPE 9 Heavy Engineering Work e.g. on Pellet Mills
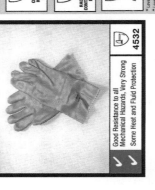
- ✓ Good Resistance to all Mechanical Hazards, Very Strong
- ✓ Some Heat and Fluid Protection
- 4532

Explanation of Protection Ratings

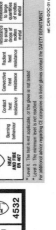

MECHANICAL HAZARDS EN 388

PERFORMANCE LEVELS*			
0 to 4	0 to 5	0 to 4	0 to 4
Abrasion resistance	Blade cut resistance	Tear resistance	Puncture resistance

CHEMICAL HAZARDS EN 374

Liquidproof test	Permeation test

MICRO ORGANISMS EN 374

Liquidproof according to the air leak test

RADIOACTIVE CONTAMINATION EN 421

COLD HAZARDS EN 511

PERFORMANCE LEVELS*		
0 to 4	0 to 4	0 to 1
Convective cold resistance	Contact cold resistance	Water permeability

HEAT AND FIRE EN 407

PERFORMANCE LEVELS*				
0 to 4	0 to 4	0 to 3	0 to 4	0 to 4
Burning behaviour	Contact heat resistance	Convective heat resistance	Radiant heat resistance	Resistance to small drops of molten metal	Resistance to large quantities of molten metal

* Level X: The test is not applicable, or the glove is not tested
* Level 0: The minimum level is not reached
For additional advice regarding CE categories on listed gloves contact the SAFETY DEPARTMENT

ref: CAN-DOC-01 rev.1

Note that types provided may vary from those illustrated above, but will give similar protection levels

CHAPTER 14

Physical and psychological health hazards and risk control

14.1 Noise ▶ 422

14.2 Vibration ▶ 428

14.3 Radiation ▶ 434

14.4 Stress ▶ 442

14.5 Further information ▶ 444

14.6 Practice revision questions ▶ 444

> **This chapter covers the following NEBOSH learning objectives:**
> 1. Outline the health effects associated with exposure to noise and appropriate control measures
> 2. Outline the health effects associated with exposure to vibration and appropriate control measures
> 3. Outline the health effects associated with ionising and non-ionising radiation and appropriate control measures
> 4. Outline the meaning, causes and effects of work-related stress and appropriate control measures

Introduction

Occupational health is concerned with physical and psychological hazards as well as chemical and biological hazards. The physical occupational hazards have been well known for many years, and recent emphasis has been on the development of lower risk workplace environments. Physical hazards include topics such as electricity, display screen equipment (DSE) and manual handling, which were covered in earlier chapters; and noise, vibration and radiation, which are discussed in this chapter.

The physical and psychological hazards discussed in this chapter are covered by the provisions of ILO Code of Practice 'Ambient factors in the workplace' and this code should be considered as the basis for eliminating or controlling exposure to ionising and non-ionising radiation, ultraviolet, infrared and (in some circumstances) visible radiation, noise, vibration, high and low temperatures and humidity. The code does not apply to other ambient factors such as shift work, ergonomic factors or psychological factors, such as workplace stress or violence.

However, it is only really in the last 20 years that psychological hazards have been included among the occupational health hazards faced by many workers. This is now the most rapidly expanding area of occupational health, and includes topics such as mental health and workplace stress (as well as violence to staff and substance abuse covered in Chapter 7).

14.1 Noise

14.1.1 Introduction

There was considerable concern for many years over the increasing cases of occupational deafness and this led to the introduction by the ILO of the Code of Practice 'Ambient factors in the workplace'. The provisions of this code should be considered as the basis for eliminating or controlling exposure to noise. The information on the assessment of noise exposure and protective and preventative measures contained in the code and provided for in the Working Environment (Air Pollution, Noise and Vibration) Convention (No. 148), and Recommendation (No. 156), 1977, and the ILO Code of Practice 'Protection of workers against noise and vibration in the working environment' (Geneva, 1984) should apply globally.

In addition, for the prediction of the amount of hearing loss expected to occur as a function of noise exposure level and duration, age and sex, when no national provisions are available, the international consensus standard ISO 1999, '*Acoustics: Determination of occupational noise exposure and estimate of noise-induced hearing impairment* (1990)', should apply.

In summary, the code requires the employer to:

▶ assess noise levels and keep records;
▶ reduce the risks from noise exposure by using engineering controls in the first instance and the provision and maintenance of hearing protection as a last resort;
▶ provide workers with health surveillance;
▶ provide workers with information and training.

Many occupations have potential noise problems including construction, manufacturing, entertainment, the uniformed services and call centres.

The main purpose of the code is to control noise levels rather than measuring them. This involves the better design of machines, equipment and work processes, and ensuring that personal protective equipment is correctly worn and workers are given adequate training and health surveillance.

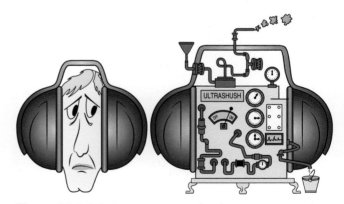

Figure 14.1 It is better to control noise at source rather than wear ear protection

Sound is transmitted through the air by sound waves which are produced by vibrating objects. The vibrations cause a pressure wave which can be detected by a receiver, such as a microphone or the human ear. The ear may detect vibrations which vary from 20 to 20 000 (typically 50–16,000) cycles each second (or Hertz – Hz). Sound travels through air at a finite speed (342 m/s at 20°C and sea level). The existence of this speed is shown by the time lag between lightning and thunder during a thunderstorm. Noise normally describes loud, sudden, harsh or irritating sounds although noise is defined as any audible sound.

Noise may be transmitted directly through the air, by reflection from surrounding walls or buildings or through the structure of a floor or building. In construction work, the noise and vibrations from a pneumatic drill will be transmitted from the drill itself, from the ground being drilled and from the walls of surrounding buildings.

14.1.2 Health effects of noise

The human ear

There are three sections of the ear: the outer (or external) ear, the middle ear and the inner (or internal) ear. The sound pressure wave passes into and through the outer ear and strikes the eardrum causing it to vibrate. The eardrum is situated approximately 25 mm inside the head. The vibration of the eardrum causes the proportional movement of three interconnected small bones in the middle ear, thus passing the sound to the cochlea situated in the inner ear.

Within the cochlea the sound is transmitted to a fluid causing it to vibrate. The motion of the fluid induces a membrane to vibrate which, in turn, causes hair cells attached to the membrane to bend. The movement of the hair cells causes a minute electrical impulse to be transmitted to the brain along the auditory nerve. Those hairs nearest to the middle ear respond to high frequency, while those at the tip of the cochlea respond

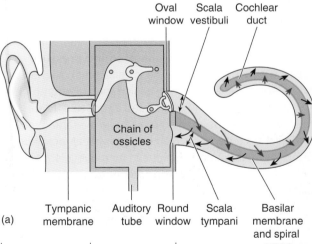

(a)

(b)

Figure 14.2 Passage of sound waves: (a) the ear with cochlea uncoiled; (b) summary of transmission

to lower frequencies. (**Frequency** is the number of occurrences of a repeating event over a given period of time and is measured in Hertz (Hz).)

There are about 30,000 hair cells within the ear and noise-induced hearing loss causes irreversible damage to these hair cells.

Ill-health effects of noise

Noise can lead to ear damage on a temporary (acute) or permanent (chronic) basis.

There are three principal **acute** effects:

▶ **temporary threshold shift** – caused by short excessive noise exposures and affects the sensory cells in the cochlea by reducing the flow of nerve impulses to the brain. The result is a slight deafness or sensitivity, particular to speech, which is reversible when the noise is removed;

▶ **tinnitus** – a ringing in the ears caused by an intense and sustained high noise level. It is caused by the over-stimulation of the hair cells. The ringing sensation continues for up to 24 hours after the noise has ceased;

▶ **acute acoustic trauma** – caused by a very loud peak noise such as an explosion. It affects either the eardrum or the bones in the middle ear and is usually reversible. Severe explosive sounds can permanently damage the eardrum.

Occupational noise can also lead to one of the following three **chronic** hearing effects:

▶ **noise-induced hearing loss** – results from permanent damage to the cochlear hair cells. It affects the ability to hear speech clearly but the ability to hear is not lost completely;

▶ **permanent threshold shift** – this results from prolonged exposure to loud noise and is irreversible due to the permanent reduction in nerve impulses to the brain. This shift is most marked at the 4000 Hz frequency, which can lead to difficulty in hearing certain consonants and some female voices;

▶ **tinnitus** – is the same as the acute form but becomes permanent. It is a very unpleasant condition, which can develop without warning.

It is important to note that, if the level of noise exposure remains unchanged, noise-induced hearing loss will lead to a permanent threshold shift affecting an increasing number of frequencies. The hearing loss caused can be temporary or permanent.

▶ **temporary deafness** can occur after leaving a noisy place. Hearing usually recovers within a couple of hours. This is a sign that continued exposure to loud noise could permanently damage your hearing;

▶ **sudden extremely loud noise** can cause instant damage; and

▶ **repeated exposure** causes gradual hearing loss due to repeated exposure. This is more common and it

14

423

can take years for a worker to realise just how deaf they have become.

Presbycusis is the term used for hearing loss in older people which may have been exacerbated by occupational noise earlier in their lives.

14.1.3 Noise assessments

Several countries specify action levels at which the hearing of employees must be protected. The conclusion as to whether any of those levels has been breached is reached after an assessment of noise levels has been made. However, before noise assessment can be discussed, noise measurement and some sample statutory action levels must be described. See Chapter 18 for a suitable noise assessment form.

Noise measurement

Sound intensity is measured by a unit known as a pascal (Pa – N/m²), which is a unit of pressure similar to that used when inflating a tyre. If noise was measured in this way, a large scale of numbers would be required ranging from 1 at one end to 1 million at the other. The sound pressure level (SPL) is a more convenient scale because:

▶ it compresses the size of the scale by using a logarithmic scale to the base 10;
▶ it measures the ratio of the measured pressure, p, to a reference standard pressure, p_0, which is the pressure at the threshold of hearing (2×10^{-5} Pa).

The unit is called a decibel (dB) and is defined as:

$$SPL = 20\log_{10}(p/p_0)\text{dB}$$

It is important to note that since a logarithmic scale to the base 10 is used, each increase of 3 dB is a doubling in the sound intensity. Thus, if a sound reading changes from 75 dB to 81 dB, the sound intensity or loudness has increased by four times.

Finally, as the human ear tends to distort its sensitivity to the sound it receives by being less sensitive to lower frequencies, the scale used by sound meters is weighted so that readings mimic the ear. This scale is known as the A scale and the readings known as dB(A). There are also three other scales known as B, C and D.

Originally the A scale was used for sound pressure levels (SPLs) up to 55 dB, the B scale for levels between 55 dB and 85 dB and the C scale for values above 85 dB. However today, the A scale is used for nearly all levels except for very high SPLs when the C scale is used. The B scale is rarely used and the D scale is mainly used to monitor jet aircraft engine noise.

Table 14.1 gives some typical decibel readings for common activities and Table 14.2 gives some values for woodworking machines published by the UK HSE.

Table 14.1 Some typical sound pressure levels (SPL) (dB(A) values)

Activity or environment	SPL [dB(A)]
Threshold of pain	140
Pneumatic drill	125
Pop group or disco	110
Heavy lorry	93
Street traffic	85
Conversational speech	65
Business office	60
Living room	40
Bedroom	25
Threshold of hearing	0

Table 14.2 Typical noise levels at wood working machines

Machine	Noise level (dB)
Beam panel saws and sanding machines	97
Boring machines	98
Band re-saws, panel planers and vertical spindle moulders	100
Portable woodworking tools	101
Bench saws and multiple ripsaws	102
High-speed routers and moulders	103
Thicknessers	104
Edge banders and multi-cutter moulding machines	105
Double-end tenoners	107

Noise is measured using a sound level meter which reads SPLs in dB(A) and the **peak sound pressure** in pascal (Pa), which is the highest noise level reached by the sound. There are two basic types of sound meter – integrated and direct reading meters. Meters which integrate the reading provide an average over a particular time period, which is an essential technique when there are large variations in sound levels. This value is known as the **continuous equivalent noise level (L_{Eq})**, which is normally measured over an 8-hour period.

Direct reading devices, which tend to be much cheaper, can be used successfully when the noise levels are continuous at a near constant value.

Another important noise measurement is the **daily personal exposure level** of the worker, $L_{EP,d}$, which is measured over an 8-hour working day. Hence, if a person was exposed to 87 dB(A) over 4 hours, this would equate to an $L_{EP,d}$ of 84 dB(A) since a reduction of 3 dB(A) represents one-half of the noise dose.

The UK HSE Guidance on the Control of Noise at Work Regulations, L108, offers some very useful advice on the implementation of the Regulations and should be read by anyone who suspects that they have a noise problem at work. The guidance covers 'equipment and procedures for noise surveys' and contains a noise exposure ready reckoner which can be used to evaluate $L_{EP,d}$ when the noise occurs during several short intervals and/or at several different levels during the 8-hour period.

The ILO Code of Practice recommends that the level of noise and/or duration of exposure should not exceed the limits established by the competent authority or other internationally recognised standards. The assessment should consider the following:

(a) the risk of hearing impairment;

(b) the degree of interference to speech communications essential for safety purposes;

(c) the risk of nervous fatigue, with due consideration to the mental and physical workload and other non-auditory hazards or effects.

To prevent the adverse effects of noise on workers, employers should:

(a) identify the sources of noise and the tasks which give rise to exposure;

(b) seek the advice of the competent authority and/or the occupational health service about exposure limits and other standards to be applied;

(c) seek the advice of the supplier of processes and equipment about expected noise emission;

(d) if this advice is incomplete or otherwise of doubtful value, arrange for measurements by persons competent to undertake these in accordance with current national and/or internationally recognised standards.

Noise action levels

Those countries that have legislation to control noise levels have introduced exposure action level values and exposure limit values.

An **exposure action value** is a level of noise at which certain action must be taken.

An **exposure limit value** is a level of noise at the ear above which an employee must not be exposed. Therefore, if the workplace noise levels are above this value, any ear protection provided to the employee must reduce the noise level to the limit value at the ear.

These exposure action and limit values in the European Union are as follows.

1. The lower exposure action levels are:
   (a) a daily or weekly personal noise exposure of **80 dB(A)**;
   (b) a peak sound pressure of **135 dB(C)**.

2. The upper exposure action levels are:
   (a) a daily or weekly personal noise exposure of **85 dB(A)**;
   (b) a peak sound pressure of **137 dB(C)**.

3. The exposure limit values are:
   (a) a daily or weekly personal noise exposure of **87 dB(A)**;
   (b) a peak sound pressure of **140 dB(C)**.

The peak exposure action and limit values are defined because high level peak noise can lead to short-term and long-term hearing loss. Explosives, guns (including nail guns), cartridge tools, hammers and stone chisels can all produce high peak sound pressures.

If the daily noise exposure exceeds the lower exposure action level, then a noise assessment should be carried out and recorded by a competent person. There is a very simple test which can be done in any workplace to determine the need for an assessment. Table 14.3 gives information on the simple test to determine the need for a noise risk assessment.

Table 14.3 Simple observations to determine the need for a noise risk assessment

Observation at the workplace	Likely noise level [dB(A)]	A noise risk assessment must be made if this noise level persists for
The noise is noticeable but does not interfere with normal conversation – equivalent to a domestic vacuum cleaner	80	6 hours
People have to shout to be heard if they are more than 2 m apart	85	2 hours
People have to shout to be heard if they are more than 1 m apart	90	45 minutes

If there is a marked variation in noise exposure levels during the working week, then a weekly rather than daily personal exposure level, $L_{EP,w}$, may be used. It is only likely to be significantly different to the daily exposure level if exposure on one or two days in the working week is 5 dB(A) higher than on the other days, or the working week has three or fewer days of exposure. The weekly exposure rate is not a simple arithmetic average of the daily rates. If an organisation is considering the use of a weekly exposure level, then the following provisions should be made:

▶ hearing protection must be provided if there are very high noise levels on any one day;

▶ the employees and their representatives must be consulted on whether weekly averaging is appropriate;

▶ an explanation must be given to the employees on the purpose and possible effects of weekly averaging.

Finally, if the working day is 12 hours, then the action levels must be reduced by 3 dB(A) because the action levels assume an 8-hour working day.

14

The ILO Code recommends that noise measurements should be used to:

(a) quantify the level and duration of exposure of workers and compare it with exposure limits as established by the competent authority or internationally recognised standards;

(b) identify and characterise the sources of noise and the exposed workers;

(c) create a noise map for the determination of risk areas;

(d) assess the need both for engineering noise prevention and control and for other appropriate measures and for their effective implementation;

(e) evaluate the effectiveness of existing noise prevention and control measures.

The UK HSE Guidance document L108 gives further detailed advice on noise assessments and surveys. The most important points are that the measurements should be taken at the working stations of the employees closest to the source of the noise and over as long a period as possible, particularly if there is a variation in noise levels during the working day. Other points to be included in a noise assessment are:

► details of the noise meter used and the date of its last calibration;

► the number of employees using the machine, time period of usage and other work activities;

► an indication of the condition of the machine and its maintenance schedule;

► the work being done on the machine at the time of the assessment;

► a schematic plan of the workplace showing the position of the machine being assessed;

► other noise sources, such as ventilation systems, that should be considered in the assessment. The control of these sources may help to reduce overall noise levels;

► recommendations for future actions, if any.

Other actions which the employer should undertake when the lower exposure action level is exceeded are to:

► inform, instruct and train employees on the hearing risks;

► supply hearing protection to those employees requesting it;

► ensure that any equipment or arrangements provided under the Regulations are correctly used or implemented.

The additional measures which the employer should take if the upper exposure action level is reached are:

► reduce and control exposure to noise by means other than hearing protection;

► establish hearing protection zones, marked by notices and ensure that anybody entering the zone is wearing hearing protection;

► supply hearing protection and ensure that it is worn.

Figure 14.3 Typical ear protection zone sign

Noise legislation usually places a duty on the employer to undertake health surveillance for workers whose exposure regularly exceeds the upper action level irrespective of whether ear protection was worn.

Health surveillance may include:

► a pre-employment or pre-assignment medical examination;

► periodical medical examinations at intervals prescribed as a function of the magnitude of the exposure hazards;

► medical examinations prior to resumption of work after a period of extended sickness.

The ILO Code also places the following obligations on workers:

► for noise levels above the lower exposure action level, they must use any control equipment (other than hearing protection), such as silencers, supplied by the employer and report any defects;

► for noise levels above the upper exposure action level, they fulfil the obligations given above and wear the hearing protection provided;

► they must see their doctor if they feel that their hearing has become damaged;

► present themselves for health surveillance.

14.1.4 Noise control techniques

In addition to **reduced time exposure** of employees to the noise source, there is a simple hierarchy of control techniques:

► reduction of noise at source;

► reduction of noise levels received by the employee (known as attenuation);

► personal protective equipment, which should only be used when the above two remedies are insufficient.

The ILO Code recommends that in the case of existing processes and equipment, employers should first consider whether the noisy process is necessary at all, or whether it could be carried out in another way without generating noise. If the elimination of noisy processes and equipment as a whole is impracticable, then their individual sources should be separated out and their relative contribution to the overall sound pressure level identified. Once the causes or sources of noise are identified, the first step in the noise control process should attempt to control noise at source.

Reduction of noise at source

There are several means by which noise could be reduced at source:

▶ change the process or equipment (e.g. replace solid tyres with rubber tyres or replace diesel engines with electric motors);
▶ change the speed of the machine;
▶ improve the maintenance regime by regular lubrication of bearings, tightening of belt drives.

If reducing the noise at source or intercepting it does not sufficiently reduce worker exposure, then the final options suggested by the ILO Code of Practice for reducing exposure are:

▶ to confine the work activity to a relatively small area where an acoustical booth or shelter may be installed;
▶ to minimise the time workers spend in the noisy environment.

Attenuation of noise levels

There are many methods of attenuating or reducing noise levels and these are covered in detail in the guide to the ILO Code of Practice. The more common ones will be summarised here.

▶ **Orientation or relocation of the equipment** – turn the noisy equipment away from the workforce or locate it away in separate and isolated areas.
▶ **Enclosure** – surrounding the equipment with a good sound-insulating material can reduce sound levels by up to 30 dB(A). Care will need to be taken to ensure that the machine does not become overheated.
▶ **Screens or absorption walls** – can be used effectively in areas where the sound is reflected from walls. The walls of the rooms housing the noisy equipment are lined with sound absorbent material, such as foam or mineral wool, or sound absorbent (acoustic) screens are placed around the equipment.
▶ **Damping** – the use of insulating floor mountings to remove or reduce the transmission of noise and vibrations through the structure of the building such as girders, wall panels and flooring.
▶ **Lagging** – the insulation of pipes and other fluid containers to reduce sound transmission (and, incidentally, heat loss).

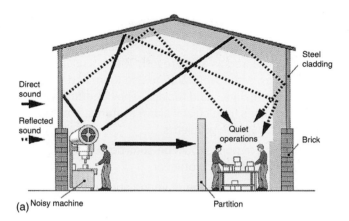

(a) Noisy machine

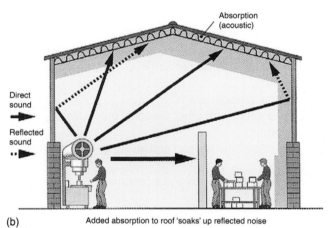

(b) Added absorption to roof 'soaks' up reflected noise

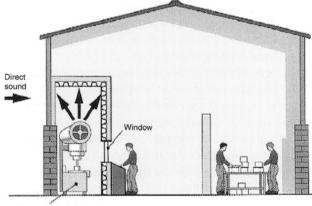

(c) Noisy machine enclosed with internal absorption

Figure 14.4 Noise paths found in a workplace: (a) the quiet area is subjected to reflected noise from a machine somewhere else in the building; (b) the correct use of absorption in the roof will reduce the reflected noise reaching the quiet area; (c) segregation of the noisy operation will benefit the whole workplace

▶ **Silencers** – normally fitted to engines which are exhausting gases to atmosphere. Silencers consist of absorbent material or baffles.
▶ **Isolation of the workers** – the provision of sound-proofed workrooms or enclosures isolated away from noisy equipment (a power station control room is an example of worker isolation).

427

14.1.5 Personal ear protection

The provision of personal hearing protection should only be considered as a last resort. There is usually resistance from the workforce to use them and they are costly to maintain and replace. They interfere with communications, particularly alarm systems, and they can present hygiene problems.

The following factors should be considered when selecting personal ear protection:

▶ suitability for the range of sound spectrum of frequencies to be encountered;
▶ noise reduction (attenuation) offered by the ear protection;
▶ pattern of the noise exposure;
▶ acceptability and comfort of the wearer, particularly if there are medical problems;
▶ durability;
▶ hygiene considerations;
▶ compatibility with other personal protective equipment;
▶ ease of communication and ability to hear warning alarms;
▶ maintenance and storage arrangements;
▶ cost.

There are two main types of ear protection: earplugs and ear defenders (earmuffs).

Earplugs are made of sound absorbent material and fit into the ear. They can be reusable or disposable, and are able to fit most people and can easily be used with safety glasses and other personal protective equipment. Their effectiveness depends on the quality of the fit in the ear which, in turn, depends on the level of training given to the wearer. Permanent earplugs come in a range of sizes so that a good fit is obtained. The effectiveness of earplugs decreases with age and they should be replaced at the intervals specified by the supplier. A useful simple rule to ensure that the selected ear plug reduces the noise level at the ear to 87 dB(A) is to choose one with a manufacturer's rating of 83 dB(A). This should compensate for any fitting problems. The main disadvantage of ear plugs is that they do not reduce the sound transmitted through the bone structure which surrounds the ear and they often work loose with time.

Ear defenders (earmuffs) offer a far better reduction of all sound frequencies. They are generally more acceptable to workers because they are more comfortable to wear and they are easy to monitor as they are clearly visible. They also reduce the sound intensity transmitted through the bone structure surrounding the ear. A communication system can be built into earmuffs. However, they may be less effective if the user has long hair or is wearing spectacles or large ear rings. They may also be less effective if worn with helmets or face shields and uncomfortable in warm conditions. Maintenance is an important factor with

earmuffs and should include checks for wear and tear and general cleanliness.

Selection of suitable ear protection is very important as it should not just reduce sound intensities below the statutory action levels but also reduce those intensities at particular frequencies. Normally advice should be sought from a competent supplier who will be able to advise on ear protection to suit a given spectrum of noise using 'octave band analysis'.

Finally, it is important to stress that the use of ear protection must be well supervised to ensure that not only is it being worn correctly, but that it is, in fact, actually being worn. This is a particular problem where lone workers are required to wear personal protective equipment, such as ear defenders. The requirement can be emphasised by the use of signs and warning notices on equipment and in company vehicles. Unannounced monitoring visits to the work site also encourage the use of personal protective equipment. If there is a persistent refusal to wear personal protective equipment, then disciplinary procedures must be initiated.

The issue of ear defenders to the workforce is a last resort after measures outlined in 14.1.4 have been considered.

The ILO Code recommends that employers should ensure that workers who may be exposed to significant levels of noise are trained:

(a) in the effective use of hearing protection devices;
(b) to identify and report any new or unusual sources of noise;
(c) to understand the role of audiometric examination.

Employers should also ensure that workers in noisy environments are informed of:

(a) the factors leading to noise-induced hearing loss and the consequences for the victim, especially young workers;
(b) the necessary precautions, especially the use of hearing protection devices;
(c) the effects that a noisy environment may have on their general safety;
(d) the symptoms of adverse effects of exposure to high levels of noise.

14.2 Vibration

14.2.1 Ill-health due to vibration

Hand-held vibrating machinery (such as pneumatic drills, sanders and grinders, powered lawn mowers and strimmers and chainsaws) can produce health risks from hand–arm or whole-body vibration (WBV).

Advice for eliminating or controlling exposure to vibration is given in the provisions of the ILO Code of Practice 'Ambient factors in the workplace'. The code recommends that exposure limits should be established according to current international knowledge. International consensus standards describe useful methods for quantifying vibration severity for whole-body vibration in ISO 2631-1:1997 and for hand-transmitted vibration in ISO 5349:1986. In addition to these standards and this code, the information on the assessment of vibration exposure and protective and preventative measures are given in the Working Environment (Air Pollution, Noise and Vibration) Convention (No. 148), and Recommendation (No. 156), 1977, and the ILO Code of Practice 'Protection of workers against noise and vibration in the working environment' (Geneva, 1984).

If workers or others are frequently exposed to hand-transmitted or whole-body vibration, and obvious steps do not eliminate the exposure, employers should assess the hazard and risk to safety and health from the conditions, and the prevention and control measures to remove the hazards or risks or to reduce them to the lowest practicable level by all appropriate means.

Hand–arm vibration syndrome

Hand–arm vibration syndrome (HAVS) describes a group of diseases caused by the exposure of the hand and arm to external vibration. Some of these have been described under WRULDs, such as carpal tunnel syndrome.

However, the best known disease is **vibration white finger (VWF)** in which the circulation of the blood, particularly in the hands, is adversely affected by the vibration. The early symptoms are tingling and numbness felt in the fingers, usually sometime after the end of the working shift. As exposure continues, the tips of the fingers go white and then the whole hand may become affected. This results in a loss of grip strength and manual dexterity. Attacks can be triggered by damp and/or cold conditions and, on warming, 'pins and needles' are experienced. If the condition is allowed to persist, more serious symptoms become apparent including discolouration and enlargement of the fingers. In very advanced cases, gangrene can develop leading to the amputation of the affected hand or finger. VWF was first detailed as an industrial disease in 1911.

The risk of developing HAVS depends on the frequency of vibration, the length of exposure and the tightness of the grip on the machine or tool. Some typical values of vibration measurements for common items of equipment used in industry are given in Table 14.4.

When assessing the risk of HAVS developing among employees, the source of the vibration, such as reciprocating, rotating and vibrating tools and equipment, needs to be considered first together with the age of the equipment, its maintenance record, its suitability for the job and any information or guidance available from the manufacturer. The number of employees using the tooling or equipment, the duration and frequency of their use and any relevant personal

Table 14.4 Examples of vibration exposure values measured by HSE on work equipment

Equipment	Condition	Vibration reading (m/s²)
Road breakers	Typical	12
	Modern design and trained operator	5
	Worst tools and operating conditions	20
Demolition hammers	Modern tools	8
	Typical	15
	Worst tools	25
Hammer drills	Typical	9
	Best tools and operating conditions	6
	Worst tools and operating conditions	25
Large-angle grinders	Modern vibration-reduced designs	4
	Other types	8
Small-angle grinders	Typical	2–6
Chainsaws	Typical	6
Brush cutter or strimmer	Typical	4
	Best	2
Sanders (orbital)	Typical	7–10

factors, such as a pre-existing circulatory problem, all form part of the assessment. Environmental factors, particularly exposure to cold and/or wet weather and the nature of the job itself, are also important factors to be considered during such a risk assessment. Finally, an examination of the existing controls and their effectiveness and the frequency, magnitude and direction of the vibration are important elements of the evaluation. Other issues could include the effectiveness of any personal protective equipment provided and any instruction or training given.

Whole-body vibration

Whole-body vibration (WBV) is caused by vibration from machinery passing into the body either through the feet of standing workers or the buttocks of sitting workers. It is the shaking or jolting of the human body through a supporting surface (usually a seat or the floor), for example when driving or riding on a vehicle along an unmade road, operating earth-moving machines or standing on a structure attached to a large, powerful, fixed machine, which is impacting or vibrating.

The most common ill-health effect is back pain which, in severe cases, may result in permanent injury. Reasons for back pain in drivers can include:

▶ poor design of controls, making it difficult for the driver to operate the machine or vehicle easily or to see properly without twisting or stretching;

▶ incorrect adjustment by the driver of the seat position and hand and foot controls, so that it is

14

Regular exposure to HAV can cause a range of permanent injuries to hands and arms, collectively known as hand–arm vibration syndrome (HAVS). The injuries can include damage to the:

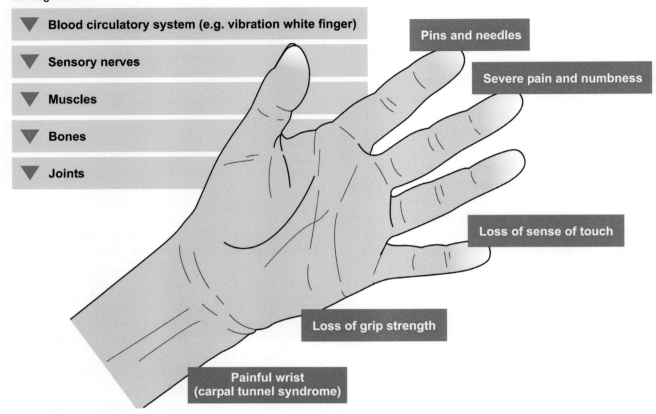

▼ Blood circulatory system (e.g. vibration white finger)

▼ Sensory nerves

▼ Muscles

▼ Bones

▼ Joints

Pins and needles

Severe pain and numbness

Loss of sense of touch

Loss of grip strength

Painful wrist (carpal tunnel syndrome)

Figure 14.5 Injuries which can be caused by hand–arm vibration

necessary to continually twist, bend, lean and stretch to operate the machine;

▶ sitting for long periods without being able to change position;

▶ poor driver posture;

▶ repeated manual handling and lifting of loads by the driver;

▶ excessive exposure to whole-body vibration, particularly to shocks and jolts;

▶ repeatedly climbing into or jumping down from a high cab or one which is difficult to get in and out of.

The risk increases where the driver or operator is exposed to two or more of these factors together.

Other acute effects include reduced visual and manual control, and increased heart rate and blood pressure. Chronic or long-term effects include permanent spinal damage, damage to the central nervous system, hearing loss and circulatory and digestive problems.

The most common occupations which generate WBV are driving fork-lift trucks, construction vehicles and agricultural or horticultural machinery and vehicles. There is growing concern throughout the European Union about this problem.

Control measures include the proper use of the equipment including correct adjustments of air or hydraulic pressures, seating and, in the case of

vehicles, correct suspension, tyre pressures and appropriate speeds to suit the terrain. Other control measures include the selection of suitable equipment with low vibration characteristics, work rotation, good maintenance and fault reporting procedures.

The UK Health and Safety Executive (HSE) commissioned measurements of WBV on several machines and some of the results are shown in Table 14.5.

Table 14.5 Machines which could produce significant whole-body vibration

Machine	Activity	Vibration reading (m/s²)
3-tonne articulated dumper truck	Removal of spoil	0.78
Dumper truck	Transport of spoil	1.13
25-tonne articulated dumper truck	Transport of spoil	0.91
Bulldozer	Dozing	1.16
4-tonne twin-drum	Finishing tarmac	0.86
80-tonne rigid dumper truck	Transport of spoil	1.03

Preventative and precautionary measures

The control strategy outlined in Chapter 4 can certainly be applied to ergonomic risks. The common measures used to control ergonomic ill-health effects are to:

▶ implement results of task analysis and identification of repetitive actions;

▶ eliminate vibration-related or hazardous tasks by performing the job in a different way;

▶ ensure that the correct equipment (properly adjusted) is always used;

▶ introduce job rotation so that workers have a reduced time exposure to the hazard;

▶ during the design of the job ensure that poor posture is avoided;

▶ undertake a risk assessment;

▶ keep reports from employees and safety representatives;

▶ keep ill-health reports and absence records;

▶ introduce a programme of health surveillance;

▶ ensure that employees are given adequate information on the hazards and develop a suitable training programme;

▶ ensure that a programme of preventative maintenance is introduced and include the regular inspection of items such as vibration isolation mountings;

▶ keep up to date with advice from equipment manufacturers, trade associations and health and safety sources (more and more low vibration equipment is becoming available).

A very useful and extensive checklist for the identification and reduction of WRULDs is given in Appendix 2 of the UK HSE *Guide to work-related upper limb disorders* HSG60.

14.2.2 European vibration exposure limits

A daily exposure limit and action values for both HAV and WBV has been introduced into the UK and the European Union. These values are as follows:

1. For HAV:

   (a) the daily exposure limit value normalised to an 8-hour reference period is 5 m/s^2

Figure 14.6 Breaker mounted on an excavator's jib to reduce vibrations

(b) the daily exposure action value normalised to an 8-hour reference period is 2.5 m/s^2.

2. For WBV:

   (a) the daily exposure limit value normalised to an 8-hour reference period is 1.15 m/s^2

   (b) the daily exposure action value normalised to an 8-hour reference period is 0.5 m/s^2.

An exposure limit value must not be exceeded. If an exposure action value is exceeded, then action must be taken to reduce the value. The term A(8) is added to the exposure limit or action value to denote that it is an average value spread over an 8-hour working day. Thus the daily exposure limit value for HAV is 5 m/s^2 A(8).

Hand–arm vibration

Many machines and processes used in industry produce HAV. Typical high-risk processes include:

▶ grinding, sanding and polishing wood and stone;

▶ cutting stone, metal and wood;

▶ riveting, caulking and hammering;

▶ compacting sand, concrete and aggregate;

▶ drilling and breaking rock, concrete and road surfaces; and

▶ surface preparation, including de-scaling and paint removal.

There are several ways to ascertain the size of the vibration generated by equipment and machines. Manufacturers must declare vibration emission values for portable hand-held and hand-guided machines and provide information on risks. Other important sources for vibration information include scientific and technical journals, trade associations and online databases. The experience of the UK HSE has shown that the vibration level is higher in practice than that quoted by many manufacturers. The reasons for this discrepancy may be that:

▶ the equipment is not well maintained;

▶ the equipment is not suitable for the material being worked;

▶ the tool has not been purchased from a reputable supplier;

▶ the accessories are not appropriate or are badly fitted;

▶ the operative is not using the tool properly.

In view of these problems, it is recommended that the declared value should be doubled when comparisons are made with exposure limits. As the exposure limit or action value is averaged over 8 hours, it is possible to work with higher values for a reduced exposure time. Table 14.6 shows the reduction in exposure time as the size of the vibration increases.

The following points summarise the important measures which should be taken to reduce the risks associated with HAV:

1. avoid, whenever possible, the need for vibration equipment;

2. undertake a risk assessment which includes a soundly based estimate of the employees' exposure to vibration;

3. develop a good maintenance regime for tools and machinery. This may involve ensuring that tools are regularly sharpened, worn components are replaced or engines are regularly tuned and adjusted;

4. introduce a work pattern that reduces the time exposure to vibration;

5. issue employees with gloves and warm clothing. There is a debate as to whether anti-vibration gloves are really effective but it is agreed that warm clothing helps with blood circulation which reduces the risk of VWF. Care must be taken so that the tool does not cool the hand of the operator;

6. introduce a reporting system for employees to use so that concerns and any symptoms can be recorded and investigated; and

7. encourage employees to check regularly for any symptoms of vibration related ill-health and to participate in any health surveillance measures provided.

It is important that drill bits and tools are kept sharp and used intact – an angle grinder with a chipped cutting disc will lead to a large increase in vibration as well as being dangerous.

Table 14.6 The change in exposure times as vibration increases

The change in exposure times as vibration increases						
Value of vibration (m/s²)	2.5	3.5	5	7	10	14	20
Exposure time to reach action value (hours)	8	4	2	1	30 min	15 min	8 min
Exposure time to reach limit value (hours)	over 24	16	8	4	2	1	30 min

Whole-body vibration

The ILO Code of Practice 'Ambient factors in the workplace' requires the control of the risks from whole-body vibration. This should be based on an assessment of the risk and exposure. In most cases it is simpler to make a broad assessment of the risk rather than try to assess exposure in detail, concentrating the main effort on introducing controls to remove the hazards or risks or to reduce them to the lowest practicable level by all appropriate means.

The requirements of the code are to assess the vibration risk to employees by taking the following actions:

▶ consider the sources of vibration and the tasks which give rise to exposure;

▶ seek the advice of the competent authority about exposure limits and other standards to be applied;

▶ seek the advice of the supplier of vehicles and equipment about their vibration emission;

▶ use vibration measurements to:

(a) quantify the level and duration of exposure of workers and compare it with exposure limits as established by the competent authority or other standards to be applied;

(b) identify and characterise the sources of vibration and the exposed workers;

(c) assess the need both for engineering vibration control and for other appropriate measures and for their effective implementation;

(d) evaluate the effectiveness of particular vibration prevention and control measures.

▶ decide if they are likely to be exposed above the national daily exposure action value (EAV) and if they are:

▷ introduce a programme of controls to eliminate or reduce their daily exposure so far as is reasonably practicable;

▷ decide if they are likely to be exposed above the recommended exposure limit value (ELV) and if they are they should take immediate action to reduce their exposure below the limit value;

▶ provide information and training on health risks and controls to employees at risk;

▶ consult the trade union safety representative or employee representative about the risks and the plans to address these risks;

▶ keep a record of the risk assessment and control actions;

▶ review and update the risk assessment regularly.

If a broad risk assessment has been made and the appropriate and reasonable control actions taken, there is no need to measure the exposure to vibration of the employee. Most machine and vehicle activities in normal use produce daily exposures below the limit value. But some off-road machinery operated for long periods in conditions that generate high levels of vibration or jolting may exceed the exposure limit value.

The measurement of WBV is very difficult and can only be done accurately by a specialist competent person. If the risk assessment has been made and the recommended control actions are in place, there is no need to measure the exposure of employees to vibration. However, the UK HSE has suggested that employers can use the following checklist to estimate whether exposure to WBV is high:

1. there is a warning in the machine manufacturer's handbook that there is a risk of WBV;

2. the task is not suitable for the machine or vehicle being used;

3. operators or drivers are using excessive speeds or operating the machine too aggressively;

4. operators or drivers are working too many hours on machines or vehicles that are prone to WBV;

Figure 14.7 (a) Vibrating roller with risk of whole-body vibration; (b) remote control vibrating plate weighing 1.2 tons with compaction in excess of a 7-ton roller which eliminates the risk of whole-body vibration. The operator is protected from vibrations, noise and dust. The machine can only be operated if line of sight is intact. In case of a loss of control the proximity recognition sensor keeps the operator safe.

5. road surfaces are too rough and potholed or floors uneven;
6. drivers are being continually jolted or when going over bumps rising from their seats;
7. vehicles designed to operate on normal roads are used on rough or poorly repaired roads;
8. operators or drivers have reported back problems.

If one or more of the above applies, then exposure to WBV may be high. The actions for controlling the risks from WBV include the following. Ensure that:

▶ the driver's seat is correctly adjusted so that all controls can be reached easily and that the driver weight setting on their suspension seat, if available, is correctly adjusted. The seat should have a back rest with lumbar support;
▶ anti-fatigue mats are used if the operator has to stand for long periods;
▶ the speed of the vehicle is such that excessive jolting is avoided. Speeding is one of the main causes of excessive WBV;
▶ all vehicle controls and attached equipment are operated smoothly;
▶ only established site roadways are used;
▶ only suitable vehicles and equipment are selected to undertake the work and cope with the ground conditions;
▶ the site roadway system is regularly maintained;
▶ all vehicles are regularly maintained with particular attention being paid to tyre condition and pressures, vehicle suspension systems and the driver's seat;
▶ work schedules are regularly reviewed so that long periods of exposure on a given day are avoided and drivers have regular breaks away from the vehicle;
▶ prolonged exposure to WBV is avoided for at risk groups (older people, young people, people with a history of back problems and pregnant women);
▶ employees are aware of the health risks from WBV, the results of the risk assessment and the ill-health reporting system. They should also be trained to drive in such a way that excessive vibration is reduced.

A simple health monitoring system that includes a questionnaire checklist should be agreed with employees or their representatives (available on the UK HSE website) to be completed once a year by employees at risk.

14.2.3 Additional recommendations from the ILO Code of Practice

Health surveillance

The ILO Code of Practice 'Ambient factors in the workplace' recommends that workers should be given a pre-employment medical examination for jobs involving hand–arm vibration to check whether Raynaud's phenomenon of non-occupational origin or hand–arm vibration syndrome (HAVS) is present from earlier employment. Where these symptoms are diagnosed, such employment should not be offered unless vibration has been satisfactorily controlled.

If a worker is exposed to hand-transmitted vibration, the occupational health professional responsible for health surveillance should:

14

(a) examine the worker periodically, as prescribed by national laws and regulations, for HAVS and ask the worker about any symptoms;

(b) examine the worker for symptoms of possible neurological effects of vibration, such as numbness and elevated sensory thresholds for temperature, pain and other factors.

If it appears that these symptoms exist and may be related to vibration exposure, the employers should be advised that control may be insufficient. Because of possible association of back disorders with whole-body vibration, workers exposed to WBV should be advised during health surveillance about the importance of posture in seated jobs, and about correct lifting technique.

Training and information

The ILO Code also recommends that employers should ensure that workers who are exposed to significant vibration are:

(a) informed about the hazards of prolonged use of vibrating tools;

(b) informed of the measures that the worker can take to minimise risk, particularly the proper adjustment of seating and working positions;

(c) instructed in the correct handling and use of hand tools;

(d) encouraged to report finger blanching, numbness or tingling, without facing unwarranted discrimination, for which there should be recourse in national law.

14.3 Radiation

14.3.1 Ionising radiation

Ionising radiation is emitted from radioactive materials, either in the form of directly ionising alpha and beta particles or indirectly ionising X-rays and gamma rays or neutrons. It has a high energy potential and an ability to penetrate, ionise and damage body tissue and organs.

All matter consists of atoms, within which is a nucleus containing protons and neutrons, and orbiting electrons. The number of protons within the atom defines the element: hydrogen has 1 proton and lead has 82 protons. Some atoms are unstable and will change into atoms of another element, emitting ionising radiation in the process. The change is called radioactive decay and the ionising radiations most commonly emitted are alpha and beta particles and gamma rays. X-rays are produced by bombarding a metal target with electrons at very high speeds using very high voltage electrical discharge equipment. Neutrons are released by nuclear fission and are not normally found in manufacturing processes.

Alpha particles consist of two protons and two neutrons and have a positive charge. They have little power to penetrate the skin and can be stopped using very flimsy material, such as paper. Their main route into the body is by ingestion.

Beta particles are high-speed electrons whose power of penetration depends on their speed, but penetration is usually restricted to 2 cm of skin and tissue. They can be stopped using aluminium foil. There are normally two routes of entry into the body: inhalation and ingestion.

Gamma rays, which are similar to *X-rays*, are electromagnetic radiation and have far greater penetrating power than alpha or beta particles. They are produced from nuclear reactions and can pass through the body.

The two most common measures of radiation are the Becquerel (Bq), which measures the activity of a radioactive substance per second, and the Sievert (Sv), which measures the biological effects of the radiation – normally measured in milliSieverts (mSv).

Thus, the Becquerel (Bq) measures the amount of radiation in a given environment and the milliSievert (mSv) measures the ionising radiation dose received by a person.

Ionising radiation occurs naturally from man-made processes and about 87% of all radiation exposure is from natural sources. In the UK, the Ionising Radiations Regulations specify a range of dose limits, some of which are given in Table 14.7.

Table 14.7 Typical radiation dose limits

	Dose (mSv)	Area of body
Employees aged 18 years	20	Whole body per year
Trainees 16–18 years	6	Whole body per year
Any other person	1	Whole body per year
Women employees of child-bearing age	13	The abdomen in any consecutive 3-month period
Pregnant employees	1	During the declared term of the pregnancy

Harmful effects of ionising radiation

Ionising radiation attacks the cells of the body by producing chemical changes in the cell DNA by ionising it (thus producing free radicals), which leads to abnormal cell growth. The effects of these ionising attacks depend on the following factors:

▶ the size of the dose – the higher the dose then the more serious will be the effect;

▶ the area or extent of the exposure of the body – the effects may be far less severe if only a part of the body (e.g. an arm) receives the dose;

▶ the duration of the exposure – a long exposure to a low dose is likely to be more harmful than a short exposure to the same quantity of radiation.

Figure 14.8 Typical ionising radiation warning sign

Acute exposure can cause, dependent on the size of the dose, blood cell changes, nausea and vomiting, skin burns and blistering, cataracts, collapse and death. **Chronic exposure** can lead to anaemia, leukaemia and other forms of cancer. It is also known that ionising radiation can have an adverse effect on the function of human reproductive organs and processes. Increases in the cases of sterility, stillbirths and malformed foetuses have also been observed.

The health effects of ionising radiation may be summarised into two groups – **somatic effects**, which refer to cell damage in the person exposed to the radiation dose and **genetic effects**, which refer to the damage done to any future children of the irradiated person.

Sources of ionising radiation

The principal workplaces which could have ionising radiation present are the nuclear industry, medical centres (hospitals and research centres), educational centres and the construction industry. Radioactive processes are used for the treatment of cancers, and radioactive isotopes are used for many different types of scientific research. X-rays are used extensively in hospitals, but they are also used in industry for non-destructive testing (e.g. crack detection in welds and testing the thickness and density of materials). Smoke detectors, used in most workplaces, also use ionising radiations.

Ionising radiations can also occur naturally – the best example being radon, which is a radioactive gas that occurs mainly at or near granite outcrops where there is a presence of uranium. It is particularly prevalent in Devon and Cornwall in the UK. The gas enters buildings normally from the substructure through cracks in flooring or around service inlets.

The International Commission on Radiological Protection (ICRP) makes recommendations on radiation dose limits and

Figure 14.9 X-ray generator cabinet

▶ is an advisory body providing recommendations and guidance on radiation protection;
▶ was founded in 1928 by the International Society of Radiology (ISR, the professional society of radiologist physicians);
▶ is an Independent Registered Charity (a 'not-for-profit organisation') in the United Kingdom; and currently has its small Scientific Secretariat in Canada.

It has recently reaffirmed that radon exposure in dwellings is an existing exposure situation. Furthermore, the Commission's protection policy for these situations continues to be based on setting a level of annual dose of around 10 mSv from radon where action would almost certainly be warranted to reduce exposure. Taking account of new research, the Commission has revised the upper value for the reference level for radon gas in dwellings from the value in the 2007 Recommendations of 600 Bq/m^3 to 300 Bq/m^3.

In the UK, the Ionising Radiation Regulations have set two action levels above which remedial action, such as fitting sumps and extraction fans, has to be taken to lower the radon level in the building. The first action level is 400 Bq/m^3 in workplaces and 200 Bq/m^3 in domestic properties. At levels above 1000 Bq/m^3, remedial action should be taken within one year. The average background radiation in the UK is 2.4 mSv per year. Background radiation levels are much higher elsewhere in the world – 260 mSv have been recorded in Iran. About half of the background radiation in the UK is caused by radon.

Radon gas is responsible for 5% of lung cancer cases. Radon is not generally a problem unless it is able to accumulate in confined spaces such as basements, cellars or lift shafts. An impermeable membrane beneath the floor of the building can be laid to prevent the gas entering the building through the floor. This latter treatment is normally only suitable for new

14

Figure 14.10 Radon monitoring equipment

buildings but does cost less than 10% of the cost of a sump and has no running costs.

About 14% of exposures to ionising radiation are due to medical exposures during diagnostic or treatment processes. The hazards are categorised as either stochastic or non-stochastic. The principal stochastic hazard is cancer in various forms. The main non-stochastic effects include radiation burns, radiation sickness, cataracts and damage to unborn children.

Personal radiation exposure can be measured using a film badge, which is worn by the employee over a fixed time interval. The badge contains a photographic film which, after the time interval, is developed and an estimate of radiation exposure is made. A similar device, known as a radiation dose meter or detector, can be positioned on a shelf in the workplace for three months, so that a mean value of radiation levels may be measured. Instantaneous radiation values can be obtained from portable hand-held instruments, known as Geiger counters, which continuously sample the air for radiation levels. Similar devices are available to measure radon levels.

14.3.2 Non-ionising radiation

Non-ionising radiation includes ultraviolet, visible light (this includes lasers which focus or concentrate visible light), infrared and microwave radiations. As the wavelength is relatively long, the energy present is too low to ionise atoms which make up matter. The action of non-ionising radiation is to heat cells rather than change their chemical composition.

Ultraviolet (UV) radiation occurs with sunlight and with electric arc welding. In both cases, the skin and the eyes are at risk from the effect of burning. The skin will burn (as in sunburn) and repeated exposure can lead to skin cancer. Skin which is exposed to strong sunlight should be protected either by clothing or sun creams. This problem has become more common with the reduction in the ozone layer (which filters out much ultraviolet light). The eyes can be affected by a form of conjunctivitis which feels like grit in the eye, and is called by a variety of names dependent on the activity causing the problem. Arc welders call it 'arc eye' or 'welder's eye' and skiers 'snow blindness'. Cataracts caused by the action of ultraviolet radiation on the eye lens are another possible outcome of exposure.

The most dangerous form of skin cancer, malignant melanoma, has increased by over 40% over the last 10 years making it the cancer with the fastest rising number of cases in the UK and one of the most common forms of cancer in the UK with over 50,000 new cases every year. Outdoor workers that could be at risk include farm or construction workers, market gardeners, outdoor activity workers and some public service workers. Those who are particularly at risk have:

▶ fair or freckled skin that does not tan, or goes red or burns before it tans;
▶ red or fair hair and light coloured eyes;
▶ a large number of moles.

Employers need to be aware of the risks their employees are taking when they work outside without adequate protection from the sun. With growing concern following the rise in skin cancers, the UK HSE has suggested the following hierarchy of controls for outdoor working:

▶ relocate some jobs inside a building or to a shady location;
▶ undertake some outdoor work earlier or later in the day;
▶ provide personal protection such as:
  ▷ wearing long sleeve shirts or loose clothing with a close weave;
  ▷ wearing hats with a wide brim;
  ▷ using a high factor sunscreen of at least SPF15 on any exposed skin;
▶ provide suitable education and training for outdoor workers;
▶ provide suitable information and supervision to instigate safe systems of work that protect workers from the sun.

UV radiation has some beneficial effects such as the accumulation of vitamin D which strengthens bones.

Where UV insectocutors are used, normally in kitchens, to control flying insects, it is important that the correct emitting labels are fitted and not ones that, for example, are used to stencil surgical instruments.

Lasers use visible light and light from the invisible wavelength spectrum (infrared and ultraviolet). As the word laser implies, they produce 'light **a**mplification by **s**timulated **e**mission of **r**adiation'. This light is highly concentrated and does not diverge or weaken with distance and the output is directly related to the chemical composition of the medium used within the particular laser. The output beam may be pulsating or continuous – the choice being dependent on the task of the laser. Lasers have a large range of applications including bar code reading at a supermarket checkout, the cutting and welding of metals and accurate measurement of distances and elevations required in land and mine surveying. They are also extensively used in surgery for cataract treatment and the sealing of blood vessels.

The International Electrotechnical Commission (IEC) and the American National Standards Institute (ANSI) have defined seven classes of laser (1, 1M, 2, 2M, 3R, 3B and 4) in ascending size of power output. Classes 1, 2 and 2M are relatively low hazard and only emit light in the visible band. Direct eye contact must be avoided with classes 3R, 3B and 4. These are more hazardous than classes 1 and 2 and the appointment of a laser safety officer is recommended. A Class 4 laser can burn the skin and cause permanent eye damage as a result of direct, diffuse or indirect exposure to the beam. Such Class 4 lasers can ignite combustible materials, and are a fire risk. All lasers should carry information stating their class and any precautions required during use.

The main hazards associated with lasers are eye and skin burns, toxic fumes, electricity and fire. The vast majority of accidents with lasers affect the eyes. Retinal damage is the most common and is irreversible. Cataract development and various forms of conjunctivitis can also result from laser accidents. Skin

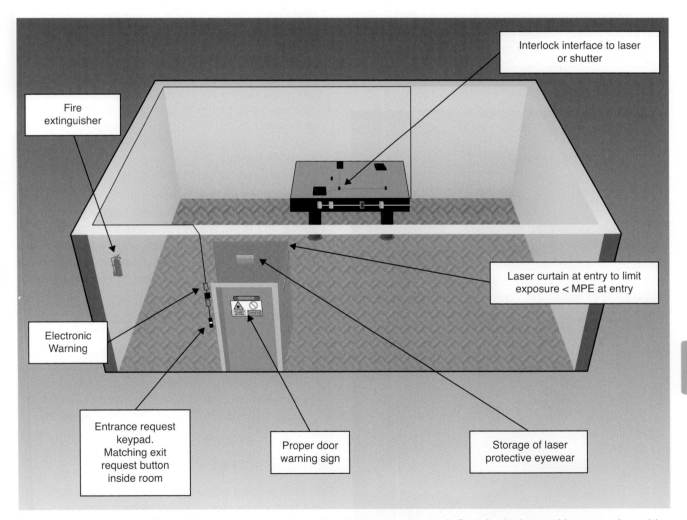

Figure 14.11 A proper Class 4 laser area setup for defeatable access control. Curtain design and layout varies with environment. Class 4 laser areas in a research or university environment usually run long-term experiments that require unattended operation. In such cases, defeatable entryway controls are appropriate. By design, persons who have been properly trained and given the keypad access code may momentarily 'defeat' the interlock to enter and exit the laser controlled area protected by a magnetically locked door

burning and reddening (erythema) are less common and are reversible.

Infrared radiation is generated by fires and hot substances and can cause eye and skin damage similar to that produced by ultraviolet radiation. It is a particular problem to fire fighters and those who work in foundries or near furnaces. Eye and skin protection are essential.

Microwaves are used extensively in cookers and mobile telephones and there are ongoing concerns about associated health hazards (and several inquiries are currently under way). The severity of any hazard is proportional to the power of the microwaves. The principal hazard is the heating of body cells, particularly those with little or no blood supply to dissipate the heat. This means that tissues such as the eye lens are most at risk from injury. However, it must be stressed that any risks are higher for items such as cookers than for low-powered devices such as mobile phones.

The measurement of non-ionising radiation normally involves the determination of the power output being received by the worker. Such surveys are best performed by specialists in the field, as the interpretation of the survey results requires considerable technical knowledge.

Figure 14.12 Metal furnace – source of infrared heat

14.3.3 Radiation protection strategies

Ionising radiation

Protection is obtained by the application of shielding, time and distance either individually or, more commonly, using a mixture of all three.

Shielding is the best method because it is an 'engineered' solution. It involves the placing of a physical shield, such as a layer of lead, steel or concrete, between the worker and the radioactive source. The thicker the shield the more effective it is.

Time involves the use of the reduced time exposure principle and thus reduces the accumulated dose.

Distance works on the principle that the effect of radiation reduces as the distance between the worker and the source increases.

Other measures include the following:

▶ effective emergency arrangements;
▶ training of employees;
▶ the prohibition of eating, drinking and smoking adjacent to exposed areas;
▶ a high standard of personal cleanliness and first-aid arrangements;
▶ strict adherence to personal protective equipment arrangements, which may include full body protection and respiratory protection equipment;
▶ procedures to deal with spillages and other accidents;
▶ prominent signs and information regarding the radiation hazards;
▶ medical surveillance of employees.

A **Radiation Protection Supervisor** should be appointed by the employer as the competent person to advise on the necessary measures for compliance with national legislation. The person appointed, who is normally an employee, must be competent to supervise the arrangements in place and have received relevant training.

The **Radiation Protection Adviser** is appointed by the employer to give advice to the Radiation Protection Supervisor and the employer on any aspect of working with ionising radiation including the appointment of the Radiation Protection Supervisor. The Radiation Protection Adviser is usually an employee of a national or international organisation with expertise in ionising radiation.

Non-ionising radiation

For ultraviolet and infrared radiation, eye protection in the form of goggles or a visor is most important, particularly when undertaking arc welding or furnace work. Skin protection is also likely to be necessary for the hands, arms and neck in the form of gloves, sleeves and a collar. For construction and other outdoor workers, protection from sunlight is required, particularly for the head and nose. Sun creams should also be used.

For laser operations, engineering controls such as fixed shielding and the use of non-reflecting surfaces around the workstation are recommended. For lasers in the higher class numbers, special eye protection is recommended. A risk assessment should be undertaken before a laser is used.

Engineering controls are primarily used for protection against microwaves. Typical controls include the enclosure of the whole microwave system in a metal surround and the use of an interlocking device that will not allow the system to operate unless the door is closed.

Intense sources of artificial light in the workplace, particularly from UV radiation and powerful lasers, can harm the eyes and skin of workers and need to be properly managed. The Control of Artificial Optical Radiation at Work Regulations came into force in 2010 in the UK and implements the Physical Agents (Artificial Optical Radiation) Directive (2006/25/EC). The aim of the Regulations is to ensure that standards are set so that all workers are protected from harm arising from exposure to hazardous sources of artificial light. As with the previous Noise and Vibration Regulations, it contains provisions on risk assessment, control of exposure, health surveillance and information, instruction and training. The Regulations are based on the limit values incorporated in the guidelines issued by the International Commission on Non-Ionising Radiation Protection (ICNIRP).

Workers should be protected from hazardous sources of light in the workplace, to ensure that the eyes and skin of workers are properly protected from intense sources of light at work that can damage the eyes and skin. Such light sources include ultraviolet, visible and infrared radiation produced by artificial sources, such as lasers and welding arcs. Examples of such hazardous light sources include:

▶ welding work (both arc and oxy-fuel) and plasma cutting causing mainly eye damage;

Figure 14.13 Specialised eye protection for work with lasers

▶ furnaces and foundries causing eye and skin damage;
▶ printing involving the UV curing of inks causing mainly skin damage;
▶ motor vehicle repairs involving the UV curing of paints causing mainly skin damage;
▶ medical and cosmetic treatments involving laser surgery, blue light and UV therapies causing both eye and skin damage;
▶ all use of Class 3B and Class 4 lasers potentially causing permanent eye and skin damage.

14.3.4 ILO recommendations for radiation protection

Ionising radiation

The ILO Code of Practice 'Ambient factors in the workplace' applies to any workplace where workers may be occupationally exposed to ionising radiation. The basic principles and framework for radiation protection of workers are described in the Radiation Protection Convention (No. 115), and Recommendation (No. 114), 1960.

Detailed guidance on ionising radiation is given in the *International basic safety standards for protection against ionising radiation and for the safety of radiation sources* (jointly sponsored by various international organisations including the ILO and WHO), and the ILO Code of Practice 'Radiation protection of workers, ionising radiations,' (Geneva, 1987).

The responsibilities of the competent national authority (or authorities) concerned with radiation protection should include:

(a) the formulation of the necessary criteria, standards and regulations for radiation protection, in consultation with the representative organisations of employers and workers concerned;

(b) the establishment of a system for notification, registration or licensing as required in the *Basic safety standards;*

(c) the provision of general guidance necessary for the implementation of the requirements;

(d) the establishment of a system of inspection to ensure that the measures taken are in compliance with the relevant requirements;

(e) the arrangements to protect the health of itinerant workers in regard to radiation safety, so that the established radiation exposure limits are not exceeded.

The ILO Code of Practice specifies the following prevention and control requirements for ionising radiations:

1. Whatever the situation, the radiation protection programme should provide the following level of detail:

(a) the documented assignment of managerial responsibilities including organisational arrangements. This may require the allocation of respective responsibilities for occupational radiation protection and safety between employers and the registrant or licensee;

(b) the designation of suitable controlled or supervised areas;

(c) the local rules for workers and the supervision of work;

(d) the arrangements for monitoring workers and the workplace;

(e) the system for recording and reporting all the relevant information related to the control of exposures and the monitoring of individuals;

(f) the education, training and information programme;

(g) the procedure for reviewing and auditing periodically the performance of the radiation protection programme;

(h) health surveillance details;

(i) the appointment of a radiation protection officer, when required by the competent authority, to oversee the application of the regulatory requirements.

2. Management, in accordance with the requirements of the *Basic safety standards* referred to earlier in this section, should designate a:

(a) **controlled area**: any area in which specific protective measures or safety provisions are or could be required for:
   (i) controlling access and normal exposures or preventing the spread of radioactive contamination during normal working conditions;
   (ii) preventing or limiting the extent of potential exposures;

(b) **supervised area** which is any area not already designated as a controlled area but where occupational exposure conditions need to be kept under review.

3. Management, in consultation with workers and/or their representatives, if appropriate, should:

(a) develop written local rules that stipulate general organisational structures and special procedures to be followed in controlled areas;

(b) include in the local rules the values of relevant exposure levels, and the procedure to be followed in the event that any such levels are exceeded;

(c) ensure that the rules, procedures, protective measures and safety provisions are known to and observed by workers and other persons to whom they apply;

(d) ensure that any work involving occupational exposure is adequately supervised.

4. When engineering and operational control measures are not sufficient to provide the required level of protection for the operational tasks, management should ensure that workers are:

(a) provided with suitable, adequate, well-maintained and tested personal protective equipment;

(b) given adequate instruction in its proper use and maintenance.

5. Management should assess the occupational exposure of workers:

(a) by individual monitoring, where appropriate. Its nature, frequency and precision should be determined based on the magnitude and possible fluctuations of exposure levels and the likelihood and magnitude of potential exposures;

(b) by monitoring of the workplace. Its nature and frequency should depend on the ambient radiological conditions of the workplace and the fluctuations thereof and be sufficient to assess exposure and review the classification of controlled areas and supervised areas. A programme for the monitoring of the workplace under the supervision of a radiation protection officer, if so required by the competent authority, should be established, maintained and kept under review.

6. Management should:

(a) maintain and retain up-to-date exposure records in accordance with national laws and internationally recognised practice for each worker for whom assessment of occupational exposure is required;

(b) maintain accurate and current records of the findings of the workplace monitoring programme and make them available to workers and/or their representatives.

7. As required by the *Basic safety standards*, management should establish a quality assurance programme, the nature and extent of which should be commensurate with the magnitude and the likelihood of potential exposures from the ionising source.

8. The radiation protection programme should be reviewed on a regular basis. Audits and/or reviews should be scheduled on the basis of the status and importance of the activity. Management should make arrangements for an independent assessment of the implementation of the radiation protection programme in order to identify and correct administrative and management problems in the achievement of its objectives.

Non-ionising radiation

The ILO Code of Practice 'Ambient factors in the workplace' applies to any workplace where workers may be occupationally exposed to non-ionising ultraviolet and infrared radiation. Detailed guidance on specific non-ionising radiations are given in *The use of lasers in the workplace,* ILO Occupational Safety and Health Series No. 68 (Geneva, 1993); and *Visual display units: Radiation protection guidance,* ILO Occupational Safety and Health Series No.70 (Geneva, 1994).

The code requires that employers assess equipment and activities likely to give rise to hazardous exposure to non-ionising radiation. The assessment should include any outdoor work which exposes workers to the sun. They should also take all necessary safety precautions and prevention and control measures to reduce the risk of exposure to hazardous levels of non-ionising radiation and to other associated hazards.

Employers should:

(a) provide specialised eye protection to workers at risk of exposure to lasers;

(b) provide effective eye and skin protection to workers exposed to UV emissions, including welding helmets, and organise work patterns and worker location to ensure the protection of non-welders;

(c) erect warning signs to prevent casual access to welding areas, high-level infrared and laser zones;

(d) where practicable, in the case of outdoor work:

(i) minimise exposure of workers to the sun by organising the work so that it can be carried out in the shade;

(ii) protect workers by appropriate clothing and personal protection, such as sunscreen ointment or lotions and eye protection, when necessary.

Employers should arrange for appropriate health surveillance by occupational health personnel who should assess the possible need for examination, including ophthalmic and skin examination, for those exposed to significant levels of non-ionising radiation and/or involved in work with lasers.

Finally, employers should provide training and information so that workers who are likely to be exposed to significant levels of non-ionising radiation and/or involved in work with lasers are made aware of:

(a) the health hazards of non-ionising radiation and the sources and activities that may pose a risk of exposure, especially the effects of the sun;

(b) the importance in outdoor work of using any available shade and personal protection including protective clothing and sunscreen ointments and lotions;

(c) the serious risks to eyesight if proper protection is not used, for example in welding and laser operations;

(d) the serious limitations of blue lenses (used in steelworks and foundries to check the temperature of the melt) in providing eye protection;

(e) the hazards of maintenance and cleaning operations and the correct use of lamp shields and enclosures;

(f) the fact that some perfumes and medicines can cause sensitisation on exposure to UV radiation.

The EU EMF Directive

Electric, magnetic and electromagnetic fields (EMFs) are a form of non-ionising radiation that arises whenever electrical energy is used. An electric field is generated by electric charges, while a magnetic field occurs around an electric current. Electromagnetic fields are mutually produced by time-varying magnetic and electric fields. When electric fields act on conductive materials (such as the human body), they influence the distribution of electric charges at their surface causing current to flow through the body to the ground. Magnetic fields induce circulating currents within the human body. The strength of these currents depends on the intensity of the outside magnetic field. If sufficiently large, these currents can stimulate nerves and muscles or cause other biological effects such as nausea.

Common sources of electromagnetic fields include work processes such as radiofrequency heating and welding; household electrical wiring and appliances; electrical motors; computer screens; telecommunications; transport and distribution of electricity; broadcasting; and security detection devices.

The EU EMF Directive covers a frequency spectrum from 0 to 300 GHz, which includes static magnetic fields and low frequency electric and magnetic fields, through to radio frequency and microwave frequencies.

14.3.5 Welding operations

Over 1,000 accidents involving welding work are reported to the UK HSE each year. There are several different types of welding operation. The most common are:

▶ manual metal arc welding;
▶ metal inert-gas welding (MIG);
▶ tungsten inert-gas welding (TIG);
▶ oxy-acetylene welding.

The non-ionising radiation hazards caused by arc welding are not the only hazards associated with welding operations. The hazards from fume inhalation, trailing cables and pipes and the manual handling of cylinders are also present. There have also been serious injuries resulting from explosions and fires during welding processes. Accidents are often caused by lack of training and faulty equipment and they are often made worse by the lack of complete personal protective equipment. Many of these accidents occur on farms

14

where welding equipment is used to make on-the-spot repairs to agricultural machinery.

14.4 Stress

It has been estimated that work-related stress costs the UK around £3.8 billion a year. The cost of work-related injuries and illnesses in the United States has been estimated at some $300 billion (£200bn) a year. In 2011, according to final fatality data from the Bureau of Labor Statistics, 4,693 workers were killed on the job – an average of 13 a day and corresponding to an incidence rate of 3.5 per 100,000 workers. This is almost six times the UK rate of 0.6 per 100,000 workers. This has been accompanied by an increase in civil law claims resulting from stress at work. Stress and the related issues of workplace bullying and harassment are issues that simply cannot be ignored.

Work-related stress

In 2001, the UK HSE defined work-related stress as the reaction people have to excessive pressure or other types of demand placed on them. An important distinction is made in this definition between pressure, which can be a positive state if managed appropriately and a normal reaction to reasonable demands, and stress which can arise in response to intense, continuous or prolonged exposure to excessive pressures and can be detrimental to health. However, a recent report commissioned by the HSE has concluded that there is no simple and universal definition of work-related stress, largely because of the complex nature of work-related stress.

The European Commission defines work-related stress as an 'emotional and psycho-physiological reaction to aversive and noxious aspects of work, work environments and work organisations. It is a state characterised by high levels of arousal and distress and often by feelings of not coping'.

However, the recent HSE report distinguished work-related stress as different from occupational stress. The former includes cases where work may have aggravated the experience of stress and associated symptoms of ill-health regardless of original cause. Here work may be a contributory factor but not necessarily the sole cause. The latter refers to cases where work is the sole cause of the experience of stress and associated symptoms of ill-health.

Stress is not a disease – it is the natural reaction to excessive pressure. It can be defined as the reaction that people have when they are unable to cope with excessive pressures and demands. Stress can lead to an improved performance but is not generally a good thing as it is likely to lead to both physical and mental ill-health, such as high blood pressure, peptic ulcers, skin disorders and depression.

Symptoms of stress

There are four groups of stress symptoms – physical, behavioural, emotional and cognitive. The physical symptoms include chest pains, nausea and frequent colds. Sufferers may eat less, smoke and drink more and develop nervous habits. They may be unable to relax, become irritable and depressed. They may also have memory problems, be unable to concentrate and become very anxious. Stress victims experience a severe lack of control.

Causes of stress

Most people experience stress at some time during their lives, e.g. during an illness or death of a close relative or friend. However, recovery normally occurs after the particular crisis has passed. The position is, however, often different in the workplace because the underlying causes of the stress, known as work-related stressors, are not relieved but continue to build up the stress levels until the employee can no longer cope.

The basic workplace stressors are:

▶ the job itself – boring or repetitive, unrealistic performance targets or insufficient training, job insecurity or fear of redundancy;
▶ individual responsibility – ill-defined roles and too much responsibility with too little power to influence the job outputs;
▶ working conditions – cramped, dirty and untidy workplace; unsafe practices; lack of privacy or security; inadequate welfare facilities; threat of violence; excessive noise, vibration or heat; poor lighting; lack of flexibility in working hours to meet domestic requirements and adverse weather conditions for those working outside;
▶ management attitudes – poor communication, consultation or supervision, negative health and safety culture, lack of support in a crisis;
▶ relationships – unhappy relationship between workers, bullying, sexual and racial harassment.

The ILO has produced a manual which includes easy-to-apply checkpoints for identifying stressors in working life and mitigating their harmful effects, as well as guidance on linking workplace risk assessment with the process of stress prevention.

Control measures

Possible solutions to all these stressors have been addressed throughout this book and involve the creation of a positive health and safety culture, effective training and consultation procedures and a set of health and safety arrangements which work on a day-to-day basis. The UK HSE has produced its own generic stress audit survey tool which is available free of charge on its website and advises the following action plan:

▶ identify the problem;

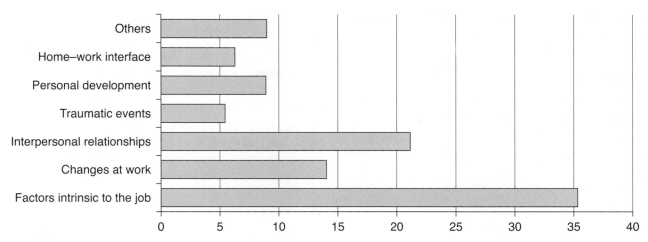

Figure 14.14 Breakdown of mental health cases by type of event which precipitated stress between 2010 and 2012

- identify the background to the problem and how it was discovered;
- identify the remedial action required and give reasons for that action;
- identify targets and reasonable target dates;
- agree a review date with employees to check that the remedial action is working.

The following additional measures have also been found to be effective by some employers:

- take a positive attitude to stress issues by becoming familiar with its causes and controls;
- take employees' concerns seriously and develop a counselling system which will allow a frank, honest and confidential discussion of stress-related problems;
- develop an effective system of communication and consultation and ensure that periods of uncertainty are kept to a minimum;
- set out a simple policy on work-related stress and include stressors in risk assessments;
- ensure that employees are given adequate and relevant training and realistic performance targets;
- develop an effective employee appraisal system which includes mutually agreed objectives;
- discourage employees from working excessive hours and/or missing break periods (this may involve a detailed job evaluation);
- introduce job rotation and increase job variety;
- develop clear job descriptions and ensure that the individual is matched to them;
- encourage employees to improve their lifestyle (e.g. many local health authorities provide smoking cessation advice sessions);
- monitor incidents of bullying, sexual and racial harassment and, where necessary, take disciplinary action;
- train supervisors to recognise stress symptoms among the workforce;
- avoid a blame culture over accidents and incidents of ill-health.

The individual can also take action if he/she feels that he/she is becoming over-stressed. Regular exercise, change of job, review of diet and talking to somebody, preferably a trained counsellor, are all possibilities.

A risk assessment will reveal whether stress is a problem in a particular workplace. The risk assessment needs to relate to the individual, such as absentee record, production performance and which of the stressors are applicable, and to the organisation, such as training programmes, communication and appraisal procedures. Many stressful conditions can be reduced by an effective 'whistle-blowing' policy that enables individuals to highlight any concerns to senior managers or directors in a confidential manner.

In the UK, there have been several successful civil actions for compensatory claims resulting from the effects of workplace stress. However, the Court of Appeal in 2002 redefined the guidelines under which workplace stress compensation claims may be made. Their full guidelines should be consulted and consist of 16 points. In summary, these guidelines are as follows:

- no occupations should be regarded as intrinsically dangerous to mental health;
- it is reasonable for the employer to assume that the employee can withstand the normal pressures of the job, unless some particular problem or vulnerability has developed;
- the employer is only in breach of duty if they have failed to take the steps that are reasonable in the circumstances;
- the size and scope of the employer's operation, its resources and the demands it faces are relevant in deciding what is reasonable, including the interests of other employees and the need to treat them fairly in, for example, the redistribution of duties;
- if a confidential counselling advice service is offered to the employee, the employer is unlikely to be found in breach of his/her duty;
- if the only reasonable and effective action is to dismiss or demote the employee, the employer is

14

not in breach of his/her duty by allowing a willing worker to continue in the same job;

▶ the assessment of damages will take account of any pre-existing disorder or vulnerability and the chance that the claimant would have succumbed to a stress-related disorder in any event.

Stress usually occurs when people feel that they are losing control of a situation. In the workplace, this means that the individual no longer feels that they can cope with the demands made on them. Many such problems can be partly solved by listening to rather than talking at people.

14.5 Further information

Ambient factors in the workplace, International Labour Organisation (ILO) Code of Practice (CoP), ISBN 92-2- 11628-X http://www.ilo.org/safework/info/standards-and-instruments/WCMS_107729/lang--en/index.htm

Radiation Protection C115 and R114, 1960 http://www.ilo.org/dyn/normlex/en/f?p=1000:12100:0::NO::P12100_ILO_CODE:C115 http://www.ilo.org/dyn/normlex/en/f?p=1000:12100:0::NO::P12100_ILO_CODE:R114

Radiation protection of workers, ionising radiations, ILO CoP, 1987, ISBN 9-22-105996-0 http://www.ilo.org/wcmsp5/groups/public/---ed_protect/---protrav/---safework/documents/normativeinstrument/wcms_107833.pdf

Working Environment (Air, Pollution, Noise and Vibration) C148 and R156, 1977 http://www.ilo.org/dyn/normlex/en/f?p=1000:12100:0::NO::P12100_ILO_CODE:C148 http://www.ilo.org/dyn/normlex/en/f?p=1000:12100:0::NO::P12100_ILO_CODE:R156

14.6 Practice revision questions

1. (a) **Outline** the mechanism by which noise is transmitted by the human ear to the brain.

   (b) **Explain** the following terms in relation to noise exposure at work in both the short and long term:
   (i) 'noise-induced hearing loss'
   (ii) 'tinnitus'
   (iii) temporary and permanent threshold shift.

   (c) **Outline** the options that might be considered to reduce the risk of hearing damage to workers who are exposed to high levels of noise.

2. (a) **Give** the meaning of the following terms used in noise measurement:
   (i) frequency;
   (ii) 'daily personal noise exposure' ($L_{EP,d}$);
   (iii) decibel (Db);
   (iv) 'A' Weighting.

   (b) **Identify TWO** types of noise measurement technique.

3. (a) **Outline** appropriate control measures to reduce the levels of noise to which the workers are exposed.

   (b) **Outline** the criteria that should be used when selecting suitable hearing protection and **identify** the limitations of such protective equipment.

   (c) **Outline** the contents of a health surveillance programme required to protect workers who are regularly exposed to high noise levels.

4. **Outline** the advantages **AND** disadvantages of:
   (a) earmuffs;
   (b) ear plugs.

5. **Explain** the meaning of the following terms in relation to noise control in the workplace:
   (a) silencing
   (b) absorption
   (c) damping
   (d) isolation
   (e) lagging.

6. (a) **Describe** hand–arm vibration (HAV) and whole-body vibration (WBV).

   (b) **Outline** the ill-health effects and the associated symptoms caused by the exposure of workers to HAV and WBV.

7. A worker operates a hammer drill for long periods of the working day.

   (a) **Outline** the issues to be considered when assessing the risk of hand–arm vibration syndrome (HAVS) being developed by the worker.

   (b) **Outline** the control measures that should be taken to minimise the risk of the worker developing HAVS.

8. A dumper truck driver, working on a construction site, spends much of the working day driving over very rough ground. After several weeks, the driver complains of back pain.

   (a) **Identify** possible reasons for the back pain associated with the truck driving.

   (b) **Outline** actions that could be taken by the employer to control the risks from whole-body vibration.

9. (a) **Identify TWO** types of ionising radiation.

   (b) **Outline** possible means of ensuring that workers are not exposed to unacceptable levels of ionising radiation.

10. (a) For each of the following types of non-ionising radiation, **identify** a source and the possible ill-health effects on exposed individuals:
    (i)   infrared radiation
    (ii) ultraviolet radiation
    (iii) lasers
    (iv) microwaves.

    (b) **Identify** the controls available to protect people against exposure to these non-ionising radiations.

11. Construction workers are often required to work outdoors in a variety of weather conditions. **Outline** the hazards and the associated control measures that should be taken to protect these workers when they are working outdoors:

    (a) in strong sunlight
    (b) in cold temperatures.

12. (a) **Explain** the meaning of 'work-related stress'.

    (b) **Identify** the ill-health effects of excessive stress to workers in the workplace.

    (c) **Identify FOUR** issues that could lead to workplace stress.

    (d) **Identify FOUR** behavioural symptoms of occupational stress.

    (e) **Outline** the control measures that could be taken to minimise the risk of occupational stress.

14

445

CHAPTER 15

Summary of ILO, OSH Conventions, legal frameworks and country examples

15.1 ILO International Conventions on OSH referenced in the NEBOSH International General Certificate syllabus ▶ 448

15.2 Typical OSH legal frameworks in the USA, EU and UK ▶ 471

15.3 National implementing legislation ▶ 476

15.4 Common themes in national legislation ▶ 527

Appendix 15.1 Seoul Declaration on Safety and Health at Work ▶ 533

Appendix 15.2 ILO – C155 Occupational Safety and Health Convention, 1981 ▶ 534

This chapter covers:

1. Summaries of ILO International Conventions, some recommendations and codes of practice on OSH referenced in the NEBOSH International General Certificate syllabus guide
2. A number of typical legal frameworks which are used in the USA, EU and UK
3. Summaries of primary health and safety legislation in over 20 countries/areas including the USA, UK, EU, Russian Federation, India, China, Brazil, United Arab Emirates and South Africa
4. Common themes in national legislation

15.1 ILO International Conventions on OSH referenced in the NEBOSH International General Certificate syllabus

15.1.1 Introduction

This chapter aims to introduce the reader to the ILO Conventions on Occupational Safety and Health (OSH) and indicate how they have set out to influence national legal frameworks. It will show how, increasingly, the national legal frameworks are being moulded to the ILO vision and the chapter has a summary, as at October 2014, of the main general OSH laws in over 20 countries across the world, taken from all continents.

The chapter includes summaries of the following ILO documents:

▶ Promotional framework for ILO–OSH Convention C187 and Recommendation R197
▶ Occupational Safety and Health Convention C155, 1981 and Recommendations R164
▶ Asbestos Convention, C162, 1986
▶ Chemicals, Convention C170, 1990
▶ Hours of Work and Rest Periods (Road Transport) C153, 1979 (No. 161)
▶ Hygiene (Commerce and Offices) Convention, C120, 1964
▶ Occupational Cancer, Convention C139, 1974
▶ Occupational Health Services, Convention, C161, 1981
▶ Radiation Protection, Convention C115, 1960
▶ Safety and Health in Construction Convention, C167, 1988
▶ Safety and health in the use of machinery, ILO Code of Practice
▶ Welfare Facilities Recommendation, R102, 1956
▶ Working Environment (Air Pollution, Noise and Vibration) Convention 1977, C148.

Occupational safety and health is a cross-disciplinary area concerned with protecting the safety, health and welfare of people engaged in work or employment. The goal of all occupational safety and health programmes is to foster a safe work environment. As a secondary effect, it may also protect co-workers, family members, employers, customers, suppliers, nearby communities and other members of the public who are impacted by the workplace environment. It may involve interactions among many subject areas, including occupational medicine, occupational (or industrial) hygiene, public health, safety engineering, chemistry, health physics, ergonomics, toxicology, epidemiology, environmental health, industrial relations, public policy, sociology and occupational health psychology.

The opening address by Guy Ryder, the Director General of ILO, to the XX World Congress on Safety and Health at Work 2014 in Frankfurt, was reported by the ILO in the following:

'Ebola and the tragedies it is causing are in the daily headlines – which is right. But work-related deaths are not. So, the task ahead is to establish a permanent culture of consciousness.

'The challenge we face is a daunting one. Work claims more victims around the globe than does war. The figures cited by the ILO are staggering. Every minute, five people die worldwide as a result of a work accident or an occupational disease: that's more than 2.3 million employees per year. In addition, some 313 million people have a (non-fatal) accident at work every year. Around 160 million workers suffer from an occupational disease. But these figures are only the tip of the iceberg,' says Guy Ryder. As reliable data on occupational health and safety are still unavailable, the ILO believes that the unofficial number of dead, injured and ill workers is much higher. In addition to all of the personal suffering involved, this situation also has quite an impact on the global economy: the ILO estimates that about 4% of the global gross domestic product – that is, around US$ 2.8 trillion – is lost due to work accidents and occupational diseases.

'So much suffering, such high costs, and yet, in these days of trade liberalization, globalization and economic crises, OSH is still regarded in many countries as a luxury, which arbitrarily falls victim to budget cuts,' says the Director-General of the ILO. This is despite the fact that, particularly during an economic recovery, labour protection could bring significant added value and sustainable development to the business world. Ryder therefore calls on political decision-makers to demonstrate even greater commitment to building national and international occupational safety and health programmes.

There are however some encouraging signs. At the G20 Leaders Summit held in St. Petersburg in 2013, the leaders decided to initiate a 'Task Force on Employment' with the ILO in order to achieve greater safety at work within the countries of the G20. Moreover, the ILO itself has also set itself the goal of advancing the cause of occupational safety and health worldwide even stronger than before. A flagship programme is currently being set up for this purpose. It does not have a name yet, but 'Vision Zero would be the perfect headline'.

Despite such developments, a great deal must still be done in order to reach the desired goal. It is essential that governments, employers and workers engage in a social dialogue. Preventive actions must in particular benefit workers with low and medium incomes, as well as the many millions of people around the world working in the informal economy without adequate work contracts or protection. 'The right to a safe and healthy working environment is a basic human right,' Guy Ryder reminded his listeners, also referring here to the United Nations Universal Declaration on Human Rights. This right is indispensable to achieving social justice and decent work. Therefore, the ILO is striving towards this goal for all working women and men the world over.

A report of a general survey concerning the Occupational Safety and Health Convention, 1981 (No. 155), the Occupational Safety and Health Recommendation, 1981 (No. R164), and the Protocol of 2002 to the Occupational Safety and Health Convention, 1981, was presented, by the Committee of Experts, entitled 'ILO standards on occupational safety and health'. The report summary includes the following.

> The promotion of decent, safe and healthy working conditions and environment has been a constant objective of International Labour Organisation (ILO) action since the Organisation was founded in 1919. A significant body of international instruments and guidance documents has been developed by the ILO over the past 90 years to assist constituents in strengthening their capacities to prevent and manage workplace hazards and risks. The present survey examines three central ILO instruments in this area: the Occupational Safety and Health Convention, 1981 (No. 155), the Occupational Safety and Health Recommendation, 1981 (No. 164), and the Protocol of 2002 to Convention No. 155. These instruments provide a blueprint for setting up and implementing comprehensive national occupational safety and health (OSH) systems based on prevention and continuous improvement.
>
> The present survey highlights the progress made by ILO member States, the continuing and increased relevance of the instruments at issue and the basic strategy they advocate. These instruments were designed to be applied progressively and their application can be adapted to specific national conditions and developments. Although further efforts should

be deployed to ensure that OSH protection is extended to all workers and all branches of economic activity, the flexibility clauses and principle of progressive application provided for in Convention No. 155 should permit an increasing number of countries to consider ratifying and giving effect to it.

The strategy advocated by Convention No. 155 and Recommendation No. 164 calls for action in essential areas pertaining to OSH, namely for the formulation, implementation and periodical review of a national OSH policy; the full participation at all levels of employers, workers and their respective organisations, as well as other stakeholders; the definition of national institutional responsibilities and of the respective responsibilities, duties and rights of employers, workers and their representatives; and the requirements regarding knowledge, education and training, and information.

A significant number of countries, particularly among developing countries, report that they are in the process of formulating or updating their national policies, and developing their regulatory and enforcement systems. Several countries are also in the process of developing, reorienting or implementing policies, focusing on and targeting emerging issues such as stress and musculoskeletal disorders (MSDs), assistance to small and medium-sized enterprises (SMEs), and the promotion of best practices.

While further information, in particular from the social partners, would have allowed it to get a more reliable global picture of the practical application of OSH requirements, the Committee of Experts concludes that a majority of ILO member States, to a large and increasing extent, give effect to the provisions not only of the Convention, but also of the Recommendation. This level of involvement is a clear indication that these instruments have a place at the heart of national action in the area of OSH. This survey also highlights the crucial importance of tracing progress in the implementation of national OSH policies through the collection and analysis of data on their practical application and statistics on occupational accidents and diseases, and that vigorous promotional efforts are called for to increase the ratification rate of the Protocol and its implementation in practice. ILO advice, assistance and technical cooperation may be crucial for many member States to enable them further to improve their national OSH systems, and efforts should be made to provide such assistance.

15

The relevance and importance of the national policy and systems approach in Convention No. 155 has been further reaffirmed through the Promotional Framework for Occupational Safety and Health Convention, 2006 (No. 187), and its Recommendation (No. 197). These instruments complement the instruments in this survey by providing further guidance on the systems approach to the management of OSH at all levels and the progressive establishment of a preventative safety and health culture based on the continuous provision of OSH information, training and education. The increasing rate of ratification of Convention No. 187 is a clear endorsement by the tripartite constituents of the ILO's timely action in the area of OSH, and indicates a renewed interest in the prevention of occupational accidents and diseases and improving working conditions and environment. In view of their close linkage, further efforts should be made to promote Convention No. 155 together with Convention No. 187. The joint support by the social partners of these efforts is an essential element in the process of achieving decent, safe and healthy working conditions and environment.'

See:

http://www.ilo.org/global/What_we_do/ Officialmeetings/ilc/ILCSessions/98thSession/ ReportssubmittedtotheConference/lang--en/ docName--WCMS_103485/index.htm

http://www.ilo.org/ilolex/english/recdisp1.htm (for ILO labour standards)

http://www.ilo.org/public/english/protection/safework/ (for the ILO SafeWork portal)

http://www.oshupdate.com/ (for general OSH updates)

http://www.ilo.org/dyn/legoshen/f?p=14100:1000:0:: NO::: (Global Database on Occupational Safety and Health Legislation).

15.1.2 Promotional framework for ILO–OSH Convention No. 187 and Recommendation 197

(a) A promotional framework for OSH

At its 91st Session (2003), the International Labour Conference adopted a Global Strategy on OSH, which was designed progressively to improve safety and health in the world of work. In response to this strategy, the International Labour Conference adopted the Promotional Framework for Occupational Safety and Health Convention (No. 187) and its accompanying Recommendation (No. 197) in 2006. The main purpose of Convention No. 187 is to ensure that a higher priority

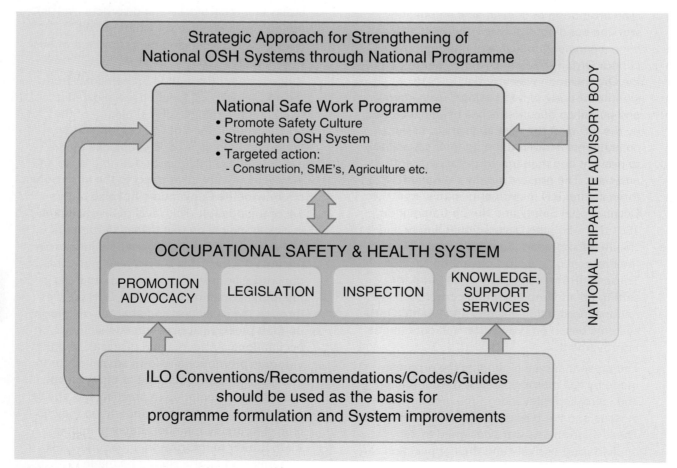

Figure 15.1 ILO's strategic approach to strengthening national OSH systems

is given to OSH in national agendas and to foster political commitments in a tripartite context for the improvement of working conditions and environment. It has a promotional rather than prescriptive content and is based on two fundamental concepts outlined in the above Global Strategy, namely to:

▸ develop a preventative safety and health culture; and
▸ apply a systems approach to managing OSH nationally.

This means the continual monitoring, evaluation and improvement of all the different 'building blocks' making up the national OSH system. The Convention defines in general terms the elements and function of the national policy, the national system and the national programme. Further specific areas of action, operational details and mechanisms such as the development and maintenance of a national OSH profile are provided in the Recommendation. See Figure 15.1.

The Promotional Framework for Occupational Safety and Health Convention 2006 (No. 187) has so far (November 2014) been ratified by thirty-two member states.

(b) OSH and safety culture

A national preventative safety and health culture is one in which the right to a safe and healthy working environment is respected at all levels. It is also one where governments, employers, workers and other interested stakeholders actively participate in securing a safe and healthy working environment through a system of defined rights, responsibilities and duties, and where the principle of prevention is accorded the highest priority. Building and maintaining such a culture requires a permanent mobilisation of all available means of action, particularly education and training, to increase general awareness, knowledge and understanding of the concepts of hazards and risks and how they may be prevented and controlled. Educational systems need to recognise the concepts of workplace hazards, risks and prevention, including them within national curricula as appropriate, thereby promoting greater continuity between public and workplace safety and health issues.

(c) OSH and management systems

In recent years, governments, enterprises and international organisations have all been giving greater attention to the need to adopt systematic models for managing OSH. The so-called OSH management systems approach (see Chapter 1) provides a promising strategy for augmenting traditional command-and-control approaches with performance improvement tools, more effective health and safety auditing concepts, and schemes for management systems.

The need for a global approach to OSH management was recognised as a logical and necessary response to increasing economic globalisation, while the benefits of systematic models of managing OSH became

apparent as a result of the impact of ISO standards for quality and the environment. Current management science theories suggest that performance is better in all areas of business, including OSH, if it is measured and continuous improvement sought in an organised fashion. Drawing from the principles defined in the *ILO Guidelines on occupational safety and health management systems, 2001*, Convention No. 187 applies a similar approach to the management of national OSH systems to ensure they are improved through a continuous cycle of policy review, evaluation and action for improvement. The different steps in the OSH Management Cycle of continuous improvement are illustrated in Chapter 1.

(d) National OSH policy

The elaboration of a national OSH policy by the social partners on a consensual basis is the most visible demonstration of the national commitment to promote a decent, safe and healthy working environment. To ensure widest support, its development, implementation and periodic review have to be carried out through a collaborative process involving government, organisations of employers and workers, and other stakeholders with OSH-related responsibilities and activities. An endorsement of the policy at the highest level of government is the most effective way to raise general awareness of the importance of OSH in achieving decent, safe and healthy working conditions and environment, and building a preventative safety and health culture.

The Promotional Framework for Occupational Safety and Health Convention 2006 (No. 187) amplifies the provisions of the Occupational Safety and Health Convention, 1981 (No. 155), calling for the formulation and periodical review of a national OSH policy by asking for an endorsement of the national programme at the highest level of government.

(e) National OSH systems

OSH is a complex subject, involving a large number of specific disciplines and requiring consideration of a wide range of workplace and environmental hazards. National OSH systems need somehow to capture such complexities if they are to function coherently and effectively, embracing a wide range of skills, knowledge and analytical capacities within appropriate organisational structures and mechanisms.

National OSH systems comprise the infrastructures, mechanisms and specialised human resources needed to translate the goals of the national policy into practice. Because they reflect the effects of socioeconomic and technological changes on working conditions and environment, national OSH systems are dynamic and need to be built through an ongoing cycle of review, performance and evaluation. Matching national OSH policy and programmes, they will also

15

Table 15.1 Essential elements of a national OSH system

▶ Legislation, and any other relevant OSH instruments;

▶ One or more authorities or bodies responsible for OSH;

▶ Regulatory compliance mechanisms, including systems of inspection;

▶ A national tripartite advisory mechanism addressing OSH issues;

▶ Arrangements to promote at the enterprise level, cooperation between employers and workers;

▶ OSH information and advisory services;

▶ Systems for the provision of OSH training;

▶ Occupational health services;

▶ Research on OSH;

▶ A mechanism for the collection and analysis of data on occupational injuries and diseases;

▶ Provisions for collaboration with relevant insurance or social security schemes covering occupational injuries and diseases; and

▶ Support mechanisms for a progressive improvement of OSH conditions in micro, small and medium-sized enterprises, and in the informal economy.

need to be readjusted from time to time to meet new perceived needs and to respond to the challenges of a continuously evolving world of work. Table 15.1 lists the essential elements of a national OSH system according to Convention No. 187.

15.1.3 Occupational Safety and Health Convention 1981, No. 155 and Recommendations R164

(a) Introduction

R164 OSH Recommendations were agreed at the 67th Session of ILO on 22 June 1981 to supplement the Occupational Safety and Health Convention 1981, No. 155 which had been agreed on 3 June. The Recommendations apply to all branches of economic activity (including public services) and to all categories of workers, including self-employed people. The workplace covers all places where workers are required to go by their work which are under the direct or indirect control of the employer. Health includes more than the absence of disease and also involves the physical and mental elements affecting workers' health which are directly related to safety and hygiene at work.

The full text of Convention 155 is given in Appendix 15.2. Here is a summary of Recommendations R164. For the full text of the Convention 187 and Recommendations go to the ILO website.

See:

http://www.ilo.org/global/What_we_do/Publications/lang--en/docName--WCMS_094524/index.htm
http://www.ilo.org/ilolex/cgi-lex/convde.pl?C187
http://www.ilo.org/ilolex/cgi-lex/convde.pl?C155
http://www.ilo.org/ilolex/cgi-lex/convde.pl?R164
http://www.ilo.org/ilolex/cgi-lex/convde.pl?R197

(b) Areas of action

Appropriate measures, taking into account the principle of giving priority to eliminating hazards at their source, should be taken in the following areas:

(a) design, siting, structural features, installation, maintenance, repair and alteration of workplaces and means of access thereto and egress therefrom;

(b) lighting, ventilation, order and cleanliness of workplaces;

(c) temperature, humidity and movement of air in the workplace;

(d) design, construction, use, maintenance, testing and inspection of machinery and equipment liable to present hazards and, as appropriate, their approval and transfer;

(e) prevention of harmful physical or mental stress due to conditions of work;

(f) handling, stacking and storage of loads and materials, manually or mechanically;

(g) use of electricity;

(h) manufacture, packing, labelling, transport, storage and use of dangerous substances and agents, disposal of their wastes and residues, and, as appropriate, their replacement by other substances or agents which are not dangerous or which are less dangerous;

(i) radiation protection;

(j) prevention and control of, and protection against, occupational hazards due to noise and vibration;

(k) control of the atmosphere and other ambient factors of workplaces;

(l) prevention and control of hazards due to high and low barometric pressures;

(m) prevention of fires and explosions and measures to be taken in case of fire or explosion;

(n) design, manufacture, supply, use, maintenance and testing of personal protective equipment and protective clothing;

(o) sanitary installations, washing facilities, facilities for changing and storing clothes, supply of drinking water, and any other welfare facilities connected with occupational safety and health;

(p) first-aid treatment;

(q) establishment of emergency plans;

(r) supervision of the health of workers.

(c) Action at national level

Taking into account the areas of action listed above, the competent authority or authorities in each country are required in the Recommendations to:

(a) issue or approve regulations, codes of practice or other suitable provisions on occupational safety and health and the working environment, account being taken of the links existing between safety and health, on the one hand, and hours of work and rest breaks, on the other;

(b) from time to time review legislative enactments concerning occupational safety and health and the working environment, and provisions issued or approved in pursuance of clause (a) of this Paragraph, in the light of experience and advances in science and technology;

(c) undertake or promote studies and research to identify hazards and find means of overcoming them;

(d) provide information and advice, in an appropriate manner, to employers and workers and promote or facilitate cooperation between them and their organisations, with a view to eliminating hazards or reducing them as far as practicable; where appropriate, a special training programme for migrant workers in their mother tongue should be provided;

(e) provide specific measures to prevent catastrophes, and to coordinate and make coherent the actions to be taken at different levels, particularly in industrial zones where undertakings with high potential risks for workers and the surrounding population are situated;

(f) secure good liaison with the International Labour Occupational Safety and Health Hazard Alert System set up within the framework of the International Labour Organisation;

(g) provide appropriate measures for handicapped workers.

National authorities should also:

▶ Organise a system for inspection;
▶ In consultation with employers and employees' organisations, promote OSH;
▶ Coordinate OSH activities at national, regional or local levels by public authorities and all other bodies concerned with OSH;
▶ Ensure that vulnerable workers, for example, those with a handicap, are covered.

(d) Action at the level of undertaking

The following obligations are placed on employers by the Recommendations:

(a) to provide and maintain workplaces, machinery and equipment, and use work methods, which are as safe and without risk to health as is reasonably practicable;

(b) to give necessary instructions and training, taking account of the functions and capacities of different categories of workers;

(c) to provide adequate supervision of work, of work practices and of application and use of occupational safety and health measures;

(d) to institute organisational arrangements regarding occupational safety and health and the working environment adapted to the size of the undertaking and the nature of its activities;

(e) to provide, without any cost to the worker, adequate personal protective clothing and equipment which are reasonably necessary when hazards cannot be otherwise prevented or controlled;

(f) to ensure that work organisation, particularly with respect to hours of work and rest breaks, does not adversely affect occupational safety and health;

(g) to take all reasonably practicable measures with a view to eliminating excessive physical and mental fatigue;

(h) to undertake studies and research or otherwise keep abreast of the scientific and technical knowledge necessary to comply with the foregoing clauses.

In addition measures should include the appointment of workers' safety delegates and of workers' safety and health committees and/or joint safety and health committees (with equal numbers of workers' and employers' representatives), all to be in accordance with national practice. The safety delegates and committees should:

(a) be given adequate information on safety and health matters, enabled to examine factors affecting safety and health, and encouraged to propose measures on the subject;

(b) be consulted when major new safety and health measures are envisaged and before they are carried out, and seek to obtain the support of the workers for such measures;

(c) be consulted in planning alterations of work processes, work content or organisation of work, which may have safety or health implications for the workers;

(d) be given protection from dismissal and other measures prejudicial to them while exercising their functions in the field of occupational safety and health as workers' representatives or as members of safety and health committees;

(e) be able to contribute to the decision-making process at the level of the undertaking regarding matters of safety and health;

(f) have access to all parts of the workplace and be able to communicate with the workers on safety and health matters during working hours at the workplace;

(g) be free to contact labour inspectors;

(h) be able to contribute to negotiations in the undertaking on occupational safety and health matters;

(i) have reasonable time during paid working hours to exercise their safety and health functions and to receive training related to these functions;

(j) have recourse to specialists to advise on particular safety and health problems.

Employers should also, having regard to the size and activities of the undertaking, make provision for:

(a) the availability of an occupational health service and a safety service, within the undertaking, jointly with other undertakings, or under arrangements with an outside body;

(b) recourse to specialists to advise on particular occupational safety or health problems or supervise the application of measures to meet them.

Employers should be required to:

(a) where the nature of the operations in their undertakings warrants it, set out in writing their policy and arrangements in the field of occupational safety and health, and the various responsibilities exercised under these arrangements, and to bring this information to the notice of every worker, in a language or medium the worker readily understands.

(b) verify the implementation of applicable standards on occupational safety and health regularly, for instance by environmental monitoring, and to undertake systematic safety audits from time to time.

(c) keep such records relevant to occupational safety and health and the working environment as are considered necessary by the competent authority or authorities; these might include records of all notifiable occupational accidents and injuries to health which arise in the course of or in connection with work, records of authorisation and exemptions under laws or regulations to supervision of the health of workers in the undertaking, and data concerning exposure to specified substances and agents.

Workers should be required to:

(a) take reasonable care for their own safety and that of other persons who may be affected by their acts or omissions at work;

(b) comply with instructions given for their own safety and health and those of others and with safety and health procedures;

(c) use safety devices and protective equipment correctly and do not render them inoperative;

(d) report forthwith to their immediate supervisor any situation which they have reason to believe could present a hazard and which they cannot themselves correct;

(e) report any accident or injury to health which arises in the course of or in connection with work.

Where a worker complains, in good faith, about what they consider is a breach of statutory requirements or a serious inadequacy in measures taken by the employer on OSH or the working environment, no measures prejudicial to the worker should be taken.

15.1.4 Asbestos Convention, C162, 1986

This Convention applies to all activities involving exposure of workers to asbestos in the course of work. It requires national laws to be formulated to prescribe how workers should be protected from exposure to asbestos. It requires enforcement of national laws and

employers to be made responsible for compliance. Employees are required to comply with prescribed safety and hygiene procedures within the limits of their responsibilities.

Part II General Principles – requirements include:

▶ Making work where exposure to asbestos may occur subject to regulations which must be reviewed periodically;

▶ Ensuring that enforcement is secured by an adequate system of inspection;

▶ Making employers responsible for compliance with the requirements;

▶ Ensuring cooperation between employers and workers.

Part III Protective and Preventative Measures – include:

▶ Ensuring that exposure to asbestos is prevented or controlled by one or more of the following:
  ▷ making work in which exposure to asbestos may occur subject to regulations prescribing adequate engineering controls and work practices, including workplace hygiene;
  ▷ prescribing special rules including authorisation for the use of asbestos or certain types of asbestos;

▶ Replacing asbestos-based products with alternative materials whenever possible;

▶ Total or partial prohibition of the use of asbestos-containing materials;

▶ The banning of crocidolite (blue asbestos) or products containing this form of asbestos;

▶ Forbidding the spraying of all forms of asbestos;

▶ Notifying the competent authorities of certain types of work involving exposure to asbestos;

▶ Producers and suppliers of asbestos and manufacturers and suppliers of products containing asbestos shall be made responsible for adequate labelling of the container and, where appropriate, the products, in a language and manner easily understood by the workers and the users concerned, as prescribed by the competent authority;

▶ The prescribing of limits for the exposure of workers to asbestos;

▶ Fixing and reviewing the exposure limits;

▶ Taking all appropriate precautions to prevent or control the release of asbestos dust into the air and that exposure limits are complied with;

▶ Where exposure exceeds exposure limits employers must provide, maintain and replace free of charge adequate respiratory protective equipment (RPE) and special protective clothing as appropriate. RPE must comply with national standards and can only be used as a supplementary, temporary, emergency or exceptional measure and not as an alternative to technical control;

▶ Demolition work involving asbestos-containing materials to be undertaken by competent contractors who must draw up a work plan for the demolition providing for:

▷ protection of workers
▷ limiting the release of asbestos dust
▷ the safe disposal of waste containing asbestos
▷ consulting with workers or their representatives;

▶ Employers must (where workers' personal clothing may become contaminated with asbestos dust):

▷ provide appropriate working clothing and
▷ make provision for handling and cleaning clothing and other special PPE
▷ prohibit workers taking home working clothes and PPE
▷ provide washing facilities including showers or bath;

▶ Employers must dispose of waste containing asbestos material without exposing workers or the nearby population.

Part IV Surveillance of the working environment and workers health

▶ Employers must:

▷ measure the concentration of airborne asbestos dust;
▷ monitor exposure of workers;
▷ keep records of exposure for periods prescribed by the competent authority;
▷ give workers and their representatives access to the records;
▷ in addition carry out monitoring of the working environment when requested by workers;

▶ Employers must provide, free of charge, medical examinations for exposed workers and inform workers of the results;

▶ The competent authority must provide a system for the notification of occupational disease caused by asbestos.

Part V Information and Education

▶ Competent authority must make arrangements for the dissemination of information and education of all concerned on the health hazards and precautionary measures and ensure that employers have written policies concerning the protection of workers;

▶ Employers must ensure that workers are informed about the health hazards of asbestos and precautionary measures.

R172 – Asbestos Recommendation, 1986 (No. 172)

Recommendation concerning Safety in the Use of Asbestos gives more detail on the Convention requirements and can be found at: http://www.ilo.org/dyn/normlex/en/f?p=NORMLEXPUB:12100:0::NO::P12100_ILO_CODE:R172

The Convention C162 can be found at: http://www.ilo.org/dyn/normlex/en/f?p=1000:12100:0::NO::P12100_ILO_CODE:C162

The ILO CoP Safety in the Use of Asbestos http://www.ilo.org/safework/info/standards-and-instruments/codes/WCMS_107843/lang--en/index.htm

15.1.5 Chemicals Convention C170, 1990

This Convention applies to all branches of economic activity in which chemicals are used. Exclusions are permitted for special problems as long as overall protection is not inferior to the application of the full Convention. Special provisions for confidential information liable to harm competition is also permitted. The Convention does not apply to organisms, but does apply to chemicals derived from organisms.

'Chemicals' means chemical elements and compounds, and mixtures thereof, whether natural or synthetic. *'Hazardous Chemical'* includes any chemical which has been classified as hazardous in accordance with Article 6 or for which information exists to indicate that the chemical is hazardous.

Part II General Principles – includes:

▶ Need to consult employers' and workers' organisations;

▶ Member states should formulate, implement and periodically review a coherent policy on safety in the use of chemicals at work;

▶ Competent authorities must have the powers to prohibit or restrict the use of certain hazardous chemicals, or require advance notification and authorisation before such chemicals are used.

Part III Classification – includes:

▶ Classification systems for all chemicals:

▷ Should be established by the competent authority or a body approved or recognised by the competent authority, in accordance with national or international standards;
▷ Such systems (plus labelling and marking) must take account of the United Nations Recommendations on transport of dangerous goods.

▶ Labelling and Marking:

▷ All chemicals shall be marked so as to indicate their identity;
▷ Hazardous chemicals must be labelled, to be easily understood by workers, so as to provide essential information on classification, hazards and safety precautions.

▶ Chemical Safety Data Sheets:

▷ Must be provided to employers for hazardous chemicals containing detailed information on their identity, supplier, classification, hazards, safety precautions and emergency procedures (more detail is given in R177 Chemicals Recommendation 1990);
▷ Criteria must be established by a competent authority or an approved body;

▷ Chemical or common name used to identify the chemical on the data sheet must be the same as that used on the label.

▶ Responsibilities of suppliers includes ensuring that:
  ▷ Chemicals have been properly classified;
  ▷ Such chemicals have been properly marked;
  ▷ Hazardous chemicals are properly labelled;
  ▷ Chemical safety data sheets are prepared for hazardous chemicals and supplied to employers.
  ▷ Revised labels and safety data sheets are supplied to employers whenever new relevant safety and health information becomes available;
  ▷ Not yet classified chemicals are assessed in order to determine whether they are hazardous chemicals.

Part IV Responsibilities of Employers – includes:

▶ Identification – employers must:
  ▷ Ensure all chemicals used at work are labelled or marked as required by this Convention and where required chemical safety data sheets are made available to workers or their representatives;
  ▷ When receiving unmarked or unlabelled chemicals, obtain the information from the supplier before the chemical is used;
  ▷ Ensure that only properly, identified, assessed and labelled or marked chemicals are used and that any necessary precautions are taken when they are used;
  ▷ Maintain a record of hazardous chemicals used, cross-referenced to the chemical data sheet and made available to employees or their representatives.

▶ Transfer of chemicals – employers must:
  ▷ Ensure when chemicals are transferred into other containers the contents are indicated in a manner which will make known to workers their identity, any hazards and safety precautions necessary.

▶ Exposure – Employers must:
  ▷ Ensure workers are not exposed beyond the permitted exposure limits or other exposure criteria for chemicals being used;
  ▷ Assess workers' exposure to hazardous chemicals;
  ▷ Monitor and record exposure to hazardous chemicals when this is necessary for safety or prescribed by the competent authority;
  ▷ Ensure that records are kept for a period prescribed by the competent authority and are accessible to the workers or their representatives.

▶ Operational control – employers must:
  1. Assess the risks arising from the use of chemicals at work and protect workers against such risks by appropriate means such as:
     (a) the choice of chemicals that eliminate or minimise the risk;
     (b) the choice of technology that eliminates or minimises the risk;
     (c) the use of adequate engineering control measures;
     (d) the adoption of working systems and practices that eliminate or minimise the risk;
     (e) the adoption of adequate occupational hygiene measures;
     (f) where recourse to the above measures does not suffice, the provision and proper maintenance of personal protective equipment and clothing at no cost to the worker, and the implementation of measures to ensure their use.
  2. Employers shall:
     (a) limit exposure to hazardous chemicals so as to protect the safety and health of workers;
     (b) provide first-aid;
     (c) make arrangements to deal with emergencies.

▶ Disposal – employers must ensure that:
  ▷ Unwanted hazardous chemicals and empty containers with residues are properly disposed of so that risks to safety, health or the environment are eliminated or minimised, in accordance with national law and practice;

▶ Information and Training – employers must:
  ▷ inform the workers of the hazards associated with exposure to chemicals used at the workplace;
  ▷ instruct the workers how to obtain and use the information provided on labels and chemical safety data sheets;
  ▷ use the chemical safety data sheets, along with information specific to the workplace, as a basis for the preparation of instructions to workers, which should be written if appropriate;
  ▷ train the workers on a continuing basis in the practices and procedures to be followed for safety in the use of chemicals at work.

▶ Cooperation – employers must:
  ▷ Cooperate as closely as possible with workers or their representatives in respect of the safe use of chemicals at work.

Part V Duties of Workers – workers must:

▶ Cooperate with employers and comply with all procedures and practices relating to the safe use of chemicals at work;

▶ Take all reasonable steps to eliminate or minimise risk to themselves and to others from the use of chemicals at work.

Part VI Rights of workers and their representatives

▶ Workers must have the right to remove themselves from danger resulting from the use of chemicals when they have reasonable justification to believe there is an imminent and serious risk to their safety or health, and shall inform their supervisor immediately;

► Workers who remove themselves in accordance with the last paragraph must be protected against undue consequences;

► Workers and their representatives must have the right to:
  ▷ information on the identity of chemicals used at work, the hazardous properties of such chemicals, precautionary measures, education and training;
  ▷ the information contained in labels and markings;
  ▷ chemical safety data sheets;
  ▷ any other information required to be kept by this Convention;

► Where disclosure of an ingredient may cause harm the identity can be protected when providing the information in a manner approved by the competent authority.

Part VII Responsibility of Exporting States

This part lays down detailed administrative requirements for exporting states under this Convention and any future changes to the Convention.

R177 – Chemicals Recommendation, 1990 (No. 177) sets out recommendations concerning safety in the use of chemicals at work which includes details on the requirements for labels, chemical safety data sheets, monitoring, operational controls and medical surveillance of workers. Details can be found at: http://www.ilo.org/dyn/normlex/en/f?p=1000:12100:0::NO::P12100_ILO_CODE:R177

The Convention C170 can be found at: http://www.ilo.org/dyn/normlex/en/f?p=1000:12100:0::NO::P12100_ILO_CODE:C170

Safety in the Use of Chemicals at Work, International Labour Office (ILO) Code of Practice (CoP), ILO, 1993, ISBN 92-2-108006-4 can be found at: http://www.ilo.org/safework/info/standards-and-instruments/codes/WCMS_107823/lang--en/index.htm

Registration, Evaluation, Authorisation and Restriction of Chemicals (REACH) http://www.hse.gov.uk/reach/

Classification, Labelling and Packaging of Substances and Mixtures Regulations EC No. 1272/2008 http://ec.europa.eu/enterprise/sectors/chemicals/documents/classification

Directive 2006/42/EC – machinery directive http://eur-lex.europa.eu/LexUriServ/LexUriServ.do?uri=OJ:L:2006:157:0024:0086:EN:PDF

15.1.6 Hours of Work and Rest Periods (Road Transport) 1979 (No. 161): C153

► This Convention concerns the Hours of Work and Rest Periods in Road Transport. It applies to wage-earning drivers working, whether for undertakings engaged in transport for third parties or for undertakings transporting goods or passengers for own account, on motor vehicles engaged professionally in the internal or international transport by road of goods or passengers.

► Except as otherwise provided herein, this Convention further applies to owners of motor vehicles engaged professionally in road transport and non-wage-earning members of their families, when they are working as drivers.

► The competent authorities can exempt, urban transport, transport for: agriculture and forestry operations; sick and injured people; rescue, salvage or fire-fighting; national defence and police services; taxis, others where speeds or routes are restricted.

► The competent authority shall lay down adequate standards for driving times and rest periods.

► Employers' and workers' organisations must be consulted.

► For the purpose of this Convention the term hours of work means the time spent by wage-earning drivers on:
  (a) driving and other work during the running time of the vehicle; and
  (b) subsidiary work in connection with the vehicle, its passengers or its load.

► Periods of mere attendance or stand-by, either on the vehicle or at the workplace and during which the drivers are not free to dispose of their time as they please, may be regarded as hours of work.

► No driver shall be allowed to drive continuously for more than four hours without a break.

► The competent authority, taking into account particular national conditions, may authorise the period referred to in the previous paragraph to be exceeded by not more than one hour.

► The length of the break and, as appropriate, the way in which the break may be split shall be determined by the competent authority.

► The competent authority may specify cases in which the provisions are inapplicable because drivers have sufficient breaks as a result of stops provided for in the timetable or as a result of the intermittent nature of the work.

► The maximum total driving time, including overtime, shall exceed neither nine hours per day nor 48 hours per week. The total driving times may be calculated as an average over a number of days or weeks to be determined by the competent authority.

► Every wage-earning driver shall be entitled to a break after a continuous period of five hours of work. The length of the break and, as appropriate, the way in which the break may be split shall be determined by the competent authority.

► The daily rest of drivers shall be at least ten consecutive hours during any 24-hour period starting from the beginning of the working day. The daily rest may be calculated as an average over periods to be

15

determined by the competent authority: Provided that the daily rest shall in no case be less than eight hours and shall not be reduced to eight hours more than twice a week.

▶ The competent authority may provide for exceptions to the provisions as regards the duration of the daily rest periods and the manner of taking such rest periods in the cases of vehicles having a crew of two drivers and of vehicles using a ferryboat or a train.

▶ During the daily rest the driver shall not be required to remain in or near the vehicle if they have taken the necessary precautions to ensure the safety of the vehicle and its load.

▶ The competent authority shall:
  (a) provide for an individual control book and prescribe the conditions of its issue, its contents and the manner in which it shall be kept by the drivers; and
  (b) lay down a procedure for notification of the hours worked in accordance with this Convention and the circumstances justifying them.

▶ Each employer shall:
  (a) keep a record, in a form approved by the competent authority, indicating the hours of work and of rest of every driver employed by him; and
  (b) place this record at the disposal of the supervisory authorities in a manner determined by the competent authority.

▶ The competent authority shall make provision for:
  (a) an adequate inspection system, with verification carried out in the undertaking and on the roads; and
  (b) appropriate penalties in the event of breaches of the requirements of this Convention.

Details of the full Convention C153 Hours of Work and Rest Periods (Road Transport) 1979 (No. 161) can be found at: http://www.ilo.org/dyn/normlex/en/f?p=NORMLEXPUB:12100:0::NO:12100:P12100_INSTRUMENT_ID:312499:NO

15.1.7 Hygiene (Commerce and Offices) Convention, C120, 1964

This Convention concerns hygiene in commerce and offices.

It applies to:

▶ trading establishments;
▶ establishments, institutions and administrative services in which the workers are mainly engaged in office work;
▶ in so far as they are not subject to national legislation or other arrangements concerning hygiene in industry, mines, transport or agriculture, any departments of other establishments, institutions, or administrative services in which departments the

workers are mainly engaged in commerce or office work.

Member states shall:

▶ give effect to the Convention through national legislation;
▶ provide adequate inspection to ensure compliance;
▶ introduce penalties to enforce the legislation.

Part II General Provisions

General requirements include the requirement to ensure that:

▶ All premises used by workers, and the equipment of such premises, shall be properly maintained and kept clean.
▶ All premises used by workers shall have sufficient and suitable ventilation, natural or artificial or both, supplying fresh or purified air.
▶ All premises used by workers shall have sufficient and suitable lighting; workplaces shall, as far as possible, have natural lighting.
▶ As comfortable and steady a temperature as circumstances permit shall be maintained in all premises used by workers.
▶ All workplaces shall be so laid out and workstations so arranged that there is no harmful effect on the health of the worker.
▶ A sufficient supply of wholesome drinking water or of some other wholesome drink shall be made available to workers.
▶ Sufficient and suitable washing facilities and sanitary conveniences shall be provided and properly maintained.
▶ Sufficient and suitable seats shall be supplied for workers and workers shall be given reasonable opportunities of using them.
▶ Suitable facilities for changing, leaving and drying clothing which is not worn at work shall be provided and properly maintained.
▶ Underground or windowless premises in which work is normally performed shall comply with appropriate standards of hygiene.
▶ Workers shall be protected by appropriate and practicable measures against substances, processes and techniques which are obnoxious, unhealthy or toxic or for any reason harmful. Where the nature of the work so requires, the competent authority shall prescribe personal protective equipment.
▶ Noise and vibrations likely to have harmful effects on workers shall be reduced as far as possible by appropriate and practicable measures.
▶ Every establishment, institution or administrative service, or department thereof, to which this Convention applies shall, having regard to its size and the possible risk:
  (a) maintain its own dispensary or first-aid post; or
  (b) maintain a dispensary or first-aid post jointly with other establishments, institutions or

administrative services, or departments thereof; or

(c) have one or more first-aid cupboards, boxes or kits.

The full text of the Hygiene (Commerce and Offices), ILO Convention, 1964 (No 120) – C120 can be found at: http://www.ilo.org/dyn/normlex/en/f?p=1000:12100:0::NO::P12100_ILO_CODE:C120

15.1.8 Occupational Cancer Convention C139, 1974

This is a Convention concerning prevention and control of occupational hazards caused by carcinogenic substances and agents.

Member states shall:

▶ periodically determine the carcinogenic substances and agents to which occupational exposure shall be prohibited or made subject to authorisation or control;

▶ make every effort to have carcinogenic substances and agents to which workers may be exposed in the course of their work replaced by non-carcinogenic substances or agents or by less harmful substances or agents; in the choice of substitute substances or agents account shall be taken of their carcinogenic, toxic and other properties;

▶ ensure that the number of workers exposed to carcinogenic substances or agents and the duration and degree of such exposure shall be reduced to the minimum compatible with safety;

▶ prescribe the measures to be taken to protect workers against the risks of exposure to carcinogenic substances or agents and shall ensure the establishment of an appropriate system of records;

▶ take steps so that workers who have been, are, or are likely to be exposed to carcinogenic substances or agents are provided with all the available information on the dangers involved and on the measures to be taken;

▶ take measures to ensure that workers are provided with such medical examinations or biological or other tests or investigations during the period of employment and thereafter as are necessary to evaluate their exposure and supervise their state of health in relation to the occupational hazards;

▶ by legislation or other method and in consultation with employers' and workers' organisations take steps to realise this Convention, specify who is responsible for compliance, and organise inspection services to supervise compliance.

More details are given in Occupational Cancer Recommendation, R147, 1974 which can be found at: http://www.ilo.org/dyn/normlex/en/f?p=1000:12100:0::NO::P12100_ILO_CODE:R147

The full text of the Occupational Cancer Convention C139, 1974 can be found at: http://www.ilo.org/dyn/normlex/en/f?p=NORMLEXPUB:12100:0::NO::P12100_INSTRUMENT_ID:312284

15.1.9 Occupational Health Services, Convention, C161, 1981

This is a convention concerning occupational health services.

Part I Principles of National Policy

▶ For the purposes of this Convention:
  ▷ the term *occupational health services* means services entrusted with essentially preventive functions and responsible for advising the employer, the workers and their representatives in the undertaking on:
    (i) the requirements for establishing and maintaining a safe and healthy working environment which will facilitate optimal physical and mental health in relation to work;
    (ii) the adaptation of work to the capabilities of workers in the light of their state of physical and mental health.

Each member state shall:

▶ formulate, implement and periodically review a coherent national policy on occupational health services in consultation with employers' and workers' organisations;

▶ develop progressively occupational health services for all workers;

▶ if services cannot be immediately established draw up plans for their establishment;

▶ ensure that the competent authority consults employers' and workers' organisations on the measures taken to give effect to this Convention.

Part II Functions

Without prejudice to employer responsibilities and having regard to the necessity for workers participation, occupational health services shall have the following functions:

▶ identification and assessment of the risks from health hazards in the workplace;

▶ surveillance of the factors in the working environment and working practices which may affect workers' health, including sanitary installations, canteens and housing where these facilities are provided by the employer;

▶ advice on planning and organisation of work, including the design of workplaces, on the choice, maintenance and condition of machinery and other equipment and on substances used in work;

▶ participation in the development of programmes for the improvement of working practices as well as testing and evaluation of health aspects of new equipment;

15

▶ advice on occupational health, safety and hygiene and on ergonomics and individual and collective protective equipment;

▶ surveillance of workers' health in relation to work;

▶ promoting the adaptation of work to the worker;

▶ contribution to measures of vocational rehabilitation;

▶ collaboration in providing information, training and education in the fields of occupational health and hygiene and ergonomics;

▶ organising of first-aid and emergency treatment;

▶ participation in analysis of occupational accidents and occupational diseases.

Part III Organisation

Provision shall be made for the establishment of occupational health services by legislation or by collective agreement or any other manner approved by the competent authority. Employers and workers shall cooperate and participate in the implementation of occupational health services on an equitable basis.

Part IV Conditions of Operations

▶ Occupational health services shall be multidiscipline and cooperate with other bodies concerned with health services. Personnel shall enjoy full professional independence from employers, workers and their representatives;

▶ Qualifications of personnel to be determined by the competent authority;

▶ Surveillance of workers shall involve no loss of earnings to them and shall be free of charge and take place as far as possible during working hours;

▶ All workers shall be informed of health hazards in their work;

▶ Occupational health services shall be informed by the employer and workers of any known and suspected factors which might affect the workers' health;

▶ Occupational health services shall be informed of the occurrences of ill-health and absence from work for health reasons, but are not required to verify the reasons for absence.

Recommendations concerning Occupational health services R171 give more details of the requirements of the convention and can be found at: http://www.ilo.org/dyn/normlex/en/f?p=1000:12100:0::NO::P12100_ILO_CODE:R171

The Convention on Occupational Health Services can be found at: http://www.ilo.org/dyn/normlex/en/f?p=NORMLEXPUB:12100:0::NO:12100:P12100_ILO_CODE:C161

15.1.10 Radiation Protection, Convention C115, 1960

This Convention concerns the protection of workers against ionising radiation.

Part I General Provisions

Member states are required to produce legislation, codes of practice or other appropriate means to put this Convention into effect. It applies to all activities involving exposure of workers to ionising radiation in the course of their work. It does not apply to radioactive substances whether sealed or unsealed, nor equipment that generates ionising radiation, which are exempted due to the limited doses of ionising radiation which can be received from them.

The Convention requires that all appropriate steps are taken to protect workers against ionising radiation. Rules and precautions shall be adopted and data essential for effective protection must be made available.

Part II Protective Measures – include the following:

▶ every effort to restrict exposure of workers to ionising radiation to the lowest practicable level, and any unnecessary exposure shall be avoided by all parties concerned;

▶ maximum permissible doses of ionising radiations which may be received from sources external to or internal to the body and maximum permissible amounts of radioactive substances which can be taken into the body shall be fixed for various categories of workers. (see Radiation protection of workers, ionising radiations, ILO CoP, 1987 for international exposure limits). Limits must be kept under constant review;

▶ appropriate levels shall be fixed for workers who are directly engaged in radiation work and are:
  (a) aged 18 and over;
  (b) under the age of 18;

▶ no worker under the age of 16 shall be engaged in work involving ionising radiations;

▶ appropriate levels shall be fixed for workers who are not directly engaged in radiation work, but who remain or pass where they may be exposed to ionising radiations or radioactive substances;

▶ appropriate warnings shall be used to indicate the presence of hazards from ionising radiations. Any information necessary in this connection shall be supplied to the workers;

▶ all workers directly engaged in radiation work shall be adequately instructed, before and during such employment, in the precautions to be taken for their protection;

▶ the notification in a prescribed manner of work involving exposure of workers to ionising radiations;

▶ monitoring of workers and places of work shall be carried out in order to measure the exposure of workers to ionising radiations and radioactive substances;

▶ all workers directly engaged in radiation work shall undergo an appropriate medical examination prior to or shortly after taking up such work and subsequently undergo further periodic medical examinations;

▶ circumstances shall be specified, in which, because of the nature and/or degree of the exposure, the following action shall be taken promptly:

▷ the worker shall undergo an appropriate medical examination;

▷ the employer shall notify the competent authority in accordance with its requirements;

▷ persons competent in radiation protection shall examine the conditions in which the worker's duties are performed;

▷ the employer shall take any necessary remedial action on the basis of the technical findings and the medical advice;

▶ no worker shall be employed or shall continue to be employed in work where they could be subject to exposure to ionising radiations contrary to qualified medical advice;

▶ provision of appropriate inspection services to supervise implementation of requirements.

15.1.11 Safety and Health in Construction Convention, C167, 1988

This Convention concerns safety and health in construction work. It applies to all construction activities, namely building, civil engineering, and erection and dismantling work, including any process, operation or transport on a construction site, from the preparation of the site to the completion of the project.

For the purpose of this Convention:

▶ The term construction covers:

(i) building, including excavation and the construction, structural alteration, renovation, repair, maintenance (including cleaning and painting) and demolition of all types of buildings or structures;

(ii) civil engineering, including excavation and the construction, structural alteration, repair, maintenance and demolition of, for example, airports, docks, harbours, inland waterways, dams, river and avalanche and sea defence works, roads and highways, railways, bridges, tunnels, viaducts and works related to the provision of services such as communications, drainage, sewerage, water and energy supplies;

(iii) the erection and dismantling of prefabricated buildings and structures, as well as the manufacturing of prefabricated elements on the construction site;

▶ The term construction site means any site at which any of the processes or operations described in the subparagraph above are carried on.

Part II General Provisions – include:

▶ consultation between employers' and workers' organisations;

▶ adoption and maintenance of legislation, practical standards or codes of practice or other appropriate method of giving effect to the Convention with due regard to recognised international standards;

▶ measures to be taken to ensure cooperation between employers and workers;

▶ responsibility for compliance to be placed on employers and the self-employed;

▶ whenever two or more employers undertake activities simultaneously at one construction site:

▷ the principal contractor, or other person or body with actual control over or primary responsibility for overall construction site activities, shall be responsible for coordinating the prescribed safety and health measures and for ensuring compliance with national legislation;

▷ where the principal contractor, or other person or body, is not present at the site, they shall nominate a competent person or body at the site with the authority and means necessary to ensure on his behalf coordination and compliance with requirements;

▷ each employer shall remain responsible for the application of the prescribed measures in respect of the workers placed under his authority;

▶ whenever employers or self-employed persons undertake activities simultaneously at one construction site they shall have the duty to cooperate in the application of the prescribed safety and health measures;

▶ those concerned with the design and planning of a construction project shall take into account the safety and health of the construction workers in accordance with national legislation and practice;

▶ workers shall have the right and the duty at any workplace to participate in ensuring safe working conditions and to express views on the working procedures adopted as they may affect safety and health;

▶ workers shall have the duty to:

▷ cooperate as closely as possible with their employer in the application of the prescribed safety and health measures;

▷ take reasonable care for their own safety and health and that of other persons who may be affected by their acts or omissions at work;

▷ use facilities placed at their disposal and not misuse anything provided for their own protection or the protection of others;

▷ report forthwith to their immediate supervisor, and to the workers' safety representative where one exists, any situation which they believe could present a risk, and which they cannot properly deal with themselves;

▷ comply with the prescribed safety and health measures;

▶ a worker shall have the right to remove himself from danger when he has good reason to believe that there is an imminent and serious danger to his safety or health, and the duty to inform his supervisor immediately;

15

461

- where there is an imminent danger to the safety of workers the employer shall take immediate steps to stop the operation and evacuate workers as appropriate.

Part III Preventive and Protective Measures – these cover:

- Safety of Workplaces
  - ▷ All appropriate precautions shall be taken to ensure that all workplaces are safe and without risk of injury to the safety and health of workers.
  - ▷ Safe means of access to and egress from all workplaces shall be provided and maintained, and indicated where appropriate.
  - ▷ All appropriate precautions shall be taken to protect persons present at or in the vicinity of a construction site from all risks which may arise from such site.

- Scaffolds and Ladders
  - ▷ Where work cannot safely be done on or from the ground or from part of a building or other permanent structure, a safe and suitable scaffold shall be provided and maintained, or other equally safe and suitable provision shall be made.
  - ▷ In the absence of alternative safe means of access to elevated working places, suitable and sound ladders shall be provided. They shall be property secured against inadvertent movement.
  - ▷ All scaffolds and ladders shall be constructed and used in accordance with national legislation.
  - ▷ Scaffolds shall be inspected by a competent person in such cases and at such times as shall be prescribed by national legislation.

- Lifting Appliances and Gear
  - ▷ Every lifting appliance and item of lifting gear, including their constituent elements, attachments, anchorages and supports, shall:
    - (a) be of good design and construction, sound material and adequate strength for the purpose for which they are used;
    - (b) be properly installed and used;
    - (c) be maintained in good working order;
    - (d) be examined and tested by a competent person at such times and in such cases as shall be prescribed by national legislation; the results of these examinations and tests shall be recorded;
    - (e) be operated by workers who have received appropriate training in accordance with national legislation.
  - ▷ No person shall be raised, lowered or carried by a lifting appliance unless it is constructed, installed and used for that purpose in accordance with national legislation, except in an emergency situation in which serious personal injury or fatality may occur, and for which the lifting appliance can be safely used.

- Transport, Earth-moving and Materials-handling Equipment

- ▷ All vehicles and earth-moving or materials-handling equipment shall:
  - (a) be of good design and construction taking into account as far as possible ergonomic principles;
  - (b) be maintained in good working order;
  - (c) be properly used;
  - (d) be operated by workers who have received appropriate training in accordance with national legislation.
- ▷ On all construction sites on which vehicles, earth-moving or materials-handling equipment are used:
  - (a) safe and suitable access ways shall be provided for them; and
  - (b) traffic shall be so organised and controlled as to secure their safe operation.

- Plant, Machinery, Equipment and Hand Tools
  - ▷ Plant, machinery and equipment, including hand tools, both manual and power driven, shall:
    - (a) be of good design and construction, taking into account as far as possible ergonomic principles;
    - (b) be maintained in good working order;
    - (c) be used only for work for which they have been designed unless a use outside the initial design purposes has been assessed by a competent person who has concluded that such use is safe;
    - (d) be operated by workers who have received appropriate training.
  - ▷ Adequate instructions for safe use shall be provided where appropriate by the manufacturer or the employer, in a form understood by the users.
  - ▷ Pressure plant and equipment shall be examined and tested by a competent person in cases and at times prescribed by national legislation.

- Work at Heights including Roofwork
  - ▷ Where necessary to guard against danger, or where the height of a structure or its slope exceeds that prescribed by national legislation, preventive measures shall be taken against the fall of workers and tools or other objects or materials.
  - ▷ Where workers are required to work on or near roofs or other places covered with fragile material, through which they are liable to fall, preventive measures shall be taken against their inadvertently stepping on or falling through the fragile material.

- Excavations, Shafts, Earthworks, Underground Works and Tunnels
  - ▷ Adequate precautions shall be taken in any excavation, shaft, earthworks, underground works or tunnel:
    - (a) by suitable shoring or otherwise to guard against danger to workers from a fall

or dislodgement of earth, rock or other material;

(b) to guard against dangers arising from the fall of persons, materials or objects or the inrush of water into the excavation, shaft, earthworks, underground works or tunnel;

(c) to secure adequate ventilation at every workplace so as to maintain an atmosphere fit for respiration and to limit any fumes, gases, vapours, dust or other impurities to levels which are not dangerous or injurious to health and are within limits laid down by national legislation;

(d) to enable the workers to reach safety in the event of fire, or an inrush of water or material;

(e) to avoid risk to workers arising from possible underground dangers such as the circulation of fluids or the presence of pockets of gas, by undertaking appropriate investigations to locate them.

▶ Cofferdams and Caissons
  ▷ Every cofferdam and caisson shall be:
    (a) of good construction and suitable and sound material and of adequate strength;
    (b) provided with adequate means for workers to reach safety in the event of an inrush of water or material.
  ▷ The construction, positioning, modification or dismantling of a cofferdam or caisson shall take place only under the immediate supervision of a competent person.
  ▷ Every cofferdam and caisson shall be inspected by a competent person at prescribed intervals.

▶ Work in Compressed Air
  ▷ Work in compressed air shall be carried out only in accordance with measures prescribed by national legislation.
  ▷ Work in compressed air shall be carried out only by workers whose physical aptitude for such work has been established by a medical examination and when a competent person is present to supervise the conduct of the operations.

▶ Structural Frames and Formwork
  ▷ The erection of structural frames and components, formwork, falsework and shoring shall be carried out only under the supervision of a competent person.
  ▷ Adequate precautions shall be taken to guard against danger to workers arising from any temporary state of weakness or instability of a structure.
  ▷ Formwork, falsework and shoring shall be so designed, constructed and maintained that it will safely support all loads that may be imposed on it.

▶ Work Over Water

▶ Where work is done over or in close proximity to water there shall be adequate provision for:
  (a) preventing workers from falling into water;
  (b) the rescue of workers in danger of drowning;
  (c) safe and sufficient transport.

▶ Demolition
  ▷ When the demolition of any building or structure might present danger to workers or to the public:
    (a) appropriate precautions, methods and procedures shall be adopted, including those for the disposal of waste or residues, in accordance with national laws or regulations;
    (b) the work shall be planned and undertaken only under the supervision of a competent person.

▶ Lighting
  ▷ Adequate and suitable lighting, including portable lighting where appropriate, shall be provided at every workplace and any other place on the construction site where a worker may have to pass.

▶ Electricity
  ▷ All electrical equipment and installations shall be constructed, installed and maintained by a competent person, and so used as to guard against danger.
  ▷ Before construction is commenced and during the progress thereof adequate steps shall be taken to ascertain the presence of and to guard against danger to workers from any live electrical cable or apparatus which is under, over or on the site.
  ▷ The laying and maintenance of electrical cables and apparatus on construction sites shall be governed by the technical rules and standards applied at the national level.

▶ Explosives
  ▷ Explosives shall not be stored, transported, handled or used except:
    (a) under conditions prescribed by national legislation; and
    (b) by a competent person, who shall take such steps as are necessary to ensure that workers and other persons are not exposed to risk of injury.

▶ Health Hazards
  ▷ Where a worker is liable to be exposed to any chemical, physical or biological hazard to such an extent as is liable to be dangerous to health, appropriate preventive measures shall be taken against such exposure.
  ▷ The preventive measures referred to in the paragraph above shall comprise:
    (a) the replacement of hazardous substances by harmless or less hazardous substances wherever possible; or
    (b) technical measures applied to the plant, machinery, equipment or process; or

15

(c) where it is not possible to comply with subparagraphs (a) or (b) above, other effective measures, including the use of personal protective equipment and protective clothing.

▷ Where workers are required to enter any area in which a toxic or harmful substance may be present, or in which there may be an oxygen deficiency, or a flammable atmosphere, adequate measures shall be taken to guard against danger.

▷ Waste shall not be destroyed or otherwise disposed of on a construction site in a manner which is liable to be injurious to health.

▶ Fire Precautions

▷ The employer shall take all appropriate measures to:
(a) avoid the risk of fire;
(b) combat quickly and efficiently any outbreak of fire;
(c) bring about a quick and safe evacuation of persons.

▷ Sufficient and suitable storage shall be provided for flammable liquids, solids and gases.

▶ Personal Protective Equipment and Protective Clothing

▷ Where adequate protection against risk of accident or injury to health, including exposure to adverse conditions, cannot be ensured by other means, suitable personal protective equipment and protective clothing, having regard to the type of work and risks, shall be provided and maintained by the employer, without cost to the workers, as may be prescribed by national legislation.

▷ The employer shall provide the workers with the appropriate means to enable them to use the individual protective equipment, and shall ensure its proper use.

▷ Protective equipment and protective clothing shall comply with standards set by the competent authority taking into account as far as possible ergonomic principles.

▷ Workers shall be required to make proper use of and to take good care of the personal protective equipment and protective clothing provided for their use.

▶ First-aid

▷ The employer shall be responsible for ensuring that first-aid, including trained personnel, is available at all times. Arrangements shall be made for ensuring the removal for medical attention of workers who have suffered an accident or sudden illness.

▶ Welfare

▷ At or within reasonable access of every construction site an adequate supply of wholesome drinking water shall be provided.

▷ At or within reasonable access of every construction site, the following facilities shall, depending on the number of workers and the duration of the work, be provided and maintained:
(a) sanitary and washing facilities;
(b) facilities for changing and for the storage and drying of clothing;
(c) accommodation for taking meals and for taking shelter during interruption of work due to adverse weather conditions.

▷ Men and women workers should be provided with separate sanitary and washing facilities.

▶ Information and Training

▷ Workers shall be adequately and suitably:
(a) informed of potential safety and health hazards to which they may be exposed at their workplace;
(b) instructed and trained in the measures available for the prevention and control of, and protection against, those hazards.

▶ Reporting of Accidents and Diseases

▷ National legislation shall provide for the reporting to the competent authority within a prescribed time of occupational accidents and diseases.

Part V Implementation

▶ Each Member state shall:
(a) take all necessary measures, including the provision of appropriate penalties and corrective measures, to ensure the effective enforcement of the provisions of the Convention;
(b) provide appropriate inspection services to supervise the application of the measures to be taken in pursuance of the Convention and provide these services with the resources necessary for the accomplishment of their task, or satisfy itself that appropriate inspection is carried out.

More detail on the requirements are given in the Safety and Health in Construction Recommendations, 1988, R175, which can be found at: http://www.ilo.org/dyn/normlex/en/f?p=1000:12100:0::NO::P12100_ILO_CODE:R175

The full Convention C167 can be found at: http://www.ilo.org/dyn/normlex/en/f?p=1000:12100:0::NO::P12100_ILO_CODE:C167

For general guidance see Safety and Health in Construction, ILO CoP, ILO Geneva, 1992, ISBN 92-2-107104-9 which can be found at: http://www.ilo.org/safework/info/standards-and-instruments/codes/WCMS_107826/lang--en/index.htm

15.1.12 Safety and health in the use of machinery, ILO Code of Practice

This 2013 ILO Code of Practice sets out principles concerning safety and health in the use of machinery and defines safety and health requirements and precautions applicable to governments, workers and employers, and also to designers, manufacturers and suppliers of machinery.

The code is not intended to replace national laws, regulations or accepted standards. Its object is to provide guidance to those who may be engaged in the framing of provisions relating to the use of machinery at work, such as competent authorities and the management of companies where machinery is supplied or used. The code also offers guidelines to designers, manufacturers, suppliers and employers' and workers' organisations.

Part I of the code sets out the scope, objectives, hierarchy of controls and definitions, as well as the general obligations, responsibilities and duties of the competent authority, designers and manufacturers, suppliers and employers, workers and their organisations. Part II deals with technical requirements and specific measures that should be taken in order to protect workers' safety and health.

This summary will cover Part I of the code only. For the full text of Part I and Part II go to Safety and health in the use of machinery, ILO CoP http://www.ilo.org/wcmsp5/groups/public/---ed_protect/---protrav/---safework/documents/normativeinstrument/wcms_164653.pdf

Also see: Safety of machinery – General principles for design – Risk assessment and risk reduction, ISO 12100:2010, ISBN 978-0-580-74262-0 http://www.iso.org/iso/catalogue_detail?csnumber=51528

ISO 12100:2010 specifies basic terminology, principles and a methodology for achieving safety in the design of machinery. It specifies principles of risk assessment and risk reduction to help designers in achieving this objective. These principles are based on knowledge and experience of the design, use, incidents, accidents and risks associated with machinery. Procedures are described for identifying hazards and estimating and evaluating risks during relevant phases of the machine life cycle, and for the elimination of hazards or sufficient risk reduction. Guidance is given on the documentation and verification of the risk assessment and risk reduction process.

Part I General Requirements

1. General provisions

▶ Scope and application:
  ▷ This code applies to any work activity in which machinery is used.
  ▷ The code is intended to apply generally to the design, manufacture, supply and use of machinery for use at work. It does not take into account the particular specificities relating to certain categories of machinery, such as weapons, pressure vessels, medical devices, seagoing vessels, vehicles and trailers solely for transportation of passengers by rail, road, air or

water, machinery for military use and household appliances for domestic use, which are typically covered by special legislation at the national level.
  ▷ This code applies to all stages of the life cycle of the machinery, including second-hand, rebuilt, modified or redeployed machinery for use at work.

▶ Objectives achieved by:
  ▷ ensuring that all machinery for use at work is designed and manufactured to eliminate or minimise the hazards associated with its use;
  ▷ ensuring that employers are provided with a mechanism for obtaining from their suppliers necessary and sufficient safety information about machinery to enable them to implement effective protective measures for workers; and
  ▷ ensuring that proper workplace safety and health measures are implemented to identify, eliminate, prevent and control risks arising from the use of machinery.

▶ Hierarchy of controls:
  ▷ The approach most commonly used is referred to as the hierarchy of controls, from preferred to least desirable, as follows:
    (a) elimination;
    (b) substitution;
    (c) engineering controls;
    (d) administrative (procedural) controls; and
    (e) personal protective equipment (PPE).

▶ Definitions include:
  ▷ *Guard:* A part of machinery specifically designed to provide protection by means of a physical barrier.
  ▷ *Life cycle*: All phases of the life of machinery, i.e.:
    (a) transport, assembly and installation;
    (b) commissioning;
    (c) use; and
    (d) decommissioning, dismantling and disposal.
  ▷ *Machinery*: An assembly fitted with, or intended to be fitted with, a drive system other than one using only directly applied human or animal effort, consisting of linked parts or components, at least one of which moves, and which are joined together for a specific application.
  ▷ *Maintenance*: Workplace activities such as constructing, installing, setting up, testing, adjusting, inspecting, modifying, and maintaining machinery on a preventive, periodic and predictive basis. These activities include lubrication, cleaning or unjamming of machinery and making adjustments or tool changes where a worker may be exposed to the unexpected energisation or startup of the machinery or release of hazardous stored energy.
  ▷ *Protective device*: A safeguard other than a guard which reduces risk, either alone or in conjunction with a guard.

15

2. General obligations, responsibilities and duties
▶ Competent authority should:
  ▷ formulate, implement and periodically review a coherent national policy on safety in the use of machinery;
  ▷ establish and from time to time review laws, regulations and standards for safety in the use of machinery;
  ▷ establish mechanisms to ensure compliance with national laws and regulations. These should include an adequate and appropriate system of risk-based inspection;
  ▷ where appropriate, require designers, manufacturers and suppliers to provide it with safety and health-related information on the assessment of the hazards and risks associated with machinery;
  ▷ ensure that guidance is provided to employers, workers and their representatives to help them comply with their legal obligations under the policy and provide assistance to all parties as necessary;
  ▷ ensure that machinery on the market satisfies the legal OSH requirements;
  ▷ endeavour to promote close cooperation between designers, manufacturers, suppliers, employers, workers and their representatives;
  ▷ establish, apply, and periodically review a system for the recording and notification by employers of occupational accidents, occupational diseases and dangerous occurrences caused by machinery;
  ▷ have a system for investigating occupational accidents, diseases and dangerous occurrences as appropriate;
  ▷ in accordance with national law and practice:
    (a) periodically carry out inspections and monitor compliance with relevant legislation;
    (b) inform employers, workers and their representatives of the findings of inspections for the implementation of required remedial action;
    (c) have the authority to issue an order to stop the use of machinery in situations where there is an imminent or serious danger to the safety or health of workers; and
    (d) produce and update safety guidance, where appropriate, in cooperation with the representative organisations of employers and workers;
  ▷ have sufficient human and financial resources to fulfil its responsibilities.
▶ Responsibilities of designers and manufacturers – they should:
  ▷ design machinery to be inherently safe so that hazards are eliminated;
  ▷ where the above is not possible, ensure that adequate technical protective measures are provided, so that safety and health risks are reduced to the lowest practical level, using the hierarchy of controls referring to section 3.4 of the code relating to ergonomics;
  ▷ ensure that machinery meets legal requirements;
  ▷ that the relevant certification, markings or documentation are available in accordance with national law and practice;
  ▷ provide instructions for installation and use of the machinery, including the information needed by the employer, operator and users of the machinery for its safe operation and maintenance;
  ▷ monitor and study any reports of malfunctions, dangerous occurrences and accidents and diseases involving the actual machinery in question or similar machinery, and any remedial measures that have been taken to control unacceptable risks that have been identified, in order to prevent recurrences. Manufacturers and designers should use the collected information on accidents and diseases to improve the safety of machinery. Manufacturers should inform designers and customers of serious defects which affect safety and health they have identified in the design or use of the machinery and the action they should take. This action could include product recall;
  ▷ carry out a frequently repeated process of risk assessment and risk reduction as part of the design process. This should cover the full range of uses and reasonably foreseeable misuses, hazards and their elimination as is reasonably practicable, estimate the risks and if they are adequately controlled or further protective measures are needed;
  ▷ ensure that machinery they produce for the workplace complies with the requirements set out in the relevant sections of Part II of the code or other corresponding international or national standards and recommendations;
  ▷ ensure that machinery is designed and constructed in such a way that it fits the purpose for which it is intended. It should be operated, adjusted and maintained without putting persons at risk during its operation under foreseeable conditions, but also taking into account any reasonably foreseeable misuse;
  ▷ ensure in the design process that consideration is given to measures for eliminating or reducing any reasonably foreseeable risk during transporting, installing, dismantling, disabling and scrapping of machinery;
  ▷ ensure that machinery is designed and constructed to take into account possible constraints to which the operator may be subject as a result of the necessary or anticipated use of personal protective equipment (PPE);

▷ ensure that machinery is supplied with all the special equipment and accessories essential to enable it to be adjusted, maintained and used safely;

▷ with any information and warnings that are essential for its safe use. Information and warnings on the machinery should preferably be provided in the form of readily understandable symbols or pictograms;

▷ provide instructions for the safe use of the machinery. The code at 2.2.14.1 lays down the detail for these instructions.

▶ General responsibilities of suppliers:

▷ Suppliers should ensure that machinery they supply:

(a) fulfils the safety requirements of the country or market in which the machinery is put into service;

(b) is accompanied by instructions for use in the language or languages of the country or market in which it is put into service; and

(c) is marked in accordance with the relevant national law and practice in force where the machinery is used;

▷ Suppliers should ensure that relevant new OSH information that becomes available for the machinery they supply is passed on to their customers where practical;

▷ Where suppliers assemble machinery prior to or on delivery, they should ensure that guards and protective devices are not damaged or missing;

▷ When second-hand machinery is sold, suppliers should ensure that the machinery is safe and meets the technical requirements set out in this code, and national legislation in the country to which it is being supplied.

▶ General responsibilities of employers who should:

▷ only select machinery after carefully taking into account all factors affecting OSH and working conditions, in addition to economic and technical criteria;

▷ buy machinery for use in the workplace only if it complies with national legislation and relevant international standards;

▷ ensure that when second-hand machinery is introduced at the workplace it is safe and meets the technical requirements determined by national legislation;

▷ ensure that, where the safety of machinery depends on the installation, it should be subject to an initial inspection (after installation and before first being put into service). It should be inspected if moved to a new site or location;

▷ ensure that machinery is safe through regular inspection by a competent person;

▷ ensure that machinery exposed to exceptional conditions such as accidents, adverse weather or prolonged periods of inactivity likely to affect the safety of the machinery, is subject to special inspections by competent persons;

▷ record the results of inspections, where appropriate, and use them for improving safety in the use of machinery. The record should be kept for a suitable period of time;

▷ ensure that when machinery is hired or moved around from one workplace to another and, where there are national requirements to that effect, it is accompanied by relevant documentation indicating that a recent inspection has been carried out;

▷ before the machinery is put into service, make sure that they understand all the instructions provided. On the basis of that information, they should assess the risks arising from actual situations in which the machinery is used, taking into account work materials, the placement of the machinery in the work area, operating procedures, organisation of work at the workplace, workers' capabilities and the overall working environment;

▷ ensure that the machinery they use complies with the requirements set out in relevant sections of Part II of the code, or other corresponding international or national standards and recommendations;

▷ reassess the risks arising from the use of existing machinery periodically, whenever modifications are made, or if work conditions change significantly, taking into account the information provided by the manufacturer and supplier;

▷ take appropriate measures to protect workers against the risks identified by the assessment by following the hierarchy of controls in the code;

▷ implement measures to ensure that PPE, where it is necessary, is available, used and stored and maintained safely and in good working order. Where PPE is necessary to protect the safety and health of workers, it should be fit-for-purpose, suited to the individual and provided at no cost to workers. Workers should be consulted in its selection and trained in its use;

▷ continuously monitor the safety of the machinery, including any changes in the working environment and organisation of work;

▷ undertake an ergonomic risk assessment (see Appendix V of the code) of machinery use to ensure that the safety and health protection of workers is optimised during the work task process;

▷ establish appropriate recording systems relating to safety and health in the use of machinery and document the relevant information on matters such as significant safety and health hazards and risks arising from machinery used in the workplace, the arrangements for prevention and control, and details of any dangerous occurrences

15

or accidents that occur. The records should be available to workers and periodically reviewed;

▷ consider creating documented work methods for machinery identified as high risk following the risk assessment. This could include, but should not be limited to:

   (a) safe operating procedures (SOPs);

   (b) job safety analysis (JSA);

   (c) safe work method statements (SWMSs);

   (d) work instruction (WI).

▷ take the measures necessary to ensure that machinery is suitable for the work to be carried out, or otherwise properly adapted for its intended purpose, and is safe for workers;

▷ ensure that machinery is correctly installed and safeguarded and that protective devices and markings are used so that workers are protected from danger to their safety and health;

▷ ensure adequate and competent supervision of work and work practices, including adherence to work procedures;

▷ take all necessary measures to ensure that, throughout its working life, machinery is maintained in a condition such that it continues to meet the relevant safety requirements;

▷ ensure the safety of machinery through a system of preventive maintenance, including regular inspections and testing, where appropriate, of protective devices and guards and emergency stops. Any defects should be rectified promptly:

   ▷ where appropriate, the maintenance systems should include written procedures and communication on how the work can be carried out safely (for example, 'permit to work' systems, procedures for working in confined spaces and lock-off procedures);

   ▷ emergency prevention, preparedness and response arrangements should be established and maintained in relation to the use of machinery in cooperation with external emergency services;

   ▷ where a machine has an inspection log it should be kept up to date;

▷ ensure that if maintenance is required while the machinery is running, it is performed by competent persons, and risk mitigation measures, such as use of 'hold to run' controls with slow running speeds, are applied;

▷ ensure that the decommissioning and disposal of machinery are carried out safely, taking into account the manufacturer's instructions and in accordance with national law and practice;

▷ ensure that workers have received the necessary training, information and instructions to perform the work competently and safely. The extent of the training and instruction received and required should be reviewed and updated.

▶ Workers' responsibilities – they should:

(a) follow safe working methods as instructed by their employers;

(b) cooperate with their employers in ensuring safety in the use of machinery;

(c) use and take care of PPE, protective clothing and any facilities made available to them, and not misuse anything provided for their own protection or the protection of others;

(d) participate actively in safety and health training,

(e) take all reasonable steps to eliminate or minimise the risk to themselves and to others resulting from their use of machinery at work; and

(f) inform their supervisor without delay of any situation which they believe could present a risk.

▶ Workers' should have the rights to consult with and be consulted by employers on all issues relating to their safety and health including training and information. They should have the right to remove themselves from danger if they have reasonable justification to believe they are in imminent and serious danger, without suffering undue consequences. They should have the right to appeal to the competent authority.

15.1.13 Welfare Facilities Recommendation, R102, 1956

These recommendations concern welfare facilities for workers. It applies to manual and non-manual workers employed in public or private undertakings, excluding agriculture and sea transport. The facilities specified in this Recommendation may be provided by means of public or voluntary action:

(a) through laws and regulations, or

(b) in any other manner approved by the competent authority after consultation with employers' and workers' organisations, or

(c) by virtue of collective agreement or as otherwise agreed upon by the employers and workers concerned.

Part III Feeding Facilities

A. Canteens

▶ Canteens providing appropriate meals should be set up and operated in or near undertakings where this is desirable, having regard to the number of workers employed by the undertaking, the demand for and prospective use of the facilities, the non-availability of other appropriate facilities for obtaining meals and any other relevant conditions and circumstances.

▶ The competent authority or some other appropriate body should make suitable arrangements to give information, advice and guidance to individual undertakings with respect to technical questions involved in the setting up and operation of canteens.

B. Buffets and Trolleys

▶ In undertakings where it is not practicable to set up canteens providing appropriate meals, and in other undertakings where such canteens already exist, buffets or trolleys should be provided, where necessary and practicable, for the sale to the workers of packed meals or snacks and tea, coffee, milk and other beverages. Trolleys should not, however, be introduced into workplaces in which dangerous or harmful processes make it undesirable that workers should partake of food and drink there.

▶ Some of these facilities should be made available not only during the midday or midshift interval but also during the recognised rest pauses and breaks.

C. Messrooms and Other Suitable Rooms

▶ In undertakings where it is not practicable to set up canteens providing appropriate meals, and, where necessary, in other undertakings where such canteens already exist, messroom facilities should be provided, where practicable and appropriate, for individual workers to prepare or heat and take meals provided by themselves.

▶ The facilities so provided should include at least:
  (a) a room in which provision suited to the climate is made for relieving discomfort from cold or heat;
  (b) adequate ventilation and lighting;
  (c) suitable tables and seating facilities in sufficient numbers;
  (d) appropriate appliances for heating food and beverages;
  (e) an adequate supply of wholesome drinking water.

D. Mobile Canteens

▶ In undertakings in which workers are dispersed over wide work areas, it is desirable, where practicable and necessary, and where other satisfactory facilities are not available, to provide mobile canteens for the sale of appropriate meals to the worker.

E. Other Facilities

▶ Special consideration should be given to providing shift workers with facilities for obtaining adequate meals and beverages at appropriate times.

▶ In localities where there are insufficient facilities for purchasing appropriate food, beverages and meals, measures should be taken to provide workers with such facilities.

F. Use of Facilities

▶ The workers should in no case be compelled, except as required by national legislation for reasons of health, to use any of the feeding facilities provided.

Part IV Rest Facilities

A. Seats

▶ In undertakings where any workers, especially women and young workers, have in the course of their work reasonable opportunities for sitting without detriment to their work, seats should be provided and maintained for their use.

▶ Seats so provided should be in adequate numbers and reasonably near the work posts of the workers concerned.

▶ In undertakings where a substantial proportion of any work can be properly done seated, seats should be provided and maintained for the workers concerned.

▶ The seat should be of a design, construction and dimensions suitable for the worker and the work; a footrest should be provided where necessary.

B. Rest Rooms

▶ In an undertaking where alternative facilities are not available for workers to take temporary rest during working hours, a rest room should be provided, where this is desirable, having regard to the nature of the work and any other relevant conditions and circumstances. In particular, rest rooms should be provided to meet the needs of women workers; of workers engaged on particularly arduous or special work requiring temporary rest during working hours; or of workers employed on broken shifts.

▶ The facilities so provided should include at least:
  (a) a room in which provision suited to the climate is made for relieving discomfort from cold or heat;
  (b) adequate ventilation and lighting;
  (c) suitable seating facilities in sufficient numbers.

Part V Recreation Facilities

▶ Appropriate measures should be taken to encourage the provision of recreation facilities for the workers in or near the undertaking in which they are employed, where suitable facilities organised by special bodies or by community action are not already available and where there is a real need for such facilities as indicated by the representatives of the workers concerned.

▶ Such measures, where necessary, should be taken by works committees or other bodies established by national legislation.

▶ Whatever may be the methods adopted for providing recreation facilities, the workers should in no case be under any obligation to participate in the utilisation of any of the facilities provided.

Parts VI and VII

Cover details of facilities management and financing using a variety of different methods.

Part VIII Transport Facilities

▶ Where, in accordance with national or local custom, workers provide their own means of transport to and from work, suitable parking or storage facilities should be provided where necessary and practicable.

▶ Where a substantial proportion of the workers experience special difficulties in travelling to and from work owing to the inadequacy of public transport services or unsuitability of transport timetables, the undertakings in which they are employed should endeavour to secure from the

15

organisations providing public transport in the locality concerned the necessary adjustments or improvements in their services.

► Where the workers' transport difficulties are primarily due to peak transport loads and traffic congestion at certain hours and where such difficulties cannot otherwise be overcome, the undertaking in which they are employed should, in consultation with the workers concerned and with the public transport and traffic authorities, and, where appropriate, with other undertakings in the same locality, endeavour to adjust or stagger times of starting and finishing work in the undertaking as a whole or in some of its departments.

► Where adequate and practicable transport facilities for the workers are necessary and cannot be provided in any other way, the undertakings in which they are employed should themselves provide the transport.

► Wherever necessary, undertakings should arrange for adequate transport facilities to be available, either through the services of public transport or otherwise, to meet the needs of shift workers at times of the day and night when ordinary public transport facilities are inadequate, impracticable or non-existent.

15.1.14 Working Environment (Air Pollution, Noise and Vibration) Convention 1977, C148

This Convention concerns the protection of workers against occupational hazards in the working environment due to air pollution, noise and vibration. For the purpose of this Convention:

► the term *air pollution* covers all air contaminated by substances, whatever their physical state, which are harmful to health or otherwise dangerous;

► the term *noise* covers all sound which can result in hearing impairment or be harmful to health or otherwise dangerous;

► the term *vibration* covers any vibration which is transmitted to the human body through solid structures and is harmful to health or otherwise dangerous.

Part II General Provisions

Include the requirement for:
► National laws to prescribe the measures to be taken for the prevention and control of, and protection against, occupational hazards in the working environment due to air pollution, noise and vibration.
► Consultation with employer and workers' organisations;
► Provision for close cooperation between employers and workers;
► Responsibilities being placed on employers for compliance with the requirements;

► Workers being required to comply with safety provisions;
► Workers and their representatives having the right to present proposals and receive information and training and appeal to appropriate bodies to ensure protection.

Part III Preventive and Protective Measures – include the requirement:

► for the competent authority to establish and revise as necessary, criteria for determining the hazards of exposure to air pollution, noise and vibration in the working environment and where appropriate specify exposure limits;

► to keep the working environment free from any hazard due to air pollution, noise or vibration by:
  ▷ technical measures applied to new plant or processes in design or installation, or added to existing plant or processes; or, where this is not possible,
  ▷ by supplementary organisational measures;

► to provide suitable PPE where the precautions have not brought air pollution, noise and vibration within the set exposure limits;

► to supervise (free of charge to workers) the health of workers exposed or liable to be exposed to these occupational hazards. This includes pre-assignment medical examination and periodic examination as determined by the competent authority;

► to ensure that all persons concerned are adequately and suitably:
  ▷ informed of potential occupational hazards in the working environment due to air pollution, noise and vibration; and
  ▷ instructed in the measures available for the prevention and control of, and protection against, those hazards;

► to promote research in the field of prevention and control of hazards;

► for the employer to appoint a competent person, or use a competent outside service, to deal with matters pertaining to the prevention and control of air pollution, noise and vibration in the working environment;

► to ensure the appropriate enforcement of the precautionary measures and inspection of workplaces.

More detailed guidance is given in R156 Working Environment (Air Pollution, Noise and Vibration) Recommendations at: http://www.ilo.org/dyn/normlex/en/f?p=1000:12100:0::NO::P12100_ILO_CODE:R156

Ambient factors in the workplace, International Labour Organisation (ILO) Code of Practice (CoP), ISBN 92-2-11628-X can be found at: http://www.ilo.org/safework/info/standards-and-instruments/WCMS_107729/lang--en/index.htm

Occupational Exposure to Airborne Substances Harmful to Health, ILO CoP, ILO Geneva, ISBN 92-2-102442-3 http://www.ilo.org/wcmsp5/groups/public/---ed_protect/---protrav/---safework/documents/normativeinstrument/wcms_107851.pdf

15.2 Typical OSH legal frameworks in the USA, EU and UK

Although the USA, EU and UK are all intimately bound up with the decisions of the ILO and regularly report back to Geneva on OSH matters, neither the USA or the UK have formerly signed up to Convention 155. The UK has ratified the more recent framework Convention 187. US and EU organisations were signatories to the Seoul Declaration of June 2008 (Appendix 15.1). The main legal frameworks of all three, however, follow the ILO concepts and in many cases were probably major contributors to the ILO standards. The UK has been picked as it is the home of NEBOSH and this book has been written around its International Certificate syllabus. Quite clearly the UK also follows and is required to follow the EU framework as a member state of the EU. This section covers the OSH frameworks only, while section 15.3 covers in more detail legislation in the USA, UK and the EU plus 20 other national legislatures.

15.2.1 USA OSH framework

(a) Introduction

In the United States the Occupational Safety and Health Act of 1979 created both the National Institute for Occupational Safety and Health (NIOSH) and the Occupational Safety and Health Administration (OSHA). OSHA, in the U.S. Department of Labor, is responsible for developing and enforcing workplace safety and health regulations. NIOSH, in the U.S. Department of Health and Human Services, is focused on research, information, education, and training in occupational safety and health.

OSHA have been regulating occupational safety and health since 1971. Occupational safety and health regulation of a limited number of specifically defined industries was in place for several decades before that, and broad regulations by some individual States was in place for many years prior to the establishment of OSHA.

The Occupational Safety and Health (OSH) Act was enacted to 'assure safe and healthful working conditions for working men and women'. The OSH Act created the Occupational Safety and Health Administration (OSHA) at the federal level and provided that States could run their own safety and health programmes as long as those programmes were at least as effective as the federal programme. Federal and State safety personnel work to ensure worker safety and health through worksite enforcement, education and compliance assistance, and cooperative and voluntary programmes. Enforcement and administration of the OSH Act in States under federal jurisdiction are handled primarily by OSHA. Safety and health standards related to field sanitation and certain temporary labour camps in the agriculture industry are enforced by the U.S. Department of Labor's (DOL) Wage and Hour Division (WHD) in States under federal jurisdiction. For more details on the OSH Act 1971 see section 15.3.23.

If a worksite is located in a **state plan** State, additional safety and health requirements may apply.

(b) State OSHA programme

Section 18 of the Occupational Safety and Health Act of 1970 (the Act) encourages States to develop and operate their own job safety and health programmes. OSHA approves and monitors State plans and provides up to 50% of an approved plan's operating costs.

States must set job safety and health standards that are 'at least as effective as' comparable federal standards. (Most States adopt standards identical to federal ones.) States have the option to promulgate standards covering hazards not addressed by federal standards.

A State must conduct inspections to enforce its standards, cover public (State and local government) employees, and operate occupational safety and health training and education programmes. In addition, most States provide free on-site consultation to help employers identify and correct workplace hazards. Such consultation may be provided either under the plan or through a special agreement under section 21(d) of the Act.

(c) How does a State establish its own programme?

To gain OSHA approval for a developmental plan – the first step in the state plan process – a State must assure OSHA that within three years it will have in place all the structural elements necessary for an effective occupational safety and health programme. These elements include: appropriate legislation; regulations and procedures for standards setting, enforcement, appeal of citations and penalties; a sufficient number of qualified enforcement personnel.

Once a State has completed and documented all its developmental steps, it is eligible for certification.

15

Certification renders no judgement as to actual State performance, but merely attests to the structural completeness of the plan.

At any time after initial plan approval, when it appears that the State is capable of independently enforcing standards, OSHA may enter into an 'operational status agreement' with the State. This commits OSHA to suspend the exercise of discretionary federal enforcement in all or certain activities covered by the State plan.

The ultimate accreditation of a State's plan is called final approval. When OSHA grants final approval to a State under section 18 (e) of the Act, it relinquishes its authority to cover occupational safety and health matters covered by the State. After at least one year following certification, the State becomes eligible for final approval if OSHA determines that it is providing, in actual operation, worker protection 'at least as effective' as the protection provided by the federal programme. The State also must meet 100% of the established compliance staffing levels (benchmarks) and participate in OSHA's computerised inspection data system before OSHA can grant final approval.

Employees finding workplace safety and health hazards may file a formal complaint with the appropriate plan State or with the appropriate OSHA regional administrator. Complaints will be investigated and should include the name of the workplace, type(s) of hazard(s) observed and any other pertinent information.

Anyone finding inadequacies or other problems in the administration of a State's programme may file a Complaint About State Program Administration (CASPA) with the appropriate OSHA regional administrator as well. The complainant's name is kept confidential. OSHA investigates all such complaints, and where complaints are found to be valid, requires appropriate corrective action on the part of the State.

Twenty-seven states (June 2015), Puerto Rico and the Virgin Islands have OSHA-approved state plans and have adopted their own standards and enforcement policies. For the most part, these States adopt standards that are identical to Federal OSHA. However, some States have adopted different standards applicable to this topic or may have different enforcement policies.

The Occupational Safety and Health State Plan Association (OSHSPA) is the organisation of officials from the 26 states and territories that operate OSHA-approved state plans. OSHSPA also serves as the link from the state plans to Congress and to federal agencies that have occupational safety and health jurisdiction. The group holds three meetings a year with Federal OSHA, giving state programmes the opportunity to address common problems and share information. It also provides information to States or territories that are considering application for state plan status.

15.2.2 EU OSH framework

(a) EU legislative process in brief

A proposal for a Directive or Regulation is presented by the European Commission and is reviewed, usually on two occasions, by the European Parliament and the Council. If the European Parliament and Council agree after the first reading (rare), the proposal will be adopted. If no agreement is reached after the second reading, the proposal goes through a conciliation process before being adopted. The Directive or Regulation comes into force after publication in the Official Journal (OJ) of the EU. References in the Calendar to '1st' or '2nd' readings are references to the European Parliament's plenary vote during 1st or 2nd reading.

Directives must be implemented through national legislation in individual member states, with the deadline usually laid down in the Directive. **Regulations** come into force in all EU member states upon publication. **Decisions** are directly binding on those to whom they are addressed. Communications are prepared by the Commission and may be followed by proposals for legislation.

For a summary of the subjects covered by EU OSH legislation see 15.3.7.

(b) Community strategy on health and safety at work (2014–2020)

(i) Summary

The evaluation of the 2007–12 EU strategy on health and safety at work confirmed its overall effectiveness and that its main objectives were achieved. However, implementation continues to be a challenge, in particular for micro and small companies, which have difficulties in complying with some regulatory requirements. When asked why an EU framework is necessary, respondents cited the need to ensure similar standards and workers' protection in all EU member states; a level playing field for Europe's businesses; and to help member states to set a focus in their workplace health and safety policies.

The EU Occupational Safety and Health (OSH) Strategic Framework 2014–2020 – the European Commission has adopted a new Strategic Framework on Health and Safety at Work 2014–2020, which identifies key challenges and strategic objectives for health and safety at work, presents key actions and identifies instruments to address these. This new framework aims at ensuring that the EU continues to play a leading role in the promotion of high standards for working conditions both within Europe and internationally, in line with the Europe 2020 Strategy.

(ii) Challenges and objectives

The strategic framework identifies the following challenges:

▶ to improve implementation of existing health and safety rules, in particular by enhancing the capacity of micro and small enterprises to put in place effective and efficient risk prevention strategies;

▶ to improve the prevention of work-related diseases by tackling new and emerging risks without neglecting existing risks;

▶ to take account of the ageing of the EU's workforce.

To meet these challenges, the strategic framework contains a series of strategic objectives:

▶ Further consolidating national health and safety strategies.

▶ Providing practical support to small and micro enterprises to help them to better comply with health and safety rules. Businesses would benefit from technical assistance and practical tools, such as the Online Interactive Risk Assessment (OiRA), a web platform providing sectoral risk assessment tools.

▶ Improving enforcement by member states.

▶ Simplifying existing legislation where appropriate to eliminate unnecessary administrative burdens, while preserving a high level of protection for workers' health and safety.

▶ Addressing the ageing of the European workforce.

▶ Improving prevention of work-related diseases to tackle existing and new risks such as nanomaterials, green technology and biotechnologies.

▶ Improving statistical data collection to have better evidence and developing monitoring tools.

▶ Reinforcing coordination with international organisations (such as the International Labour Organisation (ILO), the World Health Organisation (WHO) and the Organisation for Economic Cooperation and Development (OECD)) to contribute to reducing work accidents and occupational diseases and to improving working conditions worldwide. When all is said and done, any adaptation of the legal framework must also make that framework less complex and more effective. The Commission emphasises that simplified legislation should not lead to a reduction in existing levels of protection.

The strategic framework identifies instruments to implement these actions, highlighting for instance the EU funds, such as the European Social Fund (ESF) and the Employment and Social Innovation (EaSI) programme, that are available to support the implementation of health and safety rules.

(c) Advisory Committee on Safety and Health at Work (ACSHW)

The Advisory Committee on Safety and Health at Work (ACSH) is a tripartite body set up in 2003 by a Council Decision (2003/C 218/01) to streamline the consultation process in the field of safety and health at work and rationalise the bodies created in this area by previous Council Decisions – namely the former Advisory Committee on Safety, Hygiene and Health Protection at Work (established in 1974) and the Mines Safety and Health Commission for safety and health in coal mining and the other extractive industries (established in 1956). To ensure continuity concerning questions previously dealt with by the Mines Safety and Health Commission, a Standing Working Party (SWP) on the mining industry has been established within the Committee (Article 5 (4), of the Council Decision 2003/C 218/01).

The Committee is charged with assisting the European Commission in the preparation, the implementation and the evaluation of activities in the fields of safety and health at work. Its tasks are:

▶ To give opinions on Community initiatives in the area of occupational safety and health (such as new legislation, Community programmes);

▶ To contribute proactively to identifying Community priorities and to establish relevant policy strategies;

▶ To encourage the exchange of views and experience (be an interface between the national and European level).

The Committee comprises full members made up of one government representative, one representative of trade unions and one representative of employers' organisations for each EU Member State. Two alternate members are appointed for each full member.

The Committee, which has two plenary meetings per year, includes three interest groups, each with representatives from national governments, trade unions and employers' organisations respectively. Each interest group selects one of its members to be its spokesperson and designates a coordinator.

The Committee is chaired by the Commission and its activities are coordinated by a 'Bureau', composed of two representatives from the Commission and the spokespersons and coordinators designated by the interest groups.

The Committee's annual work programme is prepared by the Bureau for adoption by the Committee.

15

The Committee has established a number of working parties to deal with specific technical issues and organises workshops and seminars on specific topics.

(d) European Agency for Safety and Health at Work

The Agency was set up by Council Regulation (EC) No. 2062/94 of 18 July 1994 establishing a European Agency for Safety and Health at Work

The Agency's role is to:

▶ collect and analyse technical, scientific and economic information on health and safety at work in the Member States and to pass it on to the Community bodies, other Member States and interested parties;

▶ collect and analyse technical, scientific and economic information on research into safety and health at work and disseminate the results of this research;

▶ promote and support cooperation and exchange of information and experience amongst the Member States in the field of safety and health at work, including information on training programmes;

▶ organise conferences and seminars (such as the European Health and Safety at Work Week) and exchanges of national experts in the field of safety and health at work;

▶ supply the Community bodies and the Member States with the technical, scientific and economic information they require to formulate and implement judicious and effective policies designed to protect the safety and health of workers;

▶ establish an information network in cooperation with the Member States, and coordinate it, including national, Community (the European Foundation for the Improvement of Living and Working Conditions) and international bodies and organisations which provide this type of information and services;

▶ collect and make available information on safety and health matters from and to third countries and international organisations: the World Health Organisation (WHO), the International Labour Organisation (ILO), the Pan American Health Organisation (PAHO), the International Migration Office (IMO), etc.;

▶ provide technical, scientific and economic information on methods and tools for implementing preventive activities, especially for small and medium-sized enterprises, and identify good practices;

▶ contribute to the development of Community action programmes and strategies relating to the protection of safety and health at work, without prejudice to the Commission's sphere of competence;

▶ ensure that the information disseminated is easily understood by the end-users.

The Agency collaborates as closely as possible with institutions, foundations, specialist bodies and programmes at Community level in order to avoid any duplication. For example, it works together with the European Foundation for the Improvement of Living and Working Conditions.

The Agency will set up a network comprising:

▶ the main component elements of the national information networks, including the national social partner organisations in accordance with national legislation;

▶ the national focal points;

▶ European topic centres.

Member States regularly inform the Agency of the main component elements of their information networks on health and safety at work. The relevant national authorities coordinate and forward the information to be provided to the Agency at national level.

The Agency has a steering and management structure comprising a Governing Board, a Bureau and a Director.

Its Governing Board comprises 78 members, of whom 25 members represent the governments of the Member States, 25 members represent employer organisations, 25 members represent employee organisations and three members represent the Commission. The Members of the Governing Board have a three-year term of office which is renewable. The Board's headquarters is in Bilbao, Spain.

The Bureau comprises 11 members: the chairman and the three vice-chairmen of the Governing Board, one coordinator for each of the three groups of representatives (employers, workers and government), an additional representative for each of these three groups, and a Commission representative. The Bureau monitors the implementation of the Governing Board's decisions and takes all necessary steps to ensure that the Foundation is managed properly between meetings of the Governing Board.

The Agency is headed by a Director appointed by the Governing Board.

The inaugural meeting of the Governing Board took place at the Agency's headquarters in Bilbao (Spain) on 25 and 26 October 1995.

See: http://europa.eu/legislation_summaries/employment_and_social_policy/health_hygiene_safety_at_work/index_en.htm

http://osha.europa.eu/en

http://ec.europa.eu/social/main.jsp?catId=148&langId=en

15.2.3 UK OSH framework

(a) Introduction

Great Britain has a tradition of health and safety regulation going back over 150 years. The present system came into being with the Health and Safety at Work etc. Act (HSW Act) in 1974 with further significant modifications in 2008. The effect of this is to provide a

unified institutional structure and legal framework for health and safety regulation.

The Health and Safety Executive (HSE) enforces the law in many workplaces, ranging from health and safety in nuclear installations and mines, through to factories, farms, hospitals and schools, offshore gas and oil installations, the safety of the gas grid and the electricity distribution system, the movement of dangerous goods and substances and many other aspects of the protection both of workers and the public. In addition, over 400 local authorities are responsible for enforcement in a wide range of other activities, including the retail and finance sectors, and other parts of the services sector, particularly leisure.

The standards of health and safety achieved in Great Britain are delivered by the flexible regulatory system introduced by the HSW Act, and are typified by the Management of Health and Safety at Work Regulations 1999. They also reflect a long tradition of health and safety regulation going back to the 19th century. Since the HSW Act was passed, the HSE has been engaged in progressive reform of the law, seeking to replace detailed industry-specific legislation with a modern approach in which regulations, wherever possible, express goals and general principles, and detailed requirements are placed in codes and guidance. Approved codes have a special place in British health and safety law – they set out ways of achieving standards. Those who depart from a code must be prepared to show that their own approach is an equally valid way of meeting the legal requirements. In this way, flexibility is allowed for technological development within a framework set by mandatory regulations.

A fundamental principle of the British system is that responsibility for health and safety lies with those who own, manage and work in industrial and commercial undertakings. This includes the self-employed. They must assess the risks attached to their activity and take appropriate action. Workforce involvement and, in particular, the work of health and safety representatives, has made an important contribution to raising standards of health and safety.

(b) The Health and Safety Executive

The HSE consists of a governing Board of up to 12 non-executive directors and approximately 3,500 staff.

Members of the Board are appointed by the Secretary of State for Work and Pensions after consultation with organisations representing employers, employees, local authorities and others, as appropriate. The HSE's staff include inspectors, policy advisers, technologists, and scientific and medical experts.

The HSW Act and related legislation are primarily enforced by the HSE or local authorities, according to the main activity carried out at individual work premises. The Health and Safety (Enforcing Authority) Regulations 1998 allocate the enforcement of health and safety legislation at different premises between local authorities and the HSE. The HSE's statutory responsibilities under the HSW Act include proposing health and safety law and standards to ministers. In preparing its proposals, it relies on the advice of its staff and on scientific research carried out by its in-house agency the Health and Safety Laboratory (HSL) and externally. It also consults extensively with organisations representing professional interests in health and safety, business managers, trade unions, and scientific and technological experts. This is managed through a network of advisory committees and by public invitation to comment on particular proposals. Special efforts are made to seek the views of small firms, often using a range of intermediary organisations representing trade, sector or business interests.

(c) Local authorities

Local authorities enforce health and safety law mainly in the distribution, retail, office, leisure and catering sectors. The HSE liaises closely with local authorities on enforcement matters through the HSE/Local Authorities Enforcement Liaison Committee (HELA). Partnership teams (comprising HSE and local authority staff) and an enforcement liaison officer network in HSE regional offices across Britain also provide advice and support. HELA was set up in 1975 to provide effective liaison between the HSE and local authorities. Reconstituted in 2006, it provides a strategic oversight of the partnership aiming to maximise its effectiveness in improving health and safety outcomes – including enforcement priorities for local authorities. A Local Government Panel, comprising local authority councillors, was also established in 2006 and regularly meets the HSE Board for a strategic dialogue on local, central and devolved government issues that impact on health and safety regulatory functions. It also reviews the effectiveness and performance of the partnership between the two enforcing authorities.

(d) Ministerial responsibilities

Health and safety is regulated in the same way across the whole of Great Britain and a number of different Secretaries of State are responsible to Parliament at

15

Westminster for the activities of the HSE in different areas. The Secretary of State for Work and Pensions answers to Parliament on the HSE's staffing and resourcing, on matters affecting protection of workers and on all other HSE activities, except when these come within the specific area of responsibility of another Secretary of State, e.g. the Secretary of State for Energy and Climate Change on nuclear safety, the Secretary of State for Business, Enterprise and Regulatory Reform on the health and safety aspects of barriers to trade, the Secretary of State for Environment, Food and Rural Affairs on certain aspects of pesticide safety, and the Secretary of State for the Home Department on the security of explosives. In most of these matters, the HSE and local authorities act by virtue of their powers and duties under the HSW Act and its associated legislation, or European legislation. In a few, they act under agreements as the agent of the Secretary of State concerned. The HSE is required to submit to the Secretary of State such proposals as it considers appropriate for making regulations under any of the relevant statutory provisions, and to submit to the Secretary of State particulars of what it proposes to do for the purpose of performing its functions.

(e) Advisory committees

The HSE provides policy, technological and professional advice. Other expert advice comes from the HSE's network of advisory committees who deal with particular hazard areas and some with particular industries. Each includes a balance of employer and employee representatives and, where appropriate, technological and professional experts. The committees are supported by HSE staff whose main function is to recommend standards and guidance and, in some cases, to comment on policy issues or to recommend an approach to a particular new problem.

(f) Scotland

The health and safety system in Scotland is the same as in England. Inspection arrangements and accountabilities involving the HSE and 32 Scottish local authorities are similar to the rest of Great Britain. The area of greatest difference is Scotland's distinctive legal system; only the Crown Office and Procurator Fiscal Service can prosecute and the HSE prepares cases for their consideration. The Scottish government has devolved powers in relation to health, education, environment, the Fire and Rescue Service, etc., and as a consequence different arrangements are made with ministers and other stakeholders in Scotland in relation to consultation, engagement and implementation of legislative changes, and in developing and implementing inspection priorities and programmes. The HSE has established a Partnership on Health and Safety in Scotland (PHASS) with the devolved administration, business and union stakeholders and the interested professional bodies to ensure that initiatives are undertaken jointly wherever possible.

(g) Wales

The health and safety system and the legal system in Wales is the same as in England. Inspection arrangements and accountabilities involving the HSE and Welsh local authorities are the same as in England. The Welsh Assembly government has devolved powers in relation to health, education, environment, the Fire and Rescue Service, etc., and as a consequence different arrangements may need to be made with ministers and other stakeholders in Wales in relation to consultation, engagement and implementation of legislative changes, or in developing and implementing inspection priorities and programmes.

See: http://www.hse.gov.uk/pubns/web42.pdf

15.3 National implementing legislation

Different states take different approaches to legislation, regulation, and enforcement. The following summaries from 21 countries/areas are some examples from across the world. The accident and occupational disease for most areas are shown in Table 15.2.

Figures 15.2 and 15.3 show the work-related fatal accidents by world areas and the causes of work-related disease fatalities.

Figure 15.4 shows some of the progress that has been made internationally to develop OSH programmes.

Table 15.2 Estimates of work-related occupational accidents and diseases – 2001 data mainly
Source ILO – Safework programme

Country	Labour force	ILO estimate of fatal accidents 2002 data	Fatal accidents per 100,000 2002	ILO average estimate of 3 day plus lost time accidents	3 day accidents per 100,000	Deaths work related diseases	Total deaths disease + accidents + dangerous substances
Australia	9,796,300	236	2.4	180,486	1,842	6,634	8133
Bangladesh	60,200,000	14,403	23.93	8,980,000	14,917	28,661	47,886
Brazil	83,000,000	11,304	13.62	11,366,000	13,694	44,375	70,133
Canada	16,200,000	899	5.55	789,000	4,870	10,962	14,206
China	737,000,000	73,595	9.99	68,692,000	9,320	414,024	597,279
EU	224,050,000	11,369	5.07	4,340,000	1,937	50,279	79,606
Egypt	19,200,000	3,884	20.23	2,245,000	11,693	26,175	34,500
India	443,000,000	48,176	10.87	30,627,000	6,914	325,350	432,572
Indonesia	95,700,000	18,220	19.04	12,900,000	13,480	50,279	79,606
Japan	67,500,000	2,077	3.08	1,538,000	2,279	46,621	57,579
Korea (South)	22,100,000	3,148	14.24	1,542,000	6,977	11,665	16,439
Malaysia	9,600,000	1,578	16.44	920,000	9,583	5,279	7,682
Mexico	39,600,000	6,149	15.53	5,823,000	14,705		
Nigeria	51,600,000	9,631	18.66	7,167,000	13,890	68,747	92,512
Philippines	33,300,000	6,019	18.08	4,269,000	12,820	16,658	26,356
Russian Federation	63,600,000	6,974	10.97	1,786,000	2,808	60,040	78,561
Saudi Arabia	5,800,000	1,096	18.90	632,000	10,897	8,660	11,243
South Africa	11,300,000	2,643	23.39	1,455,000	12,876	15,102	20,139
Thailand	34,400,000	7,490	21.77	5,305,000	15,422	18,540	30,194
Trinidad and Tabago	572,000	92	16.08	70,999	12,412	302	462
Turkey	21,600,000	4,122	19.08	2,881,000	13,338	30,365	40,452
UK	27,200,000	225	0.83	180,000	662	20,522	24,339
USA	141,800,000	6,821	4.81	5,069,000	3,575	98,210	124,133
WORLD 2001	2,848,000,000	351,000	12.32	268,000,000	9,410	2,033,000	2,380,000
WORLD 2003	2,941,000,000	358,000	12.17	337,000,000	11,458	1,950,000	2,310,000

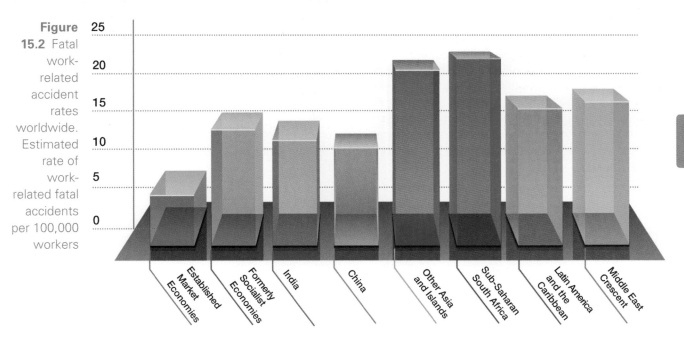

Figure 15.2 Fatal work-related accident rates worldwide. Estimated rate of work-related fatal accidents per 100,000 workers

15

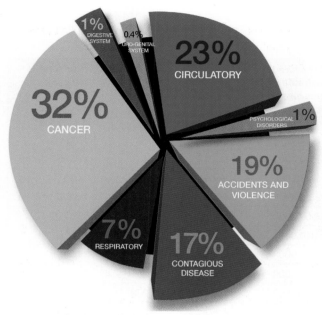

Figure 15.3 Disease fatalities attributed to work. The 2.2 million work-related deaths are broken down as shown

15.3.1 Australia

(Population in 2013: 23,639,000. Labour force: 12,440,000)

In Australia currently (2015) all states and territories are responsible for making and enforcing their own OHS laws. Although these draw on a similar approach for regulating workplaces, there are some differences in the application and detail of the laws. The Occupational

---

Major progress in the development of national OSH and inspection policies

18 countries have developed national OSH profiles based on ILO guidance, namely Azerbaijan, Benin, China, Egypt, Georgia, Guatemala, Iraq, Kazakhstan, Kenya, Kuwait, Kyrgyzstan, Mexico, Mongolia, Pakistan, Uganda, United Republic of Tanzania, Uzbekistan and Yemen, while national OSH profiles are being prepared in others: Algeria, Costa Rica, Croatia, Iran, Malaysia, Mozambique, Nicaragua, Panama, Seychelles, Sri Lanka, Tajikistan and Vietnam.

9 countries included OSH aspects in their national plans of action for Decent Work:

Bangladesh, Ethiopia, Guatemala, India, Morocco, Nepal, Nicaragua, Panama and Sri Lanka, while Panama, Nicaragua and Guatemala are developing the OSH component of the Decent Work national programmes.

35 countries made progress in this area, including:

▶ the adoption of a new OSH Act or revision of present legislation in Ethiopia, Ireland, Kazakhstan, Kenya, Kyrgyzstan, Lebanon, Morocco, Nigeria, Saudi Arabia, Uganda, United Arab Emirates and United Republic of Tanzania.

▶ setting up National Safety and Health Committees: Algeria has set up a new NIOSH and revitalised its tripartite Conseil supérieur de la Prévention, Argentina, Botswana and Columbia, in the United Arab Emirates, the OSH Unit of the Ministry of Labour is being restructured. Serbia is revitalising its National Council on OSH.

▶ signing agreements/accords with a key component on OSH: Azerbaijan, Kazakhstan, Russia, South Africa, Tajikistan and Uzbekistan. An agreement signed with the Gulf Co-operation Council resulted with ILO assistance in the revision of legislation in the field of OSH: and the preparation of OSH guides for the oil and petrochemical industries. A collaboration agreement signed by the IAPRP (Inter-Africaine de Prévention des risques Professionnels) with particular focus on the development of a specialised university degree for occupational physicians. Protocols of Co-operation have been signed between the Polish National Labour Inspection and counterparts in Bulgaria, Serbia and Ukraine.

▶ implementation of legislative reform on integrated labour inspection: Armenia, Bulgaria, Chile, Costa Rica, Vietnam and Serbia. Laos included the development of an integrated labour inspection into its 5-year national action plan.

Elsewhere, projects to strengthen the capacity of labour inspectorates have been sponsored by the ILO, its partners and donor organisations. In the Republic of Serbia, for example, a 2-year long project to improve the effectiveness of the labour inspectorate and social partners has almost been finished. The Nordic countries (Finland, Iceland, Norway, Sweden and Denmark) developed the so-called 'scoreboard', a tool to measure and compare performances of their labour inspectorates but intended for use in other EU countries as well. The idea has been taken up in Africa, where Ministers in ARLAC approved an action plan for strengthening labour inspection and an integrated system, with an African version of the Scoreboard.

Figure 15.4 Some national and international OSH development and activities, 2004–2005

Health and Safety (Commonwealth Employment) Act 1991 follows many of the general requirements laid down in the UK HSW Act 1974. However, it only applies to Commonwealth workplaces. A range of national standards have been produced including standards for asbestos, atmospheric contaminants, carcinogenic substances, competences, confined spaces, construction, dangerous goods, education, environmental tobacco smoke, hazardous substances, injury and disease reporting, lead, ionising radiation, labelling, major hazards, manual handling, noise, occupational overuse, plant and design.

The Australian government has identified occupational health and safety (OHS) as a priority area for reform. One of the key elements of the OHS reform agenda is harmonisation – moving towards one set of national OHS laws. The harmonisation of OHS legislation aims to reduce the incidence of workplace death, injury and disease right across Australia. On 7 September 2009 the Senate passed without amendments the Safe Work Australia Act 2008 [No. 2]. Safe Work Australia became operational in November 2009 as an independent statutory agency with primary responsibility to improve occupational health and safety and workers' compensation arrangements across Australia. This also gives effect to the Intergovernmental Agreement for Regulatory and Operational Reform in Occupational Health and Safety, agreed by the Council of Australian Governments (COAG) on 3 July 2008. Safe Work Australia operates under the Commonwealth Government's accountability and governance frameworks. Under the revised arrangements, the Safe Work Australia Council is known as Safe Work Australia, with former Council members being reappointed. Safe Work Australia still has 15 members, including an independent Chair, nine members representing the Commonwealth and each State and Territory, two members representing the interests of workers, two representing the interests of employers, and the CEO of Safe Work Australia. The functions of the Council remain the same.

Harmonised OHS laws are essential to the government's aim of a streamlined national economy to drive productivity in the future. The creation of Safe Work Australia is an important step towards achieving harmonised occupational health and safety laws across Australia. The government is working cooperatively with all jurisdictions to develop uniform, equitable and effective safety standards and protections for all Australians.

The functions of Safe Work Australia include, but are not limited to:

▶ developing national policy relating to OHS and workers' compensation;

▶ preparing model OHS legislation, model regulations, model codes of practice and other material relating to OHS;

▶ developing a policy, for approval by the Ministerial Council, dealing with the compliance and enforcement of the Australian laws that adopt the approved model OHS legislation;

▶ monitoring the adoption of model OHS legislation, model regulations, model codes of practice by the Commonwealth, States and Territories;

▶ developing proposals for harmonising workers' compensation arrangements across the Commonwealth, States and Territories and national workers' compensation arrangements for employers with workers in more than one of those jurisdictions;

▶ advising the Ministerial Council on matters relating to OHS or workers' compensation; and

▶ such other functions that are conferred on it by any other Commonwealth Act.

See: http://safeworkaustralia.gov.au/

The Work Health and Safety Act 2011 No. 137 came into force on 1 January 2012.

With the latest amendments it is available at: http://www.comlaw.gov.au/Details/C2014C00471

The objectives are given in Section 3 of the Act as follows:

1. The main object of this Act is to provide for a balanced and nationally consistent framework to secure the health and safety of workers and workplaces by:

    (a) protecting workers and other persons against harm to their health, safety and welfare through the elimination or minimisation of risks arising from work; and

    (b) providing for fair and effective workplace representation, consultation, cooperation and issue resolution in relation to work health and safety; and

    (c) encouraging unions and employer organisations to take a constructive role in promoting improvements in work health and safety practices, and assisting persons conducting businesses or undertakings and workers to achieve a healthier and safer working environment; and

    (d) promoting the provision of advice, information, education and training in relation to work health and safety; and

    (e) securing compliance with this Act through effective and appropriate compliance and enforcement measures; and

    (f) ensuring appropriate scrutiny and review of actions taken by persons exercising powers and performing functions under this Act; and

    (g) providing a framework for continuous improvement and progressively higher standards of work health and safety; and

15

(h) maintaining and strengthening the national harmonisation of laws relating to work health and safety and to facilitate a consistent national approach to work health and safety in this jurisdiction.

2. In furthering subsection (1)(a), regard must be had to the principle that workers and other persons should be given the highest level of protection against harm to their health, safety and welfare from hazards and risks arising from work as is reasonably practicable.

Each of the six state and two territory governments under the Australian Commonwealth government has its own laws on occupational health and safety. The core of these are the Work Health and Safety Act 2011(NSW) or (Queensland). These individual State acts have exactly the same objectives as the federal act.

See for example http://www.deir.qld.gov.au/workplace/resources/pdfs/guide-whs-act-2011.pdf

Most workers in Australia are protected by nationally uniform work health and safety laws. This includes employees, contractors, sub-contractors, outworkers, apprentices and trainees, work experience students, volunteers and employers who perform work.

The WHS Act also provides protection for the general public so that their health and safety is not placed at risk by work activities.

15.3.2 Bangladesh

(Population in 2014: 157,187,000. Labour force: 78,620,000)

The Bangladesh Labour Act 2006 (BLA 2006) both consolidated the law that existed in 25 Acts and Regulations (including the Factories Act 1965, and Industrial Relations Ordinance 1969 – all of which were repealed) whilst at the same time making some significant amendments. This has now been further significantly amended by the Bangladesh Labour (Amendment) Act 2013. The amendments to the Bangladesh Labour Act 2006, adopted on 15 July 2013, will hopefully prove to be the first step towards fulfilling the government's obligation to respect fully the fundamental rights to freedom of

association and collective bargaining and to address the critical need to bolster occupational safety and health.

The 2006 Act imposes obligations in the following areas of Labour law:

– conditions of service and employment including wages and payment (and establishment of Wages Boards), employment of young people, maternity benefits, working hours and leave;
– health, safety, hygiene, and welfare, and compensation for injury;
– trade unions and industrial relations.

Although the Act imposes certain obligations upon trade unions and the government, most duties in the Act are imposed upon 'the employer' of an establishment. The Act defines 'employer' to include:

▶ 'any person responsible for the management, supervision and control of the establishment';
▶ 'every director, manager, secretary, agent or other officer or person concerned with the management of the affairs' of the establishment; and the
▶ 'owner … or the person who controls [the establishment] absolutely or manager or any other effective officer of the management'.

Some obligations in the Act do not state explicitly upon whom the duties have been imposed – this is the case in relation, for example, to many of the health, safety and welfare obligations. However, it is clear, implicitly, that these duties are imposed upon employers – since there is no other way to construe the Act.

The health, safety and welfare duties and obligations that were contained in the old Factories Act – and which have now been transposed to chapters 5 (health and hygiene), 6 (safety), 7 (special provisions with regard to health and safety), and 8 (welfare measures) – only applied to 'factories'. The 2006 Act applies to a much wider category of premises than covered by previous legislation. It applies to all 'establishments' which are defined widely to include shops, hotels, restaurants, factories (though these must employ more than five workers), plantations, docks, transport services, construction sites and 'any premises in which workers are employed for the purposes of carrying on any industry'. It does not apply to the agricultural sector – though it does apply to tea plantations, and certain obligations apply to tea gardens.

Health, safety and welfare duties: the key obligations are summarised in Table 15.3. They are almost identical to the duties that were set out in the Factories Act 1965 and the UK Factories Act 1961.

See: http://www.corporateaccountability.org/international/bangladesh/law/main.htm

For the 2013 Amendment see: http://www.mole.gov.bd/index.php?option=com_content&task=view&id=475&Itemid=543

Table 15.3 Summary of obligations (with relevant section)

SAFETY	HEALTH	WELFARE
Precautions in case of fire (62)	Obligations relating to cleanliness (51)	First-aid appliances (89)
Fencing of machinery (63)	Adequate ventilation and reasonable temperature (52)	Safety Record Book (90)
Safe working next to machinery (64)	Action to prevent exposure to dust and fume (53)	Washing facilities (91)
Suitable striking gear (65)	Arrangements for disposal of waste (54)	Canteens (92)
Safe 'self-acting machines' (66)	Safe humidification (55)	Rest rooms (93)
Casing and guards for new machines (67)	Prevention of overcrowding (56)	Rooms for children (94)
Precautions in relation to cranes, lifts and hoists (68, 69)	Sufficient and suitable lighting (57)	
Precautions relating to revolving machinery (70)	Provision of pure drinking water (58)	
Safe use of pressure plants (71)	Access to latrines and urinals (59)	
Safe means of access (72)	Clean and hygienic spittoons (60)	
Covering and fencing of dangerous spaces (73)		
Precautions relating to carrying of weights (74)		
Precautions against exposure to dangerous fumes in confined space (77)		
Safety measures relating to explosive or flammable gas (78)		

15.3.3 Brazil

(Population in 2014: 203,332,000. Labour force: 107,300,000)

(a) Awakening of OSH issues in Brazil

In the middle of the 1970s, the military rulers in Brazil were horrified to discover that the country was not only the world champion at soccer but was also the world champion for the number of work accidents. In 1976 one worker in six suffered a registered lost-time accident. This grim realisation emerged because the consolidation of the Social Security system had, at least partly, resulted in the production of national statistics for the entire registered workforce.

A series of measures were taken to remedy the situation including the amendment of laws to include: the compulsory employment of OSH professionals in workplaces of particular size and/or with certain levels of risk; an increase to 14 days for the minimum lost time necessary to receive government assistance. Over a 10-year period, more than 100,000 professionals and technicians were trained. Some universities introduced OSH in medical and engineering curricula. Fundacentro was also formed during this period and made an important contribution to the training of professionals. Fundacentro has since become the leading occupational safety and health research institute in Latin America. The role of the state as an agent of social control was also reinforced. It would be true to say that the careers of most researchers and government inspectors contributing to this special issue have largely resulted from the important changes that occurred during this period.

From the 1980s onwards there was a gradual return to full democratic rule and many technicians, researchers and civil servants embraced the workers' cause. The OSH question became highly politicised but nearly all research and prevention efforts were concentrated on the formal labour market and unionised workers. The accident rate among these workers has significantly declined since the mid-1970s. In 1970 the fatal accident rate per 100,000 workers was 31, reducing to 11 by the year 2000. The rate of lost-time accidents in 1970 was 16,600 per 100,000 workers and this declined to 1,100 in 2000. The rate of work-related illness declined far less, from 83 in 1970 to 70 in 2000.

The economically active population in Brazil has grown significantly in recent years – from 89 million in 2003 to 107 million in 2013. During this same period, the

15

481

number of commercial establishments increased 50% and more than 12.5 million new jobs with work permits were created (formal jobs). Despite these advances, the country still faces serious problems in the world of work. Of the 92 million workers inserted in the labour market in 2009, about 50% were informal and did not enjoy many of their constitutional labour rights, making their work an uncertain and risky activity. Some of the informal employers do not pay taxes, significantly shrinking the country's tax base. Finally, despite improvements in combating forced labour and child labour, these are still persistent in the country.

A series of actions were carried out in order to build up a 'Planning System' to strengthen the inspectorate and its social partners. The planning system included: analyses of accident statistics, ergonomics; evaluation of OSH management systems; evaluation of results from inspectorate interventions.

(b) Enhancement of Occupational Health and Safety in Brazil (EOHSBI)

According to the ILO, the need for Occupational Health and Safety is particularly acute in developing and newly industrialised countries such as Brazil, where more than 80% of the world's workers live. In the poorer regions of Brazil, the health, well-being and often the survival of the entire family are critically dependent on the health and productivity of its working members. As in other newly industrialised countries, many Brazilian women, men and youth work in poor and hazardous conditions.

This attitude towards OHS is slowly changing among the better-informed and more-enlightened enterprises and is being replaced by the realisation that workers' health, safety and well-being are integral components of social responsibility, improved productivity and economic sustainability. Yet many industries (especially SMEs) do not know the real benefits of improved OHS performance, or how to achieve these standards, and need the support of institutions like SESI (Serviço Social da Indústria) to guide them.

(c) Brazilian OHS legislation and mandatory programmes

In 1994, the Federal Ministry of Labour and Employment (MTE) introduced a series of mandatory Brazilian OHS programmes, with the objective of reducing occupational accidents and illnesses. These include:

▶ The Medical Control Programme in Occupational Health, *Programa de Controle Médico de Saúde Ocupacional (PCMSO)*;
▶ The Prevention Programme in Environmental Risks, *Programa de Prevenção dos Riscos Ambientais (PPRA, NR-9)*; and
▶ The Work Conditions and Environmental Programme for Construction Industries, *Programa de Condições*

e Meio Ambiente do Trabalho na Indústria da Construção (PCMAT).

The Brazilian regulation, NR-9, requires employers to develop and implement programmes which comply with the standards outlined in the Environmental Risks Prevention Programme (PPRA). The objective of the programme is to identify potential OHS and environmental risks in the workplace, and to provide a framework for the development of effective policies and practices that minimise these risks in enterprises.

These regulations and mandatory programmes that provide the regulatory context for OHS policy and practice in Brazilian industry have not, for various reasons, produced the desired results. Factors that contribute to poor OHS standards in Brazilian industry include the following:

▶ a lack of awareness among enterprises, especially SMEs, of the social and economic value of sustainable OHS practices;
▶ limited access to relevant OHS information, and understanding of how to use this information to develop and implement practices that reduce OHS-related illnesses, accidents and deaths;
▶ the absence of quality management systems to monitor and support the control of OHS risks and hazards;
▶ low capacity for the development and implementation of continuous improvement in OHS standards in the workplace;
▶ superficial compliance with regulations and mandatory programmes;
▶ a poor system for the reporting of work-related accidents, illnesses and deaths resulting in inaccurate and unreliable data;
▶ low capacity for the development of programmes which respond to the needs of specific vulnerable groups;
▶ a national worker accident insurance system that does not provide reimbursement for medical services delivered to employees;
▶ the absence of a forum for impartial arbitration of OHS cases under civil and labour laws;
▶ an insignificant differential in penalties or premium payments between the companies with high and low OHS risk;
▶ an artificial division between policies and programmes for occupational health and general health care services; and
▶ the lack of integration between the regulations and programmes, and the public and private institutions responsible for their enforcement.

In 2010 the ILO reported on labour inspection in Brazil for the promotion of decent work called 'The Good Practices of Labour Inspection in Brazil' http://www.ilo.org/wcmsp5/groups/public/---ed_norm/---declaration/documents/publication/wcms_155946.pdf

15.3.4 Canada

(Population in 2014: 35,540,000. Labour force: 19,080,000)

(a) Occupational health and safety agencies

There are 14 jurisdictions in Canada – one federal, ten provincial and three territorial – each having its own occupational health and safety legislation. For most people in Canada, the provincial or territorial agency in the area where they work should be their point of contact. There are some exceptions to this. Federal legislation covers employees of the federal government and Crown agencies and corporations across Canada. Approximately 10% of the Canadian workforce falls under the OHS jurisdiction of the federal government. The remaining 90% of Canadian workers fall under the legislation of the province or territory where they work.

Occupational health and safety legislation in Canada outlines the general rights and responsibilities of the employer, the supervisor and the worker. There is special 'right-to-know' legislation that applies to hazardous products. It actually comprises several pieces of legislation collectively called WHMIS – the Workplace Hazardous Materials Information System. It is a comprehensive plan for providing information on hazardous materials intended for use in workplaces. WHMIS applies in all Canadian workplaces which are covered by occupational health and safety legislation and where WHMIS-controlled products are used.

The federal health and safety legislation is commonly referred to as the Canada Labour Code Part II and regulations. The Canada Labour Code also applies to employees of companies or sectors that operate across provincial or international borders. These businesses include:

▶ airports;
▶ banks;
▶ canals;
▶ exploration and development of petroleum on lands subject to federal jurisdiction;
▶ ferries, tunnels and bridges;
▶ grain elevators licensed by the Canadian Grain Commission, and certain feed mills and feed warehouses, flour mills and grain seed cleaning plants;

▶ highway transport;
▶ pipelines;
▶ radio and television broadcasting and cable systems;
▶ railways;
▶ shipping and shipping services; and
▶ telephone and telegraph systems.

(b) Provincial and territorial jurisdictions

In each province or territory, there is an Act (typically called the Occupational Health and Safety Act or something similar) which applies to most workplaces in that region. The Act usually applies to all workplaces except private homes where work is done by the owner, occupant or servants. Generally, it does not apply to farming operations unless made to do so by a specific regulation. The legislation should be consulted to find out who is or is not covered.

At the provincial and territorial level, the name of the government department responsible for OHS varies with each jurisdiction. Usually it is called a ministry or department of labour. In some jurisdictions, it is a workers' compensation board or commission that has the responsibility for occupational health and safety. Each provincial or territorial department is responsible for the administration and enforcement of its Occupational Health and Safety Act and regulations. A list of Canadian government departments with chief responsibility for occupational health and safety is available at: http://www.ccohs.ca

Many basic elements (e.g. rights and responsibilities of workers, responsibilities of employers, supervisors, etc.) are similar in all the jurisdictions across Canada. However, the details of the OHS legislation and how the laws are enforced vary from one jurisdiction to another. In addition, provisions in the regulations may be 'mandatory', 'discretionary' or 'as directed by the Minister'.

General responsibilities of governments for occupational health and safety include:

▶ enforcement of occupational health and safety legislation;
▶ workplace inspections;
▶ dissemination of information;
▶ promotion of training, education and research;
▶ resolution of OHS disputes.

(c) Employees' rights and responsibilities

Employees' responsibilities include the following:

▶ responsibility to work in compliance with OHS acts and regulations;
▶ responsibility to use personal protective equipment and clothing as directed by the employer;
▶ responsibility to report workplace hazards and dangers;
▶ responsibility to work in a manner as required by the employer and use the prescribed safety equipment.

15

Employees have the following three basic rights:

▶ right to refuse unsafe work;
▶ right to participate in the workplace health and safety activities through a joint health and safety committee (JHSC) or as a worker health and safety representative;
▶ right to know, or the right to be informed about, actual and potential dangers in the workplace.

(d) Manager or supervisor's responsibilities

As a manager or supervisor, he or she:

▶ must ensure that workers use prescribed protective equipment devices;
▶ must advise workers of potential and actual hazards;
▶ must take every reasonable precaution in the circumstances for the protection of workers.

Managers and supervisors act on behalf of the employer, and hence have the responsibility to meet the duties of the employer as specified in the Act.

(e) Employer's responsibilities

An employer must:

▶ establish and maintain a joint health and safety committee, or cause workers to select at least one health and safety representative;
▶ take every reasonable precaution to ensure the workplace is safe;
▶ train employees about any potential hazards and in how to safely use, handle, store and dispose of hazardous substances and how to handle emergencies;
▶ supply personal protective equipment and ensure workers know how to use the equipment safely and properly;
▶ immediately report all critical injuries to the government department responsible for OHS;
▶ appoint a competent supervisor who sets the standards for performance, and who ensures safe working conditions are always observed.

(f) Health and safety committees

Generally, legislation in different jurisdictions across Canada state that health and safety committees or joint health and safety committees:

▶ must be composed of one-half management and at least one-half labour representatives;
▶ must meet regularly – some jurisdictions require committee meetings at least once every three months while others require monthly meetings;
▶ must be co-chaired by one management chairperson and worker chairperson;
▶ employee representatives are elected or selected by the workers or their union.

More details about these committees are in the Health and Safety Committees Section on the website (www.ccohs.ca).

The role of health and safety committees or joint health and safety committees include:

▶ act as an advisory body;
▶ identify hazards and obtain information about them;
▶ recommend corrective actions;
▶ assist in resolving work refusal cases;
▶ participate in accident investigations and workplace inspections;
▶ make recommendations to the management regarding actions required to resolve health and safety concerns.

An employee can refuse work if he/she believes that the situation is unsafe to either himself/herself or his/her co-workers. When a worker believes that a work refusal should be initiated, then:

▶ the employee must report to his/her supervisor that he/she is refusing to work and state why he/she believes the situation is unsafe;
▶ the employee, supervisor, and a JHSC member or employee representative will investigate;
▶ the employee returns to work if the problem is resolved with mutual agreement;
▶ if the problem is not resolved, a government health and safety inspector is called;
▶ inspector investigates and gives decision in writing.

(g) Enforcement of legislation

The legislation holds employers responsible to protect employee health and safety. Enforcement is carried out by inspectors from the government department responsible for health and safety in each jurisdiction. In some serious cases, charges may also be laid by police or Crown attorneys under Section 217.1 of the Canada Criminal Code (also known as 'Bill C-45'). This section imposes a legal duty on employers and those who direct work to take reasonable measures to protect employees and public safety. If this duty is 'wantonly' or recklessly disregarded and bodily harm or death results, an organisation or individual could be charged with criminal negligence.

(h) Due diligence

Due diligence is the level of judgement, care, prudence, determination and activity that a person would reasonably be expected to do under particular circumstances.

Applied to occupational health and safety, due diligence means that employers shall take all reasonable precautions, under the particular circumstances, to prevent injuries or accidents in the workplace. This duty also applies to situations that are not addressed elsewhere in the occupational health and safety legislation.

To exercise due diligence, an employer must implement a plan to identify possible workplace hazards and carry out the appropriate corrective action to prevent accidents or injuries arising from these hazards.

'Due diligence' is important as a legal defence for a person charged under occupational health and safety legislation. If charged, a defendant may be found not guilty if he or she can prove that due diligence was exercised. In other words, the defendant must be able to prove that all precautions, reasonable under the circumstances, were taken to protect the health and safety of workers.

(i) How does an employer establish a due diligence programme?

The conditions for establishing due diligence include several criteria:

▶ The employer must have in place written OHS policies, practices and procedures. These policies, etc. would demonstrate and document that the employer carried out workplace safety audits, identified hazardous practices and hazardous conditions and made necessary changes to correct these conditions, and provided employees with information to enable them to work safely.

▶ The employer must provide the appropriate training and education to the employees so that they understand and carry out their work according to the established polices, practices and procedures.

▶ The employer must train the supervisors to ensure they are competent persons, as defined in legislation.

▶ The employer must monitor the workplace and ensure that employees are following the policies, practices and procedures. Written documentation of progressive disciplining for breaches of safety rules is considered due diligence.

▶ There are obviously many requirements for the employer but workers also have responsibilities. They have a duty to take reasonable care to ensure the safety of themselves and their co-workers – this includes following safe work practices and complying with regulations.

▶ The employer should have an accident investigation and reporting system in place. Employees should be encouraged to report 'near misses' and these should be investigated also. Incorporating information from these investigations into revised, improved policies, practices and procedures will also establish the employer is practising due diligence.

▶ The employer should document, in writing, all of the above steps: this will give the employer a history of how the company's occupational health and safety programme has progressed over time. Second, it will provide up-to-date documentation that can be used as a defence to charges in case an accident occurs despite an employer's due diligence efforts.

All of the elements of a 'due diligence programme' must be in effect before any accident or injury occurs. If employers have questions about due diligence, they should seek legal advice for their jurisdiction to ensure that all appropriate due diligence requirements are in place.

Due diligence is demonstrated by employer's actions before an event occurs, not after.

See the following for the latest version of the Occupational Health and Safety Act current to 2014:

http://laws-lois.justice.gc.ca/PDF/C-13.pdf

15.3.5 China

(Population in 2014: 1,367,460,000. Labour force: 797,000,000)

(a) Introduction

In China workers are covered by the Law of the People's Republic of China on Work Safety which came into effect in November 2002. The law covers workers in production and business activities and is aimed at enhancing supervision and control over work safety, preventing accidents and keeping their occurrence at a low level, ensuring the safety of people's lives and property and promoting the development of the economy.

The principle of giving first priority to safety issues and laying stress on prevention is upheld. Responsibilities are laid on principal leading members of production and business units who are in full charge of work safety of their own units. Participation of workers in the democratic management of working safely is upheld. The law is enforced by the relevant department under the State Council which must formulate national standards for work safety.

Principal leading members of production and business units are charged with setting up responsibilities for work safety, making arrangements for rules and operating regulations, guaranteeing an effective personal input, supervising and inspecting work safety, setting up rescue plans and reporting accidents to

15

485

the authorities. The decision-making bodies in each unit must make sufficient funds available. Mines, construction units and units manufacturing, storing or marketing dangerous articles must have full-time people for the control of work safety. Other units where there are in excess of 300 people employed must have full-time people for the control of work safety.

The responsibility for work safety still remains with the unit whose managers must have appropriate knowledge of work safety and competence for its control. Education and training is required for all employees in units who must pass the qualification tests to be assigned to posts. Safety equipment must be properly designed, tested, installed, maintained and used. The sources of grievous danger must be assessed and records kept. Protective equipment must be provided as necessary and workplace inspections carried out. Employees have the right to complain about unsafe conditions and practices and to stop work where necessary. They have a right to compensation but must abide by safety rules and operating instructions.

The main laws covering OSH issues include:

▶ People's Republic of China Labour Law (1995)
▶ Code of Safety and Health in Factories (1956)
▶ The Production Safety Law of the People's Republic of China (November 2002)
▶ Trade Union Law (1992, reviewed in 2001)
▶ Factory Safety and Sanitation Regulations.

(b) Summary of health and safety requirements

(i) Management and training

1. The factory must have a health and safety management system.
2. Workers must receive appropriate health and safety training including fire training, production safety, the correct use of protective equipment and first-aid for workers exposed to dangers. Workers should be retrained if there are any new techniques, materials or equipment introduced.
3. A factory with more than 300 employees must establish a production safety committee or should appoint a full-time Safety Officer. If there are less than 300 employees a part-time Safety Officer should be appointed.
4. The employer must provide healthcare facilities for all employees and provide regular check-ups for workers in hazardous jobs.
5. Pregnant women must not work with hazardous machinery or hazardous chemicals.
6. Women who are pregnant or menstruating must not work in low temperatures or do heavy labour.
7. The employer must pay employment injury insurance premiums according to the law.

8. The main production site must ensure that any sub-contracting units have suitable conditions for safe production. A special agreement with the contractor or leaseholder must be entered into, specifying the duties and functions of each party in the administration of production safety (November 2002).
9. Each worker's contract of employment must list any possible occupational diseases associated with the job. If the employer does not do this, the worker is entitled to refuse to perform hazardous tasks, and the employer cannot dismiss the worker on these grounds.
10. Workers are allowed to stop work without penalty if the working conditions are unsafe.
11. The factory must truthfully report any production safety accidents.

(ii) Fire precautions

1. Factories must comply with the detailed regulations on fire precautions including the number of fire extinguishers, fire safety signs, fire exits, etc.
2. Sufficient fire exits must be provided from all areas of the production site. These must be kept unlocked and clear of obstructions.
3. Smoking must be strictly prohibited at worksites where inflammable and explosive substances are used.

(iii) Safe use of machinery and chemicals

1. The employer must ensure that machinery and other equipment is safe.
2. The employer must provide appropriate personal protective equipment (PPE) free of charge and must regularly check the equipment (appropriate PPE is specified in the Code of Safety and Health).
3. Electrical equipment and wires must be safely insulated, equipped with safety fuses and regularly inspected and repaired.
4. Clear safety warning signs should be displayed on relevant equipment.
5. The layout of machines, worktables and equipment must facilitate safe operation and must have a passage of at least one metre between them.
6. Equipment that produces harmful steam, gas or dust must be sealed tightly or sufficiently ventilated.
7. Raw materials and products must not obstruct work or free passage.
8. Where hazardous substances are used, relevant supervisions, controls and emergency plans should be provided. Employees must be informed of the relevant emergency measures.
9. Hazardous substances should be stored in a separate area to the production site.
10. Washing facilities should be provided in areas where acids or other corrosive substances are used.

(iv) Hygiene and factory environment

1. The worksite should be kept clean and neat.
2. Workers must have access to clean drinking water. The containers and drinking vessels should be cleaned and sterilised every day.
3. Toilet facilities must be provided near the worksite and must be segregated by sex.
4. Toilets without drainage systems must have covers on the holes.
5. Showers should be provided in the bathrooms.
6. Female sanitation rooms should be provided near the worksite with hot water, washing tub and waste disposal bins.
7. The worksite should be provided with hand-washing facilities if needed, with soap available.
8. The worksite shall be provided with covered spittoons, which must be cleaned at least once a day.
9. Locker rooms and rest rooms should be equipped with lockers or clothes-hangers.
10. Number of toilets:
    ▷ Men's: if there are less than 100 male workers, the factory should provide one toilet for every 25 workers. If more than 100 male workers, the factory must provide an additional toilet for every 50 workers. There should also be the same number of urinals as toilets.
    ▷ Women's: if there are less than 100 female workers, the factory should provide one toilet for every 20 workers. If more than 100 female workers, the factory should provide an additional toilet for every 35 workers.

(v) Accommodation

1. Each worker must have sufficient living space. (Some provinces have specific laws on this point, for example in Shenzhen the minimum requirement is 2.0 square metres per worker.)
2. Dormitories and toilet facilities must be segregated by sex.
3. Workers must have access to clean, running drinking water in dormitories.
4. Dormitories must not be connected with production or warehouse areas.
5. Dormitories must be a safe distance from areas where hazardous chemicals are used or stored.
6. Fire regulations specify that:
    ▷ there must be enough exits to allow people to leave in an emergency (two fire exits from each floor)
    ▷ exits must be marked, unlocked and clear of obstructions
    ▷ there must be an audible fire alarm.

(vi) Age of workers

1. The minimum working age is 16 throughout China.
2. If a worker under the age of 16 is employed at a factory, the employer is responsible for returning the child to his/her parent or guardian's place of residence and must cover all costs associated with this.
3. Young workers (between the ages of 16 and 18) must:
    ▷ be registered with the local labour department
    ▷ not do hazardous work
    ▷ have regular physical examinations.

(vii) Hours of work

1. Standard hours per day = 8 hours.
2. Standard hours per week = 40 hours.
3. Overtime:
    ▷ Workers may work a maximum of 3 hours overtime on any one day and 36 hours per month
    ▷ Overtime premiums (150% of normal hourly pay)
    ▷ Women who are seven or more months pregnant or breast-feeding must not work overtime or at night
    ▷ Local waivers for extended hours are permitted if in writing from the Department of Labour.

(viii) Time off

1. Workers are entitled to two days off per week, on average. Any work on these days off should be paid as overtime.
2. Workers must have at least 11 days statutory holiday per year.
    ▷ 1 March – 1 day
    ▷ Chinese New Year (Spring Festival) – 3 days
    ▷ Ching Ming Festival – 1 day
    ▷ Labour Day (1 May) – 1 day
    ▷ Dragon Boat Festival – 1 day
    ▷ Mid-Autumn Festival – 1 day
    ▷ National Day (1 October) – 3 days.
3. Maternity leave: 90 days paid leave includes 15 days before and 75 days after the birth, with 15 additional days for difficult labour or twins.
4. After giving birth, women are entitled to one paid working hour per day for baby feeding until the baby is aged 12 months.
5. Workers' entitlement to additional paid leave is set out in local regulations. These cover:
6. Annual leave: workers should generally have 5 days paid annual leave after 1 year but less than 10 years of accumulative employment; 10 days paid leave after 10 years but less than 20 years of accumulative employment; 15 days after 20 years of accumulative employment.
7. Sick leave: depending on length of service, between 3 and 24 months. Time during prescribed medical treatment and recuperation period must be paid at not less than 80% of the local minimum wage (unless overridden by local law which may prescribe a different level of pay). This applies even if the illness or injury is non-work-related.
8. Other paid leave, provisions vary between localities but in general:

15

▷ Marriage leave: 3 days. 10 days if groom and bride are at least 25 and 23 years old respectively.

▷ Maternity leave: 90 days leave includes 15 days before and 75 days after the birth, with 15 additional days for difficult labour or twins.

▷ Conjugal leave: 30 days annually (if spouses living separately).

▷ Parental leave: 20 days every 4 years (for couples with parents in another province).

▷ Filial leave: 20 days per year or 45 days every 2 years (for single workers whose parents live in another province).

▷ Bereavement leave: 1–3 days.

Workers are entitled to normal pay when taking this leave above.

15.3.6 Egypt

(Population in 2014: 87,365,000. Labour force: 27,690,000)

In Egypt occupational health and safety is covered by the Labour Code No. 12 of 2003, Labour Law (as amended by Law No. 90/2005).

The preface to the code sets out new features of the Labour Code, including the balance in duties and responsibilities between workers and employers; due consideration to the social dimensions of the economy; new provisions that reflect global and local economic changes; reflection of fundamental principles contained in international labour standards; safeguarding formerly acquired rights; and recognition of workers' right to a periodic annual wage increase of at least 7%.

The Code consists of six parts:

Part I: Definitions and general provisions.

Part II: Individual labour relations (chapters and sections on the employment of Egyptians at home and abroad, the employment of foreigners, labour contracts, wages, leave, workers' duties, supervision of workers and accountability, hours of work, employment of women and children and the termination of the labour relationship).

Part III: Vocational training and guidance (the licensing and exercise of vocational training, skill assessment and trade practice licensing).

Part IV: Collective labour relations (consultation and cooperation, collective bargaining, collective labour agreements, labour disputes).

Part V: Occupational safety and health and the work environment (definitions, worksites, structures and permits, safeguarding the work environment, social and health services, occupational safety and health and environment inspection, regulation of occupational safety and health and environment equipment, research bodies and advisory services).

Part VI: Labour inspection and sanctions (labour inspection, judicial authority and sanctions).

The Code excludes from its scope of application the following categories: public servants including those working in local government units and public authorities, domestic servants and members of the employer's family who are also his dependents. Debts due to workers under the provisions of this Code have priority over all other debts. Provisions of the Code apply to foreign workers provided there is reciprocal treatment.

Part V applies to all worksites and establishments including offshore and means of transport. The following issues are covered in Part V:

▶ hazards in the working environment including heat and cold, noise and vibration, lighting, harmful and dangerous radiation, atmospheric pressure changes, static and dynamic electricity, explosion risks;

▶ dangers from work tools and machines, lifting equipment, articles, apparatuses, means of transport, handling and power transmission;

▶ dangers from construction, building, digging and risk of collapse;

▶ dangers from infection with bacteria, viruses, fungi, parasites and any other biological risks; this includes dealing with sick people and infected animals;

▶ dangers from chemicals whether solid, liquid or gaseous substances including concentrations in work areas, stock levels, precautions for transporting, storing, handling and using dangerous chemicals. Also covered is the requirement to keep a register with all the data concerning each chemical (obtained from suppliers) and a register to record exposures, ensuring that containers are properly labelled and training workers to deal with dangerous chemicals;

▶ must supply first-aid, means of rescue, clean-up of the workplace, safe organisation of the workplace;

▶ areas providing food and drink must have health certificates indicating that they are free of epidemic and contagious diseases;

▶ fire risks including the provision of latest fire-fighting and protection equipment, alarms, early warning, protective insulation, and automatic fire extinguishing equipment whenever necessary;

▶ evaluation of risks of expected industrial and natural disasters, and prepare an emergency plan;

▶ medical pre-employment examinations;

▶ training workers, informing them of any risks and compelling them to use the protective measures and PPE which must be provided free;

▶ workers are required to use protective measures and PPE and to care for them;

▶ carrying out daily or every shift workplace inspections and where necessary medical examination and any complaints of sickness;

▶ providing means of transport to work where there is no suitable public transport;

▶ providing suitable food and dwellings in remote areas (including suitable dwellings for married workers);

▶ where over 50 workers are employed must provide the necessary social and cultural services to workers;

▶ in industrial establishments where 15 workers or more are employed or 50 or more in non-industrial ones a six-monthly report on diseases and injuries must be sent in to the manpower directorate in the first half of July and January. For grave accidents notice must be given within 24 hours;

▶ administrative authorities have the obligation to maintain a specialised authority to carry out inspections.

See: http://www.egypt.gov.eg/english/laws/labour/default.aspx

15.3.7 European Union

(Population in 2014: 507,416,000. Labour force: 228,600,000)

In the European Union, member states have enforcing authorities to ensure that the basic legal requirements relating to occupational safety and health are met. In many EU countries, there is strong cooperation between employer and worker organisations (e.g. unions) to ensure good OSH performance as it is recognised this has benefits for both the worker (through maintenance of health) and the enterprise (through improved productivity and quality). In 1996 the European Agency for Safety and Health at Work was founded.

Member states of the European Union have all transposed into their national legislation a series of Directives that establish minimum standards on occupational safety and health. These Directives (of which there are about 20 on a variety of topics) follow a similar structure requiring the employer to assess the workplace risks and put in place preventative measures based on a hierarchy of control. This hierarchy starts with elimination of the hazard and ends with personal protective equipment.

(a) Main principles

The key principles relating to the prevention and protection of the health and safety of workers are defined in the 1989 Framework Directive (89/391/EEC). It constitutes the basis for all subsequent individual Directives.

The basic objective of the Framework Directive is to encourage improvements in occupational health and safety and it covers all sectors of activity, both public and private.

It establishes the principle that the employer has a duty to ensure the safety and health of workers in every aspect related to their work. The employer is obliged to develop an overall health and safety policy, namely by:

▶ assessing the safety and health risks which cannot be avoided, updating these assessments in the light of changing circumstances, and taking the appropriate preventative and protective measures;

▶ making a record of the risk assessment and of the list of accidents at work;

▶ informing workers and/or their representatives about potential risks and preventative measures taken;

▶ consulting workers and/or their representatives on all health and safety matters and ensuring their participation;

▶ providing job-specific health and safety training;

▶ designating workers to carry out activities related to the prevention of occupational risks;

▶ implementing measures on first-aid, fire-fighting and the evacuation of workers.

The worker, on the other hand, also has several obligations to, inter alia, follow employers' health and safety instructions or to report potential dangers.

The Framework Directive also promotes the workers' right to make proposals relating to health and safety, to appeal to the competent authority and to halt work in the event of serious danger, as part of the participative approach laid down by the Directive.

The underlying goal is to adequately protect the health and safety of workers and ensure that at the end of his/her working day, the worker will return to his/her family in good health.

15

(b) Minimum safety and health requirements for the workplace

Whilst the general responsibilities of employers for the safety and health of their workers are covered under the Framework Directive (89/391/EEC), special provisions have been introduced according to the type of location where work is being undertaken. These are covered under a Directive (89/654/EEC) on the minimum safety and health requirements for the workplace, and under subsequent individual location-specific Directives.

This covers, in particular, the construction sector, which has one of the worst occupational safety and health records in Europe. Workers in this sector have greater exposure to biological, chemical and ergonomic risk factors, as well as noise and temperature.

The Directive requires health and safety considerations to be taken on board during the design and organisation of projects. It also provides for the establishment of a chain of responsibility, linking all the players involved to minimise any risks.

Besides being obliged to inform, consult and seek the participation of workers on the matters covered by the Directive, employers must comply with other general requirements such as the regular cleaning of workplaces.

(c) Temporary and mobile worksites

Temporary and mobile worksites are covered by a Directive (92/57/EEC) which sets minimum safety and health requirements. Coordinators for health and safety have to be appointed and a plan prepared before construction begins, and procedures must be implemented to ensure that risks are adequately managed.

The Directive applies to all sectors of activity, both public and private (e.g. industrial, agricultural, commercial, administrative, service, educational, cultural, leisure). The Directive does not apply to drilling and extraction in the extractive industries.

(d) Work equipment

Providing workers with the right equipment is an important factor in ensuring their health and safety. The minimum safety and health requirements for work equipment were laid down by a 1989 EU Directive (89/655/EEC) which has since then been updated by amending acts (95/63/EC).

This Directive, which came into force in December 1992, obliges employers to choose equipment specific to the working conditions and the known hazards.

It furthermore requires that workers be given written instructions, adequate health and safety information and training. In 2001 special provisions were introduced relating to working at height (ladders, scaffolding and ropes, etc.) where minimum requirements were introduced (Directive 2001/45/EC).

Specifically, the EU works to ensure work equipment is correctly adapted to workers' safety needs in the following areas:

(e) Personal protective equipment (PPE)

EU legislation was introduced in 1989 (89/656/EEC) establishing the minimum requirements for the assessment, selection and correct use of personal protective equipment. This is defined as equipment designed to be worn or held by the worker to protect him against hazards encountered at work.

The employer must provide the appropriate equipment free of charge and ensure that it is in good working order and in a hygienic condition.

Before choosing personal protective equipment, the employer is required to assess the extent to which it complies with the conditions set out in the Directive to analyse the risks to see if they cannot be avoided by other means.

It is worth noting that another Directive exists which sets conditions for placing safe personal protective equipment on the market (89/686/EEC).

(f) Manual handling of loads involving risk

In 1990, EU legislation was introduced (Directive 90/269/EEC) covering the manual handling of loads to avoid potential back injuries to workers.

Under this legislation employers must take appropriate steps to avoid the need for manual handling of loads, or, where this cannot be avoided, to take the appropriate organisational measures to reduce the risk involved. Workers need to be informed about the weight of a load and its characteristics and be properly trained in correct handling.

(g) Display screens

Computer screens used at work are covered by the 1990 EU legislation (Directive 90/270/EEC) on the minimum safety and health requirements for work with display screen equipment.

Employers are obliged to analyse their workstations, evaluate the safety and health conditions, and remedy any impact on eyesight, physical problems or mental stress.

The daily work routines of workers must be planned in such a way as to provide periodic breaks or changes of activity.

In addition, workers are entitled to have their eyesight tested before using display screens and at regular intervals thereafter if they experience visual difficulties. They must be provided with special corrective appliances, if required, at no additional cost to them.

(h) Health and safety signs at work

Safety signs are an important method of avoiding accidents at work and, in 1992, EU legislation was introduced (Directive 92/58/EEC) laying down the minimum requirements.

Signs must be provided by employers where hazards cannot be avoided or adequately reduced by other preventative measures.

'Health and/or safety signs' can in practice be a signboard, a colour, an illuminated sign or acoustic signal, a hand signal or a verbal communication.

A safe and healthy working environment is an essential element of an EU citizen's quality of work. It is also a collective concern and the social and economic benefits of improving health and safety at work are recognised by national governments across the EU. The EU's work in the area of the prevention and protection of the health and safety of workers at work began as far back as 1952 under the European Coal and Steel Community. In line with the Treaty, the EU defines at European level the minimum requirements in the field of health and safety at work.

In this regard, the European Commission is constantly monitoring developments and presents legislative proposals to the Council and the European Parliament with a view to addressing new risks or to adapt continuously the EU legislative framework to take into account the state of art and changes in the workplace.

The main principles of prevention and protection of health and safety of workers, that are applicable to all sectors of activity, are laid down in the 1989 Framework Directive (89/391/EEC).

(i) Chemical agents

Chemicals play an important part in many aspects of everyday life. In the workplace too, chemical agents are encountered in a wide range of sectors and circumstances. So in the EU's workplaces, the first need is to determine whether hazardous chemical agents are present. Risks to workers' safety and health have to be assessed, and risk management measures should be established with the aim of reducing workers' exposure to those risks.

A Directive (98/24/EC) lays down minimum requirements which apply to all hazardous chemical agents that are or may be present in EU workplaces. These minimum requirements comprise:

▶ Indicative and binding Occupational Exposure Limit values (OELs) and biological limit values (employers and workers are to be kept informed of these).

▶ Determination and risk assessment of hazardous chemical agents.

▶ General principles for prevention of risks associated with hazardous chemical agents.

▶ Specific protection and prevention measures.

▶ Arrangements to deal with accidents, incidents and emergencies.

▶ Information and training for workers.

▶ Prohibition of the production, manufacture or use of certain chemicals to prevent workers' exposure.

▶ Health surveillance.

▶ Consultation and participation of workers.

▶ Preparation and adoption of technical guidance.

The European Commission, Directorate General for Employment, Social Affairs and Equal Opportunities, has published practical, non-binding guidelines on this. The aim is to help EU member states in drawing up national guidelines to facilitate compliance of the national measures that implement the Directive on chemical agents (98/24/EC).

(j) Physical agents

A number of individual Directives have been enacted controlling the exposure of workers to potentially damaging physical agents in the workplace such as explosive atmospheres, vibration, noise, electromagnetic fields, optical radiation and ionising radiation. These Directives lay down minimum requirements for worker protection.

(i) Risk of explosive atmospheres

An 'explosive atmosphere' is a mixture with air, under atmospheric conditions, of flammable substances in the form of gases, vapours, mists or dusts in which, after ignition has occurred, combustion spreads to the entire unburned mixture.

Under a Directive (1999/92/EC) on minimum requirements for improving the safety and health protection of workers potentially at risk from explosive atmospheres, the employer must take technical and/or organisational measures to prevent the formation of explosive atmospheres, prevent the ignition of explosive atmospheres, and reduce the effects of an explosion in such a way that workers are not at risk.

The employer must ensure that a health and safety protection document, describing explosion protection measures, is prepared and kept up to date.

See also 'Workplaces – Equipment and protective systems used in potentially explosive atmospheres'.

(ii) Exposure to mechanical vibration

Mechanical vibration poses a potential risk to workers as it may give rise to musculoskeletal, neurological and vascular disorders. A specific Directive (2002/44/EC) sets out to improve protection of workers against the risks involved and lays down minimum health and safety requirements.

The Directive specifies two different types of vibration: vibration which, when transmitted to the human hand–arm system, entails risks to the health and

15

safety of workers (in particular vascular, bone or joint, neurological or muscular disorders), and vibration which, when transmitted to the whole body, entails risks to the health and safety of workers (in particular lower-back morbidity and trauma of the spine).

The Directive lays down exposure limit values and exposure 'action values' above which employers must take measures.

(iii) Exposure to noise

Exposure to noise and, notably, risks to hearing are dealt with under a Directive (2003/10/EC) where the exposure limit value is fixed at 87 decibels (taking into account the attenuation provided by the individual hearing protectors worn by the workers) and the exposure action values are fixed at 80 decibels (lower value) and 85 decibels (upper value).

The employer must assess and, if necessary, measure the levels of noise to which workers are exposed. The results of this assessment must be recorded on a suitable medium and kept up to date on a regular basis.

If the risks arising from exposure to noise cannot be prevented by other means, properly fitting individual hearing protectors must be made available to workers and used by them in accordance with a Directive (89/656/EEC) on the use of personal protective equipment.

(iv) Electromagnetic fields

Exposure to electromagnetic fields was covered under a Directive (2004/40/EC) which was repealed and replaced with Directive 2013/35/EU in June 2013. It lays down two types of value for exposure of workers: 'exposure limit values' (frequencies that are recognised as having harmful effects on the human cardiovascular system or the central nervous system) and 'action values', or values above which employers must take the measures specified in the Directive. Member states are required to bring into force any laws, regulations and/or administrative provisions necessary to comply with the Directive by 1 July 2016.

The EU will publish a Practical Guide early in 2016.

(v) Exposure to artificial optical radiation

In certain activities workers may be exposed to artificial optical radiation from, for example, laser equipment or UVA which can have chronic adverse effects on the eyes and skin.

A Directive (2006/25/EC) on the minimum health and safety requirements regarding the exposure of workers to risks arising from physical agents (artificial optical radiation) reduces the level of exposure to this radiation both in the design of workstations to reduce the risks at source and also fixes exposure limit values for workers exposed to non-coherent radiation and laser radiation.

The workers or their representatives must receive the necessary information and training, for example in the use of protective equipment.

(vi) Dangers arising from ionising radiation

This area falls under the provision of the European Atomic Energy Community (Euratom) Treaty. Issues relating to the protection of the health of workers and the public against the dangers arising from ionising radiation are dealt with under a specific Directive (96/29/Euratom).

Other issues

Other key issues affecting health and safety at work adopted by the EU include:

► Categories of workers
► Carcinogenic agents
► Asbestos
► Biological agents
► Psychosocial factors
► Ergonomics
► Occupational diseases.

15.3.8 India

(Population in 2014: 1,261,370,000. Labour force: 487,300,000)

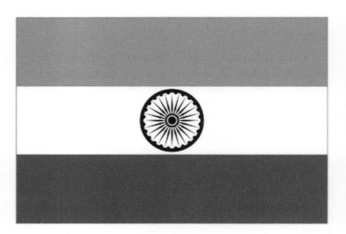

(a) Factories Act 1948 and 1987 amendments

In India workers are covered by the Factories Act 1948 (Act No. 63 of 1948), as amended by the Factories (amendment) Act, 1987 (Act No. 20 of 1987). It covers 10 or more workers (20 or more where there is no power used) working or who were working on any day in the last 12 months where a manufacturing process is carried on. Workers include employees and agency workers. Enforcement is by state and local authority inspectors who have significant powers of entry and inspection. Most responsibilities are placed on the factory occupier who must inform the authorities if a new manager is appointed.

The general duties were updated by the 1987 Act and are in line with more recent health and safety legislation such as the UK HSW Act 1974. It includes requirements that are so far as is reasonably practicable and covers health, safety and welfare. It covers safe plant and equipment, safe use, handling, storing and transport of articles and substances, provision of information, instruction, training and supervision, provision of safe places of work throughout the factory and safe access and egress and the provision of a safe working environment without risks to the health of workers. The Act also requires adequate facilities and arrangements for welfare at work.

Every occupier must provide a written statement of their general policy and their organisation and arrangements. Manufacturers, importers or suppliers have to ensure that articles are properly designed and constructed. Specific prescriptive requirements are laid down for health and cleanliness, ventilation and temperatures, dust and fumes, humidification, overcrowding, lighting, drinking water and toilets.

Safety requirements include the fencing of machinery, cranes, hoists and lifting equipment, pressure plant, floors, stairs, pits and openings and the lifting of excessive weights. Requirements also cover dangerous fumes, gases and explosive or inflammable dusts, etc. Fire precautions are included together with appropriate building maintenance. In any factory where 1,000 or more people are employed or where the State government thinks there are risks, a safety officer must be employed if so required by the State government.

There are special provisions for detailed notification of hazardous processes with emergency plans and measures to control the hazards. Workers are given the right to complain and participate in safety management where there are hazardous processes. The employment hours and requirements for adults, women, young people and children are laid down in detail. The reporting of accidents and cases of disease are also covered by the Act.

For full text see ILO NATELEX data base.

Table 15.4 Summary of obligations under the Factory Acts

Section 7(1)	Notification of factory occupation – 15 days before.
Section 7(4)	Notification of a new manager being appointed – within 7 days.
Section 7A(2)	(a) the provision and maintenance of plant and systems of work in the factory that are safe and without risks to health; (b) the arrangement in the factory for ensuring safety and absence of risks to health in connection with the use, handling, storage and transport of articles and substances; (c) the provision of such information, instruction, training and supervisions as are necessary to ensure the health and safety of all workers at work; (d) the maintenance of all places of work in the factory in a condition that is safe and without risks to health and the provision and maintenance of such means of access to, and egress from, such place as are safe and without such risks; (e) the provision, maintenance or monitoring of such working environment in the factory for the workers that is safe, without risks to health and adequate as regards facilities and arrangements for their welfare at work.
Section 7A(3)	(3) Except in such cases as may be prescribed, every occupier shall prepare, and, as often as may be appropriate, revise, a written statement of his general policy with respect to the health and safety of the workers at work and the organisation and arrangements for the time being in force for carrying out that policy, and to bring the statement and any revision thereof to the notice of all the workers in such manner as may be prescribed.

Section 7B(1)	(1) Every person who designs, manufactures, imports or supplies any article for use in any factory shall:
	 (a) ensure, so far as is reasonably practicable, that the article is so designed and constructed as to be safe and without risks to the health of the workers when properly used;
	 (b) carry out or arrange for the carrying out of such tests and examination as may be considered necessary for the effective implementation of the provisions of clause (a);
	 (c) take such steps as may be necessary to ensure that adequate information will be available:
	    (i) in connection with the use of the article in any factory;
	    (ii) about the use for which it is designed and tested; and
	    (iii) about any conditions necessary to ensure that the article, when put to such use, will be safe, and without risks to the health of the workers:
	
	 Provided that where an article is designed or manufactured outside India, it shall be obligatory on the part of the importer to see:
	   (a) that the article conforms to the same standards if such article is manufactured in India, or
	   (b) if the standards adopted in the country outside for the manufacture of such article is above the standards adopted in India, that the article conforms to such standards.

Section 8	Appointment of inspectors.
Section 9	Powers of inspectors.
Section 10	Appointment of certifying surgeons who may appoint other medical practitioners to carry out medical examinations under the Factories Acts.
Section 11	Cleanliness and removal of dirt and refuse, repainting walls, drainage of floors.
Section 12	Disposal of waste and effluents.

15

Section 13	Adequate ventilation and fresh air circulation. Maintain reasonable comfort conditions – insulation of roofs and walls – separating high temperature processes.
Section 14	(1) In every factory in which, by reason of the manufacturing process carried on, there is given off any dust or fume or other impurity of such a nature and to such an extent as is likely to be injurious or offensive to the workers employed therein, or any dust in substantial quantities, effective measures shall be taken to prevent its inhalation and accumulation in any workroom, and if any exhaust appliance is necessary for this purpose, it shall be applied as near as possible to the point of origin of the dust, fume or other impurity, and such point shall be enclosed so far as possible.
Section 15	Must provide means of regulating humidity.
Section 16	No room in any factory shall be overcrowded to an extent injurious to the health of the workers employed therein.
Section 17	In every part of a factory where workers are working or passing, there shall be provided and maintained sufficient and suitable lighting, natural or artificial, or both.
Section 18	In every factory effective arrangements shall be made to provide and maintain at suitable points conveniently situated for all workers employed therein a sufficient supply of wholesome drinking water.
Section 19	Provision of separate and sufficient latrines and urinals (males).
Section 20	Provision of spittoons.

Section 21	Fencing of machinery – in every factory the following, namely:
	(i) every moving part of a prime-mover and every flywheel connected to a prime-mover, whether the prime-mover or flywheel is in the engine-house or not;
	(ii) the headrace and tailrace of every water-wheel and water-turbine;
	(iii) any part of a stock bar which projects beyond the head stock of a lathe; and
	(iv) unless they are in such position or of such construction as to be safe to every person employed in the factory as they would be if they were securely fenced, the following, namely:
	    (a) every part of an electric generator, a motor or rotary convertor;
	    (b) every part of transmission machinery; and
	    (c) every dangerous part of any other machinery; shall be securely fenced by safeguards of a substantial construction which shall be constantly maintained and kept in position while the parts of machinery they are fencing, are in motion or in use.

Section 22	Work on or near machinery in motion only by specially trained adult males with special clothing – restrictions on examination and subsequent necessary lubrication and adjustment of machines in motion.
Section 23	Employment restrictions for young persons on dangerous machines.
Section 24	Provision of special striking gear for cutting off power and other means of cutting off power in emergencies.
Section 25	Safety around self-acting machinery.
Section 26	Casing of new machinery to make it safe.
Section 27	Prohibition of employment of women and children near cotton-openers.
Section 28	Every hoist and lift shall be: (i) of good mechanical construction, sound material and adequate strength; (ii) properly maintained, and shall be thoroughly examined by a competent person at least once in every period of six months, and a register shall be kept containing the prescribed particulars of every such examination; (iii) properly protected with enclosures and interlocked gates. When used for carrying persons: Fitted with additional ropes or chains, automatic holding devices and brakes.

Section 29	Safety of lifting machinery, chains, ropes and lifting tackle.
	(a) all parts, including the working gear, whether fixed or movable, of every lifting machine and every chain, rope or lifting tackle shall be:
	    (i) of good construction, sound material and adequate strength and free from defects;
	    (ii) properly maintained; and
	    (iii) thoroughly examined by a competent person at least once in every period of 12 months, or at such intervals as the Chief Inspector may specify in writing, and a register shall be kept containing the prescribed particulars of every such examination.
	Plus requirement for marking of safe working load.

Section 30	Safety of revolving machinery like grinding wheels.
Section 31	Safety and examination of pressure plant.
Section 32	Safety of floors stairs and means of access.
Section 33	Safety at pits, sumps, openings in floors, etc.
Section 34	Lifting of excessive weights.
Section 35	Protection of eyes.
Section 36	Precautions against dangerous fumes, gases, etc.
Section 36A	Precautions regarding the use of portable electric light.
Section 37	Explosive or inflammable dust, gases, etc.

Section 38	Precautions in case of fire including: (a) safe means of escape for all persons in the event of a fire, and (b) the necessary equipment and facilities for extinguishing fire.
Section 39	Powers to require specifications of defective parts or test of stability.
Section 40	Safety of buildings and machinery.
Section 40A	Maintenance of buildings.

Section 40B	Safety Officers
	(1) In every factory:
	  (i) wherein 1,000 or more workers are ordinarily employed, or
	  (ii) wherein, in the opinion of the State government, any manufacturing process or operation is carried on, which process or operation involves any risk of bodily injury, poisoning or disease or any other hazard to health, to the person employed in the factory, the occupier shall, if so required by the State government by notification in Official Gazette, employ such number of Safety Officers as may be specified in that notification.
	(2) The duties, qualifications and conditions of service of Safety Officers shall be such as may be prescribed by the State government.

Section 41A	Provisions for hazardous processes.
Section 41B	Compulsory disclosure of information by the factory occupier.
Section 41C	Specific responsibilities of the occupier in relation to hazardous processes.
Section 41D	Powers of Central government to appoint Inquiry Committee.
Section 41E	Emergency standards.
Section 41F	Permissible limits of exposure of chemical and toxic substances.
Section 41G	Workers' participation in safety management.
Section 41H	Right of workers to warn about imminent danger.
Section 42	Washing facilities.
Section 43	Facilities for storing and drying clothing.
Section 44	Facilities for seating.
Section 45	First-aid provision.
Section 46	Canteens – over 250 workers.
Section 47	Shelters, rest rooms and lunch rooms.
Section 48	Crèches.
Section 49	Welfare Officers.
Chapter VI	Working hours of adults.
Chapter VII	Employment of young persons.
Chapter VIII	Annual leave with wages.
Chapter IX	Special powers.
Chapter X	Penalties and procedures.

(b) National Policy on Safety, Health and Environment at Work Place

In 2010 the Government of India issued a new National Policy on Safety, Health and Environment at Work Place. In this policy they state that:

'The fundamental purpose of this National Policy on Safety, Health and Environment at work place, is not only to eliminate the incidence of work related injuries, diseases, fatalities, disaster and loss of national assets and ensuring achievement of a high level of occupational safety, health and environment performance through proactive approaches but also to enhance the well-being of the employee and society, at large. The necessary changes in this area will be based on a coordinated national effort focused on clear national goals and objectives.'

Every Ministry or Department may work out their detailed policy relevant to their working environment as per the guidelines on the National Policy.

GOALS:

The Government firmly believes that building and maintaining national preventive safety and health culture is the need of the hour. With a view to develop such a culture and to improve the safety, health and environment at work place, it is essential to meet the following requirements:

▶ providing a statutory framework on Occupational Safety and Health in respect of all sectors of industrial activities including the construction sector, designing suitable control systems of compliance, enforcement and incentives for better compliance;

▶ providing administrative and technical support services;

▶ providing a system of incentives to employers and employees to achieve higher health and safety standards;

▶ providing for a system of non-financial incentives for improvement in safety and health;

15

▶ establishing and developing the research and development capability in emerging areas of risk and providing for effective control measures;

▶ focusing on prevention strategies and monitoring performance through an improved data collection system on work related injuries and diseases;

▶ developing and providing required technical manpower and knowledge in the areas of safety, health and environment at work places in different sectors;

▶ promoting inclusion of safety, health and environment, improvement at work places as an important component in other relevant national policy documents;

▶ including safety and occupational health as an integral part of every operation.

The action programme set up includes:

(i) Enforcement

With the provision of effective enforcement mechanisms for enforcing all laws and regulations as well as the provision of compensation and rehabilitation of affected persons. It also includes subsidies and loans to help enforcement. The amendment of existing laws and the adoption of national standards, including sharing best practice and developing new innovative enforcement methods including financial incentives. Lastly it is concerned with the setting up of safety and health committees.

(ii) National standards

Development of appropriate standards, codes of practice and manuals consistent with international standards.

(iii) Compliance

Includes:

▶ encouraging government to assume the fullest responsibility for the administration and enforcement of work place standards;

▶ adopting a systems approach with continuous improvement to include self-regulation and auditing mechanisms, which can authenticate management systems;

▶ measures to prevent catastrophes;

▶ recognising best practice;

▶ provision of adequate penal deterrents;

▶ encouraging 'responsible care';

▶ specifically focusing on certain occupational diseases like pneumoconiosis and silicosis;

▶ promoting safe and clean technology.

(iv) Awareness

The policy encourages safety, health and environment awareness activities and consultation between employers' and employees' representatives including the local and national communities. It also encourages

schools, professional and vocational courses to include safety health and environment in their teaching.

(v) Research and Development

The policy encourages research and requires the establishment of priorities and coordination nationally.

(vi) Occupational safety and health skills development

The policy requires the provision of training programmes to increase the number and competence of people engaged in OSH and environment at the work place. (The NEBOSH International Certificate is part of fulfilling this policy.) Also included are: the provision of information and advice to employers' and employees' organisations; the establishing of occupational health services.

(vi) Data collection

The policy includes the provision of workplace and national statistics on work-related injury and disease plus the self-employed. Developing the means of access to the data and other information is part of the policy.

(vii) Review

The policy requires that:

> An initial review and analysis shall be carried out to ascertain the current status of safety, health and environment at workplace and building a national Occupational Safety and Health profile; and

> National Policy and the action programme shall be reviewed at least once in five years or earlier if felt necessary to assess relevance of the national goals and objectives.

This policy is a very far-reaching commitment by the Indian government which will take considerable time, effort and resources to achieve. Time will tell how much resources and effort are actually expended on its implementation.

(c) Summary of health and safety requirements in India

(i) Management and training

▶ Workers must receive training in what they should do if there is a fire emergency and some workers must be trained in first-aid.

▶ A factory employing 500 or more workers should employ a Welfare Officer.

▶ A factory employing 1,000 or more workers is required to appoint one or more Safety Officers.

▶ The employer must set up a Safety Committee consisting of equal number of representatives of workers and management.

▶ The employer must provide workers with information relating to health and safety at work.

The worker has the right to be sponsored by the employer for health and safety training at a training centre or institute.

- The employer must appoint competent persons to supervise the hazardous activities within the factory.
- The employer must produce a site emergency plan and also disaster control measures for the factory.
- Pregnant women may request light duties.

(ii) Health and first-aid

- The factory must provide sufficient first-aid boxes for employees (at least one first-aid box for every 150 workers).
- If there are more than 20 employees there must be a staff member who is trained in first-aid, either by St John's Ambulance or the Red Cross of India.
- An accident register should be maintained. If any accident occurs in the factory premises, which results in the death of the worker, the employer must (within 7 days of the death) send a report to the Commissioner for Employee's Compensation giving the circumstances attending the death.
- The employer must maintain accurate and up-to-date health records for all the workers.
- The employer shall pay compensation to employees for injuries and illness resulting from employment. If the injury results in death, the amount of compensation payable to his dependents would be equal to 50% of the monthly wages of the employee multiplied by a figure ranging from 228.54 to 99.37 (depending upon the age of the employee) or an amount of Rs.80,000, whichever is more. If the Employees' State Insurance (ESI) scheme applies, these issues are addressed by ESI.

(iii) Canteen/crèche

- If there are more than 250 workers, canteen(s) shall be provided and maintained by the employer.
- Food served should be provided at a no-profit basis (in practice it is often subsidised by the employer).
- In a factory with more than 150 workers employed, adequate and suitable rest rooms and a lunch room, with provision of drinking water, shall be provided and maintained by the employer.
- In a factory with more than 30 women workers employed, the employer shall provide and maintain suitable room(s) for the use of children under the age of 6 years.

(iv) Fire precautions

- The employer shall take necessary measures to prevent outbreak and spread of fire.
- The employer shall provide and maintain safe means of escape for all workers in the event of fire.
- The employer shall provide and maintain the necessary equipment and facilities for extinguishing fire.

- The employer shall ensure that the workers are familiar with the means of escape in case of fire and have been adequately trained.

(v) Safe use of machinery and chemicals

- The employer must provide personal protective equipment where necessary.
- The employer must ensure that machinery is safe and properly guarded.
- The employer must protect workers who are repairing the machinery while in motion.
- The employer shall not allow young persons to work on dangerous machines unless the worker has received sufficient training or is under adequate supervision.
- The employer shall provide and maintain suitable devices for cutting off power in emergencies from running machinery.
- All electrical machinery shall be encased or guarded.
- The employer shall protect workers from injury to their eyes and also from dangerous dust, gas, fumes and vapours.
- The employer shall maintain the building and machinery in such a condition that it does not cause any danger to human life or safety.
- Exposure of chemical and toxic substances in manufacturing processes (whether hazardous or otherwise) should not exceed the maximum permissible threshold limits specified by the law.
- Adequate and suitable facilities for washing should be maintained in each factory, near to areas where potentially hazardous chemicals are used.

(vi) Hygiene and factory environment

- The factory should be kept clean and tidy.
- Every worker should have at least 14.2 cubic metres of space.
- Adequate lighting, ventilation and temperature control must be provided.
- Effective measures shall be taken to prevent dust inhalation and accumulation at the work place.
- All wastes and effluents should be treated and disposed of in a safe manner.
- Workers must have access to clean drinking water. If there are more than 250 workers, the factory must provide cooled drinking water in hot weather.
- Adequate toilet facilities should be provided at the work place and should be maintained in a clean and sanitary state. There should be at least one toilet for 20 employees.
- Separate enclosed accommodation must be provided for male and female workers.

See the full text of the 1948 Factories Act at: http://www.ilo.org/dyn/natlex/docs/ELECTRONIC/32063/76807/F2079761311/IND32063.pdf

15

15.3.9 Indonesia

(Population in 2014: 252,164,000. Labour force: 120,000,000)

In Indonesia there is an extensive legal OSH framework. The main law on OSH is the Work Safety Act (Law No. 1, 1970). This law covers all workplaces and emphasises primary prevention.

The Health Act (Law No. 23, 1992) dedicates its Article 23 to occupational health, stating that occupational health is carried out so that all workers can work in good health without endangering themselves or their community, and to gain optimal work productivity along with the labour protection programme (Department of Health, 2002).

The Act covers accident and fire prevention, explosion potential, first-aid, PPE, temperatures, humidity, dust, dirt, smoke, gas, radiation, sound and vibration, occupational disease both physical and psychological, illumination, air circulation, cleanliness, unison of workers and work tools/environment (ergonomics), construction work, loading, unloading, handling and storage, safe transport, shock from electric current, improving the situation if accidents increase. The issues covered include pre-employment or change of employment health examinations.

Managers are responsible to explain to every worker:

(a) the conditions and dangers which may occur in his workplace;

(b) all safety devices and protective equipment which it is obligatory to provide at the workplace,

(c) the personal protective equipment provided for the personnel concerned;

(d) the safety system and conduct in connection with carrying out the work.

Managers have a responsibility to report certain accidents to the Minister of Manpower. They must provide PPE free of charge to workers and others entering the workplace such as visitors and contractors.

Legislative regulations shall lay down the obligations and rights of workers to:

(a) provide accurate information upon request by a Safety Inspector or Safety Expert;

(b) use obligatory personal protective equipment;

(c) fulfil and obey obligatory safety and health conditions;

(d) request the manager to carry out all obligatory safety and health conditions;

(e) raise objection to work regarding which, in his opinion, doubts exist concerning obligatory safety, health and personal protective equipment requirements otherwise, within the limits of his responsibility.

Indonesia has adopted one of the most comprehensive laws on OSH management systems (OSH-MS) at large or high-risk enterprises. The regulation stipulates that 'Any company employing 100 employees or more, or containing harmful potential substances issued due to process characteristic or production material which may cause occupational accidents such as explosion, fire, contamination and occupational disease is obligated to implement an OSH-MS.'

A systematic audit, endorsed by the government, is necessary to measure the OSH-MS practice. A company is awarded an OSH-MS certificate if it complies with at least 60% of 12 main elements, or in 166 criteria. Currently, an agency, called PT Sucofindo, is authorised by the Department of Manpower and Transmigration (DEPNAKER) for auditing and certifying the companies for OSH-MS. An institution, called Patra Nirbaya, has been approved by the Department of Mining and Energy to carry out similar activities in oil companies.

Moreover, the more recently passed Manpower Act (Law No. 13, 2003) refers to OSH-MS (Articles 86 and 87). First, the Act stipulates that every worker has the right to receive protection against safety and health hazards, protection against immorality and indecency, and treatment that shows respect to human dignity and religious values. Second, it states that every enterprise must apply an OSH-MS, to be integrated into the enterprise's general management system. The Director of OSH Standards of DEPNAKER identifies two particular priorities: (i) establishment of a more coherent national OSH administration, and (ii) promotion of OSH-MS.

The establishment of OSH committees is meant to improve the OSH enforcement and implementation at the enterprise level. All companies hiring more than 50 workers must have an OSH committee and register it at the local DEPNAKER office.

See: http://www.ilo.org/public/english/region/asro/manila/downloads/wp9.pdf

http://www.ilo.org/public/english/region/asro/bangkok/asiaosh/std_leg/national/indonsia/idsafety.htm

http://www.ilo.org/dyn/natlex/docs/SERIAL/64764/56412/F861503702/idn64764.PDF

http://www.nakertrans.go.id/?

15.3.10 Japan

(Population in 2014: 127,090,000. Labour force: 65,620,000)

In Japan workers are covered by the Industrial Safety and Health Act (Law No. 57 of June 1972) with the latest amendment shown in 2009 Ordinance No. 47 of March 2007. The purpose is to secure the health and safety of workers in workplaces and facilitate comfortable working environments. The Act not only prescribes minimum standards that employers are obliged to observe but also requires employers to positively take measures for ensuring workers' safety and health. Responsibilities are primarily placed on employers and also on designers, manufacturers, constructors and importers. Workers have responsibilities to observe requirements and to cooperate with employers. Japan's Industrial Safety and Health Law and related regulations were enacted in 1972 and they prescribe detailed safety and health measures to be taken by employers. The 1972 Act covers specific safety requirements such as machinery, dangerous substances, dust, fumes, pressure vessels, lifting machinery, examination of equipment. The government has also made an announcement of detailed guidelines for employers to implement measures in compliance with the Industrial Safety and Health Law and Regulations.

The 12th Industrial Accident Plan was introduced in 2013 and lasts to 2018. It is published by the Ministry of Health, Labour and Welfare. It is available in Japanese and English (and can be downloaded at: http://www.jisha.or.jp/english/index.html). The Japan Industrial Safety and Health Association (JISHA), which was established in 1964 under the Industrial Accident Prevention Organisations Act, is a legal entity whose membership consists of employers' associations. JISHA's overall objective is to help prevent work-related accidents and injuries and protect the health of workers by promoting safety and health efforts undertaken by employers and employers' associations, and by offering safety and health guidance and services. The JISHA Standards were developed in line with the OSHMS Guideline by the Ministry of Labour, Health and Welfare

(first released in 1991, revised in 2006) and ILO's Guidelines on OSHMS (ILO-OSH 2001), with some of JISHA' s own concepts added. This standard requires procedures to be established for the risk assessment arising from machinery, equipment, chemical substances or working behaviour. It is possible to get Certification to this standard.

(a) Guidelines on Occupational Safety and Health Management Systems Ministry of Labour Notification No. 53, 30 April 1999

Amendment: Ministry of Health, Labour and Welfare Notification No. 113, 10 March 2006.

Responsibilities in the Guidelines are placed on employers who must set up responsibilities and consult with employees. Here is a summary of the main articles.

Article 1. The purpose of these Guidelines is to raise the level of safety and health in workplaces. To achieve these objectives, these Guidelines have been designed to encourage employers to adopt, with the cooperation of their workers, a series of processes, and to engage in continuous and voluntary safety and health activities, thereby preventing industrial accidents, promoting workers' health and facilitating the establishment of a comfortable working environment.

Article 2. These Guidelines do not stipulate specific measures that employers are required to adopt in accordance with the Industrial Safety and Health Act (Act No. 57, 1972, hereinafter referred to as the 'Act'), in order to reduce or eliminate hazards or health impairments associated with machinery, equipment, chemical substances, etc.

Article 3. The terms that appear in these Guidelines are used in accordance with the following definitions.

1. Occupational Safety and Health Management System (OSHMS)

An Occupational Safety and Health Management System (OSHMS) constitutes a framework for a series of voluntary safety and health management activities, based on which the following measures are implemented systematically and continuously in workplaces. These measures are implemented as part of overall business management activities including production management.

(a) Release of a safety and health policy.
(b) Risk assessments and control measures based on the results.
(c) Establishment of safety and health objectives.
(d) Formulation, implementation, evaluation and improvement of safety and health plans.

2. System audit

A system audit is a review and assessment carried out by an employer in order to determine whether

15

the measures to be taken in accordance with its OSHMS are being properly implemented, taking into consideration the period of its safety and health plan.

Article 6. Employers shall establish specific procedures for incorporating workers' opinions in setting safety and health objectives and in formulating, implementing, evaluating and improving safety and health plans. Such procedures may include, for example, the use of a safety and health committee, etc. (meaning a safety and health committee, a safety committee or a health committee; hereinafter the same). Employers shall incorporate workers' opinions in their safety and health measures in accordance with these procedures.

Article 7. In order to establish a structure for properly implementing measures to be taken in accordance with the OSHMS, employers shall implement the following measures.

▶ Employers shall clearly stipulate the roles, responsibilities and authority of workers engaged in system management at each organisational level (referring to a general manager supervising the overall business operations of a workplace, and managerial or supervisory personnel in production, safety and health, and related departments, such as senior managers, managers, section chiefs, and foremen who are in charge of the OSHMS; hereinafter the same). Employers shall ensure that all workers, contractors and other persons concerned are fully informed of such roles, responsibilities and authority of said parties.

▶ Employers shall designate workers who engage in system management at each organisational level.

▶ Employers shall make a reasonable effort to provide sufficient personnel and budget for the OSHMS.

▶ Employers shall provide their workers with education and training about the OSHMS.

▶ Employers shall use a safety and health committee, etc. for the implementation of measures to be taken in accordance with the OSHMS.

Article 8. Employers shall specify the following items in written form.

1. Safety and health policy.
2. Roles, responsibilities and authority of workers engaged in system management at each organisational level.
3. Safety and health objectives.
4. Safety and health plans.
5. Procedures established in accordance with the provisions of Article 6, Paragraph 2 of this Article, Article 10, Article 13, Paragraph 1 of Article 15, Article 16 and Paragraph 1 of Article 17.

Article 10. Employers shall establish specific procedures for risk assessment in accordance with the Guidelines adopted pursuant to Paragraph 2,

Article 28-2 of the Act, and shall implement risk assessment in accordance with these procedures.

Article 13. Employers shall establish specific procedures to properly and continuously implement their safety and health plans, and shall implement the plans in accordance with these procedures.

Employers shall establish specific procedures to fully inform all workers, related contractors and other persons concerned of measures necessary to properly and continuously implement their safety and health plans. Employers shall inform them in accordance with these procedures.

Article 15. Employers shall establish specific procedures for conducting routine monitoring and making improvements with respect to the implementation of a safety and health plan. Employers shall monitor and improve the plan in accordance with these procedures.

When employers formulate a new safety and health plan, they shall incorporate into the plan the results of the monitoring and improvements specified in the preceding paragraph and the results of investigation specified in Article 16.

Article 16. Employers shall establish specific procedures to determine a cause, identify a problem and to take corrective actions if an industrial accident or any accident occurs. Employers shall determine the cause, identify any problems and take corrective actions in accordance with these procedures if such an accident occurs.

Article 17. Employers shall formulate a plan for periodic system audits, and establish specific procedures to properly implement system audits on the matters specified in Articles 5 through 16. Employers shall conduct the audits in accordance with these procedures.

15.3.11 Korea, South

(Population in 2014: 50,423,000. Labour force: 25,860,000)

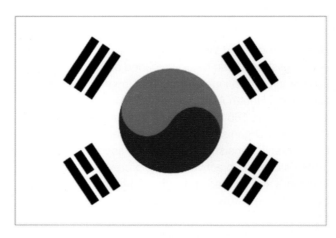

In South Korea OSH is covered by the Occupational Safety and Health Act 1990 (Act No. 4220, 13 January

1990) as amended by 18 subsequent Acts, the last being Degree 23248, 25 October 2011. The purpose of this Act is to maintain and promote the safety and health of workers by preventing industrial accidents and diseases through establishing standards on occupational safety and health and clarifying where the responsibility lies, and by creating a comfortable work environment. The Act covers workers and responsibilities are primarily laid on employers. It covers all businesses or workplaces including state and local authority premises.

Employers must:

▶ observe the standards and provide workers with information;
▶ prevent health problems caused by physical fatigue and mental stress, safeguard lives and maintain and promote the safety and health of workers;
▶ report as required to the Ministry of Labor;
▶ record prescribed accidents and diseases;
▶ post the major contents of the Act for workers;
▶ cooperate with the representatives of workers;
▶ put a person in charge of the safety and health management system (safety and health manager) which must include:
  ▷ accident and disease prevention plan;
  ▷ safety and health regulations;
  ▷ education of workers;
  ▷ inspection and improvement of the work environment;
  ▷ management of health;
  ▷ investigation of accidents and diseases;
  ▷ recording of statistics;
  ▷ whether safety devices meet the standards;
  ▷ prevention of hazards as prescribed;
▶ appoint a workplace supervisor to carry out safety and health measures such as inspections;
▶ appoint a safety manager at the workplace;
▶ appoint a health manager at the workplace;
▶ establish a safety and health committee composed of equal numbers of workers and employers.

Employers must take measures to prevent hazards and dangers from:

▶ machines, tools and other equipment;
▶ explosive, combustible or inflammable substances;
▶ electricity, heat or other energy;
▶ improper work method in areas such as excavation, timbering, quarrying, handling heavy objects, dismantling and operating;
▶ workers falling down or the collapsing of structures;
▶ gases, vapours, dust, fumes, etc.;
▶ radiation rays, high temperature, low temperature, ultrasonic waves, noises, vibration, pressures;
▶ health problems from display equipment;
▶ repetitive work;
▶ poor ventilation, lighting, thermal insulation, cleaning, etc.;

Designers, manufacturers, importers of machinery, tools and other equipment, raw materials, builders of any construction must comply with the standards.

Workers must observe the standards and are subject to measures for the prevention of industrial accidents and diseases taken by the employer or other related organisations.

The Korean Occupational Safety and Health Agency has its own website which states its mission as follows.

The Korea Occupational Safety and Health Agency was founded with the end in view of contributing to the national economy. As such, KOSHA shall effectively carry out projects aimed at preventing occupational accidents, promoting workers' safety and health, and encouraging owners to launch accident prevention activities.

See: http://english.kosha.or.kr/english/main.do

http://www.koilaf.org/KFeng/engLaborlaw/laborlaw/OCCUPATIONALSAFETYANDHEALTHACT.pdf

http://www.ilo.org/dyn/natlex/natlex_browse.details?p_lang=en&p_country=KOR&p_classification=14&p_origin=SUBJECT

15.3.12 Kuwait

(Population in 2014: 3,065,000. Labour force: 2,380,000. 80% non-Kuwatis)

(a) Introduction

In Kuwait the principle national laws and OSH regulations are:

▶ Act No. 38 of 1964, concerning Labour in the Private Sector (**Labor Act**);
▶ Ministerial Order No. 43 of 1979, concerning the conditions which must be fulfilled in areas and places of work to protect workers, undertakings and materials handled by workers, from the dangers of work, injury and occupational diseases;
▶ Ministerial Order No. 57 of 1982, concerning protection against risks due to exposure to benzene;
▶ Ministerial Decision No. 56 of 1982, concerning protection from machinery;
▶ Ministerial Order No. 45 of 1979, concerning the tables of measuring the levels and standards of safety in the places and areas of work;

15

- Ministerial Decree No. 164/2006, amending several provisions from Ministerial Decree No. 114/1996, regarding the Requirements and Conditions that should be provided in workplaces to protect the employees from work hazards;
- The Law of Labor in the Private Sector No. 6 of 2010.

(b) Labor Act Chapter 9 – Working conditions

Most of the Labor Act concerns terms of employment and trade union rights. However, chapters 9 and 12 (see (c)) cover working conditions and reporting of incidents. The ninth chapter of the Labor Act No. 38 includes the following issues:

- **Article 40.** The employer shall provide adequate means of protection for labourers at work against accidents resulting from the use of machines, cogwheels, cranes, conveyors, etc. It shall also be incumbent upon him to take necessary precautions to protect labourers from falls, sharp and falling bodies and splinters as well as from inflammable, explosive, caustic and poisonous material, electric currents, reflected light, etc.
- **Article 41.** Preventative measures and necessary precautions against accidents shall be organised in conformity with whatever decision/s the Ministry of Social Affairs and Labor may take.
- **Article 42.** Without breach of the decisions of the Ministry of Public Health and of the Municipality of Kuwait regarding licences for public, commercial, industrial and other undertakings, it shall be incumbent upon the employer to take necessary precautions to ensure perfect cleanliness, ventilation, sufficient lighting and sanitation in compliance with the detailed instructions emanating from the Ministry of Social Affairs and Labor.
- **Article 43.** The employer shall take the necessary precautions to protect his labourers from the trades' diseases inasmuch as the industries and businesses in respect of which a schedule shall issue from the Ministry of Social Affairs and Labor are concerned. The Ministry of Social Affairs and Labor shall regulate the necessary measures of protection in every industry.
- **Article 44.** It shall be incumbent upon every employer to provide a first-aid kit stocked with medicaments, bandages and antiseptics. The kit shall be placed in a conspicuous spot on the working site and shall be of easy access to the labourers. One first-aid kit in the charge of a trained attendant shall be allotted to every 100 labourers.
- **Article 45.** The employer shall provide adequate means of transport for those of his labourers working in areas not served by ordinary means of transport.
- **Article 46.** The employer employing labourers in districts distant from built-up areas shall provide

them with suitable accommodation, good potable water and means of supply in conformity with the terms agreed upon by both parties. The Ministry of Social Affairs and Labor shall determine the regions to which the provisions of this article shall be applicable.

(c) Labor Act Chapter 12 – Concerning compensation for labour injuries and occupational disease

Chapter 12 includes the following issues:

Article 61. *First: Labor Injuries*: Where a labourer sustains an injury in the course of and arising from his employment the employer shall immediately report the accident to: (a) the Police Station in charge of the area of the working site; (b) the Ministry of Social Affairs and Labor or the branch thereof covering the area of the working site. It shall be permissible for the labourer to report the accident himself where his condition permits him to do so.

Article 62. The report shall include the name of the labourer, his occupation, address and nationality with a brief account of the accident and whatever measures were taken in first-aid or treatment.

Article 63. The injured labourer shall be entitled to medical treatment in a government hospital or in a private clinic at the employer's discretion. The medical practitioner treating shall determine in his medical report the period of treatment, the infirmity resulting from the injury and the labourer's capacity to pursue work. Where a dissension arises thereon the matter shall be referred to the Ministry of Public Health for arbitration whose opinion shall be final. The employer shall be bound to meet treatment expenses fully, medicaments and transport included.

Article 64. The injured labourer shall receive full wages during the period of treatment determined by the medical practitioner. Where the period of treatment exceeds six months he shall be paid only half-wage until he is cured, proved incapacitated or dies.

Article 65. Where injured in an accident arising from the employment or in the course thereof the labourer or his legal heirs after him shall be entitled to compensation for the injury in accordance with the Schedule issuing by Order of the Minister of Social Affairs and Labor in conformity with this article provided that compensation shall not be due where it is established from the examination: (a) that the labourer deliberately injured himself; (b) that the injury occurred by reason of gross and premeditated misconduct on the part of the labourer. The foregoing shall be applicable unless there results from the injury the death of the labourer or there ensues therefrom a permanent disability the order of which exceeds (25%) of total disability.

Article 66. *Second: Occupational Diseases*: The Ministry of Social Affairs and Labor shall prepare a schedule of occupational diseases and of the trades and works inducing them. New occupational diseases may be added to the schedule.

Article 67. The provisions of articles (61), (62), (63), (64) and (65) of this Ordinance shall become applicable where a labourer is smitten by any of the diseases listed in the schedule or develops symptoms thereof.

Article 68. The responsibility of former employers towards the last employer by whom the injured labourer is employed shall be determined in the light of the report of the medical practitioner treating. Those (former employers) shall be required to pay compensation, each in proportion to the period spent by the labourer in his service. In the application of the provisions of this article, it shall be a condition that the trades and professions exercised by labourers be such as to have originated the disease befalling them.

(d) Labor Law No. 6 of 2010

Articles 80 to 83 set out the types of files and records an employer must keep for each worker. These indicate different types of leave taken, occupational accidents and diseases, occupational safety and health data, and start and end of employment. The records must contain data required by these articles. An employer must post in a visible location a schedule of hours of work and weekly rest, dates of public holidays as well as penalties (sic), he shall provide workers with means for securing their occupational safety and health and to protect them from work hazards. A worker shall not assume the costs for such equipment or suffer any wage deductions to cover costs.

Article 84 requires the employer to place, in a visible location, at the place of work, safety instructions and measures to protect from occupational diseases. The article entrusts to the competent Minister the task of issuing instructions regarding safety instructions, warnings and equipment.

Article 85 states that the Minister in charge, based on the opinion of the competent authorities, determines by a decision the type of activities for which occupational safety and health equipment is required and names the technicians who are to operate them.

Article 86 requires the employer to take all necessary precautions for the protection of workers.

Article 87 requires workers to use protective gear with due care and apply instructions regarding them.

Article 88. Due to the legislator's desire to protect workers, in this article the employer is required, subject to the terms of the Social Security Legislation, to provide insurance coverage for his workers against occupational injuries and diseases.

In **Article 89** the legislator has added a new provision stating that the following provisions do not apply to workers subject to insurance clauses for occupational accidents as covered by the terms of the Social Security Law.

Article 90 sets out the measures in the case of an occupational injury to be taken by the employer and worker, if his condition permits, and the authorities to whom it must be reported.

Article 91. Without prejudice to the provisions of Law No. 1, of 1999, the employer shall bear all costs incurred for the treatment of a worker for occupational diseases or injury, including the cost of medicines and transportation. Both worker and employer may oppose the findings of the medical report before the Medical Tribunal of the Ministry of Health.

Article 92 requires each employer to periodically provide the competent Ministry with statistics of occupational injuries and diseases in his establishment.

Article 93 gives the injured person the right to receive full pay throughout his treatment. If treatment lasts for more than six months he shall be paid half his wages until he recovers, or until he is declared disabled, or till his death.

Article 94 refers to a decision by the Minister, who after consulting the Minister of Health, sets the amount of compensation due to a worker or his beneficiaries.

Article 95 sets out those cases where an injured worker my not receive compensation.

Article 96 extends the protection granted to the worker under articles 93 to 95 of this Law if he suffers any symptoms of an occupational disease within one year of leaving service.

Article 97 sets the liability of former employers for whom the worker has worked, each in proportion to the worker's period of employment, for payment of compensation due for injury and as set out in this Law. The article provides that an insurance company or the Public Institution for Social Services may, after having paid compensation to the worker, claim from the employer(s) the compensation paid as per the provisions of paragraph one of this Article.

See: http://www.ilo.org/dyn/travail/docs/656/Explanatory%20Memorandum.pdf

15.3.13 Malaysia

(Population in 2014: 30,377,000. Labour force: 13,200,000)

In Malaysia the Department of Occupational Safety and Health (DOSH) under the Ministry of Human Resources is responsible to ensure that the safety, health and welfare of workers in both the public and private sector

15

is upheld. DOSH is responsible to enforce the Factory and Machinery Act 1969 and the Occupational Safety and Health Act 1994.

(a) Occupational Safety and Health Act

- ▶ The Occupational Safety and Health Act is an Act which provides the legislative framework to secure the safety, health and welfare among all Malaysian workforces and to protect others against risks to safety or health in connection with the activities of persons at work.
- ▶ This Act was gazetted on 24 February 1994 and may be cited as the Occupational Safety and Health Act 1994. This Act is a practical tool superimposed on existing safety and health legislation.
- ▶ The aims of this Act are:
  - ▷ to secure the safety, health and welfare of persons at work against risks to safety or health arising out of the activities of persons at work
  - ▷ to protect persons at a place of work other than persons at work against risks to safety or health arising out of the activities of persons at work
  - ▷ to promote an occupational environment for persons at work which is adapted to their physiological and psychological needs
  - ▷ to provide the means whereby the associated occupational safety and health legislation may be progressively replaced by a system of regulations and approved industry codes of practice operating in combination with the provisions of this Act designed to maintain or improve the standards of safety and health.
- ▶ The provision of the Occupational Safety and Health Act 1994 are based on the self-regulation scheme. Its primary responsibility is to ensure safety and health of work lies with those who create the risks and those who work with the risks.
- ▶ Through a self-regulating scheme that is designed to suit the particular industry or organisation, this Act also aims to establish effective safety and health organisation and performance.
- ▶ The concept of self-regulation encourages cooperation, consultation and participation of employees and management in efforts to upgrade the standards of safety and health at the workplace.
- ▶ The Occupational Safety and Health Act 1994 is enforced by the Department of Occupational Safety

and Health (DOSH), a government department under the Ministry of Human Resources Malaysia.
- ▶ Department of Occupational Safety and Health (DOSH) will ensure through enforcement and promotional works that employers, self-employed persons, manufacturers, designers, importers, suppliers and employees always practise safe and health work culture, and always comply with existing legislation, guidelines and codes of practice.
- ▶ Department of Occupational Safety and Health (DOSH) will also formulate and review legislation, policies, guidelines and codes of practice pertaining to occupational safety, health and welfare as a basis in ensuring safety and health at work.
- ▶ Department of Occupational Safety and Health (DOSH) is also the secretariat to the National Council for Occupational Safety and Health, a council established under section 8 of the Occupational Safety and Health Act 1994.
- ▶ The National Council for Occupational Safety and Health shall have power to do all things expedient or reasonably necessary for or incidental to the carrying out of the objects of this Act.

(i) The Act affection

- ▶ All employers with more than five employees are required by the legislation to formulate a written Safety and Health Policy.
- ▶ The object of the Safety and Health Policy is to demonstrate the company's commitment and concern to ensure safety and health at the place of work. When making decisions or performing work activities of the organisation, issues on safety and health stated in the policy must be taken into account.
- ▶ The Occupational Safety and Health Act 1994 specifies the general duties of employers, self-employed persons, manufacturers, designers, suppliers and employees.
- ▶ Among the provisions of the Act is the establishment of the safety and health committee, the appointment of a safety and health officer and the enforcement, investigation and offences.

(ii) The written Safety and Health Policy

- ▶ The following describes the essential ingredients for the written Safety and Health Policy as required by law.
- ▶ The written policy is divided into three main parts, namely:
  - ▷ General Policy Statement
  - ▷ Organisation
  - ▷ Arrangements.
- ▶ The General Policy Statement concerns the overall intent of the employer to look after the safety and health of the workforce. This statement can be simple and brief. Essentially it should:

- ▷ point out that the management accept responsibility for safety and health of the employees and others who may be affected by the work activities;
- ▷ present a summary of the policy's goals;
- ▷ emphasise the importance of safety and health to overall business performance;
- ▷ include a reference to other parts of the policy document which go into more details; and
- ▷ be dated and signed by the person at the top management in the organisation such as the Chairman or Managing Director.
- ▶ The second part of the policy on Organisation should describe the safety and health responsibilities. This is primarily about the role of each person. Among others it should include:
  - ▷ the list of safety and health responsibilities of all levels of management;
  - ▷ the role of employees in the implementation of the policy. It is the duty of each employee not to endanger himself or others by his actions or omissions, and to cooperate in all measures provided for his safety and health;
  - ▷ the structure and role of safety and health committees and other in-house safety and health organisation, if any.

The Arrangements or final part of the written policy concerns practical systems and procedures.

It deals mainly with potential hazards and measures to be taken to solve the problem. Essentially it should specify detailed arrangements for ensuring that the policy is being implemented, including:

- ▶ the arrangement for training and instructions;
- ▶ information about hazards that may be in certain processes, the control measures and the ways in which employees should cooperate for their own safety and health;
- ▶ explain the company's safe system of work including procedures and rules;
- ▶ a scheme for the issuance, use and maintenance of personal protective equipment (PPE);
- ▶ the procedure for investigation and reporting of accidents; and
- ▶ emergency measures such as first-aid and fire arrangements.

It is important that contents of the policy be made known to employees during induction course and job training. The policy statement should be displayed at strategic locations in the workplace.

(iii) Duties of an employee

- ▶ It shall be your duty as an employee while at work:
  - ▷ to take reasonable care at work for the safety of yourself and other persons;
  - ▷ to cooperate with your employer or any other person in the discharge of any duty, under the Act or Regulations;
  - ▷ to wear or use at all times any protective equipment and clothing provided by your employer for the purpose of preventing risks to your safety and health;
  - ▷ to comply with any instruction or measure on occupational safety and health as required under the Act or Regulations.
- ▶ If you contravene this provision of the Act, you shall be guilty of an offence and shall, on conviction, be liable to a fine not exceeding RM1,000 or to imprisonment for a term not exceeding three months or to both.
- ▶ If you intentionally, recklessly or negligently interfere with or misuse anything provided or done in the interests of safety, health and welfare in pursuance of the Act, you shall be guilty of an offence and shall, on conviction, be liable to a fine not exceeding RM20,000 or to imprisonment for a term not exceeding two years or to both.

(iv) Duties of a safety and health officer

- ▶ A safety and health officer shall advise the employer on the measures to be taken in the interests of safety and health at the place of work.
- ▶ A safety and health officer shall inspect the place of work to determine any hazard liable to cause bodily injury and to investigate any accident, near miss, dangerous occurrence, occupational poisoning or disease.
- ▶ It is also the duty of a safety and health officer to assist the employer or safety and health committee in organising and implementing an Occupational Safety and Health programme.
- ▶ Other duties of a safety and health officer include:
  - ▷ to become the secretary of a safety and health committee;
  - ▷ to assist the safety and health committee in inspections;
  - ▷ to collect, analyse and maintain statistics;
  - ▷ to assist any officer in carrying out his duty under the Act and regulations; and
  - ▷ to carry out any other instruction made by the employer on any matters pertaining to safety and health at the workplace.

(v) Safety and health committee

- ▶ Pursuant to section 30 of the Occupational Safety and Health Act 1994, every employer shall establish a safety and health committee at the workplace if there are 40 or more persons employed at the place of work.
- ▶ A person who contravenes the above provisions shall be guilty of an offence and shall, on conviction, be liable to a fine not exceeding RM5,000 or imprisonment for a term not exceeding six months or to both.
- ▶ The functions of the safety and health committee include:

15

▷ to keep under review the measures taken to ensure the safety and health of persons at the place of work;

▷ to investigate any matter at the place of work which a member of the committee or a person employed thereat considers is not safe or is a risk to health and which has been brought to the attention of the employer;

▷ to attempt to resolve any matter referred to it and, if it is unable to do so, shall request the Director General of Occupational Safety and Health to undertake an inspection of the place of work for that purpose.

See: http://www.jobsdb.com.my/MY/EN/V6HTML/ JobSeeker/handbook/regulation-of-employment/ occupational-safety_3.htm

15.3.14 Mexico

(Population in 2014: 119,713,000. Labour force: 51,500,000)

The North American Free Trade Area (NAFTA) signed by Canada, Mexico and the USA, in 1993, has two side agreements: the Agreement on Environmental Cooperation; and the Agreement on Labour Cooperation. In the case of Mexico, NAFTA's side agreements introduced an interesting (and innovative) approach to the development of occupational safety and health. The Agreement on Labour Cooperation considers that the national legislation of the three NAFTA member countries (Canada, Mexico and the USA) contains the basic principles of worker protection. Therefore, the Agreement does not oblige the parties to modify their national legislation in this regard, but requires them to guarantee the effective application of their legislation. In addition, it creates an institutional scheme to ensure compliance with these obligations, and it also establishes a special system for dispute resolution in cases related to the lack of effective enforcement of OSH, child labour and minimum wage labour standards (NAFTA).

These general regulations on OSH came into effect (except for parts of Title II) on 21 April 1997. They cover:

I – General OSH measures: definitions; responsibilities of employers and employees; scope and role of Standards issued by the *Secretaría*; research,

authorisation, evaluation and advisory activities of the *Secretaría*.

II – Safety requirements in the workplace: buildings and workplaces; fire prevention and protection; equipment, machinery, pressure vessels, steam generators; electrical installations; tools; handling, transportation and storage of materials in general, and of dangerous chemical substances in particular.

III – Occupational hygiene: noise and vibration; ionising and non-ionising radiation; biological hazards; work in abnormal air pressure; work under abnormal thermal conditions; lighting; ventilation; personal protective equipment; ergonomic considerations; welfare facilities; order and cleanliness.

IV – Safety and health organisation in the workplace: generalities; safety and health committees; notification of and statistics on occupational accidents and diseases; OSH programmes in the workplace; OSH training; preventative activity in occupational medicine and in OSH.

V – Protection of minors and of pregnant and nursing women.

VI – Enforcement, inspection and administrative sanctions.

Federal Labour Law 1995

This Law entered into force on 1 October 1995. Topics covered include: conditions of work; law; Mexico; plant safety and health organisation; responsibilities of employees; responsibilities of employers; schedule of occupational diseases; women; workmen's compensation; young persons.

Employer obligations

The constitution establishes a universal employer responsibility to recompense for work-related death and disability. It also establishes a general obligation to look after employees. Specific duties arise from numerous laws.

The LFT (Ley Federal del Trabajo), LSS (Logistics Support and Supply), RFSHMAT (Federal Regulation of Occupational Health and Safety) and Standards Act require employers to:

▶ Abide by standards;

▶ Maintain safety and health programmes;

▶ Maintain acquiescence and compliance-verification systems;

▶ Ensure accurate equipment and perilous substance controls;

▶ Assist venture of joint committees;

▶ Provide worker training and information about risks;

▶ Allow labor inspections;

▶ Provide information and reports to authorities;

▶ Protect pregnant and lactating women.

Under Mexican law, it is required that workers be given training in Spanish with materials in Spanish such as labels that come on the containers of the chemicals they are using.

Failure to abide by regulations may result in fines ranging from 15 to 315 times the minimum wage, which may be doubled for repeat offenders. The Secretariat of Labor and Social Welfare (STPS) may also order partial or total temporary shutdown of a workplace until health and safety records are met.

Employee duties

Several duties are imposed on employees. They must:

- Obey standards;
- Aid imperilled co-workers;
- Cooperate with joint committees;
- Partake in training;
- Use required protective equipment;
- Endure medical exams;
- Notify employers of contagious diseases contracted;
- Report safety breaches to employer.

Employees have the right to have the joint committee notify them of the workplace safety and health trace, to be in attendance and speak liberally during inspections, to acquire copies of inspection results, and to contribute to legal proceedings.

See: http://offshoregroup.com/2012/06/22/mexicos-industrial-health-and-safety-regulations/

See for useful information in English on Safety and Health issues in Mexico: http://mhssn.igc.org

15.3.15 Nigeria

(Population in 2014: 178,517,000. Labour force: 51,540,000)

In Nigeria the principal law regulating safety and health in industrial establishments is the Factories Act No. 16 of 1987. This Act covers exactly the same grounds as the UK Factories Act 1961 which has now been largely repealed. The Nigerian government has ratified the Occupational Safety and Health Convention, 1981 (No. 155).

The Nigerian Factories Act 1987 covers the following:

(a) Part I – Registration of factories

1. Register of factories.
2. Registration of existing factories.
3. Registration of new factories.
4. Notification of change in particulars furnished.
5. Appointment of Factories Appeal Board.
6. Appeal to Board from decision of Director of Factories.

(b) Part II – Health (general provisions)

7. Cleanliness.
8. Overcrowding.
9. Ventilation.
10. Lighting.
11. Drainage of floors.
12. Sanitary conveniences.
13. Duty of inspector as to sanitary defects remediable by local authority.

(c) Part III – Safety (general provisions)

14. Prime movers.
15. Transmission machinery.
16. Powered machinery.
17. Other machinery.
18. Provisions as to unfenced machinery.
19. Construction and maintenance of fencing.
20. Construction and disposal of new machinery.
21. Vessels containing dangerous liquids.
22. Self-acting machines.
23. Training and supervision of inexperienced workers.
24. Hoists and lifts.
25. Chains, ropes and lifting tackle.
26. Cranes and other lifting machines.
27. Register of chains, etc. and other lifting machines.
28. Safe means of access and safe place of employment.
29. Precautions in places where dangerous fumes are likely to be present.
30. Precautions with respect to explosives or other inflammable dust, gas, vapour or substance.
31. Steam boilers.
32. Steam receivers and steam containers.
33. Air receivers.
34. Exception as to steam boilers, steam receivers and steam containers and air receivers.
35. Prevention of fire.
36. Safety provisions in case of fire.
37. Power of inspector to issue improvement notice.
38. Power of inspector to issue prohibition notice as to dangerous factory.
39. Appeal against notice.

(d) Part IV – Welfare (general provisions)

40. Supply of drinking water.
41. Washing facilities.
42. Accommodation for clothing.

15

43. First-aid.

44. Exemption if ambulance room is provided.

(e) Part V – Health, safety and welfare (special provisions and regulations)

45. Removal of dust or fumes.

46. Meals in certain dangerous trades.

47. Protective clothing and appliances.

48. Protection of eyes in certain processes.

49. Power to make regulations for certain health, safety and welfare.

50. Power to take samples.

(f) Part VI – Notification and investigation of accidents and industrial diseases

51. Notification of accidents.

52. Power to extend dangerous occurrences provisions as notice of accidents.

53. Notification of industrial disease.

(g) Part VII – Special applications, extension and miscellaneous provisions

See http://www.ilo.org/dyn/natlex/country_profiles. nationalLaw?p_lang=en&p_country=NGA for detailed analysis of the OSH legislative coverage in Nigeria.

See: http://www.ilo.org/dyn/natlex/natlex_browse. details?p_lang=en&p_country=NGA&p_ classification=14&p_origin=COUNTRY&p_ sortby=SORTBY_COUNTRY

15.3.16 Oman

(Population in 2014: 4,067,000. Labour force: 968,800)

(a) The main Oman health and safety legislation

Oman Labour Law, 1973 (Royal Decree No. 35/2003 and published in the Official Gazette No. 742, dated 3 May 2003) (http://www.manpower.gov.om/en/law_safety.asp)

This covers:

▶ Definitions and general provisions;

▶ Employment of citizens and regulation of foreigners' work;

▶ Contracts of work;

▶ Wages, leaves and working hours;

▶ Employment of juveniles and women;

▶ Industrial safety (in the original wording):

Article 87 requires every employer to acquaint the worker before employing them with the hazards of his occupation and the protective measures taken and take the necessary precautions during the course of their work against health hazards and the dangers of the work and machinery and shall:

1. *Endeavour to provide whatever is necessary for safety in health in the workplace to enable the workers to perform their duties;*

2. *Ensure the places of work are always clean and satisfy the conditions of comfort, safety and occupational health;*

3. *Ensure that machinery, equipment and tools are installed and maintained in the best conditions of safety.*

The employer shall not charge the workers or deduct any amount from their wages for providing such protection.

Article 88. The workers shall refrain from committing any act which is intended to prevent any instructions being carried out, misuse or causing of damage, or destruction, of the equipment which are placed for the protection, safety and health of the workers, who work for them, and shall use such means of protection and maintain the means of protection which are in their custody with due care, and shall carry out the instructions laid down to safeguard their personal health and safety.

Article 89. The following shall be determined by a decision of the Minister in coordination with the relevant governmental authorities:

1. *The general measures of safety and occupational health which shall be applied in all places of work especially in connection with lighting, ventilation, air circulation, drinking water, water closets, carrying away dust and smoke, workers' lodging, as well as, preventative measures against fire;*

2. *The measures relating to some types of work.*

Article 90. The Ministry shall appoint inspectors to ensure that the employers carry out the instructions set out by the Minister's decisions in respect of the measures specified in Section 89. Such inspectors shall have the right to enter the places of the work, inspect the records relating to the workers, address queries to anyone as they may decide and make records. On the basis of such record, the relevant directorate shall send a warning in writing to the employer who is in violation of such instructions to eliminate such violations within such period as they may specify.

In the event of the existence of any danger which threatens the safety and health of the workers, the Ministry shall take the necessary measures to close down the place of work wholly or partially, or to stop the operation of one or more machinery until the elimination of the causes of such danger. The Ministry may also request the assistance of the Royal Oman Police if necessary.

▶ Employment of workers in mines and quarries;
▶ Labour disputes;
▶ Representative committees;
▶ Penalties.

(b) Occupational Safety Regulations issued by Royal decree No. 32/2003; also governed by the Labour Code (Ministerial Decision 286/2008)

This covers:

▶ Definitions and general provisions (Article 1)

Includes definitions such as:

(OSH) supervisor: any person who performs a protective role in establishments through supervising the application of the conditions and standards of safekeeping workers against the various hazards caused by work environment on an assignment by the administrative body of the establishment.

Serious accident: an accident that causes a permanent total or partial disability or death to a worker, or causes injury to more than one worker, or leads to losses in production and equipment.

▶ General provisions (Articles 2–14)

These cover:

▶ How inspectors perform their tasks and take immediate steps to prevent serious danger.
▶ The need for employers to facilitate the work of inspectors.
▶ Employer must inform the workers of all the hazards they may be exposed to and all the protective procedures in place. Instructions must be prominently posted and appropriate warning signs and procedures are available. The employer must make periodic assessments and record the results.
▶ Workers must use the means of protection and carry out instructions.
▶ Protective equipment must be provided free of charge.
▶ If an employer employs (10) or more workers, they shall set up an OSH programme adequate to the nature and size of the establishment to incorporate the following:

1. The policy and goals of the OSH in the establishment.
2. The duties and commitments of the employer and the worker.
3. The organisation and management of the OSH.
4. The specified authorities and responsibilities of the establishment's management who

are authorised to develop and implement the OSH policies and goals.

5. The specific work hazards that result from work, the methods of their assessment and the mechanisms for analysing them.
6. The specified protective arrangements, the emergency plans.
7. Specify training programmes on the OSH procedures.
8. Identify specifications regarding the purchase or rent of OSH equipment for work.
9. The mechanism for monitoring the performance of OSH in the activities of the contractors working with the establishment.
10. A timetable for testing the equipment or materials that may expose workers to hazards.
11. A timetable for conducting medical examinations for workers.
12. Investigation into the work accidents and taking the necessary actions to prevent the repetition of such accidents.
13. The procedures that have to be carried out by workers in cases of serious hazards.
14. The procedures that have to be carried out by the workers who are exposed to occupational hazards before leaving the worksite.
15. Prohibitions related to accident site.
16. A method of submitting or receiving workers' complaints regarding work hazards and the means of handling them.

▶ The employer who employs (50) or more workers shall assign a qualified supervisor to handle the OSH tasks.
▶ The employer who employs (50) or more workers shall also be committed to providing the department or section with periodical statistics on serious accidents, work injuries and occupational diseases.
▶ The owner of the establishment shall be committed to notifying the department or section in writing within (24) hours of any serious accident, work injury or occupational disease.
▶ General arrangements (Articles 15–16)
  ▷ Worksite safety is covered including storage of materials, movement of materials, manual handling, mechanical handling, fire persistence of buildings, surfaces, space for workers, quality of floors, drainage and suitable handrails for ditches, space and layout of machinery, even floor with suitable circulation and working space, arrangements for storage, access and exits and emergency facilities, construction and use of ladders and staircases, size of exits doorways, suitable seating, protection from falling or flying objects, protection of open water and toilets and washing facilities.
  ▷ The employer has to make sure that the conditions prevailing in the workplace are

509

sufficiently safe for the workers' health particularly in terms of: lighting, ventilation, heat and cold, noise and drinking water.

▶ Utilities (Article 17)

The employer shall take all necessary actions to assure that conditions prevailing in the utilities of the workplace are sufficient for safekeeping the safety and health of workers. The following particularly are covered: lighting, workers' sleeping places, areas designated for serving food, clothes changing places and rest rooms.

▶ Specifications of work uniform and personal protection equipment (Article 18)

The employer must provide work clothes and personal protection equipment which suits the nature of work performed. Details are laid down.

▶ Medical care (Articles 19–25)

This covers medical examinations, periodic medical test and requirements in detail for exposure to an occupational disease; requirements for first-aid to include:

1. The kit must contain the requirements of first-aid such as the medications and equipment needed for first-aid.
2. The kit must be close to the water source in a reasonable temperature with nothing in it but the medical requirements. A crescent should be drawn on the kit.
3. The inspector may ask for more first-aid requirements or their quantity if necessary according to the size of the establishment or its nature of work.

▶ Health-friendly workplaces (Articles 26–27)

This includes promoting healthy eating, physical exercise, prohibition of smoking and consideration of psychological health.

▶ Protection of women (Article 28)

Protection of women to ensure they are not exposed to conditions which do not suit women's physiological capabilities or have a negative impact of children or unborn foetuses.

▶ Protection of the disabled (Articles 29–30)

People with a disability should not be required to perform unsuitable tasks and suitable working conditions and measures should be introduced to facilitate their disability.

▶ Precautions against hazards (Articles 31–43)

Fire – The employer must provide the necessary methods and equipment for fire-fighting for the kind of work being practised, considering the following:

▷ Buildings must be constructed of fire-resistant materials.
▷ Provision of lighted emergency exists and these must be kept free of obstructions and open in the direction of escape.
▷ Fire-fighting equipment to be tested every six months.

▷ Fire extinguishers must be kept available and people trained to use them.
▷ Safe use and handling of dangerous gases, vapours or liquids.
▷ Adequate distance, determined by the authorities, between work sections must be strictly observed.
▷ Insulating hot pipes to prevent ignition.
▷ Fire protection for flammable material stores.
▷ Prohibition of matches or cigarette lighters.
▷ Warning systems for areas where explosive or combustible substances are used or stored.
▷ Storage of flammable material away from artificial light (heat sources). Safe storage of air and gas cylinders.
▷ Getting rid of flammable or combustible waste.
▷ Using safe containers, tanks or storage.
▷ Classifying stored materials according to nature and characteristics.

▶ Precautions against mechanical and electrical risks including:

▷ Tools and protective means to ensure safety during their use.
▷ Fencing of moving parts.
▷ Precautions against vibration.
▷ Worker must use tools provided and must be trained to use them.
▷ Technical issues must be addressed and warning signs posted at power generating plant and transmission grids.
▷ Measures to ensure protection of non-electrical personnel.
▷ Installing equipment and cables to the correct standards.
▷ Provide and use additional protection equipment for maintenance including indicators, rubber insulators, etc.
▷ Take precautions against damp, wet or flammable atmospheres.
▷ Control static electrical charges efficiently.
▷ Conduct regular inspections of the electrical installations, cables, etc.

▶ Take precautions against the hazards of lift tools, heavy duty machinery and workers' buses including:

▷ Proper licence for driver of a lift tool.
▷ All parts of lift tool to be properly made and fixed and of durable substances.
▷ Lift tools not to be used unless tested and retested: lift ropes and joints within last three months; lifters and lifting wheels within last six months; crane and weight lifters within last 14 months.
▷ Signs showing safe weights allowed.
▷ Lifters only allowed to carry persons in a properly designed compartment.
▷ Proper sizing of pulleys and construction of ropes or chains.

- ▷ Efficient breaks to stop falling of loads.
- ▷ Proper securing of hooks or ropes.
- ▷ Protection of moving parts and extent of movement of the whole lifter.
- ▷ Warning signs for lifters used on the roads.
- ▷ Support of a trained lifter guide or indicator equipment needed for the operator.
- ▶ Heavy tools precautionary measures to include:
  - ▷ Training of drivers.
  - ▷ Before use safety checks.
  - ▷ Not lift persons with the machine.
  - ▷ Driver may not leave the equipment operating and must return the moving parts to the ground, put on the handbrake and remove the keys before leaving.
  - ▷ Warning equipment and flashing lights to be used close to intersections or blind corners.
  - ▷ Driving speeds, reversing and other operations to be done safely.
  - ▷ Charging of batteries to be done in safe ventilated areas.
  - ▷ Access to the machine should be safe to use.
  - ▷ Seat should be designed to absorb vibrations.
- ▶ Workers' transport buses to be safe with adequate ventilation, air conditioning and a licensed driver.
- ▶ Precautionary measures for manual work tools.
- ▶ Precautionary measures for boilers and pressure vessels including air receivers.
- ▶ Precautionary measures to protect workers against the risk of exposure to chemicals such as gases, dusts, hazardous liquids and acids.
- ▶ Provision of adequate protection against harmful rays, substances which cause occupational cancer and asbestos.
- ▶ There are also specialist precautions for:
  - ▷ Construction drilling demolition and civil engineering works.
  - ▷ Agriculture and animal breeding works.
  - ▷ Sea port works.

(c) Other Decree

Royal Decree No. 46/95 Issuing the Law of Handling and Use of Chemicals.

See: http://www.ilo.org/dyn/natlex/natlex_browse. details?p_lang=en&p_isn=83520

15.3.17 Russian Federation

(Population in 2014: 146,149,000. Labour force: 75,290,000)

(a) Introduction

In the Russian Federation health and safety is governed by the Labor Code of the Russian Federation of 31 December 2001 (Federal Law No. 197-FZ of 2001). This was last amended in 2013.

Quoting from article 1:

The main objectives of the labor law shall be creating the necessary legal conditions for achieving an optimal harmonisation of the parties to labor relations' interests, the state's interests as well as legal regulation of labor relations and other relations directly linked to them as for:

- ▶ organisation of labor and management of labor;
- ▶ job placement with a specific employer;
- ▶ professional training, re-training and skill improvement of employees directly at a certain employer's facilities;
- ▶ social partnership, collective bargaining, concluding collective contracts and agreements;
- ▶ participation of employees and of labor unions in determining working conditions and in applying the labor law in the cases stipulated by the law;
- ▶ material liability of employers and employees in the sphere of labor;
- ▶ surveillance and control (including control by labor unions) of compliance with the labor law (including the law on occupational safety);
- ▶ settlement of labor disputes.

The Code places responsibilities primarily on employers in relation to employees and covers legal regulation of the work of certain categories of employees (organisation heads, part-timers, women, persons with family liabilities, young people, civil servants and others).

(b) Chapter 2: Agreements

Chapter 2 covers: Labour relations which are the relations based on an agreement between an employee and an employer on the personal performance by the employee of a work function for payment.

Under the Code employees are entitled to:

- ▶ conclude, amend and terminate the labour contract in the manner and under the terms and conditions set by this Code, other federal laws;
- ▶ be provided with the job specified in the labour contract;

15

▶ a workplace meeting the conditions set by state labour organisation and occupational safety standards and by the collective contract;

▶ timely and complete payment of wages in accordance with his/her qualification, the job complexity, the quantity and quality of the work performed;

▶ rest and leisure ensured by setting the normal working time duration, by shorter working time for certain jobs and categories of employees, by providing weekly days-off, non-working holidays, paid annual leaves;

▶ complete true information concerning the working conditions and occupational safety requirements in the workplace;

▶ professional training, re-training and skill improvement in the manner set by this Code, other federal laws;

▶ association, including the right to organise labour unions and join them in order to protect his/her labour rights, freedoms and legitimate interests;

▶ participation in managing the organisation in the forms stipulated by this Code, other federal laws and the collective contract;

▶ engage in collective bargaining and conclude collective contracts and agreements through his/her representatives as well as to be informed on implementation of the collective contract, agreements;

▶ protection of his/her rights, freedoms and legitimate interests in any ways not proscribed by laws;

▶ settlement of individual and collective labour disputes, including the right to strike in the manner set by this Code, other federal laws;

▶ redress of the damage inflicted on the employee in connection with his/her performance of the work duties and compensation of the moral damage in the manner set by this Code, other federal laws;

▶ mandatory social insurance in the cases stipulated by federal laws.

The employee shall:

▶ perform his/her work duties vested by the labour contract in good faith;

▶ comply with the internal working regulations of the organisation;

▶ maintain the work discipline;

▶ fulfil the set work norms;

▶ meet occupational safety requirements;

▶ take due care of the employer's and other employees' property;

▶ immediately inform the employer or his/her direct superior about emerging situations hazardous to human life and health, to the employer's property.

(c) Chapters 3–5: Social partnerships

Chapters 3 and 4 of the Code cover social partnerships, representatives of employees (both unionised and non-unionised) and employers. This covers consultation over OSH matters at the workplace. Chapter 5 covers the formation of tripartite commissions at Russian federation and territorial levels. Also commissions at industry and organisation levels.

(d) Chapters 6–13: Labour contracts and agreements and working time

Chapters 6–13 cover the requirements for labour contracts and agreements, some of which can relate to agreements over OSH issues. Chapter 14 covers the protection and use of personal information about employees. Employees' rights and privacies are well documented and most information outside legal requirements needs written permission from the employee concerned.

Section IV, Chapter 15 covers working time which is a period of time during which an employee has to perform their job duties according to internal rules of an organisation and conditions of a labour agreement and other periods of time that are considered working time according to laws and other legislative standard acts.

Normal length of working time cannot exceed 40 hours in a week.

An employer is responsible for keeping a record of actual working time of each employee.

Normal length of working time is reduced for:

▶ young people below the age of 16 by 16 hours;
▶ invalids of certain categories by 5 hours;
▶ employees between 16 and 18 by 4 hours;
▶ employees working in dangerous or harmful environments by 4 hours.

An employer must establish an incomplete working day or an incomplete working week on request of a pregnant woman, one of the parents who have a child below the age of 14 (or an invalid child below the age of 18) and of a person who is nursing a disabled family member according to results of a medical examination.

The maximum length of a working day is also specified for certain vulnerable people. Night working hours are also regulated. Some of the restrictions can be overridden by working agreements. Chapter 17 covers rest days and break-times during each work day.

(e) Section X, Chapters 33–36 cover labour safety requirements

The duties in respect of ensuring labour safety in the organisation are imposed on the employer.

According to the Code the employer must ensure the following:

- employees' safety at exploiting buildings, edifices, equipment, instruments, raw produce and materials used in the production, and at executing technological processes;
- employees using means of individual and collective protection;
- working conditions that meet labour safety requirements at every working place;
- employees' schedule in accordance with the laws of the Russian Federation and the laws of subjects of the Russian Federation;
- purchase at their own expense and distribution among employees working under harmful or dangerous conditions, doing work under specific temperature conditions, or doing work that inflicts contamination, of special working clothes, boots, and other means of individual protection, rinsing and neutralising substances, according to the statutory norms; teaching employees safe methods and ways of work and rendering first-aid after industrial accidents; instructing employees on labour safety; organising for employees practical study at the workplace and examination of labour safety requirements and safe methods and ways of work knowledge;
- non-admission to work of persons who have not undergone, in the statutory order, training and instruction on labour safety, and have not done practical study and examination of labour safety requirements knowledge;
- control over labour conditions at working places and employees' correct usage of means of individual and collective protection;
- attestation of working places in respect of labour conditions with following certification of labour safety measures taken in the organisation;
- organisation at their own expense, in cases stated in this Code, of employees' compulsory preliminary (when hiring employees) and periodical (during employees' work time) medical examinations (surveys), as well as urgent medical examinations (surveys) upon employees' request and according to the medical comment, preserving employees' position and average earnings for the time of the aforenamed medical examinations (surveys); non-admission to work of employees who have not undergone compulsory medical examinations (surveys), and/or in case there are any medical contraindications;
- informing employees of labour conditions and labour safety at their working place, of any existing risk to their health, and of all compensations and means of individual protection they are eligible to receive;
- providing state bodies of labour safety administration, state supervising and controlling bodies, trade unions bodies of control over labour law and labour safety abidance with information and documentation allowing them to exercise their powers;
- taking measures of precaution against emergencies; preserving employees' life and health in emergency situations, including rendering first-aid to the injured;
- investigation and registration of industrial accidents and professional diseases as established in this Code and other normative legal acts;
- providing sanitary and treatment-and-prophylactic services for employees in accordance with labour safety requirements;
- unconstrained admission to the enterprise of state bodies of labour safety administration officers, state bodies of supervision and control over labour law and other normative legal acts containing labour law propositions abidance officers, Russian Federation Fund of Social Insurance bodies officers, and public controlling bodies representatives with a view to inspect the enterprise labour conditions and labour safety and to investigate industrial accidents and professional diseases;
- complying with orders of state bodies of supervision and control over labour law and other normative legal acts containing labour law propositions abidance officers, and considering representations of public controlling bodies within time periods stated in this Code and in other federal laws;
- employees' compulsory social insurance against industrial accidents and professional diseases;
- informing employees of labour safety requirements;
- working out and sanctioning labour safety instructions for employees, taking into account the opinion of an elected trade union body or other employees' authorised bodies;
- presence of normative legal acts assembly containing labour safety requirements in accordance with the specific character of an enterprise's activity.

Employees performing hard work, working in hazardous environments and certain other categories such as the food industry are required to have periodic medical examinations (yearly if under 21).

(f) Employees' duties

According to the Code employees must:

- abide with the labour protection requirements established by the law and other legal regulations as well as by labour protection rules and instructions;
- use the means of individual and shared protection correctly;
- be trained how to use safe methods and techniques for execution of works on labour protection and how to give first-aid in the event of production accidents, be instructed about labour protection,

513

take a trainee course at the workstation, and be checked for the knowledge of the labour protection requirements;

▶ notify his direct or higher-level supervisor immediately of any situation, which may be hazardous for people's life and health, of any accident, which occurred in production, or about deterioration of his health, including symptoms of an acute occupational disease/poisoning;

▶ take preliminary (prior to starting employment) and regular (in the course of employment) medical checkups/examinations.

(g) Compliance of production facilities and products to labour protection requirements

The Code requires production facilities and products to comply with requirements as follows:

▶ production facilities construction and reconstruction projects as well as machinery, gear as well as other production equipment and processes shall be in line with the labour protection requirements;

▶ construction, reconstruction, technical retooling of production facilities, manufacture and introduction of new equipment or introduction of new techniques shall be not allowed, unless findings are available after government examination of labour conditions stating that the projects indicated in the first part of this Article comply with the labour protection requirements;

▶ new production facilities or those under reconstruction cannot be commissioned, unless findings of the relevant government bodies, which exercise supervision and control over compliance with the labour protection requirements, are available;

▶ the use of harmful or dangerous substances, materials, products or goods and/or rendering of services, for which no metrological methods and means have been devised yet and no toxicological evaluation (in terms of health protection or medical and biological assessment) has been carried out, shall be not allowed in production;

▶ if harmful or dangerous substances are to be applied, which are new or have been not used in the organisation previously, the employer must agree the ways of preserving the employees' lives and health with the relevant bodies of government supervision and control over the compliance with the labour protection requirements, before he begins using the above mentioned substances;

▶ machinery, gear and other production equipment, vehicles, processes, materials and chemical agents, individual and shared means of employees' protection, including those of

foreign make, shall comply with the health protection requirements established in the Russian Federation and be provided with certificates of compliance.

(h) Chapters 35 and 36

Chapters 35 and 36 cover the organisation of labour protection with the requirement to set up committees where there are over 100 employees and where there is no labour protection/experts in the organisation to contract with outside experts to provide labour protection services.

Every employee has the right to:

▶ a working place which meets the labour protection requirements;

▶ mandatory social insurance against accidents in production and occupational diseases according to the federal law;

▶ obtain from their employer, the relevant government agencies and public organisations plausible information about the labour conditions and labour protection at their working place, the existing risk of damage to their health as well as about the protective measures against the impact of harmful and/or hazardous production factors;

▶ refuse to do the work, if there is a danger for their life and health due to the violation of the labour protection requirements, except for the cases envisioned by federal laws, until such a danger is eliminated;

▶ be provided with individual and shared protection means according to the labour protection requirements on the employer's account;

▶ be trained to use safe labour methods and techniques on the employer's account;

▶ be re-trained on the employer's account in the case that their workplace is liquidated due to the violation of labour protection requirements;

▶ request for an inspection of labour conditions and labour protection at their working place by government supervision and control agencies overseeing the compliance with the labour and labour protection law, by employees carrying out public examination of labour conditions and/or by the trade union bodies exercising control over the compliance with the labour and labour protection law;

▶ apply on labour protection issues to Russian Federation state power bodies, Russian Federation's subjects' state power bodies, their employer, employers' associations and/or trade unions, their associations and other representative bodies commissioned by the employees;

▶ participate in person or their proxies in consideration of the issues relating to the assurance of safe working conditions at their workplace and in the investigation of an accident in production, which has happened to them, or of their occupational disease;

▶ extraordinary medical check-up (examination) according to medical recommendations, while their workplace (position) and average wage during said medical check-up (examination) shall be kept safe;

▶ compensations laid down by the law, collective bargain agreement or their employment contract, if they are engaged in hard work or work with harmful and/or dangerous labour conditions.

Adequate welfare provision must be made including sanitary and utility rooms, rooms for meals, rooms for giving medical aid, rooms for recreation during the working time and psychic relaxation.

(i) Training and vocational/professional education in the field of labour protection

All the organisation's employees, including its manager, shall take training courses in labour protection and have their knowledge of the labour protection requirements tested in line with the procedure established by the government of the Russian Federation.

The employer or a person authorised by him shall instruct in labour protection all persons taking up a job and the employees being transferred to another job. Arrangements shall be made to teach them to use safe work execution methods and techniques and to give first-aid to victims.

The employer shall provide for training of the persons taking up jobs involving harmful and/or hazardous labour conditions to teach them to use safe work execution methods and techniques. The persons in question shall then take a trainee course in their working places and examinations. They shall be trained in labour protection on a regular basis and checked for knowledge of the labour protection requirements in the course of work.

The government shall promote the organisation of labour protection education in educational establishments of general primary, general basic, general secondary (full) education and those of general vocational, secondary vocational, higher professional and postgraduate professional education.

The government shall provide for vocational/professional education of labour protection specialists in educational establishments of secondary vocational and higher professional education.

(j) Employer's duties in case of an accident in production

If there is an accident in production, the employer or his representative must do the following:

▶ immediately arrange for first-aid to the victim and his delivery to a healthcare establishment, if necessary;

▶ take urgent action to prevent the emergency situation from developing and involving other persons by way of injuring factors;

▶ until the investigation of the accident in production begins, secure the situation the way it was at the time of the accident, unless this endangers other people's life and health and leads to an emergency; and if it is not possible to secure it, take a record of the actual situation (draw up sketches, take photographs and other actions);

▶ provide for well-timed investigation of the accident in production and its registration according to this chapter;

▶ immediately inform the victim's relations about the accident in production, also send a message to agencies and organisations specified in this Code and other regulatory legal acts.

If an accident in production involves a group of people (two or more) or there is a heavy (serious) accident in production, or a fatal accident in production, the employer or his representative shall advise within a day, respectively, of an accident, which took place in the organisation:

▶ the relevant government labour inspectorate;

▶ the prosecutor's office serving the area, where the accident took place;

▶ the federal executive power body, to which authority the organisation belongs;

▶ the Russian Federation's subject's executive power body;

▶ the organisation, which had sent the employee, to whom the accident happened;

▶ the territorial associations of trade union organisations;

▶ the territorial government supervision agency, if the accident took place in an organisation or facility reporting to that agency;

▶ the underwriter for mandatory social insurance against accidents in production and occupational diseases.

The Code lays down detailed requirements for accident investigation which involves the setting up of a commission of at least three people (more if it is serious or there are multiple fatalities and to then include a government labour protection representative).

See: http://www.ilo.org/dyn/legosh/en/f?p=14100:1100:0::NO::P1100_ISO_CODE3,P1100_YEAR:RUS,2013

15.3.18 South Africa

(Population 2014: 54,002,000. Labour force: 18,540,000)

15

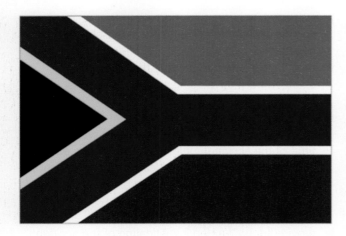

The Occupational Health and Safety Act No. 85 of 1993 as amended by Occupational Health and Safety Amendment Act No. 181 of 1993 and 2005 No. R859 (Concerning Written Policies) aims to provide for the health and safety of persons at work and for the health and safety of persons in connection with the activities of persons at work and to establish an advisory council for occupational health and safety.

The Occupational Health and Safety Act applies to all employers and workers, but not to:

▶ mines, mining areas or any mining works (as defined in the Minerals Act);

▶ load line ships, fishing boats, sealing boats, whaling boats (as defined in the Merchant Shipping Act) and floating cranes; whether in or out of the water; and

▶ people in or on these areas or vessels.

Employers' duties include:

1. To prepare a written policy concerning the protection of the health and safety of his employees at work, including a description of his organisation and the arrangements for carrying out and reviewing the policy;

2. the provision and maintenance of systems of work, plant and machinery that, as far as is reasonably practicable, are safe and without risks to health;

3. taking such steps as may be reasonably practicable to eliminate or mitigate any hazard or potential hazard to the safety or health of employees, before resorting to personal protective equipment;

4. making arrangements for ensuring, as far as is reasonably practicable, the safety and absence of risks to health in connection with the production, processing, use, handling, storage or transport of articles or substances;

5. establishing, as far as is reasonably practicable, what hazards to the health or safety of persons are attached to any work which is performed, any article or substance which is produced, processed, used, handled, stored or transported and any plant or machinery which is used in his business, and he shall, as far as is reasonably practicable, further

establish what precautionary measures should be taken with respect to such work, article, substance, plant or machinery in order to protect the health and safety of persons, and he shall provide the necessary means to apply such precautionary measures;

6. providing such information, instructions, training and supervision as may be necessary to ensure, as far as is reasonably practicable, the health and safety at work of his employees;

7. as far as is reasonably practicable, not permitting any employee to do any work or to produce, process, use, handle, store or transport any article or substance or to operate any plant or machinery, unless the precautionary measures contemplated in paragraphs (b) and (d), or any other precautionary measures which may be prescribed, have been taken;

8. taking all necessary measures to ensure that the requirements of this Act are complied with by every person in his employment or on premises under his control where plant or machinery is used;

9. enforcing such measures as may be necessary in the interest of health and safety;

10. ensuring that work is performed and that plant or machinery is used under the general supervision of a person trained to understand the hazards associated with it and who have the authority to ensure that precautionary measures taken by the employer are implemented; and

11. every employer shall conduct his undertaking in such a manner as to ensure, as far as is reasonably practicable, that persons other than those in his employment who may be directly affected by his activities are not thereby exposed to hazards to their health or safety.

12. Every self-employed person shall conduct his undertaking in such a manner as to ensure, as far as is reasonably practicable, that he and other persons who may be directly affected by his activities are not thereby exposed to hazards to their health or safety.

13. Any person who designs, manufactures, imports, sells or supplies any article for use at work shall ensure, as far as is reasonably practicable, that the article is safe and without risks to health when properly used and that it complies with all prescribed requirements.

14. Any person who erects or installs any article for use at work on or in any premises shall ensure, as far as is reasonably practicable, that nothing about the manner in which it is erected or installed makes it unsafe or creates a risk to health when properly used. Any person who manufactures, imports, sells or supplies any substance for use at work shall:
(a) ensure, as far as is reasonably practicable,

that the substance is safe and without risks to health when properly used; and

(b) take such steps as may be necessary to ensure that information is available with regard to the use of the substance at work, the risks to health and safety associated with such substance, the conditions necessary to ensure that the substance will be safe and without risks to health when properly used and the procedures to be followed in the case of an accident involving such substance.

15. Every employer shall:

(a) as far as is reasonably practicable, cause every employee to be made conversant with the hazards to his health and safety attached to any work which he has to perform, any article or substance which he has to produce, process, use, handle, store or transport and any plant or machinery which he is required or permitted to use, as well as with the precautionary measures which should be taken and observed with respect to those hazards;

(b) inform the health and safety representatives concerned beforehand of inspections, investigations or formal inquiries of which he has been notified by an inspector, and of any application for exemption made by him in terms of section 40; and

(c) inform a health and safety representative as soon as reasonably practicable of the occurrence of an incident in the workplace or section of the workplace for which such representative has been designated.

16. Every employee shall at work:

(a) take reasonable care for the health and safety of himself and of other persons who may be affected by his acts or omissions;

(b) as regards any duty or requirement imposed on his employer or any other person by this Act, cooperate with such employer or person to enable that duty or requirement to be performed or complied with;

(c) carry out any lawful order given to him, and obey the health and safety rules and procedures laid down by his employer or by anyone authorised thereto by his employer, in the interest of health or safety;

(d) if any situation which is unsafe or unhealthy comes to his attention, as soon as practicable report such situation to his employer or to the health and safety representative for his workplace or section thereof, as the case may be, who shall report it to the employer; and

(e) if he is involved in any incident which may affect his health or which has caused an injury to himself, report such incident to his employer or to anyone authorised thereto by the employer, or to his health and safety representative, as soon as practicable but not later than the end of the particular shift during which the incident occurred, unless the circumstances were such that the reporting of the incident was not possible, in which

case he shall report the incident as soon as practicable thereafter.

17. No person shall intentionally or recklessly interfere with or misuse anything which is provided in the interest of health or safety.

18. Chief executive officer charged with certain duties

(1) Every chief executive officer shall as far as is reasonably practicable ensure that the duties of his employer as contemplated in this Act, are properly discharged.

(2) Without derogating from his responsibility or liability in terms of subsection (1), a chief executive officer may assign any duty contemplated in the said subsection, to any person under his control, which person shall act subject to the control and directions of the chief executive officer.

(3) The provisions of subsection (1) shall not, subject to the provisions of section 37, relieve an employer of any responsibility or liability under this Act.

(4) For the purpose of subsection (1), the head of department of any department of State shall be deemed to be the chief executive officer of that department.

19. Health and safety representatives

(1) Subject to the provisions of subsection (2), every employer who has more than 20 employees in his employment at any workplace, shall, within four months after the commencement of this Act or after commencing business, or from such time as the number of employees exceeds 20, as the case may be, designate in writing for a specified period health and safety representatives for such workplace, or for different sections thereof.

(2) An employer and his employees or their representatives shall consult in good faith regarding the arrangements and procedures for the nomination or election, period of office and subsequent designation of health and safety representatives in terms of subsection (1): Provided that if such consultation fails, the matter shall be referred for arbitration to an inspector, whose decision shall be final.

(3) Arbitration by an inspector in terms of subsection (2) shall not be subject to the provisions of the Arbitration Act, 1965 (Act No. 42 of 1965), and a failure of the consultation contemplated in that subsection shall not be deemed to be a dispute in terms of the Labour Relations Act, 1956 (Act No. 28 of 1956).

(4) Only those employees employed in a full-time capacity at a specific workplace and who are acquainted with conditions and activities at that workplace or section thereof, as the case may be, shall be eligible for designation as health and safety representatives for that workplace or section.

(5) The number of health and safety representatives for a workplace or section thereof shall in the case of

15

shops and offices be at least one health and safety representative for every 100 employees or part thereof, and in the case of all other workplaces at least one health and safety representative for every 50 employees or part thereof: Provided that those employees performing work at a workplace other than that where they ordinarily report for duty, shall be deemed to be working at the workplace where they so report for duty.

(6) If an inspector is of the opinion that the number of health and safety representatives for any workplace or section thereof, including a workplace or section with 20 or fewer employees, is inadequate, he may by notice in writing direct the employer to designate such number of employees as the inspector may determine as health and safety representatives for that workplace or section thereof in accordance with the arrangements and procedures referred to in subsection (2).

(7) All activities in connection with the designation, functions and training of health and safety representatives shall be performed during ordinary working hours, and any time reasonably spent by any employee in this regard shall for all purposes be deemed to be time spent by him in the carrying out of his duties as an employee.

20. Functions of health and safety representatives

(1) A health and safety representative may perform the following functions in respect of the workplace or section of the workplace for which he has been designated, namely:

(a) review the effectiveness of health and safety measures;

(b) identify potential hazards and potential major incidents at the workplace;

(c) in collaboration with his employer, examine the causes of incidents at the workplace;

(d) investigate complaints by any employee relating to that employee's health or safety at work;

(e) make representations to the employer or a health and safety committee on matters arising from paragraphs (a), (b), (c) or (d), or where such representations are unsuccessful, to an inspector;

(f) make representations to the employer on general matters affecting the health or safety of the employees at the workplace;

(g) inspect the workplace, including any article, substance, plant, machinery or health and safety equipment at that workplace with a view to the health and safety of employees, at such intervals as may be agreed upon with the employer: Provided that the health and safety representative shall give reasonable notice of his intention to carry out such an inspection to the employer, who may be present during the inspection;

(h) participate in consultations with inspectors at the workplace and accompany inspectors on inspections of the workplace;

(i) receive information from inspectors as contemplated in section 36; and

(j) in his capacity as a health and safety representative attend meetings of the health and safety committee of which he is a member, in connection with any of the above functions.

(2) A health and safety representative may, in respect of the workplace or section of the workplace for which he has been designated:

(a) visit the site of an incident and attend any inspection in loco;

(b) attend any investigation or formal inquiry held in terms of this Act;

(c) in so far as is reasonably necessary to perform his functions, inspect any document which the employer is required to keep in terms of this Act;

(d) accompany an inspector on any inspection;

(e) with the approval of the employer (which approval shall not be unreasonably withheld), be accompanied by a technical adviser, on any inspection; and

(f) participate in any internal health or safety audit.

(3) An employer shall provide such facilities, assistance and training as a health and safety representative may reasonably require and as have been agreed upon for the carrying out of his functions.

(4) A health and safety representative shall not incur any civil liability by reason of the fact only that he failed to do anything which he may do or is required to do in terms of this Act.

21. Health and safety committees

(1) An employer shall in respect of each workplace where two or more health and safety representatives have been designated, establish one or more health and safety committees and, at every meeting of such a committee as contemplated in subsection (4), consult with the committee with a view to initiating, developing, promoting, maintaining and reviewing measures to ensure the health and safety of his employees at work.

(2) A health and safety committee shall consist of such number of members as the employer may from time to time determine: Provided that:

(a) if one health and safety committee has been established in respect of a workplace, all the health and safety representatives for that workplace shall be members of the committee;

(b) if two or more health and safety committees have been established in respect of a workplace, each health and safety representative for that workplace shall be a member of at least one of those committees; and

(c) the number of persons nominated by an employer on any health and safety committee established in terms of this section shall not exceed the number of health and safety representatives on that committee.

(3) The persons nominated by an employer on a health and safety committee shall be designated in writing by the employer for such period as may be determined by him, while the health and safety representatives shall be members of the committee for the period of their designation in terms of section 17(1).

(4) A health and safety committee shall hold meetings as often as may be necessary, but at least once every three months, at a time and place determined by the committee: Provided that an inspector may by notice in writing direct the members of a health and safety committee to hold a meeting at a time and place determined by him: Provided further that, if more than 10% of the employees at a specific workplace has handed a written request to an inspector, the inspector may by written notice direct that such a meeting be held.

(5) The procedure at meetings of a health and safety committee shall be determined by the committee.

(6) (a) A health and safety committee may co-opt one or more persons by reason of his or their particular knowledge of health or safety matters as an advisory member or as advisory members of the committee.

(b) An advisory member shall not be entitled to vote on any matter before the committee.

(7) If an inspector is of the opinion that the number of health and safety committees established for any particular workplace is inadequate, he may in writing direct the employer to establish for such workplace such number of health and safety committees as the inspector may determine.

22. Functions of health and safety committees

(1) A health and safety committee:

(a) may make recommendations to the employer or, where the recommendations fail to resolve the matter, to an inspector regarding any matter affecting the health or safety of persons at the workplace or any section thereof for which such committee has been established;

i. shall discuss any incident at the workplace or section thereof in which or in consequence of which any person was injured, became ill or died, and may in writing report on the incident to an inspector; and

ii. shall perform such other functions as may be prescribed.

(2) A health and safety committee shall keep record of each recommendation made to an employer in terms of subsection (1)(a) and of any report made to an inspector in terms of subsection (1)(b).

(3) A health and safety committee or a member thereof shall not incur any civil liability by reason of the fact only that it or he failed to do anything which it or he may or is required to do in terms of this Act.

(4) An employer shall take the prescribed steps to ensure that a health and safety committee complies with the provisions of section 19(4) and performs the duties assigned to it by subsections (1) and (2).

See: http://www.labour.gov.za/downloads/legislation/acts/occupational-health-and-safety/amendments/Amended%20Act%20-%20Occupational%20Health%20and%20Safety.pdf

or: http://www.labour.gov.za/DOL/legislation/acts/occupational-health-and-safety/read-online/amended-occupational-health-and-safety-act

15.3.19 Trinidad and Tobago

(Population in 2014: 1,328,019. Labour force: 621,000)

In Trinidad and Tobago a new era for occupational safety and health began with the assenting by Parliament of the Occupational Safety and Health Act (OSHA), 2004, in January of that year. The OSHA was amended by the Occupational Safety and Health (Amendment) Act, 2006, proclaimed on 17 February 2006, and operational since that date. This Act is of significance because as of 17 August 2007 it repealed and replaced the Factories Ordinance Chapter 30, No. 2, which had governed health and safety in Trinidad and Tobago since 1948; it was left operational for a few months while the arrangements for the implementation arm of the Occupational Safety and Health Authority, provided for in OSHA, were put in place. The OSHA brings the legislation in step with the country's rapid industrialisation spurred by increased activities in the construction and petrochemical sectors.

One of the key changes brought about by the enactment of this legislation is the widening of the scope of categories of protected workers. While the Factories

Ordinance made provisions for persons employed in **factories only,** the OSHA covers most workers in all aspects of work undertaken in an industrial establishment (defined as a factory, shop, office, place of work or other premises excluding residential premises) that may have significant impacts on the health and safety of the employees, with the exception of workers in private homes (domestic workers).

The OSHA promotes voluntary compliance by facilitating a shift to stronger self-governance by the employer and the workers and a more regulatory role by the government. It seeks to ensure that the promotion of high safety and health standards does not hinder business performance, productivity and efficiency while, on the other hand, it addresses employee rights issues such as sufficient protection and avenues for redress in accordance with the law and good industrial relations practices.

Other key features of the OSHA are that it addresses the roles and responsibilities of employers, employees, occupiers, suppliers and manufacturers as follows:

▶ **Employers** have the duty of care to ensure the safety, health and welfare at work of all employees and any third party that can be affected by his/her undertaking.

▶ **Employees** must assume personal responsibility for their own safety and actions at work and must be aware of and respect all potential hazards. They are required to take reasonable care for their own safety and health and for the safety and health of other persons who may be affected by their acts or omissions at work. They must comply with the employer's occupational safety and health and welfare rules, instructions, safe work practices, established procedures, permit requirements and codes.

▶ **Occupiers** or those with ultimate control over the affairs of an industrial establishment must ensure that the employer discharges his duty of care for the safety and health of employees and members of the public on or in the vicinity of the premises. This will include the provision and maintenance of safe means of access and egress and emergency response planning for those using their premises.

▶ A duty is imposed on anyone in the supply chain who **designs, manufactures, imports or supplies** any technology, machinery, plant, equipment or material for use at work to ensure that such items are safe and without risk to the safety and health of employees when properly used;

▶ It promotes consultation between **employers** and **workers' organisations** on occupational safety and health issues. For example, the OSHA requires employers or occupiers of industrial establishments of 25 or more employees to prepare in consultation with the worker representatives, a general policy with respect to occupational safety and health. Such

an employer/occupier is also expected to establish a joint (trade union/employee and employer) Safety and Health Committee to review health and safety measures as well as investigate matters considered to be unsafe or a risk to health at the industrial establishment; and

▶ It empowers **inspectors** to take enforcement and legal actions. Inspectors are permitted to enter, inspect, examine and take samples from any industrial establishment at all reasonable times in their conduct of inspection duties. Inspectors can also serve prohibition or improvement notices and initiate legal proceedings against persons in breach of the Act.

With respect to administration of the legislation, the OSHA provides for the establishment of two entities, the Occupational Safety and Health Authority (referred to as the Authority) and the Occupational Safety and Health Agency (referred to as the Agency).

The Occupational Safety and Health Authority is the regulatory body that is responsible to the government for the implementation of the provisions of the OSHA. It has a tripartite composition. The Authority was established in October 2006 to, inter alia:

▶ act as an Advisory Body to the Minister of Labour on policy, standards and matters related to occupational safety and health;

▶ make recommendations to the Minister of Labour in respect of regulations under the Act and seek his approval for the issuance of codes of practice;

▶ guide the Minister of Labour on the organisational structure, staff requirements and operations for the proper and efficient functioning of the Agency;

▶ provide leadership, guidance, direction and control in the execution and implementation of the National Occupational Safety and Health Policy;

▶ ensure that occupational safety and health standards are established, communicated and enforced;

▶ reach out to employers and employees through research, education and awareness, technical assistance and consultation programmes; and

▶ ensure that all stakeholders are kept informed and adequately advised on occupational safety and health.

The Occupational Safety and Health Agency is established as an administrative and occupational safety and health inspection and enforcement group. Whereas the Authority has the primary function of policy formulation, the Agency is responsible for the implementation of those policies. The Agency is also responsible for initiating consultation with government entities performing various inspection functions, with the objective of formulating memoranda of understanding, establishing mechanisms for coordination across jurisdictional lines and the provision for the implementation of integrated occupational safety and health programmes. Enforcement of the provisions of the Act and related regulations is the responsibility of the Agency, and this is achieved through suitably qualified inspectors.

See: http://www.ttparliament.org/legislations/a2006-03.pdf

http://sta.uwi.edu/hse/OSHA2004.pdf

15.3.20 Turkey

(Population in 2014: 76,667,000. Labour force: 27,910,000)

Turkey's Occupational Health and Safety Act 2012 (No. 6331) (OHS Act) was enacted by Parliament on 26 June 2012, and entered into force on 30 June 2012. However, articles 6 and 7 will enter into force gradually, on 1 January 2014 (for moderate and very high hazard workplaces with less than 50 employees), and on 1 June 2016 (for low hazard workplaces with less than 50 employees, and government establishments).

The purpose of the OHS Act is to provide a framework within which employees and employers can interact with respect to occupational health and safety. The OHS Act extends the scope of safety regulations, and enforces the regulations for all workplaces including the public sector, and workers – including apprentices and interns. The OHS Act removes various regulations that were mainly in Section V of Labor Act No. 4857, pertaining to the occupational health and safety duties for employers and employees.

While workplaces are divided by the OHS Act into three general hazard categories (low, moderate and very high) according to their activities, the Act also introduces financial support for the OHS expenses of workplaces with fewer than 10 employees.

General responsibility of the employer

The employer shall:

▶ have a duty to ensure the safety and health of workers in every aspect related to the work;

▶ take measures necessary for the safety and health protection of workers, including prevention of occupational risks and the provision of information, training and necessary organisation;

▶ monitor and check whether occupational health and safety measures that have been taken in the workplace are followed and ensure that nonconforming situations are eliminated;

▶ carry out a risk assessment or get one carried out;

▶ take into consideration the worker's capabilities as regards health and safety where he entrusts tasks to a worker;

▶ take appropriate measures to ensure that workers other than those who have received adequate information and instructions are denied access to areas where there is life-threatening and special hazard.

Principles of protection from risks – Article 5

The employer shall fulfil this on the following principles:
(a) avoiding risks;
(b) evaluating the risks which cannot be avoided;
(c) combating the risks at source;
(d) adapting the work to the individual, especially as regards the design of workplaces, the choice of work equipment and the choice of work and production methods, with a view, in particular, to avoiding or minimising if cannot be avoided, the adverse effects of monotonous work and work at a predetermined work-rate on health and safety;
(e) adapting to technical progress;
(f) replacing the dangerous by the non-dangerous or the less dangerous;
(g) developing a coherent overall prevention policy which covers technology, organisation of work, working conditions, social relationships and the influence of factors related to the working environment;
(h) giving collective protective measures priority over individual protective measures;
(i) giving appropriate instructions to the workers.

Occupational health and safety services – Article 6

In order to provide occupational health and safety services including activities related to the protection and prevention of occupational risks, the employer shall:

designate workers as occupational safety specialist, occupational physician and other health staff. In case there is lack of personnel in the undertaking competent enough to be designated, the employer shall enlist a joint health and safety unit to partially or fully provide these services. Provided that the employer has the required qualifications and documents, these services can be offered by the

15

employer considering the hazard class and the number of workers;

implement measures in accordance with legislation, relating to safety and health, notified in writing by the designated person;

keep the designated persons or external consultants informed of issues affecting the safety and health of workers.

Article 7 concerns state subsidies for occupational health and safety services.

Article 8 concerns occupational physicians and safety specialists.

Article 9 relates to determining the hazard class on the basis of a circular from the Director General.

Article 10 concerns risk assessment, control, measurement and research.

Article 11 concerns emergency plans, fire-fighting and first-aid.

Article 12 concerns evacuation.

Article 13 concerns workers' right to abstain from work and the procedure to be followed.

Article 14 concerns recording and notification of occupational accidents and diseases.

Article 15 concerns health surveillance.

Article 16 concerns information to be provided by the employer to workers.

Article 17 concerns the training of workers in working time and at no financial cost to workers.

Article 18 concerns consultation with and participation of workers.

Article 19 concerns workers' obligations:

It shall be the responsibility of each worker to take care as far as possible of his own safety and health and that of other persons affected by his acts or commissions at work in accordance with his training and the instructions related to occupational health and safety given by his employer.

To this end, workers must in particular, in accordance with their training and the instructions given by their employer:

(a) make correct use of machinery, apparatus, tools, dangerous substances, transport equipment and other means of production; use such safety devices correctly and refrain from changing or removing arbitrarily safety devices fitted;

(b) make correct use of the personal protective equipment supplied to them and protect themselves;

(c) immediately inform the employer and/or the workers' representative of any work situation they have reasonable grounds for considering represents a serious and immediate danger to safety and health and of any shortcomings in the machinery, apparatus, tools, facilities and buildings;

(d) cooperate with the employer and/or workers' representative to enable any tasks or requirements imposed by the competent authority to protect the safety and health of workers at work to be carried out;

(e) cooperate with the employer and/or workers' representative for occupational health and safety of workers within their field of activity.

Article 20 concerns workers' representatives.

Article 21 concerns the National Occupational Health and Safety Council.

Article 22 concerns the organisation-based Occupational Health and Safety Committee.

Article 23 concerns the coordination of occupational health and safety services where there is more than one employer in the same working environment.

Articles 24–27 concern inspection and penalties for non-compliance.

Article 28 concerns prohibition and influence of drugs and alcohol in the workplace.

Article 29 concerns safety reports or accident prevention policy documents for workplaces where a serious industrial accident can take place.

See: http://www.ilo.org/dyn/natlex/docs/ MONOGRAPH/92011/106963/F1028231731/ TUR92011%20Eng.pdf

15.3.21 **United Arab Emirates**

(Population in 2014: 9,446,000. Labour force: 4,588,000. 85% expatriates)

Federal Law to Regulate Employment Relationships (No. 8 of 1980) as amended up to 2007. This applies to most non-governmental establishments except in agriculture and domestic service. There are special provisions to protect women (cannot be employed at night or in hazardous jobs), pregnant women, nursing mothers and young persons (under 15).

The main provisions with respect to occupational health and safety include the following:

1. Every employer must provide adequate means of protection for the employee from the hazards of injuries and vocational diseases that may occur during work as well as the hazards of fire and other hazards arising from use of machines and other tools, and he must apply all other means

of protection as approved by the Ministry of Labour and Social Affairs, and the employee must use protective equipment and clothing provided to him for such purpose and he must abide by all instructions of the employer aiming at his protection from dangers and must not act in a way that may obstruct the application of said instruction.

2. Every employer must display at a conspicuous point in the place of business detailed instructions concerning methods to prevent fire and protect employees from dangers while they perform their duties. Said instructions shall read in Arabic and, if necessary, in another language understood by the employees.

3. Without prejudice to the provisions of by-laws and regulations issued by concerned government authorities the employer must provide proper cleanliness and ventilation in each place of business and must provide such places with adequate illumination, potable water and toilets.

4. The employer must provide employees with means of medical care according to the standards decided by the Minister of Labour and Social Affairs in collaboration with the Minister of Health.

5. The employer or his representative at the time of appointment must keep employees informed of the dangers related to their profession and preventative measures they have to take. Moreover, the employer must display detailed written instructions in this respect at places of business.

6. The employee shall abide by instructions and orders related to business safety and precautions, and adopt precautionary methods and pledge to care for items thereof in his possession. It is prohibited for an employee to act in any way that may contravene enforcement of said instructions or misuse methods placed for health and safety protection of employees or which may cause loss or damage to the same.

7. Each employer who employs employees in areas that are remote from cities where there is no access to normal means of transportation shall provide employees with the following facilities:

   1. Adequate means of transport;
   2. Adequate accommodation;
   3. Drinking water;
   4. Proper foodstuff;
   5. Medical aid equipment;
   6. Entertainment and sports amenities.
      Areas to which all or part of the provisions of this Article apply shall be stated by decision of the Minister of Labour and Social Affairs. With exception of foodstuff, all services referred to in this Article shall be at the expense of the employer and nothing hereof is to be borne by the employee.

8. If the employee sustains a labour injury or occupational disease as enumerated in Schedule (1) and (2) attached to the Law, the employer or its representative must report the accident instantly to the Police and Labour Department or any of its branches having jurisdiction over the place of business. The report must include the employee's name, age, vocation, address, and nationality in addition to a brief description of the accident, its circumstances and the arrangements made for the employee's medical aid or treatment.

9. The police shall carry out necessary investigation, upon receipt of the report which contains statements of witnesses and employer or his representative and statement of injured if his condition so allows, and the report must indicate in particular if the accident is related to work, and whether it was deliberate or a result of gross misconduct on the part of the employee.

10. Following the investigation, the police must send a copy of the report to the Labour Department and another to the employer. The Labour Department may request that the investigation be completed or otherwise it shall have the investigation directly completed if it is deemed necessary.

11. In cases of Labour accidents and occupational diseases the employer shall pay the employee's treatment expenses at government or private hospitals until he recovers or his disability is proven. Treatment includes admission in hospitals or sanatorium, and surgical operation, X-ray and laboratory fees in addition to medicines and rehabilitation equipment purchased, artificial limbs and apparatus provided to the disabled persons. Moreover, the employer must pay the transport expenses arising from the treatment of the employee.

12. If the injury prevents the employee from carrying out his duties, the employer must pay him a financial subsidy equal to full pay throughout the period of treatment or for a period of six months, whichever is shorter. If treatment lasts for more than six months, said subsidy shall be reduced to the half for another period of six months or until the employee recovers from illness or his disability becomes certain or he dies, whichever occurs first.

13. Neither the injured employee nor the members of his family shall be entitled to indemnity in respect of injury or disability if it has not caused death and if the investigations by the competent authorities have established that the employee has deliberately caused injury to himself with intention of committing suicide or to obtain indemnity or sick leave or otherwise, or if the employee was at the time of the incident under the influence of drug or alcoholic drinks, or if he has wilfully violated safety instructions displayed conspicuously at the place of business or if his injury or disability

15

resulted from serious premeditated misdemeanour on his part or if he has refused unreasonably the medical check-up or treatment as prescribed by the medical board. In any of the cases hereinabove, the employer shall not be under obligation to provide treatment or any financial subsidy to the employee.

14. Labour inspection shall be undertaken by specialised inspectors attached to the Ministry of Labour and Social Affairs, and having the prerogatives and powers provided for in this Law. Labour inspectors shall carry cards issued by the Ministry of Labour and Social Affairs certifying their capacity.

15. Labour inspectors shall have the following powers:
    (a) Control the proper implementation of the provisions of the Labour Law particularly in respect of the conditions of work, remuneration and protection and safety of employees during the performance of their duties and such other matters related to the health and safety of employees and the employment of juveniles and women.
    (b) Supply employers and employees with information and technical guidance to enable them to adopt the best methods for the implementation of the provisions of this Law.
    (c) Report to competent authorities, any problems which the existing rules cannot remedy and to propose whatever is necessary to this.
    (d) Make report of cases found in violation of the provisions of the Labour Law, regulations and decisions issued for their implementation.

16. The labour inspector shall have the following powers:
    1. Enter any establishment governed by the provisions of this Law at any time of the day or night without prior notice provided he performs that during working hours.
    2. Carry out any inspection or investigation as may be necessary to ensure the proper implementation of the Law, and in particular he shall:
       (a) Interrogate the employer or employees either alone or in presence of witnesses in respect of any matter related to implementation of the provisions of the Law.
       (b) Inspect all documents required to be kept in accordance with the provisions of the Labour Law and decisions made in execution thereof and to obtain photocopies and extracts therefrom.
       (c) Take one or more samples of materials used or related to the industrial operations or in other operations subject to inspection if such materials are believed to be harmful to the health or safety of employees, in

order to have them analysed in government laboratories to determine the extent of harm and inform the employer or his representative of the results and to take appropriate measures in this regard.
    (d) Ensure that notices and publications are displayed conspicuously at the place of business in accordance with the provisions of the Law.

For more information see the ILO 'Overview of the Occupational Safety and Health Situation in the Arab Region', a study prepared for discussion at the Inter-Regional Tripartite Meeting on Occupational Safety and Health, Damascus, 18–20 November 2007 at: www.ilo.org/public/english/region/arpro/beirut/.../osh_habib.pdf http://www.gulftalent.com/repository/ext/UAE_Labour_Law.pdf

15.3.22 United Kingdom

(Population in 2014: 64,105,000. Labour force: 31,760,000)

(a) Introduction

In the UK, health and safety legislation is drawn up and enforced by the Health and Safety Executive and local authorities (the local council) under the Health and Safety at Work etc. Act 1974. Increasingly in the UK the regulatory trend is away from prescriptive rules, and towards risk assessment. Recent major changes to the laws governing asbestos and fire safety management embrace the concept of risk assessment.

(b) Duties imposed by the Health and Safety at Work etc. Act 1974

The HSW Act 1974 is based on the principle that those who create risks to employees or others in the course of carrying out work activities are responsible for controlling those risks. The Act places specific responsibilities on employers, the self-employed, employees, designers, manufacturers, importers and suppliers. The Act and associated legislation also place duties in certain circumstances on others, including landlords, licensees and those in control of work activities, equipment or premises. Under the main provisions of the Act, employers have legal responsibilities in respect of the health and safety of their employees and other people who may be affected by their undertaking and exposed to risks as a result. Employees are required to take reasonable care for the health and safety of themselves and others.

Most duties are expressed as goals or targets which are to be met 'so far as is reasonably practicable', or through exercising 'adequate control' or taking 'appropriate' (or 'reasonable') steps. Qualifications such as these involve making judgements as to whether existing control measures are sufficient and, if not, what else should be done to eliminate or reduce the risk. The main duties placed on employers and the self-employed under sections 2 and 3 of the Act, for example, are qualified by the phrase 'so far as is reasonably practicable'. This means that the extent of the risk must be balanced against the difficulty involved (in terms of time, money or trouble) in controlling the risk further; additional controls are not necessary if the difficulty in implementing them would be grossly disproportionate to the risk, or to the reduction in risk that would be achieved. This judgement is an essential part of the risk assessment process and will be informed by approved codes of practice, published standards and HSE or industry guidance on good practice where available. The size of the business and its financial strength do not determine the health and safety standards to be achieved.

(c) Regulations, codes of practice and guidance

The Act states that legislation passed before 1974 should be 'progressively replaced by a system of regulations and approved codes of practice'. At the time the Act came into force there were some 30 statutes and 500 sets of regulations. In carrying out the reform of the law, the general principle has been that regulations, like the Act itself, should, so far as possible, express general duties, principles and goals and that subordinate detail should be set out in approved codes and guidance. Following a review of health and safety regulation in 2012 the process of reform continues. Further change results from the European legislative process, which sometimes imposes more detailed and specific requirements than would be envisaged under the Act. Regulations are made by the appropriate government minister, normally on the basis of proposals submitted by the HSE after consultation, as previously explained. They have to be laid before Parliament and become law 21 days after being submitted to Parliament, unless an objection is made.

Approved Codes of Practice (ACoPs) are approved by the HSE with the consent of the appropriate Secretary of State – they do not require agreement from Parliament. ACoPs have a special status in law. Failure to comply with the provisions of an ACoP may be taken by a court in criminal proceedings as evidence of a failure to comply with the requirements of the Act or of regulations to which the ACoP relates, unless it can be shown that those requirements were complied with in some other equally effective way. ACoPs (which can be updated more easily) provide flexibility to cope with innovation and technological change without a lowering of standards.

During 2013 and 2014 a major review resulted in a reduction in the number of ACoPs and a huge revamp of the guidance available from the HSE. All guidance was made free to download from the HSE's website, which is a good resource of material at a variety of levels. The HSE have also started to charge for interventions where breaches of health and safety legislation are discovered.

See: http://www.hse.gov.uk

15.3.23 United States

(Population in 2014: 318,958,000. Labour force: 155,400,000)

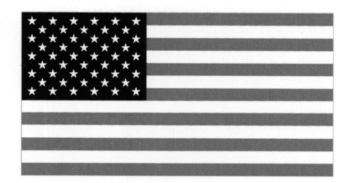

In the USA the Occupational Safety and Health Act is the primary federal law which governs occupational health and safety in the private sector and federal government in the United States. It was enacted by Congress in 1970 and was signed by President Richard Nixon on 29 December 1970. Its main goal is to ensure that employers provide employees with an environment free from recognised hazards, such as exposure to toxic chemicals, excessive noise levels, mechanical dangers, heat or cold stress, or unsanitary conditions.

The Act can be found in the United States Code at title 29, chapter 15.

(a) History of federal workplace safety legislation

Efforts by the US Federal Government to ensure workplace health and safety were minimal until the passage of OSHA. The American system of mass production encouraged the use of machinery, while the statutory regime did nothing to protect workplace safety. For most employers, it was cheaper to replace a dead or injured worker than it was to introduce safety measures. Tort law provided little recourse for relief for

15

the survivors of dead workers or for injured employees. After the Civil War, some improvements were made through the establishment of state railroad and factory commissions, the adoption of new technology (such as the air brake), and more widespread availability of life insurance. But the overall impact of these improvements was minimal.

The first federal safety legislation was enacted in the Progressive period. In 1893, Congress passed the Safety Appliance Act, the first federal statute to require safety equipment in the workplace (the law applied only to railroad equipment, however). In 1910, in response to a series of highly publicised and deadly mine explosions and collapses, Congress established the federal Bureau of Mines to conduct research into mine safety (although the Bureau had no authority to regulate mine safety). Backed by trade unions, many states also enacted workers' compensation laws which discouraged employers from permitting unsafe workplaces. These laws, as well as the growing power of labour unions and public anger toward poor workplace safety, led to significant reductions in worker accidents for a time.

Industrial production increased significantly in the United States during the Second World War, and industrial accidents soared. Winning the war took precedence over safety, and most labour unions were more concerned with maintaining wages in the face of severe inflation than with workplace health and safety. After the war ended, however, workplace accident rates remained high and began to rise. In the two years preceding OSHA's enactment, 14,000 workers died each year from workplace hazards, and another 2 million were disabled or harmed. Additionally, the 'chemical revolution' introduced a vast array of new chemical compounds to the manufacturing environment. The health effects of these chemicals were poorly understood, and workers received few protections against prolonged or high levels of exposure. While a few states, such as California and New York, had enacted workplace safety as well as workplace health legislation, most states had not changed their workplace protection laws since the turn of the century.

(b) Passage of the OSH Act

In the mid-1960s, growing awareness of the environmental impact of many chemicals had led to a politically powerful environmental movement. Some labour leaders seized on the public's growing unease over chemicals in the environment, arguing that the effect of these compounds on worker health was even worse than the low-level exposure plants and animals received in the wild. On 23 January 1968, President Lyndon B. Johnson submitted a comprehensive occupational health and safety bill to Congress. Led by the United States Chamber of Commerce and the National Association of Manufacturers, the legislation was widely opposed by business. Many labour leaders,

including the leadership of the American Federation of Labor and Congress of Industrial Organizations (AFL–CIO), did not fight for the legislation, claiming workers had little interest in the bill. The legislation died in committee.

On 14 April 1969, President Richard Nixon introduced two bills into Congress which would have also protected worker health and safety. The Nixon legislation was much less prescriptive than the Johnson bill, and workplace health and safety regulation would be advisory rather than mandatory. However, Representative James G. O'Hara and Senator Harrison A. Williams introduced a much stricter bill similar to the Johnson legislation of the year before. Companion legislation introduced in the House also imposed an all-purpose 'general duty' clause on the enforcing agency as well. With the stricter approach of the Democratic bill apparently favoured by a majority of both chambers, and unions now strongly supporting a bill, Republicans introduced a new, competing bill. The compromise bill established the independent research and standard-setting board favoured by Nixon, while creating a new enforcement agency. The compromise bill also gave the Department of Labor the power to litigate on the enforcement agency's behalf (as in the Democratic bill). In November 1970, both chambers acted: the House passed the Republican compromise bill, while the Senate passed the stricter Democratic bill (which now included the general duty clause). A conference committee considered the final bill in early December 1970. Union leaders pressured members of the conference committee to place the standard-setting function in the Department of Labor rather than an independent board. In return, unions agreed to let an independent review commission have veto power over enforcement actions. Unions also agreed to removal of a provision in the legislation which would have let the Secretary of Labor shut down plants or stop manufacturing procedures which put workers in 'imminent danger' of harm. In exchange for a Republican proposal to establish an independent occupational health and safety research agency, Democrats won inclusion of the 'general duty' clause and the right for union representatives to accompany a federal inspector during inspections. The conference committee bill passed both chambers on 17 December 1970, and President Nixon signed the bill on 20 December 1970.

The Act went into effect on 28 April 1971 (now celebrated as Workers' Memorial Day by American labour unions).

(c) Description of the OSH Act

The Act created the Occupational Safety and Health Administration (OSHA), an agency of the Department of Labor. OSHA was given the authority to both set and enforce workplace health and safety standards. The

Act also created the independent Occupational Health and Safety Review Commission to review enforcement priorities, actions and cases.

The Act also established the National Institute of Occupational Safety and Health (NIOSH), an independent research institute in the, then, Centers for Disease Control.

The Act defines an employer to be any 'person engaged in a business affecting commerce who has employees, but does not include the United States or any state or political subdivision of a State'. The Act applies to employers as diverse as manufacturers, construction companies, law firms, hospitals, charities, labour unions and private schools. Churches and other religious organisations are covered if they employ workers for secular purposes. The Act excludes the self-employed, family farms, workplaces covered by other federal laws (such as mining, nuclear weapons manufacture, railroads and airlines) and state and local governments (unless state law permits otherwise). The Act does cover federal agencies, as well as the United States Postal Service.

Section 5 of the Act contains the 'general duty clause'. The 'general duty clause' requires employers to: (1) maintain conditions or adopt practices reasonably necessary and appropriate to protect workers on the job; (2) be familiar with and comply with standards applicable to their establishments; and (3) ensure that employees have and use personal protective equipment when required for safety and health. OSHA has established regulations for when it may act under the 'general duty clause'. The four criteria are: (1) there must be a hazard; (2) the hazard must be a recognised hazard (e.g. the employer knew or should have known about the hazard, the hazard is obvious, or the hazard is a recognised one within the industry); (3) the hazard could cause or is likely to cause serious harm or death; and (4) the hazard must be correctable (OSHA recognises not all hazards are correctable). Although theoretically a powerful tool against workplace hazards, it is difficult to meet all four criteria. Therefore, OSHA has engaged in extensive regulatory rule-making to meet its obligations under the law.

Due to the difficulty of the rule-making process (which is governed by the Administrative Procedures Act), OSHA has focused on basic mechanical and chemical hazards rather than procedures. Major areas which its standards currently cover are: toxic substances, harmful physical agents, electrical hazards, fall hazards, hazards associated with trenches and digging, hazardous waste, infectious disease, fire and explosion dangers, dangerous atmospheres, machine hazards and confined spaces.

Section 8 of the Act covers reporting requirements. All employers must report to OSHA within eight hours if an employee dies from a work-related incident, or three or more employees are hospitalised as a result of a work-related incident. Additionally, all fatal on-the-job heart attacks must also be reported. Section 8 permits OSHA inspectors to enter, inspect and investigate, during regular working hours, any workplace covered by the Act. Employers must also communicate with employees about hazards in the workplace. By regulation, OSHA requires that employers keep a record of every non-consumer chemical product used in the workplace. Detailed technical bulletins called material safety data sheets (MSDSs) must be posted and available for employees to read and use to avoid chemical hazards. OSHA also requires employers to report on every injury or job-related illness requiring medical treatment (other than first-aid) on OSHA Form 300, 'Log of Work-Related Injuries and Illnesses' (known as an 'OSHA Log' or 'Form 300'). An annual summary is also required and must be posted for three months, and records must be kept for at least five years.

Section 11(c) of the Act prohibits any employer from discharging, retaliating or discriminating against any employee because the worker has exercised rights under the Act. These rights include complaining to OSHA and seeking an OSHA inspection, participating in an OSHA inspection, and participating or testifying in any proceeding related to an OSHA inspection.

Section 18 of the Act permits and encourages states to adopt their own occupational safety and health plans, so long as the state standards and enforcement 'are or will be at least as effective in providing safe and healthful employment' as the federal OSH Act. States which have such plans are known as 'OSHA States'. As of 2007, 22 states and territories operated complete plans and four others had plans which covered only the public sector.

See: https://www.osha.gov/index.html

15.4 Common themes in national legislation

15.4.1 Prevention of occupational accidents and disease

Most countries regard prevention of occupational accidents and ill health, or the equivalent aim of minimisation of occupational risks to workers' health and safety, as a national objective (whether or not it is explicitly stated as such).

In a good number of the countries that opt for an 'open' definition of prevention, legislation sets down principles of prevention to act as a bridge between the general duty to prevent and specific obligations. Such principles are the criteria by which an employer should make choices between different options, and include:

▶ Giving preference to prevention measures at the source of risk versus collective protection measures,

15

while the latter are to be preferred to individual protection measures.

▶ Analysing risks and their possible solutions by looking at all the factors (material and organisational) that contribute towards creating the risks.

▶ Endeavouring to adapt the job and the working environment to the worker (rather than the other way around).

▶ Organising prevention by integrating it with all hierarchical levels and functional units of the enterprise, etc.

15.4.2 General duties

In a large majority of countries, the law sets forth in one form or another employers' general duty to look after workers' health and safety (employer's duty of prevention).

There should be a single basic regulatory framework at a high level that can serve as a context for the remaining regulations, setting down employers' general obligations and the rights and obligations of workers. This same basic regulatory framework, or supplementary regulations, may also set down employers' additional obligations with respect to certain worker groups or categories that require special consideration due to the nature of their personal circumstances (pregnant women, the disabled, young persons, special-susceptible workers, etc.), working conditions (that may require a shorter working day, for instance) or contract terms (including temporary or seasonal workers).

It is then advisable to group together the provisions regulating work environment. As this essentially depends on the sector in which the enterprise operates (industry and services, construction, mining, fisheries, transport, etc.), regulation is generally established by sector, and these regulations as a whole are the so-called regulatory framework (although the rules applicable to industry and services are generally regarded as the 'general framework'). This regulatory platform governs the safety conditions that must be extant in workplaces (structural stability, space, means of access and evacuation routes, signage, auxiliary services, etc.). Moreover, sectoral or cross-cutting regulations also govern the service and protection facilities (electrical installations, lighting, ventilation and air conditioning, gas, heating, elevation, fire safety, etc.) in workplaces.

Finally, systematic regulation should be put in place relating to work environment, resources and special processes through a suitably ordered set of rules dealing with:

▶ Safety, selection and acquisition, installation, use, maintenance, storage and disposal (where applicable) of work equipment and facilities, materials and products and collective or personal protection equipment (i.e. 'work resources').

▶ The conditions applicable to the physical, chemical and biological work environment (noise, vibrations, thermal stress, ionising and non-ionising radiation, etc.).

▶ Procedures for performing potentially hazardous operations (handling of heavy loads, work with live electrical facilities, in explosive atmospheres, in high- or low-pressure environments, etc.).

15.4.3 Prevention activities

An employer's duty to look after workers' health and safety comes from the fact that essentially the employer determines the conditions under which they work. Accordingly, the employer must take into account this duty to protect workers when conceiving of, choosing, conditioning, introducing and using and/or maintaining:

▶ work premises or locations and their environment (physical, chemical and biological);

▶ work resources (facilities, materials, substances and equipment), including means for collective or personal protection required for 'safe' work; and

▶ work organisation and procedures, including safety precautions.

Apart from the above, countries' regulations often call for employers to carry out certain specific preventative measures, some of the more common being:

15.4.4 Risk assessment

In a fair number of countries, risk assessment is not compulsory. In others, assessment is compulsory only where, in order to verify that a legal requirement is being met (for instance, the air-borne maximum permissible concentration of a chemical), measurements and testing need to be performed.

15.4.5 Training and information for workers

Training and information for workers in connection with the safety and health aspects (both inherent risks and preventative measures) of their jobs are addressed by most bodies of legislation on occupational safety and health, be it as a duty placed on employers and/or as a right vested in workers.

15.4.6 Control of working conditions

This concept basically covers two different types of activity. On one hand, some countries' legislation contemplates regular general inspection of a company's job positions, not as a way of making or updating a risk assessment but in order to check, by direct observation or through contacts with managers and workers through specifically planned visits, that the situation is under control and that no changes, anomalies or incidents requiring corrective action have arisen. The second activity under this heading is regular control of critical elements: most countries require that the employer maintain and regularly monitor certain facilities or equipment where a fault could seriously endanger workers' health or safety, for example cranes, hoists, pressure systems and extraction systems.

15.4.7 Surveillance of workers' health

Legislation on this matter varies considerably from one country to another. In some countries, surveillance of workers' health is an obligation pertaining to the employer, while in others it is the responsibility of the national health system.

15.4.8 Action in the face of emergencies

In almost all countries there is some sort of rule addressing protective action and conditions in the face of emergencies, especially with a view to fire risks. It is also a common feature of most bodies of legislation to require employers to take steps to provide, as quickly as possible, first-aid and emergency health care to accident victims (including required links with external emergency healthcare services).

15.4.9 Investigation and notification of damage to health

In most countries, an employer is required to notify the competent authorities and/or insurance bodies of any occupational accidents and professional diseases involving more than a given number of days of sick leave.

15.4.10 Recording and documentation of information

Employers' duties regarding recording and documentation of information on preventative issues vary appreciably between countries. In some, employers are barely required to record accidents at all; in others, however, the opposite extreme is the case – employers are required to document and have available for the competent authority virtually all preventative measures performed (risk assessment, training, health surveillance, etc.), the procedures used and the results obtained, planned activities and emergency plans, the enterprise's preventative organisation and its prevention policy.

15.4.11 Organisation and management of preventative OSH services

Like any other activity, prevention should be suitably managed and, as it is performed within a certain organisation – the enterprise – it must be integrated with all hierarchical levels and functional units, without prejudice to there also being a unit specifically engaged in prevention. Regulations are very variable regarding prevention management and integration. Some countries do not regulate these issues, on the basis that they are exclusively the job of employers. An increasing number of countries set certain legal requirements for prevention management. Depending on specific cases, such requirements may apply to prevention policy, to preventative organisation, to planning and performance, to assessment of results or to actions towards improvement. These five fields are the basic elements of what is often called a system of occupational health and safety management.

15.4.12 Cooperation and coordination between enterprises

It is often the case that the relationship – or the lack of one – between two enterprises affects the standards of protection of the workers in one or both of the enterprises. When this happens, it is appropriate to introduce regulations governing coordination between employers in order to enable or reinforce compliance with their respective preventative obligations. A number of countries have introduced this requirement into regulations.

15.4.13 Workers' rights, duties and participation

An employer's duty to protect workers' health and safety is obviously of a piece with workers' right to protection. Hence, in some countries regulations are expressed as workers' rights, while in others they take the form of employers' duties. Such is the case, for instance, of a worker's right to:

▶ training and information on the risks involved in their job and the preventative measures adopted;
▶ to leave their workplace in the event of serious and imminent risk; and
▶ surveillance of their health (where necessary) and knowledge of the results.

Legislation in a fair number of countries also includes, besides these 'individual' rights (directly relating to a job position), a range of certain 'collective' rights of workers. This enables and reinforces workers' participation in the general field of prevention in the enterprise. By their collective nature, these rights must be exercised through workers' representatives or participation bodies vested in certain powers and competences. Although the possibilities are wide-ranging, such representatives or participation bodies can have five basic types of 'rights':

▶ the right to be informed on risks present in the enterprise, potential emergencies, accidents and health hazards that have arisen (while keeping

15

medical histories confidential) and the preventative measures adopted;

▶ the right to access (while respecting the confidentiality of trade secrets) to workplaces, enterprise documentation on prevention, personnel and entities (in-house, external or government) relevant to the field, and the information provided by the latter;

▶ the right to be consulted (sufficiently in advance) regarding any significant action or decision on prevention, specifically in order to enable their participation in the preparation, implementation and assessment of preventative plans and programmes;

▶ the right to receive training to enable them to perform their functions adequately;

▶ the right to have reasonable (paid) time to carry out their functions and not be dismissed or penalised for taking such time.

Workers' representation and participation can be articulated in two basic ways (or as a combination of both). The earliest and most traditional form is to have a Joint Committee made up of equal numbers of representatives for the employer and for the workers. In this case, the rights listed above are vested in the Committee itself and not in Committee members. The second form is to have workers' representatives (generally known as 'safety delegates' or 'prevention delegates') specifically vested in (individually or as a group) the rights outlined above, who deal directly with the employer (being informed and consulted by the employer or its representatives) without the need of using a 'joint-based' forum. Under either of the two systems, workers' representatives must be chosen by workers, directly or through their 'general' representatives.

As for workers' obligations, legislation in many countries sets the workers' general duty to cooperate with their employer in protecting their own and third parties' safety. Frequently this general duty is replaced or developed through a range of specific duties that aim for the worker to:

▶ comply with the rules of working procedures (including safe use of means for work and adequate use of the means for protection);

▶ comply with the rules set down for emergencies;

▶ advise on risks that they detect (specifically, serious and/or imminent risk).

15.4.14 Product safety and occupational safety

Regulations protecting users and consumers are also often called product safety regulations, at least when they concern products that can be imported and exported: equipment, materials, chemicals, etc. The target parties of such regulations are, in essence, product manufacturers (or, where applicable, product importers or suppliers), and the goal is that only 'safe' products be present on the market. A product is deemed safe if it does not endanger the health and safety of a user who installs, maintains and uses it in accordance with the manufacturer's instructions.

Product safety legislation is growing for two reasons. First, its development is in line with a process that is unfolding spontaneously in society as a whole. It is increasingly common, at least in developed countries, for consumers to demand a certain standard of quality when buying a particular product. There are currently thousands of product safety standards set down by national, regional and international standardisation bodies. These standards create a convergence of technical knowledge with legal stipulations. In addition to this convergence, this pool of standards provides indispensable information for carrying out training and informational activities.

One of the aspects that product safety regulations address with special care is the content of the instruction manual (which must obviously be in a language that the user can understand), because for a product subject to these regulations the expressions 'safe use' and 'use according to manufacturer's instructions' are equivalent in principle.

15.4.15 Certification and marking

The existence of safety requirements does not guarantee compliance. Products are normally manufactured on a mass scale. A manufacturer must always start with a design that meets requirements; then they may choose to make a prototype to check that requirements are met. In any event, the manufacturer must monitor production in order to ensure, within an acceptable margin of error, that each unit produced adheres to the design and the prototype. For its part, the competent authority must also endeavour to ensure compliance with safety requirements and, as far as possible, do so before the products are put on the market and distributed. For that reason it is frequent to demand that products compulsorily be subject to certification, or homologation, in accordance with a standard.

15.4.16 Monitoring compliance with regulations

To enforce regulations effectively it is necessary to have in place an inspection system covering all businesses in order to identify non-compliances and rectify and/or penalise them.

Table 15.5 Common themes in national OSH legislation

Action at national level	Should be single basic regulatory framework at high level setting general requirements and obligations with detailed regulations on specific subjects	
	Set up a system for producing regulations, codes of practice and enforcing them at workplaces including inspection, advice and penalties	
	Cover all workplaces	
	All workers including self-employed	
	Special consideration of the vulnerable	
	Act on principles of prevention giving preference to prevention at the source of the risk	
	Put responsibilities on employers	
	Coordinate OSH organisations	
	Require cooperation between employers on common sites	
	Collect and disseminate OSH statistics	
	Promote OSH research and information and consult with employers' and employees' organisations	
	Set up procedures to control major risks and the surrounding populations	
	Review legislation and modify as necessary	
	Requirements for designers, suppliers, importers and manufacturers	
Action by employers at level of the undertaking	OSH management	Requirement to produce a written policy (where appropriate) with arrangements for OSH and responsibilities and inform workers in a language they understand
		Operate with an OSH management system in place appropriate to the size and activities at the workplace
		Carry out risk assessment and follow principles of prevention
		Training employees and provide instruction, information and supervision
		Consult employees and set up joint OSH committees
		Provide safe systems of work
		Inspect workplaces and take remedial action
		Investigate and report accidents and cases of ill-health to the authorities
		Keep records on OSH as required by the national authority
		Provide expertise on OSH either in house or through external OSH experts
		Provide adequate resources
	Workplace requirements to cover	Structural stability and safe access and egress from workplaces
		Adequate space and evacuation routes in case of emergencies
		Maintaining clean and properly decorated workplaces as appropriate to the activity
		Lighting, suitable temperatures, ventilation
		Seating and general ergonomics of workstations
		Provison of PPE (at no cost to worker)
		Electrical and gas installations
		First-aid and emergency arrangements
		Fire precautions and equipment
		Proper welfare facilities including toilets, washing, storage of personal items of clothing, etc.
		Suitable signs and warning information
		Special provisions for the vulnerable such as young people, pregnant women, nursing mothers and people with a disability
	Work equipment requirements to cover	Safe provision and use of machinery and other work equipment
		Control of critical plant like cranes, lifts, pressure vessels, ventilation equipment
		Procedures such as permits to work for potentially hazardous operations, e.g. explosive atmospheres, live electrical work, confined spaces
		Safe movement of vehicles and mobile equipment both on the premises and insofar as it is affected by the employer on public roads
		Safe equipment for working at height including access, working platforms and specialised access equipment

15

	Chemical agents or substances to cover	Assessment of hazardous substances, their elimination and/or control in the workplace
		Provision of occupational health services, working environment monitoring, health monitoring of employees
		Control of dangerous substances which are explosive, highly flammable, oxidising
		Keeping of adequate health records
		Lead and asbestos
	Physical agents to cover	Noise, vibrations, thermal stress, optical radiation such as lasers, ionising and non-ionising radiation
		Repetitive strain and ergonomics
		Manual handling
		Computer workstations
Action by employees at the level of the undertaking	Take reasonable care for themselves and others	
	Comply with OSH instructions and procedures	
	Use safety devices and PPE correctly	
	Report any hazardous situations and any accidents or injury to health	

APPENDIX 15.1 Seoul Declaration on Safety and Health at Work

The Safety and Health Summit

Having met in Seoul, Republic of Korea, on 29 June 2008, on the occasion of the XVIII World Congress on Safety and Health at Work, jointly organised by the International Labour Office, the International Social Security Association (ISSA) and the Korea Occupational Safety and Health Agency (KOSHA), with the participation of senior professionals, employers' and workers' representatives, social security representatives, policy makers and administrators.

Recognising the serious consequences of work-related accidents and diseases, which the International Labour Office estimates lead to 2.3 million fatalities per year world-wide and an economic loss of 4% of global Gross Domestic Product (GDP),

Recognising that improving safety and health at work has a positive impact on working conditions, productivity and economic and social development,

Recalling that the right to a safe and healthy working environment should be recognised as a fundamental human right and that globalisation must go hand in hand with preventative measures to ensure the safety and health of all at work,

Recognising the importance of the instruments on safety and health at work of the International Labour Organisation (ILO) and the substantial role of the ISSA and its members' contribution in implementing these instruments,

Recalling that the promotion of occupational safety and health and the prevention of accidents and diseases at work is a core element of the ILO's founding mission and of the Decent Work Agenda,

Recalling that the prevention of occupational risks and the promotion of workers' health constitute an essential part of the ISSA's mandate and of its Conceptual Framework of Dynamic Social Security,

Recognising the importance of education, training, consultation and the exchange of information and good practices on prevention and the promotion of preventative measures,

Recognising the important role played by governments and the social partners, professional safety and health organisations and social security institutions in promoting prevention and in providing treatment, support and rehabilitation services,

Recognising the importance of cooperation among international organisations and institutions,

Welcoming progress achieved through international and national efforts to improve safety and health at work,

Declares that

1. Promoting high levels of safety and health at work is the responsibility of society as a whole and all members of society must contribute to achieving this goal by ensuring that priority is given to occupational safety and health in national agendas and by building and maintaining a national preventative safety and health culture.

2. A national preventative safety and health culture is one in which the right to a safe and healthy working environment is respected at all levels, where governments, employers and workers actively participate in securing a safe and healthy working environment through a system of defined rights, responsibilities and duties, and where the principle of prevention is accorded the highest priority.

3. The continuous improvement of occupational safety and health should be promoted by a systems approach to the management of occupational safety and health, including the development of a national policy taking into consideration the principles in Part II of the ILO Occupational Safety and Health Convention, 1981 (No. 155).

4. Governments should
   ▷ Consider the ratification of the ILO Promotional Framework for Occupational Safety and Health Convention, 2006 (No. 187) as a priority, as well as other relevant ILO Conventions on safety and health at work and ensure the implementation of their provisions, as a means to improve national performance on safety and health at work in a systematic way.
   ▷ Ensure that continued actions are taken to create and enhance a national preventative safety and health culture.
   ▷ Ensure that the occupational safety and health of workers is protected through an adequate and appropriate system of enforcement of safety and health standards, including a strong and effective labour inspection system.

5. Employers should ensure that
   ▷ Prevention is an integral part of their activities, as high safety and health standards at work go hand and hand with good business performance.
   ▷ Occupational safety and health management systems are established in an effective way to improve workplace safety and health.
   ▷ Workers and their representatives are consulted, trained, informed and involved in all measures related to their safety and health at work.

6. Affirming the workers' right to a safe and healthy working environment, workers should be consulted on safety and health matters and should:

> Follow safety and health instructions and procedures, including on the use of personal protective equipment.

> Participate in safety and health training and awareness-raising activities.

> Cooperate with their employer in measures related to their safety and health at work.

7. The World Congress on Safety and Health at Work is an ideal forum to share knowledge

and experiences in achieving safe, healthy and productive workplaces.

8. Progress made on achieving safety and health at work should be reviewed on the occasion of the XIX World Congress on Safety and Health at Work in 2011.

9. The Summit participants commit to taking the lead in promoting a preventative safety and health culture, placing occupational safety and health high on national agendas.

APPENDIX 15.2 ILO – C155 Occupational Safety and Health Convention, 1981

The General Conference of the International Labour Organisation,

Having been convened at Geneva by the Governing Body of the International Labour Office, and having met in its Sixty-seventh Session on 3 June 1981, and

Having decided upon the adoption of certain proposals with regard to safety and health and the working environment, which is the sixth item on the agenda of the session, and

Having determined that these proposals shall take the form of an international Convention,

adopts this twenty-second day of June of the year one thousand nine hundred and eighty-one the following Convention, which may be cited as the Occupational Safety and Health Convention, 1981:

PART I. SCOPE AND DEFINITIONS

Article 1

1. This Convention applies to all branches of economic activity.

2. A Member ratifying this Convention may, after consultation at the earliest possible stage with the representative organisations of employers and workers concerned, exclude from its application, in part or in whole, particular branches of economic activity, such as maritime shipping or fishing, in respect of which special problems of a substantial nature arise.

3. Each Member which ratifies this Convention shall list, in the first report on the application of the Convention submitted under Article 22 of the Constitution of the International Labour Organisation, any branches which may have been excluded in pursuance of paragraph 2 of this Article, giving the reasons for such exclusion and describing the measures taken to give adequate protection to workers in excluded branches, and shall indicate in subsequent reports any progress towards wider application.

Article 2

1. This Convention applies to all workers in the branches of economic activity covered.

2. A Member ratifying this Convention may, after consultation at the earliest possible stage with the representative organisations of employers and workers concerned, exclude from its application, in part or in whole, limited categories of workers in respect of which there are particular difficulties.

3. Each Member which ratifies this Convention shall list, in the first report on the application of the Convention submitted under Article 22 of the Constitution of the International Labour Organisation, any limited categories of workers which may have been excluded in pursuance of paragraph 2 of this Article, giving the reasons for such exclusion, and shall indicate in subsequent reports any progress towards wider application.

Article 3

For the purpose of this Convention–

(a) the term branches of economic activity covers all branches in which workers are employed, including the public service;

(b) the term workers covers all employed persons, including public employees;

(c) the term workplace covers all places where workers need to be or to go by reason of their work and which are under the direct or indirect control of the employer;

(d) the term regulations covers all provisions given force of law by the competent authority or authorities;

(e) the term health, in relation to work, indicates not merely the absence of disease or infirmity; it also includes the physical and mental elements affecting health which are directly related to safety and hygiene at work.

PART II. PRINCIPLES OF NATIONAL POLICY

Article 4

1. Each Member shall, in the light of national conditions and practice, and in consultation with the most representative organisations of employers and workers, formulate, implement and periodically review a coherent national policy on occupational safety, occupational health and the working environment.

2. The aim of the policy shall be to prevent accidents and injury to health arising out of, linked with or occurring in the course of work, by minimising, so far as is reasonably practicable, the causes of hazards inherent in the working environment.

Article 5

The policy referred to in Article 4 of this Convention shall take account of the following main spheres of action in so far as they affect occupational safety and health and the working environment:

(a) design, testing, choice, substitution, installation, arrangement, use and maintenance of the material elements of work (workplaces, working environment, tools, machinery and equipment, chemical, physical and biological substances and agents, work processes);

(b) relationships between the material elements of work and the persons who carry out or supervise the work, and adaptation of machinery, equipment, working time, organisation of work and work processes to the physical and mental capacities of the workers;

(c) training, including necessary further training, qualifications and motivations of persons involved, in one capacity or another, in the achievement of adequate levels of safety and health;

(d) communication and cooperation at the levels of the working group and the undertaking and at all other appropriate levels up to and including the national level;

(e) the protection of workers and their representatives from disciplinary measures as a result of actions properly taken by them in conformity with the policy referred to in Article 4 of this Convention.

Article 6

The formulation of the policy referred to in Article 4 of this Convention shall indicate the respective functions and responsibilities in respect of occupational safety and health and the working environment of public authorities, employers, workers and others, taking account both of the complementary character of such responsibilities and of national conditions and practice.

Article 7

The situation regarding occupational safety and health and the working environment shall be reviewed at appropriate intervals, either over-all or in respect of particular areas, with a view to identifying major problems, evolving effective methods for dealing with them and priorities of action, and evaluating results.

PART III. ACTION AT THE NATIONAL LEVEL

Article 8

Each Member shall, by laws or regulations or any other method consistent with national conditions and practice and in consultation with the representative organisations of employers and workers concerned, take such steps as may be necessary to give effect to Article 4 of this Convention.

Article 9

1. The enforcement of laws and regulations concerning occupational safety and health and the working environment shall be secured by an adequate and appropriate system of inspection.

2. The enforcement system shall provide for adequate penalties for violations of the laws and regulations.

Article 10

Measures shall be taken to provide guidance to employers and workers so as to help them to comply with legal obligations.

Article 11

To give effect to the policy referred to in Article 4 of this Convention, the competent authority or authorities shall ensure that the following functions are progressively carried out:

(a) the determination, where the nature and degree of hazards so require, of conditions governing the design, construction and layout of undertakings, the commencement of their operations, major alterations affecting them and changes in their purposes, the safety of technical equipment used at work, as well as the application of procedures defined by the competent authorities;

(b) the determination of work processes and of substances and agents the exposure to which is to be prohibited, limited or made subject to authorisation or control by the competent authority or authorities; health hazards due to the simultaneous exposure to several substances or agents shall be taken into consideration;

(c) the establishment and application of procedures for the notification of occupational accidents and diseases, by employers and, when appropriate, insurance institutions and others directly concerned, and the production of annual statistics on occupational accidents and diseases;

(d) the holding of inquiries, where cases of occupational accidents, occupational diseases or any other injuries to health which arise in the course of or in connection with work appear to reflect situations which are serious;

(e) the publication, annually, of information on measures taken in pursuance of the policy referred to in Article 4 of this Convention and on occupational accidents, occupational diseases and other injuries to health which arise in the course of or in connection with work;

(f) the introduction or extension of systems, taking into account national conditions and possibilities, to examine chemical, physical and biological agents in respect of the risk to the health of workers.

Article 12

Measures shall be taken, in accordance with national law and practice, with a view to ensuring that those who design, manufacture, import, provide or transfer machinery, equipment or substances for occupational use–

(a) satisfy themselves that, so far as is reasonably practicable, the machinery, equipment or substance does not entail dangers for the safety and health of those using it correctly;

(b) make available information concerning the correct installation and use of machinery and equipment and the correct use of substances, and information on hazards of machinery and equipment and dangerous properties of chemical substances and physical and biological agents or products, as well as instructions on how known hazards are to be avoided;

(c) undertake studies and research or otherwise keep abreast of the scientific and technical knowledge necessary to comply with subparagraphs (a) and (b) of this Article.

Article 13

A worker who has removed himself from a work situation which he has reasonable justification to believe presents an imminent and serious danger to his life or health shall be protected from undue consequences in accordance with national conditions and practice.

Article 14

Measures shall be taken with a view to promoting in a manner appropriate to national conditions and practice, the inclusion of questions of occupational safety and health and the working environment at all levels of education and training, including higher technical, medical and professional education, in a manner meeting the training needs of all workers.

Article 15

1. With a view to ensuring the coherence of the policy referred to in Article 4 of this Convention and of measures for its application, each Member shall, after consultation at the earliest possible stage with the most representative organisations of employers and workers, and with other bodies as appropriate, make arrangements appropriate to national conditions and practice to ensure the necessary coordination between various authorities and bodies called upon to give effect to Parts II and III of this Convention.

2. Whenever circumstances so require and national conditions and practice permit, these arrangements shall include the establishment of a central body.

PART IV. ACTION AT THE LEVEL OF THE UNDERTAKING

Article 16

1. Employers shall be required to ensure that, so far as is reasonably practicable, the workplaces, machinery, equipment and processes under their control are safe and without risk to health.

2. Employers shall be required to ensure that, so far as is reasonably practicable, the chemical, physical and biological substances and agents under their control are without risk to health when the appropriate measures of protection are taken.

3. Employers shall be required to provide, where necessary, adequate protective clothing and protective equipment to prevent, so far as is reasonably practicable, risk of accidents or of adverse effects on health.

Article 17

Whenever two or more undertakings engage in activities simultaneously at one workplace, they shall collaborate in applying the requirements of this Convention.

Article 18

Employers shall be required to provide, where necessary, for measures to deal with emergencies and accidents, including adequate first-aid arrangements.

Article 19

There shall be arrangements at the level of the undertaking under which–

(a) workers, in the course of performing their work, cooperate in the fulfilment by their employer of the obligations placed upon him;

(b) representatives of workers in the undertaking cooperate with the employer in the field of occupational safety and health;

(c) representatives of workers in an undertaking are given adequate information on measures taken by the employer to secure occupational safety and health and may consult their representative organisations about such information provided they do not disclose commercial secrets;

(d) workers and their representatives in the undertaking are given appropriate training in occupational safety and health;

(e) workers or their representatives and, as the case may be, their representative organisations in an undertaking, in accordance with national law and practice, are enabled to enquire into, and are consulted by the employer on, all aspects of occupational safety and health associated with their work; for this purpose technical advisers may, by mutual agreement, be brought in from outside the undertaking;

(f) a worker reports forthwith to his immediate supervisor any situation which he has reasonable justification to believe presents an imminent and serious danger to his life or health; until the employer has taken remedial action, if necessary, the employer cannot require workers to return to a work situation where there is continuing imminent and serious danger to life or health.

Article 20

Cooperation between management and workers and/or their representatives within the undertaking shall be an essential element of organisational and other measures taken in pursuance of Articles 16 to 19 of this Convention.

Article 21

Occupational safety and health measures shall not involve any expenditure for the workers.

PART V. FINAL PROVISIONS

Article 22

This Convention does not revise any international labour Conventions or Recommendations.

Article 23

The formal ratifications of this Convention shall be communicated to the Director-General of the International Labour Office for registration.

Article 24

1. This Convention shall be binding only upon those Members of the International Labour Organisation whose ratifications have been registered with the Director-General.

2. It shall come into force twelve months after the date on which the ratifications of two Members have been registered with the Director-General.

3. Thereafter, this Convention shall come into force for any Member twelve months after the date on which its ratification has been registered.

Article 25

1. A Member which has ratified this Convention may denounce it after the expiration of ten years from the date on which the Convention first comes into force, by an act communicated to the Director-General of the International Labour Office for registration. Such denunciation shall not take effect until one year after the date on which it is registered.

2. Each Member which has ratified this Convention and which does not, within the year following the expiration of the period of ten years mentioned in the preceding paragraph, exercise the right of denunciation provided for in this Article, will be bound for another period of ten years and, thereafter, may denounce this Convention at the expiration of each period of ten years under the terms provided for in this Article.

Article 26

1. The Director-General of the International Labour Office shall notify all Members of the International Labour Organisation of the registration of all ratifications and denunciations communicated to him by the Members of the Organisation.

2. When notifying the Members of the Organisation of the registration of the second ratification communicated to him, the Director-General shall draw the attention of the Members of the Organisation to the date upon which the Convention will come into force.

Article 27

The Director-General of the International Labour Office shall communicate to the Secretary-General of the United Nations for registration in accordance with Article 102 of the Charter of the United Nations full particulars of all ratifications and acts of denunciation registered by him in accordance with the provisions of the preceding Articles.

Article 28

At such times as it may consider necessary the Governing Body of the International Labour Office shall present to the General Conference a report on the working of this Convention and shall examine the desirability of placing on the agenda of the Conference the question of its revision in whole or in part.

Article 29

1. Should the Conference adopt a new Convention revising this Convention in whole or in part, then, unless the new Convention otherwise provides:

(a) the ratification by a Member of the new revising Convention shall ipso jure involve the immediate denunciation of this Convention, notwithstanding the provisions of Article 25 above, if and when the new revising Convention shall have come into force;

(b) as from the date when the new revising Convention comes into force this Convention shall cease to be open to ratification by the Members.

2. This Convention shall in any case remain in force in its actual form and content for those Members which have ratified it but have not ratified the revising Convention.

Article 30

The English and French versions of the text of this Convention are equally authoritative.

CHAPTER 16

Study skills

16.1 Introduction ▶ 540

16.2 Find a place to study ▶ 540

16.3 Time management ▶ 540

16.4 Blocked thinking ▶ 541

16.5 Taking notes ▶ 541

16.6 Reading for study ▶ 541

16.7 Free learning resources from the Open University ▶ 541

16.8 Organising for revision ▶ 542

16.9 Organising information ▶ 542

16.10 Being aware of your learning style ▶ 544

16.11 How does memory work? ▶ 544

16.12 How to deal with exams ▶ 545

16.13 The examiners' reports ▶ 546

16.14 Conclusion ▶ 547

16.15 Further information ▶ 547

> **This chapter covers:**
> 1. How to plan and organise self-study work
> 2. How to manage study and revision time
> 3. Understanding and organising revision
> 4. Understanding the concept of memory techniques
> 5. Understanding how to tackle examinations in the exam room and afterwards
> 6. Understanding about examiners' reports and what some of the latest reports are saying
> 7. Understanding the marks allocated to each NEBOSH question

16.1 Introduction

This chapter speaks directly to you, the student. For the NEBOSH certificate you need study skills and exam techniques and this means careful planning from the beginning of the course.

Think in terms of:

▶ clear and realistic goals, both short and long term;
▶ techniques for studying and passing exams;
▶ a well-organised approach;
▶ strong motivation.

People say that genius is 99% application and only 1% inspiration so good organisation and exam technique are what really matters when it comes to passing exams. Studying is an activity in its own right and there are ways in which you can make your study more effective and give yourself the greatest possible chance of success. The place where you study, the way you plan your study time, how you adapt your study to suit your lifestyle, the way that you take notes, memory and concentration skills, revision techniques, and being clear about the contents of the syllabus and what is expected by the examiner – attention to all these details will give you the best chance of passing the exam.

16.2 Find a place to study

The skill of studying usually has to be learned, so it's worth thinking carefully about the basics. If you can find somewhere quiet and free from distractions, and keep that place just for your studies, it will be easier to get down to work. You need a place with good ventilation to help you to stay alert, and somewhere that has a comfortable temperature. Make sure there is enough light – a reading lamp helps to prevent eyestrain and tiredness.

An upright chair is better than an armchair. Make sure that the workspace is large enough. The size recommended by Warwick University, for example, is a minimum of 60 cm × 1.5 m. If you can, find a place where your study materials do not have to be cleared away – it's easy to put off starting work if they are not instantly accessible.

16.2.1 Make a study plan

Start by making a study timetable. First of all, put aside some time for study so that it does not get squeezed out of your working week. Your timetable will need to include time for carrying out assignments set by tutors, time for going over lecture notes and materials, any further reading rated as 'essential' and, if possible, reading rated as 'desirable'. You should allocate time for revision as the course progresses because the more regularly you revise, the more firmly the information will become fixed in your memory. This will make it much easier to remember when you take the exams.

After every hour of study, take a short break and, if possible, have a change of atmosphere. Physical exercise will help to increase your concentration. Even a short, brisk walk round the block will improve your attention span.

Vary your activities while you are studying. For example, you could spend some time gathering information from books, the internet and so on, then maybe work on a diagram or a graph, followed by writing, reading and so on.

16.3 Time management

Finding time to study and take exams when your life is already filled with work, family and social commitments is clearly not going to be easy. The Open University (OU) has been working with adult learners since 1969 and they provide a lot of useful, free information about this at: www.open.edu/openlearn/free-course. Here are some ideas from 'manage your time effectively' to help you to plan.

The aim is to develop good work habits and time-management practices. These are often a matter of developing the right attitudes towards your work and towards your time.

Ten tips to help you manage your time:

1. establish goals and targets;
2. work smarter, not harder – value your time;
3. avoid attempting too much;
4. schedule activities, allow time for emergencies, minimise interruption, identify and use your peak energy times, put similar tasks together;
5. stay in control of paper work and electronic work;
6. minimise day dreaming;
7. use a log to track time wasting;
8. put a time limit on some tasks;

9. be decisive, finish things;
10. work at a steady pace and review your progress regularly.

Keep a log of your learning so that you can reflect on how you use your time – are you managing it effectively? Think about how you deal with the unexpected, with emergencies. How do you prioritise your tasks and your time? Can you set deadlines? Can you meet them? If not, how do you deal with the situation?

There is a section on time management in the Guide to the NEBOSH International General Certificate. This subject is also mentioned in 16.12.1 Planning and revision.

16.4 Blocked thinking

Sometimes your thinking can become blocked. If this happens, try leaving the problem area of study alone for a few days and tackle a different part of the subject, since if you concentrate too hard on something that is too difficult you are likely to lose confidence. Usually, after a break, the difficulty vanishes and the problem clears. This could be because the solution has emerged from another perspective, or because you have learnt something new that has supplied the answer. Psychologists think it's possible that some problems can be solved during sleep when your mind has a chance to wander and think laterally.

16.5 Taking notes

The most efficient way to store notes is in a loose-leaf folder, because you can easily add extra information. Write only on one side of the sheet and use margined paper. The facing pages can be used to make summaries or extra points. If notes are clearly written and well spaced out, they are much more straightforward to work from and more attractive to return to later. Use colour, highlighting and underlining too; notes that look good are more appealing and easier to revise.

During revision it will be important to be able to identify subjects quickly, so use headings, numbering, lettering, bullet points, indentation and so on. It is worth spending a bit of extra time to make revision easier.

Key words and phrases are better than continuous prose when you are taking notes. While you are writing down information it is easy to miss essential points that are being made by the tutor. Try to read through your notes within 36 hours to make sure that they are completely clear, so that when you revise you will not be puzzling over what you have written. At this point you may be able to add more information, while it is fresh in your mind. Reading through notes in this way also helps to fix the information in your memory and the

level of recall will be further improved by reading them again a week later. Although it seems time-consuming, this technique will save you a lot of time in the long run.

Many people find that they can understand and remember more effectively by making 'mind maps' (also known as 'pattern notes'). These will be discussed in more detail in the revision section.

16.6 Reading for study

Most people will be working as well as studying, so you will need techniques to help you make the best use of your time. For example, you don't need to read a whole book for study purposes, apart from books written specifically to a syllabus, such as this one. Use the index and be selective. In this book an example would be Chapter 15 on legal frameworks and ILO conventions.

Where a book is not specifically written to a syllabus, time-saving techniques such as skim reading can be used. First of all, flip through the book to see what it contains, looking at the contents, summaries, introduction, tables, diagrams and so on. It should be possible to identify areas that are necessary to the syllabus. These can be marked with removable adhesive strips or 'Post-it' notes.

Second, quickly read through the parts of the book that you have identified as useful. Two ways of doing this are:

(a) run a finger slowly down the centre of a page, watching it as it moves. You will pick up relevant words and phrases as your finger moves down the page;
(b) read in phrases, rather than word by word, which will increase your reading speed.

Both these methods need practice but they save a lot of time and effort. It is better to read through a piece two or three times, quickly, than to read it once slowly; but if there is a really difficult piece of information that is absolutely essential to understand, read it aloud, slowly.

Finally, at the end of each section or chapter, make a few brief notes giving the essential points.

It is worth learning to speed-read if a course involves a lot of reading. Tony Buzan's series The Mind Set explains the technique.

16.7 Free learning resources from the Open University

The OpenLearn section of the Open University is a very useful resource which is free and is available worldwide. There is an extensive study skills section containing 30 units which range from 3 to 50 hours of study. Have a look at 'The Learning Space'. Here are a few examples of what is available:

16

- EAL_1. Am I ready to study in English? (5 hours)
  This unit provides a series of exercises to help you to reflect on the use of English at an academic (as opposed to everyday) level. It is designed for people who have been educated in a language other than English, or for those who have studied in a country where the conventions are different from those used in the British educational system.
- LDT 101_1. Revision and examinations (6 hours)
  This unit is for people who are unsure about exams. Perhaps you have not taken an exam for a long time, or have had a bad experience with exams in the past; or maybe you have never taken an exam? This unit aims to help you to develop techniques for revision and exam taking and to reassure you.
- LDT 101_3. Learning how to learn (6 hours)
  This unit looks at learning as an active process and provides activities that will help you to learn more effectively.

You can access these free resources on www.open.edu/openlearn/free-course

16.8 Organising for revision

As the exam approaches, you will need to organise all the information from the course in preparation for revision. Ideally, of course, you should have been doing this throughout, but it is not too late to start. It will just mean extra work. Information from lectures, reading, practical experience and assignments will have to be organised into a form that makes it easier for you to remember. As well as this, you will already possess some knowledge that is relevant to the syllabus. Don't

underestimate the importance of this. It is worth spending some time thinking about how this store of information relates to and can be used to expand and back up the new information that you have learnt on the course. Examiners' Reports have, in the past, referred specifically to the advantages to be gained from this. For example:

> *Some candidates, perhaps from their work experience, showed a good knowledge of a permit-to-work system and produced reasonable answers.*

But be warned. The examiners also pointed out that:

> *Answers based solely on 'what we do in our organisation' … do not earn high marks.*

16.9 Organising information

You can organise the information in preparation for an exam in several ways:

- read back through the notes on a regular basis while the course is still in progress; this will help to fix the information in your memory;
- make revision cards: condense the information down so that it fits on a set of postcards. This makes it possible to carry a lot of information around in your pocket. The activity of condensing the information is revision in itself. This way you can keep your revision material available and spend a few spare moments, from time to time, reading it through.

For example: 'Outline the factors that may increase risks to pregnant employees.' On a card, the information could be condensed as shown in Figure 16.1.

Pregnant Employees – Risks

Exposure to chemicals
{ pesticides
lead
mutagens (inter-cellular change)
teratogens (affect embryo)

biological (e.g. hepatitis)

Physical (ionizing radiations extremes of temperature)

manual handling

ergonomic (prolonged standing awkward body movements)

Stress

issues relating to use/wearing of protective equipment

Figure 16.1 Revision notes

Make mind maps (also known as pattern notes): like revision cards, the act of making them is revision in itself. Because they are based on visual images some people find them easier to remember and they can be made as complex or as simple as you like. Use colour and imagery. It will be easier to follow and to recall the information (Buzan). Figure 16.2 shows a mind map based on the report writing section of this book.

Use key words as an aid to memory. For example, a list of key words for 'ways of reducing the risk of a fire starting in the workplace' could be reduced to an eight-letter nonsense word, as follows:

FLICSHEV:

Flammable liquids – provide proper storage facilities
Lubrication – regular lubrication of machinery
Incompatible chemicals need segregation
Control of hot work
Smoking – of cigarettes and materials, to be controlled
Housekeeping (good) – prevent accumulation of waste
Electrical equipment – needs frequent inspection for damage
Ventilation – outlets should not be obstructed.

The effort involved in making the list and inventing the word helps to fix the information in your memory.

Learning by rote is a popular learning method in many countries and is very useful for specific learning tasks, such as tables and formulae. But it is not appropriate to this type of syllabus. Rote learning can prevent you from applying what you have learnt to a wide range of situations. If you limit yourself in this way you will be in danger of not having sufficient understanding of the depth of the subject, particularly if you learn set answers to exam questions. The NEBOSH examiners' reports specifically warn against rote learning of set answers. They point out that 'Candidates should prepare themselves for this vocational examination by ensuring their understanding'. Working through old exam papers is very helpful and you can learn a lot by doing revision in this way, but don't be tempted to learn a 'perfect' version of an answer as they are all designed to make you think about specific problems that you might encounter in your working life.

It is useful to have a range of techniques for memorising, since what works for one person may not

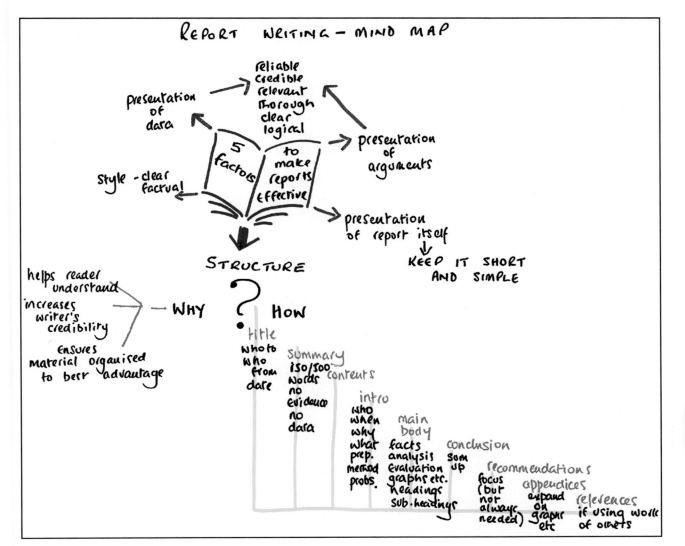

Figure 16.2 Mind map – report writing

work for another (see 16.10). There are several other memory techniques available, but the ones described in this paragraph are the most commonly used.

16.10 Being aware of your learning style

All of us are different and we learn in different ways, so it makes sense to be able to identify your own preferred style of learning. The main learning styles are **visual, auditory** and **kinaesthetic** and it helps if you can choose ways of learning and remembering that fit in with your own personal style.

For example, if you are a person who learns in a **visual** way, you may find it easier to understand and recall information when it is presented in the form of diagrams, the use of flip charts, handouts, demonstrations and videos. When you are taking notes and later in revision, you will perhaps find it more effective to:

Use colour
Organise in columns and categories
Use mnemonics in a visual way (see the 'key word' tip in 16.9)
Use mind mapping
Use index cards
Highlight, circle and underline
Make posters and/or use Post-it notes.

If you are a person who leans towards learning in an **auditory** way, you will probably find lectures, discussions, dialogue and music are useful, so here are a few ideas to help you to learn more effectively:

Discussion of new ideas and information
Recording information as well as taking notes
Using word association
Repeating facts with your eyes closed
Watching videos
Using rhyme, rhythm and music to help you to retain information
Using mnemonics aurally (see 16.9).

Maybe you are a person who works best with a **kinaesthetic** approach to learning, and find that touching, feeling and actively doing are the best ways for you to learn. Hands-on activity, role play, simulation and practising skills are likely to be the methods you are most comfortable with. Here are a few possibilities:

Copying notes repeatedly to organise them in your mind
Making on-site visits
Using colour and highlighting
Making posters and models to aid memory
Skim reading, then reading in detail
Taking study breaks
Revise on an exercise bike.

(Arp, 2014)

This is not to say that people only conform to the visual, the auditory or the kinaesthetic model.

Most people will be on a spectrum and use more than one type of learning, but you have probably already identified the one that is prevalent in your learning style and you can use that information to make both the learning process and revision a bit easier.

16.11 How does memory work?

Understanding how your memory works will help you to use it more effectively. In common with other scientific subjects, there is still a lot to be discovered about the workings of the mind, and a considerable amount of disagreement about what has been discovered so far; but psychologists generally seem to agree about the way that memory works.

The process of remembering is divided, roughly, into two sections – short-term and long-term memory. Items that go into short-term memory, if no further attention is paid to them, will fade away and be forgotten. If they are rehearsed they will stay in short-term memory for a while. Think, for example, of how you remember a phone number when you make a call. Storing something in long-term memory requires a greater 'depth of processing', that is to say, more mental activity is required so that:

(a) information stays in the memory (storage); and
(b) information can be found when it is needed (retrieval).

If you are studying on the NEBOSH course, it is likely that you are a mature student. This will give you some advantages when it comes to learning. Because your life and work experience will almost always be more comprehensive and of a higher complexity than that of younger students, you are likely to possess more 'schemas' (areas of knowledge) to which new information can be attached, and you will have more experiences already stored in your brain that can provide an explanation for the new pieces of knowledge that you are acquiring. Add to this a very high level of motivation, a stronger level of incentive towards success and more determined application, and it becomes clear that you, as a mature student, have many advantages.

Research shows that, as learners, we take in 10% of what we read; 20% of what we hear; 30% of what we see; 50% of what we see and hear; 70% of what we ourselves say; and 90% of what we ourselves do (Northedge, 2012). You can see from this that people who are simultaneously working at and studying a subject have a marked advantage.

But, it is also true that mature students tend to be very nervous about exams and assignments and often need a good deal of reassurance and support to enable them to realise their full potential.

16.12 How to deal with exams

There are three stages to taking an exam:

- planning and revision;
- the exam room;
- after the exam.

For people who require *special consideration*, there are provisions within NEBOSH to allow extra time or the use of special equipment. If you think you may be eligible for reasonable adjustments/special consideration, you should apply to NEBOSH several weeks before the date of the exam. Have a look at the NEBOSH Guide if you think any of this may apply to you.

16.12.1 Planning and revision

1. It is absolutely essential that you know what you are going to be examined on and what form the exam will take. You should read through the syllabus and if you are concerned about any area of it, this should be raised with your course tutor well before the date of the exam.

2. Read the Guide to the NEBOSH International General Certificate in Occupational Health and Safety and the examiners' reports. After the exams every year, the examiners highlight the most common mistakes made by students (see Section 16.13 on examiners' reports). They also provide useful information about, for example, pass rates, levels attained (distinction, credit, pass, fail), time management and other hints on exam technique.

3. You should work through some recent past papers, against the clock, to get used to the 'feel' of the exam. If possible, ask your tutor to set up mock exams and make sure that the papers are marked. Some of the shorter courses will not be able to provide this service, and if this is the case, do try to do at least one or two timed questions as part of your revision. You may find it difficult at first to write an answer as a first draft. Now that everyone

is used to the facilities provided by word processing, you may need to practice writing an answer without being able to edit the text. Past papers, examiners' reports and syllabuses are available from NEBOSH. The website is at www.nebosh.org.uk

4. You will need to know where the exam is to be held and the date and time. If possible, visit the building beforehand to help build confidence about the location, availability of parking and so on.

5. Make a chart of the time leading up to the exam. Include all activities, work, leisure, and social, as well as the time to be used for revision, so that the schedule is realistic.

6. Try to eat and sleep well and take some exercise.

7. Revision techniques were covered in the sections on revision and memory earlier in this chapter.

Set realistic targets, then achieve them. Make plans and stick to them.

16.12.2 In the examination room

- Read through the exam paper very carefully.
- Check the instructions – how many questions have to be answered? From which section(s)?
- Make a time plan.
- Underline **command words**, e.g. 'identify' (see Table 16.1). Using the words in the question when writing the answers will help to keep the answer on track.
- Stick to the instructions given in the question. A working knowledge of the command verbs is vital to your success. NEBOSH no longer uses the command verb 'sketch', but candidates may use sketches to illustrate their answers where appropriate.
- Write clearly. Illegible answers don't get marked.
- Look at the number of marks allocated to a question to pick up clues as to how much time should be spent on it.
- Mark questions which look possible and identify any that look impossible.

Table 16.1 Terminology used in NEBOSH exams

Command word	Definition
Identify	To give reference to an item, which could be its name or title. NB: normally a word or phrase is sufficient provided the reference is clear.
Give	To offer for consideration, acceptance, or use of another. NB: give an example of; give the meaning of.
Outline	To indicate the principal features or different parts of. NB: an exhaustive description is not required. What is sought is a brief summary of the major aspects of whatever is stated in the question.
Describe	To give a detailed written account of the distinctive features of a subject. The account should be factual, without any attempt to explain. When describing a subject (or object) a test of sufficient detail would be that another person would be able to visualise what you are describing.
Explain	To provide an understanding. To make an idea or relationship clear. NB: this command word is testing the candidate's ability to know or understand why or how something happens. Is often associated with the words 'how' or 'why'.

16

- It is rarely necessary to answer exam questions in a particular order. Start with the question that you feel most comfortable with since it will help to boost your confidence. Make sure it is clearly identified by number for the examiner.
- Answer the question that is set, not the one you wish was on the paper.
- If ideas for other answers spring to mind while you are writing, jot down a reminder on a separate piece of paper. It is easy to forget that bit of information when you are concentrating on something else.
- Plan the use of time and plan the answers. Include some time to check over each answer.
- Stick to the time plan; stick to the point; make points quickly and clearly.

Early marks in an exam question are easier to pick up than the last one or two, so make sure that all the questions are attempted within the time plan. No marks are given for correct information that is not relevant to the question.

Don't be distracted by the behaviour of other students. Someone who is requesting more paper has not necessarily written a better answer; they may simply have larger handwriting. People who start to scribble madly as soon as they turn over the question sheet are not in possession of some extra ability – they simply haven't planned their exam paper properly.

Keep calm, plan carefully, don't panic.

16.12.3 After the exam

If there are several exams to be taken, you need to stay calm, relaxed and confident. It is not a good idea to get into discussion about other people's experiences of the exam. After one exam, focus on the next. If something went wrong during the exam (for example, illness or severe family problems), the tutor and the examining board should be alerted immediately.

16.13 The examiners' reports

16.13.1 A few points from the examiners' reports

The examiners' reports provide information to help candidates and tutors in future exams. They aim to be constructive and to help you towards a better understanding of the syllabus and how you are assessed.

The examiners note that many candidates are well prepared and that their answers are relevant and comprehensive and show that they not only have the necessary knowledge, but that they are able to apply it to the workplace situation.

On the other hand, there are some candidates who appear to be unprepared and the examiners emphasise that good preparation is essential. You will need to study the content of the syllabus, while at the same time understanding how the concepts contained in it apply in the workplace. The examiners also stress the importance of applying the information that you have learned, to the question that is asked on the paper. It is especially important to understand that rote-learning will not lead to success in this type of exam.

Here is a quote from the latest examiners' reports:

'In order to meet the pass standard for this assessment, acquisition of knowledge and understanding across the syllabus are prerequisites. However candidates need to demonstrate their knowledge and understanding in answering the questions set. Referral of candidates in this unit is invariably because they are unable to write a full, well-informed answer to one or more of the questions asked.

Some candidates find it difficult to relate their learning to the questions and as a result offer responses reliant on recalled knowledge and conjecture and fail to demonstrate a sufficient degree of knowledge and understanding. Candidates should prepare themselves for this vocational examination by ensuring their understanding, not rote-learning pre-prepared answers.'

The section on 'common pitfalls' warns about common mistakes that are made by people taking these exams. Here are seven things which often cause people to fail to reach their full potential:

- Failure to apply the basic principles of exam technique can make the difference between passing the exam and being referred.
- Some candidates do not attempt all the required questions or they fail to provide complete answers. Even if your mind goes blank, do make an attempt to answer a question that is compulsory. Applying basic health and safety management principles can generate points that will add to your mark.
- If you provide information that is relevant to the topic but not relevant to the question that is set, that information cannot gain any marks. Always answer the question that is set.
- Knowledge of the command words is essential (see Table 16.1). They are the instructions that guide you on the depth of answer that is required. For example, if a question asks you to 'describe' something, you will not get many marks if you 'outline' it.
- Where a question is divided into sub-sections, you need to indicate clearly which part of the question you are answering. Use the numbering from the question in your answer. If your answer is structured to address the different parts of the question, that will help to draw out the points you need to make in your answer.

- Time planning is absolutely vital. If you spend a lot of time on one answer and give a lot of unnecessary information, you may be left with insufficient time to address all the questions.
- If you have doubts about the legibility of your handwriting, practise writing at speed. The examiner has to be able to read it to mark it.

The examiners also point out that you do not need to start a new page for each section of a question.

16.13.2 Marks for practice revision questions

We have been asked about the allocation of marks to the practice revision questions shown at the end of each chapter. However, since the marks awarded to each part of a question can vary, only general guidance can be given.

All one- or two-part questions are given 8 marks with a minimum of 2 marks awarded to any part.
Questions with three or more parts are given 20 marks with a minimum of 4 marks for each part.

NEBOSH exam questions have a similar format and allocation of marks.

16.14 Conclusion

Passing health and safety exams and assessments has a lot in common with any other subject being examined. Read the questions carefully and plan your answers. The old carpenter's saying 'Measure twice, think twice, cut once' can be applied to exam technique:

Read twice, think twice and write once.

16.15 Further information

Arp, D. Standards Manager NEBOSH (2014) Presentation to course providers 'Preparing Students for Assessment'.

Buzan, T. A wide range of study skills books by this author can be found at www.thinkbuzan.com

Guide to the NEBOSH International General Certificate.

Leicester, Coventry and Nottingham Universities Study Skills booklets.

NEBOSH Examiners' Reports.

NEBOSH Guidance on command words used in learning outcomes and question papers (March 2013) online.

Northedge, A. 2012 The Good Study Guide. The Open University. ISBN 978 0 7492 5974 7.

Free learning resources from the Open University www.open.edu/openlearn/free-course

16

CHAPTER 17

Specimen answers to practice questions

17.1 Introduction ▶ 550

17.2 The written examinations ▶ 550

17.3 GC3 – the practical application ▶ 555

Appendix 17.1 Practical application report ▶ 559

Appendix 17.2 Practical application observation sheets ▶ 563

> **This chapter covers:**
> 1. The depth of answer needed for long and short NEBOSH questions
> 2. The different types of questions used by NEBOSH
> 3. The requirements for the NEBOSH practical inspection and the management report

17.1 Introduction

The NEBOSH International General Certificate is assessed by two written examination papers (IGC1 – Management of international health and safety and GC2 – Control of workplace hazards) and a practical assessment (GC3 – Health and safety practical application). Both examination papers and the practical application must be passed within a five-year period so that the NEBOSH International General Certificate may be awarded.

NEBOSH is concerned about malpractice problems if published NEBOSH questions are used in publications so the questions used in this book are not used in NEBOSH examinations. In this chapter, specimen answers are given to one long and two short questions for each of the two written papers. A specimen practical assessment and management report is also given for the practical application. It is important to stress that there are no unique answers to these questions but the answers presented should provide a useful guide to the depth and breadth expected by the NEBOSH examiners. Candidates are strongly advised to read past examiners' reports which are published by NEBOSH and available on their website. These are very useful documents because they indicate some common errors made by candidates. Finally, learning by 'rote' of certain key words should be discouraged since examiners always expect such key words to be put into a context that is relevant to the particular examination question.

17.2 The written examinations

The previous chapter on study skills gives some useful advice on tackling examinations and should be read in conjunction with this chapter. NEBOSH has a commendably thorough system for question paper preparation to ensure that no candidates are disadvantaged by question ambiguity. Candidates should pay particular attention to the meaning of command verbs, such as 'outline', used in the questions.

Candidates have always had most difficulty with the command verb '**outline**' and, for this reason, several of the specimen questions chosen use '**outline**' so that some guidance on the depth of answer expected by the examiners can be given.

NEBOSH no longer uses the command word 'sketch'; however, candidates may use sketches to illustrate their answers, when appropriate.

It is difficult to give a definitive guide on the exact length of answer to the examination questions because some expected answers will be longer than others and candidates answer in different ways. As a general guide, for the long answer question on the examination paper, it should take about 25 minutes to write about one and a half pages (550–620 words). Each of the 10 short answer questions require about half a page of writing (170–210 words).

17.2.1 IGC1 – Management of international health and safety

Paper 1 – question 1

(a) **Outline** the factors that should be considered when selecting individuals to assist in carrying out risk assessments in the workplace. **(5)**
(b) **Describe** the key stages of a general risk assessment. **(5)**
(c) **Outline** a hierarchy of measures for controlling exposures to hazardous substances. **(10)**

Figure 17.1 Select a competent and experienced person to carry out a risk assessment

Answer:

(a) The most important factor is the competence and experience of the individuals in hazard identification and risk assessment. Some training in these areas should offer evidence of the required competence. They should be experienced in the process or activity under assessment and have technical knowledge of any plant or equipment used. They should have knowledge of any relevant standards relating to the activity or process.

They must be keen and committed but also aware of their own limitations. They need good communication skills and to be able to write interesting and accurate reports based on evidence and the detail found in health and safety standards and codes of practice. Some IT skills would also be advantageous. Finally, the views of their immediate supervisor should be sought before they are selected as team members.

(b) There are five key stages to a risk assessment suggested as follows:

The first stage is hazard identification, which involves looking at significant hazards which could result in serious harm to people. Trivial hazards should be ignored. This will involve touring the workplace concerned looking for the hazards in consultation with workers themselves and also reviewing any accidents, ill-health or incidents that have occurred.

Stage 2 is to identify the persons who could be harmed – this may be workers, visitors, contractors, neighbours and even the general public. Special groups at risk, like young persons, nursing or expectant mothers and people with a disability, should also be identified.

Stage 3 is the evaluation of the risks and deciding if existing precautions or control measures are adequate. The purpose is to reduce all residual risks after controls have been put in to as low as is reasonably practicable. It is usual to have a qualitative approach and rank risks as high, medium or low after looking at the severity of likely harm and the likelihood of it happening. A simple risk matrix can be used to get a level of risk.

The team should then consider whether the existing controls are adequate and meet any guidance or legal standards using the hierarchy of controls.

Stage 4 of the risk assessment is to record the significant findings which should be done. The findings should include any action that is necessary to reduce risks and improve existing controls, preferably set against a timescale. The information contained in the risk assessment must be disseminated to workers and discussed at the next health and safety committee meeting.

Stage 5 is a timescale set to review and possibly revise the assessment which must also be done if there are significant changes in the workplace or the equipment and materials being used.

(c) The various stages of the usual hierarchy of risk controls are in **bold** in this answer.

Elimination or **substitution** is the best and most effective way of avoiding a hazard and its associated risks. Elimination occurs when a process or activity is totally abandoned because the associated risk is too high. Substitution describes the use of a less hazardous form of the substance. There are many examples of substitution, such as the use of

water-based rather than oil-based paints and the use of asbestos substitutes.

In some cases it is possible to **change the method of working** so that exposures are reduced, such as the use of rods to clear drains instead of strong chemicals. It may be possible to use the substance in a safer form; for example, in liquid or pellets to prevent dust from powders. Sometimes the pattern of work can be changed so that people can do things in a more natural way; for example, by encouraging people in offices to take breaks from computer screens by getting up to photocopy or fetch documents.

Reduced or limited time exposure involves reducing the time that the employee is exposed to the hazardous substance by giving the worker either other work or rest periods.

If the above measures cannot be applied, then the next stage in the hierarchy is the introduction of **engineering controls**, such as isolation (using an enclosure, a barrier or guard), insulation (used on any electrical or temperature hazard) or ventilation (exhausting any hazardous fumes or gases either naturally or by the use of extractor fans and hoods). If ventilation is to be used, it must reduce the exposure level for employees to below the occupational exposure limit (OEL).

Housekeeping is a very cheap and effective means of controlling risks. It involves keeping the workplace clean and tidy at all times and maintaining good storage systems for hazardous substances.

A **safe system of work** describes the safe method of performing the job.

Training and information are important but should not be used in isolation. Information includes such items as signs, posters, systems of work and general health and safety arrangements.

Personal protective equipment (PPE) should only be used as a last resort. There are many reasons for this. It relies on people wearing the equipment at all times and it must be used properly.

Where necessary **health surveillance** should be introduced to monitor the effects on people and air quality may need to be monitored to check exposure.

Welfare facilities, which include general workplace ventilation, lighting and heating and the provision of drinking water, sanitation and washing facilities, are the next stage in the hierarchy.

All risk control measures, including **training and supervision** must be **monitored** by competent people to check on their continuing effectiveness. Periodically the risk control measures should be **reviewed**. Monitoring and other reports are crucial for the review to be useful. Reviews often take place at safety committee and/or at management meetings. A serious accident or incident should lead to an immediate review of the risk control measures in place.

Finally, special control requirements are needed for carcinogens.

Paper 1 – question 2

Outline ways in which employers may motivate their workers to comply with health and safety procedures. **(8)**

Answer:

Motivation is the driving force behind the way a person acts or the way in which people are stimulated to act. The best way to motivate workers to comply with health and safety procedures is to improve their understanding of the consequences of not working safely, their knowledge of good safety practices and the promotion of their ownership of health and safety. This can be done by effective training (induction, refresher and continuous) and the provision of information showing the commitment of the organisation to safety and by the encouragement of a positive health and safety culture with good communications systems. Managers should set a good example by encouraging safe behaviour and obeying all the health and safety rules themselves even when there is a difficult conflict between production schedules and health and safety standards. A good working environment and welfare facilities will also encourage motivation. Involvement in the decision-making process in a meaningful way, such as regular team briefings, the development of risk assessments and safe systems of work, health and safety meetings and effective joint consultation arrangements, will also improve motivation as will the use of incentive schemes. However, there are other important influences on motivation such as recognition and promotion opportunities, job security and job satisfaction. Self-interest, in all its forms, is a significant motivator.

Although somewhat negative, it is necessary sometimes to resort to disciplinary procedures to get people to behave in a safe way.

Paper 1 – question 3

(a) **Explain** why young persons may be at a greater risk from accidents at work. **(4)**
(b) **Outline** the measures that could be taken to minimise the risks to young workers. **(4)**

Answer:

(a) Young workers have a lack of experience, knowledge and awareness of risks in the workplace. They are often willing to work hard and want to

please their supervisor and can become over-enthusiastic. This can lead to the taking of risks without the realisation of the consequences. Some younger workers have underdeveloped communication skills and a limited attention span. Their physical strength and capabilities may not be fully developed and so they may be more vulnerable to injury when manually handling equipment and materials. They are also more susceptible to physical, biological and chemical agents such as temperature extremes, noise, vibration, radiation and hazardous substances.

(b) A specific risk assessment should be made before a young person is employed. This should help to identify the measures which should be taken to minimise the risks to young people.

Measures should include:

▶ additional supervision to ensure that they are closely looked after, particularly in the early stages of their employment;

▶ induction and other training to help them understand the hazards and risk at their workplace;

▶ not allowing them to be exposed to extremes of temperature, noise or vibration;

▶ not allowing them to be exposed to radiation, or compressed air and diving work;

▶ carefully controlling levels of exposure to hazardous materials so that exposure to carcinogens is as near zero as possible and other exposure is below the occupational exposure limits which are set for adults;

▶ not allowing them to use highly dangerous machinery like power presses and circular saws, explosives and mechanical lifting equipment such as fork-lift trucks;

▶ restricting the weight that young persons manually lift to well below any weights permitted for adults.

There should be clear lines of communication and regular appraisals. A health surveillance programme should also be in place.

17.2.2 GC2 – Control of workplace hazards

Paper 2 – question 1

A glassworks produces covers for streetlights and industrial lighting. The process involves molten glass being blown by hand and shaped in moulds.

(a) **Identify FOUR** health effects that may be caused by working in the hot conditions of the glass factory. **(4)**
(b) **Describe** measures that could be taken in order to minimise the health effects of working in such hot environments. **(6)**

Figure 17.2 Glass-blowing factory

(c) **Outline** the factors relating to the *task* and the *load* that may affect the risk of injury to a worker engaged in stacking the finished product onto racking. **(10)**

Answer:

(a)

1. heat exhaustion due to high ambient temperature;
2. dehydration due to excessive sweating;
3. heart stress and, in extreme cases, heat stroke due to prolonged exposure to high ambient temperatures;
4. burns from handling hot molten glass;
5. the eyes can also be affected by high-intensity light from looking at molten glass (additional answer).

(b) The health effects of working in a hot environment can be reduced by the gradual acclimatisation of new workers. Even after the initial acclimatisation, frequent rest periods will be necessary to allow the body to acclimatise to the hot conditions on a daily basis. Rest should be in cool areas which in summer may need to be artificially cooled. If, in addition, the humidity is high, a good supply of ventilation air will be needed to help control sweating. An adequate supply of cold drinking water is essential to avoid dehydration.

Workers in hot conditions should wear appropriate clothes, which must be a compromise between lighter garments to promote evaporation of perspiration, and protective clothes to prevent burns. It will be necessary to provide protective leather or fire-resistant aprons and gloves, and appropriate eye and face protection such as eye visors. Visors may need to be supplied with cooling air to keep people cool and permit proper vision. Screens could also be provided to protect workers from radiant heat. Periodic health surveillance should be provided.

Finally, workers should be trained to recognise ill-health effects on others.

(c) The **task** should be analysed in detail so that all aspects of manual handling are covered including the use of mechanical assistance. This will involve a manual handling risk assessment. The number of people involved and personal factors, such as age and health, should also be considered. A satisfactory body posture must be adopted with the feet firmly on the ground and slightly apart. To avoid work-related upper limb disorders (WRULDs) there should be no stooping or twisting of the trunk; it should not be necessary to reach upwards as this will place additional stresses on the arms, back and shoulders. The further the load is held or manipulated from the trunk, the more difficult it is to control and the greater the stress imposed on the back. These risk factors are significantly increased if several of them are present at the same time. The load should not be carried over excessive distances (greater than 10 m). The frequency of lifting, and the vertical and horizontal distances the load needs to be carried (particularly if it has to be lifted from the ground and/or placed on a high shelf) can lead to fatigue and a greater risk of injury. If the loads are handled whilst the individual is seated, the legs are not used during lifting and stress is placed on the arms and back.

There should not be excessive pulling, pushing or sudden movements of the load. The state of floor surfaces and the footwear of the individual should ensure that slips and trips are avoided.

There should be sufficient rest or recovery periods and/or the changing of tasks, particularly in the hot ambient temperatures of the glassworks. This enables the body to recover more easily from strenuous activity.

The imposition of a high rate of work is a particular problem with some automated production lines and can be addressed by spells on other work away from the line.

The handling capability of an individual is approximately halved when he/she becomes a member of a team. Visibility, obstructions and the roughness of the ground must all be considered when team handling takes place.

The **load** must also be carefully considered during the assessment – it may be too heavy. The maximum load that an individual can lift will depend on the capability of the individual and the position of the load relative to the body. There is therefore no safe load but guidance is available from health and safety literature, which does give some advice on loading levels. If the load is too bulky or unwieldy, its handling is likely to pose a risk of injury. Visibility around the load is important, as is awareness that it may hit obstructions or become unstable in windy conditions. The position of the centre of gravity is

17

important for stable lifting – it should be as close to the body as possible; however, this may be difficult if the load is hot, such as in boxes or trays of recently blown glass. They should be allowed to cool sufficiently.

The load becomes difficult to grasp when it is carried over slippery surfaces, has rounded corners or there is a lack of foot room. Sometimes the contents of the load are likely to shift. This is a particular problem when the load is a container full of smaller items, such as small glass covers. These are glass components which may shatter if dropped and leave shards of glass to be carefully cleared up.

The load is likely to be hot and could be sharp as well in places or when broken so that PPE, such as leather gloves and aprons along with eye protection, may be required.

Paper 2 – question 2

Outline the precautions that may be needed when carrying out repairs to the flat roof of a building. **(8)**

Answer:

Roof work is hazardous and requires a specific risk assessment and method statement prior to the commencement of work so that the required

Figure 17.3 Flat roof protection: (a) using handrails, and toe boards; (b) using a harness and proprietary anchor

precautions may be identified. The particular hazards are fragile roofing materials, including those materials which deteriorate and become more brittle with age and exposure to sunlight, exposed edges, unsafe access equipment and falls.

There must be suitable means of access such as scaffolding, ladders and crawling boards. Suitable edge protection will be needed in the form of guard rails to prevent the fall of people or materials, and access must be restricted to the area below the work using visible barriers. Warning signs indicating that the roof is fragile should be displayed at ground level. Protection should be provided in the form of covers where people work near to fragile materials and roof lights. The means of transporting materials to and from the roof may require netting under the roof and even weather protection.

Precautions will be required for other hazards associated with roof work, such as overhead services and obstructions, the use of equipment such as gas cylinders and bitumen boilers and manual handling.

Finally, only trained and competent persons must be allowed to work on roofs and they must wear footwear having a good grip. It is good practice to ensure that a person does not work alone on a roof.

Paper 2 – question 3

For **EACH** of the following agents, **outline** the principal health effects **AND identify** a typical workplace situation in which a person might be exposed:

(a) carbon monoxide **(2)**
(b) asbestos **(2)**
(c) legionella bacteria **(2)**
(d) hepatitis virus. **(2)**

Answer:

(a) Carbon monoxide is a colourless, tasteless and odourless gas. It causes headaches and breathlessness and, at higher concentrations, unconsciousness and death. The most common occurrence of carbon monoxide is in exhaust gas from a vehicle engine. Working in a vehicle repair garage without proper ventilation to exhaust gases would expose a person to carbon monoxide fumes.

(b) Asbestos produces fine fibres which can become lodged in the lungs. This can lead to asbestosis (scaring of the lungs), lung cancer or mesothelioma – cancer of the lining of the lung. Asbestos can be found in buildings, in ceiling tiles and as lagging around heating pipes. When these sites are disturbed, the asbestos fibres become airborne and inhalable affecting those engaged in maintenance or demolition work.

(c) Legionella is an airborne bacterium and is found in a variety of warm water sources between 20°C and 45°C. It produces a form of pneumonia caused by the bacteria penetrating the alveoli in the lungs. The disease is known as Legionnaires' disease and has symptoms similar to influenza. The three most common systems at risk from the bacteria are water systems that incorporate a cooling tower, air conditioning units, and showers. People working on these systems or working in the area of infected systems are at risk, particularly if they are over 45 years of age, and it affects men more than women.

(d) Hepatitis is a disease of the liver and can cause high temperatures, nausea, jaundice and liver failure. The virus can be transmitted from infected faeces (Hepatitis A) or by infected blood (Hepatitis B and C). Hospital workers and first-aiders who come into contact with blood products are at risk of hepatitis.

17.3 Unit GC3 – Health and Safety Practical Application

Guidance and information for accredited course providers and candidates is essential reading before attempting the practical application. The guide contains details on the aims of the application, the rules to be followed, the breadth and depth of observations and management report expected and the mark scheme. The candidates undertake the practical application in their own workplace. There is also a greater emphasis on the standard and content of the report and recommendations following the observations made in the workplace.

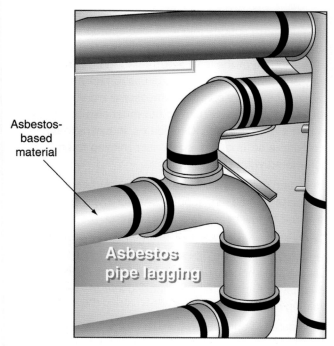

Asbestos-based material

Figure 17.4 Asbestos pipe lagging

The NEBOSH Practical application Guidance V1 (Aug 2013) (under revision at Aug 2015) is reproduced in 17.3.1. A specimen practical application, with associated observations and management report, is given in 17.3.2. It is important to emphasise that this specimen practical application (or any parts of it) must not be copied as this is malpractice.

17.3.1 Requirements of the practical application

The aim of the practical application, Unit GC3, is to assess a candidate's ability to complete successfully two activities:

▶ To carry out unaided a safety inspection of a workplace, identifying the more common hazards, deciding whether they are adequately controlled and, where necessary, suggesting appropriate and cost-effective control measures.

▶ To prepare a report that persuasively urges management to take appropriate action, explaining why such action is needed (including reference to possible breaches of international standards) and identifying, with due consideration of reasonable practicability, the remedial measures that should be implemented.

This will require candidates to apply the knowledge and understanding gained from their studies of elements of Units IGC1 and GC2 in a practical environment and to carry out an evaluation of information gathered during the inspection. The time allowed to complete the assessment is not restricted but candidates should aim to complete the inspection and the report in two hours.

The practical application may be submitted in the candidate's own handwriting or be word processed.

The submission must include:

▶ completed observation sheets covering a number and range of hazards and good practice, identifying suitable control measures and timescales;
▶ an introduction and executive summary;
▶ main findings of the completed inspection;
▶ conclusions which summarise the main issues identified in the candidate's workplace;
▶ completed recommendations table.

The practical application must be carried out in the candidate's own workplace. The workplace should be large enough to provide a sufficient range of hazards in the areas covered to provide an adequate range for identification. If the workplace is very large, in order that the practical application is manageable the candidate should limit the area considered. Where the candidate does not have access to a workplace, the course provider should be consulted to help in making arrangements for the candidate to carry out the practical application at suitable premises. Providers seeking to run the practical in this way should contact NEBOSH for advice and approval.

17

Date of assessment

Assessment of the practical unit (GC3) must normally take place **within 10 working days of** (before or after) the date of the IGC1 and/or GC2 written papers (the 'date of the examination'). Completion of study for both IGC1 and GC2 is recommended in order to undertake the practical application unit GC3.

Completion of observation sheets

Candidates will be supplied with a sufficient number of observation sheets from a course provider which may be photocopied for the purpose. An example observation sheet is given in Appendix 17.2. The observation sheets must be completed during the inspection. Only brief details of each hazard are required including where the hazard was located and the nature of the hazard. For example, 'housekeeping could be better' does not give enough information about the particular hazard. Candidates should avoid the tendency to simply identify the tasks being undertaken, or the equipment that is being used, instead of providing an indication of how the tasks or items of equipment pose a risk.

The observation sheets should be completed by:

▶ identifying, in the left-hand column, any hazards, unsafe work practices, consequences and examples of good practice observed during the inspection;

▶ commenting, in the next column, on the adequacy of existing controls and identifying any immediate, medium-term and long-term remedial actions needed;

▶ stating, in the right-hand column, a reasonable timescale for the actions identified.

There should be sufficient information on the observation sheets to enable the candidate to complete a report to management on their findings. Candidates are also advised to make notes on the area inspected, including activities taking place, in order to complete the introduction to their report. Whilst poor spelling and grammar will not be marked or penalised, if the assessor is unable to read or to understand the notes made by candidates during their inspection then invariably fewer marks will be awarded than would otherwise have been the case. The practical application may be submitted in the candidate's own handwriting or word processed.

For most practical assessments, the examiner will be looking for between 20 and 30 observations.

Marking of observation sheets

Range and outline of hazards and consequences (15 marks)

Candidates should outline 20 uncontrolled hazards to gain maximum marks, but are strongly recommended to outline more than 20 (but no more than 30) in case of duplication or inappropriate hazards being outlined. Candidates must demonstrate their understanding of how identified hazards have the potential to cause harm, for example boxes stored on the floor may cause obstruction of access, egress routes and/or a potential of musculoskeletal injury if lifted. This information must be recorded in the 'Hazards and consequences' column. Candidates are expected to identify different types of hazards such as hazardous substances, fire, electrical, work equipment, ergonomic, housekeeping, noise, vibration, transport, manual handling and health hazards and should also consider if there are any welfare and environmental problems. At least four different types of hazard must be included for maximum marks. In addition candidates are expected to comment on situations where there is adequate control of hazards and where good practice is being observed, although candidates should place the emphasis on uncontrolled hazards. A maximum of 1 mark is available for good practice.

Candidates who repeat identical hazards will only be awarded 1 mark for that hazard (e.g. inadequate labelling of an unknown chemical substance). Candidates should focus on physical conditions and not on poor policies and procedures.

It is important to note that credit can only be gained in this section for clearly identified hazards.

Identification of suitable control measures and timescales (15 marks)

Candidates are expected to give thought to what is required to immediately control the risk from each outlined hazard **AND** to identify the need for long-term actions to control the risk. This requires candidates to distinguish between the symptoms and the root causes of hazards.

For example, the immediate action on a spillage may be 'clean up spillage and inform supervisor' whereas medium-term actions might include appropriate supervisor training, regular inspections and investigation of the source of the leak. A further longer term action may be to modify the work process to tackle the root cause. The proposed control measures must not only remove or control the hazard but must also be realistic in terms of timescales. Candidates should indicate a measure of time, e.g. supervisor training to be completed within three months.

If existing controls are in place and considered adequate, candidates should consider any measures required to maintain this level of control.

Candidates should avoid generic phrases being repeatedly used, e.g. 'monitor' and 'train staff'. Candidates should give appropriate clarification by giving examples of appropriate monitoring and the type of training required.

Candidates should be aware that if unsuitable control measures are suggested full marks cannot be awarded.

Short-term measures to improve housekeeping will do little to improve the lack of safety management systems and procedures evidenced by materials and equipment left lying around.

Candidates should also note that where the hazard is not clearly outlined full credit cannot be gained for control measures as assessors will be unable to determine appropriateness of the measures proposed.

Completion of report

Candidates should use the 'Candidate report template' used in Appendix 17.1 to structure their report. The length of the report should be between 700 and 1,000 words and should not simply duplicate the observation sheets. Candidates can consult reference books when preparing the report, but plagiarism will be dealt with as malpractice.

The report should not contain photographs, printed text or any other extraneous material.

The report should be written in such terms that a manager would be able to take reasonable action based on facts. Reports based on unsupported generalities and those that simply reiterate the contents of the observation sheets will be awarded low marks.

Candidates should aim to complete their report in one hour.

If none of the criteria to award marks is met, then candidates should be aware that zero marks will be awarded.

Marking of report

Candidates are advised to refer to the marking sheet and the marking matrix given in the NEBOSH Guidance document for NEBOSH International General Certificate Health and Safety Practical Application (GC3).

Report – Introduction and executive summary

Introduction providing an overview of the chosen area (5 marks)

Candidates should start with the details of the inspection, stating where and when the inspection took place. A clear and appropriate description of the chosen area and of the activities occurring in the area should be given.

Executive summary (5 marks)

The executive summary should be written after the candidate has completed the rest of the report but it should be inserted at the beginning of the report. The purpose of the executive summary is to provide a **concise** overview of the important points arising from the work and **summarise** the main conclusions and recommendations arising from it.

An executive summary should provide sufficient information to enable a busy manager to make a decision as to whether or not to read the full report and to provide a persuasive case for implementation of recommendations made.

Report – Main findings of the inspection

The main findings of the inspection should form the main body of the report and include the following aspects:

The report should be well structured, the appropriate length and not duplicate observation sheets. The report should be concise, readable and highly selective in terms of action required by management. Candidates should include balanced arguments on why action is needed and explain the effect it would have on the standards of health and safety at the workplace and the possible effects on the business overall.

Quality of interpretation of findings and clear references to strengths and weaknesses (15 marks)

There should be a logical progression from the issues identified on the observation sheets.

The key issues should be appropriately selected and discussed, and should not include any hazards not outlined on the observation sheets and does not duplicate the observation sheets.

Identification of possible breaches of international standards (5 marks)

Candidates should be able to identify international standards and conventions listed in the syllabus that may have been infringed, e.g. Guidelines on Occupational Safety and Health Management Systems (ILO-OSH) 2001. Credit will also be given for reference to appropriate local standards.

Candidates must ensure that any legislation, standards or convention referenced is relevant to the area/location inspected and must demonstrate a clear understanding of the reasons for the breaches. Candidates should identify a minimum of five breaches.

Persuasiveness/conciseness/technical content (10 marks)

The candidate must explain why management need to take action including clear legal, moral and financial arguments why such action is needed. Reference can be made to the list of observations and recommended actions, calling particular attention to any recommendations which could have a high cost in terms of finance, inconvenience or time.

Financial benefits, such as increased productivity, may also accrue from making appropriate changes to safety systems. The possible costs of not taking action should be included.

17

Report – Conclusions and recommendations

Clear and concise conclusions which are clearly related to report findings and are effective in convincing management to take action (15 marks)

This section should provide a concise summary of the findings identified in the main body of the candidate's report. The conclusions should not introduce new issues or additional factors.

Recommendations which present realistic actions to improve health and safety in the chosen area (15 marks)

Candidates should include recommendations based upon their conclusions.

Recommendations must be presented using the recommendations table included in Appendix 17.1. The recommendations must be realistic, appropriately prioritised and have appropriate resource implications. Candidates are not expected either to know or to estimate actual costs but should demonstrate that they are aware of cost implications. For example, candidates recommending the complete resurfacing of a site roadway because of a small pothole, without commenting on its overall condition, will receive low marks. If training is recommended as a solution to a problem, candidates should indicate if this is likely to require a few hours of work-based instruction or

several days of more costly off-the-job training. It is the assessment of magnitude of the cost that is important, rather than precise figures, e.g. candidates may refer to the number of worker hours as a measure of cost.

Recommendations should be prioritised. The most pressing issues, those which present the highest risk levels and those that can be done immediately at little or no cost, should be addressed first. Target dates should be included, for example, 'plus one week' or 'plus three months'.

If none of the criteria to award marks is met, then zero marks will be awarded.

Important note

The guidelines outlined above and the following specimen practical application are completely valid at the time of publication of this book. However, NEBOSH reviews these guidelines and the format used in the two appendices from time to time so that it is important for course providers and others to consult the NEBOSH website periodically to check whether the guidelines and the report format have been amended. The specimen practical application is given as an illustration only.

17.3.2 Specimen practical application

The specimen practical assessment took place in a small joinery (woodworking) facility which is situated in a self-contained building. Details of the inspection report are given in Appendix 17.1 and the observations made in Appendix 17.2.

APPENDIX 17.1 Practical application report

UNIT GC3 – THE HEALTH AND
SAFETY PRACTICAL APPLICATION

Candidate report template

Student number: <u>C9887</u>

Location: <u>S & J Joinery Training Workshops</u>

Date of review: <u>27/January/2015</u>

Introduction including overview of area inspected and activities taking place

This report follows a health and safety inspection of the workshops and welfare facilities at S & J Joinery Training Workshops which is a small joinery (woodworking) facility situated in a self-contained building.

The facility is used to train 20 carpentry and joinery trainees and produce window frames for use on various building projects for local construction firms. The average age of the trainees is 24, although a minority of them is under 18 years. There are five employees, two supervisors, two technicians and an administrator, who administers the training scheme and liaises with the customers.

Within the building are two workshops – a bench and a machine workshop – and an administration office, welfare facilities, a rest room, two lecture rooms and a storeroom. The inspection only took place in the workshops and included the welfare facilities. Therefore, the assessment excluded the administration office, rest room and lecture rooms.

Executive summary

There have been no serious accidents since records began some five years ago although there have been several minor ones. This record appears to have led to a relaxation in the monitoring of standards. Several inexpensive improvements in the health and safety management systems and safeguards are recommended. These improvements will lower the risk of injury to the workforce. This will not only lead to better protection of individuals but also satisfy many international standards and conventions such as the Guidelines on Occupational Safety and Health Management Systems (ILO-OSH 2001) and the ILO Occupational Safety and Health Convention (C155) and the Recommendation (R164).

The most serious concern is the lack of appropriate guarding on the jig saw and circular saw. Neither of these machines must be used until adequate guarding is in place. Other areas of concern are:

1. A lack of effective trainee supervision on health and safety matters.

2. Inadequate fire and emergency arrangements.
3. The lack of a specific risk assessment that covers the activities of the trainees.
4. The presence of several slip, trip and manual handling hazards.
5. The lack of a review of all risk assessments, particularly those concerning substances hazardous to health.

A closer involvement by the trainee supervisors in the monitoring of health and safety issues in a proactive manner in both workshops will produce considerable improvements in overall health and safety performance. This monitoring should cover specific issues such as the wearing of personal protective equipment (PPE) to general housekeeping issues ensuring clean and tidy workshops.

These improvements, together with those listed below as recommendations, will bring long term benefits to the company by reducing the risk of injury and ill-health to the workforce and improvements to the quality of the training.

Main findings of the inspection

It was noted during the inspection that the company is very successful in attracting trainees and in the supply of quality window frames to local construction firms. There were several examples of good practice and these are to be commended (the effective use of health and safety posters is an example). However, it is apparent from the observations made during the inspection that there are a number of hazards in the two workshops that present significant safety and health risks which need to be addressed by the company.

The key areas of concern are:

1. **Management of health and safety**
   Many of the observed weaknesses result from a poor management system which could be improved if the technicians and administrator were given specific health and safety duties and, if necessary, relevant training of up to half-a-day in duration. One member of staff should be given the responsibility for monitoring health and safety throughout the building after receiving the appropriate off-site training. At least one member

17

of staff should also be trained in first-aid.

In particular, the repair and testing of electrical equipment, the inadequately guarded machines, and the problems with the use of personal protective equipment, can be addressed by assigning specific health and safety duties to staff. Supervisors also need to be more proactive in ensuring that benches are kept tidy and trainees adhere to health and safety rules.

Many observations indicate that risk assessments, if they have been made, are no longer valid and need to be reviewed urgently. The hazardous substance assessments may need to be revised and a specific risk assessment to be made for persons under the age of 18 years.

The health and safety committee must assess the quality and contents of the induction course given to trainees because of the large number of hazards found.

Many of these concerns are in breach of the standards required by the Guidelines on Occupational Safety and Health Management Systems (ILO-OSH 2001) and the ILO Occupational Safety and Health Convention (C155) and the Recommendation (R164).

An improvement in the level of supervision is essential and specific training in the supervision of young trainees should be considered. A satisfactory level of training could be achieved by an in-house course of 1 day duration.

2. Fire and emergency arrangements

The arrangements for fire and other emergencies are inadequate and would not meet the requirements of the Guidelines on Occupational Safety and Health Management Systems (ILO-OSH 2001) and the ILO Occupational Safety and Health Convention (C155) and the Recommendation (R164).

Detailed fire risk assessments together with fire-fighting arrangements, fire information and fire training (including fire drills) are required. Therefore, there is an urgent need to install 'Fire Exit' signs, and two fire extinguishers in the machine shop. This last point should have been noted by the extinguisher service contractor.

All emergency fire exit doors should remain unobstructed at all times and unlocked whenever the building is in use. Security could be improved by fitting an alarm system which is activated whenever the door is opened or by the use of a quick release mechanism.

3. The two workshops

There are serious concerns regarding the guarding of the jig saw and the circular saw. The operation of these machines **must** cease immediately until they are properly guarded. The local exhaust ventilation (LEV) system must be examined and maintained on a regular basis with records being kept and the broom replaced by an industrial-grade vacuum cleaner. Wood dust causes health problems which can be very serious. These problems are in breach of several international standards including 'Ambient factors in the workplace' (ILO Code of Practice), 'Safety in the Use of Chemicals at Work' (ILO Code of Practice) and 'Working Environment (Air Pollution, Noise and Vibration) ILO Convention (C148) and Recommendation (R156)'.

The use of personal protective equipment, particularly ear defenders, gloves and safety glasses, needs to be monitored by the supervisors. However, the rules concerning the use of the machines by trainees are commendable.

4. Manual handling issues

A mechanical lift truck should be purchased and used for the movement of window frames and timber. A manually operated truck should be sufficient and much cheaper than a fork-lift truck with its considerable training costs.

Manual handling and trip hazards could be eased considerably by the erection of suitable shelving on the walls of the two workshops. Only one day's supply of timber should be kept in the workshops. Any excess timber stocks should remain in the storeroom. The use of shelving is a very cheap option.

5. Other issues

All personal protective equipment must be properly maintained and stored.

The heating system should be checked by a competent person because the workshops felt cold on the day of the inspection. Low temperatures can reduce workers' performance and awareness of risks with an increased risk of injury. The ILO Code of Practice 'Ambient factors in the workplace' requires employers to assess the risks to health and safety of workers if work with bare hands is carried out for extended periods of time at temperatures below 15°C.

An improvement in housekeeping would ensure that floors and benches are kept clear and unobstructed.

Conclusions

The workshops were found to be clean, well maintained and well lit. The trainees clearly enjoyed working there. There were, however, a number of issues which need to be addressed urgently to avoid serious accidents and compensation claims following the resultant injuries. These issues are listed in the table of recommendations below.

Work on the jig saw and circular saw must cease immediately until both machines are properly guarded. The lack of reasonable arrangements for fire and other emergencies are also a serious concern. The whole of the workforce including the trainees must be aware of the emergency arrangements and be able to reach a place of safety in the event of such an emergency.

Health and safety management systems must be planned and implemented to include proactive and reactive monitoring programmes, regular audits of the system and both specific and general training for all levels of personnel within the company including supervisors. It is recommended that the technicians and administrator are given specific health and safety duties and, if necessary, relevant training of up to half-a-day in duration. At least one member of staff should also be trained in first-aid. A realistic timetable for these changes will also be required.

The vast majority of recommendations made in this report are low in terms of time, effort and money. Their implementation will lead to a considerable improvement in the health and safety of the workforce and a consequent lower level of risk of injury, ill-health and prosecution. There are, therefore, clear legal, moral and financial arguments in favour of the recommendations. The resultant increase in morale and well-being amongst the workforce could result in an increase in productivity and profitability for the company.

Recommendations

– include as a table in the following format:

Recommendation	Likely resource implications	Priority	Target date
The operation of the jig saw and the circular saw must cease immediately until they are properly guarded.	The guards are not expensive and may be fitted in 2 man hours. In the case of the circular saw, an adjustment is required and this should only be done by a competent person.	High	Immediate
Install 'Fire Exit' signs at all emergency exits and two suitable fire extinguishers in the machine shop. Emergency exit directional signs should also be erected.	This is a low cost and, if necessary, the fire extinguisher maintenance contractor should be able to help – 1 man hour.	High	1 week
The emergency fire exit door should remain unobstructed at all times and unlocked whenever the building is in use. All emergency exit doors should be monitored daily to ensure that they are unlocked and unobstructed.	An appropriate notice should be erected adjacent to the doors and a specialist security device purchased, possibly using advice from the fire extinguisher maintenance contractor. This should not be a high cost item and take 2 man hours to organise.	High	1 week
Suitable shelving should be erected on the walls of the two workshops for timber storage. Only one day's supply of timber should be kept in the workshops and the store room used for longer term storage.	Sturdy shelving will be required with strong wall fixings. This job could be easily done within the company. The costs will be low and take a maximum of 1 man day.	Medium	1 month
One member of staff should be trained to monitor the health and safety arrangements throughout the building. This will involve attendance on a recognised health and safety course off-site.	There are a variety of health and safety courses available and it is important that a reputable one is selected. Although this will be more costly than many of the recommendations, it need not be too high.	High	1 month

17

Some technicians and supervisors should be given specific health and safety duties and, if necessary, relevant training in general health and safety. At least one member of staff should also be trained in first-aid.	The first-aid training will probably be the most expensive and last for 3/4 days. General health and safety training should be done in-house by a suitably qualified member of staff. The general course should last for 1/2 day.	Medium	3 months
The local exhaust ventilation (LEV) system must be examined and maintained on a regular basis with records being kept. This must be done by a competent person.	This will involve a thorough examination by an independent competent person and should take about an hour. This may be one of the more costly items but these costs could be reduced if the company insurer could provide the service.	High	1 week then every 14 months
Review all risk assessments, including a young person assessment. This review should take place under the direction of the health and safety committee.	This will take one person up to 2 days to complete.	Medium	3 months
The broom should be replaced by an industrial-grade vacuum cleaner.	This type of vacuum cleaner is more expensive that a domestic one but is not too costly.	Medium	2 months
A mechanical lift truck should be purchased to reduce manual handling of loads.	A manually operated truck is cheaper than a fork-lift truck and less expensive to maintain. It will be adequate for the tasks in the workshops.	Medium	3 months
The heating system should be inspected by a competent person and improved. A repair and maintenance programme may be possible. The resource implications will only be known after the inspection.	Possibly a new boiler will be required making this the most expensive recommendation. However, repairs to the existing heating system may well be possible.	Medium priority if auxiliary heating is employed in the meantime.	3 months
Health and safety should be included in the induction training for trainees.	The induction training should last for 2 hours per trainee and include health and safety rules and relevant procedures.	High	Immediate

APPENDIX 17.2 Practical application observation sheets

GC3 – THE HEALTH AND
SAFETY PRACTICAL APPLICATION

Candidate's observation sheet

Sheet number 1 of 5

Student name: J Brown

Student number: C 9887

Place inspected: S & J Joinery Training Workshops

Date of inspection: 27/January/2015

Observations	Control measures	Timescale
Hazards and consequences	Immediate and longer-term actions	
GENERAL OBSERVATIONS		
1. Several risk assessments have been completed but no specific risk assessment has been done for the young trainees. This could expose the young trainees to injuries caused by their unfamiliarity with the hazards present and their general inexperience.	The completed risk assessments cover many of the physical and health hazards.	
	These risk assessments are to an acceptable standard. However, a specific risk assessment to cover the hazards faced by the young trainees must be completed as soon as possible.	Immediate
	A review timetable for all risk assessments must be established.	1 month
	All risk assessments should be reviewed by the Health and Safety committee.	Each year on an ongoing basis
2. No fire exit signs are posted in the building. This will lead to confusion (and possible panic) in the event of a fire or other emergency.	Obtain and install fire exit signs and review the fire risk assessment.	Immediate
	Inform employees and trainees of fire escape routes.	1 week
	Organise some basic fire training.	1 month
3. Emergency exit door is locked to protect building against unauthorised entry and the door was obstructed by several lengths of timber. This would cause confusion (and possibly serious injury) in the event of a fire. The timber also presents a trip hazard and could also lead to injury.	This door must be unlocked immediately and other means found to maintain the security of the building (such as special bolts that can be broken in an emergency).	Immediate
	Remove timber and any other obstacles from the door area.	Immediate
	Monitor doors regularly and include this problem in the fire training.	Weekly
4. No fire drills have been undertaken for some time. In the event of a fire, this will cause confusion amongst the trainees and others who will be unsure of the fire procedure.	Undertake an immediate trial evacuation of the building to designated assembly point.	1 week
	Arrange 1 fire drill for all new trainees.	1 week after enrolment
	Arrange at least 2 fire drills for employees per year and record evacuation time.	6 months
5. All electrical equipment, other than the electric fire in the office and the sander in the workshop, has been PAT tested. Such testing should ensure that all electrical equipment is safe to use and unlikely to lead to electric shocks or fires	In general this is good practice although a report should be prepared for the Health and Safety committee into the reasons for the two omissions.	For the next meeting
	Arrange for the fire and the sander to be PAT tested and the results recorded. All hand-held tools should be checked each week and PAT tested every 6 months.	1 week
		6 months
	Ensure that records are kept and equipment re-tested at regular intervals.	Each year

17

GC3 – THE HEALTH AND SAFETY PRACTICAL APPLICATION

Student name: J Brown
Place inspected: S & J Joinery Training Workshops

Student number: C 9887
Date of inspection: 27/January/2015

Observations	Control measures	Timescale
Hazards and consequences	Immediate and longer-term actions	
GENERAL OBSERVATIONS (Cont.)		
6. Risk of slips, trips and falls due to trailing cables in both workshops. Slips, trips and falls are the most common causes of accidents in the workplace. The trailing cables also present a risk of electric shock.	Fix covers over cables and re-route them away from busy walkways.	Immediate
	Reorganise workshop layout and/or work activities to reduce risk of tripping.	1 week
	Install additional power sockets to reduce the need for long trailing cables.	6 months
7. Timber left lying on floor of both workshops – slip, trip and fall hazard. This could cause injuries to individuals who do not see the obstructions on the floor.	All unused timber must be stacked in the corner of workshop or in the adjacent empty room.	Immediate
	Suitable shelving should be erected to store timber safely. Only store a day's requirement in the workshops. Include this arrangement in future training and monitor.	2 weeks
	Decide on a future strategy for timber storage and implement.	6 months
8. First-aid box is fully and properly stocked and appropriate notices displayed in each workshop.	This is commendable and no action is required other than monitoring the contents of the first-aid box to ensure that it is refilled.	Each week
9. No trained first-aider in the building. If required, help is obtained from next building. Help may not be available when needed.	Arrange for at least 1 member of staff (preferably 2) to be trained as a first-aider. Also an Appointed Person must be nominated to contact the emergency services when necessary.	3 months
10. Risk of back injuries from lifting heavy lengths of timber and manufactured window frames – likely to cause back and shoulder injuries leading to sickness leave.	Ensure that timber lengths are always lifted by at least 2 persons and window frames by 4 persons.	Immediate
	Undertake a manual handling risk assessment to see whether mechanical assistance can be used.	Immediate
	Give everyone a basic manual handling training course including supervisors.	2 weeks
	Monitor activities regularly.	1 month
11. Several useful posters on manual handling, noise and the correct use of tools and machines, displayed in both workshops.	This is to be commended. May need updating in the longer term. Keep under review at the Health and Safety committee.	Every 3 months
12. Lighting levels throughout were good.	No action is required but keep under review.	

GC3 – THE HEALTH AND SAFETY PRACTICAL APPLICATION

Candidate's observation sheet

Sheet number **3 of 5**

Student name: _J Brown_   Student number: C **9887**

Place inspected: _S & J Joinery Training Workshops_   Date of inspection: **27**/**January**/**2015**

Observations	Control measures	Timescale
Hazards and consequences	Immediate and longer-term actions	
GENERAL OBSERVATIONS (Cont.)		
13. There were several instances of torn gloves and scratched safety glasses on the benches in the 2 workshops. The gloves will be ineffective in protecting workers from injury and the scratched safety glasses may cause workers to make serious mistakes.	Replace gloves and glasses.	Immediate
	Ensure that all PPE is properly stored and maintained. Prominent notices should be displayed and all infringements recorded.	1 month
	Review all PPE risk assessments and monitor that it is being used.	1 month
	Ensure that trainees receive instruction on the correct use of PPE.	At induction
14. Toilet and washing facilities were satisfactory and cleaned on a regular basis. Barrier cream was provided and trainees instructed to use it. 1 toilet door lock was broken.	A good cleaning regime is in place.	
	Repair toilet door lock.	Immediate
	Monitor cleanliness of toilets.	Daily
15. Heating levels were low throughout the building and the workshops were cold. Low temperatures can reduce workers' performance and awareness of risks – increased risk of injury.	Check all room temperatures.	Immediate
	Have boiler and heating system checked and, if necessary, provide auxiliary heating.	1 week
	Monitor to ensure that a reasonable temperature is maintained.	Weekly from Sept to March
	This could be a costly problem in terms of finance, time and inconvenience.	
16. Despite several notices forbidding the consumption of food and drink, several trainees had bottles of drinking water on their work benches. This could lead to serious health risks to the trainees if the food becomes contaminated with hazardous substances. The water could also pose a problem if electrical equipment is nearby.	Supervisors must enforce ban on food and drink in the workshops.	Immediate
	Install a fresh drinking water facility in the trainee rest room with cups provided.	1 month
17. All fire extinguishers regularly serviced.	Commendable and inspect regularly to maintain good practice.	Immediate
18. Several cupboard doors under the benches throughout the building were left open leading to risk of minor injury – likely to cause knee or leg injury.	Inform staff and trainees to keep all cupboards shut and post appropriate notice on notice board.	Ongoing
	Ensure that supervisors Monitor compliance.	

17

GC3 – THE HEALTH AND SAFETY PRACTICAL APPLICATION

Student name: J Brown

Place inspected: S & J Joinery Training Workshops

Student number: C 9887

Date of inspection: 27/January/2015

Observations	Control measures	Timescale
Hazards and consequences	Immediate and longer-term actions	
BENCH WORKSHOP		
19. There was an unmarked glass bottle containing a liquid (solvent?) on one of the benches. This could be a substance hazardous to health that could cause injury and/or health problems.	Identify, label the bottle and place it in a secure store. If it cannot be identified then dispose of it safely.	Immediate
	If it is identified, ensure that a risk assessment has been completed.	Immediate
	Supervisor must inform trainees of the unsafe practice of leaving unmarked bottles on benches.	Immediate
	Ensure that the use of hazardous substances is covered during induction training.	Each induction
20. Frayed lead to hand-held sander – risk of electric shock to the user or anyone who touches it.	Remove it from use immediately.	Immediate
	Replace lead, PAT test sander and place it on the PAT register.	1/2 weeks
	Include in regular future PAT inspection schedule.	Every 3 months
	Include electrical hazards and precautions in induction training for trainees.	Each induction
21. Chisels and a Stanley knife left on benches – could lead to serious injuries (deep cuts) to hands/fingers if touched accidentally.	Remove from benches and place in tool rack.	Immediate
	Check to ascertain whether Stanley knife necessary. If it is, use one with a retractable blade.	Immediate
	Report finding to supervisors and inform trainees of dangers.	Ongoing
	Monitor.	
MACHINE WORKSHOP		
22. Jig saw unguarded – guard has been removed. This is very serious and could lead to a serious hand injury particularly to inexperienced trainees.	Prohibit use of machine until guard installed.	Immediate
	Supervisors to monitor future use of jig saw and encourage everybody to report any absences of guards in the future.	Immediate and on-going
	Include need for correct guards on machines in trainee training sessions.	Each induction

GC3 – THE HEALTH AND SAFETY PRACTICAL APPLICATION

Student name: _J Brown_   **Student number: C 9887**

Place inspected: _S & J Joinery Training Workshops_   **Date of inspection: 27/January/2015**

Observations	Control measures	Timescale
Hazards and consequences	Immediate and longer-term actions	
<u>MACHINE WORKSHOP (Cont.)</u>		
23. Top guard on circular saw set too high – another serious problem with a significant risk of hand/arm injury to the operator, such as the amputation of a limb.	Prohibit use of machine until guard is correctly set. Inform supervisor of dangerous hazard.	Immediate
	This indicates another serious weakness in the supervision process. An urgent meeting of all supervisors is required to discuss this and other problems.	Immediate
	Set guard to correct gap.	Immediate
	Ensure that inspection of all machine guards is part of maintenance schedule and monitor compliance.	Each week
24. Noise assessment has been done but several persons not wearing ear defenders even though noise levels are in excess of the upper exposure action level and mandatory signs are in place – indicates poor trainee supervision. High noise levels could lead to hearing damage.	Noise assessments are made every year and available in the machine shop. This is commendable. All persons in the machine shop must be trained to fit ear defenders properly and wear them when any machine is running.	Immediate
	Include use of ear defenders in the trainee induction training.	Each induction
25. Only a water fire extinguisher available in the machine shop. All machines are electrically powered – could produce an electric shock if used on an electric motor.	Replace with 2 fire extinguishers – a carbon dioxide and a powder one.	Immediate
	Investigate the reasons for this occurrence with the servicing contractor.	3 months
	Again review fire risk assessment.	1 week
26. LEV not been inspected for 3 years. Possible reasons for dust?	Arrange for an immediate inspection by a competent person to check the efficiency of LEV.	Immediate
	Set up an ongoing contract for future inspections.	3 months
	Train staff and monitor compliance.	3 months
27. Only a broom available for sweeping up excess wood dust. Inhalation hazard.	Remove the broom and replace with an industrial grade vacuum cleaner.	Immediate
	Monitor that it is being used.	3 months
28. Only trainees over the age of 18 years are allowed to use the machines and then only with close supervision.	This is very good practice and is commendable.	

17

CHAPTER 18

International sources of information and guidance

18.1 Introduction ▶ 570

18.2 How to search the internet effectively ▶ 570

18.3 Some useful websites ▶ 572

18.4 Health and safety forms ▶ 574

This chapter covers:

1. Guidance on searching the internet
2. Websites for principal sources of information and guidance on health and safety issues

It will also give you access to:

3. A range of forms for use in the management of health, safety and fire issues which are illustrated at the end of the chapter

---

18.1 Introduction

The authors and publishers are not responsible for the contents or reliability of the linked websites and do not necessarily endorse the views expressed within them. Listing should not be taken as endorsement of any kind. We cannot guarantee that these links will work all the time and we have no control over the availability of the linked pages. For this chapter of the book, as in Chapter 16, a less formal style has been chosen, speaking directly to you, the student.

Today using the internet for access to legislation, standards and guidance is universal and of great benefit to the international occupational health and safety (OH&S) practitioner or person seeking information. Clearly there are many organisations who give small amounts of information free in order to gain your membership or custom. As internet users know, documents, guidance, reports, etc. can remain sitting on websites for years. So it is very important to ensure that they are up to date and authoritative. Guidance from good reliable sources is usually kept up to date and often states the date on which the site was last updated.

Fortunately in the OH&S field there are many organisations that freely allow the download of information because they are either governments, government funded or have a strong wish to make good guidance widely available. These include IOSH, RoSPA and many similar OSH organisations around the world; government departments in the USA, UK, Europe and beyond; universities and local authorities. In the UK, the HSE now allows free download of most of their extensive, authoritative and well-produced guidance. This is a significant change of policy in the last few years and provides a wealth of information for the OH&S practitioner and others.

Finding these sources of information and getting the best results from your searches can be a daunting task. The following general guidance may help when searching for occupational, safety, health, fire, chemical and environmental information on the internet. The websites are mainly in English but some are also in several other languages or a local language and English.

There is also a very useful free, UK Open University, 9-hour introductory course, Finding information in information technology and computing,

LIB 5 at: http://openlearn.open.ac.uk/course/view. php?id=2370

18.2 How to search the internet effectively

18.2.1 The key to successful searching

Remember, you are smarter than a computer. Use your intelligence. Search engines are fast, but dumb.

A search engine's ability to understand what you want is very limited. It will obediently look for occurrences of your keywords all over the Web, but it doesn't understand what your keywords mean or why they're important to you. To a search engine, a keyword is just a string of characters. It doesn't know the difference between cancer the crab and cancer the disease … and it doesn't care.

But **you** know what your query means (at least, we hope you do!). Therefore, you must supply the brains. The search engine will supply the raw computing power. While searching, bookmark any other sites of interest. Here are a few tips.

18.2.2 Use unique, specific terms

It is simply amazing how many web pages are returned when performing a search. You might guess that the terms "hard hat" are relatively specialised. A Google search of those terms returned 65,700,000 results! To reduce the number of pages returned, use *unique* terms that are *specific* to the subject you are researching. "Safety hard hat" reduces this to 93,000. Enter a short question, such as "what time is it?"

18.2.3 Use the minus sign (-) to narrow the search

How many times have you searched for a term and had the search engine return something totally unexpected? Terms with multiple meanings can return a lot of unwanted results. The rarely used but powerful minus sign, equivalent to a Boolean NOT, can remove many unwanted results. For example, when searching for the insect caterpillar, references to the company Caterpillar, Inc. will also be returned. Use Caterpillar -Inc to exclude references to the company or <u>Caterpillar -Inc -Cat</u> to further refine the search.

18.2.4 Use of + sign

Place a **plus sign** (+) prior to each word to view each "word separately" within your search results, such as +writer+grammar+punctuation.

18.2.5 Use quotation marks for exact phrases

I often remember parts of phrases I have seen on a web page or part of a quotation I want to track down. Using quotation marks around a phrase will return only those exact words in that order. It's one of the best ways to limit the pages returned. Example: "<u>Safe as houses</u>". Of course, you must have the phrase exactly right – and if your memory is as good as mine, that can be problematic.

18.2.6 Don't use common words and punctuation

Common terms like *a* and *the* are called <u>stop words</u> and are usually ignored. Punctuation is also typically ignored. But there are exceptions. Common words and punctuation marks *should* be used when searching for a specific phrase inside quotes. There are cases when common words like *the* are significant. For instance, <u>Raven</u> and <u>The Raven</u> return entirely different results.

18.2.7 Capitalisation

Most search engines do not distinguish between uppercase and lowercase, even within quotation marks. The following are all equivalent:

▶ technology
▶ Technology
▶ TECHNOLOGY
▶ "technology"
▶ "Technology"

18.2.8 Drop the suffixes

It's usually best to enter the base word so that you don't exclude relevant pages. For example, *bird* and not *birds*, *walk* and not *walked*. One exception is if you are looking for sites that focus on the act of walking, enter the whole term *walking*.

18.2.9 Maximise AutoComplete

Ordering search terms from general to specific in the search box will display helpful results in a drop-down list and is the most efficient way to use AutoComplete. Selecting the appropriate item as it appears will save time typing.

The standard <u>Google start page</u> will display a drop-down list of suggestions supplied by the Google search engine. This option can be a handy way to discover similar, related searches. For example, typing in *health and safety* will not only bring up the suggestion *health and safety at work* but also *health and safety legislation* and *health and safety jobs*.

18.2.10 Specific file type search

In Google you can search by specific type of file by writing **filetype:<name_of_type>**, for example: "filetype:pdf" and write the subject that you are going to search then you will only get pdf files on that subject in the results.

18.2.11 Authors, institutions and other sources

Do you know the name of any author(s) working in this subject? By using the author's name, you may retrieve other references to similar work on the subject of your choice.

Is there an institution or competent authority(ies) known to have done some work in this area? Again try using the name; you may retrieve even more references.

Are any journals/indexing/abstracting service(s) specialising in the subject which are known to you? Again you can add these to your search.

Are there any information centre(s) specialising in the subject? This is similar to author searches because these information centres may well have produced a publication on the subject.

18.2.12 Don't forget the phone book

Telephone directories (white pages and yellow pages) are widely available now on the Web. This means that if you know someone's name and what town they live in, you can access their address, phone number, even their age. There are also databases ('reverse look-up') that allow you to type in a phone number and get the name and address of the person who owns the phone.

If you know the address of the person you are seeking, you can easily get a map of their town, street, and neighbourhood on the many web map sites. Some maps are precise enough to show the exact location of the person's home.

18.2.13 Privacy issues

The same things that you can find out about other people, other people can find out about you.

Here's a list of some of the databases someone might access when researching you:

▶ Property sales and ownership;
▶ Land Registry;
▶ Postal service change-of-address records;
▶ Telephone books, past and present;
▶ Automobile registration records;

18

▶ Other vehicle registration, i.e. boats, private airplanes;

▶ Subscription lists for magazines and periodicals;

▶ National marketing databases;

▶ National email directories;

▶ Voter registration lists;

▶ Bankruptcy filings;

▶ Court proceedings/judgements;

▶ Professional licences;

▶ Tax records;

▶ Credit bureau records;

▶ Domain name registration records.

If you are concerned about your privacy, you can ask to have your personal information removed from web databases. It is difficult to remove all trace of yourself, though. Some events and transactions are legitimately matters of public record, and more of these public records are becoming available every year via the Web.

18.3 Some useful websites

1. **British Institute for Occupational Hygienists: www.bohs.org/**
BIOH is the professional body for Occupational Hygienists in the UK and organises examinations and awards in Occupational Hygiene.

2. **CIS: http://www.ilo.org/safework/cis/WCMS_CON_TXT_SAF_INF_STT_EN/lang--en/index.htm**
International Occupational Safety and Health Information Centres (CIS) and its National and Collaborating Centres which are located all over the world. The goal of this Network is to help its members and the rest of the world to find information from the participating countries on subjects related to occupational safety and health (OSH). The CIS Centres Information Network can be accessed in three languages (English, French and Spanish).

3. **Department for Communities and Local Government (DCLG): https://www.gov.uk/government/collections/fire-safety-law-and-guidance-documents-for-business**
DCLG has responsibility for Fire and has published a series of guides on the Regulatory Reform Fire Safety Order 2005 which can be downloaded free from their website.

4. **European Agency for Safety and Health at Work: http://osha.europa.eu/en**
This Agency was set up by the European Union to promote the exchange of technical, scientific and economic information between member states. Has links to the European Commission legislation and also to each of the member states' legislation. Also contains links to Australia, Canada, Iceland, Switzerland, Norway, USA legislation. Note that the legislation appears in the language of that country.

5. **European Commission EUR-LEX portal: http://eur-lex.europa.eu/en/index.htm**
Details of European legislation, publications including the Official Journal of the European Union which is updated daily, links to other Commission sites.

6. **FIRA International Ltd: http://www.fira.co.uk/**
FIRA has driven the need for higher standards through testing, research and innovation for the furniture and allied industries. A non-government funded organisation, FIRA is supported by all sections of the industry through the Furniture Industry Research Association, ensuring ongoing research programmes which bring benefits to all. With unrivalled support from across the whole industry, FIRA also has the influence and capability to help shape legislation and regulations.

7. **Fire Protection Association: www.thefpa.co.uk/**
The Fire Protection Association is the UK's national fire safety organisation. The association works to identify and draw attention to the dangers of fire and the necessary fire prevention measures.

8. **Glossary: www.ilo.org/public/english/protection/safework/cis/products/safetytm/glossary.htm**
Glossary of Terms on Chemical Safety from the International Programme on Chemical Safety (IPCS) which is a joint venture between the United Nations Environment Programme (UNEP), the International Labour Organisation (ILO), and the World Health Organisation (WHO). The language of chemical safety is drawn from many sources. These include medicine, toxicology, pharmacology, epidemiology, ecotoxicity and environmental pollution. Its terminology has developed in an unstructured manner with proliferation into multiple terms, some with overlapping, alternative or even ambiguous meanings. To facilitate international communication and comprehension this Glossary of Terms on Chemical Safety has been produced. It does not claim to be an exhaustive compilation nor a definitive list of approved terms, but because language and terminology are not static the compilers welcome suggestions for additions and improvements.

9. **Hazards Forum: http://hazardsforum.org.uk**
The Hazards Forum was established in 1989 by the four major engineering institutions, the Institutions of Civil, Electrical, Mechanical and Chemical Engineers. The Forum was set up because of concern about several major disasters, both technological and natural. The Forum believes that there is widespread public misunderstanding of the nature of hazards and the risks they pose. The public reaction to the risks from genetically modified foods and those from smoking are two examples of differing attitudes. These attitudes may result from media attention; suspicion of

assurances of safety bodies with a vested interest; or just ignorance. Whatever the reason the Forum believes there is a need to fill an educated but unbiased role and this the Forum aspires to do.

10. **Health and Safety Executive (HSE): www.hse. gov.uk**

    The HSE's main website containing general information on their areas of work and the latest news.

11. **Health Development Agency (HDA): www.hda-online.org.uk**

    The HDA is a special health authority that aims to improve the health of people in the UK – in particular to reduce inequalities in health between those who are well off and those on low incomes or reliant on state benefits.

12. **Chartered Institute of Waste Management (CIWM): http://www.ciwm.co.uk**

    The IWM is the professional body which represents over 4,000 waste management professionals – predominantly in the UK but also overseas. The IWM sets the professional standards for individuals working in the waste management industry and has various grades of membership.

13. **Institution of Occupational Safety and Health (IOSH): http://www.iosh.org.uk**

    IOSH is the chartered organisation representing occupational safety and health practitioners predominately in the UK but elsewhere also. IOSH has a valuable website with technical services for members, training courses and many local branch meetings for practitioners.

14. **IOSH International Links: http://www.iosh. org.uk**

    IOSH has a comprehensive International site. International development is an integral part of IOSH's vision: 'a world of work which is safe, healthy and sustainable'. We need to improve safety and health at work not only within the UK, but also in developing and newly industrialised countries. The site gives access to linked sites like the ILO, WHO and local OSH organisations in many countries on all continents.

15. **International Labour Organisation (ILO): http://www.ilo.org/global/lang--en/index.htm**

    The ILO is the global body responsible for drawing up and overseeing international labour standards. Working with its member states, the ILO seeks to ensure that labour standards are respected in practice as well as principle. The International Labour Organisation's site contains the Conventions and Recommendations which are continually being developed. Contains the full text of these documents.

    ILO SafeWork bookshelf: http://www.ilo.org/safework_bookshelf/english?d&nd=170000102&nh=0

The SafeWork Bookshelf is a collection of key OSH documents produced, in whole or in part, by the ILO. It was compiled by CIS, the information arm of the SafeWork Programme of the ILO.

16. **ISO (International Organisation for Standardisation): http://www.iso.org/iso/home.htm**

    ISO is the world's largest developer and publisher of International Standards.

    ISO is a network of the national standards institutes of 163 countries, one member per country, with a Central Secretariat in Geneva, Switzerland, that coordinates the system.

17. **Legislation – UK: http://www.legislation.gov.uk**

    This website is managed by The National Archives on behalf of HM Government. Publishing all UK legislation is a core part of the remit of Her Majesty's Stationery Office (HMSO), part of The National Archives, and the Office of the Queen's Printer for Scotland. These are freely available in full text. All acts from 1987 and Statutory Instruments from 1988 are available on the website. There are separate pages for Scottish, Northern Ireland and Welsh legislation.

18. **Legislation – Other countries: www.gksoft. com/govt/en**

    Governments around the world on the Web. Lists government websites from over 220 countries. Classified under Worldwide Governments, European Governments, African Governments, American Governments, Asian Governments, Oceanian Governments. There are some multi-government institutions also listed.

19. **National Examination Board in Occupational Safety and Health (NEBOSH): http://www. nebosh.org.uk**

    NEBOSH is the main accredited awarding body for Level 6 and Level 3 health and safety qualifications. Their awards are run by over 400 training organisations.

20. **The National Safety Council USA: http://www. nsc.org/Pages/Home.aspx**

    The National Safety Council saves lives by preventing injuries and deaths at work, in homes and communities, and on the roads, through leadership, research, education and advocacy.

21. **OSH World: www.oshworld.com**

    List of organisations, governments, associations, trade and technical associations, and trade unions listed under country and then alphabetically by name. Major organisations listed in each country including the ILO Health and Safety Information Centres.

22. **OSH Links** (formerly known as the Internet Directory): **http://www.ccohs.ca/oshlinks/about. html**

    An extensive resource that can be used to find primarily Canadian websites that

18

cover occupational health and safety subjects.

23. **Royal Society for the Prevention of Accidents (RoSPA): http://www.rospa.com/**
The Royal Society for the Prevention of Accidents is a registered charity established more than 90 years ago that aims to campaign for change, influence opinion, contribute to debate, educate and inform – for the good of all. By providing information, advice, resources and training, RoSPA is actively involved in the promotion of safety and the prevention of accidents in all areas of life – at work, in the home, on the roads, in schools, at leisure and on (or near) water. RoSPA's mission is to save lives and reduce injuries.

24. **Safety Health and Environment Intra Industry Benchmarking Association (SHEiiBA): www. sheiiba.com**
The Safety Health and Environmental Intra Industry Benchmarking Association offers web-based benchmarking tools designed for intra company knowledge sharing and performance.

25. **Specifications and Standards**
American National Standards Institute: www. ansi.org
British Standards Institution: www.bsi-global. com
European Committee for Standardisation: www.cenorm.be

ILI – Informe London Information: www.ili.co.uk International Standards Organisation: www.iso.ch

26. **WorkSafe Western Australia: http://www. commerce.wa.gov.au/WorkSafe/**
WorkSafe is a division of the Department of Commerce, the Western Australian State Government agency responsible for the administration of the Occupational Safety and Health Act 1984. The principal objective of the Act is to promote and secure the safety and health of people in the workplace.

27. **World Health Organisation (WHO): http://www. who.int/about/en**
WHO is the directing and coordinating authority for health within the United Nations system. It is responsible for providing leadership on global health matters, shaping the health research agenda, setting norms and standards, articulating evidence-based policy options, providing technical support to countries and monitoring and assessing health trends.

18.4 Health and safety forms

All the forms illustrated are available on the companion website in Microsoft Word format for readers to download and use.

18.4.1 Management

M1 GENERAL HEALTH & SAFETY RISK ASSESSMENT EXAMPLE 1	No.	

Firm/Company		Department	
Contact name		Nature of Business	
Telephone no.			

Principal Hazards

1.
2.
3.
4.
5.

Persons at Risk

Main Legal Requirements

1.
2.
3.
4.
5.

Significant Risks	**Consequences**
1.	
2.	
3.	
4.	
5.	

Existing Control Measures

1.
2.
3.
4.
5.

Residual risk, i.e. after controls are in place

Severity		Likelihood		Residual Risk = SxL	

Information - relevant HSE and trade publications

Comments from Line Manager	Comments from Risk Assessor

Signed		Date		Signed		Date	
Review Date							

M1 – Risk assessment 1

18

M2 RISK ASSESSMENT REPORT FORM EXAMPLE 2

| Firm/Company | | Date of Assessment | |
| Assessor | | Date of review | |

Hazards	Persons Affected	Risks	Initial Risk Level	Existing Controls	Additional Controls	Action by whom?	Action by when?	Done

M2 – Risk assessment 2

576

M3 CONTRACTORS' RISK ASSESSMENT EXAMPLE FOR CONFINED SPACES

INITIAL RISK ASSESSMENT DATE

	Significant Hazards	Persons at Risk	Low	Medium	High	Residual risk
1	Poisoning from toxic gases	Employees			√	Low
2	Asphyxiation – lack of oxygen	Employees			√	Low
3	Explosion	Employees			√	Low
4	Fire	Employees			√	Low
5	Excessive heat	Employees		√		Medium
6	Drowning	Employees			√	Low
7						
8						

ACTION ALREADY TAKEN TO REDUCE THE RISKS

Compliance with:

HSE Guidance L101 - Entry into Confined Spaces
Local Authority/client safety standards, e.g. on sewer entry.
Entry into Confined Spaces Regulations 1997
The Construction (Design and Management) Regulations 2007
Dangerous Substances and Explosive Atmospheres Regulations 2002

Planning:

Eliminate need for entry or use of hazardous materials by selection of alternative methods of work or materials.
Assessment of ventilation available and possible local exhaust ventilation requirements, potential presence of hazardous gases/atmosphere, process by-products, need for improved hygiene/welfare facility.
Prepare a method statement for the work.

Physical:

Documented entry system will apply, with a Permit to Work and method statement. Adequate ventilation will be present or arranged. Detection equipment will be present before entry to check on levels of oxygen and presence of toxic or explosive substances. The area will be tested before entry and continually during the presence of persons in the confined space. Breathing apparatus or airlines will be provided if local ventilation is not possible. Where no breathing apparatus is assessed as being required, emergency BA and rescue harnesses will be provided. Rescue equipment including lifting equipment, resuscitation facilities, safety lines and harnesses will be provided. A communication system with those in confined space will be established. Air will not be sweetened with pure oxygen. Precautions for safe use of any plant or heavier-than-air gases in the confined space must be established before entry. Necessary PPE and hygiene facilities will be provided for those entering sewers.

Managerial/Supervisory:

The management role is to decide on the nature of the confined space and to put a safe system into operation, including checking the above and preparing a method statement. Flood potential and isolations must be checked.

Training:

Full training required for all entering and managing confined spaces. Rescue surface party to be trained, including first aid and operation of testing equipment. All operatives must be certified as trained and supervisory staff trained to the same standard. All persons entering a confined space should be physically fit.

Site Manager's Comments

RISK RE-ASSESSMENT DATE	

M3 – Risk assessment – example confined space

18

M4 CONTRACTORS' RISK ASSESSMENT EXAMPLE FOR WORK ON FRAGILE ROOFS

INITIAL RISK ASSESSMENT

	Significant Hazards	Persons at Risk	Low	Medium	High	Residual risk
1	Falls of persons through materials	Employees and Public below			√	Medium
2	Access across fragile material	Employees		√		Low
3						
4						
5						
6						

ACTION ALREADY TAKEN TO REDUCE THE RISKS

Compliance with:

Lifting Operations & Lifting Equipment Regulations 1998 (LOLER)
Provision and Use of Work Equipment Regulations 1998 (PUWER)
Work at Height Regulations 2005
HSE Guidance HSG 150 – Health and Safety in Construction
HSE Guidance HSG 33 – Safety in Roof work
Construction (Design and Management) Regulations 2015.

Planning:

Fragile materials will be identified before work begins. In each case, a risk assessment will be made to provide a safe system of work taking account of the work to be done, access/egress requirements and protection of the area beneath the work area. A method statement will be prepared.

Physical:

Suitable means of access will be provided, such as roof ladder, crawling boards, scaffolding or mobile elevating work platform. Handrails will be fitted. Where access is possible alongside fragile materials such as roof lights, covers will be provided or the fragile material will be fenced off. Soft landing airbags or catch nets will be provided as appropriate. Barriers and signs will be provided so as to isolate the area below fragile materials while work is in progress. No person is permitted to walk on suspected fragile materials for any purpose, including access and surveying.

Managerial/Supervisory:

The role of management is to prepare a method statement prior to commencement of the work, and to arrange for provision of suitable access equipment and trained personnel as required by the safe system devised. Managers must check risk assessments and method statements supplied by subcontractors and others, including the self-employed, to ensure that the proposed work method is safe.

Training:

All operatives must be given specific instructions on the system of work to be used in each case. Selection may be required of operatives who have experience of the work and are physically fit. MEWP operators must be fully trained with a suitable certificate of competence.

Site Manager's Comments

RISK RE-ASSESSMENT DATE	

M4 – Risk assessment – example fragile roof

M5 WORKPLACE INSPECTION REPORT FORM

Company/Organisation	
Work area covered by this inspection	
Activity carried out in work place	
Person carrying out inspection	Date of inspection

Observations	Priority/risk (H,M,L)	Actions to be taken (if any)	Time Scale
List hazards, unsafe practices and good practices		List all immediate and longer term actions required	Immediate, 1 week etc.

M5 – Workplace inspection

18

M6 WORKPLACE INSPECTION CHECKLIST

PREMISES

1	**Work at Height**	Ladders & Step ladders	Right equipment for the job? Level base? Correct angle? Secured at top and bottom? Equipment in good condition?
		Working platforms/ temporary scaffolds	Regularly inspected? Suitable for the task? Properly erected? Maintained and inspected?
		Use of mobile elevating work platforms (MEWPs)	Suitable for task? Operators properly trained? Properly maintained?
2	**Access**	Access ways	Adequate for people, machinery and work in progress? Unobstructed? Properly marked? Stairs in good condition? Handrails provided?
3	**Working Environment**	Housekeeping	Tidy, clean, well organised?
		Flooring	Even and in good condition? Non-slippery? Crowded?
		Comfort /health	Too hot/cold? Ventilation? Humidity? Dusty? Lighting? Slip risk controlled?
		Cleaning	Hygienic conditions? Normal conversation possible?
		Noise	Noise assessment needed/not needed? Noise areas designated? Tasks require uncomfortable postures or actions?
		Ergonomics	Frequent repetitive actions accompanying muscular strain? Workstation assessments needed/not needed?
		Visual display units	Chairs adjustable/comfortable/maintained properly? Cables properly controlled? Lighting OK? No glare? Washing and toilet facilities satisfactory?
4	**Welfare**	Toilets/Washing	Kept clean with soap and towels/ Adequate changing facilities? Clean and adequate/Means of heating food?
		Eating facilities	For pregnant or nursing mothers?
		Rest room	Kept clean? Suitably placed and provisioned?
		First aid	Appointed person? Trained first aider? Correct signs and notices? Eye wash bottles as necessary? Portable equipment tested?
5	**Services**	Electrical equipment	Leads tidy, not damaged? Fixed installation inspected? Equipment serviced annually?
		Gas	Hot and cold water provided?
		Water	Drinking water provided?
6	**Fire Precautions**	Fire extinguishers	In place? Full? Correct type? Maintenance contract?
		Fire instructions	Posted up? Not defaced or damaged?
		Fire alarms	Fitted and tested regularly?
		Means of escape/Fire exits	Adequate for the numbers involved? Unobstructed? Easily Opened? Properly Signed?
7	**Work Equipment**	Lifting equipment	Thoroughly examined? Properly maintained? Slings etc. properly maintained? Operators properly trained? Written schemes for inspection?
		Pressure systems	Safe working pressure marked? Properly maintained?

M6 – Workplace inspection checklist

M6 WORKPLACE INSPECTION CHECKLIST (Cont.)

PREMISES

7	Work Equipment (cont.)	Sharps	Safety knives used? Knives/needles/glass properly used/disposed of?
		Vibration	Any vibration problems with hand held machinery or with whole body from vehicle seats etc?
		Tools and equipment	Right tool for the job? In good condition?
8	Manual & Mechanical Handling	Manual handling	Moving excessive weight? Assessments carried out? Using correct technique? Could it be eliminated or reduced?
		Mechanical handling	Forklifts and other trucks properly maintained? Drivers authorised and properly trained? Passengers only where specifically intended with suitable seat?
9	Vehicles	On site	Speeding limits? Following correct route? Properly serviced? Drivers authorised?
		Road risks	Suitable vehicles used? No use of mobile phones when driving? Properly serviced? Schedules managed properly?
10	Dangerous substances	Flammable liquids and gases	Stored properly? Used properly/minimum quantities in workplace? Sources of ignition? Correct signs used?
11	Hazardous substances	Chemicals	COSHH assessments OK? Exposures adequately controlled? Data sheet information available? Spillage procedure available? Properly stored and separated as necessary?
		Exhaust ventilation	Properly disposed of? Suitable and sufficient? Properly maintained? Inspected regularly?

PROCEDURES

12	Risk assessments		Carried out? General and fire? Suitable and sufficient?
13	Safe systems of Work		Provided as necessary? Kept up to date? Followed?
14	Permits to work		Used for high risk maintenance? Procedure OK? Properly followed?
15	Personal Protective Equipment		Correct type? Worn correctly? Good condition?
16	Contractors		Is their competence checked thoroughly? Are there control rules and procedures? Are they followed?
17	Notices, Signs and Posters	Employers' liability insurance	Notice displayed? In date?
		Health and Safety law poster	Displayed? Latest design?
		Safety Signs	Correct type of sign used? Signs in place and maintained?

PEOPLE

18	Health surveillance		Specific surveillance required by law? Stress or fatigue?
19	People's behaviour		Are behaviour audits carried out? Is behaviour considered in the safety programme?
20	Training and supervision		Suitable and sufficient? Induction training? Refresher training?
21	Appropriate authorised person		Is there a system for authorising people for certain special tasks like permits to work, dangerous machinery, entry into confined spaces?
22	Violence		Any violence likely in workplace? Is it controlled? Are there policies in place?
23	Especially at risk categories	Young persons	Employed? Special risk assessments?
		New or expectant mothers	Employed? Special risk assessments?

M6 – (*Continued*)

18

M7 JOB SAFETY ANALYSIS

Job		Date	
Department		Carried out by	

Description of job

Legal requirements and guidance

Task steps	Hazards	Likelihood	Severity	Risk L x S	Controls

Safe systems of work

Job Instruction

Training requirements

Review Date	

M7 – Job safety analysis

M8 ESSENTIAL ELEMENTS — PERMIT TO WORK

1. Permit Title		2. Permit No.	

3. Job Title

4. Plant Identification

5. Description of Work

6. Hazard Identification

7. Precautions Necessary	7. Signatures

8. Protective Equipment

9. Authorisation	
10. Acceptance	
11. Extension/Shift handover	
12. Hand back	
13. Cancellation	

NOTES REFER TO BOX NO ON FORM

2. Should have any reference to other relevant permits or isolation certificates.
5. Description of work to be done and its limitations.
6. Hazard identification including residual hazards and hazards introduced by the work.
7. Precautions necessary – person(s) who carry out precautions, e.g. isolations, should sign that precautions have been taken.
8. Protective equipment needed for the task.
9. Authorisation signature confirming that isolations have been made and precautions taken, except where these can only be taken during the work. Date and time duration of permit.
10. Acceptance signature confirming understanding of work to be done, hazards involved and precautions required. Also confirming permit information has been explained to all workers involved.
11. Extension/Shift handover signature confirming checks have been made that plant remains safe to be worked on, and new acceptance/workers made fully aware of hazards/precautions. New time expiry given.
12. Hand back signed by acceptor certifying work completed. Signed by issuer certifying work completed and plant ready for testing and re-commissioning.
13. Cancellation certifying work tested and plant satisfactorily recommissioned.

Source HSE

M8 – Permit to work

18

M9 WITNESS STATEMENT FORM

Type of event		Date of occurrence	
Location of event			
Surname			
Forename(s)			
Address			
Telephone No.			
Employer			
Position (if applicable)			

Please provide an account of the event, as you saw or heard it arise, that may be of use in the incident/near miss investigation

Signed		Date	

M9 – Witness statement

M10 ACCIDENT/INCIDENT REPORT

Injured Person		Date		Time	
Position		Place of Accident			
Department		Details of Injury			
Investigation carried out					
Position		Date		Estimated absence	

Brief details of Accident (Diagrams, photographs and any witness statements should be attached where necessary)

Immediate Causes	Underlying or Root causes

Conclusions (How can this kind of incident/accident be prevented?)

Action to be taken		Completion Date	

Please ensure that an accident investigation and report is completed and forwarded to Personnel within 48 hours of the accident occurring. Some Incidents e.g. involving death or major injuries or dangerous occurrences may have to be notified to the Enforcing Authority. (Mandatory in the UK)

Signature of Manager making Report		Date	
Copies to go to Personnel Manager	Health & Safety Manager	Payroll Controller	

INJURED PERSON

Surname		Forenames	

Home address

Age		Male		Female		Occupation	

By ticking this box I give consent to my employer and Best Practice Co (if not employer) to disclose my personal information and details of the incident which appear on this form to safety representatives and representatives of employee safety for them to carry out the health and safety functions given to them by law. (May not apply outside Europe)

Signature of injured person		Date	

Employee		Agency Temp		Contractor		Visitor		Youth Trainee	

M10 – Accident report 1

18

KIND OF ACCIDENT Indicate what kind of accident led to injury or condition (tick one box)

Contact with moving machinery or material being machined	1	Injured whilst handling or lifting	5	Drowning or Asphyxiation	9	Contact with electricity or an electrical discharge	13
Struck by moving, including flying, or falling object	2	Slip, trip or fall on same level	6	Exposure to or contact with harmful substance	10	Injured by an animal	14
Struck by moving vehicle	3	Fall from height. Indicate approx. distance of fall……mtrs	7	Exposure to fire	11	Violence	15
Struck against something fixed or stationary	4	Trapped by something collapsing or overturning	8	Exposure to an explosion	12	Other kind of accident	16

Detail any machinery, chemicals, tools etc. involved

Accident first reported to:

Position & Dept.		Name	
First Aid/medical attention by	First Aider Name		Dept.
	Doctor Name	Medical Centre	Hospital

WITNESSES

Name	Position and Department	Statement obtained (attach all statements)	
		Yes	No
		Yes	No
		Yes	No
		Yes	No
		Yes	No

For Office Use Only (if relevant)

Date reported to Enforcing Authority	by telephone		On form F2508		by email	
Date reported to Company Insurers		Were the recommendations effective?				

If No say what further action should be taken:

M10 – (*Continued*)

First Aid Treatment and Accident Record

BEST PRACTICE COMPANY *logo goes here*

Date: _____ FIRST AIDER AND/OR INJURED PERSON **No:** _____

Part A: FIRST AID TREATMENT/EXAMINATION RECORD

1. About the person who was seen/treated

Name _____ Address _____

Postcode _____

Occupation _____ Employer _____

2. About you, the person filling in this record

▼ If the incident *DID NOT* happen to you fill in your name, address and occupation

Name _____ Address _____

Postcode _____

Occupation _____

3. Treatment given (if any)

please provide details _____

_____

_____

_____

_____

Part B: INCIDENT AND INJURY DETAILS (if believed to be a work related incident complete the following)

4. Details of incident

▼ Say when it happened: Date ____ / ____ / ____ Time _____

▼ Say where it happened. State which room or area _____

_____

▼ Say how the incident happened – give the cause if you can _____

_____

_____

▼ If the person who had the incident suffered an injury, say what it was _____

_____

_____

▼ Please sign the record and date it

Signature _____ Date ____ / ____ / ____

5. Further details of injury (if any)

☐ Amputation	☐ Burn - Heat	☐ Electric shock	☐ Puncture
☐ Break/Fracture	☐ Concussion/headache	☐ Foreign body	☐ Scald
☐ Bruises (contusions)	☐ Crush	☐ Graze	☐ Splinters and blisters
☐ Burn - Chemical	☐ Cuts (laceration)	☐ Multiple Injury	☐ Sprain
			☐ Strain

M11 – First aid treatment and accident record 2, revised Jan 2015

18

right hand   **left hand**

right foot   **front view**   **left foot**

6. For the Person Involved in the Incident Only

By ticking this box I give consent to my employer and Best Practice CO (if not employer) to disclose my personal information and details of the incident which appear on this form to safety representatives and representatives of employee safety for them to carry out health and safety functions given to them by law. ☐

Signature _____ Date _____ / _____ / _____

FIRST AIDER PLEASE *IMMEDIATELY* INFORM RELEVANT TEAM LEADER/LINE MANAGER AFTER FIRST AID TREATMENT IS COMPLETED

name of person notified _____ date _____

7. BP Co Team Leader/Line manager Information - Please complete immediately and inform your line manager

Was a risk assessment carried out? YES/NO   If yes, who did it? _____

If no, why not? _____

Was correct protective clothing worn?   YES/NO

Why did the incident occur? What was its cause? _____

What action has been taken to prevent a recurrence? _____

Type of incident:
* must also notify Factory Manager/Health & Safety Manager immediately

First Aid Injury ☐	*HSE Reportable - 7 days ☐	Other (give details) ☐
*Hospital Visit ☐	*HSE Specified Serious Injury ☐	
*Lost time injury ☐		

Level of Investigation required (tick box)   NONE ☐   MINIMAL ☐   LOW ☐   MEDIUM ☐   HIGH ☐

Above must be completed within 24 hours of incident

8. BP Co Management Review — by Line Manager & Health and Safety Manager or Deputy

Comments _____

Name _____ Signed _____ Date _____

Type of Incident is correct ☐   Level of Investigation is correct ☐

Name _____ Signed _____ Date _____

9. H&S Management System Entry - ONLY FOLLOWING COMPLETION OF 1-8 ABOVE

Entry by _____ Number _____

Action Allocated to _____ Date _____ / _____ / _____

M11 – (*Continued*)

18.4.2 Safety

S1 MACHINERY RISK ASSESSMENT							
MACHINE NO	Model		Manufacturer		Risk Assessment No		
	Model no.		Other		Department ID		
HAZARDS	Hazard	Yes	No	Hazard		Yes	No
	Trapping/drawing in			Stored energy/vacuum/kinetic			
	Impact/crushing/mobility			Electrical/radiation			
	Cutting/severing/sharp edges			Thermal/noise/vibration			
	Entanglement/ejection			Material/sunstance			
	Rough/slippery surface/falling			Environment/ergonomic			
WHO MIGHT BE HARMED?	Operatives		Cleaners	Maintenance		Visitors	Others

GUARDING	Fixed Guards		Interlocked Movable Guards		Adjustable Guards		Fixed Distance Guard	
	Adequate Enclosure of drives and motors		Fitted to machine		Fitted to machine		Fitted to machine	
			Closed to run		Readily Adjustable		Securely Fixed	
	Securely fixed in position		Design OK with positive switches and robust		Prevent Ejection		Prevents access to danger zone	
	Tools required to remove		Safe to open		Maintained OK		Tool to remove	
			Maintained OK				Maintained OK	

CONTROLS/ WARNINGS/ INSTRUCTIONS/ TRAINING	Controls				Instruction Training		Area	
	Clearly identified		Shrouded start		Warning signs		Lighting OK	
	Function Clear		Warning device		Signs clear		Stability OK	
	Easy to use		Isolator nearby		Safety sheet OK		Ventilation OK	
	Emergency stop		Isolator lockable		SOP available		LEV needed	
	Machine stops OK		Permit required		Training OK		Access operators	
	Safe at stop		Seating needed				Access maintenance	

ACTION REQUIRED

Severity		Frequency		Residual Risk (SxE)	

S1 – Risk assessment machinery, revised Jan 2015

18

ASSOCIATED PERMITS (specify type, if none state NA)		ON THE JOB COPY	BEST PRACTICE COMPANY logo goes here	GENERAL WORK PERMIT	Number GWP 00000000
Type	Prefix & No.				
Type	Prefix & No.				
Type	Prefix & No.				

SECTION 1 ISSUE   **PLEASE PRINT DETAILS**

Permit Receiver/Competent Person in charge of work	Location of work/equipment to be worked on
Name of person detailed to carry out work	Details of work to be done

Risk Assessment attached? Yes ☐ No ☐ Reference	Safety method Statement or Safe System of Work attached? Yes ☐ No ☐ NA ☐ Reference

SECTION 2 ISOLATION of electrical or mechanical plant, liquid or gas pipeline or other energy source – give details

Item	Lock location & reference	Isolated by: Print Name

Attached Isolation sheet? Yes ☐ No ☐ Reference	Location of Keys

SECTION 3 PREPARATIONS/PRECAUTIONS tick box

	Yes	No	NA		Yes	No	NA
1. Has every source of energy been isolated?				5. Are vessels/pipes free of toxic/flammable gas			
2. Have all isolations been tagged?				6. Any asbestos-containing materials present?			
3. Have all isolations been tested?				7. Are the risk assessment control measures in place?			
4a. Does the standard of pipeline isolation meet minimum Corporate standards?				8. Other precautions/control measures required (specify)			
4b. If not, specify additional precautions							

SECTION 4 PERSONAL PROTECTIVE EQUIPMENT tick box

Site Standard	Other specify
Safety Goggles	
Hearing Protection	
Respiratory Protection Specify Type	

SECTION 5 TIME LIMITS

From   hrs   On   /   /   To   hrs   On   /   /
Max 24 hours   See overleaf for details of time extensions allowed

SECTION 6 AUTHORISATION Permit Issuer/Authorised Person	**SECTION 7 RECEIPT** Permit Receiver/Competent Person in charge
I certify that it is safe to work in the area/on the equipment detailed in Section 1 above and that all safety measures detailed in Section 2-4 have been carried out/complied with. ALL OTHER PARTS ARE DANGEROUS Print name Date Time	I have read, understood and accept the requirements of this permit. I will ensure that everyone working under my supervision will strictly follow the requirements of this permit. I have checked the isolations. Print name Date Time

SECTION 8 SUSPENSION OF GENERAL WORK PERMIT this is an exception and must be signed on the front of 'on the job'

I certify that the task for which this Permit was issued has now been suspended. We have agreed and implemented a procedure which complies with the criteria noted in the checklist in Section 13 overleaf

Signed Permit receiver/Competent person in charge	Signed Authorised person
Date Time	Date Time

SECTION 9 CLEARANCE Permit Receiver/Competent person in charge	**SECTION 10 CANCELLATION** Permit Issuer/Authorised Person
I certify that the work for which the permit was issued is now COMPLETED and that all persons at risk have been WITHDRAWN and WARNED that it is NO LONGER SAFE to work on the plant specified on this permit and that GEAR TOOLS and EQUIPMENT are all CLEAR Print name Date Time	This permit to work is hereby CANCELLED, all plant is restored to safe operating conditions, including the replacement of guards. Print name Date Time
Signed	Signed

S2(a) – General work permit

PERMIT TIME EXTENSION/TRANSFER

Time Extensions

General Work Permits are normally only valid for a maximum of 24 hours.
When a permit is extended beyond 24 hours a Permit Issuer/Authorised Person must check ALL the conditions in the Permit. When satisfied there have been no changes and it is safe to continue the work specified in Section 1, Time Extension/Transfer Section 11 must be completed. The permit may be continued beyond a shift and be transferred to another Permit Receiver/Competent Person, who must check that all the precautions are in place and that they understand all of the requirements of the permit. Section 11 must be signed by the new Permit Receiver.

General Note

Except for Section 8 Suspension, the front sheet must not be altered or changed in any way. If there is a need to change the permit conditions (except Section 8 overleaf) then a new permit must be used. Extension sheet Mxxx must be used when Section 11 below is full.

SECTION 11 TIME EXTENSION/TRANSFER OPERATIONAL PERIODS

Permit is extended from	Print Name	Date as FRONT of Permit
_____ hrs to _____ hrs	Permit receiver	Signed
_____ hrs to _____ hrs	Permit receiver	Signed
Permit is extended from		**Date**
_____ hrs to _____ hrs	Authorised Person	Signed
_____ hrs to _____ hrs	Permit receiver	Signed
_____ hrs to _____ hrs	Permit receiver	Signed
_____ hrs to _____ hrs	Permit receiver	Signed
Permit is extended from		**Date**
_____ hrs to _____ hrs	Authorised Person	Signed
_____ hrs to _____ hrs	Permit receiver	Signed
_____ hrs to _____ hrs	Permit receiver	Signed
_____ hrs to _____ hrs	Permit receiver	Signed
Permit is extended from		**Date**
_____ hrs to _____ hrs	Authorised Person	Signed
_____ hrs to _____ hrs	Permit receiver	Signed
_____ hrs to _____ hrs	Permit receiver	Signed
_____ hrs to _____ hrs	Permit receiver	Signed
Permit is extended from		**Date**
_____ hrs to _____ hrs	Authorised Person	Signed
_____ hrs to _____ hrs	Permit receiver	Signed
_____ hrs to _____ hrs	Permit receiver	Signed
_____ hrs to _____ hrs	Permit receiver	Signed
Permit is extended from		**Date**
_____ hrs to _____ hrs	Authorised Person	Signed
_____ hrs to _____ hrs	Permit receiver	Signed
_____ hrs to _____ hrs	Permit receiver	Signed
_____ hrs to _____ hrs	Permit receiver	Signed

SECTION 12 ADDITIONAL PEOPLE WORKING ON PLANT

SECTION 13 SUSPENSION OF GENERAL WORK PERMIT check list

Operation of machinery or plant with safeguards removed must be considered a measure of last resort and may only be carried out where it is not practicable to perform tasks, such as examinations or adjustments, in any other manner with the safeguards in place. Should it be necessary to re-energise part of the machinery to which this permit applies before the task is complete a Permit Issuer/Authorised Person must complete the following checklist & sign Section 8.

CHECK LIST	tick box	Yes	NA	CHECK LIST	tick box	Yes	NA
Is number of people involved as small as possible?				Is area free of obstructions & slipping hazards?			
Have precautions been taken to keep people away?				Are all other parts suitably protected?			
Has loose clothing been avoided?				Are suitable precautions against hazards maintained?			
Is safe access with foothold & handhold provided?				Are people competent, briefed and supervised as necessary?			
Is there a means of stopping equipment?							
Check list completed by: print name				Signed		Date	

S2(b) – General work permit: back of form

18

18.4.3 Health

BEST PRACTICE COMPANY	Site	Best Site	Work Area	Laboratory	COSHH Assessment No.	001

Substance / Material:	FORMALDEHYDE	No of Personnel involved in Activity:	One Person
Task / Activity :	Laboratory Work	Frequency of Operation:	Once per day
Hazardous Content:	Formaldehyde* (40%), Methanol (10%).	Duration of Exposure:	Up to ½ Hour Daily
		Quantity / Amount Exposed To:	Use of small quantities under laboratory conditions.
Health Hazard Groups: A / B / C / D / E	Group "C"	Specific Work Area:	Inside well ventilated Area

Procedures Reference: Corporate Minimum Standard, H&SMan-002 Section 2 H&SMan-02 Section 3	Routes of Entry to Body:	INHALATION	✓	SKIN	✓
		INGESTION	✓	EYES	✓

Hazard Classification:	Residual Risk: **HIGH HAZARD** **LIQUID**	(WEL) Workplace Exposure Limits: **Formaldehyde *2.5mg/m3TWA, 2.5mg/m3STEL WEL [15mins]**

Monitoring Arrangements

If using engineering controls, ensure maintenance (Regulation 9.)

RECORDS   MAINTENANCE

Health Surveillance Requirements

Material contains MEL and skin sensitising assigned substance, consider monitoring and skin checks (Regulation 10&11)

BLOOD SAMPLING   LUNG FUNCTION   SKIN CHECK

Hazards from Activity / Health Risks

AVOID CONTACT WITH SKIN & EYES
CAUSES BURNS
TOXIC IF SWALLOWED, BY INHALATION AND IN CONTACT WITH SKIN
LIMITED EVIDENCE OF A CARCINOGENIC EFFECT
MAY CAUSE SENSITIZATION BY SKIN CONTACT
RISK OF SERIOUS DAMAGE TO EYES
Do not breathe in vapour
When using do not eat, drink or smoke

First Aid

INHALATION - REMOVE TO FRESH AIR AND REST
IN THE EVENT OF **SIGNIFICANT EXPOSURE** CALL FOR MEDICAL ASSISTANCE IMMEDIATELY
INGESTION - DO NOT INDUCE VOMITING
INGESTION - GIVE PLENTY OF WATER IN SIPS
INGESTION - GET IMMEDIATE MEDICAL ATTENTION
EYE - IRRIGATE WITH WATER FOR AT LEAST 15 MINUTES
EYE CONTACT - GET IMMEDIATE MEDICAL ATTENTION
SKIN - REMOVE CLOTHING & WASH CONTAMINATED AREA WITH PLENTY OF WATER
SKIN CONTACT - GET IMMEDIATE MEDICAL ATTENTION

Control Measures / Storage

Personnel must wear lab coat or protective overalls when working with substance
Chemical Resistant Gloves must be worn [Nitrile]
Eye Protection to BS EN 166 Grade 3
Wear rubber boots
General or local exhaust ventilation should be used to control airborne levels
Store away from oxidising agents, stored at a temperature sufficiently high to avoid the formation of polymers

PPE / Workplace Environment Control Measure Symbols

Chemical Overalls   Glasses   Chemical   Dust Mask

Spillage and Disposal

VENTILATE AREA
DO NOT ALLOW SPILLAGE TO ENTER MAINS DRAIN/SEWERS/WATER COURSES
MARK THE AREA AND WARN ALL PERSONNEL
WEAR NITRILE GLOVES
WEAR EYE GOGGLES (GRADE 3)
WEAR RPE WITH ORGANIC FILTER (A)
WEAR PROTECTIVE OVERALLS & CHEMICAL/SAFETY FOOTWEAR
ABSORB IN SAND OR INERT ABSORBENT MATERIAL
COLLECT INTO A CONTAINER, CLOSE LID
DISPOSE OF FOLLOWING SUITABLE SITE PROCEDURE OR SEEK L.A. GUIDANCE

Fire and Emergency

ISOLATED SMALL SCALE FIRE:
DRY POWDER – WATER SPRAY OR MIST – FOAM – CARBON DIOXIDE (CO2)
LARGE FIRE: EVACUATE AREA, CALL FIRE BRIGADE OR FOLLOW SITE PROCEDURE
WEAR SELF-CONTAINED BREATHING APPARATUS AND PROTECTIVE CLOTHING
TOXIC FUMES ARE PRODUCED WHEN SUBSTANCE IS INVOLVED IN A FIRE
KEEP CONTAINERS COOL WITH WATER SPRAY

Supplier: **Chemical distribution Ltd**
Address: **Bridge Way, Anytown**

Post Code: **AN19 1SW**
Telephone: **+44 (0) 1206 543210**
Emergency Tel: **01234 567891**
MSDS Reference:
Link to Manufactures Safety Data Sheet

Assessor Name	A.Another
Date of Assessment	10.01.11
Approved By	A.Another
Date Re-Assessment Due	10.01.13
Assessment Printed Date	10.01.12
Assessment Revision No	01

H1a – COSHH assessment example

BEST PRACTICE COMPANY logo goes here	Site		Work Area		COSHH Assessment No.	

| **Substance / Material:**
Task / Activity :
Hazardous Content:

Health Hazard Groups:
A / B / C / D / E | | **No of Personnel involved in Activity:**
Frequency of Operation:
Duration of Exposure:
Quantity / Amount Exposed To:
Specific Work Area: | |

Procedures Reference:	Routes of Entry to Body:	INHALATION		SKIN	
		INGESTION		EYES	

Hazard Classification:	Residual Risk:	(WEL) Workplace Exposure Limits:

Monitoring Arrangements:

Hazards from Activity / Health Risks

Control Measures / Storage

First Aid

PPE / Workplace Environment Control Measure Symbols

Spillage and Disposal

Supplier:
Address:

Post Code:
Telephone:
Emergency Tel:
MSDS Reference:
Link to Manufactures Safety Data Sheet

Fire and Emergency

Assessor Name	
Date of Assessment	
Approved By	
Date Re-Assessment Due	
Assessment Printed Date	
Assessment Revision No	

H1 – COSHH assessment 1 record form example

18

COSHH 1 - DETAILS OF SUBSTANCES USED OR STORED

Name of Manager:

Name of Department Area:

SUBSTANCE DETAILS

1. Information from the label

Trade name:	Manufacturer's name:

Name of any chemical constituents listed:

Hazard marking...	corrosive		irritant		harmful		toxic		very toxic	

Risk phrases or Hazard Statements noted on label (e.g. Harmful in contact with skin)

Safety phrases or Precautionary Statements noted on label (e.g. Avoid contact with skin)

PRECAUTIONS noted on label (e.g. Use in well ventilated area)

2. Have you got Safety Data Sheet or MSDS for this product	YES		NO	

DETAILS OF USE

3. What is it used for?

4. By whom?

5. How often?

6. Where?

7. What CONTROL measures (precautions) are used?
(e.g. local ventilation, goggles, respirator, protective gloves etc.)

8. Is it ABSOLUTELY ESSENTIAL to keep/use this substance?	YES		NO	
9. Can it be DISPOSED OF NOW?	YES		NO	

H2 – COSHH assessment 2 record form example

COSHH 2 - ASSESSMENT OF A SUBSTANCE

1. Name of Substance:

2. The process or description of job where the substance is used:

3. Location of the process where the substance is used:

4. Health & Safety information on substance:

a) Hazards to health:

b) Precautions required:

5. Number of persons exposed:

6. Frequency and duration of exposure:

7. Control measures that are in use:

8. The assessment, an evaluation of the risks to health:

9. Details of steps to be taken to reduce the exposure:

10. Action to be taken (name):	**Date:**

11. Date of next assessment/review:

12. Name and position of person making this statement:

13. Date of assessment:

H2 – (*Continued*)

18

EXAMPLE OF A WORKSTATION SELF ASSESSMENT CHECKLIST

Name:	Department:	Date:

The completion of this checklist will enable you to carry out a self-assessment of your own workstation. Your views are essential in order to enable us to achieve our objective of ensuring your comfort and safety at work. Please circle the answer that best describes your opinion, for each of the questions listed.

The form should be returned to……………………………………………………………as soon as it has been completed.

ENVIRONMENT

1. Lighting

Describe the lighting at your workstation.	About right		Too bright		Too dark
Do you get distracting reflections on your screen?	Never		Sometimes		Constantly
What control do you have over local lighting?	Full control		Some control		No control

2. Temperature and Humidity

At your workstation, is it usually:	Comfortable		Too warm		Too cold
Is the air around your workstation:	Comfortable				Too dry

3. Noise

Are you distracted by noise from work equipment?	Never		Occasionally		Constantly

4. Space

Describe the amount of space around your workstation.	Adequate		Inadequate

FURNITURE

5. Chair

Can you adjust the height of the seat?		Yes		No
Can you adjust the height and angle of the backrest? Is the chair stable?		Yes		No
Is the chair stable?		Yes		No
Does it allow movement?		Yes		No
Is the chair in a good state of repair?		Yes		No
If your chair has arms, do they get in the way?		Yes		No

6. Desk

Is the desk surface large enough to allow you to place all your equipment where you want it?		Yes		No
Is the height of the desk suitable?	Yes		Too high	Too low
Does the desk have a matt surface (non reflectant)?		Yes		No

7. Footrest

If you cannot place your feet flat on the floor whilst keying, has a footrest been supplied?		Yes		No

H3 – Workstation self assessment checklist example, page 1

EXAMPLE OF A WORKSTATION SELF ASSESSMENT CHECKLIST (cont.)

8. Document Holder

If it would be of benefit to use a document holder, has one been supplied?	Yes		No	
If you have a document holder, is it adjustable to suit your needs?	Yes		No	

DISPLAY SCREEN EQUIPMENT

9. Display Screen

Can you easily adjust the brightness and the contrast between the characters on screen and the background	Yes		No	
Does the screen tilt and swivel freely?	Yes		No	
Is the screen image stable and free from flicker?	Yes		No	
Is the screen at a height, which is comfortable for you?	Yes		No	

10. Keyboard

Is the keyboard separate from the screen?	Yes		No	
Can you raise and lower the keyboard height?	Yes		No	
Can you easily see the symbols on the keys?	Yes		No	
Is there enough space to rest your hands in front of the keyboard?	Yes		No	

11. Software

Do you understand how to use the software?	Yes		No	

12. Training

Have you been trained in the use of your workstation?	Yes		No	
Have you been trained in the use of software?	Yes		No	
If you were to have a problem relating to display screen work, would you know the correct procedures to follow?	Yes		No	
Do you understand the arrangements for eye and eyesight tests?	Yes		No	

Any other comments

H3 – Workstation self assessment checklist example, page 2

18

EXAMPLE OF A NOISE ASSESSMENT RECORD FORM

Name of Department			Date of Survey			
Lower Exposure Action Level: 80 dBA daily or weekly			**Upper Action Level:** 85 dBA daily or weekly		**Peak Pressure:** 135 dB(C) / 137 dB(C)	
Workplace	Number of persons exposed	Noise level (leg) dB (A)	Daily exposure period	$L_{EP,d}$ dB (A)	Peak pressure dB (C)	Comments

General Comments:

Instrument used		Date of last calibration			
Signature		Position		Date	

H4 – Noise assessment record form example

MANUAL HANDLING OF LOADS: ASSESSMENT CHECKLIST

Section A - Preliminary	Department:	Date:

TASK DESCRIPTION: Factors beyond the limits of the guidelines?	IS AN ASSESSMENT NEEDED? (i.e. is there a potential risk of injury, and are the factors beyond the limits of the guidelines?) YES/NO

If 'YES 'continue. If 'NO' the assessment need go no further.

Tasks covered by this assessment (detailed description): Locations: People involved: Date of assessment:	Diagrams and other information:

SECTION B – See separate sheet for detailed analysis

SECTION C – Overall assessment of the risk of injury	Low	Med	High

SECTION D – Remedial action needed

Remedial steps that should be taken, in priority order:

a

b

c

d

e

f

g

h

Date by which action should be taken:
Date for reassessment:

Assessor's name:	Signature

H5 – Manual handling of loads assessment record form

18

MANUAL HANDLING RISK ASSESSMENT: EMPLOYEE CHECKLIST					
Task Description			**Employee's ID no.**		
RISK FACTORS					
A. Task characteristics	**Yes/No**	**Risk Level**			**Current Controls**
1. Loads held away from trunk?		H	M	L	
2. Twisting?					
3. Stooping?					
4. Reaching upwards?					
5. Extensive vertical movements?					
6. Long carrying distances?					
7. Strenuous pushing or pulling?					
8. Unpredictable movements of loads?					
9. Repetitive handling operations?					
10. Insufficient periods of rest/recovery?					
11. High work rate imposed?					
B. Load characteristics					
1. Heavy?					
2. Bulky?					
3. Difficult to grasp?					
4. Unstable/unpredictable?					
5. Harmful (sharp/hot)?					
C. Work environment characteristics					
1. Postural constraints?					
2. Floor suitability?					
3. Even surface?					
4. Thermal/humidity suitability?					
5. Lighting suitability?					
D. Individual characteristics					
1. Unusual capability required?					
2. Hazard to those with health problems?					
3. Hazard to pregnant workers?					
4. Special information/training required?					
Any further action needed?					
Details:					

H6 – Manual handling risk assessment employee checklist

18.4.4 Fire

F1 FIRE SAFETY MAINTENANCE CHECKLIST

A fire safety maintenance checklist can be used as a means of supporting your fire safety policy. This example list is not intended to be comprehensive and should not be used as a substitute for carrying out a fire risk assessment. You can modify the example where necessary, to fit your premises and may need to incorporate the recommendations of manufacturers and installers of the fire safety equipment/ systems that you may have installed in your premises. Any ticks in the grey boxes should result in further investigation and appropriate action as necessary. In larger and more complex premises you may need to seek the assistance of a competent person to carry out some of the checks.

Source: Department for Communities and Local Government Fire Safety Guides

Fire Safety Maintenance Checklist	Yes	No	N/A	Comments
DAILY CHECKS (not normally recorded)				
Escape routes				
Can all fire exits be opened immediately and easily?				
Are fire doors clear of obstructions?				
Are escape routes clear?				
Fire warning systems				
Is the indicator panel showing 'normal'?				
Are whistles, gongs or air horns in place?				
Escape lighting				
Are luminaires and exit signs in good condition and undamaged?				
Is emergency lighting and sign lighting working correctly?				
Fire-fighting equipment				
Are all fire extinguishers in place?				
Are fire extinguishers clearly visible?				
Are vehicles blocking fire hydrants or access to them?				
WEEKLY CHECKS				
Escape routes				
Do all emergency fastening devices to fire exits (push bars and pads, etc.) work correctly?				
Are external routes clear and safe?				
Fire-warning systems				
Does testing a manual call point send a signal to the indicator panel? (Disconnect the link to the receiving centre or tell them you are doing a test.)				
Did the alarm system work correctly when tested?				
Did staff and other people hear the fire alarm?				
Did any linked fire protection systems operate correctly? (e.g. magnetic door holder released, smoke curtains drop)				
Do all visual alarms and/or vibrating alarms and pagers (as applicable) work?				
Do voice alarm systems work correctly? Was the message understood?				
Escape lighting				
Are charging indicators (if fitted) visible?				
Fire-fighting equipment				
Is all equipment in good condition?				
Additional items from manufacturer's recommendations.				
MONTHLY CHECKS				
Escape routes				
Do all electronic release mechanisms on escape doors work correctly? Do they 'fail safe' in the open position?				

F1 – Fire safety maintenance checklist, page 1

18

Fire Safety Maintenance Checklist	Yes	No	N/A	Comments
Do all automatic opening doors on escape routes 'fail safe' in the open position?				
Are fire door seals and self-closing devices in good condition?				
Do all roller shutters provided for fire compartmentation work correctly?				
Are external escape stairs safe?				
Do all internal self-closing fire doors work correctly?				
Escape lighting				
Do all luminaires and exit signs function correctly when tested?				
Have all emergency generators been tested? (Normally run for one hour.)				
Fire-fighting equipment				
Is the pressure in 'stored pressure' fire extinguishers correct?				
Additional items from manufacturer's recommendations.				
THREE-MONTHLY CHECKS				
General				
Are any emergency water tanks/ponds at their normal capacity?				
Are vehicles blocking fire hydrants or access to them?				
Additional items from manufacturer's recommendations.				
SIX-MONTHLY CHECKS				
General				
Has any fire-fighting or emergency evacuation lift been tested by a competent person?				
Has any sprinkler system been tested by a competent person?				
Have the release and closing mechanisms of any fire-resisting compartment doors and shutters been tested by a competent person?				
Fire warning system				
Has the system been checked by a competent person?				
Escape lighting				
Do all luminaires operate on test for one third of their rated value?				
Additional items from manufacturer's recommendations.				
ANNUAL CHECKS				
Escape routes				
Do all self-closing fire doors fit correctly?				
Is escape route compartmentation in good repair?				
Escape lighting				
Do all luminaires operate on test for their full rated duration?				
Has the system been checked by a competent person?				
Fire-fighting equipment				
Has all fire-fighting equipment been checked by a competent person?				
MISCELLANOUS				
Has any dry/wet rising fire main been tested by a competent person?				
Has the smoke and heat ventilation system been tested by a competent person?				
Has external access for the fire service been checked for ongoing availability?				
Have any firefighters' switches been tested?				
Has the fire hydrant bypass flow valve control been tested by a competent person?				
Are any necessary fire engine direction signs in place?				

F1 – Fire safety maintenance checklist, page 2

F2 FIRE RISK ASSESSMENT RECORD – SIGNIFICANT FINDINGS

Risk Assessment for		
Assessment undertaken by		

Company	Date	
Address	Completed by	
	Sign	
Sheet no.	Floor/Area	Use

Step 1 – Identify Fire Hazards

Sources of ignition	Sources of fuel	Sources of oxygen

Step 2 – People at risk

Step 3 – Evaluate, remove, reduce and protect from risk

Evaluate the risk of the fire occurring

Evaluate the risk to people from a fire starting in the premises

Remove and reduce the hazards that may cause a fire

Remove and reduce the risks to people from a fire

Assessment review

Assessment review date	Completed by	Sign

Review outcome (where substantial changes have occurred a new record sheet should be used)

F2 – Fire risk assessment record form

18.4.5 Construction

C1 CONSTRUCTION INSPECTION REPORT

1. Name and address of person for whom inspection was carried out.

2. Site address.	3. Date and time of inspection.

4. Location and description of place of work or work equipment inspected.

5. Matters which give rise to any health and safety risks.

6. Can work be carried out safely?	Yes		No	

7. If not, name of person informed.

8. Details of any other action taken as a result of matters identified in 5 above.

9. Details of any further action considered necessary.

10. Name and position of person making the report.	
11. Date and time report handed over.	
12. Name and position of person receiving report.	

C1 – Construction inspection report form

C2 EXAMPLE RISK ASSESSMENT FOR CONTRACT BRICKLAYERS

Company name:
TVW Contract Bricklayers

Date of risk assessment:
6/3/2006

IMPORTANT REMINDER

This example risk assessment shows the kind of approach a small business might take. Use it as a guide to think through some of the hazards in your business and the steps you need to take to control the risks. Please note that it is not a generic risk assessment that you can just put your company name on and adopt wholesale without any thought. This would not satisfy the law – and would not be effective in protecting people.

Every business is different – you need to think through the hazards and controls required in your business for yourself. Source HSE

What are the hazards?	Who might be harmed and how?	What are you already doing?	What further action is necessary?	Action by who?	Action by when?	Done
Falling from height	Serious injury or even fatal injury could occur if a worker falls.	▪ Agree scaffolding requirements at contract stage, including appropriate load rating and provision of loading bays. ▪ Bricklayers' supervisor to check with the site manager that the correct scaffold is provided and inspected. ▪ Workers instructed not to interfere with or misuse scaffold – supervisor to keep an eye out for problems. ▪ Ladders in good condition, adequately secured (lashed) and placed on firm surface. ▪ Band stands with handrails to be used for work on internal walls. ▪ Workers trained to put up bandstands.	▪ Scaffold requirements agreed, including loading bays and appropriate load rating.	TB	20/3/06	20/3/06
			▪ Supervisor to speak regularly to site manager to arrange scaffold alterations and ensure that weekly inspections have been carried out.	LG	from 1/5/06	
Collapse of scaffold	Operatives on scaffold may incur crush injuries, or worse, if the scaffold collapses on top of them.	▪ Agree scaffolding requirements at contact stage, including appropriate loading bays. ▪ Bricklayers' supervisor to check with the site manager that the correct scaffold is provided and inspected.	▪ Supervisor to keep a check to make sure that scaffold is not overloaded with materials.	LG	from 1/5/06	
Falling objects hitting head or body, including feet	Serious head and other injuries to workers, others on site and members of the public.	▪ Brick guards kept in position on scaffold lifts. ▪ Waste materials removed from scaffolding and placed in skip. ▪ Safety helmets and protective footwear (with steel toecaps and mid-soles) supplied and worn at all times.	▪ Supervisor to monitor use of safety hats and protective footwear.	LG	from 1/5/06	
Manual handling	All workers could suffer from back injury and long term pain if regularly lifting/ carrying heavy or awkward objects.	▪ Bricks, mortar etc. to be transported and lifted to scaffolding using telehandler provided by principal contractor. ▪ Provision of lifting bay agreed with principal contractor. ▪ Bricks/blocks to be covered with tarpaulin when stored on site to prevent taking up water. ▪ Spot boards to be raised with blocks to easy working height. ▪ Trolley to be used for moving loads of bricks around the scaffold lift. ▪ Check at tender stage for any blocks or lintels over 20kg and make arrangements.	▪ Heaviest blocks are 15kg no special arrangements necessary. ▪ Concrete lintels are well over 20kg, to be positioned using telehandler (all are accessible). ▪ All workers to be instructed not to carry materials up by hand.	VP VP VP	20/3/06 from 1/5/06 from 1/5/06	
Workers struck or crushed by moving vehicles on site	Workers could suffer serious or even fatal injuries from vehicles and machines on site when reversing.	▪ Manager to agree safe route to work area with principal contractor based upon the construction phase health and safety plan. ▪ Induction to each site to be carried out for all workers on first day.	▪ Safe route agreed with principal contractor based. ▪ Supervisor to liaise with site manager to ensure safe route stays clear. ▪ Instruct staff that they must never drive vehicles and plant on this site. ▪ High-visibility vests to be provided. ▪ Supervisor to check vests are worn on all sites where the principal contractor requires them.	TB LG LG LG LG	20/3/06 from 1/5/06 from 1/5/06 from 1/5/06 from 1/5/06	20/3/06

C2 – Risk assessment bricklayer example

18

What are the hazards?	Who might be harmed and how?	What are you already doing?	What further action is necessary?	Action by who?	Action by when?	Done
Slips and trips	All workers may suffer sprains or fractures if they trip over waste including brick bands and pallet debris. Slips at height could result in a serious fall.	■ Good housekeeping maintained at all times. ■ Waste including brick bands and pallet debris disposed of in skip. ■ Safety footwear provided to all workers. ■ Safe route to workplace agreed with principal contractor based upon the construction phase health and safety plan.	■ Temporary storage locations to be agreed with site manager. ■ Supervisor to ensure that workers wear safety footwear whenever on site.	TB LG	20/2/06 from 1/5/06	20/3/06
Stepping on nails and sharp objects	All workers could suffer foot injuries.	■ Safety boots with steel toecaps and mid-soles provided to all workers. ■ Waste disposed of in skips.	■ Explain the need to wear safety boots and dispose of waste in skips, repeat annually. ■ Supervisor to check that safety boots are always worn and waste disposed of properly.	LG LG	1/5/06 from 1/5/06	1/5/06
Hazard to eyes, cutting bricks	Bricklayers could suffer eye injury through flying brick fragments.	■ Safety goggles (EN 166 B standard) worn when breaking bicks.	■ Use of goggles to be monitored by supervisor.	LG	from 1/5/06	
Hazardous substances, mortar	Direct skin contact with the mortar could also cause bricklayer contact dermatitis and burns.	■ Risk of dermatitis or cement burns and precautions explained to all workers. ■ Use cement or cement-containing products within the use-by date. ■ Direct skin contact to be avoided, CE marked PVC gloves used when handling mortar. ■ Good washing facilities on site, with hot and cold water, soap and basins large enough to wash forearms. ■ Principal contractor's first aid includes emergency eyewash.	■ Training on how to treat exposure to be given to all operatives. ■ Supervisor to be aware of anyone with early signs of dermatitis.	TB LG	17/4/06 from 1/5/06	26/4/06
Dust from cutting bricks	Dust exposure could cause silicosis.	■ Angle grinders replaced with block splitter, removing the risk of significant dust exposure. ■ The use of a grinder for chasing etc. is not needed on this job.				
Operating cement mixer	Workers could be crushed or cut if the mixer topples or they get caught in moving parts. Damage to electrics could result in a shock.	■ Cement mixer located on firm, level ground. ■ Mixer is fully guarded and guards in place during operation. ■ Mixer is 110 volt and PAT tested every three months.	■ Supervisor to check mixer daily for obvious damage.	LG	from 1/5/06	
Noise from use of equipment, e.g. angle grinder	Workers using grinders or working near people who may suffer hearing loss.	■ Angle grinders replaced with block splitter, removing high noise levels from our work. ■ Construction phase plan shows other trades using grinders etc. should not be working close enough to cause problems.	■ Supervisor to monitor and talk to site manager if noisy work does start close by.	LG	from 1/5/06	
Vibration from use of equipment such as angle grinder	Exposure to vibration can lead to the development of 'vibration white finger' (VWF)	■ Angle grinders replaced with block splitter. No significant vibration left.				
Fire/explosion	All operatives in the vicinity could suffer from smoke inhalation or burns.	■ Suitable fire extinguisher kept in site office and welfare block. ■ Good housekeeping monitored by supervisor.	■ Supervisor to brief all workers on first day on emergency arrangements agreed with principal contractor.	LG	1/5/06	1/5/06
Welfare/ first aid	Good facilities help prevent dermatitis etc.	■ Principal contractor will have facilities on site by the time bricklaying starts, including: o flushing toilet; o hot and cold running water, soap, towels and full-size washbasins; o heated canteen with kettle etc; o first-aid equipment; o principal contractor will arrange clearing and ensure the necessary electrical and heating safety checks are made; o and site agent is appointed person for first aid.				

C2 – (*Continued*)

C3 EXAMPLE RISK ASSESSMENT FOR WOODWORK

IMPORTANT REMINDER

Company name:
The Woodworking Company

Date of risk assessment:
28/09/07

This example risk assessment shows the kind of approach a small business might take. Use it as a guide to think through some of the hazards in your business and the steps you need to take to control the risks. Please note that it is not a generic risk assessment that you can just put your company name on and adopt wholesale without any thought. This would not satisfy the law – and would not be effective in protecting people.
Every business is different – you need to think through the hazards and controls required in your business for yourself. Source HSE

What are the hazards?	Who might be harmed and how?	What are you already doing?	What further action is necessary?	Action by who?	Action by when?	Done
Exposure to wood dust	Staff risk lung diseases, such as asthma, from inhaling wood dust. Hardwood dust can cause cancer, particularly of the nose.	■ Local exhaust ventilation (LEV) provided at machines and staff are trained in using it properly. Bricklayers' supervisor to check with the site manager that the correct scaffold is provided and inspected. ■ LEV maintained to keep it in good condition and working effectively. LEV inspected every 14 months by a competent person. ■ Wood dust cleared up using a suitable vacuum cleaner, fitted with an appropriate filter. ■ Suitable respiratory protective equipment (RPE) as well as LEV for very dusty jobs, and staff trained in how to use it. ■ Staff do health surveillance questionnaire before starting, then annually. ■ Any affected staff referred to a medical professional.	■ Remind staff of the risks of wood dust, and why these controls are necessary. ■ Remind staff never to dry sweep wood dust, which just spreads the dust around.	Manager Manager	7/10/07 7/10/07	1/10/07 1/10/07
Machinery	Staff risk serious and possibly fatal cut injuries following contact with moving parts of machinery, particularly saw blades.	■ All machines guarded according to manufacturers' instructions. ■ Guards inspected regularly and maintained as necessary to ensure their good condition. ■ Staff have sufficient space at machines to work safely. ■ Staff monitored by manager to ensure guards always used. ■ All staff trained in safe use of machines by a competent person. ■ All machines braked and fitted with necessary safety features, e.g. chip limited tooling etc.	■ Download information sheets on the safe use of the machines used in the workshop from HSE website and pin them up in mess room.	Manager	4/10/07	1/1/07
Manual handling	Staff may suffer musculoskeletal disorders, such as back pain, from handling heavy/bulky objects, e.g. timber boards and machinery parts. Also risk cuts when handling tooling, or splinters when handling pallets.	■ Staff trained in manual handling. ■ Workbenches and machine tables set at a comfortable height. ■ Strong, thick gloves provided for handling tooling and pallets. ■ Panel trolley and lifting hooks available for moving boards. ■ Systems of work in place for the safe and careful handling of assembled furniture.	■ Where possible, store tooling next to the machine to reduce carrying distance. ■ Remind staff to ask for a new set of gloves when old ones show wear and tear, and not to try to lift objects that appear too heavy.	Manager Manager	30/9/07 4/4/07	29/9/07 4/4/07

C3 – Risk assessment woodwork example

18

What are the hazards?	Who might be harmed and how?	What are you already doing?	What further action is necessary?	Action by who?	Action by when?	Done
Noise	Staff and others may suffer temporary or permanent hearing damage from exposure to noise from woodworking machinery.	■ Noise enclosures used where practicable, and maintained in good condition. ■ Low-noise tooling used where possible. ■ Planned maintenance programme for machinery and LEV systems. ■ Suitable hearing protectors provided for staff and staff trained how to use them. Check and maintain them according to advice given by supplier. ■ Staff trained in risks of noise exposure. ■ Staff trained in systems of work to reduce noise exposure (e.g. suitable feed rates for certain jobs, timber control etc.).	■ Consider if certain machines could be safely mounted on anti-vibration mountings. ■ Include noise emission in specification for new vertical spindle moulder, to be purchased next year.	Manager Manager	30/9/07 30/9/07	29/9/07 29/9/07
Vehicles	Staff may suffer serious, possibly fatal, injuries if struck by a vehicle such as a lift truck or a delivery lorry.	■ Fork-lift truck maintained and inspected as per lease contract. ■ Lift truck operated only by staff who have been trained to use it. ■ Pedestrian walkways marked. ■ Only authorised people allowed in yard for deliveries/dispatch.	■ Ensure drivers get out of their vehicle and stand in a safe area while it is being loaded/unloaded.	Manager and all other staff	31/9/07	15/9/07
Slips, trips and falls	Staff could suffer injuries such as bruising or fractures if they trip over objects, or slip, e.g. on spillages, and fall.	■ Generally good housekeeping – off-cuts cleared away promptly, dust cleared regularly etc. ■ Staff wear strong safety shoes that have a good grip. ■ Good lighting in all areas.	■ Remind staff to clear up spillages of wax or polish immediately, even very minor spillages.	Manager	31/9/07	15/9/07
Electrical	Staff could get electrical shocks or burns from using faulty electrical equipment, e.g. machinery, or a faulty installation. Electrical faults can also lead to fires.	■ Residual current device (RCD) built into main switchboard. ■ Staff trained to spot and report any defective plugs, discoloured sockets or damaged cable/equipment to manager. ■ No personal electrical appliances, e.g. toasters or fans, allowed.	■ Ask landlord when the next safety check of the electrical installation will be done. ■ Confirm with landlord the system for making safe any damage to building installation electrics, e.g. broken light switches or sockets.	Company secretary Company secretary	31/9/07 31/9/07	15/9/07 15/9/07
Work at height	Falls from any height can cause bruising and fractures.	■ Strong stepladder, in good condition, provided. ■ Only trained, authorised staff allowed to work at height.	■ Condition of stepladder to be checked periodically.	Manager	31/9/07	as required
Fire	If trapped, staff could suffer fatal injuries from smoke inhalation/burns.	■ Fire risk assessment done, see www.communities.gov.uk/fire and necessary action taken.	■ Ensure the actions identified as necessary by the fire risk assessment are completed.	Manager	from now	

C3 – (Continued)

Index

Note: Page numbers in **bold** are for figures, those in *italics* are for tables.

abbreviations xxiv–xxvi
abrasion hazards **291**
abrasive wheels 280, 293, 301–2
absolute duty 12, 108
acceleration/deceleration hazards 290
access control 190
Accident Compensation Corporation (ACC) (New Zealand) 9
accident triangles 152–4
    F.E. Bird **93**, 93, 152, **153**
    H.W. Heinrich's accidents/incidents ratios **62**, 62, 152
    limitations of 152–4
accidents
    accident classification 94
    accident rates 4–5, 61, 159
    common causes of 182
    commuting accidents 4, 5, 159, 160, 230
    compensation 9
    construction industry 182, 212
    costs of *see* costs of accidents
    definition 3–4, 92–3, 152
    electrical 312
    fatal 2, 4–5, *7*, 105, 160, 182, 193, 212, 230, 312, *477*
    female workers 6
    frequency rate 61, 159
    and hot and cold environments 187
    and human factors 62
    incidence rate 61, 159
    investigations *see* incident/accident investigation
    main (preventable) factors for 4
    maintenance-related 194
    manual handling 244
    new workers 75
    reduction in number of 31, 52
    reports *see* incident/accident recording and reporting
    vehicle 193, 230, 236
    welding operations 441–2
    women workers 6
    work equipment-related 274
    young workers 105
acid rain 412
ACT stage of health and safety management system 21, 22, *23*, 25, 27, 29
    see also audits; performance review
action for improvement, ILO-OSH 2001 24
active assessment 22
active/proactive monitoring 25, 26, 140, 143–4
    see also audits; safety inspections; safety sampling; safety surveys; safety tours
acute health risks/effects 94, 377
acute toxicity 376

adapting the work to the individual 108
Advisory Committee for Roof Work (UK) 200
Advisory Committee on Safety and Health at Work (ACSHW) (EU) 473–4
aerosols 350
Africa 246
    fatal occupational injuries and diseases *5*, *7*
    fire safety legislation and standards 334
    hazardous substances classification and labelling 378
    see also Egypt; Niger; Nigeria; South Africa
agency workers 45–6, 86
agricultural/horticultural hazards and safeguards 293–4, 302–4
air bags 201
air pollution 411–12, 470
air sampling
    direct reading instruments 387
    dust observation lamps 387
    grab or spot sampling 386–7
    passive sampling 387
    qualitative monitoring 387
    records 387
    sampling pumps and heads 387
    smoke tubes 387
    stain tube detectors 386
    vane anemometers 387
alarm systems
    cold store 185
    fire 254–6
    security 190
    voice alarms 355
alcohol abuse 187, 191, 238
allergic contact dermatitis 382, 408
allergies 376
alternating current (AC) 312, 314
American Conference of Governmental Industrial Hygienists (ACGIH) 387, 388
American National Standards Institute (ANSI) 275, 333
American Society of Mechanical Engineers (ASME) 275
ammonia 406–7
amps (A) 312
anthropometry 245
appointed persons 124
architects 218
arrangements for health and safety 34–5
asbestos 10, 210, 374, 375, 380, 400–6
    asbestos-containing materials (ACMs) 400, 403, 404
    deaths 400–1, 402
    diseases 400, 402–3
    health risks 400–1
    ILO requirements 401–2, 403, 405, 454–5

main forms of 400
management and control 402–6
  accidental exposure 406
  asbestos risk register 404
  assessment survey 404
  awareness training 400, 403, 405–6
  disposal of waste 406
  exposure limits 403
  health surveillance 405, 455
  identification of presence 404
  management survey 403
  refurbishment or demolition survey 403–4
  removal work 403, 405
  respiratory protective equipment 403
safe systems of work 133–5
asbestosis 400
Asia 4, 5, *7*, 14, 135, 246
asphyxiation 380
asthma 6, 374, 379, 380, 388, 406
asthmagens 399
ATEX Directive 353–4
atmosphere tests, confined spaces 118
attitude, individual 65
auditors 175
audits 22, 27, 29, 49, 50, 143, 172–7
  actions taken after 176
  definition, scope and purpose 172–3
  external v. internal 175–6
  ILO recommendations for 24, 173–4
  pre-audit preparations 174–5
  proprietary v. inhouse systems 175
  scoring systems 164–5
Australia 9, 15
  director duties and responsibilities 47
  electrical wiring 320
  emergency call numbers 136
  fire extinguishers *358*
  fire safety legislation and standards 334
  occupational accidents and diseases *477*
  occupational exposure limits 388
  OSH legal framework 478–80
  Safe Work Australia 479
  Work Health and Safety Act (2011) 479–80
Australian Safety and Compensation Council, Hazardous Substances Information System (HSIS) 388

back pain 6, 182, 374, 429–30
bacteria 375
balance trucks 254
Bangladesh 136
  occupational accidents and diseases *477*
  OSH legal framework 480, *481*
banksmen 232, 235, 236, 260
barrier creams 397
base plates (scaffolds) 205
battery (cordless)-operated hand tools 323
bench-mounted circular saws 294, **306**, 306
bench-top grinding machines 280, 293, 301–2, **302**
benchmarking 26, 33, 35, 50
Bhopal chemical disaster 3
biological agents, forms of
  bacteria 375
  fungi 375
  moulds 375
  prions 375
  viruses 375

blood-borne viruses 410–11
Boards of Directors
  health and safety training 76
  leadership of health and safety 82–3
  and performance review 177–8
  responsibilities 46, 47, 63
boredom 65
bracings (scaffolds) 205
Brazil 15, 137
  hazardous substances classification and labelling 378
  occupational accidents and diseases *477*, 481
  OSH legal framework 481–2
bronchitis 380
brush cutters 294, **303**, 303
BS 5588: Part 8 Fire precautions in the design, construction and use of buildings 366
BS 7671: Requirements for Electrical Installation 320–1
BS EN 12453:2001 Electrically powered gates 197
buildings, principles of fire protection in
  fire compartmentation 352–3
  fire loading 351
  fire resistance of structural elements 351–2
  insulating materials 352
  surface spread of fire 351
bullying 189
buried power lines 324, 326
burning 338
burning equipment 59
burns
  electrical contact 314, 316
  fire 340
buying problems 53–4

cables (electrical) 317, 319
  buried power lines 324, 326
  overhead power lines 324, **325**
caissons 463
Canada 15
  electrical wiring 320
  emergency call number 137
  hazardous substances
    classification and labelling 378
    occupational exposure limits 388
  no-fault compensation schemes 9
  occupational accidents and diseases *477*
  OSH legal framework 483–5
Canadian Standards Association (CSA) 275
cancers 3, 6, 374, 380, 400, 406, 436, 459
carbon dioxide 407
carbon monoxide 407
carcinogenic substances 376, 388–9, 391, 398–9, 459
cardiopulmonary resuscitation (CPR) 122
cardiovascular system **381**, 381
carpal tunnel syndrome 6, 245, 429, **430**
CCTV (closed circuit television) 188, 189, 190
CE marking 52, 53, 276, 277
  limitations of 277
  requirements 277
cement dust/wet cement 216, 374, 408
cement/concrete mixers 216, 294–5, **305**, 305–6
CENELEC (Comité Européen de Normalisation Électrotechnique) 275
Central America 137
  fire protection legislation 334
certification 530
chainsaws 294, 303–4, **304**

chair design 64
changing facilities 183
CHECK stage of health and safety management system 21, 22, *23*, 25, 26
    see also incident/accident investigation; measuring health and safety performance; monitoring; safety inspections
checkout conveyor systems 294, **305**, 305
chemical agents, forms of 374–5
    dusts 374
    fibres 374–5
    fumes 375
    gases 375
    liquids 375
    mists 375
    vapours 375
chemical and biological health hazards and risk control *see* hazardous substances
Chemicals (Hazard Information and Packaging for Supply) 2009 Regulations (CHIP 4) (UK) 376, 377
Chief Executive Officers (CEOs) 30, 48, 82, 84
child labour 10, 13
children, work-related deaths *3*
China 15
    emergency numbers 135
    fire safety legislation and standards 334
    hazardous substances classification and labelling 378
    legal framework 485–8
    occupational accidents and diseases *5*, *7*, *477*
CHIP 4 *see* Chemicals (Hazard Information and Packaging for Supply) 2009 Regulations
chlorine 407
chronic health risks/effects 94, 377
chronic obstructive pulmonary disease 6
chutes 254–5, **271**
circuit breakers 322
    see also residual current devices (RCDs)
circular saws 294, **306**, 306
Classification, Labelling and Packaging of Substances and Mixtures Regulations (CLP) 376, 378
cleaning
    equipment and materials 196
    floors and stairways 194, 195, 196
clothing
    accommodation for 183
    changing facilities 183
    contaminated 398
    high-visibility 196
    protective 13, 44, 395, 397–8, 464
    and static electricity 354
CLP Regulation *see* Classification, Labelling and Packaging of Substances and Mixtures Regulations
codes of practice 14
    see also ILO codes of practice
cofferdams 463
cold environments 185, 187
collisions
    with fixed or stationary objects 193–4, 196
    with moving vehicles 193, 195, 230
combined testing and inspection 328
Comité Européen de Normalisation (CEN) 275
Comité Européen de Normalisation Électrotechnique (CENELEC) 275
commercial stakeholders 79
commitment, management 62, 63, 69, 76–7
communication 22, 24, 39, 60, 65
    with contractors 55, 56, 58

email 71, 72
graphic 72
    see also signs and symbols
and improvement of health and safety behaviour 71–2, 77
language issues 54, 65, 71, 72, 76
memos 71, 72
notice boards 71–2
safe systems of work 113
and temporary work projects 210
toolbox talks 72
verbal 71
written 71
commuting accidents 4, 5, 159, 160, 230
compactors 294, **305**, 305
compensation/compensation claims 9, 14, 161–3
competence 55, 70–1, 77
    driver 237
    electrical systems 320–1
    health and safety practitioners 49, 70
competent authorities 70
competent electrical person 314
competent institutions 70
competent persons 49, 70–1, 111–12, 120, 314
compressed air work 463
computer workstations *see* display screen equipment (DSE)
concussion 194
conduction 338
confined spaces
    atmosphere tests 118
    heat exhaustion as hazard in 185
    and permit-to-work 117–18
    personal protective equipment 118
    precautions 117–18
    rescue equipment 118–19
    risk assessment 114
    safe systems of work 113–14
    ventilation equipment 118
construction designers 218
construction inspection report form 604
construction work 54, 55, 212–21
    accidents 182, 212
    hazards and controls 211, 212–16, 294–5
        demolition 213–14, 463
        drowning prevention 214
        electricity 215, 463
        emergency procedures 215
        excavations 214, 219–21, 227–8, 462–3
        explosives 463
        falling objects 193, 195, 215
        health hazards 216, 463–4
        ILO recommendations 462–4
        noise 215–16
        safe place of work 212–13
        safety inspections and reports 208–9, 221, 224–5
        vehicles and traffic routes 214–15
        waste disposal 216
        welfare facilities 215, 464
        work at height 213, 462
    management of activities
        cooperation and coordination 217–18
        duties of designers, engineers, architects and clients 218–19
        duties of employers 216–17
        selection and control of contractors 219
        workers right and duties 218, 461–2

safe systems of work 112
scope of 212
consultation with workers 27, 30, 35, 43, 44, 60, 69
benefits of 74
and positive health and safety culture 62
and safe systems of work 112
to improve health and safety behaviour 72–4, 77
continual improvement, of health and safety management
systems 22, 24, 25, 26, 27, 29, 30, 44, 178–9
contractors 45
checklist for health and safety management 86–7
and client relationship 54–5
communication with 55, 56, 58
and construction projects 55
international contract work 56–7
legal considerations when using 54–5
management and authorisation of 57–8
and permit-to-work system 120
safety rules for 58–9
scale of contractor use 55–6
selection of 24, 57, 219
sub-contractors 45, 58, 182
supervision of 24, 58
Control of Artificial Optical Radiation at Work Regulations (2010)
(UK) 439
control of the health and safety organisation 43
control measures 108–9
engineered measures 108
management supervision 108
operational measures 108
for safe movement of people in the workplace 194–7
see also risk control measures
Control of Substances Hazardous to Health (COSHH)
regulations (UK) 388–9
assessment examples 592–5
controlled waste 413
controls (work equipment) 274
emergency stop controls 282–3
hold-to-run 298
lifting appliances 265
start controls 282
stop controls 282
two-handed devices 297–8, **299**
convection 337
conveyor systems **257**, 257–8, **270**
checkout 294, **305**, 305
cooking processes 346
cooperation/coordination between enterprises 217–18, 529
corporate responsibility 2
corrosive substances 376
COSHH Regulations see Control of Substances Hazardous to
Health (COSHH) regulations (UK)
cost-benefit analysis, risk control 105
Costa Rica 15
costs, fire-related 333
costs of accidents 7–8, 161, 442
direct costs 7–8, 93
indirect costs 8, 93
insured direct costs 7, 93
insured indirect costs 8, 93
uninsured direct costs 7–8, 93
uninsured indirect costs 8, 93
cranes 260–2
safe operation of 261
signallers 260
crushing hazards **291**

CTIF (Center of Fire Statistics) 332, 333
Cuba 451
cultural diversity/differences 54
customers
manufacturer/supplier information for 52–3
as stakeholders 28
cutting hazards 290, **291**
cylinder mowers 293–4, 302–3, **303**
Cyprus 451
Czech Republic 15, 451

damping (noise) 427
dangerous occurrence
definition 4, 93, 152
recording/reporting 159, 160
dangerous substances see flammable liquids and gases
deafness, noise-induced 423–4
deaths, work-related 442
accidents 2, 4–5, 7, 105, 160, 182, 193, 212, 230, 312, 477
asbestos-related 400–1, 402
diseases 4, 5, 6, 7, 477
fire-related **332**, 332, 340
worldwide estimates 477
decibels (dB) 424
Declaration of Conformity 276, 277
demolition 210, 213–14
deliberate controlled collapse 214
ILO recommendations 463
method statement 213–14
piecemeal 214
risk assessment 213
training 214
Denmark 451
departmental managers 48, 84–5
dermatitis 382–3, 408
allergic contact 382, 408
irritant contact 382, 408
developing countries
and international contracts 56
work-related deaths and injuries 4
dichloromethane (DCM) 407
diesel-engine exhaust emissions 407
diffuse pleural thickening 400
dilution (or general) ventilation 390, 393
direct current (DC) 312, 314
directors
and health and safety policy 33
health and safety training 61
leadership of health and safety 82–3
organisational health and safety responsibilities 46–7, 63
and performance monitoring 144
and performance review 177–8
as stakeholders 28
disabled workers/members of the public 107, 182, 185, 365–6
discrimination 13
disease, occupational see occupational diseases
display screen equipment (DSE)
compensation claims 163
EU legislation 490
and musculoskeletal problems 244, 247–8
and psychological problems 248
risk assessment 246–7
visual problems 248
workstation design 247
DO stage of health and safety management system 20, 22, 23,
25, 26, 42

see also organising health and safety management; risk assessment; risk reduction and control
document shredders 293, 301
documentation 529
  health and safety management systems 25, 29
  safe systems of work 113
doors and windows 185
drawing-in hazards **291**
drinking water 183, 186
driving at work 115, 230, 236–42
  driver competency 237
  driver fitness and health 238
  driver training 230, 237–8
  Driving CPC 237
  driving time and rest periods 240, 457–8
  health and safety rules 239–40
  ILO recommendations 240, 457–8
  IOSH advice on 242
  journey distance 239
  journey routes 239
  journey scheduling 239
  journey time 239
  risk assessment and evaluation 237–9
  road safety 236, 237
  vehicles
    condition and safety equipment 238–9
    ergonomic considerations 239
    load 239
    suitability 238
  weather conditions 239
  see also vehicles in the workplace
drowning prevention 214
drug abuse 187, 191–2, 238
  investigating size of the problem 191
  monitoring policy on 192
  planning actions on 191
  taking action on 191
due diligence 47, 484–5
dumper trucks 214, 230
dust observation lamps 387
dusts 216, 347, 354, 374, 384, 408, 409
  inhalable 374
  respirable 374, 380
duties and responsibilities
  of directors 46–7, 63
  of managers 42, 43, 47–9, 84–5
  of manufacturers and others in the supply chain 51–4
  of supervisors 48–9, 85–6
  see also employer responsibilities; workers, rights and responsibilities
duty, levels of 12–13, 108
duty of care 6

ear defenders (earmuffs) 428
earplugs 428
earth leakage circuit breakers *see* residual current devices (RCDs)
earth-moving equipment 233–5
earthing (or grounding) 313
economic reasons for OSH standards 6–9
economics, and health and safety standards 79
Egypt
  emergency call numbers 135
  legal framework 488–9
  occupational accidents and diseases *477*
ejection hazards **291**

electric drills 288
  stroboscopic effect 66
electric, magnetic and electromagnetic fields (EMFs) 441, 492
electrical safety 59, 312–30
  construction sites 215, 463
  control measures
    circuit breakers 322
    competence 320–1
    double insulation 324
    emergency procedures 326
    fuses 313, 317, 322
    inspection strategies 327–9
    insulation 322
    isolation 322–3
    maintenance strategies 327
    management systems 321
    protection against buried power lines 324, 326
    protection against overhead power lines 324, **325**
    reduced low-voltage systems 323
    residual current devices (RCDs) 313, 323–4
    safe systems of work 320–1
    selection and suitability of equipment 321–2
    testing 327–9
    training 321
    wiring regulations 320–1
  hazards 92, 292, 314–20
    cables 317, 319
    electric arcing 318
    electric shock and burns 313, 314–16, 319, 354
    electrical equipment in potentially flammable atmospheres 353–4
    electrical fires and explosions 316–18, 319, 337, 346–7
    high risks 320
    plugs and sockets 317, 319
    portable electrical equipment 319–20
    static electricity 317, 318–19, 335, 353, 354
  signs and symbols **312**, **315**
electrical waste 415
electrically powered gates 197
electricity, basic principles and definitions 312–14
  alternating current (AC) 312, 314
  amps (A) 312
  circuits 313
  competent electrical person 314
  conductors 312, 313
  direct current (DC) 312, 314
  earthing (or grounding) 313
  electric current 312
  electrical resistance 312
  equipotential bonding 313
  examination 314
  high voltage 313
  insulators 313
  isolation 314
  low voltage 313, 316
  mains voltage 312, 313
  maintenance 313
  milliamps (mA) 312
  ohms 312
  Ohm's law 312–13
  short circuits 313, 316
  supplementary bonding 313
  testing 313
  volts (V) 312
electronic waste 415
elevators 258

email communications 71
emergency call numbers 122–3, 135–7
emergency dispatchers 122
emergency procedures 22, 24, 25, 34, 39, 54, 59, 120–3, 529
    assembly and roll-call 122, 363–4
    construction sites 215
    electrical incident 326
    fire 363
    hazardous substances 399
    supervisory duties 121–2
    testing and training 123
    violations of 68
    work at height 201–2
    see also evacuation of a workplace
emergency services, contacting 122–3
emergency stop controls 282–3
employees see workers
employer responsibilities 13, 43–6, 54–5
    asbestos exposure 455
    construction work 216–17
    duty of care 6
    hazardous substances 456
    ILO recommendations on 44
    machinery/equipment 467–8
    vicarious liability 6
employers' liability insurance 8–9, 43, 44, 78
EN 2:1992 Classification of Fires 336
EN 3–6:1996 Portable Fire Extinguishers 358
End of Life Vehicle (ELV) Directive 415
enforcement agencies, role of 14–15
enforcement regimes 78
engineered measures 108
engineering controls 98
    hazardous substances 390, 392–3
engineers 218
    and permit-to-work system 120
entanglement hazards **291**
Environment Agency (UK) 413–14
environmental protection 411
    definition 3
Equalities Act (2010) (UK) 365
ergonomics 53, 63–4, 68, 239, 292–3
    ill-health effects of poor 245
    low-cost improvements in *246*
    principles and scope of 244–5
    and vibration-related ill-health 430–1
ergonomists 49
European Agency for Safety and Health at Work 16, 474
European Atomic Energy Community (Euratom) Treaty 492
European emergency call number 122–3
European Union (EU) 472–4, 489–92
    Advisory Committee on Safety and Health at Work (ACSHW) 473–4
    CE marking 52, 53, 276, 277
    Community strategy on health and safety at work (2014Ä2020) 472–3
    cost of work-related accidents/diseases 7
    Decisions 472
    Directives 14, 472
        artificial optical radiation 439, 492
        ATEX Directive 353–4
        display screens 490
        electromagnetic fields 492
        EMF Directive 441
        End of Life Vehicle (ELV) Directive 415
        explosive atmospheres 491

        Framework Directive 489, 490
        groundwaters 412–13
        hazardous substances 384, 388, 415, 491
        health and safety signs 491
        ionising radiation 492
        manual handling of loads involving risk 490
        minimum safety and health requirements 490
        noise 492
        personal protective equipment (PPE) 490
        physical agents 491–2
        Restriction of Hazardous Substances in Electrical and Electronic Equipment (RoHS) Directive 415
        solvents emissions 411
        temporary and mobile work sites 490
        temporary workers 45–6
        Waste Electrical and Electronic Equipment (WEEE) Directive 415
        work equipment 490
    Driving CPC (Certificate of Professional Competence) 237
    fire safety legislation and standards 333–4
    occupational accidents and diseases *477*
    Regulations 472
        Classification, Labelling and Packaging of Substances and Mixtures (CLP Regulations) 376, 378
        REACH (Registration, Evaluation, Authorisation and Restriction of Chemicals) 391–2
    risk assessment and management guidance 91, 127
    role of 12
    vibration exposure limits 431–3
    work-related deaths 7
evacuation of a workplace 344, 361–7
    and building plans and specifications 366–7
    means of escape
        doors 361
        escape routes 361–2
        escape times 362–3
        lighting 362
        signs 362
    people with special needs 365–6
    procedure and emergency plans 363–6
        assembly and roll-call 363–4
        fire drills 70, 364
        fire notices 364
        fire routines 363
        supervisory duties/fire marshals 363
examination
    of electrical equipment 314
    of pressure systems 281
excavations 214, 219–21
    ILO recommendations 227–8, 462–3
    inspections and reporting requirements 221
    precautions and controls 220–1
exceptional violations 68
expectant or nursing mothers 14, 105–7, 183
explosive atmospheres 347, 353–4, 491
explosives 463
extension leads 319
external stakeholders 28
eye protection equipment *395*, 395, 439, 441
    face visors 397
    goggles *395*, **397**, 397
    safety glasses 195, *395*, 397
eye strain 248

face visors 397
Factories Acts (India) 492–5

Factories Acts (Nigeria) 507–8
Factories Ordinance, Trinidad and Tobago 519–20
fall arrest equipment 200–1
falling objects 193, 195, 200, 209, 215, 290
falling-object protective structures (FOPS) 232, 233
falls
    from height 32, 182, 193, 195, 198–9
    on the same level 192–3, 194
fatalities *see* deaths
fault-tree analysis (FTA) 112
fibres 374–5
fibrosis 380, 408
Finland 15, 451
fire
    causes 338, **339**, 340
    classification of 336–7
    consequences of 340
    costs of 333
    deaths **332**, 332, 340
    electrical fires 316–18, 319, 337, 346–7
    evacuation in event of *see* evacuation of a workplace
    fire triangle 334–5, **335**
    fuel sources 335
    heat transmission and fire spread 337–8
    ignition sources 335, **339**, 341
    and oxygen 335–6
    place of occurrence 339–40
    safety legislation and standards 333–4
    vehicular activity hazards 235, 345
fire action signs **100**, 100
fire detection and alarm systems 354–6
    automatic detection 356
    manual ('break-glass') call points 355–6
    people with hearing difficulties 355
    schematic plan 355
    voice alarms 355
fire drills 70, 364
fire-fighting arrangements
    extinguishing methods
        chemical reaction 356
        cooling 356
        smothering 356
        starving 356
    fire extinguishers 356–9
        carbon dioxide (black band) 357
        classification of *358*
        colour coding 358
        foam (cream band) 357
        marking layout **359**
        powder (blue band) 357
        use-code symbols **359**
        water (red band) 356–7
        wet chemical (class 'F') 357
    inspection, maintenance and testing of equipment 360–1
    portable equipment 357–8
    responsible person 357
    sprinkler systems 358, 360
fire marshals 33, 363
fire prevention and prevention of fire spread 59, 344–54, 464
    buildings 350–3
        fire compartmentation 352–3
        fire loading 351
        fire resistance of structural elements 351–2
        insulating materials 352
        surface spread of fire 351
    control measures

cooking processes 346
electrical safety 346–7
equipment and machinery 345
fork-lift trucks and other vehicles 345
heating appliances 345–6
housekeeping 344
safe systems of work 347
smoking 346
storage 344–5
electrical equipment in potentially flammable atmospheres 353–4
flammable liquids and gases 347–50
    aerosols 350
    control measures 347–8, 349
    ILO recommendations 347–8
    risk assessment 348
    storage 349, **350**
    substitution 348–9
fire prevention training courses 76
fire risk assessment 340–4
    Stage 1. hazard identification 340–1
        combustible materials 340–1
        heat/ignition sources 341
        unsafe acts 341
        unsafe conditions 341
    Stage 2. identification of persons at significant risk 341–3
    Stage 3. risk evaluation and reduction 343
    Stage 4. findings 343
    Stage 5. monitor and review 343
    and building plans and specifications 366–7
    checklist 369–70, 603
    fire evacuation plans 344
    structural features of workplace 343–4
    temporary workplaces, maintenance and refurbishment 344
fire safety maintenance checklist 601–2
fire safety signs **100**, 100
first-aid 3, 34, 123–4, 182, 464
    box contents *124*, 124
    provisions checklist 123
    treatment and accident record form 587–8
first-aid signs **100**, 100, **124**, 124
first-aiders 34, 123, 124–5
    in relation to risk categories and employee numbers *125*
flames 335
flammable gas sign **336**
flammable liquid sign **336**
flammable liquids and gases 209, 335, 347–50, 354
    aerosols 350
    control measures 347–8, 349
    ILO recommendations 347–8
    risk assessment 348
    storage 349, **350**
    substitution 348–9
flammable solid sign **335**
flammable solids 335
floors 185, 192, 193, 194, 196
flying objects 193, 195
food facilities 183, 468–9
food safety 59, 183
footwear 193, 194, 195, 196, 354, 397
forced labour 13
fork-lift trucks 215, 258–9
    checks and maintenance 259
    driver training 76, 259
    fire hazards 345
    hazards 258–9

safe driving of **259**
and whole-body vibration 430
formal visual inspection, portable electrical appliances 327–8
fragile roof signs *101*, 101
freedom of association 13
frozen shoulder 245
fumes 374, 375
fungi 375
fuses 313, 317, 322

gases 375
flammable *see* flammable liquids and gases
general purpose trucks 254
Germany 15, 388
glass-fibre 374, 375
Globally Harmonised System (GHS) Classification and Labelling of Chemicals (GHS) 52, 376, 377, 378
Globally Harmonised System (GHS) hazard statements 418
gloves 397, 408, 420
anti-vibration 432
goggles **397**, 397
graphic communication 72
see also signs and symbols
greenhouse effect 411
grinding machines 280, 293, 301–2
groundwater 412–13
guard rails (scaffolds) 205, 206
guards/guarding 53, 274, 277–8, 295–7, 306–7
adjustable 295–6, **297**
fixed 295, **296**
hand-held power tools 287
interlocking 296, **297**

hand—arm vibration syndrome (HAVS) 429, *430*, 431–2, 433
hand-held power tools
battery (cordless) 323
electric drills 288
guarding 287
hazards 286–7, 288–9, 319–20
operating controls and switches 287
safe operations/instructions 287–8
sanders 288–90
hand-held tools (non-powered)
hazards 284
safety considerations
inspection 286
quality and design 286
suitability 285
training 286
hand protection equipment 395, 397
barrier creams 397
gloves 397, 408, 420, 432
harmful substances 376
hazard analysis 96, 112
hazard identification 22, 24, 27, 90, 96
fire risk assessment 340–1
machinery 290
hazard and operability studies (HAZOPS) 96, 112
hazard perception 66
hazardous substances 34, 39, 374–420
assessment 216, 384–7
health surveillance 385
requirements 384–5
sources of information available for 385–6
survey techniques for health risks 386–7
biological agents 375

and body systems
cardiovascular system 381
nervous system 380–1
respiratory system 380, 381
skin 382
urinary system 381–2
buying problems 53
chemical agents 374–5
classification of 378, 392
and associated health risks 376–7
CHIP 4 classifications 376, 378
GHS classifications 52, 376, 378
compensation claims 163
contractors use of 59
deaths-related to *3*, 6
EU requirements 384, 388, 415, 491
hazard warning statements 378
health effects (acute and chronic) 377
health risks 376–7, 379–83
of specific agents 406–11
and hot and cold environments 186
ILO requirements 383–4, 455–7, 463–4
labelling of 52, 376, 378, 385
material safety data sheets 109, 383, 386
Occupational Exposure Limits (OELs) 101, 384, 387–9
precautionary statements 378
routes of entry to the human body 379
ingestion of 379
inhalation of 379
injection or skin puncture 379
skin absorption 379
signs and symbols 99, **100**, 376, **377**
storage issues 209
substances of very high concern (SVHCs) 391
waste disposal 406, 413–15
worker rights and responsibilities 383–4, 456–7
hazardous substances control measures 384, 385, 389–411
asbestos 400–6
carcinogen, mutagen or substances that cause asthma 398–9
elimination of the substance 390
emergency procedures 399
engineering controls 390, 392–3
dilution (or general) ventilation 390, 393
local exhaust ventilation 390, 392–3
health records 390, 398
health surveillance 385, 390, 398, 399
hierarchy of 390
ILO recommendations 390–1, 398–9
maintenance of equipment 399
organic solvents example 399–400
personal hygiene 398
personal protective equipment (PPE) 383, 390, 393–8
preventative measures 391–2
downstream users 391
manufacturers/importers 391
other users in the supply chain 391
substitution of the substance 98, 109, 390, 391
supervisory (people) controls 390, 393
transport by road 399
hazardous waste and spillage 406, 413, 414, 415
hazards 13, 39, 210–12
checklist 127–8
chemical/biological *see* hazardous substances
construction work 211, 212–16
definition 4, 92

electrical safety 92, 292, 314–20
elimination of 95, 98
ergonomic 245
exposure to
  control of 34
  levels of 95
  reduced time 101, 393
high hazard 4, 92
machinery/work equipment *see* machinery/work equipment
  hazards
manual handling 248–9
musculoskeletal 244–6, 247–8, 248–9
pedestrian 192–7
physical health 422–42
psychological *see* psychological health hazards
secondary 319–20
vehicle operations 193, 195, 230–1, 345
waste disposal 414
workplace 182, 187–227
head injuries 194
head protection 195, 196, 200, *395*
health, definition 3
health hazard substances 376
health records 390, 398
  asbestos exposure 405
health risks 94, 108, 374, 379
  acute 94
  chronic 94
  construction sites 216
  hazardous substances 376–7, 379–83
    asbestos 400–1
  physical and psychological 422–45
health and safety advisers *see* health and safety practitioners
Health and Safety at Work (HSW) Act (UK) 42, 91, 474–5
health and safety committees 22, 24, 34, 35, 44, 69, 72, 73
  composition of 73
  terms of reference 73
health and safety culture 22, 26, 59–62, 451
  definitions 59, 60
  and health and safety performance 60–1
  human factors affecting 62–8
  ILO perspective on 60
  important indicators of 61
  improving *see* improving health and safety behaviour
  poor 61–2
  positive 27, 49, 59, 60, 63, 69
  preventative 44
  and training 74
Health and Safety Directors 47
Health and Safety Executive (HSE) (UK) 2, 3, 7, 16, 36, 46, 244,
  475
  Accident Prevention Unit 62
  advisory committees 476
  EH40 Occupational Exposure Limits 384
  guidance on workplace temperature 183–4
  hazardous substances principles of good practice 389–90
  HSG6 *Safety in Working with Lift Trucks* 259
  HSG33 *Health and Safety in Roof Work* 200
  HSG47 *Avoiding Danger from Underground Services* 326
  HSG48 *Reducing Error and Influencing Behaviour* 63, 65, 69
  HSG57 *Seating at Work* 185
  HSG65 *Managing for Health and Safety* 20, 21, 23, 24, 25–6
    definition of accident 92–3, 152
    definition of health and safety culture 59
    risk assessment 90–1
    risk management matrix 97
  HSG168 *Fire Safety in Construction* 344
  HSG258 *Controlling Airborne Contaminants at Work* 393
  HSG264 *Asbestos: The survey guide* 403
  human factors, definition of 62
  INDG69 'Violence at Work: a Guide for Employers' 189
  NDG163 'Risk assessment — A brief guide to controlling
    risks in the workplace 94
  INDG368 'Working Together: Guidance on health and safety
    for contractors and suppliers' 86
  INDG417 'Leading health and safety at work' 46, 175
  L8 Legionnaires' disease — The control of legionella bacteria
    in water systems' 410
  L22 *Safe use of work equipment* 274
  L108 Guidance on the Control of Noise at Work Regulations
    424–5, 426
  L113 *Safe use of lifting equipment* 263
  manual handling operations 250, **251**, 251, 252
  mobile elevating work platforms (MEWPS) 208
  safety notice on electrically powered gates 197
  Violence at Work: Findings from the British Crime Survey
    188
health and safety information
  provision of 69, 102
  sources of 109
health and safety management systems 14–15, 16, 451
  benefits of 29
  continual improvement of 22, 24, 25, 26, 27, 29, 30, 44
  documentation 25, 29
  key characteristics of successful systems
    continual improvement 29
    effective audit 29
    involvement of stakeholders 27–8
    positive health and safety culture 27
  key elements of 20–3
  major systems *see* Health and Safety Executive (HSE) (UK),
    HSG65; ILO-OSH 2001; ISO 45001:2016; OHSAS 18001
  problems associated with 29
  worker participation in 22, 24, 26, 28, 30
  see also audits; incident/accident investigation; measuring
    health and safety performance; monitoring; organising
    health and safety management; planning phase of
    health and safety management; risk assessment; risk
    control
health and safety policy 20, 26, 27, 29–36, 42, 43, 44, 49, 82
  arrangements section of 34–5
  checklist 38–9
  and health and safety objectives 31–3, 43
  and ILO-OSH 2001 24, 29–30, 36
  and organisation of health and safety 33–4
  and other business activities 33
  policy statement of intent 21–2, 27, 30–1, 60
  reasons for less than successful 35–6
  review of 35–6
  standards and guidance relating to 24, 29–30, 36
health and safety practitioners 33, 49–50
  insurance cover 50
  relationships outside the organisation 50
  relationships within the organisation 50
  status and competence 49, 70
  training 49
health and safety professionals 28
health and safety records
  access to 24, 25
  establishment of 25
health and safety representatives 22, 24, 25, 28, 30, 33, 69,
  72–3, 77

health surveillance 13, 34, 398, 529
  asbestos exposure 405, 455
  extreme temperature environments 187
  hazardous substances 385, 390, 398, 399
  noise exposure 426
  radiation exposure 441
  vibration exposure 433–4
hearing loss, noise-induced 423–4
hearing protection 425, 426, 428
  selection of suitable 428
  training in use of 428
heat equation 186
heat stroke/stress/exhaustion 185, 186, 187
heat transmission principles
  conduction 338
  convection 337
  direct burning 338
  radiation 338
heating appliances, fire hazards 345–6
heating and temperature 183–4
hepatitis 410, 411
high pressure hazards 291
high voltage 313
high-voltage apparatus 117
HIV (human immunodeficiency virus) 410
hoists 262
hold-to-run controls 298
homeworkers 44
hot environments 185, 186
hot surfaces 335
hot work 117
hot work permit 117, 344
hours of work, road transport drivers 240, 457–8
HSE see Health and Safety Executive
human error 64, 65, 66, 67–8
human factors 62–8
  individual (personal) factors 62, 65–7
  jobs 62, 63–4
  organisation 62, 63
human failure
  human error 64, 65, 66, 67–8
  violations 67, 68
human immunodeficiency virus (HIV) 410
humidity 186, 248
hygiene 458–9
  personal 398
  see also welfare facilities
hypothermia 187

IEC 60445 320
IEE Wiring Regulations 320–1
ill-health see work-related ill-health
ILO 2, 3, 4, 5, 12, 16, 28, 56, 275, 401
  Declaration on Fundamental Principles and Rights at Work
    13
  on employers' duties and responsibilities 13
  Guidelines on occupational safety and health management
    systems see ILO-OSH 2001
  on risk assessment 90, 91
  role and function of 9–10
  Work Improvement in Small Enterprises (WISE)
    methodology 246
  on workers' rights and responsibilities 13–14
  see also ILO Codes of Practice; ILO Conventions and
    Recommendations
ILO Codes of Practice 11, 28

Ambient factors in the workplace 13, 274, 276, 383, 384,
    422, 432, 433–4, 439, 441
  emergency procedures 120–1
  hazardous substances 390–1, 398–9
  hot and cold environments 186–7
  risk assessment guidelines 91, 92, 93, 95, 97
classification of industrial accidents 167–9
  occupational diseases 169–70
Protection of workers against noise and vibration 422, 425,
    426–7, 428
Radiation protection of workers, ionising radiations 439–40
Recording and notification of occupational accidents and
    diseases 150–1, 152, 160–1
Safety and Health in Construction 18, 200, 216–19
  excavations 227–8
  lifting equipment 264–6
  scaffolds and ladders 223
  transport, earth moving and materials-handling equipment
    233–5
Safety and health in the use of machinery 275–6, 282, 284,
    299–300, 464–8
Safety in the use of asbestos 401–2, 403, 405
Safety in the use of chemicals at work 347–8, 383–4
ILO Conventions and Recommendations 10–12, 20, 28, 448–71
  C115: Radiation Protection (1960) and (R114) 439, 460–1
  C119: Guarding of Machinery and (R118) 277–8
  C120: Hygiene (Commerce and Offices) (1964) 458–9
  C127: Maximum Weight and (R128) 244, 253–4
  C138: Minimum Age (1973) 105
  C139: Occupational Cancer (1974) 459
  C148: Working Environment (Air Pollution, Noise and
    Vibration) and (R156) 422, 429, 470–1
  C153: Hours of Work and Rest Periods (Road Transport)
    (1979) and (R153) 240, 457–8
  C155: Occupational Safety and Health (1981) and (R164) 10,
    44, 72, 111, 120, 150, 230, 334, 383, 449–50, 452–4,
    471
    action at the level of undertaking 453–4, 536–7
    action at national level 452–3, 535–6
    areas of action 452
    final provisions 537–8
    principles of national policy 535
    scope and definitions 534
  C161: Occupational Health Services (1981) 459–60
  C162: Asbestos (1986) and (R172) 10, 401, 403, 454–5
  C167: Safety and Health in Construction (1988) and (R175)
    10, 18, 208, 212, 216, 223, 461–4
  C170: Chemicals (1990) 455–7
  C182: Worst Forms of Child Labour (1999) 105
  C183: Maternity Protection and (R191) 106
  C187: OSH Promotional Framework and (R197) 44, 60, 150,
    450–1
    national OSH policy 451
    national OSH systems 451, 452
    OSH and management systems 451
    OSH and safety culture 451
  Migrant Workers (Supplementary Provisions) (1975) 11
  R102: Welfare Facilities 468–70
  R162: Older Workers Recommendation, 1980 105
ILO-OSH 2001 3, 15, 21, 23–5, 26, 31, 36, 44, 52, 90, 451
  action for improvement element 24
  audit recommendations 24, 173–4, 175
  communication recommendations 72
  and competence requirements 70
  and contractors 55
  documentation recommendations 25

emergency preparedness and response requirements 334

evaluation element 24

investigation of work-related injuries, ill-health etc. 151

organising element 24

performance monitoring and measurement 140–1

planning and implementation element 24

policy element 24, 29–30

risk assessment and control guidelines 95, 104

worker participation 25

improving health and safety behaviour 68–79

commitment from management and 69, 76–7

competence and 70–1, 77

consultation with workers and 72–4, 77

effective communication and 71–2, 77

external influences and 78–9

health and safety training and 74–6

internal influences and 76–7

legislation and 71, 78

promotion of health and safety standards and 69–71

incident/accident investigation 27, 34, 49, 50, 60, 69, 73, 150–8, 529

appropriate levels of 154–5

benefits of 151–2

compensation claims following 161–3

forms 157–8

function of 150

high-level 155

incident causes and analysis

comparison with relevant standards 157

immediate causes 156

root causes 157

underlying causes 156

low-level 154

medium-level 154–5, 158

minimal-level 154

national/international requirements 150–1

procedures 155–6

initial action 155

interview techniques 156

investigation method 155–6

remedial actions 157

types of incident or adverse events 152

incident/accident recording and reporting 25, 27, 31, 34, 44, 49, 50, 59, 60, 70, 158–63

analysis 159

example form 585–6

organisational requirements for 159–60

to external authorities 160

types of incident/accident 158–9

typical incidents which need reporting 160–1

incidents

definition 93, 152

lessons learned from 161

ratios 62

statistics 160–1

see also accidents; near miss incidents

India 15

emergency call numbers 135

Factories Acts 492–5

National Policy on Safety, Health and Environment at Work Place 495–6

occupational accidents and diseases 5 , 7, 477

silicosis 3

individual (personal) factors 62, 65–7

attitude 65

memory 66

motivation 65–6

negative 65, 67

perception 66

Indonesia 15

emergency call numbers 135

legal framework 498

occupational accidents and diseases 477

induction training 72, 75–6, 105

infrared radiation 438

ingestion of hazardous substances 379

inhalable dusts 374

inhalation of hazardous substances 379

injuries

frequency rates 159

incidence rates 159

minor versus serious 153, 154

statistics 141

inspections see safety inspections

Institute of Directors (UK) 46

Institution of Engineering and Technology (UK) 320

Institution of Occupational Safety and Health (IOSH) (UK) 7, 56, 115, 242

instruction(s) 6, 13, 34, 44, 49, 109

general equipment 299–300

for using equipment with specific risks 279–80

insulating materials (buildings) 352

insulation (electrical appliances) 322, 324

insurance

employers' liability 8–9, 43, 44, 78

health and safety practitioners 50

and information required following incident/accident 162–3

insurance companies 28, 78

internal stakeholders 28

International Commission on Non-Ionising Radiation Protection (ICNIRP) 439

International Commission on Radiological Protection (ICRP) 435

international contract work 56–7

International Labour Conference 10

International Labour Organisation see ILO

International Monetary Fund 28

International Organisation for Standardisation see ISO

international organisations 28

International Society of Radiology (ISR) 435

internet searching

privacy issues 571–2

techniques 570–1

useful websites 572–3

investigating incidents see incident/accident investigation

investors, as stakeholders 28

ionising radiation see under radiation

Ionising Radiation Regulations (UK) 435

IOSH see Institution of Occupational Safety and Health

Ireland 7, 9

director duties and responsibilities 47

irritant contact dermatitis 382, 408

irritants 376

ISO 2, 15–16, 56, 275, 334

ISO 1999:1990 Acoustics: Determination of occupational noise exposure and estimate of noise-induced hearing impairment 422

ISO 2631–1:1997 whole-body vibration 429

ISO 3864–1:2002 Graphical symbols — Safety colours and safety signs 99

ISO 3941 Classification of Fires 336

ISO 5349:1986 hand-transmitted vibration 429

ISO 7010:2003 *Graphical symbols — Safety colours and safety signs* 99, 283, 362
ISO 7165:2009 *Fire-fighting-Portable fire extinguishers — Performance and construction* 358, 359
ISO 9000 series (quality) 15, 16, 23
ISO 12100:2010 282, 290, 292, 295, 306
ISO 14000 series (environment) 15, 16, 23
ISO 45001:2016 16, 25, 97
isocyanates 407
isolation of equipment 283
   electrical equipment 314, 322–3

Jamaica 137
Japan 15, 451
   electrical wiring 320
   emergency numbers 135
   fire safety legislation and standards 334
   Guidelines on Occupational Safety and Health Management Systems 499–500
   hazardous substances classification and labelling 378
   occupational accidents and diseases 477
jib cranes 260, 261
jigs 298
job descriptions 43, 64
job design, and violence at work 190
job safety analysis (JSA) 64, 112
   form 582
job-specific training 76
jobs 62, 63–4
   and ergonomics 63–4
   health and safety considerations checklist 64–5
   health and safety failures in 65
joint occupation of premises 59

Kazakhstan 136
Key Performance Indicators (KPI) 33
kinetic energy hazards 292
knowledge-based mistakes 67
Kuwait
   Labor Act 502–3
   Labor Law No.6 (2010) 503

labelling, of hazardous substances 52, 376, 378, 385
Labor Act (Kuwait) 501–3
labour standards 10, 11
ladders 18, 59, 202–3
   correct angle **203**
   ILO recommendations 462
   securing 202, **204**
   see also stepladders
lagging (for noise reduction) 427
language issues 54, 65, 71, 72, 76
lapses 67
lasers 437–8, 439, 492
Latin America
   fatal accidents and work-related diseases 4, *5*, *7*, 477
   fire safety legislation and standards 334
   silicosis 3, 6
   see also Brazil
Latvia 136
lead 407–8
leadership of health and safety 82–3
ledgers (scaffolds) 205
legionella (Legionnaires' disease) 409–10
legislation 14, 78
   and client—contractor relationship 54–5

fire safety 333–4
identifying and keeping up with 71
safe systems of work 111
supply side 52
see also national legislation
leptospirosis (Weil's disease) 409
lift plans 260
lift trucks 259, **271**
   see also fork-lift trucks
lifting equipment *see* mechanical handling and lifting equipment
lifting tackle 262
   hooks 254, 263, **270**
   shackles and eyebolts 262
   slings 262
lifts 262
   and fire emergencies 366
   passenger 262–3
lightening strikes 319
lighting 184–5, 196, 283, 463
   fire escape routes 362
liquids 375
   flammable *see* flammable liquids and gases
load handling *see* manual handling; manually operated load handling equipment; mechanical handling and lifting equipment
loads, safe working load (SWL) 256, 260
local exhaust ventilation (LEV) 390, **392**, 392–3
   collection hood and intake 392
   filter or other air cleaning device 392–3
   ventilation ducting 392
log tongs 254
lone workers 107
   safe systems of work 114–15
   security equipment 190
   and violence 115, 188, 189
   work-related driving 115

machinery/work equipment
   buying problems 53–4
   causes of fire in 345
   compensation claims 162
   EU legislation 490
   monitoring systems 34, 60
   risk assessment 34, 274–5, 275–6
      example form 589
machinery/work equipment hazards
   agricultural/horticultural 293–4
   construction hazards 294–5
   hand-held power tools 286–7, 288–9, 319–20
   hand-held tools (non-powered) 284
   hazard identification 290
   manually operated load handling equipment 255
   manufacturing and maintenance 194, 293
   mechanical 290, **291**, 292
   mechanical handling and lifting equipment 257–9
   mobile work equipment 231
   non-mechanical 292–3
   office hazards 293
   retail hazards 294
machinery/work equipment risk control measures 295–307
   agricultural/horticultural 302–4
   construction 305–6
   guards/guarding 53, 274, 277–8, 295–7, **296**, **297**, 306–7
   hand-held power tools 287–90
   hand-held tools (non-powered) 284–5, 285–6

hold-to-run control 298
ILO recommendations 462, 464–8
information and instructions 279, 299–300
inspection 281
isolation of equipment 283
jigs, holders and push sticks 298
lighting 283
maintenance 119, 194, 274, 279, 280–1
maintenance of equipment with specific risks 279
manually operated load handling equipment 255–6
manufacturing and maintenance 301–2, 466–7
markings and warnings 283, 284
mechanical handling and lifting equipment 256, 259, 261–6
office equipment 301
personal protective equipment 300–1, 304
retail 305
stability 283
standards and requirements 275–7
start/stop and emergency controls 282–3
supplier responsibilities 467
training for specific risks 279–80
training and supervision 286, 300
trip devices 297, **298**
two-handed control devices 297–8, **299**
user responsibilities 284
work space and operating stations 283–4
maintenance of machinery/equipment 119, 274, 279, 280–1
breakdown-based 281
condition-based 280
electrical equipment 313, 327
fire equipment 360–1
hazardous substances 399
hazards associated with 194, 293
permit-to-work 119, 281
preventative planned 280–1
safe system of work 281
safeguards 301–2
Malaysia 15, 503–6
emergency call numbers 135
occupational accidents and diseases *477*
Occupational Safety and Health Act (1994) 504–6
management commitment 62, 63, 69, 76–7
Management of Health and Safety at Work Regulations (UK) 15, 91
management supervision 108
management systems *see* health and safety management systems
managers
and health and safety policy 33
health and safety training 61
and incident/accident investigations 151–2
involvement in inspections and accident investigations 69
organisational health and safety responsibilities 42, 43, 47–9, 84–5
and performance monitoring 143–4
and performance review 177–8
and permit-to-work systems 120
and safe systems of work 112
training 61, 76
managing directors 48, 84
manual handling 34, 53, 244
assessment checklist 599
compensation claims 162
control measures 251–4
capability of the individual 252

and load 252
mechanical assistance 251
and task 251–2
and working environment 252
EU legislation 490
hazards 248–9
HSE guidance 250, **251**, 251, 252
ILO recommendations on 253–3
risk assessment 216, 250–1
capability of the individual 251
employee checklist 600
load considerations 250
task analysis 250
working environment considerations 251
training 252–3
Manual Handling Operations Regulations (UK) 244
manually operated load handling equipment 254–6
hazards 255
precautions with the use of 255–6
types of 254–5, **270–1**
manufacturers' duties and responsibilities 51–3
hazardous substances 391
manufacturing hazards and safeguards 293, 301–2
marking
of machinery 283, 284
product 530
Material Safety Data Sheets (MSDSs) 109, 383, 386
materials-handling equipment 233–5
measuring health and safety performance
effective risk control 142–3
ILO-OSH 2001 requirements 140–1
purposes of 141–2
addressing different information needs 142
answering questions 142
decision making 142
traditional approach 141
mechanical assistance 255
mechanical handling and lifting equipment 256–66
controls, control devices and cabins 265
general requirements for lifting operations 256
hazards 257–9
ILO recommendation on 264–6, 462
inspection, examination and tests 263–4, 265
installation 256, 265
operation and operatives 265
organisation of lifting operations 256
positioning and installing 256, 265
risk assessment 269
types of 256–63
conveyors 257–8
cranes 260–2, 265–6
elevators 258
fork-lift trucks 258–9745
lifting ropes 266
lifts and hoists 262
passenger lifts 262–3
slings and hooks 262
medium-density fibreboard (MDF) 409
memory problems 66
memos 71, 72
mesothelioma 400, 402
Mexico 15
emergency call numbers 137
hazardous substances classification and labelling 378
occupational accidents and diseases *477*
OSH legal framework 506–7

micro-enterprises 60
microwaves 438, 439
migrant workers 46, 54, 65
   induction training 75, 76
Mines Safety and Health Commission 473
mistakes 67
mists 375
mobile elevating work platforms (MEWPS) **207**, 207–8
mobile jib cranes 261
mobile phones 190
mobile work equipment
   control strategies 231–2
   hazards 231
   ILO recommendations 233–5
   safe driving of 235
   safeguards 232–3
Moldova 136
Mongolia 135
monitoring 27, 34, 50, 63, 83
   active *see* active/proactive monitoring
   compliance with regulations 530
   ILO guidelines 140–1
   management roles in 143–4
   plant and equipment 34, 60
   reactive 25, 26, 140–1, 143, 144, 149, 150
   risk control 102–3
   safe systems of work 113
moral reasons for OHS standards 4–6
motivation, individual 65–6
moulds 375
movement of people in the workplace
   control measures 194–7
   hazards and 192–4
moving elements, hazards from 292
moving objects, collisions with/being struck by 193, 195
multinational companies 54
musculoskeletal disorders (MSDs) 6, 245, 249
musculoskeletal hazards 244–6, 247–8, 248–9
mutagenic substances 377, 388–9, 391, 398–9

National Fire Protection Association (NFPA) (USA) 333
National Institute for Occupational Safety and Health (NIOSH)
   (USA) 388, 471
national legislation/legal frameworks 471–532
   Australia 478–80
   Bangladesh 480, *481*
   Brazil 481–2
   Canada 483–5
   China 485–8
   Egypt 488–8
   India 492–7
   Indonesia 498
   Japan 499–500
   Kuwait 501–3
   Malaysia 503–6
   Mexico 506–7
   Nigeria 507–8
   Oman 508–11
   Russian Federation 511–15
   South Africa 515–19
   South Korea 500–1
   Trinidad and Tobago 519–21
   Turkey 521–2
   United Arab Emirates 522–4
   see also under European Union (EU); United Kingdom (UK);
      United States of America (USA)

national legislation/legal frameworks, common themes in
   527–32
   certification and marking 530
   control of working conditions 528–9
   cooperation and coordination between enterprises 529
   emergency procedures 529
   general duties 528
   health surveillance 529
   investigation and notification of damage to health 529
   monitoring compliance with regulations 530
   organisation and management of preventative OSH services
      529
   prevention activities 528
   prevention of occupational accidents and disease 527–8
   product safety and occupational safety 530
   recording and documentation of information 529
   risk assessment 528
   training and information for workers 528
   workers' rights, duties and participation 529–30
near miss incidents 62, 153
   definition 4, 93, 152
   investigation of 151
   reporting of 31
NEBOSH International General Certificate 550–67
   exam terminology *545*, 550
   practical application (GC3) 555–67
      requirements of 555–8
      specimen practical application 559–67
   written examinations specimen answers
      control of workplace hazards (GC2) 552–5
      management of international health and safety (IGC1)
         550–2
neighbours, as stakeholders 28
nervous system 380–1, **381**
neurotoxins 381
new workers
   accidents 75
   supervision 75
New Zealand 15
   compensation system 9
   electrical wiring 320
   emergency call numbers 136
   fire safety legislation and standards 334
   hazardous substances
      classification and labelling 378
      occupational exposure limits 388
Niger 451
Nigeria
   emergency call numbers 135
   occupational accidents and diseases *477*
night working 45, 106
NIOSH *see* National Institute for Occupational Safety and
   Health (NIOSH) (USA)
no smoking signs **100**, 100
no-fault injury compensation scheme 9
noise 39, 59, 68, 215–16, 292, 411, 422–8
   assessment record form 598
   control techniques
      attenuation of noise levels 426, 427
      personal ear protection 426, 428
      reduced time exposure 426
      reduction of noise at source 426, 427
   EU legislation 492
   exposure levels 424, 425–6
   health surveillance 426
   ill-health effects of 423–4

acute acoustic trauma 423
  hearing loss 423–4
  temporary/permanent threshold shift 423
  tinnitus 423
ILO recommendations 422, 425, 426–7, 428, 470
repeated exposure to 423–4
risk assessment 425
sudden extremely loud 423
typical levels at wood working machines *424*
noise absorption walls/screens 427
noise assessments
  noise action levels 425–6
    exposure action value 425
    exposure limit value 425
  noise measurement 424–5
    continuous equivalent noise level 424
    daily personal exposure level 424
    peak sound pressure 424
    sound pressure level (SPL) 424
non-compliance, consequences of 15
non-mechanical machinery hazards 292–3
  electrical hazards 292
  ergonomic hazards 292–3
  material/substance hazards 292
  noise hazards 292
  radiation hazards 292
  thermal hazards 292
  vibrations hazards 292
  work environment 293
Norway 15, 136
notice boards 71–2

observation skills and techniques, safety inspections 145–6
occupational diseases 2, 6, 7
  definition 93, 152
  ILO list of 169–70
  recording/reporting/investigating 150–1, 159, 160, 529
Occupational Exposure Limits (OELs) 101, 384, 387–9
  ceiling values 387, 388
  permissible exposure limit (PEL) 387–8
  short-term exposure limit (STEL) 387
  threshold limit values (TLVs) 387
  time-weighted average s(TWA) 387, 388
occupational health professionals 49
occupational health and safety, scope and nature of 2–3
Occupational Health and Safety Act (Turkey, 2012) 521–2
occupational health services 459–60
  see also health surveillance
occupational hygienists 49
occupational ill-health 374
  definition 3, 92
Occupational Safety and Health Act (Malaysia, 1994) 504–6
Occupational Safety and Health Act (South Africa, 1993) 516–19
Occupational Safety and Health Act (South Korea, 1990) 500–1
Occupational Safety and Health Act (USA, 1970) 471, 525, 526–7
Occupational Safety and Health Administration (OSHA) (USA) 15, 16, 388, 471
Occupational Safety and Health (Amendment) Act (Trinidad and Tobago, 2006), Trinidad and Tobago 519
office hazards and safeguards 293, 301
Ohm's law 312–13
OHSAS 18001:1999 16
OHSAS 18001:2007 16, 21, 23, 24, 25, 26, 27, 44, 175
older workers 105

Oman 508–11
  emergency call numbers 135
  Labour Law (1973) 508–9
  Occupational Safety Regulations 509–11
operational measures 108
organic solvents 399–400, 407
Organisation for Economic Cooperation and Development (OECD) 12
organisational factors 62, 63
organisational structure, and health and safety management 22, 26
organising health and safety management 33–4, 42–88
  and contractors 54–9
  employers' responsibilities 43–6
  health and safety culture 59–62
  health and safety practitioners 49–50
  ILO-OSH 2001 24
  managers' responsibilities 42, 43, 47–9
  manufacturers and others in the supply chain 51–4
  persons in control of premises 50–1
  self-employed people 51
  workers' responsibilities 50
OSH legal frameworks *see* national legislation/legal frameworks
OSHA *see* Occupational Safety and Health Administration (OSHA) (USA)
overhead gantry travelling cranes 260
overhead obstacle signs **99**, 99
overhead power lines 324, **325**
oxidising agent sign **336**
oxygen 335–6
ozone depletion 411–12

pagers 190
Pakistan 135
pallet stacker trucks 258
pallet trucks **255**, 255, 258
palletised goods 209
pedestal drills 293, **302**, 302
pedestrian hazards 192–7
  control/preventative measures 194–7
  electrically powered gates 197
  falls from height 32, 182, 193, 195
  falls on the same level 192–3, 194
  fixed or stationary objects 193–4, 196
  moving, falling or flying objects 193, 195
  moving vehicles 193, 195
  slip hazards 192–3, 194, 196–7
  tripping hazards 192, 193, 194
pedestrian walkways 195, 196
peer pressure 65, 68, 74–5
perception 66
performance appraisal 43
performance assessment phase of health and safety management systems 22, 26
  see also CHECK stage of health and safety management
performance improvement phase of health and safety management systems 22
  see also ACT stage of health and safety management
performance indicators 27, 33, 140
performance measurement *see* measuring health and safety performance
performance monitoring *see* monitoring
performance phase of health and safety management systems 22
  see also DO stage of health and safety management

performance review of health and safety management system
22, 27, 47, 49, 83, 176–9
and continual improvement 178–9
items to be considered 177
people involved and planned intervals 177
purpose of 176–7
role of directors and senior managers 177–8
performance standards 26, 32, 43
performance targets
health and safety 31
organisational/production 2, 62
permit-to-work system 76, 110, 115–20
example form 583
principles applying 10 115–16
procedures 116
responsibilities
authorised persons 120
competent persons 120
contractors 120
engineers 120
operatives 120
senior authorised persons 120
site managers 120
specialists 120
work requiring a permit 116–20
confined spaces 117–19
hot work 117
machinery maintenance 119, 281
work at height 119–20
work on high-voltage apparatus 117
personal factors see individual (personal) factors
personal hygiene 398
personal protective equipment (PPE) 34, 68, 69, 76, 103, 104,
195, 196, 274, 408
benefits 103
compensation claims 162
and confined spaces 118
construction sites 216
EU legislation 490
eye protection 195, *395*, 395, **397**, 397
hand and skin protection 395, 397
hazardous substances 383, 390, 393–8
ILO recommendations 464
limitations 103
machinery/work equipment use 300–1, 304
principle requirements of 394
protective clothing 13, 44, 395, 397–8, 464
respiratory (RPE) 394, *395*, 395–7
supervision 395
training 395
persons in control of premises 50–1
Philippines
emergency call numbers 136
fire safety legislation and standards 334
hazardous substances classification and labelling 378
occupational accidents and diseases *477*
photochemical smog 412
photocopiers 293, 301
Physical Agents (Artificial Optical Radiation) Directive 439,
492
physical health hazards and risk control 422–42
noise 422–8
radiation 434–42
vibration 428–34
physiotherapists 49
Piper Alpha oil platform fire (1988) 42

Plan, Do, Check, Act approach to health and safety
management 20–2, *23*, 25
PLAN stage of health and safety management see planning
phase of health and safety management
planning phase of health and safety management 20, 21–2, *23*,
26–7, 82
ILO-OSH 2001 24
OHSAS 18001:2007 25
see also health and safety policy; risk assessment
plant 38–9
buying problems 53–4
design and selection of 69
safety inspections 145
see also machinery/work equipment
platform trucks 254, **271**
plugs and sockets 317, 319
pneumoconiosis 408
Poland 15
policy see health and safety policy
pollution 411
air 411–12, 470
water 412–13
POPMAR model 25
portable conveyors 255
portable electrical equipment
hazards 319–20
inspection strategies 327–8
Portable Appliance Testing (PAT) 327–9
see also hand-held power tools
posters 72, 101
powered load handling equipment see mechanical handling
and lifting equipment
practicable level of duty 12
pregnant women 14, 105–7, 183, 246
premises
design and selection of 69
joint occupation of 59
persons in control of 50–1
safety inspections 145
pressure vessels and systems 281
preventative health and safety culture 44
preventative measures, as theme in national legislation 527–8,
529
prions 375
proactive monitoring see active/proactive monitoring
product certification and marking 530
product safety 530
production demands 77
promotion of health and safety standards 69–71
protective clothing 13, 44, 395, 397–8, 464
Provision and Use of Equipment Regulations (UK) 275
psychological health hazards 183, 248, 422
see also substance misuse; violence at work; work-related
stress
public safety 45, 232
push sticks 298

racking systems 209
radiation 292, 434–42
of heat 338
ionising radiation 39, 434–6
acute exposure 435
alpha particles 434
beta particles 434
chronic exposure 435
controlled areas 440

distance from source 438
dose limits 434
EU legislation 492
gamma rays 434
genetic effects 435
harmful effects of 434–5
ILO recommendations 439–40, 460–1
protection 435, 438, 439–40, 460–1
reduced time exposure 438
shielding 438
somatic effects 435
sources of 435–6
supervised areas 440
non-ionising radiation 39, 436–8
electric, magnetic and electromagnetic fields (EMFs) 441
ILO recommendations 441
infrared radiation 438
lasers 437–8, 439
microwaves 438, 439
protection 438–9, 441
ultraviolet (UV) 436–7, 439, 441
welding operations 441–2
radiation protection advisers 49, 438
radiation protection supervisors 438
radio systems 190
radon 435–6
REACH (Registration, Evaluation, Authorisation and Restriction of Chemicals) Regulations 391–2
reach trucks **258**, 258
reactive assessment 22
reactive monitoring 25, 26, 140–1, 143, 144, 149, 150
reasonably practicable level of duty 12–13
records
accident/incident see incident/accident recording and reporting
air sampling 387
electrical equipment inspection and testing 329
health 390, 398
health and safety, establishment of and access to 24, 25
of preventative measures 529
violence at work 189
work-related ill-health 44
recreation facilities 469
recycling 414, 415
reduced low-voltage systems 323
regulatory authorities 14–15, 28
repetitive strain injury (RSI) 245
repetitive tasks 65, 66
report writing 146–9
aims 147
presentation 147
structure 147–8
reporting incidents see incident/accident recording and reporting
Reporting of Injuries, Diseases and Dangerous Occurrences Regulations (RIDDOR) (UK) 93, 152
reproductive toxins 377, 391
residual current devices (RCDs) 313, 323–4
respirable dust 374, 380
respiratory protective equipment (RPE) 394, *395*, 395–7, 403
compressed air-line apparatus 395
filtering face piece (FFP3) device 397
filtering half-mask 395
fresh air hose apparatus 395
full-face mask respirator 395
half-mask respirator 395

powered respirator 395
selection process 395–6
self-contained breathing apparatus 395
respiratory system **380**, 380
rest facilities 183, 469
rest periods, road transport drivers 240, 457–8
Restriction of Hazardous Substances in Electrical and Electronic Equipment (RoHS) Directive 415
retail hazards and safeguards 294, 305
reverse burden approach 47
reviews
risk assessment 95, 104
risk control measures 103
see also performance review
RIDDOR see Reporting of Injuries, Diseases and Dangerous Occurrences Regulations
risk 13
avoiding 108
combating at source 108
definition 4, 92
elimination of 95
high risk 4, 92, 96–7, 104, 343
as likelihood of potential harm 92
low risk 96, 97, 104, 343
medium risk 96, 97
residual risk 4, 91, 96
risk assessment 22, 24, 25, 27, 34, 43, 44, 54, 76, 90–108, 528
confined spaces 114
construction projects 55
dangerous occurrence 93
definition 90
demolition 213
example forms 575–8
examples
contract bricklayers 605–6
hairdressing salon 129–30
office cleaning 131–2
woodwork 607–8
fire see fire risk assessment
flammable liquids and gases 348
generic 92
hazard and risk 92
ILO approach to 95
and incident investigation findings 161
incident or near miss 93
international requirements for 91
machinery/work equipment 34, 274–5, 275–6
display screen equipment (DSE) 246–7
mechanical handling and lifting equipment 269
manual handling 216, 250–1
noise 425
objectives of 93–4
occupational disease 93
occupational or work-related ill-health 92, 94
pre-travel 107–8
process
evaluation of control measures 97–104
and groups at risk 96
hazard identification 96
monitoring and review 104
recording significant findings 104
risk evaluation 96–7
qualitative 91–2, 96–7
quantitative 91, 97
reviews 95, 104
road risks 237

special cases
  disabled workers 107
  expectant and nursing mothers 105–7
  lone workers 107
  young persons 105
suitable and sufficient 91
temporary work projects 210–11
UK approach to 94–5, 96–104
vehicle operations 231–2
violence at work 189
whole-body vibration (WBV) 432
work at height 198
work equipment 34, 274–5
risk assessors 94
risk control 90, 142–3
cost-benefit analysis 105
failures in 149
hierarchy of 97–104
measures, working at height 198–209
principles of prevention 108–9
risk control measures
action plan following incident investigation 161
adapting to technical progress 108–9
adapting the work to the individual 108
administrative controls 101–3
air control 470
avoiding risks 108
collective protective measures 109
combating risks at source 108
construction work 462–4
driving at work 237–9
electrical systems 320–9
engineering controls 98
fire hazards 344–7
flammable liquids and gases 347–8, 349
giving appropriate instruction 109
hazard elimination 98
hazardous substances 383, 384, 385, 389–411
health and safety information 102
isolation/segregation 101
machinery/work equipment *see* machinery/work equipment
  risk control measures
manual handling 251–4
monitoring and supervision 102–3
noise 426–8, 470
personal protective equipment 103, 104
prioritisation of 104
radiation 438–41, 460–1
reduced time exposure 101
review of 103
safe movement of people 194–7
safe systems of work 101
safety signs **98–101**, 98–101
stress 442–4
substitution 98, 109
temporary works 210–12
training 101–2
vehicles in the workplace 231–6
vibration-related tasks 430–4, 470
welfare facilities 102
risk control systems (RCSs) 172, 173, 176
risk perception 66
road safety 230, 236, 237
Robens report (1972) (UK) 42
roller tracks 254
rollover protective structures (FOPS) 233, 274

roof work 199–200, 462
rotating elements, hazards from 292
routine violations 68
RSI (repetitive strain injury) 245
rule-based mistakes 67
Russian Federation 15
  emergency call numbers 136
  hazardous substances classification and labelling 378
  occupational accidents and diseases *477*
  OSH framework 511–15

sack trucks 254, **271**
safe movement of people in the workplace 192–7
safe movement of vehicles in the workplace 230–6
Safe Person approach 109
Safe Place approach 109
safe systems of work 22, 64, 76, 101, 110–15
  asbestos examples 133–5
  assessment of requirements 111
  behavioural controls 112
  communication and training 113
  competent persons 111–12
  confined spaces 113–14
  construction work 112
  development of 111–12
  documentation 113
  electrical systems 320–1
  fire prevention 347
  and job/operation analysis 112
  legal requirements 111
  lone workers 114–15
  maintenance work 281
  managers and 112
  monitoring 113
  preparation checklist 112–13
  procedural controls 112
  technical controls 112
  workers/consultation and 112
  see also permit-to-work system
Safe Work Australia 479
safe working load (SWL) 256, 260
safety, definition 3
safety audits *see* audits
safety committees *see* health and safety committees
safety culture questionnaire 87–8
Safety Extra Low Voltage (SELV) 323
safety glasses 195, 397
safety harnesses 200–1
safety inspections 22, 140, 143, 144–7
  checklists 145
  electrical equipment 313, 327–9
  example report forms 579–81
  excavations 221
  exercises **165–7**
  fire equipment 360–1
  observation skills and techniques 145–6
  people 145
  plant and substances 145
  premises 145
  pressure vessels and systems 281
  procedures 145
  recording forms: construction work 224–5
  reports 146–9, 221
  scaffolding 208–9
Safety Management System (SMS) Audit 143, 172
safety nets 201

safety sampling 143, 172
safety signs *see* signs and symbols
safety surveys 143, 172
safety tours 143, 172
sampling, safety 143, 172
sanders 288–90
sanitary conveniences 182–3
Saudi Arabia
   emergency call numbers 136
   occupational accidents and diseases *477*
scaffolds 18, 59, 205–7
   components 205
   faults checklist 226
   ILO recommendations 208, 223, 462
   independent tied 205, **206**
   inspection 208–9
   pre-fabricated mobile towers **207**, 207
   putlog 205
Scotland 476
seating 64, 185
security, construction site 212–13
security equipment 190
   see also CCTV
self-employed people 51, 54
self-regulation 42
sensitisation 13, 376
Seoul Declaration on Safety and Health at Work (2008) 471,
   533–4
Serbia 136
serious health hazard substances 376
service demands 77
shear hazards **291**
signallers 234, 235, 260
signs and symbols **98–101**, 98–101
   checklist 101
   electricity **312**, **315**
   EU legislation 491
   exit 362
   fire action **100**, 100
   fire extinguisher use-code **359**
   fire safety **100**, 100
   fire triangle **335**
   first-aid **100**, 100, **124**, 124
   flammable gas **336**
   flammable liquid **336**
   flammable solid **335**
   fragile roof **101**, 101
   hazardous substances 99, **100**, 376, **377**
   no smoking **100**, 100
   overhead obstacle 99
   oxidising agent **336**
   wet floor **99,** 99
silencers 427
silica/silica dust 216, 374, 380, 408
silicosis 3, 6, 408
site managers, and permit-to-work system 120
situational violations 68
skin
   main structures in the dermis **382**
   risks and protection 3, 379, 381–3, 395, 397, 436, 438–9, 441
skin cancer 436
skips 216, 414
slip hazards 192–3, 194, 196–7
slips, due to loss of concentration 67
small and medium-sized enterprises 60, 94
SMART performance standards or objectives 32, 161, 176

smog, photochemical 412
smoke 338
smoking 341, 346, 374
so far as is (reasonably) practicable 108
social reasons for OSH standards 6
societal expectations 78
sockets and plugs 319
soft landing systems 201
sole boards (scaffolds) 205
solids, flammable 335
solvent abuse 191
sound proofing 427
South Africa 515–19
   emergency call numbers 135
   fire safety legislation and standards 334
   hazardous substances
      classification and labelling 378
      occupational exposure limits 388
   occupational accidents and diseases *477*
   Occupational Health and Safety Act 516–19
South America *see* Latin America
South Korea 15, 135, 451
   hazardous substances classification and labelling 378
   occupational accidents and diseases *477*
   OSH legal framework 500–1
Spain 451
sparks/sparking 335, 345, 353
special needs, fire emergency evacuation and people with
   365–6
specialist training 76
specialists, and permit-to-work system 120
spillages 192
   hazardous substances 406, 413, 414, 415
stability/instability of machinery 283, 292
staging 203–4
stain tube detectors 386
stairways 185, 194, 195, 196
stakeholders
   commercial 79
   involvement of 27–8
standards
   fire safety 333–4
   international 15–16
   labour 10, 11
   national 16
   performance 26, 32, 43
   promotion of 69–71
   and supply chain 52
   work equipment 275–7
   see also International Organisation for Standardisation (ISO)
standards (scaffolds) 205
start controls 282
static electricity 317, 318–19, 335, 353, 354
stationary objects, striking against 193–4, 196
statistics, incident/injury/ill-health 141, 160–1
status
   of health and safety 69
   of health and safety practitioners 49
stepladders 203–4
stop controls 282
storage issues 53, 209
   and fire hazards 344–5
   flammable liquids and gases 349, **350**
stored energy hazards 292
Strategic Forum for Construction Plant Safety Group 208
stratospheric ozone depletion 411–12

stress *see* work-related stress
strimmers 294, 303
stroboscopic effect 66
study skills 540–7
  blocked thinking 541
  examiners' reports 546–7
  exams
    after the exam 546
    in the exam room 545–6
    marks for questions 547
    NEBOSH terminology *545*
    planning and revision 545
  learning styles (visual, auditory and kinaesthetic) 544
  memory 544
  note-taking 541
  organising information 542–4
  organising for revision 542
  OU free learning resources 541–2
  place to study 540
  reading for study 541
  study plan 540
  time management 540–1
sub-contractors 45, 58, 182
substance misuse at work 187, 191–2
substances of very high concern (SVHCs) 391
substitution 109, 413
  flammable liquids and gases 348–9
  hazardous substances 98, 109, 390, 391
'suitable and sufficient' 91, 108
supervision 6, 13, 24, 34, 39, 44, 48, 62, 63
  of contractors 24, 58
  hazardous substances 390, 393
  machinery/work equipment use 300
  management 108
  of new workers 75
  personal protective equipment (PPE) 395
  of risk control measures 102–3
  of temporary work projects 211
  of young persons 280
supervisors
  and health and safety policy 33
  health and safety training 61
  organisational health and safety responsibilities 48–9, 85–6
  training 76
supply chain management 51–4
  and accident reduction 52
  checklist 86–7
  and legislation and standards 52
  and multinational companies 54
  and rapid response to changing requirements 52
  and reduction of waste 51–2
Supply of Machinery (Safety) Regulations (2008) (UK) 275
surveys, safety 143, 172
Sweden 15, 451
Switzerland 136

taxi drivers 242
technical progress, adapting to 108
temperature 183–4
  extremes of 185–7
temporary work projects 210–12
  control measures
    communication and cooperation 210
    European Union (EU) 490
    management and supervision 211
    risk assessment 210–11

fire risk assessment 344
  hazards 210
temporary workers 45–6
  see also contractors; migrant workers
tenosynovitis 245, 247–8
testing
  electrical equipment 313, 327–9
  fire equipment 360–1
  pressure vessels and systems 281
tetanus 216, 379, 409
Thailand 15
  emergency call numbers 136
  occupational accidents and diseases *477*
thermal hazards 292
ties (scaffolds) 205
tinnitus 423
tiredness 65
toe boards (scaffolds) 205
toolbox talks 72, 75, 76
tours, safety 143, 172
tower cranes 261–2, 265–6
toxic substances 376
trade unions 28, 78
traffic marshals *see* banksmen
traffic routes 185, 194, 196, 214–15, 231–2
training 6, 13, 14, 24, 27, 31, 34, 35, 39, 43, 44, 60, 74–6, 528
  asbestos awareness 400, 403, 405–6
  competence 70
  demolition 214
  driver 76, 230, 237–8, 259
  electrical systems 321
  emergency procedures 123
  environment 75
  fire prevention 76
  fork-lift truck drivers 76, 259
  hand-held tools (non-powered) 286
  health and safety culture and poor levels of 62
  of health and safety practitioners 49
  hearing protection devices 428
  and individual (personal) factors 65
  induction 72, 75–6, 105
  job-specific 76
  machinery/work equipment use 279–80, 286, 300
  of managers 61, 76
  manual handling 252–3
  measuring effectiveness of 75
  personal protective equipment (PPE) 395
  radiation awareness 441
  refresher courses 69
  as risk control measure 101–2
  safe systems of work 113
  of safety delegates 74
  specialist 76
  supervisors 76
  toolbox 72, 75, 76
  vibration awareness 434
  young persons 105, 280
transoms (scaffolds) 205
transport facilities 469–70
transport of hazardous substances 399
transport hazards *see* driving at work; vehicles in the workplace
trestles 203–4
Trinidad and Tobago
  emergency call numbers 137
  Factories Ordinance 519–20

occupational accidents and diseases *477*
Occupational Safety and Health Agency 520–1
Occupational Safety and Health (Amendment) Act (2006) 519
Occupational Safety and Health Authority (OSHA) 519, 520
trip devices 297, **298**
tripping hazards 192, 193, 194
trolleys 254
trucks 254
  balance 254
  general purpose 254
  lift 259, **271**
  pallet **255**, 255, 258
  pallet stacker 258
  platform 254, **271**
  reach **258**, 258
  sack 254, **271**
  very narrow aisle (VNA) 258
  see also fork-lift trucks
trustees, as stakeholders 28
tuberculosis 408
Turkey
  emergency call numbers 136
  occupational accidents and diseases *477*
  Occupational Health and Safety Act (2012) 521–2
two-handed control devices 297–8, **299**

Ukraine 136
ultraviolet (UV) radiation 436–7, 439, 441
underground services 324, 326
undesired circumstances 152
United Arab Emirates 136, 522–4
United Kingdom (UK) 450
  Advisory Committee for Roof Work 20
  Combined Code on Corporate Governance 28, 46
  Environment Agency 413–14
  fire extinguishers *358*
  hazardous substances occupational exposure limits 388–9
  HSE/Local Authorities Enforcement Liaison Committee (HELA) 475
  Institute of Directors 46
  Institution of Engineering and Technology 320
  Institution of Occupational Safety and Health (IOSH) 7, 56, 115, 242
  occupational accidents and diseases *477*
  regulatory and legal framework 474–6, 524–5
    Chemicals (Hazard Information and Packaging for Supply) 2009 Regulations (CHIP 4) 376
    Control of Artificial Optical Radiation at Work regulations (2010) 439
    Control of Substances Hazardous to Health (COSHH) regulations 388–8
    Equalities Act (2010) 365
    Health and Safety at Work (HSW) Act 42, 91, 474–5, 524–5
    Ionising Radiation Regulations 435
    and local authorities 475
    Management of Health and Safety at Work Regulations 15, 91
    Manual Handling Operations Regulations 244
    ministerial responsibilities 4756
    Provision and Use of Equipment Regulations 275
    Reporting of Injuries, Diseases and Dangerous Occurrences Regulations (RIDDOR) 152, 193
    Supply of Machinery (Safety) Regulations (2008) 275
    Workplace (Health, Safety and Welfare) Regulations 183
  risk assessment approach 94–5, 96–104
  Robens report (1972) 42

Work at Height Safety Association (WAHSA) 202
work-related injuries/illnesses 7
workplace violence 2
see also Health and Safety Executive (HSE)
United Nations (UN) 2, 28, 56
United States of America (USA)
  cost of work-related injuries/illnesses 7, 442
  deaths at work 442
  electrical wiring 320
  emergency call numbers 137
  fire extinguishers *358*
  fire safety legislation and standards 333
  hazardous substances
    classification and labelling 378
    occupational exposure limits 387–8
  National Institute for Occupational Safety and Health (NIOSH) 388, 471
  occupational accidents and diseases *477*
  Occupational Safety and Health Administration (OSHA) 15, 16, 388, 471
  regulatory and legal framework 14, 333, 471–2
    Code of Federal Regulations, Occupational Safety and Health Standards on Toxic and Hazardous Substances 388
    Occupational Safety and Health Act (1970) 471, 525, 526–7
  workplace violence 2
unsafe acts 144, 341
unsafe conditions 96, 144, 152, 341
urinary system **381**, 381–2
user checks, portable electrical appliances 327

vacuum hazards 292
vane anemometers 387
vapours 375
vehicles in the workplace 230–6
  construction sites 214–15
  control strategies
    ILO recommendations 233–5
    management of vehicle movements 235–6
    risk assessment 231–2
    site design 232
    systems of work for system operatives 232
    vehicle selection and maintenance 232
  hazards 193, 195, 230–1, 345
  see also driving at work
ventilation 183, **184**, 317, **394**
  dilution (or general) 390, 393
  see also local exhaust ventilation
ventilation equipment, confined spaces 118
very narrow aisle (VNA) trucks 258
very toxic substances 376
vibration 39, 59, 292, 411, 428–34
  control measures 430–4
    ergonomic risks 430–1
    exposure limits 431–3
    health surveillance 433–4
    training and information 434
  EU requirements 491–2
  ill-health effects 428–31
    back pain 429–30
    carpal tunnel syndrome 429, **430**
    hand—arm vibration syndrome (HAVS) 429, **430**, 431–2, 433
    vibration white finger (VWF) 6, 429
  ILO recommendations 428–9, 432, 433–4, 470

typical exposure values for common items of equipment *429*
whole-body vibration (WBV) 429–30, 432–3
vicarious liability 6
Vietnam 15
violations 67, 68
  exceptional 68
  routine 68
  situational 68
violence at work 2, 115, 187–92
  action plans 190–1
  classification of incidents 189
  common causes of 188
  definition 189
  and job design 190
  and operating environment design 190
  and quality of service provided 189
  records of 189
  risk assessment 189
  security equipment measures 190
  workers most at risk from 188, 189
viruses 375
  blood-borne 410–11
visitors and general public 45
visual problems 248
voltage
  high 313
  low 313, 316
  mains 312, 313
volts (V) 312

Wales 476
warnings/warning devices 283
  see also alarm systems; signs and symbols
washing facilities 102, 182–3
waste
  controlled 413
  hazardous properties of 419
waste disposal 53, 216, 411, 413–15
  asbestos 406
  hazardous substances 406, 413, 414, 415
  hazards associated with 414
  management
    prevention 413
    recovery 414
    reduction of waste 414
    responsible disposal 414
    reuse/recycling 414
  practice 414–15
Waste Electrical and Electronic Equipment (WEEE) Directive 415
waste skips 216, 414
water pollution 412–13
WBGT (Wet Bulb Globe Temperature) 186
weather conditions
  and work at height 200
  and work-related driving 239
Weil's disease 409
welding equipment 59
welding operations 441–2
welfare, definition 3
welfare facilities 43, 59, 102, 182–3, 458–9
  construction sites 215, 464
  ILO recommendations 468–70
  temporary work projects 211
wet floor signs **99**, 99
wheelbarrows 254

whole-body vibration 429–30
windows and doors 185
wiring regulations 320–1
witness statement form 584
women workers
  accident rates 6
  pregnant or nursing 14, 105–7, 183, 246
wood dust 216, 409
Woodhouse Report (1966) (New Zealand) 9
woodworking machinery 279–80
work at height 32, 197–209, 213
  emergency procedures 201–2
  fall arrest equipment 200–1
  fragile roofs and surfaces 199–200
  ILO recommendations 462
  permit-to-work 119–20
  protection against falling objects 200
  risk assessment 198
  safe working practices for access equipment
    ladders 202–3
    mobile elevating work platforms (MEWPS) **207**, 207–8
    scaffolds 205–7
    stepladders, trestles and staging 203–4
Work at Height Safety Association (WAHSA) (UK) 202
work equipment *see* machinery/work equipment
Work Health and Safety Act (2011) (Australia) 479–80, 479–90
Work Improvement in Small Enterprises (WISE) methodology 246
work permit form 590–1
work-related ill-health 25
  deaths from 4, 6
  definition 3, 92
  investigations 73
  records 44
  statistics 141
  working days lost from 6
  see also occupational diseases
work-related stress 6, 68, 248, 422, 442–4
  causes of 442
  control measures 442–5
  definition 442
  symptoms 442
work-related upper limb disorders (WRULDS) 53, 93, 245–6, 249
work—rest systems 187
workers 26, 47
  disabled 107, 182, 185
  new 75
  older 105
  and positive health and safety culture 59, 60
  representation/representatives 22, 24, 25, 28, 30, 33, 69, 72–3, 77
  rights and responsibilities 13–14, 33, 50, 54, 86, 218, 529–30
    construction work 461–2
    hazardous substances 383–4, 456–7
    hearing protection 426
  and safe systems of work 112
  see also consultation with workers; lone workers; migrant workers; women workers; young persons
working conditions, control of 528–9
working platforms
  mobile elevating (NEWPs) **207**, 207–8
  scaffolds 205
workplace environment
  doors and windows 185

floors, stairways and traffic routes 185
heating and temperature 183–4
lighting 184–5
and machinery hazards 293
ventilation 183, **184**
welfare arrangements 182–3
workstations and seating 185
workplace hazards and risk control 182, 187–227
construction work 212–21
pedestrian hazards 192–7
substance misuse at work 187, 191–2
temporary work projects 210–12
violence at work 2, 115, 187–92
working at height 197–209
Workplace (Health, Safety and Welfare) Regulations (UK) 183
workplace transport *see* vehicles in the workplace
Worksafe (Western Australia) 16

workstations
compensation claims 163
design of 64, 185, 247
display screen equipment (DSE) 163, 244, 246–8, 490
self assessment checklist 596–7
World Bank 28, 56
World Congress on Safety and Health at Work (Frankfurt, 2014) 448
World Health Organisation (WHO) 2, 6, 12, 28, 400, 401

X-rays 435

young persons 65, 74
risk assessment 105
training and supervision of 105, 280
work-related deaths 105
see also child labour; children